21st Century Nanoscience –

21st Century Nanoscience – A Handbook

Exotic Nanostructures and Quantum Systems (Volume Five)

Edited by

Klaus D. Sattler

CRC Press

Taylor & Francis Group

Boca Raton London New York

CRC Press is an imprint of the
Taylor & Francis Group, an **informa** business

CRC Press
Taylor & Francis Group
6000 Broken Sound Parkway NW, Suite 300
Boca Raton, FL 33487-2742

First issued in paperback 2022

© 2020 by Taylor & Francis Group, LLC
CRC Press is an imprint of Taylor & Francis Group, an Informa business

No claim to original U.S. Government works

ISBN-13: 978-0-815-35626-4 (hbk)
ISBN-13: 978-1-03-233638-1 (pbk)
DOI: 10.1201/9780429347313

Publisher's Note

The publisher has gone to great lengths to ensure the quality of this reprint but points out that some imperfections in the original copies may be apparent.

Library of Congress Cataloging-in-Publication Data

Names: Sattler, Klaus D., editor.
Title: 21st century nanoscience : a handbook / edited by Klaus D. Sattler.
Description: Boca Raton, Florida : CRC Press, [2020] | Includes bibliographical references and index. | Contents: volume 1. Nanophysics sourcebook—volume 2. Design strategies for synthesis and fabrication—volume 3. Advanced analytic methods and instrumentation—volume 5. Exotic nanostructures and quantum systems—volume 6. Nanophotonics, nanoelectronics, and nanoplasmonics—volume 7. Bioinspired systems and methods. | Summary: "This 21st Century Nanoscience Handbook will be the most comprehensive, up-to-date large reference work for the field of nanoscience. Handbook of Nanophysics, by the same editor, published in the fall of 2010, was embraced as the first comprehensive reference to consider both fundamental and applied aspects of nanophysics. This follow-up project has been conceived as a necessary expansion and full update that considers the significant advances made in the field since 2010. It goes well beyond the physics as warranted by recent developments in the field"—Provided by publisher.
Identifiers: LCCN 2019024160 (print) | LCCN 2019024161 (ebook) | ISBN 9780815384434 (v. 1 ; hardback) | ISBN 9780815392330 (v. 2 ; hardback) | ISBN 9780815384731 (v. 3 ; hardback) | ISBN 9780815355281 (v. 4 ; hardback) | ISBN 9780815356264 (v. 5 ; hardback) | ISBN 9780815356417 (v. 6 ; hardback) | ISBN 9780815357032 (v. 7 ; hardback) | ISBN 9780815357070 (v. 8 ; hardback) | ISBN 9780815357087 (v. 9 ; hardback) | ISBN 9780815357094 (v. 10 ; hardback) | ISBN 9780367333003 (v. 1 ; ebook) | ISBN 9780367341558 (v. 2 ; ebook) | ISBN 9780429340420 (v. 3 ; ebook) | ISBN 9780429347290 (v. 4 ; ebook) | ISBN 9780429347313 (v. 5 ; ebook) | ISBN 9780429351617 (v. 6 ; ebook) | ISBN 9780429351525 (v. 7 ; ebook) | ISBN 9780429351587 (v. 8 ; ebook) | ISBN 9780429351594 (v. 9 ; ebook) | ISBN 9780429351631 (v. 10 ; ebook)
Subjects: LCSH: Nanoscience—Handbooks, manuals, etc.
Classification: LCC QC176.8.N35 A22 2020 (print) | LCC QC176.8.N35 (ebook) | DDC 500—dc23
LC record available at https://lccn.loc.gov/2019024160
LC ebook record available at https://lccn.loc.gov/2019024161

Visit the Taylor & Francis Web site at
http://www.taylorandfrancis.com

and the CRC Press Web site at
http://www.crcpress.com

Contents

Editor

Klaus D. Sattler pursued his undergraduate and master's courses at the University of Karlsruhe in Germany. He earned his PhD under the guidance of Professors G. Busch and H.C. Siegmann at the Swiss Federal Institute of Technology (ETH) in Zurich. For three years he was a Heisenberg fellow at the University of California, Berkeley, where he initiated the first studies with a scanning tunneling microscope of atomic clusters on surfaces. Dr. Sattler accepted a position as professor of physics at the University of Hawaii, Honolulu, in 1988. In 1994, his group produced the first carbon nanocones. His current work focuses on novel nanomaterials and solar photocatalysis with nanoparticles for the purification of water. He is the editor of the sister references, *Carbon Nanomaterials Sourcebook* (2016) and *Silicon Nanomaterials Sourcebook* (2017), as well as *Fundamentals of Picoscience* (2014). Among his many other accomplishments, Dr. Sattler was awarded the prestigious Walter Schottky Prize from the German Physical Society in 1983. At the University of Hawaii, he teaches courses in general physics, solid state physics, and quantum mechanics.

Contributors

Saad Alafnan
Petroleum Engineering
King Fahd University of Petroleum
and Minerals
Dhahran, Saudi Arabia

I. Yucel Akkutlu
Petroleum Engineering
Texas A&M University
College Station, Texas

A.V. Andreeva
Institute of Microelectronics
Technology and High Purity
Materials
Russian Academy of Science
Moscow Region, Russia

Portonovo S. Ayyaswamy
Department of Mechanical
Engineering and Applied
Mechanics
University of Pennsylvania
Philadelphia, Pennsylvania

Balamurugan Balasubramanian
Nebraska Center for Materials and
Nanoscience
University of Nebraska
Lincoln, Nebraska
and
Department of Physics and
Astronomy
University of Nebraska
Lincoln, Nebraska

Anjan Barman
Department of Condensed Matter
Physics and Material Sciences
S. N. Bose National Centre for Basic
Sciences
Kolkata, India

Ali Beskok
Department of Mechanical
Engineering
Southern Methodist University
Dallas, Texas

Holger F. Bettinger
Institute for Organic Chemistry and
Institute for Physical and
Theoretical Chemistry
University of Tübingen
Tübingen, Germany

Joan J. Carvajal
Departament Química Física i
Inorgànica
Física i Cristallografía de Materials i
Nanomaterials (FiCMA-FiCNA) -
EMaS
Universitat Rovira i Virgili
Tarragona, Spain

Alper Tunga Celebi
Department of Mechanical
Engineering
Southern Methodist University
Dallas, Texas

Javier Cervera
Faculty of Physics
Department of Thermodynamics
University of Valencia
Burjasot, Spain

Huan-Cheng Chang
Institute of Atomic and Molecular
Sciences
Academia Sinica
Taipei, Republic of China

Oliver Y. Chen
Institute of Atomic and Molecular
Sciences
Academic Sinica
Taipei, Republic of China

Samiran Choudhury
Department of Condensed Matter
Physics and Material Sciences
S. N. Bose National Centre for Basic
Sciences
Kolkata, India

Anulekha De
Department of Condensed Matter
Physics and Material Sciences
S. N. Bose National Centre for Basic
Sciences
Kolkata, India

Blanca del Rosal
Faculty of Science, Engineering and
Technology
Centre for Micro-Photonics
Swinburne University of Technology
Hawthorn, Australia

A.L. Despotuli
Institute of Microelectronics
Technology and High Purity
Materials
Russian Academy of Science
Moscow Region, Russia

A. Domaracka
ENSICAEN, UNICAEN, CEA,
CNRS, CIMAP
Normandie Université
Caen, France

M.-A. Durán-Olivencia
Department of Chemical Engineering
Imperial College London
London, United Kingdom

David M. Eckmann
Department of Bioengineering
University of Pennsylvania
Philadelphia, Pennsylvania
and
Department of Anesthesiology and
Critical Care
University of Pennsylvania
Philadelphia, Pennsylvania

Andrew M. Ellis
Department of Chemistry
University of Leicester
Leicester, United Kingdom

Vladimir García-Morales
Faculty of Physics
Department of Thermodynamics
University of Valencia
Burjasot, Spain

Peter Grüninger
Institute for Organic Chemistry and
    Institute for Physical and
    Theoretical Chemistry
University of Tübingen
Tübingen, Germany

Rodaina Sayed Hassan
Multidisciplinary Physics Laboratory
    (MPLAB)
Faculty of Sciences Section I
Lebanese University
Beirut, Lebanon

Elda Hegmann
Liquid Crystal Institute
Kent State University
Kent, Ohio
and
Department of Biological Sciences
Kent State University
Kent, Ohio

Torsten Hegmann
Liquid Crystal Institute
Kent State University
Kent, Ohio
and
Department of Chemistry and
    Biochemistry
Kent State University
Kent, Ohio

B.A. Huber
ENSICAEN, UNICAEN, CEA,
    CNRS, CIMAP
Normandie Université
Caen, France

Yuen Yung Hui
Institute of Atomic and Molecular
    Sciences
Academic Sinica
Taipei, Republic of China

Subhra Jana
Department of Chemical, Biological &
    Macro-Molecular Sciences
S. N. Bose National Centre for Basic
    Sciences
Kolkata, India

S. Kalliadasis
Department of Chemical Engineering
Imperial College London
London, United Kingdom

Mohammad Ehsan Khaled
Laboratory for Precision and Nano
    Processing Technologies
School of Mechanical and
    Manufacturing Engineering
The University of New South Wales
Sydney, Australia

Roopali Kukreja
University of California Davis
Davis, California

Weidong Liu
Laboratory for Precision and Nano
    Processing Technologies
School of Mechanical and
    Manufacturing Engineering
The University of New South Wales
Sydney, Australia

Joachim Maier
Max Planck Institute for Solid State
    Research
Stuttgart, Germany

José A. Manzanares
Faculty of Physics
Department of Thermodynamics
University of Valencia
Burjasot, Spain

Fabian Meder
Center for Micro-BioRobotics
Istituto Italiano di Tecnologia
Pontedera, Italy

A. Mika
ENSICAEN, UNICAEN, CEA,
    CNRS, CIMAP
Normandie Université
Caen, France

Arin Mizouri
Department of Chemistry
University of Leicester
Leicester, United Kingdom

Yann Molard
Institut des Sciences Chimiques de
    Rennes
Université de Rennes 1, CNRS
Rennes, France

Sucheta Mondal
Department of Condensed Matter
    Physics and Material Sciences
S. N. Bose National Centre for Basic
    Sciences
Kolkata, India

Alireza Moridi
Laboratory for Precision and Nano
    Processing Technologies
School of Mechanical and
    Manufacturing Engineering
The University of New South Wales
Sydney, Australia

Chinh Thanh Nguyen
Department of Mechanical
    Engineering
Southern Methodist University
Dallas, Texas

Hendrik Ohldag
Lawrence Berkeley National
    Laboratory
University of California Santa Cruz
Santa Cruz, California
and
Stanford University
Stanford, California

Dirk H. Ortgies
Nanobiology Group
Instituto Ramón y Cajal de
    Investigación Sanitaria IRYCIS
Madrid, Spain
and
Fluorescence Imaging Group
Departamento de Física de Materiales
Facultad de Ciencias
Universidad Autónoma de Madrid
Madrid, Spain

Julia Pérez-Prieto
Instituto de Ciencia Molecular
    (ICMol)
University of Valencia
Paterna, Spain

Marianne E. Prévôt
Liquid Crystal Institute
Kent State University
Kent, Ohio

Charlotte Pughe
Department of Chemistry
University of Leicester
Leicester, United Kingdom

Ravi Radhakrishnan
Department of Bioengineering
University of Pennsylvania
Philadelphia, Pennsylvania
and
Department of Chemical and
    Biomolecular Engineering
University of Pennsylvania
Philadelphia, Pennsylvania

N. Ramakrishnan
Department of Bioengineering
University of Pennsylvania
Philadelphia, Pennsylvania

P. Rousseau
ENSICAEN, UNICAEN, CEA,
    CNRS, CIMAP
Normandie Université
Caen, France

Arnab Samanta
Department of Chemical, Biological &
    Macro-Molecular Sciences
S. N. Bose National Centre for Basic
    Sciences
Kolkata, India

Oleksandr A. Savchuk
Nanophotonics Department
Ultrafast Bio- and Nanophotonics
    Group
International Iberian Nanotechnology
    Laboratory (INL)
Braga, Portugal

D. J. Sellmyer
Nebraska Center for Materials and
    Nanoscience
University of Nebraska
Lincoln, Nebraska
and
Department of Physics and
    Astronomy
University of Nebraska
Lincoln, Nebraska

Seungha Shin
Mechanical, Aerospace, and
    Biomedical Engineering
The University of Tennessee
Knoxville, Tennessee

Berlian Sitorus
Department of Chemistry
University of Leicester
Leicester, United Kingdom
and
Department of Chemistry
Tanjungpura University
Pontianak, Indonesia

Ralph Skomski
Nebraska Center for Materials and
    Nanoscience
University of Nebraska
Lincoln, Nebraska
and
Department of Physics and
    Astronomy
University of Nebraska
Lincoln, Nebraska

Dmytro Solonenko
Institut für Physik
Technische Universität Chemnitz
Chemnitz, Germany

Meng-Chih Su
Department of Chemistry
Sonoma State University
Rohnert Park, California

Kuan Vai Tam
Lunar and Planetary Science
    Laboratory
Macau University of Science and
    Technology
Macau, China

Chi Pui Tang
Lunar and Planetary Science
    Laboratory
Macau University of Science and
    Technology
Macau, China

Steffi S. Thomas
School of Chemistry
Trinity College Dublin
Dublin, Ireland

Marek W. Urban
Department of Materials Science and
    Engineering
Clemson University
Clemson, South Carolina

Julie P. Vanegas
Liquid Crystal Institute
Kent State University
Kent, Ohio

Patrick Vogt
Institut für Physik
Technische Universität Chemnitz
Chemnitz, Germany

Siyang Wang
Department of Materials Science and
    Engineering
Clemson University
Clemson, South Carolina

Yan Xu
Department of Chemical Engineering
Graduate School of Engineering
Osaka Prefecture University
Sakai, Japan
and
Japan Science and Technology Agency
    (JST), PRESTO
Kawaguchi, Japan
and
NanoSquare Research Institute
Osaka Prefecture Universit
Sakai, Japan

Nader Yaacoub
Institute of Molecules and Materials
    of Le Mans (IMMM) CNRS
    UMR-6283
Le Mans University
Le Mans, France

Shengfu Yang
Department of Chemistry
University of Leicester
Leicester, United Kingdom

P. Yatsyshin
Department of Chemical Engineering
Imperial College London
London, United Kingdom

Liangchi Zhang
Laboratory for Precision and Nano
    Processing Technologies
School of Mechanical and
    Manufacturing Engineering
The University of New South Wales
Sydney, Australia

1

Novel Nanoscience in Superfluid Helium

Arin Mizouri, Charlotte Pughe,
Berlian Sitorus, Andrew M.
Ellis, and Shengfu Yang
University of Leicester

Berlian Sitorus
Tanjungpura University

Nanoscience and nanotechnology are broad interdisciplinary areas of research that span chemistry, physics, materials science and biology. The motivation for studying these areas is driven partly by the fascinating properties offered by new materials at the nanoscale, which can differ drastically from their bulk counterparts, and can be exploited for applications in medicine,[1,2] electronics and opto-electronics,[3,4] catalysis,[5−7] and nowadays in consumer products such as clothing and cosmetics.[8,9] The properties of nanomaterials are strongly dependent on their shapes, sizes, morphology, structures and chemical composition, and it is critical to have control over these parameters in order to obtain novel materials with desired properties.

To date, the vast majority of nanomaterials have been synthesized at the room temperature or above, but low-temperature techniques have also emerged in recent years, e.g., by employing solid matrices or superfluid helium as the solvents. In the solid matrix method, hot atoms collide with a cold solid matrix formed by the condensation of rare gases, nitrogen or carbon dioxide,[10−12] which then melts and allows atoms to aggregate. This is the first step occurring in the formation of nanomaterials in the solid matrix. While the clusters are being formed, the binding energy between atoms is released, stimulating further melting and thereby allowing subsequent migration and clustering of the added materials. In contrast, the formation of nanomaterials in superfluid helium occurs in the liquid phase, where the migration and aggregation is easier than in solid matrices. The extraordinary properties of superfluid helium, such as near-zero viscosity, ultrahigh thermal conductivity and quantized vortices,[13] allow the creation of a variety of nanostructures, many being inaccessible by conventional synthetic methods. However, superfluid helium is intrinsically a poor solvent as helium atoms interact very weakly with other atoms and molecules. As a result, materials added tend to locate at the walls of the container, rather than reside within the fluid, in bulk superfluid helium.

Problems associated with the condensation of samples on the container walls can be circumvented by using liquid helium droplets, which are isolated, self-contained large clusters of helium typically composed of 10^3–10^{11} helium atoms.[14,15] Unlike bulk liquid helium, helium droplets have no container wall; so foreign species can be trapped and then confined by the surface of the droplet. Toennies et al. were the first to demonstrate that the superfluidity of bulk liquid helium also occurs on the nanoscale,[16,17] to which atoms/molecules can be added by a pickup technique,[18] i.e., by collisions between the droplets and gas-phase atoms/molecules. Helium droplets are highly "sticky", with a near unity pickup probability. In other words, the number of atoms/molecules being added to the droplets is approximately the number of collision events. By controlling the pickup conditions, formation of nanoparticles containing a few dozens to millions of atoms are possible, which depends on the size of the droplets and the number density of atoms/molecules in the pickup region. Although still in its infancy, a number of metallic nanoparticles have been synthesised using this technique. In 2011, Loginov et al. performed the first transmission electron microscopy (TEM) imaging of spherical metallic nanoparticles formed in helium droplets.[19]

One of the unique advantages of using helium droplets as nano–reactors to grow nanomaterials is the ease of forming core–shell structures, which can be achieved simply by pickup of different types of materials when they pass through two sequential pickup regions. Such a scenario was first demonstrated by the formation of molecular clusters, where a core–shell structure was deduced using mass spectrometry.[20] Evidence for the formation of a core–shell bimetallic structure was obtained in Ni/Au nanoparticles formed by sequential addition of Ni and Au to helium droplets, for which the core–shell structure was evident from X-ray photoelectron spectroscopy (XPS).[21] In this process the inter-diffusion between layers is also minimized as the reactions occur in very cold superfluid helium, giving rise to core–shell structures with clear boundary at the interface, which is difficult to achieve using conventional "hot" approaches.

This chapter is divided into three sections: (i) the properties of superfluid helium (in bulk helium and droplets), and how and why each property is useful in the synthesis of nanomaterials; (ii) fabrication of nanostructures in bulk superfluid helium; and (iii) the new possibilities for the synthesis of nanomaterials using superfluid helium droplets, and the recent advances in this emerging field of nanoscience. Finally, we will close by considering some of the remaining challenges and the prospects for superfluid helium being employed in the development of next-generation materials.

1.1 The Properties of Superfluid Helium

Helium is the only element that remains as a liquid at 0 K. This is a consequence of the very weak van der Waals interaction between helium atoms (5 cm^{-1}) and the low atomic mass (4 amu). This results in a zero-point vibrational energy being comparable to the interatomic attraction, preventing helium atoms from being pinned to a solid lattice unless a very high pressure is imposed. This very weak interaction also means a very low temperature is needed to liquefy helium. Liquid helium was first discovered in 1908 by Kammerlingh-Onnes, who cooled the element to 4.2 K. At this temperature, helium behaves as a normal liquid (He I phase)[22] but upon further cooling below 2.17 K it enters a superfluid phase (He II, see Figure 1.1 for the phase diagram of ^{4}He). Superfluidity is the consequence of Bose-Einstein condensation (BEC), in which all of the bosonic particles occupy the same quantum state.[23] In contrast, ^{3}He is a fermion and has a distinctly different phase diagram when compared with ^{4}He. Although ^{3}He can become a superfluid, the transition temperature is far lower (3 mK)[24] in order for the fermionic ^{3}He atoms to pair up and form quasi-bosons.

One of the major differences between bulk helium and helium droplets is that the temperature of bulk liquid helium can be varied via cooling or heating its container. In contrast, helium droplets have a steady-state temperature that is determined by the balance of the surface tension of the droplets and the kinetic energy carried by helium atoms. For ^{4}He, this steady-state temperature is 0.37 K, which is well below the "kick-in" temperature for forming superfluid helium.[25]

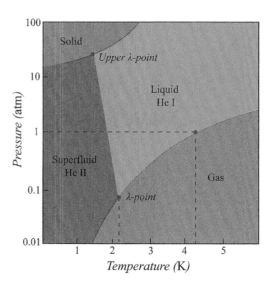

FIGURE 1.1 Phase diagram of helium. The three dots at 2.17 K, 1.76 and 4.2 K highlight the λ-point, upper λ-point and boiling point of helium, respectively.

1.1.1 Viscosity

Kapitza, Allen and Misener performed the first experimental measurements showing the superfluidity of liquid helium below 2.17 K. In these experiments, performed in 1938, they investigated the viscous flow of helium through narrow channels.[26,27] Above 2.17 K, the liquid helium was found to be viscous, just like any normal liquids but below the λ-point the viscosity was found to decrease by a factor of 10^6, allowing liquid helium to flow through microchannels at driven by a tiny pressure difference. From this observation, Kapitza coined the name "superfluid" to describe frictionless flow in the He II phase.

Some of the unusual properties of bulk superfluid helium irrelevant to the growth of nanomaterials, such as the fountain effect, film flow and creeping, will not be covered here. Instead, we will focus only on properties that are possessed by both bulk superfluid helium and helium droplets and that are important for the growth of nanomaterials, such as superfluidity (which allows the free motion of dopants and thus easy aggregation), the ultrahigh cooling rate (allowing different species to be incorporated) and quantized vortices (allowing the formation of 1-D nanostructures).

Not only does the superfluid flow unhindered, but also species contained within the fluid are able to move freely, almost unperturbed by the surrounding helium atoms. Due to the poor solubility of dopants in liquid helium, frictionless motion of impurities makes it difficult to suspend a species inside the bulk superfluid. This difficulty can be circumvented by using helium droplets. In 1998, Toennies et al. showed for the first time that atoms and molecules can be confined in very small droplets of liquid helium,[16] which can be retained for further investigations, e.g., by mass spectrometry and/or optical spectroscopy. The first experimental evidence for the superfluidity of helium droplets was

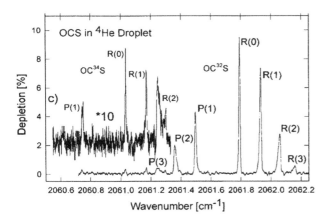

FIGURE 1.2 Absorption spectrum of OCS in ^{4}He droplets (He_N = <6000>). The factor 10 enlarged spectrum shows the results for the OC^{34}S isotope measured with an improved time constant. (Reprinted with permission from Ref. [28], AIP Publishing.)

obtained by Grebenev et al., who measured the rotationally resolved infrared spectrum of the OCS molecule.[28] The spectrum obtained showed well-resolved P and R branches (see Figure 1.2), suggesting that the OCS molecule is able to rotate in a way similar to gas phase molecules. However, liquid helium does have some significant influence on rotational spectra, e.g., by increasing the rotational moment of inertia and thus reducing the rotational constant.[29] The reduced rotational constant was interpreted as the attachment of a layer of helium atoms which follows the rotation of the OCS molecule in the droplet. The relative intensities of the vibration-rotation transitions allowed the determination of the temperature of helium droplets, which was found to be ~0.37 K. In the same study, OCS was also added to ^{3}He droplets for comparison. In this case no rotational structure can be resolved, suggesting that ^{3}He droplets are in the normal liquid phase. Indeed, this is consistent with the very low onset temperature of superfluidity for ^{3}He (3 mK) and a higher steady-state temperature of ^{3}He droplets (~0.15 K).

1.1.2 Thermal Conductivity

Superfluid helium has very unusual thermal properties, which were first observed by Allen.[30] When a torch was shone on a superfluid liquid helium reservoir, helium was found to spout from the capillary, an effect known as the fountain effect. This can be explained by the *so-called* two-fluid model in the He II phase, in which the fluid is assumed to consist of both a normal fluid fraction and a superfluid fraction. Upon heating, the superfluid fraction decreases in the region being heated, creating a concentration gradient. As a result, superfluid helium rapidly flows through the porous membrane in order to minimize the chemical potential, leading to spray of the fluid out of the capillary. Keesom et al. quantified the heat conductivity of He II as 800 times greater than that of copper at room temperature and a magnitude of 1.3×10^7 times greater

than the He I phase.[31,32] This ultrahigh thermal conductivity leads to bubble-free evaporation of superfluid helium when excess energy is introduced to the system.

A similar rapid cooling seems to occur when atoms or molecules are added to helium droplets: a cooling rate for the dopants as high as 10^{16} K/s has been proposed.[33,34] As a result, cluster formation in helium droplets involves successive binding of atoms/molecules to a pre-cooled cluster while the energy released from each binding event is instantly dissipated by the evaporative loss of helium atoms at the surface. As the helium-helium binding energy is only 5 cm^{-1}, each 1 eV of energy stimulates the evaporative loss of ~1,600 atoms.[35] Clusters grown under such conditions may have unusual structures as thermal energy is being rapidly removed; so they might localize in a shallow potential well along the potential energy surface. This has been demonstrated in several experiments, such as the formation of linear HCN clusters and cyclic H_2O hexamers,[36,37] and more recently in metallic nanoclusters.[38] New properties can be envisaged from entities with new structures.

1.1.3 Quantized Vortices

The quantization of circulation in superfluids and superconductors is one of the most remarkable macroscopic illustrations of quantum mechanics. In superfluids, the circulation is carried by the continuant atoms, each possessing an angular momentum equal to an integer multiple of $\hbar$. The investigation of the rotational behavior of superfluid ^{4}He can be traced back to the 1950s when Osborne,[39] Andronikashvili, and Kaverkin[40] found that the liquid rotates uniformly as a whole in rotating containers, and the shape of the meniscus generated in the containers is independent of the temperature. This was puzzling until Onsager proposed quantized vortices and modeled rotation in a bucket of superfluid helium as an array of concentric vortices rotating around a fixed point with quantized circulation[41]:

$$\text{Quantized circulation} = n\frac{h}{m_4} \qquad (1.1)$$

where h/m_4 represents a quantum of circulation, h is Planck's constant, m_4 is the mass of a ^{4}He atom, and n represents the quantum number ($n = 1, 2\ldots$) of the vortex. Later, Feynman proposed a slightly different theory, in which rotation of superfluid helium is manifested in the form of many vortex filaments (also called vortex lines).[42] He predicted that the radius of a vortex filament is on the order of the atomic scale (~0.5 Å) and that the centrifugal force associated with it would be balanced by the Bernoulli force due to a pressure gradient.[43,44]

In 1976, Packard et al. provided the first visual evidence supporting Feynman's theory.[45,46] This was achieved by dissolving electrons in a bath of superfluid helium over which a fluorescent screen was placed. By applying an electric potential, the electrons were accelerated from the superfluid towards the phosphor screen. The resulting image showed electrons arranged into circular arrays, with each spot representing a vortex.[47] Analogous experiments by Bewley et al. in 2006 used light scattering of H_2 clusters pinned to vortex lines to determine the vortex density of bulk He II.[48] By injecting a premixed gaseous solution of hydrogen, which was highly diluted by helium gas, into liquid helium in a steadily rotating container, micron-sized hydrogen clusters were formed. Below 2.17 K, the otherwise randomly distributed particles above the phase transition temperature were found to collect into slender filaments, and arrange themselves along uniformly spaced lines parallel to the axis of rotation, forming a rectilinear array.[48] The vortex density depends on the angular velocity of the container, Ω, and in this case was found to be 2,000 Ω vortex lines per cm^2. Imaging of pinned particles has shown that vortex lines do not have to be linear: they can also adopt circular, closed ring or curved shapes and can even tangle together.[49−51]

Quantized vortices have recently been found to be present in large helium droplets with diameters >300 nm, which allow the growth of nanostructure in one dimension, i.e., the nanowires. The first evidence for the presence of quantized vortices in helium droplets was obtained by Gomez et al.,[52] who added Ag atoms to helium droplets and then investigated the deposits obtained by allowing the doped helium droplets to collide with a solid target. By using TEM, track-like silver deposits were observed, which were attributed to the preferential growth of silver chains along the quantized vortices. In superfluid helium, the weak Bernoulli force is sufficient to trap dopants in the vortex so quantized vortices can serve as intrinsic 1-D templates that can be employed to fabricate ultrathin 1-D nanowires.[53] Both the length and diameter of these wires can be controlled by varying the helium droplet size and the rate of doping.

1.2 Growth of Nanomaterials in Bulk Superfluid Helium

Although bulk superfluid helium is intrinsically a poor solvent and it does not appear to be a good medium to grow nanomaterials, effort has been made recently to grow nanostructures, the vast majority of which are 1-D nanostructures formed by pinning metal atoms to the vortex lines instead of forming spherical particles. Due to the high thermal conductivity and the very low temperature of superfluid helium, atoms/molecules, once entering a quantized vortex, can no longer escape as helium rapidly removes all the kinetic energy; hence, the addition of materials to bulk liquid helium containing quantized vortices can lead to the formation of nanostructures in the shape of vortices – nanowires. By this route, long and thin metallic nanowires have been recently produced using bulk superfluid helium.[54,55]

1.2.1 Mechanism for Nanowire Formation

It had been thought that all processes following the laser ablation of impurities in superfluid helium are controlled by

diffusion and thus are slow and strictly isothermal, forming loose products. However, recent studies by Gordon et al. have convincingly shown that the fast coagulation process in superfluid helium is in fact not controlled by diffusion.[56] Instead, it is the concentration of impurities within the cores of quantized vortices that make thin and long wires the predominant products. In this process the coagulation is not isothermal in the initial stages and leads to vast local overheating. It is believed that this process results in melting of the coagulation products, forming densely packed structures.

Figure 1.3 illustrates the process.[56] The signal decay is different in the superfluid phase and the normal liquid phase, which provides evidence for the difference in mechanisms. In superfluid helium, two stages are proposed in the coagulation process, the first being the "hot" stage where atoms and nanoclusters coalesce and form molten nanospheres, releasing heat as a result of the condensation, which is sufficient to evaporate the surrounding liquid helium. The second stage is the aggregation of nanospheres into nanowires. The behavior is remarkably different in the normal liquid phase, which has three stages: the growth begins with a "hot" stage of spherical cluster growth, followed by clusters coalescing, eventually leading to a lump. These experiments suggest that each metal has a critical cluster size, beyond which nanoclusters can no longer melt and form spherical particles. Instead, they stick to each other in the vortex to form nanowires, where the critical cluster size determines the diameter of the wires.

Gordon et al. have recently proposed a mechanism for the condensation of atoms and nanoparticles in quantized vortices of superfluid helium.[57] In their model, vortex formation is caused by the impact of a laser pulse on the metal target, which generates a vortex near the focus of the laser. Small metal clusters and metal atoms generated by the laser ablation of a metal target then migrate into the vortex core. They can move freely along the vortex line, where there is a higher probability of collision compared with particles outside the vortex, and can aggregate into larger clusters. The coagulation releases energy that is sufficient to melt the clusters, resulting in a dense structure, which acquires a spherical shape due to the surface tension.[57,58] Meanwhile, the liquid helium surrounding the vortex line boils, and the cavitation by the gas bubbles in the normal liquid fraction then leads to the formation of further vortices. This process continues until the cluster size increases to an extent that the heat released during coagulation can no longer melt the cluster; so the aggregate starts to grow along the vortex core. Eventually, this results in elongated wires. Since the Bernoulli force is very short-ranged, the formation of larger particles during condensation leads to an increase in the particle-vortex attraction, and thus extending their lifetime in the trapped state, making it the prevalent process.[59] As a result, the main products of condensation are not spherical particles but long and thin filaments.

It was the structure and thicknesses of wires grown in bulk superfluid helium that led to the conclusion that the

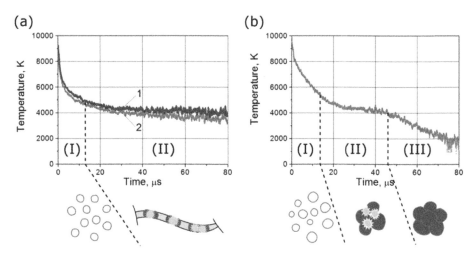

FIGURE 1.3 Temporal dependence of the temperature developed during tungsten coagulation in (a) superfluid and (b) normal helium. (a) The first stage is the hot stage of spherical cluster growth followed by the nanowire growth (stage II). (b) The first stage is the hot stage of spherical clusters growth; in the second stage, clusters coalesce. (Reprinted with permission from Ref. [56], American Chemical Society.)

nanowires pass through a stage in which molten nanoclusters coalesce. Such a suggestion seems counterintuitive, since the very high thermal conductivity of superfluid helium should eliminate any local overheating.[60] In a recent experiment, Gordon et al. investigated time-dependent thermal emission accompanying metal coagulation in He II, which showed huge local overheating of several thousand degrees.[56,61] Although superfluid helium possesses a high thermal conductivity, it can only maintain this for low heat flow. Otherwise, turbulence kicks in and suppresses the heat transfer, converting the superfluid helium into the normal fluid. The normal fluid has a much lower thermal conductivity, which then evaporates and forms an insulating envelope filled with low-pressure helium gas.

Time-resolved shadowgraph photography has been employed recently to study the laser ablation of copper and silver targets submerged in bulk normal and superfluid ^{4}He, allowing the dynamics of the vortex-assisted metal condensation to be investigated.[62] Following the laser ablation pulse, the plasma produced by ablation is confined by the surrounding solvent, causing a transient increase of the local temperature and pressure. The cavitation bubble surrounding the plasma then expands due to the high initial internal pressure, leading to a rapid reduction in both pressure and temperature inside the bubble. Since the expansion dynamics of the cavitation bubble is highly non-linear, there are shockwaves that pass into the liquid, which occur in both normal liquid helium and in superfluid helium. Therefore, when the bubble cools towards thermal equilibrium, two stages are involved, including both the gas-normal liquid and normal-superfluid. The sudden formation of large clusters occurs at the latter stage, where the fluid reaches the superfluid lambda transition.

In the experiments by Zurek and Kibble, the temperature of liquid helium was quickly reduced by a sudden pressure drop.[63,64] The rapid phase transition between the normal liquid and the superfluid provides an ideal system for observing vortex formation, which explains the formation of a thermally excited vortex near a superfluid phase transition. For laser ablation of metals in bulk superfluid helium, essentially a similar scenario occurs as the short laser pulse produces instant heating, and the system then undergoes rapid cooling by the surrounding superfluid helium. Hence, the vorticity observed during laser ablation of metal targets in superfluid helium can be interpreted by the Zurek-Kibble mechanism.

A recent review provides a thorough summary on the successes of density functional theory (DFT) in modeling the structure and dynamics of doped liquid helium droplets.[65] However, despite the success with DFT, a detailed theory of nanowire or filament growth in superfluid helium vortices does not exist. One of the remaining puzzles is the filament formation at the later stages, where individual nanowires become entangled and twisted together to form "ropes" and networks.[66] Many of the observed filaments and nanowires have rich structures that could not form from a single vortex. Furthermore, nanowires also form in normal fluid of helium at temperatures above 2.17 K. However, these are only small structures without networks or filaments, and the wires have a less smooth structure when compared with those formed in superfluid helium.[69]

1.2.2 Typical Bulk Superfluid Helium Apparatus

Figure 1.4 shows a typical bulk superfluid helium apparatus for the formation of nanowires by Gordon et al.[67] A metal target is submerged in superfluid helium, which is ablated by a pulsed laser through a pair of sapphire windows. At the base of the cell sits a TEM grid for the collection of metallic deposits, which can subsequently be removed from the liquid helium cryostat for imaging. In addition,

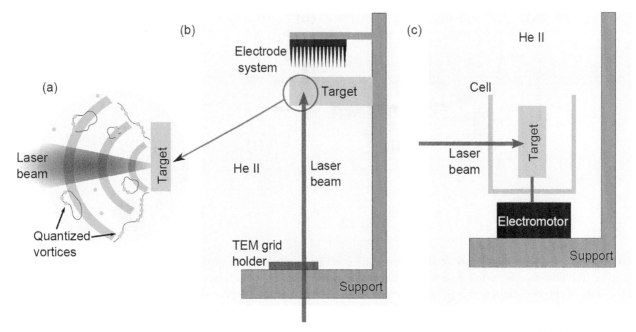

FIGURE 1.4 Apparatus for the formation of nanowires in superfluid helium by laser ablation of a metal target.[67,69] (a) A target onto which a pulsed laser is focused. The expansion of the gas and rapid cooling leads to the formation of quantized vortices. (b) A typical experimental cell immersed in liquid He II showing the target, electrode system, and a holder of TEM grids. (c) A new cell design.[69] The cell can be rotated by a motor.

an electrode system is positioned near the target which can be used to study the electrical conductivity of nanowires in liquid helium.[68]

A major drawback of using the "standard" laser ablation technique in the synthesis of nanowires is that the coagulation of the metal occurs very close to the plasma bubbles, adjacent to which the hydrodynamic instabilities can lead to structural distortion of the nanowires and thus a relatively short length. Consequently, a different cell design used mechanical rotation in superfluid helium to overcome this problem.[69] By using a miniaturized low-power electric motor in superfluid helium (see Figure 1.5c), an organized array of parallel quantized vortices can be generated by the rotation of superfluid helium inside the cryostat. The number of vortices generated by the rotation of the liquid adjacent to the target will exceed that by the laser pulse, provided that the rotation is sufficiently fast.[69] Unfortunately, this new design only worked for the duration of one set of measurements, as the motor behavior was unstable at 1.6 K, which led to the deformation of the components of the motor.

1.2.3 Nanostructures Formed in Bulk Superfluid Helium

Nanowires grown in bulk superfluid helium are generally densely packed due to the melting process, and can have crystalline, polycrystalline, and amorphous structures. Although relatively large spherical nanoparticles are often seen, the vast majority of the nanostructures formed are nanowires, which often entangle each other. Theoretical modeling by Volk et al. suggests that the large particles do not exist initially; instead, they are formed during the heating of the samples when they were removed from cold helium to ambient conditions prior to imaging,[70] in which nanoscale wires can fission into particles because of Rayleigh instability. Figure 1.5 shows typical nanostructures formed by laser ablation of metals, e.g., Au, Cu, and their alloys, in bulk superfluid helium.[71] For the latter, the growth mechanism is expected to be identical to that of pure metals due to the universal nature of processes occurring during the growth of nanostructures in superfluid helium.

Formation of ultrathin nanowires from refractory metals (e.g., Nb, Re, W, Mo) by laser ablation in bulk superfluid helium has also been explored. For these metals, the diameters of the wires are found to be different, e.g., 4.0, 2.0, and 2.5 nm for niobium, molybdenum, and tungsten, respectively,[67] which, clearly, is due to the critical particle sizes of different metals. Nanowebs have also been produced for other metals such as gold, platinum, and mercury, where the difference in diameters has also been observed. For instance, gold and platinum nanowires have diameters of 4.5 and 3.0 nm, respectively.[72] The observation of metal nanowires with different diameters provides circumstantial evidence for the mechanism proposed by Gordon et al.[56] As all the metal nanowire webs are electrically interconnected, they can potentially be used as electrodes in electrochemistry with no support.

Recent experiments have found that nanowires made from binary metal alloys may have more diverse structures than pure metals, such as periodic alternation of elemental composition and core–shell nanostructures, produced by laser ablation of metal alloy targets.[73] The growth of

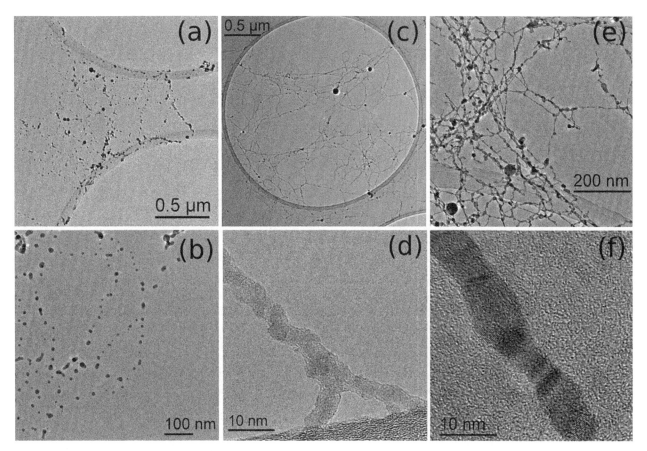

FIGURE 1.5 TEM images of sediments formed following the laser ablation of metal targets submerged in superfluid helium for (a, b) silver; (c, d) copper; (e, f) their alloys. (Reprinted with permission from Ref. [71], Royal Society of Chemistry (Great Britain).)

nanowires starts with the formation of spherical "proto-clusters" with diameters larger than 2 nm, which is large enough for spatial phase separation to occur. This is due to the fact that the phase diagram of a nanocluster correlates with surface tension rather than with free energy. Before the protoclusters stick together to form nanowires, they have a core–shell structure, with the shell material being more fusible than that of the core. When these clusters stick together in the vortex, the shell material can then fuse to allow interconnection for the core and shell respectively, leading to the formation of core–shell structures or periodic structures, respectively.

1.3 Synthesis of Nanomaterials in Superfluid Helium Droplets

Helium droplets possess important properties for the formation of nanostructures, including (i) the superfluidity, which allows the free motion of atoms and molecules inside the droplets and subsequent aggregation into nanoparticles; (ii) a very low temperature (0.37 K), which helps in the incorporation of volatile substances into the nanomaterials; (iii) the chemical inertness of helium, which allows the formation of nanomaterials with highly reactive species; (iv) the sequential pickup of different types of materials,

easing the formation of core–shell nanostructures. Other important properties include quantized vortices, which, as in the case of bulk superfluid helium, allow the formation of 1-D nanostructures. The following sections discuss how these properties can be exploited for the formation of both spherical and 1-D nanostructures.

1.3.1 Basics of Helium Droplets

The formation of superfluid helium droplets requires a low temperature in order to reach the He II phase. Helium droplets can be formed from bulk superfluid helium, for example, by agitation using a piezoelectric transducer,[74] laser levitation,[75] or magnetic field levitation.[76] However, in practice these methods are rarely used because the droplets formed are very large and are hard to control. In the context of nanomaterial synthesis, it is extremely challenging to separate and dope them with atoms and molecules. Instead, the commonly used method for forming superfluid helium droplets is expansion of pre-cooled and pressurized helium into the vacuum (see Figure 1.6 for a typical helium droplet apparatus), which is more cost-effective and is also a method consuming less helium. This generally involves forcing helium gas or liquid through a pinhole nozzle, often with a diameter of 2–10 μm. The droplets are then formed either by the condensation of helium gas or by the breakup

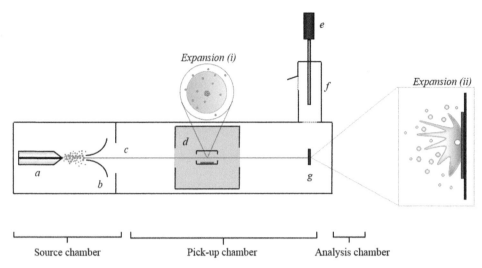

FIGURE 1.6   Typical helium droplet apparatus for the fabrication of nanoparticles. (a) Nozzle (5 μm diameter), (b) skimmer (with an aperture of 0.5 mm diameter), (c) helium droplet beam, (d) metal evaporator, (e) translation arm, (f) load-lock chamber, and (g) deposition target to align the helium droplet beam. The expanded view (i) illustrates addition of metal atoms to the helium droplets as they pass through the metal vapor. The expanded view (ii) shows the collision of nanoclusters formed in helium droplets with a surface.

of a liquid droplet beam. After their formation, the helium droplets cool rapidly by evaporative loss of helium atoms until a steady-state temperature of 0.37 K is reached.[25] They are then collimated into a beam by passing through a skimmer and enter the pickup region where dopants can be added.

Depending on the nozzle temperature and the stagnation pressure, the formation of helium droplets undergoes one of three different expansion regimes,[35,77] each having different average droplet sizes and different size distributions. Between 10 and 30 K, it is the subcritical expansion zone, where helium gas ejects from the nozzle as a gas and expands along the isentropes,[78] which crosses the gas-liquid phase separation line from the gas phase side (see Figure 1.7). The droplets are then formed by the condensation of atoms via three body collisions, with typical mean sizes $<N_{\mathrm{He}}> = 10^3$–10^4, where N_{He} is the number of helium atoms within a droplet of helium. From 6 to 10 K, the expansion is in the supercritical region, where the isentrope crosses the phase separation line from the liquid side. In this case, liquid helium emerges in the nozzle, which breaks into helium droplets. The droplets formed in this regime are much larger, with $<N_{\mathrm{He}}> = 10^4$–10^7. With further cooling of helium gas to a lower temperature, e.g., below 6 K, the curve plateaus and the expansion is subsonic. The phase separation line has already been crossed before the liquid leaves the nozzle. Rather than helium gas, liquid helium appears from the nozzle as a continuous liquid beam, which then breaks up due to Rayleigh instability and forms very large but mono-dispersed droplets containing over 10^{10} atoms.[78] Droplets nearly visible to the naked eye can be made at temperatures near 0 K, with diameters close to the diameter of the nozzle orifice. These very large droplets are important for the formation of sizeable nanoclusters as the aggregation of dopants in helium droplets releases

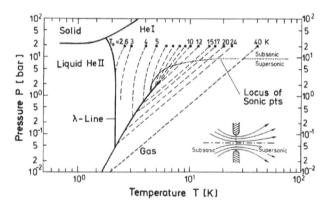

FIGURE 1.7   Expansion isentropes (dotted lines) of ^{4}He at the various expansion regimes. The stagnation pressure is 20 bar. (Reprinted with permission from Ref. [14], American Institute of Physics.)

substantial energy that needs to be dissipated, i.e., by evaporation of helium atoms.

In the pickup zone, helium droplets can acquire gas-phase dopants, as illustrated in Figure 1.6. No evaporation is necessary for gas samples and while the vapor above liquids can also be used. For solid samples, a resistively heated oven evaporator is often needed to produce vapors. Compared with bulk superfluid helium, helium droplets are far more efficient at retaining impurities because they have a boundary that can confine the dopants. After being captured by helium droplets, most species will migrate and reside in the interior of the droplets, somewhere near the center of the droplets where the potential energy is the lowest.[79] The exceptions are alkali metal atoms and small alkali clusters, as well as some alkaline earth metals, which tend to reside on the surface of helium droplet due to short-range repulsion between the diffuse valence electrons and helium atoms.[80] However, as the size of the clusters increases, the London force between the cluster and helium droplets will increase, which might surmount the short-range repulsion. Once such a critical size is reached, the metal clusters will submerge and be located inside the droplets, e.g., $n \geq 21$ for Na_n clusters[81] and $n \geq 78$ for K_n clusters.[82]

The pickup probability is near unity on collision between a dopant atom/molecule and a helium droplet. Consequently, the total number of atoms/molecules added to the droplets is determined by the number of collisions, which follows a Poisson distribution:

$$P_k(z) = \frac{z^k}{k!} \exp(-z) \qquad (1.2)$$

Here, k is the number of species picked up by the droplet; z is the collision parameter defined by $z = \sigma n l$, where σ is the geometric cross-section of the droplets, n is the number density of dopants, and l is the length of the pickup region. The number of dopants picked up by the droplets can be controlled by the size of the droplets and the number density of gas-phase dopants in the pickup region. Clearly, the former depends on the stagnation pressure and the nozzle temperature of the helium droplet source while the latter can be controlled through fine tuning of the partial pressure(s) of gas-phase atoms/molecules in the pickup region.

1.3.2 Formation of Nanoparticles in Helium Droplets

The nanocluster growth process starts with the addition of a single atom/molecule to the droplets. As a dopant enters the droplet, its kinetic energy will be instantly removed and its internal energy, such as vibrational and rotational energy, can also be removed. Experiments showed that the cooling rate can be as high as 10^{16} K/s.[34] Hauser et al. modeled the formation of metal clusters inside helium droplets by using DFT and molecular dynamic simulations.[83] The simulations suggested that a metal atom will follow a planar rosette-like trajectory, rather than a straight-line path to the center of

the droplet, until aggregation occurs. Continuous addition of atoms causes the evaporation of helium droplets in order to dissipate the kinetic energy of the metal atoms and the binding energy released in the aggregation. Meantime, the droplets decrease in size, accounting for energy released at a rate of 5 cm^{-1} energy/atom.

Formation of Large Clusters in Helium Droplets

Formation of metallic clusters in helium droplets was first carried out in 1996. By passing helium droplets through a heated oven containing silver, indium, or europium, Bartelt et al. showed that metal clusters with masses up to 2,000 amu could be formed by using mass spectrometry.[84] UV-depletion spectroscopy of silver clusters formed in helium droplets was then measured, which was very similar to the spectrum of silver clusters in bulk liquid helium. This suggested that these particles reside in the interior of the droplets. Later on, Diederich et al. demonstrated that larger magnesium clusters consisting of over 1,000 atoms can be formed.[85] Following these early experiments, very large molecular clusters of $(NH_3)_n$ ($n = 2 - 10^4$), $(CH_4)_n$ ($n = 1 - 4 \times 10^3$), and $(C_3H_6)_n$ ($n = 10 - 10^4$) were produced in helium droplets and were investigated by spectroscopy.[86,87]

The ability to form sizeable clusters has stimulated novel nanoscience by using superfluid helium droplets as the nanoreactors. The first experiment intending to form nanoparticles in helium droplets was performed by Mozhayskiy et al., who added both Ag and Au to helium droplets and formed binary metal clusters containing $\sim$500 metal atoms and investigated these clusters by spectroscopy.[88] The first deposition experiment was performed by Loginov et al., who employed TEM to image silver clusters containing 300–6,000 atoms after they were deposited on a carbon surface (Figure 1.8).[19] In a different experiment, photoabsorption spectra were recorded for Ag_n clusters with the average number $n = 6, 60, 300, 2,000,$ and $6,000$.[89] All of the spectra featured a plasmon peak around 3.6–3.8 eV, which is consistent with the formation of spherical Ag particles. However, an additional broad absorption band extending

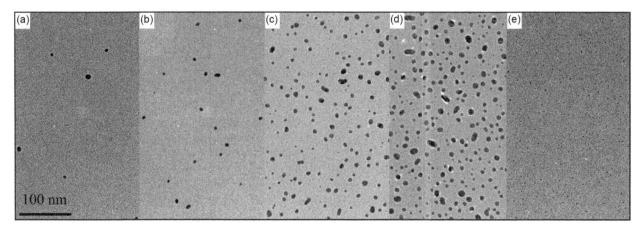

FIGURE 1.8 TEM images of Ag nanoparticles formed in superfluid helium droplets. The nanoparticle density increases from image (a) to (d) as the deposition time is increased. (Reprinted (adapted) with permission Ref. [19], American Chemical Society.)

down to 0.5 eV was observed for large clusters. This additional peak was unexpected and was interpreted by a multi-center growth mechanism, i.e., the growth of multiple small clusters at different sites within the droplets before they migrate and coagulate at the center of the droplet. Multi-center growth occurs if the time between the successive pickup events is shorter than the time required for a silver atom to reach the center of the droplets and aggregate with the existing cluster. Eventually, these small particles migrate to the center of the droplets and form a single loosely packed cluster, exhibiting broadband absorption at low frequency region. As each small cluster can have its own lattice orientation, multi-center growth can result in polycrystalline structure. This was confirmed by Ernst's group, who investigated Ag and Au nanoparticles by use of high-resolution TEM and observed polycrystalline structures.[90,91]

Formation of Core–Shell Nanoparticles in Helium Droplets

The synthesis of nanomaterials using superfluid helium droplets is not limited to the formation of spherical nanoparticles containing only one type of materials. Other possibilities are core–shell nanoparticles, alloy nanoparticles, and nanowires. Besides, it is also possible to synthesize nanoparticles with new properties, e.g., magnetic properties and catalytic properties. Core–shell nanoparticles consisting of a core surrounded by a shell made from a different material are desirable in many applications because they offer new electronic, optical, magnetic, and chemical properties.[92] By varying the diameter of the core and the thickness of the shell, the properties of core–shell nanoparticles can be tuned,[93] which is an added advantage of the core–shell structure. In addition, the surface coating can also be crafted so that it can be functionalized, making the nanoparticles compatible with different media. For example, coating nanoparticles with materials such as silica or gold is a common strategy for applications of nanoparticles in medical science.[94]

Fabrication of core–shell nanoparticles with a well-defined core and shell structure is often challenging using conventional synthetic methods, such as wet-chemical synthesis and high-temperature techniques. With these methods, reactions generally take place at room temperature or above, and so inter-diffusion between the core and the shell is difficult to avoid. Furthermore, in wet-chemical synthesis the overall structure is often loosely packed, which requires further annealing in order to produce particles with reproducible properties.[95] Such treatments can increase the inter-diffusion between core and shell materials. In the helium droplet, the strategy for forming core–shell structures is straightforward: when helium droplets sequentially pass through two or more ovens containing different types of materials, these materials will be added to the droplets following the geometric order. A core will be first grown in the droplets, and these then travel onwards to the next pickup cell and acquire a different material. Naturally, this

atom-by-atom growth allow new materials to grow on the surface of the existing clusters, resulting in core–shell structures. As the entire procedure takes place at the very low temperature of superfluid helium and the deposition is often at room temperature, inter-diffusion is minimized. In addition, the growth of core and shell(s) can be independently controlled, i.e., through the partial pressure in different pickup cells, and therefore their thicknesses can be independently varied. Moreover, nearly any material can be used as the core due to the very low temperature of superfluid helium, at which any other substances become solid. These include solid-phase materials, as well as substances that are gases or liquids at room temperature.

A proof-of-concept experiment for the formation of core–shell nanostructures was first carried out by Shepperson et al.,[20] who formed H_2O/X (X = N_2, O_2, CO_2 and CO, etc.) by sequential addition of water and another molecule to helium droplets, and the core–shell structures were deduced from mass spectrometric measurements. The same group also obtained the first evidence for the formation of core–shell nanoparticles in helium droplets in 2013.[21] By sequentially passing helium droplets through a pickup cell containing Ni vapor and a second cell with gold vapor, Ni/Au core–shell nanoparticles were formed, which were investigated by XPS. No shift of the photon energy was observed for the Ni 2p and Au 4f core levels in the XPS spectra when compared to pure metals, which excluded the possibility of alloying between Ni and Au. This confirmed a well-defined core and shell with a boundary between the two metals. Later on, Ag/Si core–shell nanoparticles with a distinct core and shell structure were reported by Ellis and Yang.[96] On TEM images a dark Ag core and a transparent Si shell are apparently visible due to the difference in electron scattering cross-sections between Ag and Si. Essentially, the materials added later grow on the surface of existing nanoclusters inside the droplets: by this means even core-multiple shell nanoparticles can be formed, and each layer can be independently controlled.

Formation of Magnetic Nanoparticles

The unique properties possessed by superfluid helium droplets make it possible to develop novel nanomaterials with unusual structures and perhaps unexpected properties. In contrast to noble metals such as Ag and Au, strong exchange interactions exist between high-spin transition metal atoms in addition to metallic bonding. Although the exchange interactions are still weaker than the metallic bonding, it is found to play a critical role in the growth of magnetic nanoparticles inside superfluid helium droplets.

Yang et al. recently synthesized Cr nanoparticles in superfluid helium droplets and investigated the magnetic properties for the first time. Cr is an antiferromagnetic material, which favors anti-parallel alignment of the atomic spins for adjacent atoms by its nature, leading to near-zero magnetization in the bulk phase. When grown in superfluid helium droplets, however, a surprisingly high magnetic moment was observed.[38] For example, Cr nanoparticles with a diameter of 2.4 nm possess a magnetic moment as high as 1.29 μ_B/atom at 5 K, and 0.45 μ_B/atom at room temperature. The magnetization saturates at a field of ~0.5 tesla at room temperature, clearly indicating that the Cr nanoparticles formed are ferromagnetic (see the hysteresis curves in Figure 1.9).

It is well-known that the magnetic properties of nanoclusters vary with their structures, for example, by purposely introducing defects and/or lattice strain.[97] For Cr nanoparticles grown in superfluid helium, the high ferromagnetic moments were attributed to defects, as evident from high-resolution scanning transmission electron microscopy (STEM). Unlike noble metal nanoparticles grown in helium droplets,[21] Cr nanoparticles were found to be extensively disordered at the atomic level. This is because of the rapid continuous cooling offered by superfluid helium, which instantly removes the excess energy released during the aggregation of Cr atoms, making the exchange interactions predominant in the absence of thermal effects. The Cr nanoparticles built atom by atom then have uneven interatomic distances. In the same study, DFT and Monte Carlo simulations were also performed and a frustrated aggregation mechanism was proposed.

Formation of Catalytic Nanoparticles

Helium droplets can be used to produce nanoparticles with diameters <5 nm, which are ideal for catalysis. Under subsonic expansion conditions, large mono-dispersed helium droplets can be obtained, which can be used to produce nanoparticles with a narrow size distribution. This is particularly important for catalysis, where the particle sizes drastically impact on the catalytic activities.

Wu et al. have recently shown that the size of Au nanoparticles on a titanium dioxide support can be tuned over a very narrow range of 0.75–3 nm by controlling the helium droplet size.[98] The $Au_N@TiO_2$ nanoparticles studied were shown to be stable up to 473 K and could potentially be used to catalyze CO oxidation reactions. Stable gold catalysts have been of increasing interest in recent years as

gold nanoparticles have displayed superior catalytic activity compared to the conventional platinum group, especially when the diameter of the nanoparticles is less than 5 nm.[99] Using helium droplets the nanoparticles can be prepared without surfactants or capping ligands, which might otherwise impede the catalytic performance.[100] Therefore, helium droplets hold promise as a route to prepare stable and tunable nanoparticles for catalysis.

1.3.3 Quantized Vortices and the Growth of Nanowires Using Superfluid Helium Droplets

Given the presence of quantized vortices in bulk superfluid helium, the question arises as to whether these might exist in helium droplets. Theoretical modeling has suggested that quantized vortices can occur in helium droplets[101–103] but the first evidence for their existence was only obtained in 2012.[52] Gomez et al. added silver atoms to helium droplets with mean diameters of 100, 300, and 1,000 nm (composed of 10^7–10^{10} helium atoms). Subsequently, the Ag-doped droplets collided with a thin carbon film substrate, and Ag nanomaterials were deposited onto the substrate for TEM imaging. Silver nanoparticles obtained from helium droplets with a diameter of 100 nm or below are mostly spherical, whereas larger droplets produced track-like deposits consisting of elongated Ag segments (nanorods). The results showed that the shapes and overall lengths of the tracks varied with the helium droplet size, suggesting that Ag atoms are pinned to quantized vortices spanning the helium droplets. In a similar experiment, Spence et al. added Ag to helium droplets at a lower doping rate and observed spherical nanoparticles aligned in a chain.[104] Later on, Volk et al. performed both experiments and theoretical modeling on the silver nanowires, suggesting that the ultrathin nanowires can fragment due to Rayleigh instability, which then melt into spheres as in the case of bulk superfluid helium.[70]

Direct imaging of quantized vortices in helium droplets was recently achieved by Gomez et al. using ultrafast X-ray diffraction.[105] In this experiment, xenon atoms were added to large helium droplets with radius ranging from 100 to 1,000 nm, and these were then illuminated by a free electron laser source. Bragg patterns from xenon clusters trapped in the vortex cores are consistent with the presence of an array of quantum vortices in the droplets. Surprisingly, the vortex densities in helium droplets appear to be ~5 orders of magnitude larger than those observed in bulk liquid helium.

Quantized vortices offer the prospect of creating a range of 1-D nanostructures in helium droplets. In a recent experiment by Latimer et al., quantized vortices were shown to lead to ultrathin 1-D nanowires made from elements such as Au, Ag, Si, Ni, and Cr. Both the length and the diameter of the nanowires can be controlled, i.e., by the size of the droplets and the partial pressure in the pickup region.[53] The diameters of the nanowires are generally in the region of 3–5 nm, and the length can be increased at higher doping

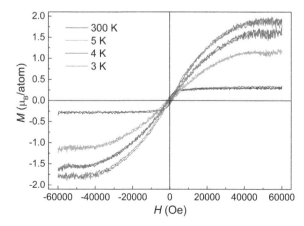

FIGURE 1.9 Hysteresis curves of Cr nanoparticles.[38]

rates. This technique has also been employed by Thaler et al. to grow Au/Ag bimetallic core–shell nanowires.[106] Both Au and Ag were evaporated by resistive heating and were then added sequentially to the helium droplets. The resulting nanowires were then investigated by energy-dispersive X-ray spectroscopy (EDX) to confirm the core–shell structure.

1.4 Perspectives and Challenges

Superfluid helium, with its unique properties, has opened up many exciting opportunities for novel nanoscience. Compared with bulk superfluid helium, helium droplets offer a much higher degree of control and flexibility in the selection of materials, and also ease the formation of core–shell and core-multiple-shell nanoparticles. Many of these are highly challenging and are hard to access by conventional synthetic methods, making superfluid helium a uniquely powerful tool for nanoscience and nanotechnologies.

Despite the fascinating opportunities opened up by superfluid helium, there exist some significant challenges for experiments in both bulk superfluid helium and helium droplets. Bulk superfluid helium has so far been used to develop nanowires and nanowebs by laser ablation of metal targets, yet the underlying physics for the generation of quantized vortices and subtle reactions for the formation of 1-D nanostructures remains debatable. The prerequisite for the formation of quantized vortices is the generation of angular momentum but it is unclear how laser ablation can generate this for helium atoms near the focus point of the laser light, and how the phase transition at the normal liquid and superfluid interface may assist the process. These are the key questions that need to be answered from fundamental aspects, which require both experimental and theoretical effort. Another major weakness of using superfluid helium in the fabrication of nanomaterials is associated with the relatively low production rates of nanoparticles and nanowires, and the high costs of both bulk liquid helium and helium droplet apparatus. For helium droplets, the production rate is restricted by the flux of the droplets; while for bulk superfluid helium, the intrinsic limitation is the evaporation of metals in superfluid helium without significant disturbance to dismiss the superfluidity. Both approaches are currently restricted to a production of sub-milligram quantities per day, and they all require sophisticated high-vacuum and low-temperature apparatuses. These, in addition to the scarceness of helium, make the approaches non-economical for many possible applications. However, this can be resolved if scale-up approaches can be developed and thus the practical impact of superfluid helium in nanoscience and nanotechnology can be promoted.

To close this chapter, we will highlight some new possibilities offered by superfluid helium. The first is the fabrication of core-multiple-shell nanostructures with distinct boundaries between layers using helium droplets. These nanostructures are extremely challenging to make with conventional techniques, and yet may offer novel properties that are important for applications spanning spintronics, photonics, electronics, and biomedical science. From a fundamental point of view, the fabrication of such new nanostructures and investigation of the associated properties are important to develop new theoretical models, which in turn will stimulate the design of new nanomaterials and the development of next-generation technologies. Another possibility is to form nanostructures with structures different from the bulk counterparts, as well as those made from conventional synthetic approaches, as manifested in the formation of ferromagnetic Cr nanoparticles. In this process the ultrahigh thermal conductivity of the very cold superfluid helium plays a critical role to remove thermal energy released during the growth of nanomaterials, hence accentuates other effects which may often be overwhelmed by thermal reactions. The resulting meta-stable structures may offer new opportunities for applications. Consequently, superfluid helium technology can impact on the fundamentals of nanoscience and can potentially open up new possibilities in developing next-generation nanotechnologies.

References

1. S. K. Murthy, *Intl. J. Nanomed.*, 2007, 2, 129.
2. X. Xia, Y. Xia, *Front. Phys.*, 2013, 9, 378.
3. A. N. Shipway, E. Katz, I. Willner, *Chem. Phys. Chem.*, 2000, 1, 18.
4. T. J. Kempa, R. W. Day, S. Kim, H. Park, C. M. Lieber, *Energy Environ. Sci.*, 2013, 6, 719.
5. E. C. Corbos, P. R. Ellis, J. Cookson, V. Briois, T. I. Hyde, G. Sanka, P. T. Bishop, *Catal. Sci. Tech.*, 2013, 3, 2934.
6. A. J. Gellman, N. Shukla, *Nat. Mater.*, 2009, 8, 87.
7. N. A. Mashayekhi, Y. Y. Wu, M. C. Kung, H. H. Kung, *Chem. Commun.*, 2012, 48, 10096.
8. A. Popov, J. Lademann, A. Priezzhev, R. Myllyla, *J. Biomed. Opt.*, 2005, 10, 064037.
9. R. Kessler, *Environ. Health Perspect.*, 2011, 119, A120.
10. S. Fedrigo, W. Harbich, J. Buttet, *Phys. Rev. B*, 1998, 58, 7428.
11. H. Jödicke, R. Schaub, A. Bhowmick, R. Monot, J. Buttet, W. Harbich, *Rev. Sci. Instrum.*, 2000, 71, 2818.
12. V. T. A. Oiko, T. Mathieu, L. Cao, J. Liu, R. E. Palmer, *J. Chem. Phys.*, 2016, 145, 166101.
13. R. J. Donnelly, *J. Phys. Condens. Matter*, 1999, 11, 7783.
14. L. F. Gomez, E. Loginov, R. Sliter, A. F. Vilesov, *J. Chem. Phys.*, 2011, 135, 154201.
15. R. Sliter, L. F. Gomez, J. Kwok, A. Vilesov, *Chem. Phys. Lett.*, 2014, 600, 29–33.
16. S. Grebenev, J. P. Toennies, A. F. Vilesov, *Science*, 1998, 279, 2083–2086.
17. M. Hartmann, R. E. Miller, J. P. Toennies, A. F. Vilesov, *Science*, 1996, 272, 1631.

18. S. Goyal, D. L. Schutt, G. Scoles, *Phys. Rev. Lett.* 1992, 69, 933–936.

19. E. Loginov, L. F. Gomez, A. F. Vilesov, *J. Phys. Chem. A*, 2011, 115, 7199–7204.

20. J. Liu, B. Shepperson, A. Ellis, S. Yang, *Phys. Chem. Chem. Phys.*, 2011, 13, 13920–13925.

21. A. Boatwright, C. Feng, D. Spence, E. Latimer, C. Binns, A. M. Ellis, S. Yang, *Faraday Discuss.*, 2013, 162, 113–124.

22. H. K. Onnes, K. N. A. V. Wetenschappen, *Proc. Ser. B Phys. Sci.*, 1908, 11, 168.

23. L. Pitaevshii, *Bose-Einstein Condensation*, Clarendon Press, Oxford, 2003, pp. 44–49, 82–86, and 118–122.

24. A. Schmitt, *Introduction to Superfluidity*, Springer International Publishing, Switzerland, 2015, p. 64.

25. M. Hartmann, R. E. Miller, J. P. Toennies, A. Vilesov, *Phys. Rev. Lett.*, 1995, 75, 1566.

26. P. Kapitza, *Nature*, 1938, 141, 74.

27. J. F. Allen, A. D. Misener, *Nature*, 1938, 141, 75.

28. S. Grebenev, M. Hartmann, M. Havenith, B. Sartakov, J. P. Toennies, A. F. Vilesov, *J. Chem. Phys.*, 2000, 112(10), 4485–4495.

29. M. V. Patel, A. Viel, F. Paesani, P. Huang, K. B. Whaley, *J. Chem. Phys.*, 2003, 118, 5011.

30. J. F. Allen, H. Jones, *Nature*, 1938, 141, 234–244.

31. W. H. Keesom, A. P. Keesom, B. F. Saris, *Physica*, 1938, 5, 281.

32. W. H. Keesom, A. P. Keesom, *Physica*, 1936, 3, 359.

33. M. Lewerenz, B. Schilling, J. P. Toennies, *J. Chem. Phys.*, 1995, 102, 8191–8207.

34. W. K. William, B. E. Applegate, J. Sztáray, B. Sztáray, T. Baer, R. J. Bemish, R. E. Miller, *J. Am. Chem. Soc.*, 2004, 126, 11283–11292.

35. J. P. Toennies, A. F. Vilesov, *Angew. Chem. Int. Ed. Engl.*, 2004, 43, 2622–2648.

36. K. Nauta, R. E. Miller, *Science*, 1999, 285, 1895.

37. K. Nauta, R. E. Miller, *Science*, 2000, 287, 293–295.

38. S. Yang, C. Feng, D. Spence, A. M. A. A. Al Hindawi, E. Latimer, A. M. Ellis, C. Binns, D. Peddis, S. S. Dhesi, L. Zhang, Y. Zhang, K. N. Trohidou, M. Vasilakaki, N. Ntallis, I. Maclaren, F. M. F. De Groot, *Adv. Matt.*, 2017, 29, 1604277.

39. D. V. Osborne, *Proc. Phys. Soc. Lond. A,* 1950, 63, 909–912.

40. E. L. Andronikashvili, and I. P. Kaverkin, *Sov. Phys. JETP* 1955, 1, 174–176.

41. L. Onsager, *Nuovo Cimento Suppl.*, 1949, 6, 249–250.

42. R. P. Feynman, Application of quantum mechanics to liquid helium, Chapter 11. in: *Progress in Low Temperature Physics*, Vol. 1, C. J. Gorter ed., North-Holland Publishing Co., Amsterdam, 1955, pp. 17–53.

43. R. J. Donnelly, Discovery and properties of quantised vortices in helium II. in: *Advances in Cryogenic Engineering*, Vol. 35, R. W. Fast ed., Springer Science and Business Media, New York, 1990, pp. 25–34.

44. R. J. Donnelly, Onsager's quantisation of circulation in superfluid helium. in: *Lars Onsager*, P. C. Hemmer, H. Holden, S. Kjelstrup Ratkje ed., World Scientific Publishing Co., Singapore, 1996, pp. 693–696.

45. R. E. Packard, T. M. Sanders, *J. Phys. Rev. A*, 1972, 6, 799.

46. R. E. Packard, E. J. Yarmchuk, *J. Phys. Rev. Lett.*, 1979, 43, 214.

47. R. J. Donnelly, *Quantised Vortices in Helium II*, Cambridge University Press, New York, 2005, pp. 1–15.

48. G. P. Bewley, D. P. Lathrop, K. R. Sreenivasan, *Nature*, 2006, 441, 588.

49. G. W. Rayfield, F. Reif, *Phys. Rev.*, 1965, 137, A1194.

50. P. Rosenbusch, V. Bretin, J. Dalibard, *Phys. Rev. Lett.*, 2002, 89, 200403.

51. D. Kivotides, Y. A. Sergeev, C. F. Barenghi, *Phys. Fluids*, 2008, 20, 055105.

52. L. F. Gomez, E. Loginov, A. F. Vilesov, *Phys. Rev. Lett.*, 2012, 108, 155302.

53. E. Latimer, D. Spence, C. Feng, A. Boatwright, A. M. Ellis, S. Yang, *Nano Lett.*, 2014, 14, 2902–2906.

54. E. B. Gordon, A. V. Karabulin, V. I. Matyushenko, V. D. Sizov, I. I. Khodos, *Chem. Phys. Lett.,* 2012, 519–520, 64–68.

55. V. Lebedev, P. Moroshkin, B. Grobety, E. B. Gordon, A. Weis, *J. Low Temp. Phys.* 2011, 165, 166–176.

56. E. B. Gordon, A. V. Karabulin, M. I. Kulish, V. I. Matyushenko, M. E. Stepanov, *J. Phys. Chem.*, 2017, *121* (48), 9185–9190.

57. E. B. Gordon, A. V. Karabulin, V. I. Matyushenko, V. D. Sizov, I. I. Khodos, *J. Exp. Theor. Phys.*, 2011, 112, 1061.

58. E. B. Gordon, A. V. Karabulin, V. I. Matyushenko, V. D. Sizov, I. I Khodos, *High Energy Chem.*, 2014, 48, 206.

59. Y. A. Sergeev, C. F. Barenghi, *J. Low Temp. Phys.*, 2009, 157, 429–475.

60. R. J. Donnelly, *Quantized Vortices in Helium II*, Cambridge University Press, Cambridge, 1991.

61. E. B. Gordon, M. I. Kulish, A. V. Karabulin, V. I. Matyushenko, *J. Low Temp. Phys.*, 2017, 43, 1086–1093.

62. E. Popov, M. Mammetkuliyev, J. Elorantaa, *J. Chem. Phys.*, 2013, 138, 204307.

63. W. H. Zurek, *Nature (London)*, 1985, 317, 505.

64. T. W. B. Kibble, *Nature (London)*, 1985, 317, 472.

65. F. Ancilotto, M. Barranco, F. Coppens, J. Eloranta, N. Halberstadt, A. Hernando, D. Mateo, M. Pi, *Int. Rev. Phys. Chem.*, 2017, 36, 621.

66. P. Moroshkin, V. Lebedev, B. Grobety, C. Neururer, E. B. Gordon, A. Weis, *Europhys. Lett.*, 2010, 90, 34002.

67. E. B. Gordon, A. V. Karabulin, V. I. Matyushenko, V. D. Sizov, I. I. Khodos, *Laser Phys. Lett.*, 2015, 12, 096002.

68. E. B. Gordon, A. V. Karabulin, V. I. Matyushenko, V. D. Sizov, I. I. Khodos, *Appl. Phys. Lett.*, 2012, 101, 052605.

69. E. B. Gordon, M. I. Kulish, A. V. Karabulin, V. I. Matyushenko, E. V. Dyatlova, A. S. Gordienko, M. E. Stepanov, *Low Temp. Phys.*, 2017, 43, 1055–1061.

70. A. Volk, D. Knez, P. Thaler, A. W. Hauser, W. Grogger, F. Hofer, W. E. Ernst, *Phys. Chem. Chem. Phys.*, 2015, 17, 24570.

71. E. B. Gordon, A. Karabulin, V. I. Matyushenko, V. Sizovc, I. I. Khodos, *Phys. Chem. Chem. Phys.*, 2014, 16, 25229–25233.

72. E. B. Gordon, A. V. Karabulin, A. A. Morozov, V. I. Matyushenko, V. D. Sizov, I. I. Khodos, *J. Phys. Chem. Lett.*, 2014, 5, 1072–1076.

73. E. B. Gordon, A. V. Karabulin, V. I. Matyushenko, S. A. Nikolaev, I. I. Khodos, *Mendeleev Commun.*, 2017, 27, 4, 385–386.

74. H. Kim, K. Seo, B. Tabbert, G. A. Williams, *Europhys. Lett.*, 2002, 58, 395.

75. M. A. Weilert, D. L. Whitaker, H. J. Maris, G. M. Seidel, *J. Low Temp. Phys.*, 1995, 98, 17.

76. M. A. Weilert, D. L. Whitaker, H. J. Maris, G. M. Seidel, *Phys. Rev. Lett.*, 1996, 77, 4840.

77. H. Buchenau, E. L. Knuth, J. Northby, J. P. Toennies, C. Winkler, *J. Chem. Phys.*, 1990, 92, 6875–6889.

78. R. E. Grisenti, J. P. Toennies, *Phys. Rev. Lett.*, 2003, 90, 234501.

79. S. Yang, A. Ellis, Clusters and nanoparticles in superfluid helium droplets: fundamentals, challenges and perspectives. in: *Nanodroplets*, vol. 18, Z. M. Wang ed., Springer, New York, 2014, pp. 237–264.

80. S. Yang, A. M. Ellis, *Chem. Soc. Rev.*, 2013, 42, 472–484.

81. C. Stark, V. V. Kresin, *Phys. Rev. B*, 2010, 81, 085401.

82. L. An der Lan, P. Bartl, C. Leidlmair, H. Schöbel, R. Jochum, S. Denifl, T. D. Märk, A. M. Ellis, P. Scheier, *Phys. Rev. B*, 2012, 85, 115414.

83. A. W. Hauser, A. Volk, P. Thaler, W. E. Ernst, *Phys. Chem. Chem. Phys.*, 2015, 17, 10805–10812.

84. A. Bartelt, J. Close, F. Federmann, N. Quaas, J. Toennies, *Phys. Rev. Lett.*, 1996, 77, 3525–3528.

85. T. Diederich, T. Döppner, T. Fennel, J. Tiggesbäumker, K. H. Meiwes-Broer, *Phys. Rev. A*, 2005, 72, 023203.

86. M. N. Slipchenko, B. G. Sartakov, A. F. Vilesov, *J. Chem. Phys.*, 2008, 128, 134509.

87. M. N. Slipchenko, H. Hoshina, D. Stolyarov, B. G. Sartakov, A. F. Vilesov, *J. Phys. Chem. Lett.*, 2016, 7, 47–50.

88. V. Mozhayskiy, M. N. Slipchenko, V. K. Adamchuk, A. F. Vilesov, *J. Chem. Phys.*, 2007, 127, 094701.

89. E. Loginov, L. F. Gomez, N. Chiang, A. Halder, N. Guggemos, V. V. Kresin, A. F. Vilesov, *Phys. Rev. Lett.*, 2011, 106, 233401.

90. P. Thaler, A. Volk, D. Knez, F. Lackner, G. Haberfehlner, J. Steurer, M. Schnedlitz, W. Ernst, *J. Chem. Phys.*, 2015, 143, 134201.

91. A. Volk, P. Thaler, M. Koch, E. Fisslthaler, W. Grogger, W. E. Ernst, *J. Chem. Phys.*, 2013, 138, 214312.

92. R. G. Chaudhuri, S. Paria, *Chem. Rev.*, 2012, 112, 2373–2433.

93. J. Kossut, *Nat. Mater.*, 2009, 8, 8–9.

94. S. Sabale, P. Kandesar, V. Jadhav, R. Komorek, R. K. Motkuri, X. Yu, *Biomater. Sci.* 2017, 5, 2122–2225.

95. J. Hughey, NNIN REU Research Accomplishments, 2005, 56.

96. A. Ellis, and S. Yang, *Sci. Lett. J.* 2015, 5, 225.

97. S. H. Baker, M. Roy, S. C. Thornton, M. Qureshi, C. Binns, *J. Phys. Condens. Matter*, 2010, 22, 385301.

98. Q. Wu, C. J. Ridge, S. Zhao, D. Zakharov, J. Cen, X. Tong, E. Connors, D. Su, E. A. Stach, C. M. Lindsay, A. Orlov, *J. Phys. Chem. Lett.*, 2016, 7, 2910–2914.

99. M. Haruta, *Chem. Rec.*, 2003, 3, 75–87.

100. X. Hu, S. Dong, *J. Mater. Chem.*, 2008, 18, 1279–1295.

101. G. H. Bauer, R. J. Donnelly, W. Vinen, *J. Low Temp. Phys.*, 1995, 98 (1), 47–65.

102. K. K. Lehmann, R. Schmied, *Phys. Rev. B*, 2003, 68 (22), 224520.

103. M. Barranco, R. Guardiola, S. Hernández, R. Mayol, J. Navarro, M. Pi, *J. Low Temp. Phys.*, 2006, 142 (1–2), 1.

104. D. Spence, E. Latimer, C. Feng, A. Boatwright, A. M. Ellis, S. Yang, *Phys. Chem. Chem. Phys.*, 2014, 16(15), 6903–6906.

105. L. F. Gomez, K. R. Ferguson, J. P. Cryan, C. Bacellar, R. M. P. Tanyag, C. Jones, S. Schorb, D. Anielski, A. Belkacem, C. Bernando, R. Boll, *Science*, 2014, 345(6199), 906–909.

106. P. Thaler, A. Volk, F. Lackner, J. Steurer, D. Knez, W. Grogger, F. Hofer, W. E. Ernst, *Phys. Rev. B*, 2014, 90(15), 155442.

Stimuli-Responsive Polymeric Nanomaterials

Siyang Wang and
Marek W. Urban
Clemson University

2.1 Introduction

Stimuli-responsive nanomaterials play an important role in modern materials' design and offer numerous opportunities for future technological advances. Although continuously increasing efforts have been devoted to the development of stimuli-responsive polymers that primarily include synthesis of stimuli-responsive building blocks and their polymerization, the formation of higher-order structures, from nano- to micro- and above, remains to be challenging. Recently published reviews summarized previous and current efforts in this field along with potential technological opportunities.[1-4] The main prerequisite for developing higher-order responsive systems is to control the placement of the stimuli-responsive building blocks, as their location may enable or disable responsiveness at higher-order hierarchical structures. The primary motivation for developing stimuli-responsive nano- and higher-order assemblies is the ability to locally or globally alter physical (thermal, electromagnetic radiations, electric/magnetic fields) and chemical (pH, ionic strength, covalent bonding, or supramolecular chemistry) responses, while empowering macroscale assemblies. Precise control of stimuli-responsive building blocks in a copolymer can be facilitated by controlled polymerization (CP) approaches, among which nitroxide-mediated radical polymerization (NMP)[5], atomic transfer radical polymerization (ATRP)[6], reversible addition-fragmentation chain transfer polymerization (RAFT)[7] along with the recently developed heterogeneous radical polymerization (HRP)[8] capable of producing ultrahigh molecular weight block copolymers that contain built-in stimuli-responsive nano-assemblies play an important role.

Numerous studies published as early as 1950s have reported the synthesis of colloidal nanomaterials, which primarily focused on spherical colloidal nanoparticles known as latexes.[9,10] While today these technologies represent significant and mature portion of polymer industries,[11,12] every once in a while, this field is rejuvenated by new advances. For example, one of the questions raised a while back was how colloidal nanoparticles form films.[13,14] The significance of this question had important implications for several technological advances because film formation significantly impacts adhesion, mechanical, and optical properties of polymeric coatings.[3,15,16] Also, hollow polymeric nanoparticles can effectively alter opacity of polymeric films.[17-20] During film formation, there are processes that may have constructive or adverse influence on interfacial properties; film–air or film–substrate interfaces may be hydrophobic or hydrophilic, depending upon the stratification of individual components.[15,16,21-25] Today, as these technologies are mature, and spherical, core–shell, and hollow colloidal nanoparticles containing hydrophobic fluoro-monomers can be produced by emulsion polymerization in water,[26] new opportunities have arisen, which is reflected by the formation of more complex anisotropic 2-3D morphologies with stimuli-responsive attributes.[27] To place these advances in prospective, Figure 2.1 illustrates an array of nanostructures produced over the years, which range from isotropic spherical,[28-29] core–shell,[30-34] hollow,[35-37] and concentric nanoparticles[38-40] to Janus particles[41-43], gibbous/inverse-gibbous,[4,44,45] or cocklebur[46,47] nanoparticles, nanowires[48] and nanotubes[49]. The advantage of introducing heterogeneities in stimuli-responsive nanomaterials offers diverse properties brought about by each individual phase/component into one nano-object, thus enabling the creation of larger scale micro- and macro-devices with a broad spectrum of properties and a small volume. Therefore, despite maturity of some of these technologies, the exploration of stimuli-responsive polymeric nanomaterials continues to be of great interest.

Numerous colloidal nano-objects with various morphologies, functionalities, and responsiveness were developed. Figure 2.1a–c depicts isotropic spherical, core–shell, and hollow nanoparticles capable of uniformly altering size, shape, color as well as other properties, which were summarized in several review articles.[50-53] Figure 2.1d–g

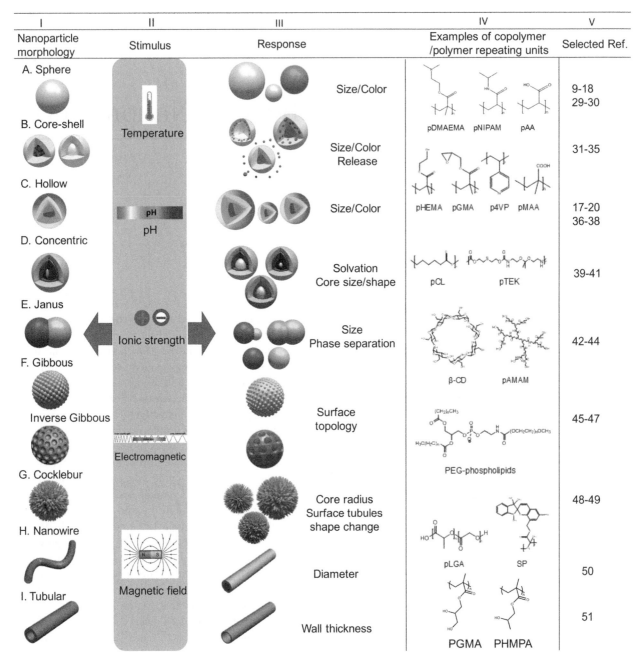

FIGURE 2.1 Representative examples of morphologies of nanomaterials (I), physical/chemical stimuli (II), responses manifested by shape, color, and dimensional changes (III), selective examples of polymers/copolymers responsible for responsiveness (IV), and references (V). Abbreviations: poly(dimethylaminoethyl methacrylate) (pDMAEMA), poly(N-isopropylacrylamide) (pNIPAM), poly-acrylic acid (pAA), polyhydroxyethylmethacrylate (pHEMA), poly(glycidyl methacrylate) (pGMA), poly(4- vinylpyridine) (p4VP), poly(methacrylic acid) (pMAA), polycaprolactone (pCL), polythioether ketal (pTEK), β-CD (β-cyclodextrin), poly(amidoamine) (PAMAM), PEGylated phospholipids (PEGphospholipids), poly(lactic-co-glycolic acid) (pLGA), spiropyran (SP).

illustrates anisotropic nano-objects that are capable of changing size and morphologies asymmetrically in 3D, thus enabling orientation or self-assembly of heterogeneous and hierarchical structures. Stimuli-responsive anisotropic nanoparticles, such as gibbous and Janus nanoparticles, nanowires, and nanotubes, are of particular interest not only in the context of mimicking biological organisms but also due to potential lithographic and biomedical applications. It should be noted that limited analytical tools are available to elucidate the origin of molecular processes responsible for their behavior and properties, thus making the measurements of asymmetric responses to external stimuli troublesome. Another challenge is to imbed stimuli-responsive components, facilitating directional interactions and multi-dimensional encoded signaling capable of interacting with an environment.

For isotropic shape nanoparticles (spherical, core–shell, and hollow) shown in Figure 2.1A–D, the property changes in response of external stimuli are typically dimensionally isotropic, but the overall size and porosity of nanoparticles can be locally altered.[50,51,54] For anisotropic nanoparticles, which typically exhibit asymmetrical distribution of chemical and/or physical properties, the responsiveness will be directional. Another example is the Janus nanoparticles shown in Figure 2.1E. The size, color, and physical or chemical properties of each hemisphere can be modulated by external stimuli.[44,55,56] As shown in Figure 2.1F, stimuli-responsive bulges on the surface of gibbous nanoparticles may alter their individual shape, while maintaining the shape of the spherical core, thus offering the adjustment of topography and surface roughness.[45] Cocklebur shaped nanoparticles[46] prepared from a combination of a nano-extruder applied to the aqueous solution containing methyl methacrylate (MMA) and n-butyl acrylate (n-BA) with azo-bis-isobutyronitrile (AIBN) or potassium persulfate (KPS) initiators and stabilized by a mixture of sodium dioctyl sulfosuccinate (SDOSS) and 1,2-bis(10,12-tricosadiynoyl)-sn-glycero-3-phosphocholine (DCPC) phospholipid may offer a number of sensing properties. Nanowires shown in Figure 2.1H are flexible[57] and can transform to spherical nanoparticles under certain conditions. As shown in Figure 2.1I, nanotubes that are constructed from stimuli-responsive walls, which expand or shrink at a physiological temperature, may offer numerous applications in biomedical technologies.[58] This chapter summarizes selective advances of stimuli-response nanomaterials and consists of two sections: responsive nanoparticles and responsive nanowires and nanotubes.

2.2   Nanoparticles

For thermal and pH-responsive nano-objects, morphological changes of these nano-objects are driven by conformational changes, hydrogen bonding (H-bonding), and/or protonation–deprotonation induced rearrangements. Thermo-responsive polymers have lower critical solution temperature (LCST) in aqueous environments, and typical examples include poly(N-isopropyl arylamide) (PNIPAM), poly(ethylene oxide) (PEO), poly(propylene oxide) (PPO), and poly(vinylcaprolactone) (PVCL). Above the LCST, homopolymers or copolymers containing these monomer units will experience the coil-to-globule transitions due to loss of hydrogen bonding between polymers and water molecules. The pH-responsive polymers usually realize on the reversibly ionizable functional groups within the side chains that induce electrostatic repulsions between charged polymer units upon environmental pH changes. As a result, polymer segments are able to extend or collapse, depending on the extent of ionization. Typical pH-responsive polymers include poly(acrylic acid) (PAA), poly(methacrylic acid) (PMAA), poly(N,N'-dimethyl aminoethyl methacrylate) (PDMAEMA), and poly(vinyl pyridine) (PVP).

There is also an increasing demand for developing photo-chromic nanoparticles that are capable of changing morphologies and color/fluorescence in response of electromagnetic irradiation. This is achieved by attaching or copolymerizing photo-responsive chromophores on the surface of inorganic nanoparticles, or along the polymer chains with soft-matter nanoparticles. The most common chromophores[59] capable of changing dimensions and absorption/emission of light in response to electromagnetic irradiation are azobenzene,[60] spiropyran,[61,62] and triphenylmethane.[54]

pH-responsive nanoparticles usually exhibit ionic strength responsiveness, typically due to manipulation of electrostatic interactions between polymer units and ions.[50] Biologically responsive nanomaterials are usually triggered by the presence of biologically active enzymes, which are often utilized in targeted drug delivery systems.[63,64] For example, liposomes consisting of polypeptide-functionalized polymers demonstrated targeted release of drugs in the presence of cancer-associated protease.[65] Colloidal nanoparticles containing conductive polymers[66,67] or gold nanoparticles[68] have been utilized for controlled drug-releasing devices in response of weak electric fields. Due to "remote" control of responses by external magnetic fields and other unique properties, magnetic nanoparticles have been widely studied and utilized in biomedical applications.[69–73] Electrochemical-responsive colloidal systems typically consist of gold nanoparticles, enabling the redox reactions of the media, which developed potential applications as electrochemical sensors.[74] Several literature reviews are available.[75–78] Among various stimuli, thermal responsive nanomaterials have been of particular interest due to numerous applications in drug delivery,[61] for example, most pathological sites dealing with inflammation can act as internal stimuli, in which temperature variations trigger responsiveness. The key criterion of thermal responsiveness is LCST in aqueous environments.

Anisotropic colloidal particles can be synthesized by several methods,[79–81] but the most common approaches are seeded growth,[82–84] self-assembly,[85–88] controlled fusion,[89] designed phase-separation,[90] selective deposition,[91,92] partial surface modification,[93–95] and seeded emulsion polymerization.[42,45,96] Of particular interests are the Janus particles with two sides of different chemistry/polarity and directional assemblies. The synthesis of Janus particles has been achieved using several approaches, which are categorized into four major strategies: masking,[97] phase-separation,[98] seeded growth,[99] and self-assembly.[100] The most versatile strategy is the selective chemical modification of exposed surfaces on temporarily immobilized spherical particles on 2D planar substrates, or at the interface of the Pickering emulsion droplets. The top surfaces of particles immobilized on 2D substrates can be chemically modified by metal deposition,[91,101,102] plasma treatments,[103] ligand exchanges,[104–106] chemical reactions,[107] electrostatic binding,[108] electrochemical growth,[109] and other means.[81] The particles can be also immobilized on the surface of electrospun fibers[110] or the Pickering emulsion droplets,[95,111] followed by chemical modifications and release of the resulting asymmetric particles. Another widely utilized synthetic procedure is the phase separation of two components mixture in one single particle, which can be realized through electrohydrodynamic co-jetting,[90] microfluidic co-flow,[112,113] and solvent assisted phase separation in polymer solution droplets.[99,114] Janus particles can be also obtained through the phase separation between the growing secondary components and seed particles. For the synthesis of inorganic Janus nanoparticles, seed particles stabilized by ligands facilitate the growth of the secondary phase on one side to form the Janus morphologies. Polymeric Janus nanoparticles can be obtained by seeded emulsion polymerization of phase-separated copolymers. In addition, the Janus nanoparticles can be synthesized *via* the self-assembly of triblock copolymers into various multi-compartment micelles or films, followed by cross-linking of the middle block and dissolution of the assembled structure.[88,115,116]

Several methods of producing nanoparticles were proposed, but perhaps emulsion polymerization is the most versatile.[27,46,117] Particularly recently developed synthetic methods of obtaining Janus nanoparticles by using a step-wise seeded emulsion polymerization, pentafluorostyrene (PFS) and nBA can be copolymerized on the p(MMA/nBA) core and the phase separation between the two copolymers results in acorn-shape Janus nanoparticles consisting of p(MMA/nBA) and p(PFS/nBA) hemispheres.[42] By tuning the T_g of the seed particles

via copolymerization of MMA and nBA with various ratios, followed by copolymerization of PFS/nBA in the presence of p(MMA/nBA) seed particles, heterogeneous nanoparticles with various morphologies were obtained due to glass transition temperature (T_g) difference and interfacial surface tension between fluorinated and acrylic copolymer phases.[98] As shown in Figure 2.2a, when $T_{\text{reaction}} > T_g$ of the seed, core–shell nanoparticles are obtained. As T_{reaction} is close to or smaller than the T_g of the seed, and the interfacial surface tension (V) between the two phases increase, acorn-shape or inverse acorn-shape morphologies can be obtained. Furthermore, upon incorporation of pH-responsive azobenzene compounds (AZO) during synthesis (Figure 2.2b), the Janus nanoparticles exhibit color responses depending upon pH changes, as demonstrated by the UV–vis absorbance spectra shown in Figure 2.2c and illustrated in Figure 2.2d.[55] Furthermore, when the p(MMA/nBA)-p(PFS/nBA) Janus nanoparticles were utilized as seed particles, and DMAEMA and nBA were copolymerized semicontinuously, triphasic Janus nanoparticles with a stimuli-responsive hemispherical shell were synthesized.[44] These triphasic Janus nanoparticles are capable of changing shape by varying temperature and/or pH.

Further structural complexity of nanoparticles can be obtained by the formation of gibbous (raspberry-like) morphologies nanoparticles. General structure of anisotropic gibbous nanoparticles is with spherical core and multidirectional bulges. The unique structure allows nanoparticles to retain high surface roughness as well

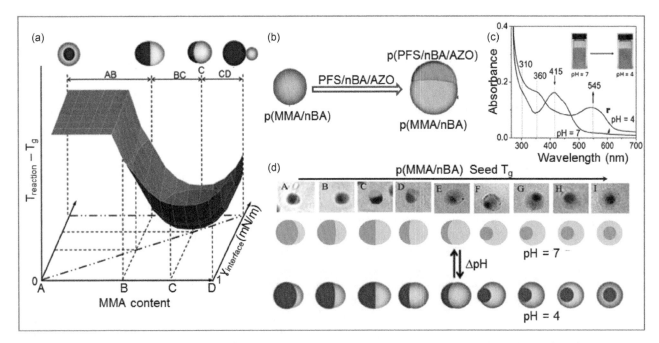

FIGURE 2.2   (a) Control of heterogeneous nanoparticle morphologies by tuning MMA content of p(MMA/nBA) seed particles; (b) synthesis of color changing Janus nanoparticles; (c) UV–vis spectra of the colloidal dispersion; and (d) nanoparticles with various morphologies exhibiting light and dark gray color at neutral and acid conditions, respectively. (Reproduced with permission from Ref. [55,98].)

as the core, bulges, and/or dimples can be consisted of stimuli-responsive entities. Therefore, gibbous nanoparticles can be equipped with localized stimuli responsiveness. Key-and-lock structures can also be obtained with suitable conditions where the size of bulges and dimples are comparable. More recently developed approach of synthesizing gibbous nanoparticles depicted in Figure 2.3a is to attach large amount of smaller size nanoparticles onto the surface of a spherical seed via chemical reactions[118], electrostatic attraction,[119] or hydrogen bonding,[120] while the stimuli responsiveness are obtained by introducing thermal or pH-responsive monomers/copolymers. For example, PMMA, PMMA/nBA, polystyrene (PSt), PMMA/PMAA, or SiO2-PMMA can be acted as seeds, followed by localized

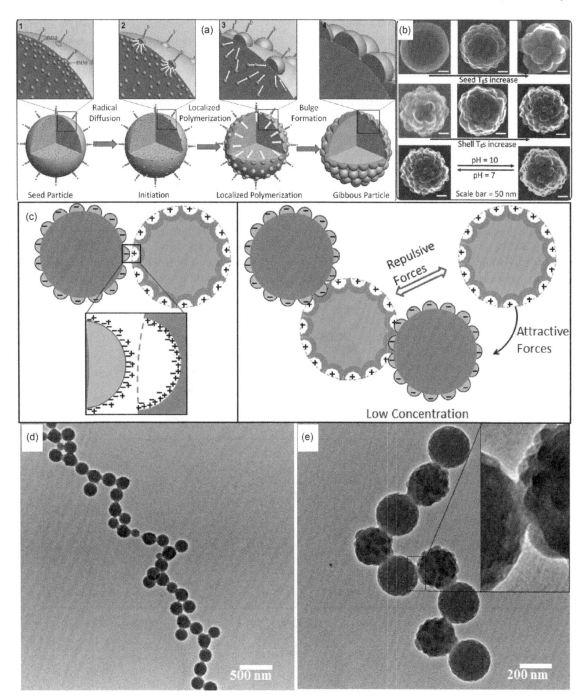

FIGURE 2.3 Schematic representation of the mechanism responsible for the formation of gibbous nanoparticles using seeded emulsion polymerization (a), TEM images for each step of the formation of gibbous nanoparticles and pH responsiveness of the shell (b) (Reproduced with permission from Ref. [45].), schematic representation of gibbous and inverse-gibbous particles with surface matching geometries and electrostatic double layers and their necklace-like assembly (c), TEM images of the assemblies of gibbous (−)/inverse-gibbous (+) particles with different concentrations: 10^{-7}g/mL (d), 10^{-8}g/mL (e), respectively. (Reproduced with permission from Ref. [1].)

copolymerization of monomers, for example, PFS, MMA, and nBA to form bulges.

In the example shown in Figure 2.3b, p(MMA-PFS/MMA) gibbous nanoparticles morphologies can be tuned by different monomer ratios, while at the same time the existence of PMAA in the bulges enables the size change of nanoparticles. The p(MMA-PFS/MMA) gibbous nanoparticles shrink at low pH and expand under high pH environment.[45] A similar approach in the synthesis of gibbous nanoparticles can be utilized with hydrophobic monomers to fabricate inverse-gibbous nanoparticles.[121] The primarily driving force in inverse-gibbous nanoparticle synthesis is to attain key-and-lock structures with shape recognition by using gibbous nanoparticles as counterpart. As shown in Figure 2.3c–e, "gear-like" assemblies can be formed via combining gibbous and inverse-gibbous nanoparticles. The gibbous nanoparticles can be synthesized from PMMA core, followed by the formation of PFS/nBA bulges. Inverse-gibbous nanoparticles may consist of PSt core and poly(heptafluoro-n-butyl acrylate(HFBA)/nBA) dimples. Upon incorporating 2,2-azobis[2-(2-imidazolin-2-yl)propane] dihydrochloride (AIPD) (positive) and KPS, opposite charges can be obtained. This unique gibbous and inverse-gibbous combination results in the formation of both size/shape matching and opposite charges compatible

surface topologies. This key-and-lock combination can be easily assembled or disassembled through changing the surface charges.

Using two-phase Janus nanoparticles (Figure 2.4), triphasic Janus nanoparticles can be synthesized using p(MMA/nBA)-p(PFS/nBA) Janus nanoparticles seeds by further copolymerization of DMAEMA and nBA, thus facilitating Janus nanoparticles with dual temperature and pH responsiveness. The control of the shape is achieved by variation of the seed T_g and copolymerization compositions of the shell. As shown in Figure 2.4A,B, triphasic stimuli-responsive Janus nanoparticles consisting of phase-separated p(MMA/nBA), p(PFS/nBA), and p(DMAEMA/nBA) copolymers can be prepared by a two-step process. Janus balance (JB), defined as the ratio of hydrophilic and hydrophobic components, can be used to quantify the geometry of Janus nanoparticles.[95] As the temperature increases, the size of Janus nanoparticles decreases with the increasing of temperature. At 25°C, the size of the particle is 147 nm, and when the temperature increases to 45°C the particle size decreases to 131 nm. Meanwhile, morphology changes occur when temperature increases to 45°C and the particle been spherical with a convex p(PFS/nBA) phase changes to ellipsoidal morphology with a concave p(PFS/nBA) phase and the

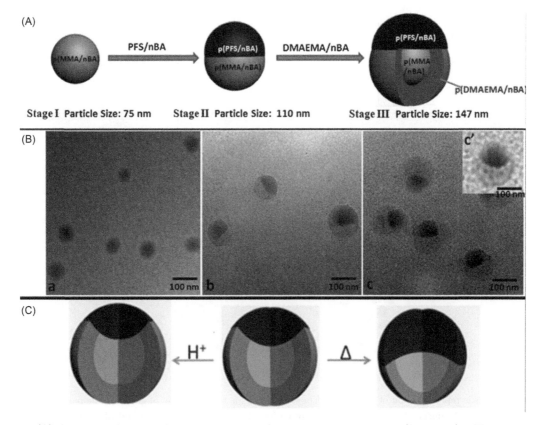

FIGURE 2.4 (A) Schematic diagram of copolymerization of stimuli-responsive JNPs. (B-a, b, c) TEM images of p(MMA-nBA), p(MMA-nBA/PFS-nBA), and p(MMA-nBA/PFS-nBA/DMAEMA-nBA) particles, respectively. The inset picture is the stimuli-responsive JNP stained by KMnO4 aqueous solution for 5 min. (C) Dimensional changes of JNPs as a function of pH and temperature. (Reproduced with permission from Ref. [43].)

JB decreases from 3.78 to 4.52. When the T_g of the seed decreases, it is easier to control the shape of the nanoparticles ranging from acorn to ellipsoidal and inverse core–shell morphologies. Also, the size of the particle increases from 145 nm to 163 nm when pH changes from 10 to 4 and the JB increases from 3.42 to 4.52 (Figure 2.4C).[43] These dual responsiveness Janus nanoparticles with tunable sizes and JB offer a broad range of applications from the stabilization of oil in water at high pH condition to multi-drug delivery systems.

Interfacial polymerization between hydrophilic acrylic acid (AA) and oleophilic styrene/divinyl benzene (St/DVB) was also utilized to synthesize Janus particles with subsequently selective electrostatic assembly of Fe_3O_4 nanoparticles on the convex surfaces resulting in the formation of hydrophilic/oleophilic magnetic Janus particles. Introducing magnetic Janus particles into surfactant-free oil-stained toluene-in-water emulsion, rapid and efficient oil droplet separation occurs under magnetic fields, while no significant separation without magnetic fields occurs.[122] Many nanotechnologies utilize a combination of organic and inorganic phases to produce ceramic or metallic nanoparticles. One can envision the development of new properties by combining inorganic (metals, metal oxides) and organic (polymer) phases into one nanoparticle designated as "ceramers" (inorganics) and "metamers" (metallic).[1]

One of many useful applications of these technologies is cancer therapy. For example, thermo-sensitive amphiphilic copolymer poly(2-(2-methoxyethoxy) ethyl methacrylate-co-oligo(ethylene glycol)methacrylate)-co-2-(dimethylamino) ethyl methacrylate-b-poly(D, L-lactide-co-glycolide) can be assembled into core–shell structured nanoparticles with co-encapsulation of two cytotoxic drugs doxorubicin and paclitaxel (NP-DT), and absorption of small interfering RNAs (siRNA) against surviving (NP-DTS).[123] This copolymer exhibits low toxicity, low immunogenicity, and thermal sensitivity owing to the presence of polyethylene glycol(PEG) and PNIPAM.[124] The LCST can be precisely tuned during polymerization by changing the amount of chain transfer agent and MEO2MA and OEGMA ratio. The LCST was designed slightly above physiological temperature which upon body contact triggers the collapse of nanoparticles to release the drug. The outer layers of the drug-loaded nanoparticles can be modified with polydopamine (PDA) (named NP-DTS-PDA), preventing premature burst release and NIR laser-triggered nanoparticle collapse. This process is shown in Figure 2.5a–c.

Polymeric nanoparticles responsive to magnetic fields usually consist of nanocarrier with magnetite (Fe_3O_4) or maghemite (γ-Fe_2O_3) cores, which are referred to superparamagnetic iron oxide nanoparticles (SPIONs).[125] SPIONs are being utilized in biomedical field applications including drug delivery and magnetic resonance imaging (MRI).[126] Magnetically induced nanocarriers also play an important role in nucleic acids delivery, siRNA, and genes deliveries. Higher transfection efficiencies can be obtained through the condensation of nucleic acids and iron oxide

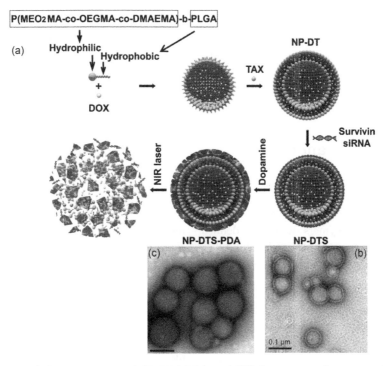

FIGURE 2.5 (a) Illustration of the construction of NP-DTS-PDA and NIR laser-triggered nanoparticle collapse for drug release. DOX represents for doxorubicin; TAX represents for paclitaxel. (b) and (c) are TEM images of NP-DTS and NP-DTS-PDA showing presence of spherical core–shell nanostructures and a thin PDA shell on the surface of NP-DTS-PDA. (Reproduced with permission from Ref. [123].)

nanoparticles with cationic coating nano-assemblies under permanent magnetic field.[127] Cationic PDMA-block-poly (e-caprolactone) (PDMS-b-PCL) micelles comprised of SN-38 (7-ethyl-10-hydroxycamptothecin) were synthesized, and human vascular endothelial growth factor (VEGF) targeting siRNA with SPIONs were introduced.[128] These complex micelles can suppress tumor growth while serving as contrast agents in MRI diagnosis. One of the obstacles utilizing nanoparticles in biological systems is their dispersion and aggregation control. These properties can be controlled using amino acid DOPA in stealth liposomes consisting of self-assembled SPIONs, which are individually stabilized with palmityl-nitro/DOPA incorporated in the lipid membrane. Upon application of alternating magnetic fields, controlled timing and dose of repeatedly released cargo from these vesicles can be accomplished by locally heating the membrane.[129]

Grafting-from and grafting-to are the most commonly used methods to modify nanoparticles in order to achieve core–shell structure with functionalized polymer shells. The key factor is the grafting densities.[130] For example, SPIONs with the grafting density less than 0.5 chains/nm^2 of PEG, ~3,000–10,000 molecular weight is reported to meet the criteria of colloidal stability.[131] The fabrication of nitrodopamine attached Fe_3O_4 surface with end-grafted PNIPAM brushes, where nitrodopamine as a key attachment point enables the modified nanoparticles with grafting density independent LCST. Such nanoparticles may exhibit excellent colloidal stability and magnetic properties, which can be applied to manipulate the reversible solubility of nanoparticles. Introducing a combination of magnetic and thermally responsive properties, a variety of applications can be envisioned; for example, separation and extraction of molecules in magnetic fields. Further coupling of dopamine affinity towards iron oxide by pH or electromagnetic-responsive monomers brings other possibilities for remote stimuli-responsive system.[132]

2.3   Nanowires and Nanotubes

Formation of anisotropic stimuli-responsive nanomaterials is accomplished by molecular design of copolymer blocks that in specific environments will self-assemble to form anisotropic shapes. Block copolymers with well-defined architectures (diblock, triblock, star-like, etc.) and block lengths are usually synthesized using living cationic/anionic, ring-opening metathesis[133] and controlled radical polymerizations (CRPs).[134] If designed properly, they may self-assemble in a solution to form dispersed nano-objects. Block copolymers consist of two or more blocks, and each block exhibits specific characteristic properties; for example, non-compatible hydrophobic–hydrophobic,[135] hydrophilic–hydrophobic,[85] cationic–anionic,[136] rod–coil,[137,138] or crystalline–amorphous blocks.[139–141] These block copolymers should have an affinity to phase-separate forming colloidal assemblies upon the manipulation of solution

conditions, such as co-solvent ratios, pH, temperature, ionic strength, the presence of organic counter ions, or inorganic nanoparticles. Several approaches have been utilized to facilitate self-assembly of block copolymers, but the most common one is to dissolve block copolymers in a common solvent, followed by slow addition of a poor solvent (solvent displacement), and removal of the common solvent upon evaporation or dialysis. As a result, various heterogeneous morphologies can be obtained.

Although self-assembly of block copolymers through post-polymerization solvent displacement in dilute solutions offers control over various morphologies, the time-consuming procedure as well as dilute conditions (usually <1 wt%) represent a significant drawback for a large-scale synthesis. Considerable progress has been made in preparing in situ assembly block copolymers and continues to be of significant interests.[142–151] Amphiphilic block copolymers can undergo self-assembly to form various nanostructures in solvent environments that are selective for one of the blocks.[85] Although nanowires or worm-like block copolymer micelles have been prepared couple of decades ago,[152] there is still strong interest to produce nanowires other than spherical morphologies with longer circulation times in therapy.[153] Figure 2.6a illustrates an example of RAFT dispersion polymerization of hydrophobic core-forming monomer hydroxypropyl methacrylate (HPMA) using poly(glycerol monomethacrylate) (PGMA) macro-chain transfer agent (CTA) and result in various intermediate morphologies as polymerization progresses (Figure 2.6b).[154] For the synthesis targeting at PGMA47-PHPMA200 diblock composition, spherical micelles ~20–30 nm were formed at conversion ~46% due to micellar nucleation of the resulting PGMA47-PHPMA92 diblocks. As polymerization continues, the spherical micelles undergo 1D fusion to form nanowires, and then become branched nanowires, which transform to 2D bilayers. At ~70% conversion, bilayers begin to wrap-up to form "jelly-fish" with hemi-vesicles and nanowires, which eventually lead to the formation of vesicles at conversions >80%. This morphology transformation from free-flowing spherical to free-flowing worm-like micelles is attributed to the increase of packing parameter as DP of hydrophobic block increases and temperature changes (Figure 2.6c).

Taking advantage of incompatibility of fluoropolymers and non-fluorinated polymers, triblock copolymers containing hydrophilic, hydrophobic, and fluorophilic blocks, capable of forming various multi-compartment micellar morphologies through self-assembly driven by the interfacial tension,[86] can be obtained. Assembled from ABC triblock copolymers, where A represents hydrophilic block, B is the hydrophobic block, and C is the fluorophilic block, the micellar structures produced by self-assembly of these block copolymers have a hydrophilic corona resulting from the solvation of hydrophilic blocks in water, and a heterogeneous core that may exhibit morphologies ranging from core–shell, gibbous, segmented worm-like, hamburger-like, disc-like, and others. ABC triblock copolymers primarily

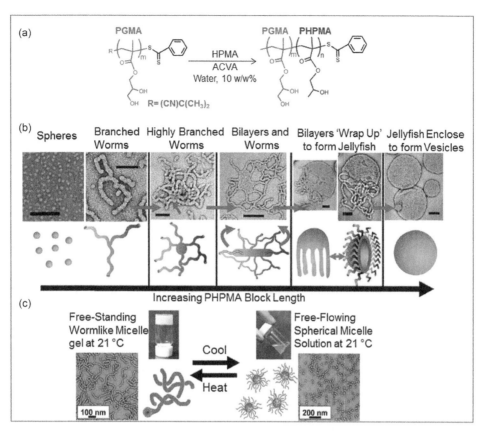

FIGURE 2.6 (a) RAFT aqueous dispersion polymerization of hydroxypropyl methacrylate (HPMA) using poly(glycerol monomethacrylate) (PGMA) macro-CTA at 10 w/w % and 70°C; (b) Suggested mechanism for the sphere-to-worm-to-vesicle transformation during the in situ synthesis; (c) Thermo-responsive aqueous solution behavior of a 10% w/w aqueous dispersion of diblock copolymer particles. TEM studies of grids prepared from a dilute aqueous dispersion of G_{54}-H_{140} dried at either 21°C or 4°C showing the reversible worm-to-sphere transition. (Reproduced with permission from Ref. [58, 154].)

form the core–shell–corona structures, capable of minimizing interfacial energy, whereas ABC triblock copolymers tend to form gibbous core–corona morphologies.

Except for conventional spherical structures, polymer nanowires that consist of ultrahigh molecular weight block copolymers can be synthesized by one-step surfactant-free HRP. When DMAEMA and styrene (St) were copolymerized (Figure 2.7a), nanowires consisting of ultrahigh molecular weight ($>10^6$) PDMAEMA-block-St with thermal-responsive PDMAEMA blocks were produced. These materials are able to change diameter upon temperature changes. Schematic illustration of nanowire directional growth is shown in Figure 2.7b. Within heterogeneous and initiator-starvation environment during the initial stage of the polymerization, spherical nanoparticle is formed. As the polymerization continues, directional growth or the nanoparticles starts resulting from the localized repulsive forces between hydrophilic blocks and the confined hydrophobic blocks, favorable high aspect ratio nanowire morphology is obtained finally. When the nanowires are heated to 70°C, the diameter of the nanowire decreases from ~70 to ~63 nm. This simple one-step synthetic method allows fabrication of hundreds of microns' scale length nanowires with components, which are hydrophilic and hydrophobic monomers, water-soluble initiator, and water as solvent.[8,48]

Polymeric nanotubes can be also produced from block copolymer self-assembly in nonpolar solvents[141,155,156] as well as aqueous environments.[157–159] Such block copolymer nanotubes incorporated with stimuli-responsive monomer units can alter their morphologies in response to environmental adjustments. Using this approach, pH- and CO_2-responsive nanotubes can be prepared from poly(ethylene oxide)-b-poly((N-amidine) dodecylacrylamide)-b-polystyrene triblock copolymers (PEO-PADA-PS), which are dissolved in THF and self-assembled by solvent displacement with water to form microtubules.[160] In the presence of CO_2 gas as stimulus, amidine-containing blocks will be protonated, thus resulting in morphology changes, from microtubules to vesicles to spheres.

Multilayers of nanotube-forming phospholipid templates that consist of thermally responsive polymers can serve as templates with tunable diameter and wall thickness. A general method to fabricate these nanotubules is to incorporate polymerizable thermal-responsive monomers into the hydrophobic area of phospholipid (PL) template. As shown in Figure 2.8, NIPAM monomers are dispersed in a hydrophilic

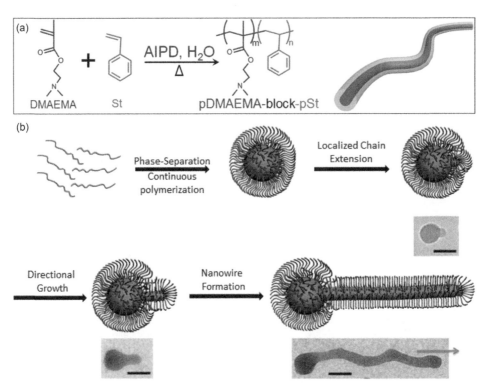

FIGURE 2.7 (a) Synthesis route of block copolymer nanowires; (b) schematic representation of block copolymer nanowire directional growth from block polymer formation, spherical micelle formation, localized chain extension, directional growth to nanowire formation. Scale bar for TEM images is 100 nm. (Reproduced with permission from Ref. [48].)

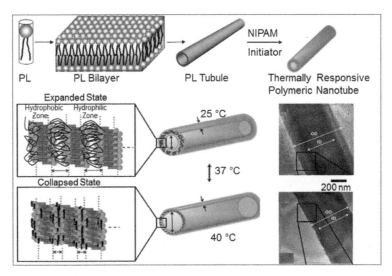

FIGURE 2.8 Schematic illustration of thermally responsive polymeric nanotubes. (Reproduced with permission from Ref. [49].)

part of the phospholipid bilayers and in situ polymerized to form thermally responsive layers (~40°C). The expanded and collapsed state can be altered when temperature is above 37°C. Under these conditions, the wall and the outer diameter of the nanotubules shrink. With another certain situation, PL can form bi-layered and tubule morphologies, when PL tubule is exposed to NIPAM and initiator, increased temperature initiates polymerization resulting in thermally responsive polymeric nanotubes. When temperature increases from 25°C to 37°C, the wall collapses due to LCST of PNIPAM. Notably,

the transition temperature of the nanotubules is tunable by the copolymerization of other monomers.[49]

Conductive polymers, in particular in the form of nanotubes or nanowires, are of a particular interest because these materials may offer many applications, including chemical, optical and bio-sensors, nano-diodes, field effect transistors, field emission and electrochromic displays, super capacitors and energy storage, actuators, drug delivery, neural interfaces, or protein purification. Although this is beyond the scope of this chapter, preparation methods and physical properties of conjugated polyaniline (PANI), polypyrrole (PPY), and poly (3,4-ethylenedioxythiophene) (PEDOT) were summarized in the literature.[161] The primary preparation approaches are hard physical template method, soft chemical template method, electrospinning, and lithography techniques. Of particular interest are electronic transport and optical properties of individual nanotubes as a function of size.[162] Notably, polymeric materials can be utilized as sacrificial phase in the formation of inorganic, stable at high temperatures nanotubes. One example is the synthesis of highly ferromagnetic, concentric iron oxide–carbon–iron oxide nanotubes (FMNTs) using Fe^{2+}/H_2O_2 redox reactions from biologically active 1,2- bis(10,12-tricosadiynoyl)-sn-glycero-3-phosphocholine ($DC_{8,9}PC$) nanotube-forming PL.[162]

2.4 Summary

From simple spherical to more complex morphologies, polymeric stimuli-responsive nanomaterials offer significant contributions to modern materials science and technological advances. While core–shell structures enable the combination of different chemical and physical properties into the same material, hollow nanoparticles allow the storage and delivery of cargo molecules with targeted destinations. Increase in the complexity of morphologies leads to the formation of Janus, gibbous/inverse-gibbous nanoparticles with controllable and heterogeneous distribution of individual components and high surface area. Localized components enable controllable responsiveness, and surface modifications provide the possibility of sensing physical and chemical environments. For example, surface receptors may sense and recognize specific bioactive environments while humidity and other environmental changes can lead to the development of physical sensors. Thus, stimuli-responsive polymers will offer new applications in medicine, transportation, or homeland security. Innovations in fabricating new devices require bulk and surface responsiveness of nanomaterials, precise control of nanomaterial architectures, and site-specific responsiveness. Various applications and impact that nanotechnology has already made on daily lives are attributed to the higher strength and light-weight materials, multiple functions, and programmed behaviors. Inorganic nanoparticles, nanorods, and nanofibers with high strength and low density have been widely used in reinforcing polymer matrices.[163] Nanoparticles produced from silicate, silica, titanium dioxide, zinc oxide, silver, and others are utilized as gas barriers, fillers, UV protectors, UV blockers, anti-microbial agents, and fillers.[164] Semiconducting quantum dots (QDs) are applied in manufacture of ion batteries, fuel cells, LEDs, diode lasers, solar cells, and imaging sensors.[165,166] Metal nanoparticles are used in catalysis, conductive components of circuit boards, transistors, sensors, and many other applications.[167] Gold[168] and iron oxide magnetic nanoparticles[169] found applications in bio-imaging and biosensors, plasmonic, and hyperthermia

therapy, and targeted delivery carriers. Combining polymeric nanomaterials with inorganics will lead to the development of smart composites with high mechanical strength, sustainability, and robust responsiveness. With respect to organic/polymeric nanoparticle, latexes and colloidal dispersions have been vastly manufactured in large-scale synthesis with direct applications as coatings, paints, and pressure-sensitive adhesives.[170] Polymeric nanoparticles with various morphologies have many untapped uses in biomedical applications, drug delivery carriers,[171] self-healing materials,[172] sensing devices, and soft robotics.

Acknowledgments

This work was supported by the National Science Foundation under Awards DMR 1744306 and OIA 1655740, and partially by the J.E. Sirrine Foundation Endowment at Clemson University.

References

1. Liu, X.; Yang, Y.; Urban, M. W., Stimuli-responsive polymeric nanoparticles. *Macromolecular Rapid Communications* **2017**, *38* (13), 1–20.
2. Liu, F.; Urban, M. W., Recent advances and challenges in designing stimuli-responsive polymers. *Progress in Polymer Science* **2010**, *35* (1–2), 3–23.
3. Urban, M. W., Stratification, stimuli-responsiveness, self-healing, and signaling in polymer networks. *Progress in Polymer Science* **2009**, *34* (8), 679–687.
4. Lu, C. L.; Urban, M. W., Stimuli-responsive polymer nano-science: Shape anisotropy, responsiveness, applications. *Progress in Polymer Science* **2018**, *78*, 24–46.
5. Zhao, B.; Li, D. J.; Hua, F. J.; Green, D. R., Synthesis of thermosensitive water-soluble polystyrenics with pendant methoxyoligo(ethylene glycol) groups by nitroxide-mediated radical polymerization. *Macromolecules* **2005**, *38* (23), 9509–9517.
6. Vasquez, E. S.; Chu, I. W.; Walters, K. B., Janus magnetic nanoparticles with a bicompartmental polymer brush prepared using electrostatic adsorption to facilitate toposelective surface-initiated ATRP. *Langmuir* **2014**, *30* (23), 6858–6866.
7. Smith, A. E.; Xu, X.; McCormick, C. L., Stimuli-responsive amphiphilic (co) polymers via RAFT polymerization. *Progress in Polymer Science* **2010**, *35* (1–2), 45–93.
8. Lu, C. L.; Urban, M. W., One-step synthesis of amphiphilic ultrahigh molecular weight block copolymers by surfactant-free heterogeneous radical polymerization. *ACS Macro Letters* **2015**, *4* (12), 1317–1320.
9. Crouch, W. W., Shortstopping an emulsion polymerization reaction with alkyl polysulfide. Google Patents, **1951**.

10. Antlfinger, G. J., Method of terminating the polymerization of monomeric materials in aqueous emulsion and a nonstaining, nondiscoloring composition for use therein. Google Patents, **1953**.

11. Provder, T.; Urban, M. W., *Film Formation in Coatings: Mechanisms, Properties, and Morphology*. ACS Publications, Washington, DC, **2001**.

12. Provder, T.; Winnik, M. A.; Urban, M. W., *Film Formation in Waterborne Coatings*. ACS Publications, Washington, DC, **1996**.

13. Chevalier, Y.; Pichot, C.; Graillat, C.; Joanicot, M.; Wong, K.; Maquet, J.; Lindner, P.; Cabane, B., Film formation with latex particles. *Colloid and Polymer Science* **1992**, *270* (8), 806–821.

14. Winnik, M. A., Latex film formation. *Current Opinion in Colloid & Interface Science* **1997**, *2* (2), 192–199.

15. Zhao, Y.; Urban, M. W., Phase separation and surfactant stratification in styrene/n-butyl acrylate copolymer and latex blend films. 17. A spectroscopic study. *Macromolecules* **2000**, *33* (6), 2184–2191.

16. Urban, M. W.; Allison, C. L.; Johnson, G. L.; Di Stefano, F., Stratification of butyl acrylate/polyurethane (BA/PUR) latexes: ATR and step-scan photoacoustic studies. *Applied Spectroscopy* **1999**, *53* (12), 1520–1527.

17. Blankenship, R.; Bardman, J., Process for preparing polymer emulsions and polymers formed therefrom. Google Patents, **2001**.

18. Kowalski, A.; Vogel, M., Sequential heteropolymer dispersion and a particulate material obtainable therefrom, useful in coating compositions as an opacifying agent. Google Patents, **1984**.

19. Tiarks, F.; Landfester, K.; Antonietti, M., Preparation of polymeric nanocapsules by miniemulsion polymerization. *Langmuir* **2001**, *17* (3), 908–918.

20. zu Putlitz, B.; Landfester, K.; Fischer, H.; Antonietti, M., The generation of "armored latexes" and hollow inorganic shells made of clay sheets by templating cationic miniemulsions and latexes. *Advanced Materials* **2001**, *13* (7), 500–503.

21. Rhudy, K. L.; Su, S.; Howell, H. R.; Urban, M. W., Self-stratified films obtained from poly(methyl methacrylate/n-butyl acrylate) colloidal dispersions containing poly(vinyl alcohol): A spectroscopic study. *Langmuir* **2008**, *24* (5), 1808–1813.

22. Stegge, J. M.; Urban, M. W., Stratification processes in thermoplastic olefins monitored by step-scan photoacoustic FT-IR spectroscopy. *Polymer* **2001**, *42* (12), 5479–5484.

23. Thorstenson, T. A.; Tebelius, L. K.; Urban, M. W., Surface and interfacial FT-IR spectroscopic studies of latexes. 7. EA/MAA latex suspension stability and surfactant migration. *Journal of Applied Polymer Science* **1993**, *50* (7), 1207–1215.

24. Zhao, Y.; Urban, M. W., Polystyrene/poly (n-butyl acrylate) latex blend coalescence, particle size effect, and surfactant stratification: A spectroscopic study. *Macromolecules* **2000**, *33* (20), 7573–7581.

25. Urban, M. W., Stimuli-responsive colloids: From stratified to self-repairing polymeric films and beyond. *Current Opinion in Colloid & Interface Science* **2014**, *19* (2), 66–75.

26. Dreher, W. R.; Singh, A.; Urban, M. W., Effect of perfluoroalkyl chain length on synthesis and film formation of fluorine-containing colloidal dispersions. *Macromolecules* **2005**, *38* (11), 4666–4672.

27. Dreher, W. R.; Jarrett, W. L.; Urban, M. W., Stable nonspherical fluorine-containing colloidal dispersions: Synthesis and film formation. *Macromolecules* **2005**, *38* (6), 2205–2212.

28. Mastrotto, F.; Salmaso, S.; Alexander, C.; Mantovani, G.; Caliceti, P., Novel pH-responsive nanovectors for controlled release of ionisable drugs. *Journal of Materials Chemistry B* **2013**, *1* (39), 5335–5346.

29. Yassine, O.; Zaher, A.; Li, E. Q.; Alfadhel, A.; Perez, J. E.; Kavaldzhiev, M.; Contreras, M. F.; Thoroddsen, S. T.; Khashab, N. M.; Kosel, J., Highly efficient thermoresponsive nanocomposite for controlled release applications. *Scientific Reports* **2016**, *6*.doi: 10.1038/srep28539.

30. Majewski, A. P.; Schallon, A.; Jerome, V.; Freitag, R.; Muller, A. H. E.; Schmalz, H., Dual-responsive magnetic core-shell nanoparticles for nonviral gene delivery and cell separation. *Biomacromolecules* **2012**, *13* (3), 857–866.

31. Ruhland, T. M.; Reichstein, P. M.; Majewski, A. P.; Walther, A.; Muller, A. H. E., Superparamagnetic and fluorescent thermo-responsive core-shell-corona hybrid nanogels with a protective silica shell. *Journal of Colloid and Interface Science* **2012**, *374*, 45–53.

32. Xu, J.; Luo, S. Z.; Shi, W. F.; Liu, S. Y., Two-stage collapse of unimolecular micelles with double thermoresponsive coronas. *Langmuir* **2006**, *22* (3), 989–997.

33. Li, X.; Fu, X. N.; Yang, H., Preparation and photocatalytic activity of eccentric Au-titania core-shell nanoparticles by block copolymer templates. *Physical Chemistry Chemical Physics* **2011**, *13* (7), 2809–2814.

34. Zhao, Y.; Urban, M. W., Novel STY/nBA/GMA and STY/nBA/MAA core–shell latex blends: Film formation, particle morphology, and cross-linking. 20. A spectroscopic study. *Macromolecules* **2000**, *33* (22), 8426–8434.

35. Shin, J. M.; Anisur, R. M.; Ko, M. K.; Im, G. H.; Lee, J. H.; Lee, I. S., Hollow manganese oxide nanoparticles as multifunctional agents for magnetic resonance imaging and drug delivery. *Angewandte Chemie-International Edition* **2009**, *48* (2), 321–324.

36. Li, G. L.; Xu, L. Q.; Tang, X. Z.; Neoh, K. G.; Kang, E. T., Hairy hollow microspheres of fluorescent shell and temperature-responsive brushes via combined

distillation-precipitation polymerization and thiol-ene click chemistry. *Macromolecules* **2010**, *43* (13), 5797–5803.

37. Fu, G. D.; Li, G. L.; Neoh, K. G.; Kang, E. T., Hollow polymeric nanostructures-synthesis, morphology and function. *Progress in Polymer Science* **2011**, *36* (1), 127–167.

38. Kamata, K.; Lu, Y.; Xia, Y. N., Synthesis and characterization of monodispersed core-shell spherical colloids with movable cores. *Journal of the American Chemical Society* **2003**, *125* (9), 2384–2385.

39. Kim, S. H.; Weitz, D. A., One-step emulsification of multiple concentric shells with capillary microfluidic devices. *Angewandte Chemie-International Edition* **2011**, *50* (37), 8731–8734.

40. Radloff, C.; Halas, N. J., Plasmonic properties of concentric nanoshells. *Nano Letters* **2004**, *4* (7), 1323–1327.

41. Misra, A.; Urban, M. W., Acorn-shape polymeric nano-colloids: Synthesis and self-assembled films. *Macromolecular Rapid Communications* **2010**, *31* (2), 119–127.

42. Hwang, S.; Roh, K. H.; Lim, D. W.; Wang, G. Y.; Uher, C.; Lahann, J., Anisotropic hybrid particles based on electrohydrodynamic co-jetting of nanoparticle suspensions. *Physical Chemistry Chemical Physics* **2010**, *12* (38), 11894–11899.

43. Lu, C. L.; Urban, M. W., Tri-phasic size-and janus balance-tunable colloidal nanoparticles (JNPs). *ACS Macro Letters* **2014**, *3* (4), 346–352.

44. Pi, M. W.; Yang, T. T.; Yuan, J. J.; Fujii, S.; Kakigi, Y.; Nakamura, Y.; Cheng, S. Y. A., Biomimetic synthesis of raspberry-like hybrid polymer-silica core-shell nanoparticles by templating colloidal particles with hairy polyamine shell. *Colloids and Surfaces B-Biointerfaces* **2010**, *78* (2), 193–199.

45. Lu, C.; Urban, M., Rationally designed gibbous stimuli-responsive colloidal nanoparticles. *ACS Nano* **2015**, *9* (3), 3119–3124.

46. Lestage, D. J.; Urban, M. W., Cocklebur-shaped colloidal dispersions. *Langmuir* **2005**, *21* (23), 10253–10255.

47. Tian, Y.; Zhang, J. P.; Tang, S. W.; Zhou, L.; Yang, W. L., Polypyrrole composite nanoparticles with morphology-dependent photothermal effect and immunological responses. *Small* **2016**, *12* (6), 721–726.

48. Lu, C.; Urban, M. W., Instantaneous directional growth of block copolymer nanowires during heterogeneous radical polymerization (HRP). *Nano Letters* **2016**, *16* (4), 2873–2877.

49. Kawano, S.; Urban, M. W., Expandable temperature-responsive polymeric nanotubes. *ACS Macro Letters* **2012**, *1* (1), 232–235.

50. Motornov, M.; Roiter, Y.; Tokarev, I.; Minko, S., Stimuli-responsive nanoparticles, nanogels and capsules for integrated multifunctional intelligent systems.

Progress in Polymer Science **2010**, *35* (1), 174–211.

51. Ganta, S.; Devalapally, H.; Shahiwala, A.; Amiji, M., A review of stimuli-responsive nanocarriers for drug and gene delivery. *Journal of Controlled Release* **2008**, *126* (3), 187–204.

52. Cayre, O. J.; Chagneux, N.; Biggs, S., Stimulus responsive core-shell nanoparticles: synthesis and applications of polymer based aqueous systems. *Soft Matter* **2011**, *7* (6), 2211–2234.

53. Mura, S.; Nicolas, J.; Couvreur, P., Stimuli-responsive nanocarriers for drug delivery. *Nature Materials* **2013**, *12* (11), 991–1003.

54. Li, M.-H.; Keller, P., Stimuli-responsive polymer vesicles. *Soft Matter* **2009**, *5* (5), 927–937.

55. Ramachandran, D.; Corten, C. C.; Urban, M. W., Color- and shape-tunable colloidal nanoparticles capable of nanopatterning. *RSC Advances* **2013**, *3* (24), 9357–9364.

56. Tu, F.; Lee, D., Shape-changing and amphiphilicity-reversing Janus particles with pH-responsive surfactant properties. *Journal of the American Chemical Society* **2014**, *136* (28), 9999–10006.

57. Yan, Q.; Zhao, Y., CO_2-stimulated diversiform deformations of polymer assemblies. *Journal of the American Chemical Society* **2013**, *135* (44), 16300–16303.

58. Blanazs, A.; Verber, R.; Mykhaylyk, O. O.; Ryan, A. J.; Heath, J. Z.; Douglas, C. I.; Armes, S. P., Sterilizable gels from thermoresponsive block copolymer worms. *Journal of the American Chemical Society* **2012**, *134* (23), 9741–9748.

59. Irie, M., Photoresponsive polymers. In: *New Polymer Materials*, Springer, Berlin, **1990**, pp 27–67.

60. Kumar, G. S.; Neckers, D., Photochemistry of azobenzene-containing polymers. *Chemical Reviews* **1989**, *89* (8), 1915–1925.

61. Ramachandran, D.; Liu, F.; Urban, M. W., Self-repairable copolymers that change color. *RSC Advances* **2012**, *2* (1), 135–143.

62. Klajn, R., Spiropyran-based dynamic materials. *Chemical Society Reviews* **2014**, *43* (1), 148–184.

63. De La Rica, R.; Aili, D.; Stevens, M. M., Enzyme-responsive nanoparticles for drug release and diagnostics. *Advanced Drug Delivery Reviews* **2012**, *64* (11), 967–978.

64. Colson, Y. L.; Grinstaff, M. W., Biologically responsive polymeric nanoparticles for drug delivery. *Advanced Materials* **2012**, *24* (28), 3878–3886.

65. Basel, M. T.; Shrestha, T. B.; Troyer, D. L.; Bossmann, S. H., Protease-sensitive, polymer- caged liposomes: A method for making highly targeted liposomes using triggered release. *ACS Nano* **2011**, *5* (3), 2162–2175.

66. Ge, J.; Neofytou, E.; Cahill III, T. J.; Beygui, R. E.; Zare, R. N., Drug release from electric-field-responsive nanoparticles. *ACS Nano* **2011**, *6* (1), 227–233.

67. Jeon, G.; Yang, S. Y.; Byun, J.; Kim, J. K., Electrically actuatable smart nanoporous membrane for

pulsatile drug release. *Nano Letters* **2011**, *11* (3), 1284–1288.

68. Balogh, D.; Tel-Vered, R.; Freeman, R.; Willner, I., Photochemically and electrochemically triggered Au nanoparticles "sponges". *Journal of the American Chemical Society* **2011**, *133* (17), 6533–6536.

69. Pankhurst, Q. A.; Connolly, J.; Jones, S.; Dobson, J., Applications of magnetic nanoparticles in biomedicine. *Journal of Physics D: Applied Physics* **2003**, *36* (13), R167.

70. Lu, A. H.; Salabas, E. e. L.; Schüth, F., Magnetic nanoparticles: Synthesis, protection, functionalization, and application. *Angewandte Chemie International Edition* **2007**, *46* (8), 1222–1244.

71. Lee, J.-H.; Huh, Y.-M.; Jun, Y.-W.; Seo, J.-W.; Jang, J.-T.; Song, H.-T.; Kim, S.; Cho, E.-J.; Yoon, H.-G.; Suh, J.-S., Artificially engineered magnetic nanoparticles for ultra-sensitive molecular imaging. *Nature Medicine* **2007**, *13* (1), 95–99.

72. Reddy, L. H.; Arias, J. L.; Nicolas, J.; Couvreur, P., Magnetic nanoparticles: Design and characterization, toxicity and biocompatibility, pharmaceutical and biomedical applications. *Chemical Reviews* **2012**, *112* (11), 5818–5878.

73. Colombo, M.; Carregal-Romero, S.; Casula, M. F.; Gutiérrez, L.; Morales, M. P.; Böhm, I. B.; Heverhagen, J. T.; Prosperi, D.; Parak, W. J., Biological applications of magnetic nanoparticles. *Chemical Society Reviews* **2012**, *41* (11), 4306–4334.

74. Guo, S.; Wang, E., Synthesis and electrochemical applications of gold nanoparticles. *Analytica Chimica Acta* **2007**, *598* (2), 181–192.

75. Luo, X.; Morrin, A.; Killard, A. J.; Smyth, M. R., Application of nanoparticles in electrochemical sensors and biosensors. *Electroanalysis* **2006**, *18* (4), 319–326.

76. Drummond, T. G.; Hill, M. G.; Barton, J. K., Electrochemical DNA sensors. *Nature Biotechnology* **2003**, *21* (10), 1192–1199.

77. Suni, I. I., Impedance methods for electrochemical sensors using nanomaterials. *TrAC Trends in Analytical Chemistry* **2008**, *27* (7), 604–611.

78. Shipway, A. N.; Katz, E.; Willner, I., Nanoparticle arrays on surfaces for electronic, optical, and sensor applications. *ChemPhysChem* **2000**, *1* (1), 18–52.

79. Jiang, S.; Chen, Q.; Tripathy, M.; Luijten, E.; Schweizer, K. S.; Granick, S., Janus particle synthesis and assembly. *Advanced Materials* **2010**, *22* (10), 1060–1071.

80. Glotzer, S. C.; Solomon, M. J., Anisotropy of building blocks and their assembly into complex structures. *Nature Materials* **2007**, *6* (8), 557–562.

81. Walther, A.; Müller, A. H., Janus particles: Synthesis, self-assembly, physical properties, and applications. *Chemical Reviews* **2013**, *113* (7), 5194–5261.

82. Lu, Y.; Xiong, H.; Jiang, X.; Xia, Y.; Prentiss, M.; Whitesides, G. M., Asymmetric dimers can be formed by dewetting half-shells of gold deposited on the surfaces of spherical oxide colloids. *Journal of the American Chemical Society* **2003**, *125* (42), 12724–12725.

83. Yu, H.; Chen, M.; Rice, P. M.; Wang, S. X.; White, R.; Sun, S., Dumbbell-like bifunctional Au-Fe$_3$O$_4$ nanoparticles. *Nano Letters* **2005**, *5* (2), 379–382.

84. Gu, H.; Zheng, R.; Zhang, X.; Xu, B., Facile one-pot synthesis of bifunctional heterodimers of nanoparticles: A conjugate of quantum dot and magnetic nanoparticles. *Journal of the American Chemical Society* **2004**, *126* (18), 5664–5665.

85. Zhang, L.; Eisenberg, A., Multiple morphologies of "crew-cut" aggregates of polystyrene-b-poly (acrylic acid) block copolymers. *Science* **1995**, *268* (5218), 1728–1731.

86. Li, Z.; Kesselman, E.; Talmon, Y.; Hillmyer, M. A.; Lodge, T. P., Multicompartment micelles from ABC miktoarm stars in water. *Science* **2004**, *306* (5693), 98–101.

87. Cui, H.; Chen, Z.; Zhong, S.; Wooley, K. L.; Pochan, D. J., Block copolymer assembly via kinetic control. *Science* **2007**, *317* (5838), 647–650.

88. Erhardt, R.; Böker, A.; Zettl, H.; Kaya, H.; Pyckhout-Hintzen, W.; Krausch, G.; Abetz, V.; Müller, A. H., Janus micelles. *Macromolecules* **2001**, *34* (4), 1069–1075.

89. Wang, Y.; Wang, Y.; Breed, D. R.; Manoharan, V. N.; Feng, L.; Hollingsworth, A. D.; Weck, M.; Pine, D. J., Colloids with valence and specific directional bonding. *Nature* **2012**, *491* (7422), 51–55.

90. Roh, K.-H.; Martin, D. C.; Lahann, J., Biphasic Janus particles with nanoscale anisotropy. *Nature Materials* **2005**, *4* (10), 759–763.

91. Takei, H.; Shimizu, N., Gradient sensitive microscopic probes prepared by gold evaporation and chemisorption on latex spheres. *Langmuir* **1997**, *13* (7), 1865–1868.

92. Smoukov, S. K.; Gangwal, S.; Marquez, M.; Velev, O. D., Reconfigurable responsive structures assembled from magnetic Janus particles. *Soft Matter* **2009**, *5* (6), 1285–1292.

93. Paunov, V. N.; Cayre, O. J., Supraparticles and "Janus" particles fabricated by replication of particle monolayers at liquid surfaces using a gel trapping technique. *Advanced Materials* **2004**, *16* (9–10), 788–791.

94. Liu, B.; Wei, W.; Qu, X.; Yang, Z., Janus colloids formed by biphasic grafting at a Pickering emulsion interface. *Angewandte Chemie* **2008**, *120* (21), 4037–4039.

95. Jiang, S.; Granick, S., Controlling the geometry (Janus balance) of amphiphilic colloidal particles. *Langmuir* **2008**, *24* (6), 2438–2445.

96. Reculusa, S.; Poncet-Legrand, C.; Ravaine, S.; Mingotaud, C.; Duguet, E.; Bourgeat-Lami, E., Syntheses of raspberrylike silica/polystyrene materials. *Chemistry of Materials* **2002**, *14* (5), 2354–2359.

97. Yake, A. M.; Snyder, C. E.; Velegol, D., Site-specific functionalization on individual colloids: Size control, stability, and multilayers. *Langmuir* **2007**, *23* (17), 9069–9075.

98. Corten, C. C.; Urban, M. W., Shape evolution control of phase-separated colloidal nanoparticles. *Polymer Chemistry* **2011**, *2* (1), 244–250.

99. Tanaka, T.; Okayama, M.; Kitayama, Y.; Kagawa, Y.; Okubo, M., Preparation of "Mushroom-like" Janus particles by site-selective surface-initiated atom transfer radical polymerization in aqueous dispersed systems. *Langmuir* **2010**, *26* (11), 7843–7847.

100. Walther, A.; Hoffmann, M.; Muller, A. H. E., Emulsion polymerization using Janus particles as stabilizers. *Angewandte Chemie-International Edition* **2008**, *47* (4), 711–714.

101. Love, J. C.; Gates, B. D.; Wolfe, D. B.; Paul, K. E.; Whitesides, G. M., Fabrication and wetting properties of metallic half-shells with submicron diameters. *Nano Letters* **2002**, *2* (8), 891–894.

102. Choi, J.; Zhao, Y.; Zhang, D.; Chien, S.; Lo, Y.-H., Patterned fluorescent particles as nanoprobes for the investigation of molecular interactions. *Nano Letters* **2003**, *3* (8), 995–1000.

103. Ling, X. Y.; Phang, I. Y.; Acikgoz, C.; Yilmaz, M. D.; Hempenius, M. A.; Vancso, G. J.; Huskens, J., Janus particles with controllable patchiness and their chemical functionalization and supramolecular assembly. *Angewandte Chemie* **2009**, *121* (41), 7813–7818.

104. Li, B.; Li, C. Y., Immobilizing Au nanoparticles with polymer single crystals, patterning and asymmetric functionalization. *Journal of the American Chemical Society* **2007**, *129* (1), 12–13.

105. Li, B.; Ni, C.; Li, C. Y., Poly (ethylene oxide) single crystals as templates for Au nanoparticle patterning and asymmetrical functionalization. *Macromolecules* **2008**, *41* (1), 149–155.

106. Wang, B.; Li, B.; Zhao, B.; Li, C. Y., Amphiphilic Janus gold nanoparticles via combining "solid-state grafting-to" and "grafting-from" methods. *Journal of the American Chemical Society* **2008**, *130* (35), 11594–11595.

107. Liu, L.; Ren, M.; Yang, W., Preparation of polymeric Janus particles by directional UV-induced reactions. *Langmuir* **2009**, *25* (18), 11048–11053.

108. McConnell, M. D.; Kraeutler, M. J.; Yang, S.; Composto, R. J., Patchy and multiregion janus particles with tunable optical properties. *Nano Letters* **2010**, *10* (2), 603–609.

109. Gong, J.; Zu, X.; Li, Y.; Mu, W.; Deng, Y., Janus particles with tunable coverage of zinc oxide nanowires. *Journal of Materials Chemistry* **2011**, *21* (7), 2067–2069.

110. Lin, C.-C.; Liao, C.-W.; Chao, Y.-C.; Kuo, C., Fabrication and characterization of asymmetric Janus and ternary particles. *ACS Applied Materials & Interfaces* **2010**, *2* (11), 3185–3191.

111. Hong, L.; Jiang, S.; Granick, S., Simple method to produce Janus colloidal particles in large quantity. *Langmuir* **2006**, *22* (23), 9495–9499.

112. Nie, Z.; Li, W.; Seo, M.; Xu, S.; Kumacheva, E., Janus and ternary particles generated by microfluidic synthesis: Design, synthesis, and self-assembly. *Journal of the American Chemical Society* **2006**, *128* (29), 9408–9412.

113. Nisisako, T.; Torii, T.; Takahashi, T.; Takizawa, Y., Synthesis of monodisperse bicolored Janus particles with electrical anisotropy using a microfluidic co-flow system. *Advanced Materials* **2006**, *18* (9), 1152–1156.

114. Tanaka, T.; Okayama, M.; Minami, H.; Okubo, M., Dual stimuli-responsive "Mushroom-like" Janus polymer particles as particulate surfactants. *Langmuir* **2010**, *26* (14), 11732–11736.

115. Gröschel, A. H.; Walther, A.; Löbling, T. I.; Schmelz, J.; Hanisch, A.; Schmalz, H.; Müller, A. H., Facile, solution-based synthesis of soft, nanoscale Janus particles with tunable Janus balance. *Journal of the American Chemical Society* **2012**, *134* (33), 13850–13860.

116. Walther, A.; André, X.; Drechsler, M.; Abetz, V.; Müller, A. H., Janus discs. *Journal of the American Chemical Society* **2007**, *129* (19), 6187–6198.

117. Singh, A.; Dreher, W. R.; Urban, M. W., Phospholipid-assisted synthesis of stable F- containing colloidal particles and their film formation. *Langmuir* **2006**, *22* (2), 524–527.

118. Wang, J. Y.; Yang, X. F., Synthesis of core-corona polymer hybrids with a raspberry-like structure by the heterocoagulated pyridinium reaction. *Langmuir* **2008**, *24* (7), 3358–3364.

119. Tsai, H. J.; Lee, Y. L., Facile method to fabricate raspberry-like particulate films for superhydrophobic surfaces. *Langmuir* **2007**, *23* (25), 12687–12692.

120. Li, G. L.; Yang, X. L.; Bai, F.; Huang, W. Q., Raspberry-like composite polymer particles by self-assemble heterocoagulation based on a charge compensation process. *Journal of Colloid and Interface Science* **2006**, *297* (2), 705–710.

121. Lu, C.; Urban, M. W., Synthesis and directional assembly of gibbous and inverse-gibbous colloidal nanoparticles. *Materials Today Communications* **2016**, *9*, 41–46.

122. Song, Y. Y.; Zhou, J. J.; Fan, J. B.; Zhai, W. Z.; Meng, J. X.; Wang, S. T., Hydrophilic/Oleophilic magnetic Janus particles for the rapid and efficient oil-water separation. *Advanced Functional Materials* **2018**, *28* (32), 1802493.

123. Ding, Y.; Su, S.; Zhang, R.; Shao, L.; Zhang, Y.; Wang, B.; Li, Y.; Chen, L.; Yu, Q.; Wu, Y.; Nie, G., Precision combination therapy for triple negative

breast cancer via biomimetic polydopamine polymer core-shell nanostructures. *Biomaterials* **2017**, *113*, 243–252.

124. Su, S. S.; Tian, Y. H.; Li, Y. Y.; Ding, Y. P.; Ji, T. J.; Wu, M. Y.; Wu, Y.; Nie, G. J., "Triple-Punch" strategy for triple negative breast cancer therapy with minimized drug dosage and improved antitumor efficacy. *Acs Nano* **2015**, *9* (2), 1367–1378.

125. Mahmoudi, M.; Sant, S.; Wang, B.; Laurent, S.; Sen, T., Superparamagnetic iron oxide nanoparticles (SPIONs): Development, surface modification and applications in chemotherapy. *Advanced Drug Delivery Reviews* **2011**, *63* (1–2), 24–46.

126. Brazel, C. S., Magnetothermally-responsive nanomaterials: Combining magnetic nanostructures and thermally-sensitive polymers for triggered drug release. *Pharmaceutical Research* **2009**, *26* (3), 644–656.

127. Lale, S. V.; Koul, V., Stimuli-responsive polymeric nanoparticles for cancer therapy. In: Thakur V., Thakur M., Voicu S. (eds.) *Polymer Gels: Perspectives and Applications.* Springer, Singapore, **2018**, pp. 27–54.

128. Lee, S.-Y.; Yang, C.-Y.; Peng, C.-L.; Wei, M.-F.; Chen, K.-C.; Yao, C.-J.; Shieh, M.-J., A theranostic micelleplex co-delivering SN-38 and VEGF siRNA for colorectal cancer therapy. *Biomaterials* **2016**, *86*, 92–105.

129. Amstad, E.; Kohlbrecher, J.; Mueller, E.; Schweizer, T.; Textor, M.; Reimhult, E., Triggered release from liposomes through magnetic actuation of iron oxide nanoparticle containing membranes. *Nano Letters* **2011**, *11* (4), 1664–1670.

130. Urban, M. W., *Stimuli-Responsive Materials: From Molecules to Nature Mimicking Materials Design.* Royal Society of Chemistry: Cambridge, UK, **2016**.

131. Zirbs, R.; Lassenberger, A.; Vonderhaid, I.; Kurzhals, S.; Reimhult, E., Melt-grafting for the synthesis of core-shell nanoparticles with ultra-high dispersant density. *Nanoscale* **2015**, *7* (25), 11216–11225.

132. Kurzhals, S.; Zirbs, R.; Reimhult, E., Synthesis and Magneto-thermal actuation of iron oxide core-PNIPAM shell nanoparticles. *Acs Applied Materials & Interfaces* **2015**, *7* (34), 19342–19352.

133. Bielawski, C. W.; Grubbs, R. H., Living ring-opening metathesis polymerization. *Progress in Polymer Science* **2007**, *32* (1), 1–29.

134. Braunecker, W. A.; Matyjaszewski, K., Controlled/living radical polymerization: Features, developments, and perspectives. *Progress in Polymer Science* **2007**, *32* (1), 93–146.

135. Erhardt, R.; Zhang, M.; Böker, A.; Zettl, H.; Abetz, C.; Frederik, P.; Krausch, G.; Abetz, V.; Müller, A. H., Amphiphilic Janus micelles with polystyrene and poly (methacrylic acid) hemispheres. *Journal of the American Chemical Society* **2003**, *125* (11), 3260–3267.

136. Liu, S.; Armes, S. P., Polymeric surfactants for the new millennium: A pH-responsive, zwitterionic, schizophrenic diblock copolymer. *Angewandte Chemie International Edition* **2002**, *41* (8), 1413–1416.

137. Jenekhe, S. A.; Chen, X. L., Self-assembled aggregates of rod-coil block copolymers and their solubilization and encapsulation of fullerenes. *Science* **1998**, *279* (5358), 1903–1907.

138. Lee, M.; Cho, B.-K.; Zin, W.-C., Supramolecular structures from rod-coil block copolymers. *Chemical Reviews* **2001**, *101* (12), 3869–3892.

139. Wang, X.; Guerin, G.; Wang, H.; Wang, Y.; Manners, I.; Winnik, M. A., Cylindrical block copolymer micelles and co-micelles of controlled length and architecture. *Science* **2007**, *317* (5838), 644–647.

140. Gädt, T.; Ieong, N. S.; Cambridge, G.; Winnik, M. A.; Manners, I., Complex and hierarchical micelle architectures from diblock copolymers using living, crystallization-driven polymerizations. *Nature Materials* **2009**, *8* (2), 144–150.

141. Raez, J.; Manners, I.; Winnik, M. A., Nanotubes from the self-assembly of asymmetric crystalline-coil poly (ferrocenylsilane-siloxane) block copolymers. *Journal of the American Chemical Society* **2002**, *124* (35), 10381–10395.

142. Wan, W.-M.; Sun, X.-L.; Pan, C.-Y., Morphology transition in RAFT polymerization for formation of vesicular morphologies in one pot. *Macromolecules* **2009**, *42* (14), 4950–4952.

143. Warren, N. J.; Armes, S. P., Polymerization-induced self-assembly of block copolymer nano-objects via RAFT aqueous dispersion polymerization. *Journal of the American Chemical Society* **2014**, *136* (29), 10174–10185.

144. Sugihara, S.; Blanazs, A.; Armes, S. P.; Ryan, A. J.; Lewis, A. L., Aqueous dispersion polymerization: a new paradigm for in situ block copolymer self-assembly in concentrated solution. *Journal of the American Chemical Society* **2011**, *133* (39), 15707–15713.

145. Boissé, S.; Rieger, J.; Belal, K.; Di-Cicco, A.; Beaunier, P.; Li, M.-H.; Charleux, B., Amphiphilic block copolymer nano-fibers via RAFT-mediated polymerization in aqueous dispersed system. *Chemical Communications* **2010**, *46* (11), 1950–1952.

146. Wan, W.-M.; Pan, C.-Y., One-pot synthesis of polymeric nanomaterials via RAFT dispersion polymerization induced self-assembly and re-organization. *Polymer Chemistry* **2010**, *1* (9), 1475–1484.

147. Wan, W.-M.; Hong, C.-Y.; Pan, C.-Y., One-pot synthesis of nanomaterials via RAFT polymerization induced self-assembly and morphology transition. *Chemical Communications* **2009**, (39), 5883–5885.

148. Chambon, P.; Blanazs, A.; Battaglia, G.; Armes, S., Facile synthesis of methacrylic ABC triblock copolymer vesicles by RAFT aqueous dispersion

polymerization. *Macromolecules* **2012**, *45* (12), 5081–5090.

149. Zhang, X.; Boissé, S.; Zhang, W.; Beaunier, P.; D'Agosto, F.; Rieger, J.; Charleux, B., Well-defined amphiphilic block copolymers and nano-objects formed in situ via RAFT-mediated aqueous emulsion polymerization. *Macromolecules* **2011**, *44* (11), 4149–4158.

150. Zhang, W.; D'Agosto, F.; Boyron, O.; Rieger, J.; Charleux, B., Toward a better understanding of the parameters that lead to the formation of nonspherical polystyrene particles via RAFT-mediated one-pot aqueous emulsion polymerization. *Macromolecules* **2012**, *45* (10), 4075–4084.

151. Delaittre, G.; Dire, C.; Rieger, J.; Putaux, J.-L.; Charleux, B., Formation of polymer vesicles by simultaneous chain growth and self-assembly of amphiphilic block copolymers. *Chemical Communications* **2009**, (20), 2887–2889.

152. Won, Y.-Y.; Davis, H. T.; Bates, F. S., Giant wormlike rubber micelles. *Science* **1999**, *283* (5404), 960–963.

153. Geng, Y.; Dalhaimer, P.; Cai, S.; Tsai, R.; Tewari, M.; Minko, T.; Discher, D. E., Shape effects of filaments versus spherical particles in flow and drug delivery. *Nature Nanotechnology* **2007**, *2* (4), 249.

154. Blanazs, A.; Madsen, J.; Battaglia, G.; Ryan, A. J.; Armes, S. P., Mechanistic insights for block copolymer morphologies: How do worms form vesicles? *Journal of the American Chemical Society* **2011**, *133* (41), 16581–16587.

155. Wang, X. S.; Winnik, M. A.; Manners, I., Swellable, redox-active shell-crosslinked organometallic nanotubes. *Angewandte Chemie International Edition* **2004**, *43* (28), 3703–3707.

156. Wang, X.; Wang, H.; Frankowski, D. J.; Lam, P. G.; Welch, P. M.; Winnik, M. A.; Hartmann, J.; Manners, I.; Spontak, R. J., Growth and Crystallization of metal-containing block copolymer nanotubes in a selective solvent. *Advanced Materials* **2007**, *19* (17), 2279–2285.

157. Yu, K.; Zhang, L.; Eisenberg, A., Novel morphologies of "crew-cut" aggregates of amphiphilic diblock copolymers in dilute solution. *Langmuir* **1996**, *12* (25), 5980–5984.

158. Yu, K.; Eisenberg, A., Bilayer morphologies of self-assembled crew-cut aggregates of amphiphilic PS-b-PEO diblock copolymers in solution. *Macromolecules* **1998**, *31* (11), 3509–3518.

159. Grumelard, J.; Taubert, A.; Meier, W., Soft nanotubes from amphiphilic ABA triblock macromonomers. *Chemical communications* **2004**, (13), 1462–1463.

160. Yan, Q.; Zhao, Y., Polymeric microtubules that breathe: CO_2-driven polymer controlled-self-assembly and shape transformation. *Angewandte Chemie* **2013**, *125* (38), 10132–10135.

161. Long, Y.-Z.; Li, M.-M.; Gu, C.; Wan, M.; Duvail, J.-L.; Liu, Z.; Fan, Z., Recent advances in synthesis, physical properties and applications of conducting polymer nanotubes and nanofibers. *Progress in Polymer Science* **2011**, *36* (10), 1415–1442.

162. Yu, M.; Urban, M. W., Formation of concentric ferromagnetic nanotubes from biologically active phospholipids. *Journal of Materials Chemistry* **2007**, *17* (44), 4644–4646.

163. Kango, S.; Kalia, S.; Celli, A.; Njuguna, J.; Habibi, Y.; Kumar, R., Surface modification of inorganic nanoparticles for development of organic–inorganic nanocomposites—A review. *Progress in Polymer Science* **2013**, *38* (8), 1232–1261.

164. Nie, Z.; Petukhova, A.; Kumacheva, E., Properties and emerging applications of self-assembled structures made from inorganic nanoparticles. *Nature Nanotechnology* **2010**, *5* (1), 15–25.

165. Kamat, P. V., Quantum dot solar cells. Semiconductor nanocrystals as light harvesters. *The Journal of Physical Chemistry C* **2008**, *112* (48), 18737–18753.

166. Medintz, I. L.; Uyeda, H. T.; Goldman, E. R.; Mattoussi, H., Quantum dot bioconjugates for imaging, labelling and sensing. *Nature Materials* **2005**, *4* (6), 435–446.

167. Fedlheim, D. L.; Foss, C. A., *Metal Nanoparticles: Synthesis, Characterization, and Applications*. CRC Press, Boca Raton, FL, **2001**.

168. Daniel, M.-C.; Astruc, D., Gold nanoparticles: Assembly, supramolecular chemistry, quantum-size-related properties, and applications toward biology, catalysis, and nanotechnology. *Chemical Reviews* **2004**, *104* (1), 293–346.

169. Gao, J.; Gu, H.; Xu, B., Multifunctional magnetic nanoparticles: Design, synthesis, and biomedical applications. *Accounts of Chemical Research* **2009**, *42* (8), 1097–1107.

170. Steward, P.; Hearn, J.; Wilkinson, M., An overview of polymer latex film formation and properties. *Advances in Colloid and Interface Science* **2000**, *86* (3), 195–267.

171. Rösler, A.; Vandermeulen, G. W.; Klok, H.-A., Advanced drug delivery devices via self-assembly of amphiphilic block copolymers. *Advanced Drug Delivery Reviews* **2012**, *64*, 270–279.

172. Urban, M. W.; Davydovich, D.; Yang, Y.; Demir, T.; Zhang, Y.; Casabianca, L., Key-and-lock commodity self-healing copolymers. *Science* **2018**, *362* (6411), 220–225.

3

Nanoparticle Superlattices

Steffi S. Thomas
Trinity College Dublin

Fabian Meder
Istituto Italiano di Tecnologia

3.1 Introduction

Nanostructured and nanopatterned materials have multiple uses in technological applications and in many research areas due to the functional properties arising from the nanoscale features. In addition, they occur in nature in many forms and are essential for all organisms. Various tools established and advanced in the last years to tailor matter at the nanoscale for numerous applications in fields spanning from electronics, optics, surface technology, towards biology and medicine. Structuring different materials in the sub-100 nm scale in a controlled way is facilitated by diverse top-down lithographical techniques. However, arranging, positioning, and combining in smaller size scales such as sub-20 nm is significantly more difficult. At this scale, a bottom-up self-assembly is more promising than top-down fabrication inspired by the assembly of atoms or molecules in nature at similar length scales. As nanoparticles can now be synthesized from many materials and in controlled shapes, they constitute ideal building blocks to assemble larger structures. Approaches based on self-assembly of nanoparticles as "pre-fabricated nano-objects" of multiple functionalities have the potential to be cost-effective and easier to use compared to lithographic methods (Gang 2016). However, nanoparticle self-assembly has limitations regarding predictability and control over structures. Current research attempts to control nanoparticle self-assembly at the nanoscale to precisely position individual components. Under certain conditions, nanoparticles assemble into crystalline lattices (nanoparticle superlattices, NSLs) (Collier et al. 1998). In other words, nanoparticles repeat in periodic distances like atoms in a crystal. The combination of nanoparticles in a crystal lattice results in interesting properties. The materials based on NSLs can have properties

greater than the expected sum of those of its individual nanoparticle components pointing towards the opportunity of new functional bottom-up materials (Heath 2007). The research on NSLs initiated in the 1990s by the works of Fendler (Kotov et al. 1994), Murray et al. (1995), Mirkin et al. (1996) and others finding that the self-assembly of NSLs creates ordering of the nanoparticle building blocks, depending on factors such as the nanoparticle material, size, shape, and in particular their surface chemistry (that ranges from simple organic ligands to engineered DNA sequences). A central perspective of NSLs over other techniques for obtaining nanostructures was proposed by Weller in a highlight article from 1996 that costly equipment for nanostructuring could be replaced by "round bottom flask and beaker" by using nanoparticles as building blocks for structural inorganic chemistry like the atoms in molecular chemistry (Weller 1996). Advances in the synthesis of nanoparticles including asymmetric particles and directional interactions have led to lattice structures that appear in similar and dissimilar structures as in atomic or molecular crystals. As a result, nanoparticles can be seen as a form of matter in their own right and controlling their self-assembly facilitates the synthesis of novel, nanostructured materials (Manoharan 2015). In summary, NSL self-assembly enables unique functional materials by a modular design comprising nanoscale precision structures with potential for advancing many areas of future technologies. This chapter gives an overview of NSLs research mainly from the last decade focusing on NSLs made from ~1 to 20 nm nanoparticles in a size range where electronic, magnetic, and optical properties are prominent. Assemblies of larger nano- and microparticles such as colloidal crystals and synthetic opals not falling into this class are given elsewhere, e.g., in Manoharan (2015). The chapter begins with an introduction of the basic driving forces for NSLs formation. We give an overview of nanoparticle building blocks and their surface chemistries as well as NSL substrates of different properties and geometry. Subsequently, we highlight the achievable control over building block ordering resulting in distinct NSL crystalline structures and give an overview of techniques to analyze and test the NSL properties. The chapter closes with examples of existing and upcoming technical applications of NSLs.

3.2 Self-assembly of NSLs

Primary components for NSLs self-assembly are nanoparticles. Figure 3.1a illustrates a spherical sub-20 nm nanoparticle as typically used in NSLs. It is important to consider that the nanoparticle is surrounded by ligands, organic molecules that stabilize the nanoparticle in a medium (e.g., an organic solvent), which contribute to the effective particle diameter and they govern and introduce specific surface properties. The ligands are essential for NSL assembly and structure formation, as detailed in Section 3.2. Under certain conditions, the nanoparticles self-assemble into an ordered structure, the NSL, a process which in many parts compares to a molecular crystallization process. Two types of monodisperse nanoparticles may assemble in binary NSLs as depicted in Figure 3.1b. In the crystal lattice, nanoparticle building blocks repeat periodically with defined interparticle distances, a key parameter for most NSL applications. Examples for binary NSLs and their corresponding crystal unit cells are shown

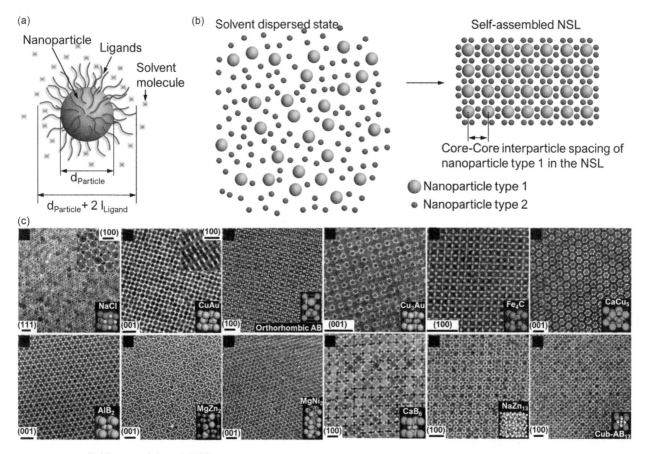

FIGURE 3.1 Self-assembly of NSLs. (a) Schematic of a spherical nanoparticle (e.g., metallic, semiconductor, metal oxide etc.) such as used in NSL self-assembly and the organic ligands grafted on the surface to stabilize and disperse the particle in the solvent. The effective particle diameter takes account of the core particle diameter (d_{Particle}) and the ligand length ($d_{\text{Particle}} + 2\ l_{\text{Ligand}}$). (b) Schematic of an NSL formed by self-assembly of a binary dispersion of monodisperse nanoparticles (two types of nanoparticles, e.g., different materials, different sizes). The self-assembly is typically induced by increasing the nanoparticle volume fraction, e.g., by solvent evaporation, leading to NSL self-assembly. Here, the {100} plane of a cubic AB_{13} lattice schematizes particle ordering with periodic interparticle distances in the NSL. The illustration does not show ligands and solvent molecules (for clarity) which are yet essential for self-assembly. (c) TEM images of binary NSLs with various structures assembled from different nanoparticles and modeled unit cells (left to right, first row: 13.4 nm γ-Fe_2O_3 and 5.0 nm Au; 7.6 nm PbSe and 5.0 nm Au; 6.2 nm PbSe and 3.0 nm Pd; 7.2 nm PbSe and 4.2 nm Ag; 6.2 nm PbSe and 3.0 nm Pd; 7.2 nm PbSe and 5.0 nm Au; second row: 6.7 nm PbS and 3.0 nm Pd; 6.2 nm PbSe and 3.0 nm Pd; 5.8 nm PbSe and 3.0 nm Pd 5.8 nm PbSe and 3.0 nm Pd; 7.2 nm PbSe and 4.2 nm Ag; 6.2 nm PbSe and 3.0 nm Pd nanoparticles. Scale bars: 20 nm, the lattice projection is labeled in each panel above the scale bar. (Adapted by permission from *Springer Nature: Nature* Shevchenko et al. 2006, copyright 2006.)

in Figure 3.1c (Shevchenko et al. 2006). The transmission electron microscopy (TEM) images give an indication of the diverse structural patterns that can be generated by NSL self-assembly in sub-10 nm periodicity, depending on types and dimensions of building blocks and their interactions. Key factors that drive NSL assembly are size and shape monodispersity of the nanoparticles, size ratio in multicomponent NSLs, particle morphology, nanoparticle concentrations, solvents, surface ligands, substrates, any molecules in the dispersant, pressure, and temperature.

Figure 3.2a–d shows the three approaches that are used or combined for NSL synthesis: (i) the controlled evaporation of a solvent in which monodisperse nanoparticles are dispersed in, (ii) the moderate destabilization of a nanoparticle solution and (iii) by biorecognition-driven reactions (e.g., DNA base-pair interactions) (Murray et al. 2000, Talapin et al. 2001, Prasad et al. 2008, Jones et al. 2015, Luo et al. 2015).

A widely used technique to obtain NSLs for research purposes is the straight-forward solvent evaporation approach that may be up-scaled, e.g., for wafer-scale production (Gaulding et al. 2015). For coating a solid substrate in the research laboratory, yet often a TEM sample grid, the nanoparticle building blocks are synthesized and dispersed in a volatile organic solvent (toluene, tetrachloroethylene, etc.). NSLs form when the particle concentration is increased by slowly evaporating the solvent at temperatures of 20°C–70°C under ambient or reduced

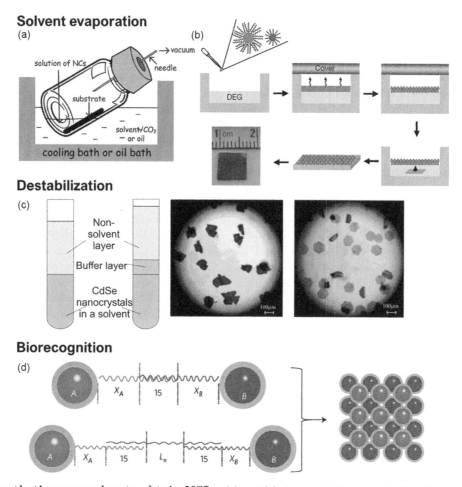

FIGURE 3.2 Synthetic approaches to obtain NSLs. (a) and (b) show synthetic setups in the solvent evaporation approach to assemble NSLs directly on solid substrates or, respectively, at the air–liquid interface from which it can be transferred on other substrates. (c) Destabilization of a nanoparticle dispersion to obtain NSLs. Left tube: A non-solvent diffuses directly into a colloidal solution of CdSe nanoparticles. Right tube: A third solvent acts as buffer layer and slows down the diffusion of the non-solvent. Optical micrographs of the obtained CdSe NSLs: Left, faster nucleation, irregular-shaped NSLs; right, slower nucleation, perfectly faceted hexagonal platelets NSLs. (d) Illustration of biorecognition-directed NSL formation by the use of DNA strands grafted on the surface of the nanoparticles. The schematic shows two kinds of assembly strategies, the upper drawing depicts a direct hybridization by base-pair interactions between different nucleic acids (X_A, X_B) grafted on two types of nanoparticles. The lower drawing shows the linker-assisted strategy in which a specific linker (L_n) is used to crosslink DNA on the nanoparticles. ((a) Adapted with permission from Bodnarchuk et al. 2010, copyright 2010 American Chemical Society. (b) Adapted by permission from *Springer Nature: Nature* Dong et al. 2010, copyright 2010. (c) Adapted by permission from John Wiley and Sons: *Advanced Materials* Talapin et al. 2001, copyright 2001. (d) Adapted by permission from *Springer Nature: Nature Nanotechnology* Zhang et al. 2013, copyright 2013.)

pressure. Assembly initiates at about a particle volume concentration of about 55% (Shevchenko et al. 2006). The NSLs eventually deposit on the substrate which is slightly tilted and immersed in the nanoparticle solution (Murray et al. 2000, Bodnarchuk et al. 2010, Luo et al. 2015). Figure 3.2a illustrates a synthetic setup. A similar procedure can be followed to assemble NSLs on liquid surfaces by dropping the nanoparticles in a hydrophobic, oily solvent on the surface of a hydrophilic liquid (Figure 3.2b). When the solvent evaporates, the NSLs form on the liquid/air interface and may subsequently be transferred on a substrate (Dong et al. 2010, Korgel 2010). NSLs can also be obtained in solution without evaporating their dispersion media by destabilization of the nanoparticle dispersion. This can be, for instance, achieved by a diffusion-controlled addition of a non-solvent into a nanoparticle solution as shown in Figure 3.2c (Talapin et al. 2001). Another commonly used approach for controlling NSL formation is exploiting the programmable nature of DNA bonds. Tethering complementary single-stranded DNA linkers to the surface of the nanoparticles with specifically designed sequences that hybridize with other nucleic acid strands grafted onto a second nanoparticle, enables to program the interactions between the particles as illustrated in Figure 3.2d (Macfarlane et al. 2011, Zhang et al. 2013, Jones et al. 2015). Upon mixing the different components, nanoparticles aggregate in a controlled manner in solution forming NSLs with structures that are plannable by designing the DNA ligand's base-pair interactions.

The assembly of nanoparticles into NSLs in the commonly used solvent evaporation and the destabilization process is a crystallization reaction, which is determined by nucleation and growth processes (Sigman et al. 2004, Bodnarchuk et al. 2011). For the formation of colloidal crystals from micron-sized particles, the structure can usually be predicted considering the packing of hard spheres (Manoharan 2015). However, for sub-20 nm containing NSLs, these models may not always apply because additional interaction forces occurring at the small nanoscale may govern the assembly and result in deviations from structures predicted by geometric packing arguments (Shevchenko et al. 2006, Batista et al. 2015).

From a thermodynamic perspective, the equilibrium configuration for a large ensemble of particles can be treated as minimization of the Helmholtz free energy (F) (Bodnarchuk et al. 2010). F is given by the internal energy (U), entropy (S), and temperature (T) in a closed system: $F = U - TS$ (Bodnarchuk et al. 2010, Manoharan 2015). The contributions to the free energy of the equilibrium nanoparticle arrangement can be subdivided into energetic and entropic contributions. Larger colloidal particles can be considered as non-interacting hard spheres and merely entropic forces govern their ordering due to the increased local free space that is available for each sphere in an ordered lattice compared to the disordered state. Small nanoparticle crystallization can also be mainly entropy-driven and then described by hard-sphere packing

(Evers et al. 2010). The entropy-driven assembly of spherical particles reveal that face-centered cubic (fcc) structures are slightly more stable compared to the hexagonal close-packed (hcp) structures with a free energy difference of about $\sim 10^{-3} k_B T$ (Bolhuis et al. 1997, Talapin et al. 2007). In agreement, fcc and hcp structures with identical packing density $\sim 74\%$ are often found in assemblies of colloidal particles whereas formation of the fcc is thermodynamically and kinetically favored (Talapin et al. 2007). Such ordering of monodisperse particles allows larger local free space available for each particle compared to the unstructured phase resulting in higher translational entropy of the spheres and drives the phase transition when the volume fraction of particles reaches a threshold of about $\sim 55\%$ (Shevchenko et al. 2006). In a binary mixture of spherical particles which have two different sizes, the packing symmetry is then often a function of the size ratio of the small and large particles ($\gamma = R_{small}/R_{large}$) (Shevchenko et al. 2006). When energetic interactions between the particles are negligible, the particles assemble in a lattice structure that maximizes the free volume available to individual spheres (Boles and Talapin 2015). Particles can translocate until volume fractions high as about $\sim 74\%$ whereas random arrangement restricts relocation already at volume fractions of about $\sim 64\%$ (Boles and Talapin 2015). Simulations have shown that entropy alone can drive the binary nanoparticle mixtures to assemble into NSLs with AB (e.g., NaCl), AB_2 (e.g., AlB_2), and AB_{13} (e.g., $NaZn_{13}$) unit cell configurations in the lattice (Shevchenko et al. 2006, Boles and Talapin 2015, Travesset 2015, 2017).

Nevertheless, often NSLs (such as those shown in Figure 3.1c) are found to occur with significantly lower packing density as the single phase face-centered cubic close packing ($\sim 74\%$ packing density) expected from entropy-driven assembly (Shevchenko et al. 2006). At sub-20 nm particle size scales, the particle dimension approaches the size ranges of molecules and surface ligands. Hence further forces more dominantly occurring at the small nanoscale influence the NSL assembly. One aspect, for instance, simply results from the smaller mass of a sub-20 nm nanoparticle, is its increased particle velocity compared to larger colloids. The sub-20 nm particle's thermal velocity is several magnitudes higher than for micron-sized colloids (Evers et al. 2010, Batista et al. 2015). Consequently, the formation of NSLs occurs relatively fast within a few minutes while still taking place at (near) equilibrium conditions and colloidal crystallization with larger colloidal takes typically much longer (Evers et al. 2010). At the sub-20 nm scale, new surface and particle properties occur (e.g., quantum confinement, plasmonic effects, higher thermal velocity, etc.) that do not arise or are less pronounced for larger nano- and microparticles (Batista et al. 2015). Specific interactions between the nanoparticles increasingly govern NSL assembly at the small nanoscale that are typically Coulombic, charge–dipole, dipole–dipole, and van der Waals interactions (Talapin et al. 2007, Bodnarchuk et al. 2010). The quantification of the energetic contributions in NSL assembly from sub-20 nm particles is often complex as different processes superimpose such that classical colloidal interaction theory can often not describe assembly (Batista et al. 2015). The effects are yet difficult to predict and often rely on empiric observations. Examples are semiconductor nanoparticles (PbSe and CdSe) that were shown to follow predictions by hard-sphere models and assembly was entropy-driven thus that the phase diagram could be controlled via assembly temperatures (Evers et al. 2010). However, when metallic nanoparticles were involved (i.e., Au), van der Waals interactions were stronger and crystal structures deviated from the models (Evers et al. 2010). Moreover, dipole–dipole interactions arising from nonlocal dipoles in nanoparticles with large dipole moments (e.g., PbSe, PbS, γ-Fe$_2$O$_3$) influence NSL structures (Talapin et al. 2007). Electrostatic interactions between charged ligands or particle surfaces further guide NSL structures (Kalsin et al. 2006, Shevchenko et al. 2006, Kostiainen et al. 2013). An essential part hence play surface ligands that are grafted on the nanoparticles and interact with each other creating specific surface properties depending on their nature, temperature, solvent, and distance (Bodnarchuk et al. 2010, Si et al. 2018). The assembly process of NSLs may also include processes like ligand-stripping (Geuchies et al. 2016) and ligand deformation (Boles and Talapin 2015), which impact structural development. Current research therefore emphases on understanding parameters controlling entropic and energetic contributions during NSL formation to create NSLs with programmable structures from a variety of building blocks (Kalsin et al. 2006, Shevchenko et al. 2006, Kostiainen et al. 2013). The next section will give insights into typical NSL building blocks, surface ligands, and substrates suitable for NSL assembly.

3.3 Building Blocks and Substrates

3.3.1 Nanoparticles

Nanoparticles are the fundamental building blocks and are essential for the functional properties of NSLs in various nanotechnology applications. Thus the choice and combination of multiple nanoparticle types in NSLs is the key when designing NSLs for certain purposes. The synthetic techniques employed for the preparation of nanoparticles is vital to ensure nanoparticles with controlled composition, surface chemistry, size, and shape that meet requirements for NSL self-assembly. Typically monodisperse, aggregate-free nanoparticles with sub-20 nm diameters and <10% size deviation are required. Such nanoparticles are made by well-controlled solution phase bottom-up syntheses followed by post-synthesis procedures that enable control over particle uniformity in size and shape such as size-selective particle separation or digestive-ripening (Murray et al. 2000, Shimpi et al. 2017). The nanoparticles used in NSLs often consist of materials such as semiconductors (PbS, PbSe, CdSe, CdTe,

CdS, Ag$_2$S, Ag$_2$Te), metals (Au, Ag, Pd, Co, Pt), and metal oxides (Fe$_3$O$_4$, Fe$_2$O$_3$, CoO, CeO$_2$) for which the solution-based synthetic routes are now well-established (Murray et al. 2000, Shevchenko et al. 2006, Zhuang et al. 2011, Si et al. 2018). Moreover, finally the processability in solution is essential to combine different nanoparticles that eventually self-assemble (Choi et al. 2016).

Nanoparticle size is a key parameter that determines particle properties (optical, electronic, magnetic), surface-volume ratio, and it is also essential for NSL packing and structure in particular in multicomponent NSLs, synthesized from nanoparticles with two (or more) different diameters. Another feature is particle shape, which can be synthetically controlled from spherical to particles with diverse anisotropic features and used to create novel structural features in NSLs. Anisotropic nanoparticles with rod-like, cubic, triangular, polyhedral, or branched tetra- and octapodal shapes also form NSLs (Figure 3.3), next to the more frequently used spherical or nearly spherical-shaped nanoparticles (e.g., icosahedral nanoparticles) to form NSLs as seen in Figure 3.1 (Shevchenko et al. 2006, Jones et al. 2010, Miszta et al. 2011, Henzie et al. 2012, Qi et al. 2012, Young et al. 2013, Si et al. 2014, Zhang et al. 2014, Lu et al. 2015). By exploiting interactions introduced via shape and surface chemistry, directional interactions in the lattices can be created that lead to complex NSL structures (Figure 3.3a). Figure 3.3b shows a NSL assembled form octapodal particles, and Figure 3.3c a binary NSL consisting of triangular and spherical nanoparticles. More complex building blocks such as quantum dot−gold heterodimer nanoparticles can form NSLs by anisotropic interactions between the two components in the nanoparticles (Figure 3.3d) (Zhu, Fan, et al. 2018). As mentioned previously, an important component during synthesis of the nanoparticles are ligands, which, e.g., guide the particle's shape development. In situ or post-synthetic surface functionalization enables to vary the surface groups on a given core particle. Such surface modifications are needed during synthesis, for maintaining particles dispersed after synthesis and to guide self-assembly.

3.3.2 Ligands

Ligands are organic molecules grafted typically during synthesis on the surface of nanoparticles (Figure 3.1a). They can be modified and exchanged, which is a possibility to guide NSL assembly by controlling the ligands. With the decrease in the particle dimension to the nanoscale, the proportion of surface-to-volume increases meaning that for same mass of material, the available surface is increased. Thus, the presence of molecules on the surface of the particles often contributes to interparticle interactions more significantly than the particle interior. These properties increase proportionally with decreasing size of the particles. Ligands are first needed during synthesis of the nanoparticles as stabilizing agents that prevent aggregation and precipitation of the nanoparticles (Murray et al. 2000).

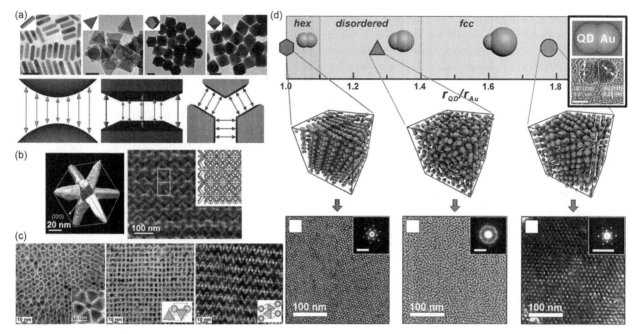

FIGURE 3.3   Anisotropic nanoparticles in NSLs. (a) Directional bonding interactions through particle shape anisotropy. Upper TEM images show differently shaped gold nanoparticles, rods, triangular prisms, rhombic dodecahedra and octahedra, scale bars 50 nm. The lower illustrations show that depending on the particle shape the interactions between ligands on the surface of the particles (here oligonucleotide interactions) varies and dictates structural details of the NSLs. (b) Left: 3D reconstruction of a single-branched nanocrystal (octapod) from STEM projections, revealing an octahedral symmetry. Right: NSL of octapodal particles matching a tetragonal structure (sketched in the inset). (c) TEM images and proposed unit cells of binary superlattices self-assembled from triangular nanoplates and spherical nanoparticles. (d) Heterodimers consisting of two different materials, here gold and CdSe-CdS quantum dots (see illustration and TEM image in the inset on the top right, scale bar 3 nm), can assemble into NSLs and its structure depends on size ratios between Au and CdSe-CdS compartments in the dimers. Lower images, TEM images of SLs assembled from heterodimers with size ratios of 1.0, 1.27, and 1.77, respectively, showing good agreement with molecular dynamic simulations (top illustrations). ((a) Adapted by permission from *Springer Nature: Nature Materials* Jones et al. 2010, copyright 2010. (b) Adapted by permission from *Springer Nature: Nature Materials* Miszta et al. 2011, copyright 2011. (c) Adapted by permission from *Springer Nature: Nature* Shevchenko et al. 2006, copyright 2006. (d) Adapted with permission from Zhu, Fan, et al. 2018. Copyright 2018 American Chemical Society.)

Thereby the molecular structure of the ligands typically includes a functional head group, which binds to the nanoparticle (e.g., $-COOH$, $-NH_2$, $-P$, $-P{=}O$, $-SH$) and an organic rest of varying chain length providing stability. Examples of molecular ligands include oleic acid, stearic acid, dodecanethiol, citrate, CTAB, trioctylphosphine, etc. and key factors during NSL assembly are their functional groups and molecular length as well as its orientation and density on the particle surface (Si et al. 2018). The ligands influence the nanoparticles physical and chemical interactions in solution such as steric hindrance, electrostatic attraction or repulsion, hydrogen bonding, and Van der Waals forces. These interactions also play a key role in the formation of NSLs giving the essential importance to ligands during NSL self-assembly such as the prevention of random aggregation of nanoparticles by steric hindrance counteracting Van der Waals attractions, guiding self-assembly by charges, and ligand–ligand interactions that maintain NSL integrity post-assembly. One factor is the ligand length that is related to the number of carbon atoms in the organic rest. The interparticle separation in NSLs increases correspondingly to the ligand's length (Chen et al. 2008, Si et al. 2018). Nevertheless, not only the ligand size is essential for interparticle interactions and self-assembly. Systematic studies on NSL formation using nanoparticles with same sizes coated by different monofunctional ligands of the same length but different functional group show that dodecanethiol- and dodecylamine-coated Au nanoparticles assemblies differed in their interparticle distance, even though both ligand molecules contain 12 carbon atom chain lengths (Prasad et al. 2003). This was not expected since the head group of the ligands would anchor to the nanoparticle surface, and the hydrophobic alkane chain is exposed to the exterior. The differences in NSLs can be partially attributed to the weaker binding of the amine group to the nanoparticles surface, which results in a higher disorder in the alkane chain and thus preventing interdigitation (Prasad et al. 2003). Furthermore, mechanisms like ligand stripping-off from the surface and ligand exchange occurring during NSL assembly should also be considered as essential driving factors during NSL structure formation (Wei et al. 2015, Geuchies et al. 2016) in addition to ligand "softness" and ligand deformability related to its molecular structure and particle surface interactions (Boles and Talapin 2015).

Furthermore, NSL assembly can be modified by using complex molecules like biomolecules such as nucleic acids and proteins that can mimic certain properties of natural systems (Kostiainen et al. 2013, Kim and Macfarlane 2016, Künzle et al. 2016). In particular DNA, opens a unique route to direct NSL assembly, and major advances have been achieved using DNA as ligands (Jones et al. 2015, Si et al. 2018). These are used to program particle binding, interparticle spacing, and structural properties of NSLs by either exploiting its specific Watson–Crick base-pairing or simply by controlling properties like DNA length (see Section 3.4.1 for further examples) (Cheng et al. 2009, Auyeung, Cutler, et al. 2012, Auyeung et al. 2015, Jones et al. 2015, Si et al. 2018).

3.3.3 Substrates

The macroscopic assembly for fabrication of NSLs requires suitable substrates next to the nanoparticle building blocks, and the substrate is often important for NSL analysis and later application. NSL assembly may occur at increasing particle volume fractions at liquid–solid, liquid–air, and liquid–liquid interfaces. Thereby the substrate might be one of the interfacing materials or NSLs are later transferred from the interface onto a specific substrate. Several approaches have been adopted to construct two- and three-dimensional self-assembled NSLs based on certain substrate geometries, or free-standing without substrate. Typical solid substrates are TEM sample grids coated by carbon, silicon nitride, silicon dioxide, or silicon wafers as larger substrates. The substrate's chemical and physical properties influence the NSL structure, such as interparticle spacing (Muralidharan et al. 2011). The NSLs containing La_3F and Au or PbSe nanoparticles shown in Figure 3.3c assemble, for example, on silicon oxide or amorphous carbon substrates in different structures, the reason remained unclear (Shevchenko et al. 2006). In the following subsections, we will give an overview of substrate effects on NSL formation.

Effect of Substrate's Chemistry and Interfaces on NSL Assembly

From the methods to synthesize NSLs mentioned in Section 3.2, the evaporation induced is the most widely used way to deposit a NSL onto a solid substrate. Nanoparticles are drop-casted on the substrate or the substrate is dipped into a solution of the nanoparticles. Next to particle interactions, packing constraints, also mass and heat transfer during drying, together with kinetic and thermodynamic factors influence the self-assembly (Bigioni et al. 2006, Weidman et al. 2016, Josten et al. 2017). Some of these factors are influenced by the substrate surface chemistry and topography and, e.g., its wettability plays a role in NSL assembly (Dong et al. 2010). If the nanoparticle dispersion does not wet the substrate, NSL growth is three-dimensional, forming facets that reflect the packing symmetry of the nanoparticles (Murray et al. 2001). Certain crystal structures on the substrate's surface may lead to an epitaxial growth of the NSLs, which means that the NSL assembles depending on substrate crystallinity (Hellstrom et al. 2013). A further role may play insulating and conducting properties of the substrate. Free electrons in a conducting carbon substrate can screen the electric field of the nonlocal dipoles of nanoparticles such as PbS, PbSe, and γ-Fe_2O_3 nanocrystals, which may stabilize NSL structures not found when assembling the same nanoparticles on insulating substrates (Talapin et al. 2007).

Even under optimized synthetic conditions, the NSLs that are grown directly on solid substrates typically exist as isolated, micrometer-sized islands scattered irregularly on the substrate, leading to a low surface coverage (Dong et al. 2010, Korgel 2010). Larger, homogeneous NSLs can be synthesized by first assembling the NSL at the liquid–air interface (Bigioni et al. 2006, Dong et al. 2010, Korgel 2010). Thereby a droplet of the nanoparticle dispersion is placed on another immiscible liquid and evaporated. The as-prepared NSL is then transferred onto the final solid substrate (as, e.g., the process depicted in Figure 3.2b). The method can be adapted to a variety of solid substrates but is limited in the choice of immiscible liquids required for the particle dispersion and for initial assembly. In order to determine the two solvents required for the monolayer formation, the following benchmarks need to be considered. First, the solvents used must be immiscible and the nanoparticles dispersion solvent must be less dense such that it does not replace the solvent acting as a substrate. Second, the nanoparticles must not disperse in the "substrate solvent" that will allow the nanoparticles to remain on top of it when the dispersant solvent evaporates. Finally, the dispersion solvent must have a higher vapor pressure than the substrate solvent, which enables a faster evaporation and also allows the nanoparticles to remain at the interface. For example, a solution of nanoparticles in an organic solvent like hexane on top of a polar solvent, for example, diethylene glycol, meets these criteria. The method is used and modified to coat substrates such as silicon wafers with large area NSLs (Gaulding et al. 2015, Wang et al. 2018).

Substrate Geometry

Typically, NSLs are synthesized on flat substrates without three-dimensional features. However, also three-dimensional substrates like spheres as shown in Figure 3.4 have been used to deposit NSLs. Figure 3.4a shows a process in which nanoparticles and the substrate, here, a larger colloidal SiO_2 particle, are encapsulated together in an oil-in-water emulsion droplet (Meder et al. 2018). Upon evaporation of the oil phase the NSLs form and deposit on the larger substrate particle. The process enables the fabrication of colloidal particles consisting of a certain core material and coated with sub-10 nm patterned NSLs (Figures 3.4b,c)

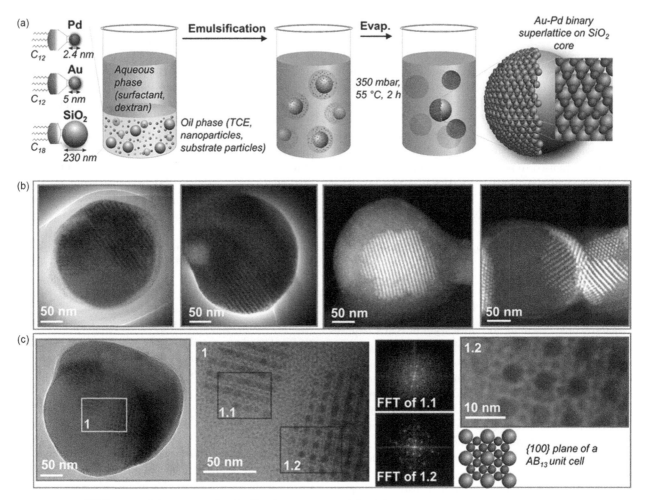

FIGURE 3.4   NSL assembly on a spherical substrate particle. (a) Illustration of the synthetic procedure for the coating process of ∼230 nm amorphous SiO₂ particles with binary Au−Pd NSLs. SiO₂, Au, and Pd building blocks are confined in an oil-in-water emulsion droplet in a controlled ratio, preserving the colloidal nature of the system during the entire process. During the evaporation of the oil phase, binary NLS form, transfer on the surface of the SiO₂ cores and finally disperse in water. (b) TEM and high-angle annular dark-field (HAADF) scanning TEM (last two images) of ∼230 nm SiO₂ particles coated with NSLs of 5 nm Au and 2.4 nm Pd nanoparticles. The electron-dense Au NPs structures appear as stripes in the lattice corresponding to the stacking perspective and their enhanced contrast over SiO₂ and the smaller interstitial 2.4 nm Pd NPs. (c) Overview and zoom-in TEM images of an Au−Pd NSL on a spherical SiO₂ substrate. The fast Fourier transforms (FFT) on the right were done of areas highlighted as panels 1.1 and 1.2. A further zooming-in on panel 1.2 shows that the Au−Pd lattice resembles the {100} plane of a AB_{13} structure with Pd NPs surrounding and spacing the Au NPs. (Adapted with permission from Meder et al. 2018. Copyright 2018 American Chemical Society.)

(Meder et al. 2018). The bottom-up design of colloidal matter might be interesting to create functional compartments for applications like micro robotics or on-particle devices. Various factors can, however, affect the formation of superlattices on spherical surfaces, which include curvature effects, formation of grain boundaries, kinetic instability, interfacial tension, etc. The assembly of matter on three-dimensional substrates is thus, in general, subject to confinements given by the substrate's geometry like curvature-based compressive and shear stress (Irvine et al. 2012). On three-dimensional soft or elastic substrates such as a droplet surface, temporary defects can enable the mobility of particle subunits and their collective rearrangement into crystalline domains with grain boundaries facilitating the self-assembly into patterned structures with a correlation

between the elasticity of the substrate surface and the achieved grain size of assembled particles (Bausch et al. 2003, Irvine et al. 2012, Meng et al. 2014). The effects of curvature are vital for analogous biological processes involving assembly of similarly sized subunits, i.e., proteins, on curved surface such as viral capsids (Roos et al. 2010) and cells (Fagan and Fairweather 2014).

Suspended or Free-Standing NSLs

Free-standing or suspended NSLs refer to "substrate-free" NSLs with minimal or no contact with a solid substrate. They are synthesized by employing the liquid-air interface, using sacrificial templates, and by tailored ligand interactions (Mueggenburg et al. 2007, Cheng et al. 2009, Estephan et al. 2013, Si et al. 2014, Rao et al. 2015, Shim et al. 2017, Yu et al. 2017). Suspended NSLs have potential applications such as soft substrates for enhanced Raman scattering (Chen, Si, et al. 2015), ionic gating (Rao et al. 2015), security label for banknotes for anticounterfeiting (Si et al. 2015), etc. Despite various applications of suspended superlattices, the fabrication technology to achieve stable structures without defects remains a challenge.

3.4 NSL Structures and Crystal Phases

The crystal structure is one of the most important characteristics of NSLs that determines the positioning of the single nanoparticle building blocks in the NSLs. It governs nanoparticle distances, interparticle spacing, nearest neighbors, etc. in multicomponent NSLs. It is consequently a major feature defining the collective interactions of the building blocks. The structures in NSLs often (though not always) resemble lattice structures also found in crystals consisting of molecular or atomic building blocks and, as in atomic crystals, features like grain boundaries, twinning, etc. also occur in NSLs. Figure 3.5a shows examples for NSL structural features that may occur on larger scales (typically μm-scale) such as terraces, steps, kinks, vacancies, grain boundaries, faceted surfaces, and so forth (Kang et al. 2013). Central emphasis of current research is yet often controlling the unit cell structures and thus nanoscale positioning of the building blocks (Boneschanscher et al. 2014). The synthetic techniques established to obtain NSLs allow controlling the NSL crystal structure to a certain degree by tailoring interaction forces during NSL self-assembly. In the following, we will give an overview of types of structures and the level of control over them in multicomponent NSLs.

Figure 3.5b shows models of commonly observed binary NSL structures (see also Figure 3.1c). The unit cells that have been found in NSLs are of AB, AB_2, AB_3, AB_5, AB_6, and AB_{13} stoichiometry as well as comprising structural motifs of Archimedean tiling (AT) and dodecagonal quasicrystal (DDQC) (Talapin et al. 2009, Ye et al. 2017) configurations (Boles and Talapin 2015). Also ternary NSLs consisting of three types of building blocks can be obtained, e.g., in ABC_4 configuration (Evers et al. 2009, Paik et al. 2015). A question to be answered for controlling the structural development is which driving forces direct self-assembly under given conditions. Entropy-driven binary assembly of micron-sized and larger nanoparticles enables to predict structures by just knowing the size ratio of two particle building blocks. However, the huge diversity of structures found in NSLs made from sub-20 nm nanoparticles are usually arrangements that fill space less densely than expected from hard-sphere models and assemblies of larger particles. Figure 3.5b shows that a majority of the crystal structures observed in NSLs have lower packing densities than 74%. It is clear that contributions particular through effects of ligands and nanoparticles properties occurring in

the small, sub-20 nm size scale (i.e., magnetic (Cheon et al. 2006, Yang et al. 2015), electronic properties (Talapin et al. 2007)) play an essential role for the establishment of the lattice structure – and provide a tool to tailor it, if rules governing the structural assemblies can be found. In case of ligands, parameters as ligand deformability (Boles and Talapin 2015), exchangeability when two types of nanoparticles are functionalized with alkane ligands of different length and group (Wei et al. 2015, Travesset 2017), stripping-off of ligands (Geuchies et al. 2016), charges (Kalsin et al. 2006, Shevchenko et al. 2006, Kostiainen et al. 2013), are ways to direct the crystal structures in NSLs (Si et al. 2018). The following two subsections give examples of techniques of controlling NSL structure during and also after their assembly.

3.4.1 Ligand Programming of NSL Structure and Stimuli-Responsive Properties

The incorporation of ligands that respond to stimuli, such as redox reactions, solvent molecules, pH, metal ions, light, and temperature, can be used to direct the self-assembling structure and to render NSLs stimuli-responsive (Grzelczak et al. 2010, Mighty linkers 2015, Si et al. 2018). It follows that upon external stimulus NSL structures can change in situ depending on varying interparticle forces. An exceptional and fascinating opportunity to tailor NSLs is given by using DNA as programmable ligand due to their unique base-pairing interactions as already mentioned in Figure 3.2d (Jones et al. 2015). It is thereby possible to disconnect the identity of the nanoparticle building block from its assembly behavior enabling a distinct synthetic control over NSLs structures. Design rules for obtaining certain NSL structures can be drawn based on DNA-driven assembly that yet cannot be achieved based on electrostatic, covalent, or non-covalent interactions (Macfarlane et al. 2011, Mighty linkers 2015). Figure 3.5d shows NSL structure engineering using DNA, which allows for an independent control of parameters like particle size, lattice parameters, and crystallographic symmetry (Macfarlane et al. 2011). A further possibility given by DNA-crosslinking of NSLs is an on-demand reprogramming of a given NSL structure by exploiting the reversibility of DNA bonds (Ross, Ku, Vaccarezza, et al. 2015, Zhang et al. 2015). A responsive change of a DNA-linked NSL structure could, for example, be exposed to a reprogramming 'input' DNA strand that selectively shifts the particle–particle interactions (Zhang et al. 2015) or by using pH-stimuli-responsive-DNA-ligands (Zhu, Kim, et al. 2018). Thus transitions from CsCl phase to various phases, including CuAu, hcp, quasi-2D, fcc, and a cluster morphology were possible (Zhang et al. 2015). By combining top-down lithography with DNA-mediated assembly, reconfigurable NSLs could be obtained by directing the layer-by-layer assembly of different plasmonic nanoparticle subunits on lithographically predefined positions to achieve reconfigurable optical properties (Lin, Mason, Li, et al. 2018).

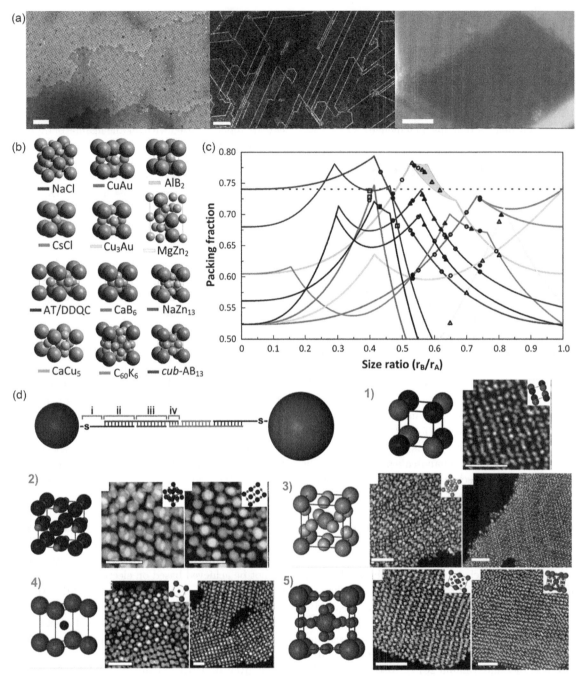

FIGURE 3.5 Micro- and nanoscale structures in NSL materials. (a) Microscale structural features in NSLs. Left SEM image: Terraces, steps, kinks, and vacancies in a binary Pt–Pd NSL, similar to those in single crystals. Middle TEM image: Grain boundaries in an Au–FeOx NSL. Right SEM image: Supercrystal of Pt nanoparticles that has faceted surfaces. (b) Models of 12 commonly observed structures in binary NSLs with unit cells of AB, AB_2, AB_3, AB_5, AB_6, and AB_{13} stoichiometry as well as the structural motifs of Archimedean tiling (AT) and dodecagonal quasicrystal (DDQC) configurations. (c) Line plots of packing density as function of nanoparticle size ratio in a binary NSL when packed in the arrangements shown in (b). The data points show phases experimentally observed in binary NSLs. Most observed NSLs appear to be open arrangements compared with close-packed monodisperse hard spheres (dotted line). (d) NSL structure engineering with DNA, which allows for separating the identity of the particle from the variables that control its assembly into NSLs. DNA strands for NSL assembly consist of (i) alkyl-thiol moiety and 10-base non-binding region, (ii) a recognition sequence that binds to a DNA linker, (iii) a spacer sequence of programmable length to control interparticle distances, and (iv) a "sticky end" sequence driving NSL assembly via DNA hybridization interactions. The model unit cells and TEM images show DNA-programmed NSLs with (1) simple cubic lattices and isostructural with (2) NaCl, (3) Cr_3Si, (4) CsCl, (5) Cs_6Cs_{60}. Scale bars are 100 nm in 1 and 2, and 200 nm in 3–5. Further examples can be found in Macfarlane et al. (2011).

(Continued)

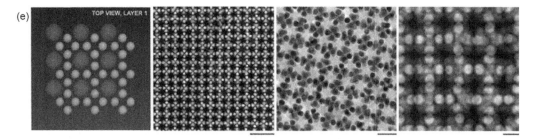

FIGURE 3.5 (CONTINUED) (e) Selective removal of one component of binary NSLs to create nanoporous materials having the same chemical composition but differing in their nanoscale architectures. Left: Structural model of a NSL, the smaller, light-gray spheres represent Au nanoparticles and the dark gray spheres represent vacancies corresponding to the positions of Fe_3O_4 nanoparticles which are later etched. HAADF-STEM and TEM images of non-close-packed NSLs after Fe_3O_4 etching, where the top layer consists of Au quartets as in the structural model. The scale bars correspond to 50 nm, 20 nm, and 10 nm. ((a) Adapted with permission from Kang et al. 2013, copyright 2013 American Chemical Society. (b) and (c) Adapted with permission from Boles and Talapin 2015, copyright 2015 American Chemical Society. (d) Adapted from the Supporting Online Material of Macfarlane et al. 2011 with permissions from the authors, copyright 2011. (e) Adapted from the Supporting Online Material of Udayabhaskararao et al. 2017 with permissions from the authors, copyright 2017.)

3.4.2 Post-assembly Processing of NSLs

Another versatile approach to tune the structure are modifications of NSLs by tailored post-assembly processing. In multicomponent NSLs consisting of different types of nanoparticles, the fact that different nanoparticle materials bear different chemical properties and a different reactivity can be used to selectively remove certain building blocks by gas or solution phase etching. This was used to etch one type of nanoparticles in binary NSLs without affecting the order of the second nanoparticle resulting in nanoallotropes with characteristic structural features (Figure 3.5e) (Udayabhaskararao et al. 2017). Another example for chemical treatments is changing the chemical composition of an NSL by a post-assembly cation exchange (Boneschanscher et al. 2014, Miszta et al. 2014). It was hence possible to transform buckled PbSe honeycomb structures, in which the PbSe nanoparticles occupy two parallel planes, into CdSe superlattices with the potential for Dirac-type electronic bands and strong spin–orbit coupling (Boneschanscher et al. 2014). To stabilize DNA-engineered NSLs, e.g., for applicability outside their assembly media (which could influence or distort the NSLs structure) a transitioning of the NSLs from solution to a solid state was performed by encapsulation in silica which preserves the initial NSLs structure (Auyeung, MacFarlane, et al. 2012). Thermal treatment is used to anneal NSLs which can induce phase transitions that, e.g., impact the magnetic properties (Sun et al. 2000, Cheon et al. 2006). In addition, by appropriate thermal annealing of an NSL, the ligands can be converted into a carbon framework, which enhances electrical conductivity and mechanical properties of the NSLs (Li, Yiliguma, et al. 2016). The various post-assembly processing tools are thus an important aspect not only for modifying NSL structure but also for creating specific properties and preparing NSLs for certain applications. Similar to doping of bulk semiconductors, doping NSLs consisting of semiconductor nanoparticles can be used to tailor electronic properties. Among different techniques, for example, the post-synthetic colloidal, atomic layer deposition (PS-cALD) process has been developed to engineer stepwise the surface stoichiometry and therefore the electronic properties of semiconductor NSLs (Oh et al. 2014). Therefore the NSLs have been exposed to metal salt solutions enabling n- or p-doping (Oh et al. 2014). By a substitutional doping of semiconductor NSLs with Au nanoparticles that occupy random positions in the lattice similar analogous to atomic doping, however, using controlled nanoparticle self-assembly in this case, the conductivity of lead selenide films has been manipulated over at six orders of magnitude (see also Section 3.6) (Cargnello et al. 2015).

3.5 Analytics of NSLs

The techniques to analyze NSLs typically focus on the final structure and structure development of NLSs. Methods to test and analyze the NSL nanoscale features and overall synthetic success are limited to those enabling sub-10 nm resolutions. One of the most applied analytical techniques is electron microscopy, mainly TEM and scanning TEM (STEM) including (high-angle annular dark-field) HAADF detection to determine assemblies, structure, and composition. Scanning electron microscopy (SEM) is used for NSLs with larger building blocks as, e.g., in Lu et al. (2015) and it is only rarely used for NSLs with features in the sub-10 nm scale (see, e.g., in Smith et al. (2009)) due to resolution limitation compared to TEM. For TEM techniques, NSLs are either directly synthesized on a TEM sample holder or transferred after assembly at, e.g., a liquid interface. Electron beam diffraction analysis gives information on the crystal structure and lattice parameters that can be used to extract the crystal structure. TEM tomography gives insights in 3D assemblies and enables 3D reconstructions that can be used to comprehend crystal structure (Auyeung, Cutler, et al. 2012, Udayabhaskararao et al. 2017). In situ measurements are possible in liquid-phase TEM, which may provide further information on assembly process. However, TEM restricts the NSLs to be observed to certain sample configurations, in particular only a limited thickness can be applied due to the necessity of adequate electron beam transmission. In addition, a limited sample size that can be imaged with sub-10 nm resolution (typically a few hundred nanometers or less) makes it difficult collecting statistically relevant data. NSLs deposited on bulk structures are due to limited electron transparency not possible to analyze via TEM.

Small and wide angle X-ray scattering (SAXS and WAXS, respectively) techniques are widely applied to study structure of NSLs. They make use of the fact that a spatial variation of electron density will scatter an X-ray beam – at nanometer length scales to low angles and in the atomic scale to high angles (Li, Senesi, et al. 2016). SAXS can be used to measure both particle size and interparticle distances with angstrom resolution and meaningful statistics over large length scales (angstroms to micro-meters) (Murray et al. 2000, Li, Senesi, et al. 2016). It can be used in situ, and in liquids as well as on dried NSL films. A full profile analysis enables to determine nanoparticle positions in unit cells of multi-building block NSLs. A combination of SAXS and WAXS permits to investigate the relation of long-range and short-range order in the NSLs. Whereas long-range order of NSLs is often lost, the NSL may maintain short-range order that includes local symmetry as indicated by peak broadening (Murray et al. 2000, Li, Senesi, et al. 2016). Grazing incidence small-angle X-ray scattering (GISAXS) instead of transmission SAXS can be useful when NLSs are deposited as a film on a substrate to extract orientation, symmetry, lattice parameter, crystallite size, and electron density (Smith et al. 2009, Senesi et al. 2013, Li, Senesi, et al. 2016, Weidman et al. 2016). The results of X-ray scattering as well as electron microscopy are typically supported by computational simulations of NSLs structures.

Different purpose-specific techniques are used to analyze electronic, magnetic, optical, or catalytic properties of NSLs which focus on collective properties and applications. For instance, to measure the conductivity of NSLs, two thin electrodes (e.g., 10–50 nm thickness) are deposited on an insulating substrate, and conduction of NSLs assembled on top of the electrodes is measured (Sahu et al. 2012, Cargnello et al. 2015). To determine optical properties such as plasmon resonances and plasmon coupling in Ag and Au containing NSLs or optical absorption polarization, extinction spectra are recorded either in solution or solid state using, for instance, common UV–Vis spectrometers or microspectrophotometers coupled to optical microscopes (Young et al. 2014, Diroll et al. 2015, Ross et al. 2016). NSL lattice symmetry-dependent reflectance are recorded as well by microscope-coupled spectrophotometers (Collier et al. 1998, Sun et al. 2018). The magnetic properties of NSLs such as saturation magnetization and coercive field are determined by

a superconducting quantum interference device (SQUID) magnetometer (Murray et al. 2000, Sun et al. 2000, Desvaux et al. 2005, Cheon et al. 2006, Chen et al. 2011, Yang et al. 2015).

3.6 Properties and Applications

The emerging and existing applications of NSLs aim to make use the unique properties of the collective interactions occurring in NSL depending on its structure and organization of selected components. The particular electronic, optical, catalytic, magnetic, and mechanical properties of NSLs show potential for applications in several fields and confirm the predicted advantages of the modular bottom-up assembly of tailored materials. In the following, we give examples of different properties occurring in NSLs and applications in optics, electronics, magnetics, and catalysis to explore which features of NSLs can be used to generate specific technological advantages (Talapin et al. 2010). Most applications are yet still in the fundamental stage and object of current research.

Among the first synthesized NSLs are those made of semiconductor quantum dots, also called quantum dot solids, with interesting electronic, optoelectronic, and thermoelectronic properties that rely on the charge transport through the NSL. Factors that control these material properties are related to the size, shape, and composition of the quantum dots and their interparticle distance and organization in the NSLs (Kagan and Murray 2015). The behavior can be compared with those in atomic solids, in which the size and composition of the atoms, the average interatomic distance, and the number of nearest neighbors direct its electronic properties, making it a metal, semiconductor, or insulator (Kagan and Murray 2015). In quantum dot-based NSLs, the interaction between the electronic wave functions of the quantum dots can be controlled by the distance of the particles and it increases for lower distances, producing tunable cooperative physical phenomena that are unique to a quantum dot solid (Kagan and Murray 2015). Wave function overlap and the electron and hole states vary as function of the interparticle distance in an ideal quantum dot NSLs. Thus ligand-engineering and post-assembly processing plays an important role to tailor interparticle distances. Often the ideal state is not possible to achieve due to distributions in particle size, material, and distances, which can be further tuned by improving NSL synthesis (Kagan and Murray 2015, Whitham et al. 2016). The advantage of quantum dot NSL semiconductors is their tunability and cheap processing, applicability to different substrates, and the opportunity to create new functionalities by the variety of nanoscale building blocks. Examples of devices under development that use these properties are quantum dot NSL field effect transistors (FETs) bearing high mobility and low-voltage, low-hysteresis, low-noise characteristics with performance metrics that exceed those of solution-processable organic semiconductors and are competitive with carbon nanotube array devices (Talapin and Murray 2005, Heath 2007, Urban et al. 2007, Hetsch et al. 2013, Oh et al. 2014, Reich et al. 2014, Kagan and Murray 2015). Further applications for quantum dot NSLs are emitting layers, in light emitting diodes (Coe et al. 2002, Kim et al. 2011, Sun et al. 2012), active layers in photovoltaic devices improving the power conversing efficacy (Talapin et al. 2010, Talgorn et al. 2011), or as active layers for light-to-electricity conversion in photodetectors (Konstantatos and Sargent 2009, Talapin et al. 2010). Analogous to atomic doping of material, doping semiconductor NSLs with other nanoparticles, e.g., Au, material properties can be varied as mentioned above. By introducing Au nanoparticles in a semiconductor NSL, one can manipulate the conductivity of NSLs over at least six orders of magnitude (Cargnello et al. 2015). Figure 3.6a shows the direct-current conductivity of PbSe NSLs doped with varying amounts of Au/Ag nanoparticles. Further processing like

carbon-coating of NSLs, for example, by thermal treatment to convert the ligands surrounding the nanoparticle building blocks into a carbon framework, can render even NSLs made from non-conductive nanoparticles into NSLs with high conductivity and, additionally, improved mechanical strength. Thus also NSLs of various building blocks that not necessarily have desired electronic properties may be modified to introduce desired electronic features by using ligand post-processing. Carbon-coated NSLs may find applications such as in batteries, fuel cells, solar cells, electrochemistry, and electrocatalysis (Li, Yiliguma, et al. 2016). The potential general electronic applications of NSLs are manifold and overlap across fields, they also include energy storage (Jiao et al. 2015, Li, Yiliguma, et al. 2016) and thermoelectric energy harvesting devices (Heath 2007) next to the devices mentioned above. All rely on the modular NSL bottom-up structure and collective interactions of their components.

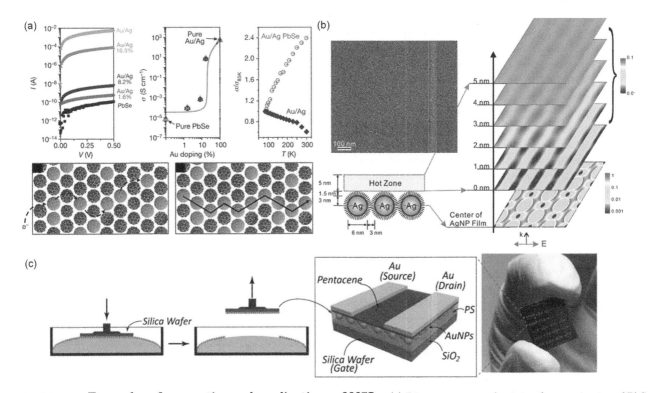

FIGURE 3.6 Examples of properties and applications of NSLs. (a) Direct-current conductivity characterization of PbSe NSLs doped with varying amounts of Au/Ag nanoparticles. Upper left graph, I–V curves; central graph, conductivity of the nanocrystal superlattices as a function of Au/Ag doping; right graph, temperature-dependent conductivity of films of PbSe superlattices doped with 16.5% Au/Ag nanocrystal, and films of pure Au/Ag nanocrystals. The lower schematic shows conductivity in systems below (left) and above (right) the percolation threshold in a honeycomb lattice showing the hopping path for electrons (dashed black arrow) and the direct electron transfer between Au nanocrystals (filled black arrow). (b) Illustration of the detection hot zone in an NSL-based SERS substrate (lower left schematic). The finite-difference time-domain simulated electric field intensity distribution is shown (right) at different vertical planes on top of an Ag nanoparticle NSL such as in the TEM image (upper left inset). The NSL structure may provide different detection zones both with strong enhancement factor for single molecule detection and high spatial detection uniformity. (c) Illustration showing the transformation process of a NSL onto a wafer and schematic configuration/photograph of a nano-floating-gate memory device. ((a) Adapted by permission from *Springer Nature: Nature* Cargnello et al. 2015, copyright 2015. (b) adapted with permission from Chen, Lin, et al. 2015, copyright 2013 American Chemical Society. (c) Adapted by permission from John Wiley and Sons: *Advanced Materials* Wang et al. 2018, copyright 2018.)

Another main property of NSLs are the collective optical phenomena that arise from the interplay of the individual nanoparticles in NSLs such as tunable reflection, optical cavity modes, and tunable photonic resonances promising various applications in sensors, photonic materials, and metamaterials (Ross, Ku, Blaber, et al. 2015, Ross et al. 2016). The optical properties of NSLs can be tuned by nanoparticle materials, sizes, shapes, interparticle distances, lattice order, and the NSL dimensions. A feature of metal nanoparticles, in particular gold and silver, are the collective oscillation of the conduction electrons upon excitation by incident light at a certain frequency, their surface plasmon resonance. When the particles are organized in an NSL, the single nanoparticle's localized surface plasmon resonance (LSPR) provides a strong oscillator and a source of pronounced scattering whereas the periodicity (determined by the interparticle spacing in the NSL) regulates which scattered wavelength will lead to constructive interference (Ross et al. 2016). Using plasmonic building blocks in NSLs hence provides the opportunity to precisely design the optical properties of a material, which could not be achieved in the meso- or macroscale bulk state (Alaeian and Dionne 2012, Ross, Ku, Vaccarezza, et al. 2015, Lin, Mason, Li, et al. 2018). For example, near-field coupling of gold nanoparticles in the lattices can be tuned by controlling interparticle separation by using ligands of different length (Chen et al. 2008). This can generate plasmonic-based subwavelength optics often challenging to create by conventional methods such as electron beam lithography (Chen et al. 2008). Systems to control distance and spacing of the plasmonic hot spots are essential in biosensing techniques as, for instance, in large area, quantitative surface enhanced Raman scattering (SERS) as shown in Figure 3.6b. NSLs can be used to introduce a homogeneous distribution of sensing hot spots and enhance the detection and performance of the sensor (Chen, Lin, et al. 2015, Xavier et al. 2018). NSLs of plasmonic nanoparticles can also be an alternative route to design photonic crystals by tuning crystal structure and interparticle distances due to the controllable spacing and the high refractive index of the lattices (Park et al. 2015, 2017, Sun et al. 2018). Photonic modes can derive from Bragg diffraction in the crystalline NSL structure (Ross et al. 2016). NSLs' structural tunability hence enables by-design plasmonic metamaterials and thus has the potential for the creation of optical circuitry and interconnects, optical cloaking materials, and data exchange (Rogach 2004, Yannopapas 2007, Alaeian and Dionne 2012, Mühlig et al. 2013, Young et al. 2014, Lewandowski et al. 2015, Ross et al. 2016). Production techniques for obtaining large-scale NSLs with controlled interparticle distance and near-field coupling such as wafer-scale technique shown in Figure 3.6c used by the authors to obtain nano-floating-gate memory devices can help to develop NSL devices for different uses in microelectronics, photoelectronics, and functional sensors (Wang et al. 2018).

Optics and electronics are promising and most advanced areas for NSL applications at current state. However, the possibilities for applying NSLs span over many further fields due to the versatility of combining various components in NSLs. Catalysis is another field that demonstrates advantages when using organized nanomaterials. NSLs provide the opportunity of combining different catalytic nanoparticles and to spatially define the location and concentration of the catalytic centers (Kang et al. 2013, Auyeung et al. 2015, Brodin et al. 2015, Li et al. 2015). Further, properties like tunable and enhanced magnetic (Cheon et al. 2006, Yannopapas 2007) and mechanical (Mueggenburg et al. 2007, Podsiadlo et al. 2010, 2011, Dreyer et al. 2016) features will lead to additional applications and many more are still being discovered.

3.7   Summary and Outlook

Using nanoparticles with tunable properties as primary building blocks of new materials requires their controlled assembly into larger structures. The different ways to obtain NSLs are versatile modular bottom-up processes to organize nanoscale objects. NSLs bear structures with specific particle positioning and periodicity and the investigation of NSL self-assembly provides a powerful tool for generating plannable materials. The great advances that have been achieved by biomolecular NSL programming, post-assembly processing and other techniques point towards clear perspectives to tailor NSL structures in cost-effective, and easy to apply self-assembly techniques. New applications and advances in various scientific and technical areas arose, are coming up, and further will arise, making use of NSL materials precisely tuned at the nanoscale. Although fascinating control over the NSL structure has been obtained in the last years, still assembly is not entirely predictable due to the various factors interplaying during NSL formation. To tailor NSLs into materials with anticipated function, it will be important to further understand how NSL properties and their composition relate and to what extent these properties depend on slight variations in the lattice structure. Further understanding of how building blocks collectively interact thus will help to create material features that enable new and advance existing technologies. The investigation of NSL assembly on wafer-scale or three-dimensional substrates will permit to realize NSLs-based devices on various platforms. This will lead to nanoscale control over materials and combine inorganic matter in a similar way nature arranges biomolecules to craft materials with fascinating functionalities.

References

Alaeian, H. and Dionne, J.A., 2012. Plasmon nanoparticle superlattices as optical-frequency magnetic metamaterials. *Optics Express*, 20 (14), 15781.

Auyeung, E., Cutler, J.I., MacFarlane, R.J., Jones, M.R., Wu, J., Liu, G., Zhang, K., Osberg, K.D., and Mirkin, C.A., 2012. Synthetically programmable nanoparticle superlattices using a hollow three-dimensional spacer approach. *Nature Nanotechnology*, 7 (1), 24–28.

Auyeung, E., MacFarlane, R.J., Choi, C.H.J., Cutler, J.I., and Mirkin, C.A., 2012. Transitioning DNA-engineered nanoparticle superlattices from solution to the solid state. *Advanced Materials*, 24 (38), 5181–5186.

Auyeung, E., Morris, W., Mondloch, J.E., Hupp, J.T., Farha, O.K., and Mirkin, C.A., 2015. Controlling structure and porosity in catalytic nanoparticle superlattices with DNA. *Journal of the American Chemical Society*, 137 (4), 1658–1662.

Batista, C.A.S., Larson, R.G., and Kotov, N.A., 2015. Nonadditivity of nanoparticle interactions. *Science*, 350 (6257), 1242477.

Bausch, A.R., Bowick, M.J., Cacciuto, A., Dinsmore, A.D., Hsu, M.F., Nelson, D.R., Nikolaides, M.G., Travesset, A., and Weitz, D.A., 2003. Grain boundary scars and spherical crystallography. *Science*, 299 (5613), 1716–1718.

Bigioni, T.P., Lin, X.M., Nguyen, T.T., Corwin, E.I., Witten, T.A., and Jaeger, H.M., 2006. Kinetically driven self assembly of highly ordered nanoparticle monolayers. *Nature Materials*, 5 (4), 265–270.

Bodnarchuk, M.I., Kovalenko, M.V, Heiss, W., and Talapin, D.V, 2010. Energetic and entropic contributions to self-assembly of binary nanocrystal superlattices: Temperature as the structure-directing factor. *Journal of the American Chemical Society*, 132 (34), 11967–11977.

Bodnarchuk, M.I., Li, L., Fok, A., Nachtergaele, S., Ismagilov, R.F., and Talapin, D.V., 2011. Three-dimensional nanocrystal superlattices grown in nanoliter microfluidic plugs. *Journal of the American Chemical Society*, 133 (23), 8956–8960.

Boles, M.A. and Talapin, D.V., 2015. Many-body effects in nanocrystal superlattices: Departure from sphere packing explains stability of binary phases. *Journal of the American Chemical Society*, 137 (13), 4494–4502.

Bolhuis, P.G., Frenkel, D., Mau, S.-C., and Huse, D.A., 1997. Entropy difference between crystal phases. *Nature*, 388, 235–237.

Boneschanscher, M.P., Evers, W.H., Geuchies, J.J., Altantzis, T., Goris, B., Rabouw, F.T., Van Rossum, S.A.P., Van Der Zant, H.S.J., Siebbeles, L.D.A., Van Tendeloo, G., Swart, I., Hilhorst, J., Petukhov, A.V., Bals, S., and Vanmaekelbergh, D., 2014. Long-range orientation and atomic attachment of nanocrystals in 2D honeycomb superlattices. *Science*, 344 (6190), 1377–1380.

Brodin, J.D., Auyeung, E., and Mirkin, C.A., 2015. DNA-mediated engineering of multicomponent enzyme crystals. *Proceedings of the National Academy of Sciences*, 112 (15), 4564–4569.

Cargnello, M., Johnston-Peck, A.C., Diroll, B.T., Wong, E., Datta, B., Damodhar, D., Doan-Nguyen, V.V.T., Herzing, A.A., Kagan, C.R., and Murray, C.B., 2015. Substitutional doping in nanocrystal superlattices. *Nature*, 524 (7566), 450–453.

Chen, C.F., Tzeng, S.D., Chen, H.Y., Lin, K.J., and Gwo, S., 2008. Tunable plasmonic response from alkanethiolate-stabilized gold nanoparticle superlattices: Evidence of near-field coupling. *Journal of the American Chemical Society*, 130 (3), 824–826.

Chen, C.J., Chiang, R.K., and Jeng, Y.R., 2011. Crystallization and magnetic properties of 3D micrometer-scale simple-cubic maghemite superlattices. *Journal of Physical Chemistry C*, 115 (37), 18142–18148.

Chen, H.Y., Lin, M.H., Wang, C.Y., Chang, Y.M., and Gwo, S., 2015. Large-scale hot spot engineering for quantitative SERS at the single-molecule scale. *Journal of the American Chemical Society*, 137 (42), 13698–13705.

Chen, Y., Si, K.J., Sikdar, D., Tang, Y., Premaratne, M., and Cheng, W., 2015. Ultrathin plasmene nanosheets as soft and surface-attachable SERS substrates with high signal uniformity. *Advanced Optical Materials*, 3 (7), 919–924.

Cheng, W., Campolongo, M.J., Cha, J.J., Tan, S.J., Umbach, C.C., Muller, D.A., and Luo, D., 2009. Freestanding nanoparticle superlattice sheets controlled by DNA. *Nature Materials*, 8 (6), 519–525.

Cheon, J., Park, J., Choi, J., Jun, Y., Kim, S., Kim, M.G., Kim, Y.-M.J., and Kim, Y.-M.J., 2006. Magnetic superlattices and their nanoscale phase transition effects. *Proceedings of the National Academy of Sciences of the United States of America*, 103 (9), 3023–3027.

Choi, J.-H., Wang, H., Oh, S.J., Paik, T., Sung, P., Sung, J., Ye, X., Zhao, T., Diroll, B.T., Murray, C.B., and Kagan, C.R., 2016. Exploiting the colloidal nanocrystal library to construct electronic devices. *Science*, 352 (6282), 205–208.

Coe, S., Woo, W.-K., Bawendi, M., and Bulović, V., 2002. Electroluminescence from single monolayers of nanocrystals in molecular organic devices. *Nature*, 420, 800–803.

Collier, C.P., Vossmeyer, T., and Heath, J.R., 1998. Nanocrystal superlattices. *Annual Review of Physical Chemistry*, 49 (1), 371–404.

Desvaux, C., Amiens, C., Fejes, P., Renaud, P., Respaud, M., Lecante, P., Snoeck, E., and Chaudret, B., 2005. Multimillimetre-large superlattices of air-stable iron-cobalt nanoparticles. *Nature Materials*, 4 (10), 750–753.

Diroll, B.T., Greybush, N.J., Kagan, C.R., and Murray, C.B., 2015. Smectic nanorod superlattices assembled on liquid subphases: Structure, orientation, defects, and optical polarization. *Chemistry of Materials*, 27 (8), 2998–3008.

Dong, A., Chen, J., Vora, P.M., Kikkawa, J.M., and Murray, C.B., 2010. Binary nanocrystal superlattice membranes self-assembled at the liquid-air interface. *Nature*, 466 (7305), 474–477.

Dreyer, A., Feld, A., Kornowski, A., Yilmaz, E.D., Noei, H., Meyer, A., Krekeler, T., Jiao, C., Stierle, A., Abetz, V., Weller, H., and Schneider, G.A., 2016. Organically linked iron oxide nanoparticle supercrystals with exceptional isotropic mechanical properties. *Nature Materials*, 15 (5), 522–528.

Estephan, Z.G., Qian, Z., Lee, D., Crocker, J.C., and Park, S.J., 2013. Responsive multidomain free-standing films of gold nanoparticles assembled by DNA-directed layer-by-layer approach. *Nano Letters*, 13 (9), 4449–4455.

Evers, W.H., Friedrich, H., Filion, L., Dijkstra, M., and Vanmaekelbergh, D., 2009. Observation of a ternary nanocrystal superlattice and its structural characterization by electron tomography. *Angewandte Chemie - International Edition*, 48 (51), 9655–9657.

Evers, W.H., De Nijs, B., Filion, L., Castillo, S., Dijkstra, M., and Vanmaekelbergh, D., 2010. Entropy-driven formation of binary semiconductor-nanocrystal superlattices. *Nano Letters*, 10 (10), 4235–4241.

Fagan, R.P. and Fairweather, N.F., 2014. Biogenesis and functions of bacterial S-layers. *Nature Reviews Microbiology*, 12 (3), 211–222.

Gang, O., 2016. Nanoparticle assembly: From fundamentals to applications: Concluding remarks. *Faraday Discussions*, 186, 529–537.

Gaulding, A., Diroll, B., Goodwin, E.D., Vrtis, Z., Kagan, C., and Murray, C., 2015. Deposition of wafer-scale single-component and binary nanocrystal superlattice thin films via Dip-Coating. *Advanced Materials 37*, 2846–2851.

Geuchies, J.J., Van Overbeek, C., Evers, W.H., Goris, B., De Backer, A., Gantapara, A.P., Rabouw, F.T., Hilhorst, J., Peters, J.L., Konovalov, O., Petukhov, A.V., Dijkstra, M., Siebbeles, L.D.A., Van Aert, S., Bals, S., and Vanmaekelbergh, D., 2016. In situ study of the formation mechanism of two-dimensional superlattices from PbSe nanocrystals. *Nature Materials*, 15 (12), 1248–1254.

Grzelczak, M., Vermant, J., Furst, E.M., and Liz-Marzán, L.M., 2010. Directed self-assembly of nanoparticles. *ACS Nano*, 4 (7), 3591–3605.

Heath, J.R., 2007. Synergy in a superlattice. *Nature*, 445, 492–493.

Hellstrom, S.L., Kim, Y., Fakonas, J.S., Senesi, A.J., Macfarlane, R.J., Mirkin, C.A., and Atwater, H.A., 2013. Epitaxial growth of DNA-assembled nanoparticle superlattices on patterned substrates. *Nano Letters*, 13 (12), 6084–6090.

Henzie, J., Grünwald, M., Widmer-Cooper, A., Geissler, P.L., and Yang, P., 2012. Self-assembly of uniform polyhedral silver nanocrystals into densest packings and exotic superlattices. *Nature Materials*, 11 (2), 131–137.

Hetsch, F., Zhao, N., Kershaw, S.V, and Rogach, A.L., 2013. Quantum dot field effect transistors. *Materials Today*, 16 (9), 312–325.

Irvine, W.T.M., Bowick, M.J., and Chaikin, P.M., 2012. Fractionalization of interstitials in curved colloidal crystals. *Nature Materials*, 11 (11), 948–951.

Jiao, Y., Han, D., Ding, Y., Zhang, X., Guo, G., Hu, J., Yang, D., and Dong, A., 2015. Fabrication of three-dimensionally interconnected nanoparticle superlattices and their lithium-ion storage properties. *Nature Communications*, 6, 1–8.

Jones, M.R., MacFarlane, R.J., Lee, B., Zhang, J., Young, K.L., Senesi, A.J., and Mirkin, C.A., 2010. DNA-nanoparticle superlattices formed from anisotropic building blocks. *Nature Materials*, 9 (11), 913–917.

Jones, M.R., Seeman, N.C., and Mirkin, C.A., 2015. Programmable materials and the nature of the DNA bond. *Science*, 347 (6224), 1260901.

Josten, E., Wetterskog, E., Glavic, A., Boesecke, P., Feoktystov, A., Brauweiler-Reuters, E., Rücker, U., Salazar-Alvarez, G., Brückel, T., and Bergstrm, L., 2017. Superlattice growth and rearrangement during evaporation-induced nanoparticle self-assembly. *Scientific Reports*, 7 (1), 1–9.

Kagan, C.R. and Murray, C.B., 2015. Charge transport in strongly coupled quantum dot solids. *Nature Nanotechnology*, 10 (12), 1013–1026.

Kalsin, A.M., Fialkowski, M., Paszewski, M., Smoukov, S.K., Bishop, K.J.M., and Grzybowski, B.A., 2006. Electrostatic self-assembly of binary nanoparticle crystals with a diamond-like lattice. *Science*, 312 (5772), 420–424.

Kang, Y., Ye, X., Chen, J., Qi, L., Diaz, R.E., Doan-Nguyen, V., Xing, G., Kagan, C.R., Li, J., Gorte, R.J., Stach, E.A., and Murray, C.B., 2013. Engineering catalytic contacts and thermal stability: Gold/iron oxide binary nanocrystal superlattices for CO oxidation. *Journal of the American Chemical Society*, 135 (4), 1499–1505.

Kim, T.H., Cho, K.S., Lee, E.K., Lee, S.J., Chae, J., Kim, J.W., Kim, D.H., Kwon, J.Y., Amaratunga, G., Lee, S.Y., Choi, B.L., Kuk, Y., Kim, J.M., and Kim, K., 2011. Full-colour quantum dot displays fabricated by transfer printing. *Nature Photonics*, 5 (3), 176–182.

Kim, Y. and Macfarlane, R.J., 2016. Transmutable nanoparticles with reconfigurable surface ligands. *Science*, 351 (6273), 579–582.

Konstantatos, G. and Sargent, E.H., 2009. Solution-processed quantum dot photodetectors. *Proceedings of the IEEE*, 97 (10), 1666–1683.

Korgel, B.A., 2010. Nanocrystal superlattices: Assembly at liquid interfaces. *Nature Materials*, 9 (9), 701–703.

Kostiainen, M.A., Hiekkataipale, P., Laiho, A., Lemieux, V., Seitsonen, J., Ruokolainen, J., and Ceci, P., 2013. Electrostatic assembly of binary nanoparticle superlattices using protein cages. *Nature Nanotechnology*, 8 (1), 52–56.

Kotov, N.A., Meldrum, F.C., Wu, C., and Fendler, J.H., 1994. Monoparticulate layer and Langmuir-Blodgett-type multiparticulate layers of size-quantized cadmium sulfide clusters: A colloid-chemical approach to superlattice construction. *Journal of Physical Chemistry*, 98 (11), 2735–2738.

Künzle, M., Eckert, T., and Beck, T., 2016. Binary protein crystals for the assembly of inorganic nNanoparticle superlattices. *Journal of the American Chemical Society*, 138 (39), 12731–12734.

Lewandowski, W., Fruhnert, M., Mieczkowski, J., Rockstuhl, C., and Górecka, E., 2015. Dynamically self-assembled silver nanoparticles as a thermally tunable metamaterial. *Nature Communications*, 6, 6590.

Li, J., Wang, Y., Zhou, T., Zhang, H., Sun, X., Tang, J., Zhang, L., Al-Enizi, A.M., Yang, Z., and Zheng, G., 2015. Nanoparticle superlattices as efficient bifunctional electrocatalysts for water splitting. *Journal of the American Chemical Society*, 137 (45), 14305–14312.

Li, J., Yiliguma, Wang, Y., and Zheng, G., 2016. Carbon-coated nanoparticle superlattices for energy applications. *Nanoscale*, 8 (30), 14359–14368.

Li, T., Senesi, A.J., and Lee, B., 2016. Small angle X-ray scattering for nanoparticle research. *Chemical Reviews*, 116 (18), 11128–11180.

Lin, Q.-Y., Mason, J.A., Li, Z., Zhou, W., O'Brien, M.N., Brown, K.A., Jones, M.R., Butun, S., Lee, B., Dravid, V.P., Aydin, K., and Mirkin, C.A., 2018. Building superlattices from individual nanoparticles via template-confined DNA-mediated assembly. *Science*, 359 (6376), 669–672.

Lu, F., Yager, K.G., Zhang, Y., Xin, H., and Gang, O., 2015. Superlattices assembled through shape-induced directional binding. *Nature Communications*, 6, 1–10.

Luo, D., Yan, C., and Wang, T., 2015. Interparticle forces underlying nanoparticle self-assemblies. *Small*, 11 (45), 5984–6008.

Macfarlane, R.J., Lee, B., Jones, M.R., Harris, N., Schatz, G.C., and Mirkin, C.A., 2011. Nanoparticle superlattice engineering with DNA. *Science*, 334 (6053), 204–208.

Manoharan, V.N., 2015. Colloidal matter: Packing, geometry, and entropy. *Science*, 349 (6251).

Meder, F., Thomas, S.S., Bollhorst, T., and Dawson, K.A., 2018. Ordered surface structuring of spherical colloids with binary nanoparticle superlattices. *Nano Letters*, 18 (4), 2511–2518.

Meng, G., Paulose, J., Nelson, D.R., and Manoharan, V.N., 2014. Elastic instability of a crystal growing on a curved surface. *Science*, 343 (6171), 634–637.

Mighty linkers, 2015. *Nature Materials*, 14 (8), 745.

Mirkin, C.A., Letsinger, R.L., Mucic, R.C., and Storhoff, J.J., 1996. A DNA-based method for rationally assembling nanoparticles into macroscopic materials. *Nature*, 382, 607–609.

Miszta, K., De Graaf, J., Bertoni, G., Dorfs, D., Brescia, R., Marras, S., Ceseracciu, L., Cingolani, R., Van Roij, R., Dijkstra, M., and Manna, L., 2011. Hierarchical self-assembly of suspended branched colloidal nanocrystals into superlattice structures. *Nature Materials*, 10 (11), 872–876.

Miszta, K., Greullet, F., Marras, S., Prato, M., Toma, A., Arciniegas, M., Manna, L., and Krahne, R., 2014. Nanocrystal film patterning by inhibiting cation exchange via electron-beam or X-ray lithography. *Nano Letters*, 14 (4), 2116–2122.

Mueggenburg, K.E., Lin, X.M., Goldsmith, R.H., and Jaeger, H.M., 2007. Elastic membranes of close-packed nanoparticle arrays. *Nature Materials*, 6 (9), 656–660.

Mühlig, S., Cunningham, A., Dintinger, J., Scharf, T., Bürgi, T., Lederer, F., and Rockstuhl, C., 2013. Self-assembled plasmonic metamaterials. *Nanophotonics*, 2 (3), 211–240.

Muralidharan, G., Sivaraman, S.K., and Santhanam, V., 2011. Effect of substrate on particle arrangement in arrays formed by self-assembly of polymer grafted nanoparticles. *Nanoscale*, 3 (5), 2138–2141.

Murray, C.B., Kagan, C.R., and Bawendi, M.G., 1995. Self-organization of CdSe nanocrystallites into three-dimensional quantum dot superlattices. *Science*, 270 (5240), 1335–1338.

Murray, C.B., Kagan, C.R., and Bawendi, M.G., 2000. Synthesis and characterization of monodisperse nanocrystals and close-packed nanocrystal assemblies. *Annual Review of Materials Science*, 30 (1), 545–610.

Murray, C.B., Sun, S., Doyle, H., and Betley, T., 2001. Monodisperse 3d transition-metal (co, ni, fe) nanoparticles and their assembly into nanoparticle superlattices. *MRS Bulletin*, 26 (12), 985–991.

Oh, S.J., Berry, N.E., Choi, J.H., Gaulding, E.A., Lin, H., Paik, T., Diroll, B.T., Muramoto, S., Murray, C.B., and Kagan, C.R., 2014. Designing high-performance PbS and PbSe nanocrystal electronic devices through stepwise, post-synthesis, colloidal atomic layer deposition. *Nano Letters*, 14 (3), 1559–1566.

Paik, T., Diroll, B.T., Kagan, C.R., and Murray, C.B., 2015. Binary and ternary superlattices self-assembled from colloidal nanodisks and nanorods. *Journal of the American Chemical Society*, 137 (20), 6662–6669.

Park, D.J., Ku, J.C., Sun, L., Lethiec, C.M., Stern, N.P., Schatz, G.C., and Mirkin, C.A., 2017. Directional emission from dye-functionalized plasmonic DNA superlattice microcavities. *Proceedings of the National Academy of Sciences*, 114 (3), 457–461.

Park, D.J., Zhang, C., Ku, J.C., Zhou, Y., Schatz, G.C., and Mirkin, C.A., 2015. Plasmonic photonic crystals realized through DNA-programmable assembly. *Proceedings of the National Academy of Sciences*, 112 (4), 977–981.

Podsiadlo, P., Krylova, G., Lee, B., Critchley, K., Gosztola, D.J., Talapin, D.V., Ashby, P.D., and Shevchenko, E.V., 2010. The role of order, nanocrystal size, and capping ligands in the collective mechanical response of three-dimensional nanocrystal solids. *Journal of the American Chemical Society*, 132 (26), 8953–8960.

Podsiadlo, P., Lee, B., Prakapenka, V.B., Krylova, G.V., Schaller, R.D., Demortière, A., and Shevchenko, E.V., 2011. High-pressure structural stability and elasticity of supercrystals self-assembled from nanocrystals. *Nano Letters*, 11 (2), 579–588.

Prasad, B.L.V, Stoeva, S.I., Sorensen, C.M., and Klabunde, K.J., 2003. Digestive-ripening agents for gold nanoparticles: Alternatives to thiols. *Chemistry of Materials*, 15 (4), 935–942.

Prasad, B.L.V., Sorensen, C.M., and Klabunde, K.J., 2008. Gold nanoparticle superlattices. *Chemical Society Reviews*, 37 (9), 1871–1883.

Qi, W., De Graaf, J., Qiao, F., Marras, S., Manna, L., and Dijkstra, M., 2012. Ordered two-dimensional superstructures of colloidal octapod-shaped nanocrystals on flat substrates. *Nano Letters*, 12 (10), 5299–5303.

Rao, S., Si, K.J., Yap, L.W., Xiang, Y., and Cheng, W., 2015. Free-standing bilayered nanoparticle superlattice nanosheets with asymmetric ionic Transport Behaviors. *ACS Nano*, 9 (11), 11218–11224.

Reich, K.V., Chen, T., and Shklovskii, B.I., 2014. Theory of a field-effect transistor based on a semiconductor nanocrystal array. *Physical Review B - Condensed Matter and Materials Physics*, 89 (23), 1–8.

Rogach, A.L., 2004. Binary superlattices of nanoparticles: Self-assembly leads to 'Metamaterials'. *Angewandte Chemie - International Edition*, 43, 148–149.

Roos, W.H., Bruinsma, R., and Wuite, G.J.L., 2010. Physical virology. *Nature Physics*, 6 (10), 733–743.

Ross, M.B., Ku, J.C., Blaber, M.G., Mirkin, C.A., and Schatz, G.C., 2015. Defect tolerance and the effect of structural inhomogeneity in plasmonic DNA-nanoparticle superlattices. *Proceedings of the National Academy of Sciences*, 112 (33), 10292–10297.

Ross, M.B., Ku, J.C., Vaccarezza, V.M., Schatz, G.C., and Mirkin, C.A., 2015. Nanoscale form dictates mesoscale function in plasmonic DNA-nanoparticle superlattices. *Nature Nanotechnology*, 10 (5), 453–458.

Ross, M.B., Mirkin, C.A., and Schatz, G.C., 2016. Optical properties of one-, two-, and three-dimensional arrays of plasmonic nanostructures. *The Journal of Physical Chemistry C*, 120 (2), 816–830.

Sahu, A., Kang, M.S., Kompch, A., Notthoff, C., Wills, A.W., Deng, D., Winterer, M., Frisbie, C.D., and Norris, D.J., 2012. Electronic impurity doping in CdSe nanocrystals. *Nano Letters*, 12 (5), 2587–2594.

Senesi, A.J., Eichelsdoerfer, D.J., Macfarlane, R.J., Jones, M.R., Auyeung, E., Lee, B., and Mirkin, C.A., 2013. Stepwise evolution of DNA-programmable nanoparticle superlattices. *Angewandte Chemie - International Edition*, 52 (26), 6624–6628.

Shevchenko, E.V., Talapin, D.V., Kotov, N.A., O'Brien, S., and Murray, C.B., 2006. Structural diversity in binary nanoparticle superlattices. *Nature*, 439 (7072), 55–59.

Shim, T.S., Estephan, Z.G., Qian, Z., Prosser, J.H., Lee, S.Y., Chenoweth, D.M., Lee, D., Park, S.J., and Crocker, J.C., 2017. Shape changing thin films powered by DNA hybridization. *Nature Nanotechnology*, 12 (1), 41–47.

Shimpi, J.R., Sidhaye, D.S., and Prasad, B.L.V., 2017. Digestive ripening: A fine chemical machining process on the nanoscale. *Langmuir*, 33 (38), 9491–9507.

Si, K.J., Chen, Y., Shi, Q., and Cheng, W., 2018. Nanoparticle superlattices: The roles of soft ligands. *Advanced Science*, 5 (1), 1–22.

Si, K.J., Sikdar, D., Chen, Y., Eftekhari, F., Xu, Z., Tang, Y., Xiong, W., Guo, P., Zhang, S., Lu, Y., Bao, Q., Zhu, W., Premaratne, M., and Cheng, W., 2014. Giant plasmene nanosheets, nanoribbons, and origami. *ACS Nano*, 8 (11), 11086–11093.

Si, K.J., Sikdar, D., Yap, L.W., Foo, J.K.K., Guo, P., Shi, Q., Premaratne, M., and Cheng, W., 2015. Dual-coded plasmene nanosheets as next-generation anticounterfeit security labels. *Advanced Optical Materials*, 3 (12), 1710–1717.

Sigman, M.B., Saunders, A.E., and Korgel, B.A., 2004. Metal nanocrystal superlattice nucleation and growth. *Langmuir*, 20 (3), 978–983.

Smith, D.K., Goodfellow, B., Smilgies, D.M., and Korgel, B.A., 2009. Self-assembled simple hexagonal AB2 binary nanocrystal superlattices: SEM, GISAXS, and defects. *Journal of the American Chemical Society*, 131 (9), 3281–3290.

Sun, L., Choi, J.J., Stachnik, D., Bartnik, A.C., Hyun, B.R., Malliaras, G.G., Hanrath, T., and Wise, F.W., 2012. Bright infrared quantum-dot light-emitting diodes through inter-dot spacing control. *Nature Nanotechnology*, 7 (6), 369–373.

Sun, L., Lin, H., Kohlstedt, K.L., Schatz, G.C., and Mirkin, C.A., 2018. Design principles for photonic crystals based on plasmonic nanoparticle superlattices. *Proceedings of the National Academy of Sciences*, 115 (28), 7242–7247.

Sun, S., Murray, C.B., Weller, D., Folks, L., and Moser, A., 2000. Monodisperse FePt nanoparticles and ferromagnetic FePt nanocrystal superlattices. *Science*, 287 (March), 1989–1992.

Talapin, D.V., Lee, J.-S., Kovalenko, M.V., and Shevchenko, E.V., 2010. Prospects of colloidal nanocrystals for electronic and optoelectronic applications. *Chemical Reviews*, 110 (1), 389–458.

Talapin, D.V., Shevchenko, E.V., Murray, C.B., Titov, A.V., and Král, P., 2007. Dipole - Dipole interactions in nanoparticle superlattices. *Nano Letters*, 7 (5), 1213–1219.

Talapin, D.V and Murray, C.B., 2005. PbSe nanocrystal solids for n-and p-channel thin film field-effect transistors. *Science*, 310 (86), 86–90.

Talapin, D.V, Shevchenko, E.V, Bodnarchuk, M.I., Ye, X., Chen, J., and Murray, C.B., 2009. Quasicrystalline order in self-assembled binary nanoparticle superlattices. *Nature*, 461 (7266), 964–967.

Talapin, D.V, Shevchenko, E.V, Kornowski, A., Gaponik, N., Haase, M., Rogach, A.L., and Weller, H., 2001. A new approach to crystallization of CdSe nanoparticles into ordered three dimensional superlattices. *Advanced Materials*, 13 (24), 1868–1871.

Talgorn, E., Gao, Y., Aerts, M., Kunneman, L.T., Schins, J.M., Savenije, T.J., Van Huis, M.A., Van Der Zant, H.S.J., Houtepen, A.J., and Siebbeles, L.D.A., 2011. Unity quantum yield of photogenerated charges and band-like transport in quantum-dot solids. *Nature Nanotechnology*, 6 (11), 733–739.

Travesset, A., 2015. Binary nanoparticle superlattices of soft-particle systems. *Proceedings of the National Academy of Sciences*, 112 (31), 9563–9567.

Travesset, A., 2017. Topological structure prediction in binary nanoparticle superlattices. *Soft Matter*, 13 (1), 147–157.

Udayabhaskararao, T., Altantzis, T., Houben, L., Coronado-Puchau, M., Langer, J., Popovitz-Biro, R., Liz-Marzán, L.M., Vukovic, L., Král, P., Bals, S., and Klajn, R., 2017. Tunable porous nanoallotropes prepared by post-assembly etching of binary nanoparticle superlattices. *Science*, 358 (6362), 514–518.

Urban, J.J., Talapin, D.V., Shevchenko, E.V., Kagan, C.R., and Murray, C.B., 2007. Synergism in binary nanocrystal superlattices leads to enhanced p-type conductivity in self-assembled PbTe/Ag2Te thin films. *Nature Materials*, 6 (2), 115–121.

Wang, K., Ling, H., Bao, Y., Yang, M., Yang, Y., Hussain, M., Wang, H., Zhang, L., Xie, L., Yi, M., Huang, W., Xie, X., and Zhu, J., 2018. A centimeter-scale inorganic nanoparticle superlattice monolayer with non-close-packing and its high performance in memory devices. *Advanced Materials*, 30 (27), 1800595.

Wei, J., Schaeffer, N., and Pileni, M.P., 2015. Ligand exchange governs the crystal structures in binary nanocrystal superlattices. *Journal of the American Chemical Society*, 137 (46), 14773–14784.

Weidman, M.C., Smilgies, D.-M., and Tisdale, W.A., 2016. Kinetics of the self-assembly of nanocrystal superlattices measured by real-time in situ X-ray scattering. *Nature Materials*, 15 (7), 775–781.

Weller, H., 1996. Self-organized superlattices of nanoparticles. *Angewandte Chemie (International Edition in English)*, 35 (10), 1079–1081.

Whitham, K., Yang, J., Savitzky, B.H., Kourkoutis, L.F., Wise, F., and Hanrath, T., 2016. Charge transport and localization in atomically coherent quantum dot solids. *Nature Materials*, 15 (5), 557–563.

Xavier, J., Vincent, S., Meder, F., and Vollmer, F., 2018. Advances in optoplasmonic sensors - Combining optical nano/microcavities and photonic crystals with plasmonic nanostructures and nanoparticles. *Nanophotonics*, 7 (1), 1–38.

Yang, Z., Wei, J., Bonville, P., and Pileni, M.P., 2015. Beyond entropy: Magnetic forces induce formation of quasicrystalline structure in binary nanocrystal superlattices. *Journal of the American Chemical Society*, 137 (13), 4487–4493.

Yannopapas, V., 2007. Artificial magnetism and negative refractive index in three-dimensional metamaterials of spherical particles at near-infrared and visible frequencies. *Applied Physics A: Materials Science and Processing*, 87 (2), 259–264.

Ye, X., Chen, J., Eric Irrgang, M., Engel, M., Dong, A., Glotzer, S.C., and Murray, C.B., 2017. Quasicrystalline nanocrystal superlattice with partial matching rules. *Nature Materials*, 16 (2), 214–219.

Young, K.L., Personick, M.L., Engel, M., Damasceno, P.F., Barnaby, S.N., Bleher, R., Li, T., Glotzer, S.C., Lee, B., and Mirkin, C.A., 2013. A directional entropic force approach to assemble anisotropic nanoparticles into superlattices. *Angewandte Chemie - International Edition*, 52 (52), 13980–13984.

Young, K.L., Ross, M.B., Blaber, M.G., Rycenga, M., Jones, M.R., Zhang, C., Senesi, A.J., Lee, B., Schatz, G.C., and Mirkin, C.A., 2014. Using DNA to design plasmonic metamaterials with tunable optical properties. *Advanced Materials*, 26 (4), 653–659.

Yu, Y., Guillaussier, A., Voggu, V.R., Houck, D.W., Smilgie, D.M., and Korgel, B.A., 2017. Bubble assemblies of nanocrystals: Superlattices without a substrate. *Journal of Physical Chemistry Letters*, 8 (19), 4865–4871.

Zhang, S.Y., Regulacio, M.D., and Han, M.Y., 2014. Self-assembly of colloidal one-dimensional nanocrystals. *Chemical Society Reviews*, 43 (7), 2301–2323.

Zhang, Y., Lu, F., Yager, K.G., Van Der Lelie, D., and Gang, O., 2013. A general strategy for the DNA-mediated self-assembly of functional nanoparticles into heterogeneous systems. *Nature Nanotechnology*, 8 (11), 865–872.

Zhang, Y., Pal, S., Srinivasan, B., Vo, T., Kumar, S., and Gang, O., 2015. Selective transformations between nanoparticle superlattices via the reprogramming of DNA-mediated interactions. *Nature Materials*, 14 (8), 840–847.

Zhu, H., Fan, Z., Yuan, Y., Wilson, M.A., Hills-Kimball, K., Wei, Z., He, J., Li, R., Grünwald, M., and Chen, O., 2018. Self-assembly of quantum dot-gold heterodimer nanocrystals with orientational order. *Nano Letters*, 18 (8), 5049–5056.

Zhu, J., Kim, Y., Lin, H., Wang, S., and Mirkin, C.A., 2018. PH-responsive nanoparticle superlattices with tunable DNA bonds. *Journal of the American Chemical Society*, 140 (15), 5061–5064.

Zhuang, Z., Peng, Q., and Li, Y., 2011. Controlled synthesis of semiconductor nanostructures in the liquid phase. *Chemical Society Reviews*, 40 (11), 5492–5513.

Holger F. Bettinger and
Peter Grüninger
University of Tübingen

4.1   Introduction

Heptacene is a member of the class of polycyclic aromatic hydrocarbons (PAH), called acenes, that is comprised of rectilinearly fused benzene rings (Figure 4.1). The smallest acene is naphthalene, followed by anthracene, tetracene (the name recommended by the International Union of Pure and Applied Chemistry is naphthacene, but the compound is often called tetracene in the literature and also in this chapter), pentacene, and hexacene. Naphthalene and anthracene have been isolated from coal tar in the first part of the 1820s and are nowadays very well-investigated fundamental organic compounds with multiple applications. The next higher homologs of the acenes, tetracene, pentacene, and hexacene were first synthesized in the 1930s (Fieser, 1931; Clar and John, 1929; Clar, 1939; Marschalk, 1939). These acenes, in particular tetracene and pentacene, are well characterized, and it is well understood how the properties such as absorption and fluorescence spectra, ionization potentials, and chemical reactivity change by successive elongation of the π systems. One particularly noteworthy property of acenes is their very rapid increase in chemical reactivity with increasing the size. For example, the bimolecular rate constants for Diels-Alder reaction with maleic anhydride, measured at $91.5 \pm 0.2\,°C$, increase by more than three orders of magnitude when going from anthracene to hexacene (Biermann and Schmidt, 1980). Likewise, the reactivity towards molecular oxygen, particularly in the presence of light, increases within the series and requires researchers to take appropriate caution when storing and handling the compounds, in particular the larger ones.

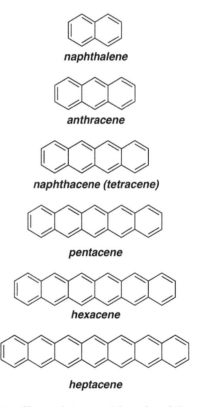

FIGURE 4.1   Chemical structural formulae of the acenes up to heptacene.

Tetracene and pentacene, along with some of their derivatives, received significant attention as organic semiconductors. There exist very good reviews on the properties of acenes and their applications as organic semiconductors, and the reader is referred to this literature for further information (Bendikov et al., 2004b; Anthony, 2006, 2008; Watanabe et al., 2013; Suzuki et al., 2014; Ye and Chi, 2014; Ortmann et al., 2015).

The next higher homolog of the acene series is heptacene. This hydrocarbon marks the edge of stability within the acene series (Zade and Bendikov, 2010). Attempts of its synthesis were first described in the 1940s by Clar (1942), but were received with some skepticism by Marschalk (1943) who worked on the heptacene synthesis at about the same time as Clar. Both jointly reported their results a few years later (Clar and Marschalk, 1950). Despite occasional reports of heptacene syntheses, the consensus within the community in the early 21st century was "that heptacene cannot be isolated in the pure state" (Bendikov et al., 2004a).

The first firm proof for the existence of heptacene was provided by the Neckers research group in 2006 (Mondal et al., 2006). The study by Neckers triggered a number of additional studies on heptacene and finally led to the observation that heptacene can exist in the solid state (Einholz et al., 2017). In addition to parent heptacene, substituted heptacenes were synthesized. The research on substituted heptacenes was pioneered by the Anthony group (Payne et al., 2005). The preparation of acenes larger than heptacene has also been reported. Under cryogenic matrix isolation conditions, octacene, nonacene, and undecacene were generated (Tönshoff and Bettinger, 2010; Shen et al., 2018), and kinetically stabilized nonacenes were synthesized by Anthony's group (Purushothaman et al., 2011). The so-called "on-surface synthesis" allowed access to heptacene (Zugermeier et al., 2017) and substituted heptacenes (Urgel et al., 2017), octacene and nonacene (Zuzak et al., 2017), decacene (Krüger et al., 2017), and undecacene (Zuzak et al., 2018). The remainder of this chapter will discuss the synthesis and properties of parent as well as substituted heptacenes taking into account the published literature up to the summer of 2018.

4.2 Synthesis of Heptacene and Substituted Heptacenes

4.2.1 Parent Heptacene

The (unsuccessful) dehydrogenation route to heptacene.

Erich Clar first obtained pentacene and hexacene by passing dihydro derivatives over hot copper powder under reduced pressure in an inert atmosphere (Clar and John, 1929; Clar, 1939). A similar dehydrogenation reaction was naturally also attempted in the synthesis of heptacene.

The controversy of Clar and Marschalk concerning the successful heptacene synthesis arose around the question whether certain dihydroheptacenes would undergo dehydrogenation in the presence of copper at high temperature. For a more detailed summary the reader is referred to reviews (Tönshoff and Bettinger, 2014; Bendikov et al., 2004b). Both researchers concluded that dihydroheptacenes cannot be dehydrogenated to heptacene. However, both isolated a highly reactive green compound and determined that this would be benzo[a]hexacene rather than the desired heptacene (Clar and Marschalk, 1950). The formation of benzo[a]hexacene was assumed to result from the angularly fused by-products **1** of reductive cyclization of the isophthalic acid derivative **2** (Figure 4.2).

Bailey and Liao reported that the dehydrogenation of a hexacosahydroheptacene over palladium/charcoal would produce heptacene in 76% yield based on the elemental analysis (Bailey et al., 1955). Presumably due to the criticism by Boggiano and Clar (1957), this work was not regarded as a successful heptacene synthesis later on.

The photochemical route to heptacene from bridged α-diketone precursor.

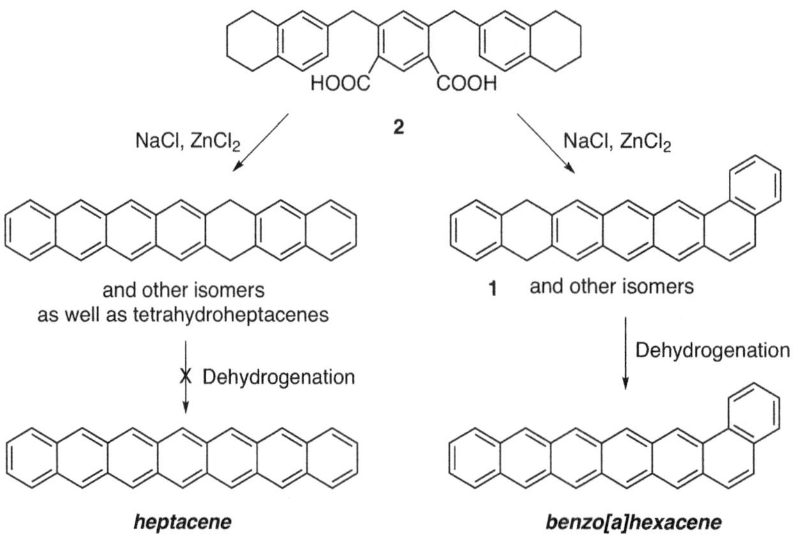

FIGURE 4.2 Possible formation of benzo[a]hexacene according to Clar and Marschalk (1950).

The investigation of the properties of heptacene only started after a successful synthetic pathway was established by the group of Neckers (Mondal et al., 2006). Neckers and co-workers synthesized a photoprecursor of heptacene 3 that incorporated an α-diketone bridge (Figure 4.3). The photo-cleavage of bridged α-diketones to yield aromatic hydro-carbons with carbon monoxide as only by-product was introduced by Strating et al. (1969) and is sometimes called the Strating-Zwanenburg reaction. The mechanism of this photobisdecarbonylation was probed by time-resolved spec-troscopy and computational methods (Mondal et al., 2008; Bettinger et al., 2013). The Strating-Zwanenburg reaction was previously employed successfully by Ono et al. for the synthesis of pentacene (Uno et al., 2005; Yamada et al., 2005) and proved to be very powerful for the synthesis of a wide variety of small and large acenes (Suzuki et al., 2014), including the largest ones known.

Neckers et al. performed the photobisdecarbonylation of 3 in a poly(methyl methacrylate) (PMMA) matrix at room temperature and identified heptacene by its characteristic optical absorption spectrum (Mondal et al., 2006). The reaction was subsequently also performed in solid noble gas matrices at cryogenic temperature (around 10 K). This technique allows investigation of a much larger part of the electromagnetic spectrum. Hence, the electronic absorption spectrum down to 200 nm and the infrared spectrum of heptacene could be measured (Mondal et al., 2009; Bettinger et al., 2007).

While pentacene can be generated from the corresponding α-diketone in toluene solution (Yamada et al., 2005), neither hexacene (Mondal et al., 2007) nor heptacene (Mondal et al., 2006) can be obtained from photobisdecarbonylation of the corresponding α-diketones in the solution phase. In the case of heptacene, the products of [π4s + π4s] photodimerization 4 are obtained instead (Figure 4.4) (Einholz and Bettinger, 2013).

The heptacenequinone reduction route to heptacene.

It is well established that the reduction of pentacene-6,13-quinone with aluminum tricyclohexoxide results in pentacene (Bruckner et al., 1960). Treliant Fang described in his Ph. D. thesis that the similar Meerwein-Ponndorf-Verley (MPV) reduction of heptacene-7,16-quinone 5 (with addi-tion to tetrabromo methane for acceleration of the reaction) resulted in diheptacenes 4 (Fang, 1986). Further research by Einholz et al. demonstrated that the dimers 4 origi-nally obtained by Fang are identical to the diheptacenes obtained from photobisdecarbonylation reaction in solution (Figure 4.4) (Einholz and Bettinger, 2013). The most impor-tant property of the dimers is that they undergo thermal cleavage: heating of the dimer mixture under the reduced pressure of a typical matrix isolation experiment (10^{-5} mbar) results in sublimation of monomeric heptacene with spectral properties that are identical to those of samples obtained from the photodecomposition of 3. In addition, heating of the mixture of 4 in a rotor used for solid-state nuclear magnetic resonance (NMR) spectroscopy caused

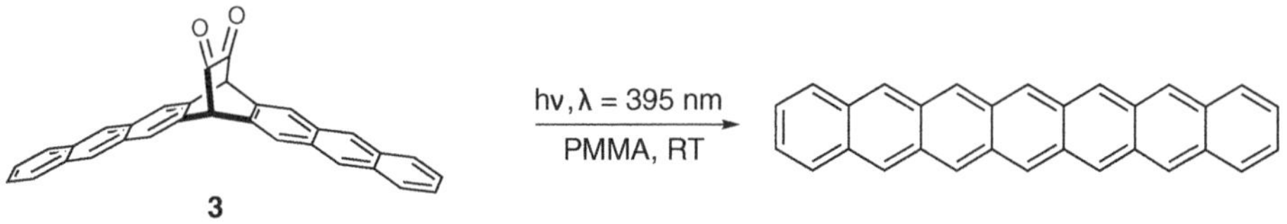

3

$$\xrightarrow{\text{hv}, \lambda = 395 \text{ nm}}{\text{PMMA, RT}}$$

FIGURE 4.3 Successful photogeneration of heptacene from photoprecursor **3** (Mondal et al., 2006).

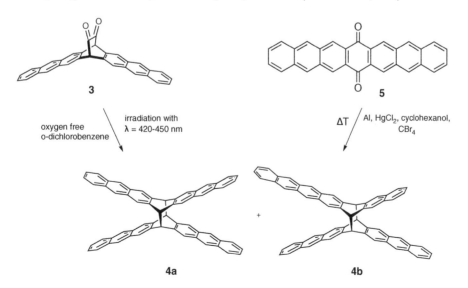

FIGURE 4.4 Photobisdecarbonylation of heptacene photoprecursor **3** in solution and MPV reduction of **5** yields isomeric forms of diheptacenes **4**.

significant changes in the solid-state ^{13}C CP-MAS-NMR spectrum: the ^{13}C signals of the aliphatic bridgehead atoms disappeared while the signal in the aromatic region sharpened. These spectral changes were considered as evidence for the formation of heptacene from **4** in the solid state (Figure 4.5) (Einholz et al., 2017).

4.2.2 Substituted Heptacenes

The syntheses of substituted heptacenes were summarized previously by Tönshoff and Bettinger (2014) as well as by Thorley and Anthony (2014). The compounds displayed in Figure 4.6 are the currently known derivatives.

The 7,16-bis-silylethynyl substituted derivatives **6a–c** were the first stabilized heptacenes, and their successful synthesis described by Anthony and co-workers marked a significant breakthrough in acene research (Payne et al., 2005). Increase in the size of the steric protecting group from iPr to Si(CH$_3$)$_3$ enhanced the stability of the compound and allowed characterization by single-crystal X-ray crystallography for **6c**. An additional increase in stability was achieved by the introduction of additional aryl groups in the 5,9,14,18 positions (Chun et al., 2008; Qu and Chi, 2010). Aryl substitution alone, as exemplified by compounds **7**, does not result in compounds of significant stability (Chun et al., 2008; Kaur et al., 2009). On the other hand, thioaryl substitution

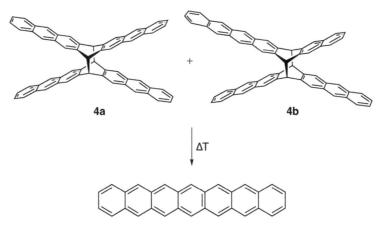

FIGURE 4.5 Formation of heptacene from the mixture of isomeric diheptacenes (**4**) under elevated temperature (∼300°C) in the solid state or in high vacuum (Einholz et al., 2017).

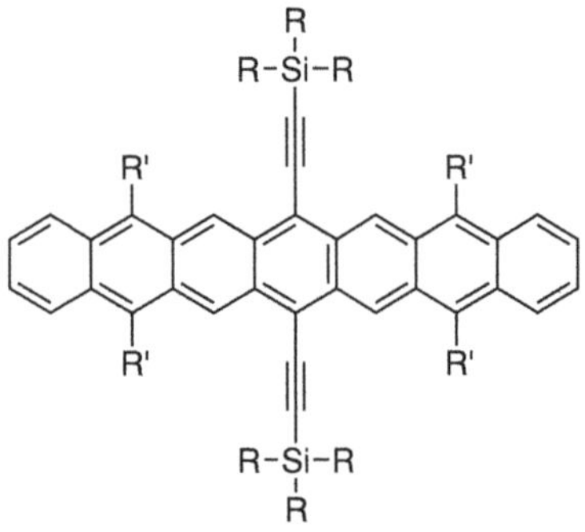

6a: R' = H, R = iPr
6b: R' = H, R = tBu
6c: R' = H, R = Si(CH$_3$)$_3$
6d: R' = C$_6$H$_5$, R = iPr
6e: R' = p(CF$_3$)C$_6$H$_4$, R = iPr

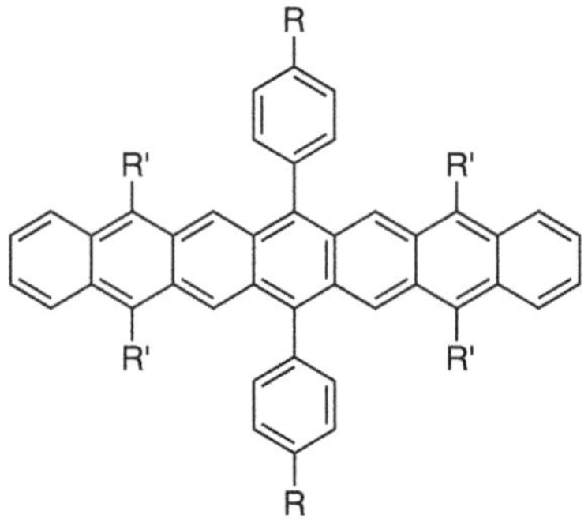

7a: R' = C$_6$H$_5$, R = H
7b: R' = C$_6$H$_5$, R = O(CH$_2$)CH(ethyl)(n-butyl)
7c: R' = H, R = tBu
7d: R' = 2,6-(CH$_3$)$_2$(C$_6$H$_3$), R = tBu

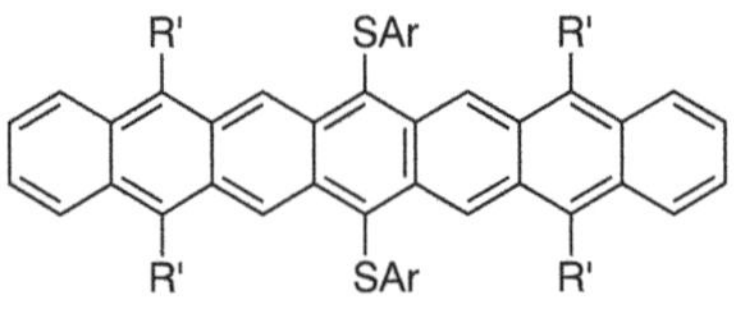

8a: R' = H, Ar = 4-tBuC$_6$H$_4$
8b: R' = 2,6-(CH$_3$)$_2$(C$_6$H$_3$), Ar = 4-tBuC$_6$H$_4$

FIGURE 4.6 Known substituted heptacenes.

as found in **8** imparts significant stabilization and indeed **8b** could be fully characterized by NMR, UV/vis, fluorescence spectroscopies and by laser desorption ionization time of flight mass spectrometry (LDI-TOF-MS) (Kaur et al., 2009).

4.3 Properties of Heptacene

4.3.1 Optical Absorption Spectrum

The optical absorption of the known stable acenes up to hexacene were investigated thoroughly (Nijegorodov et al., 1997). It is worthwhile to briefly summarize the properties of the acenes up to hexacene (Nijegorodov et al., 1997).

The absorption spectra of acenes are dominated by three major absorptions, that are $^1A \rightarrow {}^1B_b$ (β band), $^1A \rightarrow {}^1L_b$ (α band), and $^1A \rightarrow {}^1L_a$ (p band). Beginning with anthracene, the energies of the excited states decrease as $S_\beta > S_\alpha > S_p$, and the overall absorption spectrum shifts to lower energies with increasing acene length. The p band is short-axis polarized and within the visible range from tetracene onwards. Its intensity is intermediate, and the oscillator strength decreases with increasing acene size. The long-axis polarized α and β bands differ strongly in intensity: while the former is weak, the latter is the strongest band of the acene molecules investigated by Nijegorodov et al (1997). It should be noted that for acenes larger than heptacene, an additional stronger band emerges (Tönshoff and Bettinger, 2010; Bettinger et al., 2016).

The overall absorption spectrum of heptacene measured in solid argon at 10 K resembles that of the smaller members, pentacene and hexacene. The β and p bands are shifted to longer wavelengths (lower energies) as expected based on the behavior of the shorter members of the acene family. One obvious difference, however, is the shape of the p band that for all acenes shows typical vibrational fine structure. The strongest feature within the p band is the 0-0 transition that is of lowest energy. For heptacene, however, an additional weak feature at lower energy can be observed under all experimental conditions. These are (i) photogeneration from **3** within inert gas matrices at 10 K, organic glasses around 77–100 K, or within polymers at room temperature; (ii) cycloreversion of **4**, sublimation of the resulting heptacene and condensation in solid argon at 10 K; (iii) cycloreversion of **4** in 1-methylnaphthalene solution at 230°C (Figure 4.7) (Einholz et al., 2017).

The fact that from different precursors and in quite different matrices or solutions quite similar spectra are obtained makes matrix site effects an unlikely cause for this additional feature. Furthermore, a simulation of the vibrational fine structure of the p band of heptacene using quantum chemistry techniques suggests that the additional feature is not associated with the vibrational fine structure of the ^1L_a (S_p) state (Figure 4.8) (Einholz et al., 2017). One possible explanation given by Einholz et al. is that this feature is associated with an excited electronic state, $^2A_{1g}$, that falls below the ^1L_a state of heptacene according

to computations at the multiconfiguration coupled electron pair approximation (MCCEPA) ansatz (Einholz et al., 2017). A transition to such a state is electric dipole forbidden within the Franck–Condon approximation, but could gain intensity, maybe due to vibronic coupling to the ^1L_a state that is close in energy (Einholz et al., 2017).

4.3.2 Fluorescence and Photoluminescence

For the acene molecules up to hexacene fluorescence spectra, fluorescence quantum yields and life times were also determined by Nijegorodov et al. (1997). The T_1 triplet energies could be determined up to pentacene, but only be estimated by extrapolation for hexacene (Nijegorodov et al., 1997).

For parent heptacene, neither fluorescence nor phosphorescence spectra were reported in the literature. Based on the series of smaller acenes, the fluoresence quantum yield is expected to be quite small (it is 0.09 for pentacene and 0.01 for hexacene in cyclohexane solution). The fluorescence spectra of stabilized acenes **6e** and **8b** have been reported. For **6e** a strong feature at 553 nm is observed in toluene solution upon excitation at 353 nm (Qu and Chi, 2010). Likewise, **8b** shows a strong feature with vibrational fine structure at roughly 530 nm (exact value not given) upon excitation at 310 nm in chloroform (Kaur et al., 2009). These features are not in agreement with the expected $^1L_a \rightarrow {}^1A$ transition, as the ^1L_a states have their maxima around 865–870 nm (see also the discussion of nonacene fluorescence, Purushothaman et al, 2011). The latter compound **8b**, however, also shows a weak fluorescence feature with maxima at roughly 880 and 937 nm (exact value not given) upon 310 and 695 nm excitation, which could be in line with the $^1L_a \rightarrow {}^1A$ transition.

4.3.3 Electron Spin Resonance (ESR) and Nuclear Magnetic Resonance (NMR) Spectral Properties

Samples of heptacenes that were generated by photobisdecarbonylation of **3** in solid argon do not show an ESR signal in the temperature range (10–38 K) that is compatible with argon matrix isolation (Bettinger et al., 2007). These experiments show that heptacene has a singlet electronic ground state, in line with the interpretation of the optical absorption spectrum.

For various substituted heptacenes **6-8**, the solution phase ^1H NMR spectra did not show any unusual behavior, e.g., with regard to the line width, indicating that the triplet state is so high in energy that its possible thermal population, expected to follow Boltzmann statistics, does not interfere with routine ^1H NMR spectroscopy.

A solid sample of **4** and the thermal cycloreversion to heptacene was investigated by ^{13}C cross-polarized magic angle spinning (CP-MAS) NMR spectroscopy (Figure 4.9). Compound **4** shows a resonance at 54.7 ppm that is ascribed to its bridgehead carbon atoms. In addition, signals between

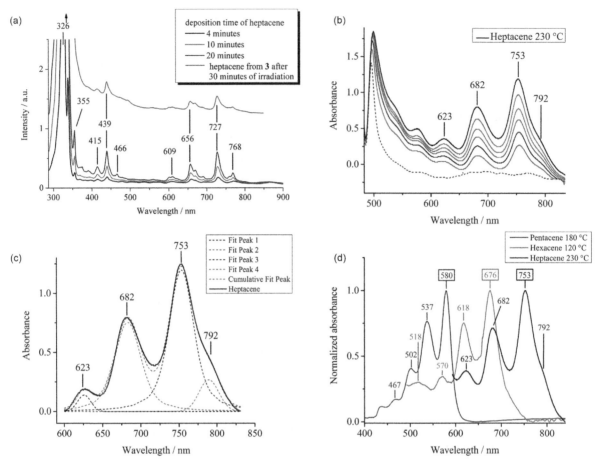

FIGURE 4.7 (a) Comparison of optical absorption spectra of heptacene (Ar, 30 K) obtained from cycloreversion of diheptacenes **4** in the presence of copper and from photobisdecarbonylation of α-diketone **3** (top trace shifted for ease of comparison). Note that there is residual intensity due to **3** in the range 450–500 nm in the top trace. (b) UV–vis spectrum of heptacene obtained by heating a solution of diheptacenes in 1-methylnaphthalene. (c) Peak fit of the *p* band region of heptacene. (d) Normalized optical spectra of pentacene, hexacene, and heptacene in the *p* band region measured in 1-methylnaphthalene at elevated temperatures. (Adapted with permission from Einholz et al., 2017. Copyright 2017 American Chemical Society.)

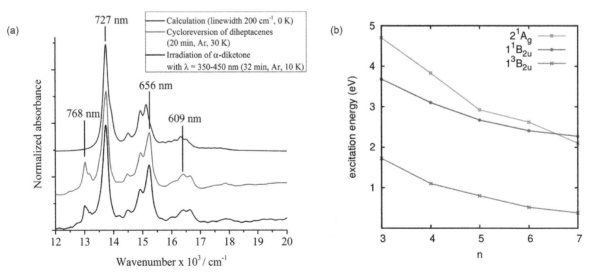

FIGURE 4.8 (a) Comparison of the *p* band region of heptacene obtained from photodecomposition of α-diketone and cycloreversion of diheptacenes with the calculated Franck–Condon profile at 0 K (Lorentzian line broadening with full width at half maximum of 200 cm^{-1} with the 0–0 transition shifted to 727 nm (13,755 cm^{-1}). (b) Energies of excited states 1^3B_{2u}, 1^1B_{2u}, and 2^1A_g with respect to the ground state of the n-acenes as computed at the MCCEPA level of theory. (Adapted with permission from Einholz et al., 2017. Copyright 2017 American Chemical Society.)

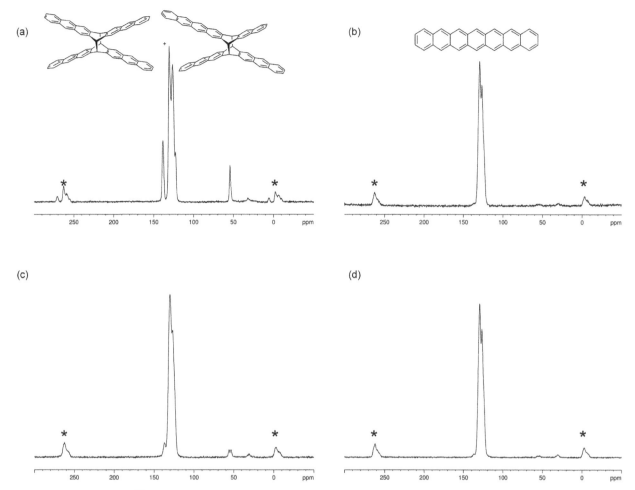

FIGURE 4.9 Solid-state ^{13}C CP-MAS NMR spectra (rotational side bands are marked by asterisks). (a) Product mixture obtained by MPV reduction of **5**. (b) Heptacene obtained from sample displayed in part (a) after heating to 300°C for 12 min. (c) Same sample as in part (b) after storage at room temperature in ambient atmosphere for 1 month. (d) Same sample as in part (c) after heating again (300°C/12 min). (Adapted with permission from Einholz et al., 2017. Copyright 2017 American Chemical Society.)

123 and 138 ppm are observed that are in the range that is typical for aromatic carbon atoms. After annealing of the sample within the rotor to 300°C for 15 min, the signal for the aromatic carbon atoms sharpens, while the bridgehead signal disappears. This indicates that the sample undergoes a chemical transformation that changes the formal hybridization of some carbon atoms from sp^3 to sp^2. This observation is compatible with the formation of heptacene, which only consists of sp^2 carbon atoms and whose computed ^{13}C NMR spectrum is in agreement with the sharpened feature observed after annealing.

4.3.4 Photoredox and Electrochemical Redox Behavior

Irradiation of heptacene that is isolated in argon with the output of a conventional low-pressure mercury lamp results in electron transfer among heptacene molecules that are randomly distributed within the matrix (Bettinger et al., 2007; Mondal et al., 2009). This photoinitiated electron transfer is caused by the 185 nm line of the mercury

spectrum and results in concomitant formation of heptacene radical anions and radical cations. Doping of the matrix with dichloromethane (1.5% by volume) as an electron trap results in selective formation of the radical cation of heptacene that has absorptions extending far into the near infrared region (Figure 4.10). The weak feature at 2,134 nm (0.58 eV) is due to an excitation of an electron from the singly occupied molecular orbital (SOMO) to the lowest unoccupied molecular orbital (LUMO) the SOMO → LUMO, while the strong feature at 1,248 nm is associated with a SOMO-1 → SOMO transition according to TD-UB3LYP/6-31G* computations (Mondal et al., 2009). Note that Mondal et al. (2009) accidentally inverted the assignments of the NIR bands to the involved orbitals.

The electrochemical properties of parent heptacene were not investigated. Cyclic voltammetry of compounds **6c–e** revealed reversible oxidation and reduction waves (Payne et al., 2005; Chun et al., 2008; Qu and Chi, 2010). The highest occupied molecular orbital (HOMO) and LUMO levels were estimated as −4.8 eV and −3.5 eV, respectively, for **6d**. These values are roughly 0.1 eV smaller due to the

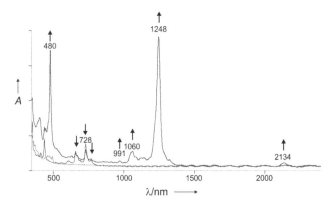

FIGURE 4.10 UV/vis-NIR absorption spectrum of the heptacene photoprecursor **3** (gray trace) in CH_2Cl_2 doped argon at 10 K, heptacene obtained after irradiation with light of wavelength 385–450 nm, and heptacene radical cation after irradiation with light of 185 nm wavelength. The arrows pointing up indicate increase, the arrows pointing downwards indicate decrease of the corresponding bands during 185 nm irradiation.

trifluoromethyl groups in **6e** (Chun et al., 2008; Qu and Chi, 2010). The band gaps estimated from the electrochemical experiments and the optical spectra range between 1.3 eV and 1.4 eV for **6c–e** (Payne et al., 2005; Chun et al., 2008; Qu and Chi, 2010).

4.3.5 Vibrational Spectroscopy

The infrared spectrum of heptacene in solid argon was also recorded (Bettinger et al., 2007; Mondal et al., 2009). The strongest absorption is in the fingerprint region at $901\ cm^{-1}$. This is due to an out-of-plane bending CH mode of b_{3u} symmetry computed at $922\ cm^{-1}$ at the RB3LYP/6-31G* level of theory. The experimental spectrum is in very good agreement with the one obtained at that level of theory. It should be noted that also the smaller acenes pentacene and hexacene isolated in argon have their strongest IR absorption at 901–$902\ cm^{-1}$ (Mondal et al., 2009).

4.3.6 Properties Determined by Computational Chemistry Techniques

As the available experimental data for heptacene is quite limited, computational chemistry methods can provide useful information. Heptacene as well as the smaller and larger acenes are of minima within D_{2h} symmetry constraints. A review discussing the electronic structure of acenes and polyacene based on computational investigations is available (Bettinger, 2010).

Many properties of the heptacene molecule can in principle be calculated and were indeed calculated (e.g., vibrational and absorption spectra, and nuclear shielding constants) using density functional theory (DFT) and other methods of computational quantum chemistry. The most relevant one with respect to understanding the electronic structure is the nature of the electronic ground state,

X^1A_g (S_0), and the energy difference between the ground state and the lowest energy triplet state, a^3B_{2u} (T_1), the so-called singlet–triplet energy gap. A number of computational treatments of the S_0 state have shown that the electronic ground state of heptacene is of multiconfigurational character (Hachmann et al., 2007; Plasser et al., 2013; Chakraborty and Shukla, 2013; Yang et al., 2016; Bettinger et al., 2016; Wu et al., 2016). This multiconfigurational character, indicative of strong static electron correlation, is due to the small energy difference between the HOMO and LUMO orbitals and is manifest by a diminished weight of the reference configuration $[\dots(HOMO)^2, LUMO^0\dots]$ of $c_0^2 = 0.74$ and an increased weight of the formally doubly excited configuration $[\dots(LUMO)^2, HOMO^0\dots]$ of $c_1^2 = 0.10$ (Bettinger et al., 2016). In Kohn-Sham density functional theory (KS-DFT), this multiconfigurational character of the wavefunction is manifested in the existence of a broken-symmetry (or spin-polarized) solution that is of lower energy than the conventional spin-restricted solution (Bendikov et al., 2004a). A similar instability of the Hartree–Fock wave function was already discovered in the 1980s, and interpreted as an anti-ferromagnetic electronic structure of the acene ground state (Baldo et al., 1983). This early prediction was basically confirmed by significantly more sophisticated electronic structure investigations (Hachmann et al., 2007).

The singlet–triplet energy difference (ΔE_{ST}) of heptacene (Table 4.1) ranges between 4 and 12 kcal/mol using polarized double-ζ basis sets of comparable quality. The smallest gap is obtained by the spin-restricted RB3LYP method that underestimates the stability of the singlet ground state, as the spin-unrestricted treatment (UB3LYP) gives a larger gap of 7 kcal/mol. The particle–particle random phase approximation (pp-RPA) and thermally assisted occupation method (TAO) are techniques that were developed for studying systems with strong static electron correlation in the context of DFT (Yang et al., 2016; Wu et al., 2016). These arrive at larger singlet–triplet gaps (9–12 kcal/mol) than broken-symmetry UB3LYP. The MCCEPA method is a true *ab initio* theory that takes into account both static

TABLE 4.1 Singlet–Triplet Energy Difference (ΔE_{ST} in kcal mol^{-1}) of Heptacene Computed at Various Levels of Theory

Level	ΔE_{ST}
RB3LYP/6-31G* [a,c]	4.4
UB3LYP/6-31G* [b,c]	7.1
pp-RPA-B3LYP/cc-pVDZ [d]	11.8
pp-RPA-B3LYP/cc-pVDZ [e]	9.0
TAO-LDA/6-31G* [f]	9.9
MC-CEPA/SV(P) [g]	8.7

[a] Spin-restricted solution for the singlet state.
[b] Spin-unrestricted solution for the singlet state.
[c] Taken from Bendikov et al. (2004a).
[d] Using the RB3LYP/6-31G* geometry of the singlet state (Yang et al., 2016).
[e] Using the UB3LYP/6-31G* geometry of the singlet state (Yang et al., 2016).
[f] Method also used for geometry optimization (Wu et al., 2016).
[g] Using the RB3LYP/SV(P) geometry (Einholz et al., 2017).

and dynamic electron correlation (Fink and Staemmler, 1993). Overall, the differences between various methods are not particularly large, and a singlet–triplet energy gap of roughly 9 kcal/mol is expected, assuming that the influence of the one-electron basis set is not particularly strong.

Yang et al. computed also a few higher lying singlet and triplet states with the pp-RPA-B3LYP method in order to judge the suitability of higher acenes for singlet fission (Yang et al., 2016). They concluded that the energetic criteria for singlet fission, $E(S_1) > 2\ E(T_1)$ and $E(T_2) > 2\ E(T_1)$, are fulfilled for tetracene and larger acenes, while the conditions $E(S_1) \approx 2$ eV and $E(T_1) \approx 1$ eV that favor high photovoltaic efficiency are not fulfilled for heptacene (Yang et al., 2016). However, based on the state energies, Yang et al. suggested that heptacene could be able to undergo singlet fission into three triplets (Yang et al., 2016). The largest acene for which singlet fission was reported is hexacene (Busby et al., 2014; Monahan et al., 2017).

4.4 Heptacene Thin Films

As the experiments have shown that heptacene can be obtained from **4**, sublimed and deposited in an argon matrix at cryogenic temperatures, films of heptacene were deposited in the matrix isolation chamber on sapphire (i.e., without argon gas) at 16 K and 298 K and studied by optical spectroscopy (Figure 4.11) (Einholz et al., 2017). These experiments revealed that heptacene forms films that can be held under vacuum for an extended period.

4.5 On-Surface Generation of Heptacene and Heptacene Derivatives

The α-diketone **2** cannot only be used as a photoprecursor of heptacene, but it can also be used to generate heptacene on a Ag(111) surface (Zugermeier et al., 2017). Heating the substance to 460 K on the surface results in bisdecarbonylation as indicated by X-ray photoelectron (XP) and near-edge X-ray absorption fine structure (NEXAFS) spectroscopy, and low-temperature scanning tunneling microscopy (STM). Annealing of a film of **2** on Ag(111) with nominal thickness of two monolayers to 460 K results in decarbonylation as evidenced by almost complete disappearance of the carbonyl oxygen component of the O 1s signal in the XP spectrum. Under these conditions, also the top monolayer is desorbed. Further annealing to 620 K results in partial desorption

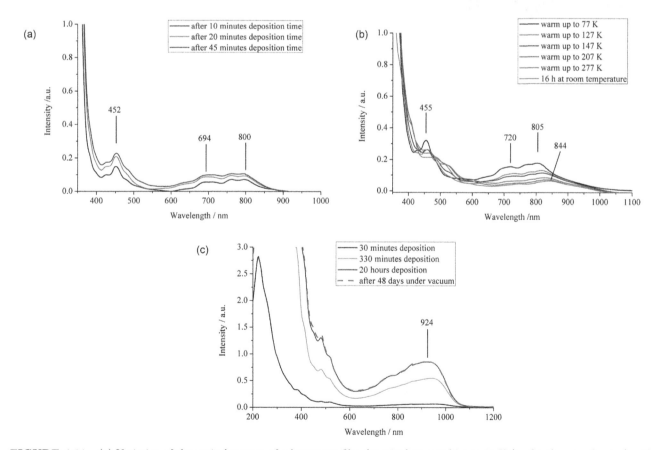

FIGURE 4.11 (a) Variation of the optical spectra of a heptacene film deposited on sapphire at 16 K (in the absence of argon) with deposition time. (b) Variation of the optical spectra of the heptacene film upon slowly warming from 16 to 295 K. (c) Optical spectra of heptacene films deposited on sapphire at 298 K. (Reprinted with permission from Einholz et al., 2017. Copyright 2017 American Chemical Society.)

of heptacene resulting in a 0.8 monolayer coverage. The NEXAFS spectra indicate that molecular plane of heptacene is oriented approximately parallel to the surface.

STM at 300 K images "fuzzy, stripe-like features" that are assigned to (tip-induced) 1D migration of **2** on the Ag(111) surface (Zugermeier et al., 2017). After annealing to 460 K, STM shows elongated molecules with lateral extensions that are compatible with heptacene. The molecules are immobile at 4 K and have their long axis aligned with the high symmetry direction of the substrate (Zugermeier et al., 2017). In addition, high-resolution images (Figure 4.12) suggest "that the center of the heptacene molecule has a lower apparent height than the two ends" (Zugermeier et al., 2017).

This distortion is in agreement with DFT computations of a heptacene molecule on Ag(111) that revealed a deviation from planarity with an angle of 3.5°. This was rationalized by a partial charge transfer from Ag to heptacene

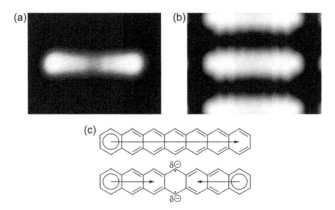

FIGURE 4.12 Comparison of an (a) experimental STM image (3.1×2.3 nm^2, -1.2 V, 2 pA, 4 K) and (b) DFT-calculated STM image ($+1.7$ V, 2 pA). (c) Neutral heptacene and partially negatively charged heptacene with their Clar sextets. (Reproduced from Zugermeier et al., 2017 with permission from The Royal Society of Chemistry.)

resulting in a system that could acquire two Clar sextets in the sense of a resonance structure (Zugermeier et al., 2017).

The bis-α-diketone **9** was employed for the generation of a heptacene–gold polymer **11** on a Au(111) surface by the Yamada and Fasel groups (Urgel et al., 2017). Upon heating of the 7,16-dibromo derivative **9** that was deposited by vacuum sublimation onto the Au(111) surface to 435 K the C-Br bonds break. A densely packed array **10** forms with C-Au distances of 2.3 ± 0.2 Å according to STM (Figure 4.13).

Further heating of the sample to 535 K results in decarbonylation with formation of heptacene–Au oligomers with two to nine heptacene molecules and individual heptacene molecules. It was assumed that residual hydrogen gas in the vacuum system resulted in hydrogenation of the formal radical centers of heptacene molecules (Urgel et al., 2017). STM and atomic force microscopy (AFM) images (Figure 4.14) of a heptacene–Au–heptacene unit reveal seven distinct lobes of the heptacene moieties and a central bright lobe that is assigned to the Au atom interconnecting the two molecules. The authors derived a C–Au distance of 2.5 ± 0.2 Å and concluded from DFT simulations that the heptacene molecule is slightly tilted (5.6°) towards the Au atom and slightly bent (6.0°) along the 7,16-direction.

Very recently, the dehydrogenation of 5,9,14,18-tetrahydroheptacene **12** was shown by STM and AFM to proceed on a Au(111) surface through annealing to 520 K for 10 min (Figure 4.15) (Zuzak et al., 2018). This transformation could also be induced by the microscope tip and a bias voltage of 2.5 V (Zuzak et al., 2018).

4.6 Conclusions and Outlook

It took more than 60 years after the first attempts of the synthesis of heptacene were reported before its existence was proven using the technique of matrix isolation in a polymer

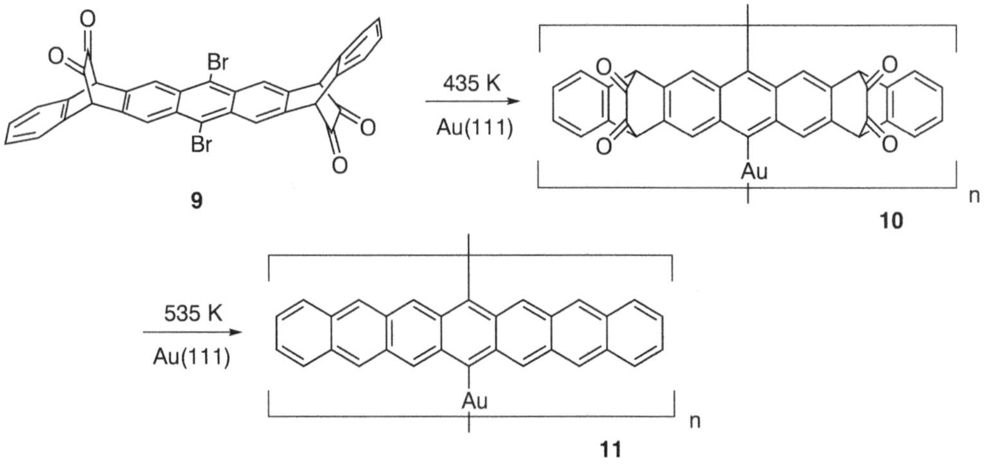

FIGURE 4.13 Stepwise on-surface generation of heptacene–gold aggregates from bis-α-diketone **9** according to Yamada and Fasel (Urgel et al., 2017).

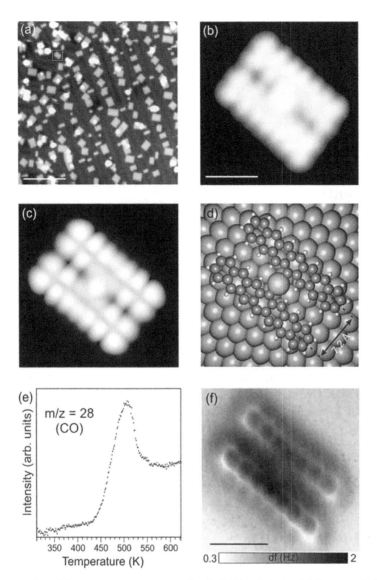

FIGURE 4.14 Thermal conversion of heptacene precursors on the Au(111) substrate. (a) Large-scale STM image of the heptacene organometallic complexes after a second annealing step at 535 K. $V_b = -2.5$ V, $I = 7$ pA. Scale bar: 15 nm. (b) High-resolution STM image of a heptacene–Au–heptacene organometallic complex (or Au-directed heptacene dimer) where intramolecular features are distinguished. $V_b = -1.5$ V, $I = 70$ pA. Scale bar: 1 nm. (c) DFT-simulated STM image of panel (b) at -1.5 eV. (d) Top view of the DFT equilibrium geometry of a heptacene–Au–heptacene organometallic complex adsorbed on a Au(111) slab. (e) Temperature-programmed desorption (TPD) graph showing the desorption of α-diketone groups as CO molecules with a maximum desorption rate at 510 K (heating rate 2 K/s). (f) Constant-height frequency-shift nc-AFM image of panel b acquired with a CO-functionalized tip (z offset -30 pm below STM set point: 5 mV, 10 pA). Scale bar = 1 nm. (Reprinted with permission from Urgel et al., 2017. Copyright 2017 American Chemical Society.)

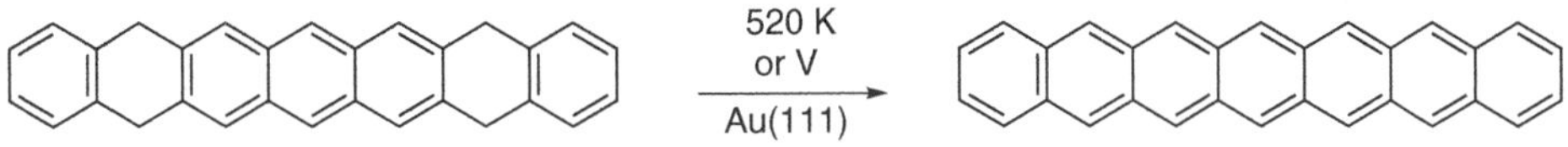

FIGURE 4.15 On-surface generation of heptacene by thermal- or voltage-induced dehydrogenation of 5,9,14,18-tetrahydroheptacene **12** according to Godlewski and Echavarren (Zuzak et al., 2018).

(PMMA). Today, kinetically stabilized heptacenes are available through conventional organic chemistry synthesis, and their characterization with standard techniques, including single crystal X-ray crystallography, was successful. Parent heptacene can nowadays be generated reliably under matrix isolation conditions and by thermal cycloreversion from diheptacenes. It remains true that heptacene is a very reactive compound and elucidation of its properties beyond those that are readily compatible with matrix isolation remains challenging. In our view, the most relevant open questions are related to the stability and electronic structure of heptacene films of variable thickness on metal single crystals, the hypothesis associated with the p band system in the optical spectrum, its fluorescence behavior, and the singlet–triplet energy gap. In a broader context, the possibility of heptacene to undergo singlet fission into three triplets is of high relevance. Our groups currently aim at further revealing the properties of heptacene by addressing the issues mentioned above.

4.7   Acknowledgments

We thank Deutsche Forschungsgemeinschaft for financial support of our acene research and Thomas Chassé for helpful discussions.

References

Anthony, J. E. 2006. Functionalized acenes and heteroacenes for organic electronics. *Chem. Rev.*, 106, 5028–5048.

Anthony, J. E. 2008. The larger acenes: Versatile organic semiconductors. *Angew. Chem. Int. Ed.*, 47, 452–483.

Bailey, W. J. & Liao, C.-W. 1955. Cyclic dienes. XI. New syntheses of hexacene and heptacene. *J. Am. Chem. Soc.*, 77, 992–993.

Baldo, M., Piccitto, G., Pucci, R. & Tomasello, P. 1983. Semiconductor-like structure of infinite linear polyacene. *Phys. Lett.*, 95A, 201–203.

Bendikov, M., Duong, H. M., Starkey, K., Houk, K. N., Carter, E. A. & Wudl, F. 2004a. Oligoacenes: Theoretical prediction of open-shell singlet diradical ground states. *J. Am. Chem. Soc.*, 126, 7416–7417.

Bendikov, M., Wudl, F. & Perepichka, D. F. 2004b. Tetrathiafulvalenes, oligoacenes, and their buckminsterfullerene derivatives: The brick and ortar of organic electronics. *Chem. Rev.*, 104, 4891–4945.

Bettinger, H. F. 2010. The electronic structure of higher acenes and polyacene: The perspective developed by theoretical analyses. *Pure Appl. Chem.*, 82, 905–915.

Bettinger, H. F., Mondal, R., Krasowska, M. & Neckers, D. C. 2013. Mechanisms for the formation of acenes from α-diketones by bisdecarbonylation. *J. Org. Chem.*, 78, 1851–1857.

Bettinger, H. F., Mondal, R. & Neckers, D. C. 2007. Stable photoinduced charge separation in heptacene. *Chem. Commun.*, 5209–5211.

Bettinger, H. F., Tönshoff, C., Doerr, M. & Sanchez-Garcia, E. 2016. Electronically excited states of higher acenes up to nonacene: A density functional theory/multireference configuration interaction study. *J. Chem. Theor. Comput.*, 12, 305–312.

Biermann, D. & Schmidt, W. 1980. Diels-alder reactivity of polycyclic aromatic hydrocarbons. 1. Acenes and benzologs. *J. Am. Chem. Soc.*, 102, 3163–3173.

Boggiano, B. & Clar, E. 1957. Four higher annellated pyrenes with acene character. *J. Chem. Soc.*, 2681–2689.

Bruckner, V., Wilhelms, A. K., Kormendy, K., Meszaros, M. & Tomasz, J. 1960. Simple synthesis of pentacene. *Tetrahedron Lett.*, 5–6.

Busby, E., Berkelbach, T. C., Kumar, B. et al. 2014. Multiphonon relaxation slows singlet fission in crystalline hexacene. *J. Am. Chem. Soc.*, 136, 10654–10660.

Chakraborty, H. & Shukla, A. 2013. Pariser-Parr-Pople model based investigation of ground and low-lying excited states of long acenes. *J. Phys. Chem. A*, 117, 14220–14229.

Chun, D., Cheng, Y. & Wudl, F. 2008. The most stable and fully characterized functionalized heptacene. *Angew. Chem. Int. Ed.*, 47, 8380–8385.

Clar, E. 1939. Hexacen, ein grüner, einfacher Kohlenwasserstoff (Aromatische Kohlenwasserstoffe, XXIV. Mitteil.). *Ber. Dtsch. Chem. Ges.*, 72B, 1817–1821.

Clar, E. 1942. Heptacen ein einfacher, "ultragrüner" Kohlenwasserstoff (Aromatische Kohlenwasserstoffe, XXXV. Mitteil.). *Chem. Ber.*, 75, 1330–1338.

Clar, E. & John, F. 1929. Polynuclear aromatic hydrocarbons and their derivatives. V. Naphthoanthracenes, their oxidation products and a new class of deeply colored hydrocarbons. *Ber. Dtsch. Chem. Ges.*, 62B, 3021–3029.

Clar, E. & Marschalk, C. 1950. Aromatic hydrocarbons. LVII. Syntheses in the heptacene series and of 1,2-benzohexacene. *Bull. Soc. Chim. Fr.*, 444–452.

Einholz, R. & Bettinger, H. F. 2013. Heptacene: Increased persistence of a 4n+2 π-Electron Polycyclic aromatic hydrocarbon by oxidation to the 4n π-electron dication. *Angew. Chem., Int. Ed.*, 52, 9818–9820.

Einholz, R., Fang, T., Berger, R. et al. 2017. Heptacene: Characterization in solution, in the solid state, and in films. *J. Am. Chem. Soc.*, 139, 4435–4442.

Fang, T. 1986. Heptacene, octacene, nonacene, supercene and related polymers. Ph. D. Thesis, University of California, Los Angeles.

Fieser, L. F. 1931. Reduction products of naphthacenequinone. *J. Am. Chem. Soc.*, 53, 2329–2341.

Fink, R. & Staemmler, V. 1993. A multi-configuration reference CEPA method based on pair natural orbitals. *Theor. Chim. Acta*, 87, 129–145.

Hachmann, J., Dorando, J. J., Avilés, M. & Chan, G. K.-L. 2007. The radical character of the acenes: A density matrix renormalization group study. *J. Chem. Phys.*, 127, 134309.

Kaur, I., Stein, N. N., Kopreski, R. P. & Miller, G. P. 2009. Exploiting substituent effects for the synthesis of a photooxidatively resistant heptacene derivative. *J. Am. Chem. Soc.*, 131, 3424–3425.

Krüger, J., García, F., Eisenhut, F. et al. 2017. Decacene: On-surface generation. *Angew. Chem. Int. Ed.*, 56, 11945–11948.

Marschalk, C. 1939. Linear hexacenes. *Bull. Soc. Chim. Fr.*, 6, 1112–1121.

Marschalk, C. 1943. Note on the article by E. Clar entitled "Heptacene, an ultragreen hydrocarbon". *Bull. Soc. Chim. Fr.*, 10, 511–512.

Monahan, N. R., Sun, D., Tamura, H. et al. 2017. Dynamics of the triplet-pair state reveals the likely coexistence of coherent and incoherent singlet fission in crystalline hexacene. *Nat. Chem.,* 9, 341–346.

Mondal, R., Adhikari, R. M., Shah, B. K. & Neckers, D. C. 2007. Revisiting the stability of hexacenes. *Org. Lett.,* 9, 2505–2508.

Mondal, R., Okhrimenko, A. N., Shah, B. K. & Neckers, D. C. 2008. Photodecarbonylation of alpha-Diketones: A mechanistic study of reactions leading to Acenes. *J. Phys. Chem. B,* 112, 11–15.

Mondal, R., Shah, B. K. & Neckers, D. C. 2006. Photogeneration of Heptacene in a Polymer Matrix. *J. Am. Chem. Soc.,* 128, 9612–9613.

Mondal, R., Tönshoff, C., Khon, D., Neckers, D. C. & Bettinger, H. F. 2009. Synthesis, stability, and photochemistry of pentacene, hexacene, and heptacene: A matrix isolation study. *J. Am. Chem. Soc.,* 131, 14281–14289.

Nijegorodov, N., Ramachandran, V. & Winkoun, D. P. 1997. The dependence of the absorption and fluorescence parameters, the intersystem crossing and internal conversion rate constants on the number of rings in polyacene molecules. *Spectrochim. Acta A,* 53, 1813–1824.

Ortmann, F., Radke, K. S., Günther, A., Kasemann, D., Leo, K. & Cuniberti, G. 2015. Materials eets concepts in molecule-based electronics. *Adv. Funct. Mater.,* 25, 1933–1954.

Payne, M. M., Parkin, S. R. & Anthony, J. E. 2005. Functionalized higher acenes: Hexacene and heptacene. *J. Am. Chem. Soc.,* 127, 8028–8029.

Plasser, F., Pašalić, H., Gerzabek, M. H. et al. 2013. The multiradical character of one- and two-dimensional graphene nanoribbons. *Angew. Chem., Int. Ed.,* 52, 2581–2584.

Purushothaman, B., Bruzek, M., Parkin, S. R., Miller, A.-F., Anthony, J. E. 2011. Synthesis and Structural Characterization of Crystalline Nonacenes. *Angew. Chem., Int. Ed.,* 50, 7013–7017.

Qu, H. & Chi, C. 2010. A stable heptacene derivative substituted with electron-deficient trifluoromethylphenyl and triisopropylsilylethynyl groups. *Org. Lett.,* 12, 3360–3363.

Shen, B., Tatchen, J., Sanchez-Garcia, E. & Bettinger, H. F. 2018. Evolution of the optical gap in the acene series: Undecacene. *Angew. Chem. Int. Ed.,* 57, 10506–10509.

Strating, J., Zwanenburg, B., Wagenaar, A. & Udding, A. C. 1969. Evidence for the expulsion of bis-CO from bridged a-diketones. *Tetrahedron Lett.,* 10, 125–128.

Suzuki, M., Aotake, T., Yamaguchi, Y. et al. 2014. Synthesis and photoreactivity of α-diketone-type precursors of acenes and their use in organic-device fabrication. *J. Photochem. Photobiol. C,* 18, 50–70.

Thorley, K. J. & Anthony, J. E. 2014. The electronic nature and reactivity of the larger acenes. *Isr. J. Chem.,* 54, 642–649.

Tönshoff, C. & Bettinger, H. F. 2010. Photogeneration of octacene and nonacene. *Angew. Chem., Int. Ed.,* 49, 4125–4128.

Tönshoff, C. & Bettinger, H. F. 2014. Beyond pentacenes: Synthesis and properties of higher acenes. *Top. Curr. Chem.,* 349, 1–30.

Uno, H., Yamashita, Y., Kikuchi, M. et al. 2005. Photo precursor for pentacene. *Tetrahedron Lett.,* 46, 1981–1983.

Urgel, J. I., Hayashi, H., Di Giovannantonio, M. et al. 2017. On-surface synthesis of heptacene organometallic complexes. *J. Am. Chem. Soc.,* 139, 11658–11661.

Watanabe, M., Chen, K.-Y., Chang, Y. J. & Chow, T. J. 2013. Acenes generated from precursors and their semiconducting properties. *Acc. Chem. Res.,* 46, 1606–1615.

Wu, C.-S., Lee, P.-Y. & Chai, J.-D. 2016. Electronic properties of cyclacenes from TAO-DFT. *Sci. Rep.,* 6, 37249.

Yamada, H., Yamashita, Y., Kikuchi, M. et al. 2005. Photochemical synthesis of pentacene and its derivatives. *Chem. Eur. J.,* 11, 6212–6220.

Yang, Y., Davidson, E. R. & Yang, W. 2016. Nature of ground and electronic excited states of higher acenes. *Proc. Natl. Acad. Sci. USA,* 113, E5098–E5107.

Ye, Q. & Chi, C. 2014. Recent highlights and perspectives on acene based molecules and materials. *Chem. Mater.,* 26, 4046–4056.

Zade, S. S. & Bendikov, M. 2010. Heptacene and beyond: The longest characterized acenes. *Angew. Chem., Int. Ed.,* 49, 4012–4015.

Zugermeier, M., Gruber, M., Schmid, M. et al. 2017. On-surface synthesis of heptacene and its interaction with a metal surface. *Nanoscale,* 9, 12461–12469.

Zuzak, R., Dorel, R., Kolmer, M., Szymonski, M., Godlewski, S. & Echavarren, A. M. 2018. Higher acenes by on-surface dehydrogenation: From heptacene to undecacene. *Angew. Chem. Int. Ed.,* 57, 10500–10505.

Zuzak, R., Dorel, R., Krawiec, M. et al. 2017. Nonacene generated by on-surface dehydrogenation. *ACS Nano,* 11, 9321–9329.

Epitaxial Silicene

Dmytro Solonenko and
Patrick Vogt
Technische Universität Chemnitz

5.1   Introduction

Since the discovery of graphene, the first elemental two-dimensional (2D) material, the interest in this family of 2D crystals has been growing over the past decade. The related search has been motivated by the expected applications of these layered structures in novel electronic devices, ranging from spintronic applications or new sensor concepts to topological insulators and beyond. The reasons for such expectations are associated with the broad range of materials that might form 2D layers and the possibility to adjust their physical properties by structural modification, external influences, or chemical functionalization. 2D materials became thus a playground for researchers to test their audacious ideas and to demonstrate breathtaking applications that might meet the future demands of our hi-tech society. However, even if some of these novel concepts are in the grasp, the translation of the fundamental innovations from research into market is a long, difficult path. This requires a detailed deep understanding of the fundamental properties of these 2D materials, a condition that cannot be circumvented. When looking at graphene, only few of the initial technological expectations have been realized so far. Even though a proof of concept could be demonstrated for some of these ideas, market-relevant applications are still awaited. This just shows that the development of fundamentally new technologies requires its time in order to understand their underlying properties.

One-atom-thin 2D materials offer such perspectives as they allow us to modify their characteristics by placing them on a substrate or into 2D heterostructures, applying an electrical or magnetic field, shining light on them, or applying an electric bias. A similar impact is much harder to achieve for conventional bulk crystals due to their rigidity and stability.

For many decades, such 2D materials were not thought to be stable. It was expected that even if a 2D crystal was formed, it would immediately break apart due to lattice fluctuations at non-zero temperatures [1]. Such small displacements of atoms within the third dimension could destroy the periodic potential of a crystal. This out-of-plane dimension provides an additional degree of freedom for the atoms to oscillate around their equilibrium positions. The only way for 2D crystals to still form is to find an extra stabilization mechanism, which will compensate the thermodynamic fluctuations. It turned out that such a stabilization can be realized by additional π bonds and the related charge delocalization, resulting from the sp^2 hybridization of carbon atoms, as found in graphene [2]. Indeed, the conjugated honeycomb lattice of graphene creates an opportunity for atoms to share their p_z orbitals restricting the atomic displacements, which stabilizes the crystal.

The strong σ bonds between sp^2 orbitals of the C atoms in the atomic layer are enhanced by the π bonds forming between the pz orbitals. It is the combination of these σ and π bonds that stabilizes the graphene lattice. In the realm of organic chemistry, it is known since the 19th century that carbon atoms can change their hybridization state depending on their chemical environment in order to become energetically more favorable [3]. In fact, C atoms can be found in four different configurations of their valence electrons as illustrated in Figure 5.1: In the non-hybridized state (left panel), the carbon atoms keep their $3s$ and $1p$ orbitals. In the case of an sp^3 hybridization, the $3s$ and $1p$ orbitals form four equivalent sp^3 orbitals, equally distanced to each

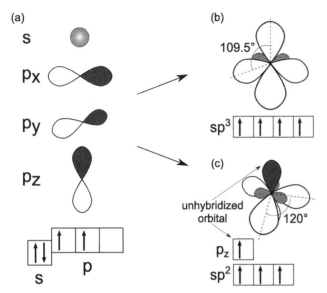

FIGURE 5.1 Schematic illustration of the different hybridization states of a carbon atom: unhybridized (a), sp^3 unhybridized (b) and sp^2, unhybridized (c). The accompanying chemical scheme demonstrates the occupation of the different orbitals by electrons according to the Pauli principle. The black-coding of p orbitals represents their dumbbell character.

other in a tetragonal orientation, which results in angles α between each two orbitals of 109.5° (Figure 5.1b). For an sp^2 configuration, $2s$ and $1p$ orbitals form $3sp^2$ hybrid orbitals, with one remaining p_z orbital perpendicular to the sp^2 hybrid orbitals (Figure 5.1c). The sp^2 hybrid orbitals are located in the same plane with angles α of 120° between them. This planar configuration of these orbitals is the prerequisite for the formation of the fully planar graphene lattice. In the sp hybridization state (not shown), only $1s$ and $1p$ orbitals hybridize with two remaining p_z orbitals which enable a C≡C triple-bond formation.

By looking at the group IV elements of the periodic table, we can safely assume that similar hybridization configurations are also found for the chemical elements of this group. However, one of the main differences of a 2D form of silicon in comparison to graphene is that it is not formed of purely sp^2 hybridized Si atoms. Generally, the physical background for the orbital hybridization is the principle of energy minimization of the atom. Depending on the chemical environment, these configurations can differ. C atoms may reside in an sp^3 hybridization, when forming a diamond crystal, or in an sp^2 hybridization, when forming graphene or graphite for better stability. Their thermodynamically most stable configuration is, however, the sp^2 hybridization (Figure 5.2a), meaning that diamond is only metastable.

Remarkably, the bond of two sp^2 hybridized C atoms has a lower bond energy in comparison to the energy of a triple bond. However, the sp hybridization is energetically unfavourable due to the bond asymmetry. For Si, the lowest bond energy is found for an sp^3 hybridization of the atoms, forming the diamond structure of bulk Si, while an

sp^2 hybridization marks only a local energy minimum for Si (Figure 5.2a).

The variable that plays a crucial role for the structure formation is the interatomic distance, illustrated in Figure 5.2b. A double bond formation implies shorter distances between the atoms. Here, the difference between C and Si atoms is due to their atomic size. The double bond length between C atoms in a honeycomb structure is 142 pm, which lies within the doubled covalent radius of a C atom (73 pm) and corresponds to 8% of a bond shortening in comparison to the bonds in diamond. A similar shortening of 8% of Si–Si bonds would imply a bond length of 216 pm, which is much shorter than the minimum possible distance of 222 pm (doubled covalent radius of Si) (Figure 5.2b). Since the reduction of the distance by moving the atoms into each other, i.e., partially overlaying the atomic spheres, is energetically not possible, the only way to slightly increase the distance between the atoms without moving them apart is to displace them "vertically", i.e., in respect to the mean line out of their initial plane which results in the corrugation. Quoting the work that has first suggested the existence of aromatic Si lattice, "...a puckering feature of the 2D Si sheet is essential and originates from the weakness of the Si skeleton's aromaticity itself" [4]. In other words, the corrugation is the only possible solution for Si atoms to form a double bond. Although the metastability of aromatic Si lattice, thus, silicene, was theoretically predicted, it has not yet been found. Due to the weakening of the Si lattice potential via buckling, the internal atomic displacements become stronger, increasing the probability of lattice collapse.

One of the possible ways to stabilize silicene is to synthesize it on a supporting substrate. The substrate is required to keep the lattice from converting into energetically more

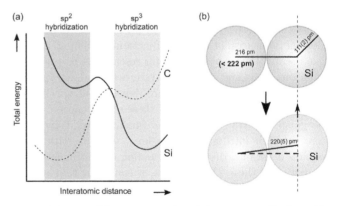

FIGURE 5.2 (a) Schematic plot of the energetically favorable hybridization states of C and Si atoms as a function of their interatomic distance. The curve only represents the local and global minima for the two elements, showing that globally for C, an sp^3 hybridization is energetically more favorable than an sp^2 hybridization, whereas it is the opposite for Si. (b) Schematic illustration of the relation between the buckling and the Si–Si bond length: If the neighboring Si atoms are buckled, their interatomic distance (Si–Si bond length) increases, making a mixed sp^2/sp^3 hybridization energetically possible.

favorable bulk Si [5]. A necessary requirement for such a substrate is to be chemically inert to the adlayer, which, otherwise, may lead to a formation of an alloy. Finally, a matching substrate has to maintain right geometrical and thermodynamical conditions for the formation of epitaxial silicene. In other words, the lattice should have a suitable lattice parameter for the epitaxy, and it has to be thermodynamically stable under the growth conditions, where especially the temperature is one of the main driving parameters for crystal growth. Such substrate was found to be silver (111), which fulfills all these requirements for the formation of silicene [6]. In the following section, we will describe the growth and the physical properties of epitaxial silicene grown on Ag(111).

5.2 Theoretical Predictions and Expectations

Silicene was first investigated (not by the name "silicene") in 1994 by Takeda and Shiraishi. At that time, the interest in low-dimensional Si structures was stimulated by the success of organic chemistry, producing Si-containing molecules where Si atoms are double-bonded, in an sp^2 hybridization state. Using density functional theory (DFT) in the local density approximation (LDA), the results for a 2D Si layer clearly showed that a slightly buckled arrangement of the Si atoms exhibits a higher stability in contrast to a fully planar arrangement (Figure 5.3a), owing to the pure sp^2-like hybridization in the latter case [4]. The energetic stability of the 2D layer was found to be a function of the bond distances and the bond angles between adjacent Si atoms. The equilibrium lattice configuration was found to have bond angles of about 10° and an interatomic distance of 0.225 nm, a configuration in which the so-called "Hellmann–Feynmann" forces, acting on the atomic nuclei, disappear. Such bond length is slightly larger than the minimum interatomic Si—Si distance, suggesting a mixed sp^2/sp^3 hybridization.

The elongated Si—Si bond length in the buckled Si layer indicates a decrease of the overlap of the sp^2 hybridized orbitals and accordingly of the p_z orbital needed to stabilize the 2D structure. Therefore, the aromaticity of the layer becomes weakened.

As a consequence of the buckling, the corrugated Si structure acquires a lattice symmetry of D_3d, which is lower in comparison to the one of a planar honeycomb sheet, having a D_6h symmetry. However, the calculations show that the electronic band structure still exhibits linear dispersions at the $\overline{K}$ points ("Dirac points") of the silicene BZ (Figure 5.3b (adapted from Ref. [7])). These linear bands stem from the honeycomb arrangement of sp^2-like hybridized Si atoms, as confirmed by tight-binding (TB) calculations [8]. For a linear electronic dispersion, where $E(k) \sim k$, the effective mass m^* is no longer given by

$$m^* = \hbar^2 \left(\frac{d^2 E(k)}{dk^2} \right)^{-1}, \qquad (5.1)$$

which is strictly restricted to parabolic bands. Applying this equation in the case of a 2D honeycomb layer would strongly disagree with "massless" charge carriers, usually expected for these lattices.

Instead, for the 2D system, the momentum p of the charge particles is described in a semi-classical approach as

$$p = \hbar k = m^* v_g, \qquad (5.2)$$

where v_g is the group velocity of the wave-packet associated with the particle and given by

$$v_g = \frac{1}{\hbar} \frac{dE(k)}{dk}. \qquad (5.3)$$

By substituting v_g in Eq. (5.2) with the expression in Eq. (5.3), we obtain for the effective mass [9]

$$m^* = \hbar \frac{1}{v_g} k = \hbar^2 k \left(\frac{dE(k)}{dk} \right)^{-1}. \qquad (5.4)$$

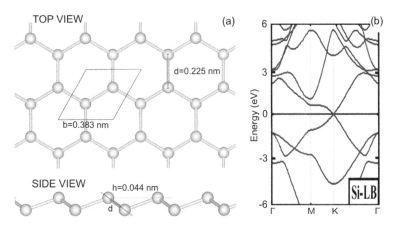

FIGURE 5.3 (a) Relaxed atomic structure model for low-buckled free-standing silicene in top and side views as derived theoretically by DFT calculations. The lattice parameters are indicated in the images: d: Si–Si interatomic distance, b: lattice constant, h: buckling distance (the values are taken from Ref. [7]). (b) Calculated electronic band structure for the low-buckled silicene structure in (a), showing linear electronic dispersions of the π and π^* bands at the $\overline{K}$ points of the Brillouin zone (BZ) (adapted from Ref. [7]).

For graphene, this yields

$$E = \hbar k v_F, \tag{5.5}$$

where $v_F \sim v_g$ is the Fermi velocity, which is independent of k in the proximity of the Dirac points, describing linear bands. Their linearity in turn suggests a zero effective mass of charge carriers for $k = 0$ moving along the 2D lattice and thus high electron mobility in the material.

DFT calculations show that a linear dispersion emerges due to its aromaticity even for a low-buckled, i.e., not planar, layer [10]. Nevertheless, it is worth to point out that some properties of these linear bands are slightly altered when compared to a planar honeycomb lattice. For instance, the slope, i.e., the Fermi velocity v_F, and the bandwidth are larger in the case of silicene compared to graphene, which can be attributed to the smaller overlap of the Si orbitals. A less steeper slope indicates a reduction in the Fermi velocity v_F of the electrons, to half the value found for graphene. Similar results are found by TB calculations for a buckled silicene lattice, provided that the interaction of the second-nearest neighbour is considered in the calculations [8].

The results also show some resemblance between silicene and a Si(111) atomic plane. The honeycomb arrangement as well as an evident buckling is similar in both cases; however, the degree of the corrugation, i.e., the buckling distance, is drastically different. For a single layer of sp^3 hybridized Si(111), a pretty large buckling distance of 0.2 nm is found within a (111) layer[11].

The Si atoms of a silicene lattice with a small buckling are neither fully sp^3 nor fully sp^2 hybridized but somewhere in between, meaning they have a mixed sp^2/sp^3 hybridization. In order to describe the degree of buckling in such a mixed hybridization state, we use the buckling ratio q, which is defined as the ratio between the displacement of a Si atom with respect to the layer plane, i.e., the height h, and the bond length d of this atom to its next neighbour ($q = h/d$) as illustrated in Figure 5.4. Takagi et al. have shown that the buckling can now be described mathematically, in analogy to the DBs of a Si(111) surface (see Figure 5.4) [12]. Assume a tetrahedron of Si atoms with the base formed of three Si atoms at the bottom, each bonded to the forth top Si atom which has a remaining DB besides the three BBs to the base atoms. We can then write the hybridization state of these orbitals explicitly in the following way, where s and p denote the initial s and p orbitals:

$$|\Psi_{DB}\rangle = \frac{|3s\rangle + \mu|3p_z\rangle}{\sqrt{1+\mu^2}},$$

$$|\Psi_{BB}\rangle = \frac{|3s\rangle + \lambda\boldsymbol{\alpha}\cdot\boldsymbol{p}}{\sqrt{1+\lambda^2}}. \tag{5.6}$$

Here, q is the buckling ratio, $\mu^2 = \frac{1}{2q^2} - \frac{3}{2}$, $\lambda^2 = \frac{2}{1-3q^2}$, $\boldsymbol{\alpha}$ is the unit vector along the BB, and $\boldsymbol{p}$ is a vector of the p-orbital. The influence of the buckling ratio is seen better when the energies of these orbitals are considered. The expectation values of the energy corresponding to these bonds are

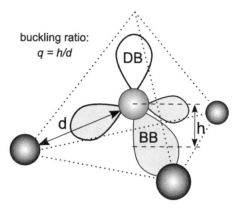

FIGURE 5.4 Schematic illustration to describe the mixed sp^2/sp^3 hybridization of Si atoms within a 2D silicene layer as suggested in ([12]). Similar to the ideal Si(111) surfaces, the Si atoms are coordinated in a tetrahedron, where the top three-fold coordinated Si atom forms back bonds (BBs) to the other tree atomy at the base and has a remaining dangling bond (DB). The buckling within such structure can be described by the buckling ratio, $q = h/d$, where h is a distance between central atom to the base of the tetrahedron and d is the Si–Si BB length. Different values of q are related to different mixed sp^2/sp^3 hybridization states.

$$E_{DB} = \frac{2q^2 E_s + (1 - 3q^2)E_p}{1 - q^2},$$

$$E_{BB} = \frac{(1 - 3q^2)E_s + 2E_p}{1 - 3q^2}, \tag{5.7}$$

where E_s and E_p are the energies of $3s$ and $3p$ orbitals of Si, respectively. Thus, when the buckling ratio, q, varies between 0 and $\frac{1}{3}$, all possible hybridization states can be described. When $q = 0$, i.e., for a fully planar layer, the energy of the BB is a mixture between $1s$ and $2p$ orbitals, while the DB has a pure p_z origin. Oppositely, $q = \frac{1}{3}$ yields the same energy for every orbital, stemming from $1\,s$ and $3\,p$ orbitals and $\theta = 109.5°$, which confirms the sp^3 hybridization state. A q value between 0 and $\frac{1}{3}$ then refers to a mixed sp^2/sp^3 hybridization of the Si atoms. This simple consideration allows describing the mixing of the orbitals in an intermediate hybridization state.

The different hybridization states, as expressed by the buckling ratio and Eqs. (5.6) and (5.7), will also have an impact on the total energy of the layer and its vibrational properties. It means that depending on q not all buckling configurations will be energetically stable. Cahangirov et al. performed DFT calculations for silicene lattices with various levels of corrugation [10]. The authors could show that the planar and the high-buckled ($\delta_{buckling} = 2$Å, $q = 0.95$) free-standing silicene lattices do not only have higher total energies compared to a layer with low buckling but also exhibit negative phonon frequencies. A negative phonon frequency is indicative of the instability of a structure, as it suggests that energy is produced rather than consumed by the vibration. On the contrary, a silicene layer with low buckling ($\delta_{buckling} = 0.44$Å) is confirmed to be stable, demonstrating no phonons with imaginary frequencies. All other structural parameters, such as the bond length, the cohesive energy, and the lattice constant agree with the ones reported by Takeda and Shiraishi [4].

Besides such structural considerations, the vibrational properties of a material allow a much deeper insight into the material properties as they reflect, for example, the atomic arrangement, the chemical composition, the electron–phonon interaction, or the crystalline quality. Free-standing silicene and diamond-like bulk Si, both have a basis consisting of two Si atoms. Accordingly, both exhibit six phonon branches, three of which are assigned to acoustical and three to optical phonons. Figure 5.5 displays the calculated phonon dispersions of both materials, along the high-symmetry axes of their BZs. While the acoustic branches are only weakly perturbed for the 2D layer in comparison to the 3D crystal, the optical phonon dispersions have a completely different character. One of the transversal optical (TO) modes acquires a lower energy at the Γ point, with an energy reduction of almost a 65%. This mode corresponds to an out-of-plane atomic motion. Due to the absence of Si atoms above and below the sheet, the energy of this vibration is reduced as a consequence of the reduction of the restoring forces, imposed by the absence of its neighbouring atoms. The other two optical phonons, the second transversal (TO) and the longitudinal (LO) phonons, are still degenerated at Γ but shifted to higher frequencies if compared to bulk-like Si. Such higher eigenfrequencies of the vibrations are related to a stronger bonding between atoms, due to the additional π bonding in the 2D layer. The stronger bonding results in a higher force constant, which is directly related to a higher frequency in the model of a harmonic oscillator. Such modifications of the 3D bulk phonon branches in the 2D counterpart show explicitly the relation between atomic structure, bonding configuration, and vibrational properties and furthermore allow identification of such 2D layers.

Summarizing this section, we point out that the physical characteristics of free-standing silicene such as the atomic, electronic, and vibrational properties have been well-studied and allow its experimental identification. In the following sections, we will give a brief overview of how silicene layers can be synthesized before discussing more deeply silicene epitaxially grown on a Ag(111) substrate.

5.3 Synthesis of Silicene on Substrate Materials

Silicene does not exist in nature and cannot be generated by exfoliation from a parent crystal. Therefore, it has to be synthesized chemically or by layer growth on a supporting substrate.

For epitaxial silicene, a few substrates were shown to host the 2D Si honeycomb lattice. At the moment, the most well-studied and understood structure that supports the growth of silicene is Ag(111) [6,12,17,18]. Figure 5.6a shows scanning tunneling microscopy (STM) topographs of the epitaxial silicene lattice. The structure exhibits a six-fold symmetry being aligned with the crystallographic directions of the silver substrate. Periodic dark spots represent the parts of the lattice with all down-buckled Si atoms, while bright protrusions are the arrangements of three up-buckled Si atoms per spot. Due to the fact that other systems do not possess a similar conformity, their main characteristics are only briefly introduced.

Shortly after the discovery of the silicene formation on Ag(111), it was reported that epitaxial 2D silicene also forms on the $ZrB_2(0001)$ surface via a segregation mechanism [15]. The actual substrate is a thin film of zirconium diboride grown on Si(111) single crystals. The top Si layer is formed when the substrate is heated and Si atoms start to diffuse through the thin ZrB_2 film to the surface, where

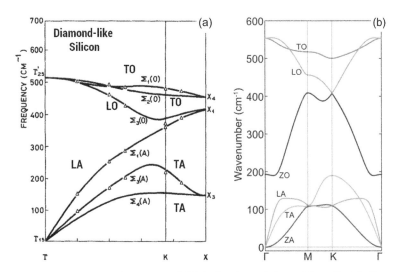

FIGURE 5.5 Theoretically calculated phonon dispersion of (a) bulk silicon (adapted from Ref. [13]) and (b) free-standing silicene (adapted from Ref. [14]). In free-standing silicene, the double degeneracy of the bulk Si L(T)O is lifted: the flexural TO mode is redshifted (ZO), while a degenerate LO/TO mode is blueshifted by about 60 cm^{-1}.

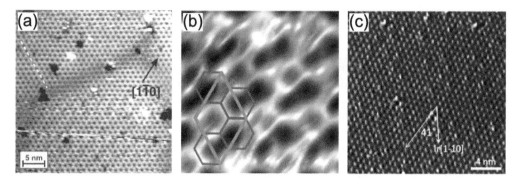

FIGURE 5.6 STM topography images of (a) epitaxial (3×3) silicene on Ag(111) (adapted from Ref. [6]), (b) $(\sqrt{3} \times \sqrt{3})$ silicene on ZrB$_2$ (adapted from Ref. [15]), and (c) $(\sqrt{7} \times \sqrt{7})$ silicene on Ir(111) (adapted from Ref. [16]).

they get arranged to form a periodic Si structure. The Si atoms assemble themselves into an interconnected 2D Si layer with a distinct $(\sqrt{3} \times \sqrt{3})$ superstructure in respect to the (2×2) unit cell of ZrB$_2$(0001). An STM image of filled states of such silicene layer can be seen in Figure 5.6b. However, photoemission data demonstrate that this reconstructed epitaxial silicene fails to exhibit the electronic properties expected for free-standing silicene [15]. The results show a substantial mixing of the electronic states of the Si layer and the ZrB$_2$ film.

Another substrate that supports the formation of a synthetic silicene lattice was reported a year later to be (111)-oriented iridium [16] (Figure 5.6c). In this case, the self-assembled epitaxial 2D Si sheets acquire a $(\sqrt{7} \times \sqrt{7})$ superstructure. Despite the superstructure formation, this layer is reported to have a negligible interaction with the substrate, based on DFT calculations [16].

For all three cases, the superstructure formation, i.e., the reconstruction of the epitaxial silicene layer, clearly points to a substantial interaction with the underlying substrate. Another common characteristic shared by all substrates mentioned is their metallicity. Besides the interaction with these substrates, the metallicity makes it more complicated to experimentally determine the electronic properties of the adlayer. Therefore, the retrieval and determination of the silicene electronic structure is possible only on an insulating substrate, the search of which is still going on. Although recent reports claim the growth of silicene on other metallic substrates, such as ZrC(111) [19] and graphite [20], these results still need to be thoroughly examined before concluding on the interaction with these substrates or the description of these Si layers as silicene-like.

5.4 Synthesis of Epitaxial Silicene on Ag(111)(1×1)

In this section, we will discuss the formation and properties of silicene formed epitaxially on the Ag(111)(1 × 1) surface. We focus on Ag(111) only because it is by far the most studied substrate for the synthesis of epitaxial silicene, in both experimental and theoretical investigations. We will see that the influence of the silver substrate on the

formation, the atomic structure, and the related properties of the silicene layer is significant. Therefore, Ag(111) acts like a model system for substrates that involve such a non-negligible interaction. The epitaxial silicene layer still shows the properties expected of a 2D system, which are, however, modified with respect to the ones of free-standing silicene. This is the reason why we will refer to it as *epitaxial silicene.*

The synthesis of epitaxial silicene on Ag(111) is usually performed in two steps, involving the preparation of the smooth substrate surface, followed by the Si deposition and the 2D layer formation. This formation is based on the self-assembly of the Si atoms in a 2D layer, which requires a substrate surface of high quality and specific preparation conditions, such as the deposition temperature. The growth mode and the role of the growth parameter will be discussed in the following.

5.4.1 Substrate Preparation and MBE-Like Growth of Si

Epitaxial growth of silicene begins with the selection of the substrate material, its orientation and finally its proper preparation. The crystalline quality of the 2D adlayer will strongly depend on the surface quality of the substrate. Although the Ag(111) surface turned out to be an ideal candidate for the epitaxial silicene growth, its lattice constant of 0.289 nm is no match for the one of free-standing silicene, having a value of 0.383 nm. Nevertheless, a commensurate Si lattice can be formed on it by matching four silicene with three Ag lattice constants. In contrast to many other metal/semiconductor systems, the silver/silicon interface is found to be abrupt without intermixing or silicide formation. This interface was already investigated decades ago in the form of Ag deposition onto the Si(111) surface, demonstrating such non-reactivity [21]. This property in connection with the almost matching lattice constants (four to three) gave the motivation for the synthesis of epitaxial silicene on Ag(111) single crystals.

For the self-assembly of the Si atoms to form silicene, the silver surface is required to be chemically clean and atomically smooth, i.e., with Ag step widths preferentially in the range of 100 nm or larger. These conditions are met when substrate preparation and layer growth are performed under ultra-high vacuum (UHV) conditions, in the same vacuum chamber. Standard, commercially available, Ag single crystals have to be transferred into this chamber and cleaned of surface contamination (water, oxygen, CO_2, etc.) present due to the transport or storage in ambient air. This unavoidable surface degradation can be recovered by surface cleaning with *in situ* Ar^+ ion bombardment and sub-sequent annealing at a temperature around 550°C just below the melting point of silver in order to "heal" the surface from roughness and defects created by the bombardment.

The steps of such substrate preparation are sketched in Figure 5.7. Usually, Ar^+ ions are used for the bombardment, which do not chemically interact with the substrate surface, accelerated, and controlled via an electric field (step I). The energy (normally in the range of 1 keV) determines the impact and thus the strength of bombardment. Higher energies can not only clean the surface more efficiently but also increase the roughness produced. The roughness is "cured" by thermal annealing (step II), where top layer Ag atoms acquire the thermal energy sufficient to break their interatomic bonds and diffuse across the substrate surface, resulting in surface flattening and formation of large terraces. The quality of such substrate surface can easily be tested by low-energy electron diffraction (LEED), which then shows a clear (1×1) diffraction pattern with sharp diffraction spots. The cycles of ion bombardment and post-annealing are repeated until satisfactory quality of the surface is achieved.

Once the substrate surface quality is sufficient, Si is deposited on the substrate by means of molecular beam epitaxy (MBE) (step III). In a very simple approach, Si atoms are thermally sublimated from a directly heated Si wafer piece facing the Ag(111) substrate. There are two most important factors that determine the quality and the atomic arrangement of the epitaxial silicene layer: the growth rate and the substrate temperature during Si deposition. Variation of both results in a complicated phase diagram for the Si formation on Ag(111) is discussed below (see Section 5.4.2).

Before looking closer on how exactly the substrate temperature influences the silicene growth, we briefly recall the fundamentals of an epitaxial growth process.

According to the theory of epitaxial growth, there are three growth regimes that determine the thin film formation on a substrate material which can basically be distinguished by the effective surface coverage (see, for example, Ref. [22]). In the layer-by-layer growth regime, single atomic layers grow subsequentially on top of each other, where the next layer starts to grow only when the previous one is completed. This growth mode is also called "Frank–van der Merwe growth" [23]. It is also referred to as the "2D layer growth", since the layers grow parallel to the substrate plane. A different growth mode is the island growth or "Vollmer–Weber growth" where 3D islands grow on the surface instead of single atomic layers. In this case, the interaction between the constituents of the film is stronger than the one with the substrate. As a result, the surface is not fully covered, and the surface coverage for Vollmer-Weber growth is the smallest. An intermediate growth regime is the so-called "Stranski–Krastanov growth", where initially a complete first atomic layer is formed, acting as a wetting layer, before island growth sets in.

For the Si/Ag(111) system, actually two of these growth modes are observed [6,24]. At first, the Si atoms deposited onto the silver surface heated at temperatures between approximately 200 and 300°C form the first atomic layer, up to the point when the substrate surface is fully covered within the diffusion length of the impinging Si atoms (Figure 5.8a). In other words, new Si atoms attach to the ones within the first layer in order to complete the 2D structure. This is an example of Frank–van der Merwe growth and a prerequisite for the formation of a 2D layer. This growth mode is illustrated in Figure 5.8b by plotting the area ratio between the Si 2p core level and the Ag 4d band emission as a function of the Si deposition time (adapted from Ref. [6]). Clearly, the Si/Ag area ratio changes linearly, indicating an initial 2D growth behavior with a corresponding deposition rate of 1 ML per hour.

After the first layer is completed, the growth mode changes to a Stranski–Krastanov growth mode, where islands composed of single layers stacked perpendicular to the surface plane form the so-called "multi-layer" structure [24]. In this case, the first epitaxial silicene layer acts as a wetting layer.

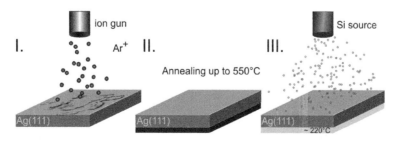

FIGURE 5.7 Schematic illustration of the synthesis of silicene on the Ag(111) surface. The synthesis involves three single steps: (I) Ar^+ ion bombardment of the Ag surface followed by an annealing step (II) in order to heal the surface from roughness produced during step (I) and (III) Si deposition from the Si source at a temperature of 220°C.

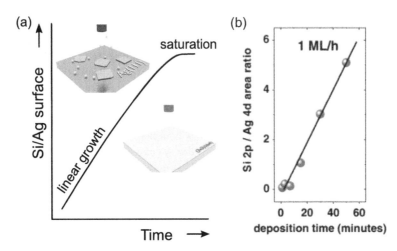

FIGURE 5.8 Growth mode of single-layer Si on Ag(111): (a) Schematic image of the growth mode of Si on Ag(111), sketched by plotting the Si/Ag surface ratio as a function of the deposition time for a constant Si deposition flux. For deposition amounts up to a complete single atomic layer (monolayer (ML)), a linear growth mode is expected for a 2D growth mode. (b) Area ratio between the Si 2p and Ag 4d core level emission lines measured by x-ray photoemission spectroscopy (XPS) as a function of the growth time and constant Si flux of 1 ML per hour. The linear fit confirms the 2D growth mode up to full coverage, expected for the synthesis of epitaxial silicene (adapted from Ref. [6]).

5.4.2 Phase Diagram

We are now ready to discuss the generic phase diagram of the Si/Ag(111) system, showing how strongly the substrate temperature during Si deposition can influence the development of different Si phases and change the growth regime between 2D and 3D [25].

According to various experimental results for the Si/Ag(111) system, it turns out that the Si atoms can get arranged in differently structured 2D layers when the substrate temperature is varied for the deposition or even by post-annealing a Si layer after deposition at room temperature [26–30]. STM topography images corresponding to the Si structures obtained at different deposition temperatures are shown in Figure 5.9. At room temperature, no ordered structure can be found, only Si clusters, which show no periodicity in LEED measurements (Figure 5.9a). The epitaxial $(3 \times 3)/(4 \times 4)$ silicene layer appears in STM

when the deposition temperature is tuned to ca. 220°C (Figure 5.9b). The domains with $(3 \times 3)/(4 \times 4)$ periodicity can be accompanied by the domains with the $(\sqrt{13} \times \sqrt{13})R13.9°$ periodicity. Their number rises with increasing deposition temperatures (Figure 5.9c). Patches of another structure with an apparent "$(2\sqrt{3} \times 2\sqrt{3})$" periodicity are also present at temperatures above 220°C. This structure will replace all other phases when Si is deposited at around 280°C (Figure 5.9d). If the temperature is increased even further, no 2D layer formation can be found any more, suggesting the end of the "2D growth" regime. This interpretation is also supported by atomic force microscopy (AFM) measurements which demonstrate the formation of 3D Si nanocrystals [14].

Figure 5.10 shows a sketch of the phase diagram related to different deposition temperatures. This phase diagram is based on STM, low-energy electron diffraction (LEED),

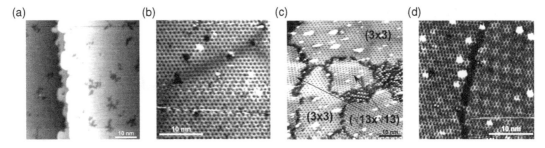

FIGURE 5.9 STM topography images of 2D Si phases formed on Ag(111) at different substrate temperatures during the Si deposition ($U_{bias} = -1.0$ V, $I = 1.08$ nA): (a) STM image of filled states after deposition of 0.1 ML of Si at room temperature. (b) 1 ML of Si deposited at 220° resulting in the formation of (3×3) epitaxial silicene. (c) 1 ML of Si deposited at 240° showing the formation of several 2D Si phases, including $(\sqrt{13} \times \sqrt{13})R13.9°$, (3×3), and "$(2\sqrt{3} \times 2\sqrt{3})R30°$" domains. (d) 1 ML of Si deposited at 280°C leading to the formation of the so-called "$(2\sqrt{3} \times 2\sqrt{3})R30°$" structure, which is characterized by the additional Moiré-like surface pattern, overlaying a more complex atomic structure (adapted from Ref. [25]).

XPS, and Raman observations, which conclusively allow determining the nature of these phases [25,26,29–31].

Si deposition on Ag(111) at room temperature (and below) causes the formation of amorphous bulk Si (aSi) only, with the Si atoms having a dominant sp^3 hybridization. Generally, it is complicated to clearly assign the hybridization state for atoms in amorphous matter due to the lack of any long-range order. However, it is safe to assume that such atomic arrangement is solely governed by the minimum energy principle, resulting for Si in an sp^3 hybridization [32]. The formation of amorphous Si is also confirmed in corresponding Raman spectra which show only a broad band centered at 480 cm^{-1} but no sharp phonon modes (Figure 5.10b) [33].

When the deposition temperature of the substrate reaches values around 180°C, the Si atoms start to locally form some ordered arrangements which decorate the step-edges of the Ag terraces (a couple of nanometers in diameter), the seeds for the growth of epitaxial silicene layers. This suggests that the substrate temperature provides enough energy for Si atoms to have sufficient diffusion length on the substrate surfaces.

For a temperature around 220°C, epitaxial (3×3) silicene layers form, which cover the entire Ag(111) terraces and the largest part of the substrate surface [34]. Epitaxial silicene on Ag(111) grows in domains, for which the numerous nucleation sites are responsible, but only laterally without second layer formation.

These epitaxial silicene layers on Ag(111) are reconstructed, forming a superstructure with a (3×3) periodicity with respect to silicene and a (4×4) periodicity with respect to the Ag(111)(1 × 1) surface. This structure is therefore often referred to as $(3 \times 3)/(4 \times 4)$. In Raman spectroscopy, the dominant $(3 \times 3)/(4 \times 4)$ silicene phase is characterized by two vibrational A modes and an E mode as seen in the related spectrum in Figure 5.10b. In Section 5.4.5, these vibrational properties are discussed in detail.

Other silicene superstructures, such as $(\sqrt{13} \times \sqrt{13})\mathrm{R} \pm 13.9°$ can also be found on the Ag(111) surface for Si deposition between 220°C and 250°C. This well-ordered $(\sqrt{13} \times \sqrt{13})\mathrm{R} \pm 13.9°$ superstructure exists in four differently rotated domains, described as differently rotated $(3 \times 3)/(4 \times 4)$ silicene domains [35].

When the substrate temperature reaches 250°C, the domains of new Si phase with the dominant $(2\sqrt{3} \times 2\sqrt{3})$ periodicity in LEED patterns appears. At a deposition temperature of 280°C, this phase becomes dominant and no traces of the $(3 \times 3)/(4 \times 4)$ structure can be found any more. The "$(2\sqrt{3} \times 2\sqrt{3})$" structure differs from the others by showing some inherent disorder in STM images. Its nature is still discussed controversially [36]. The Raman spectrum, related to the "$(2\sqrt{3} \times 2\sqrt{3})$" structure, looks very similar to one of the (3×3) phases and also exhibits an additional characteristic mode at 155 cm^{-1} which stems from its inherent disorder, therefore denoted as ID (Figure 5.10b).

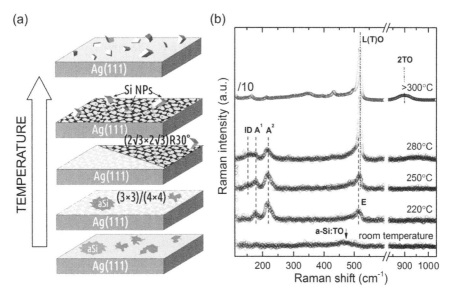

FIGURE 5.10 (a) Schematic phase diagram for the formation of different Si phases on Ag(111). For lower temperatures up to approximately 180°C, only amorphous Si forms on the Ag(111) surface. The formation of (3×3) occurs for substrate temperatures around 220°C, accompanied by patches of aSi, likely formed at surface defects. For a slightly higher temperature, multiple 2D phases are formed. From 250°C, the growth of the "$(2\sqrt{3} \times 2\sqrt{3})$" phase becomes dominant which is the only phase observed for temperature higher than 280°C. Additionally, the co-formation of 3D Si nanosized crystals sets in at these temperatures. A dominating 3D growth of Si crystallites starts from 300°C. (b) Raman spectra measured for the different preparation steps described in (a) confirming the assignment made. The preparation temperatures are varied in the range from 20 to 300°C. The spectra are stacked for clarity (adapted from Ref. [25]).

For preparation temperatures from 250°C, the beginning of the co-formation of diamond-like Si crystallites is observed, which becomes more dominant with increasing deposition temperatures. Also, the contribution of the "$(2\sqrt{3} \times 2\sqrt{3})R30°$" structure increases at higher temperatures, seen by the intensity increase of the ID mode. Consequently, the A phonon modes, being dominant in the spectrum of $(3 \times 3)/(4 \times 4)$ silicene, decline gradually.

Any multiple phase regime between these temperatures does not show any other Raman modes, as the ones described (Figure 5.10b). The related Raman spectra are simply a weighted superposition of the Raman spectra of the $(3 \times 3)/(4 \times 4)$ and the "$(2\sqrt{3} \times 2\sqrt{3})R30°$" phases. This demonstrates that all these phases are related to a very similar 2D layer, which, for example, supports the assumption that the different $(\sqrt{13} \times \sqrt{13})R\pm13.9°$ domains are differently rotated $(3 \times 3)/(4 \times 4)$ structures. Similar conclusions were drawn by a recent tip-enhanced Raman spectroscopy study, which, although naturally having higher spatial resolution due to the use of the STM tip, found only minor differences in phonon density of states [37]. The origin of the observed phonon modes and their relation to silicene is discussed in Section 5.4.5.

The last part of the phase diagram shows the formation of 3D Si crystallites at temperatures above 300°C. The higher temperature implies a higher kinetic energy of the Si atoms, which allows the system to change from the metastable 2D layer into stable bulk-like Si by dewetting from the silver surface [14,38]. In other words, the Si atoms become mobile enough to move over the step of Si monolayer and grow normally forming the well-known diamond structure, which is the most energetically favorable one for Si. The corresponding Raman spectrum shows now a very intense band around 520 cm^{-1}, which is a well-known one-phonon L(T)O mode of bulk/bulk-like Si (Figure 5.10b), unequivocally demonstrating a 2D-to-3D phase transition.

Hence, we see that epitaxial silicene can be prepared on Ag(111) for the right experimental conditions, such as the UHV environment, atomically smooth substrate surface, and a well-defined substrate temperature. In the following sections, we explicitly describe the physical properties of epitaxial $(3 \times 3)/(4 \times 4)$ silicene on Ag(111) showing that this Si phase exhibits all the properties of silicene when the interaction to the substrate is considered.

5.4.3 Atomic Structure of Epitaxial (3×3) Silicene on Ag(111)

The atomic structure of $(3 \times 3)/(4 \times 4)$ epitaxial silicene on Ag(111) can be seen as a clear hint towards the significant interaction with the underlying silver substrate surface. The formation of the superstructure with respect to the silicene (1×1) unit cell is directly related to this interaction, which leads the silicene Si atoms to change their lattice positions. But despite the lattice reconstruction, epitaxial silicene still forms a honeycomb lattice as the fundamental structural unit. The atomic model of this structure is depicted in Figure 5.11, derived from DFT calculations [39].

In this model, the Si atoms within the honeycomb grid are replaced with respect to their position in relation to the Ag(111)(1×1) surface. If the Si atoms are localized directly above an underlying Ag atom (big gray spheres), the Si atom moves slightly out of the 2D Si plane (small gray spheres), as a result of the interaction with the Ag(111) surface (see Figure 5.11a,b). If the Si atom is localized between two of these underlying Ag atoms, it moves slightly into the lattice plane (the spheres with Wyckoff positions of $(-x+y, y, z)$). In this way, the epitaxial silicene layer forms the (3×3)

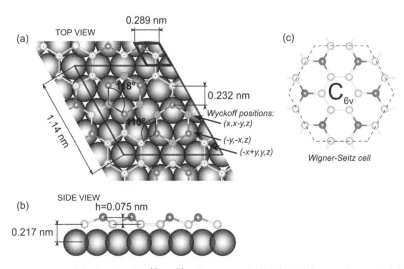

FIGURE 5.11 Atomic structure model of epitaxial (3×3) silicene on Ag(111) in (a) top view and (b) side view. The big black rhombus indicates the (3×3) unit cell, the small black rhombus the (1×1) unit cell of Ag(111). Wyckoff positions are listed for three types of Si atoms within the unit cell. (c) Wigner–Seitz unit cell of (3×3) epitaxial silicene, showing the six-fold rotational symmetry of the layer and the six mirror planes; thus it belongs to a C$_{6v}$ symmetry group.

superstructure with respect to the ideal hexagonal Si honeycomb ring. Therefore, the structure is referred to as (3×3). This means that its unit cell is 3×3 times larger than the one of free-standing silicene. At the same time, it is 4×4 times larger than the unit cell of Ag(111) surface (big black rhombus); i.e., it has a (4×4) periodicity related to Ag(111). This clearly indicates the commensurate (3:4) arrangement of the epilayer with respect to the substrate, resulting from the difference of the two lattice constants, as discussed in Section 5.4.1. This shows that a (111) plane of silver is a suitable surface for the silicene growth, since the lattice mismatch is small enough. Nevertheless, the latter also results in the change of the Si — Si bond length and bond angles.

In contrast to low-buckled free-standing silicene, where the Si atoms belong either to the upper or lower sublattices, all separated by the same distance of 0.225 nm, epitaxial silicene demonstrates a richer variation of the Si bonding configurations. As a matter of fact, there are three types of different bonding sites within the (3×3) unit cell, which belong to the following sets of Wyckoff positions: $(-y, -x, z)$, $(-x+y, y, z)$, and $(x, x-y, z)$ (also indicated in Figure 5.11a). The first set describes the six Si atoms, shifted upwards in respect to the median plane of the lattice (small dark gray in Figure 5.11). The two other sets describe the Si atoms localized at the corners of the unit cell and inside the unit cell (small light gray spheres), respectively. Depending on the actual atomic position, the Si — Si bond length varies from 0.229 to 0.232 nm.

The first bond length value is found within the flat ring of Si atoms, which coincides with corners of the unit cell and corresponds to the dark spots in the STM topography image (Figure 5.6), in the center of a so-called "flower pattern". This Si — Si bond length is clearly shorter than the one for bulk Si with a value of 0.235 nm, resulting from the sp^2/sp^3 hybridization and the related Si $=$ Si double-bond character. The second bond length corresponds to the upper-buckled Si atoms, which is closer to the one for bulk-like Si, indicative of a stronger sp^3 hybridization of these atoms. The difference of the bond length and their values therefore provide an indication for the mixed sp^2/sp^3 hybridization of the Si atoms, as also found for free-standing silicene. The same conclusion can be drawn when looking at the bond angles between the Si atoms. These bond angles have values of $110°C$ and $118°$ with only negligible deviations. These values are between and close to the angles for a fully sp^3 and sp^2 hybridization, with angles of $109.5°$ and $120°$, respectively. Lastly, the buckling distance h is another parameter, which supports the sp^2/sp^3 hybridization character in epitaxial silicene. The distance between up-buckled and down-buckled Si atoms was determined in the calculations to be 0.75 Å. This value is twice as big as the buckling distance in free-standing silicene (0.44 Å) and almost half as big as the one found for the Si(111) plane (1.5 Å). This comparison shows that the atomic arrangement of epitaxial silicene on Ag(111) effectively resembles the properties of free-standing silicene.

Despite the superstructure formation, the (3×3) layer is still highly symmetric. This can be seen by drawing the Wigner–Seitz (WS) cell, as shown in Figure 5.11c. One can nicely see the six-fold rotation symmetry of this structure and the six mirror planes within the unit cell, which assign the structure to belong to a C_{6v} symmetry group.

The described model also explains nicely the appearance of these epitaxial silicene layers in STM imaging (Figure 5.6). The "flower pattern" observed in the image is composed of six triangular structures, each consisting of three bright protrusions. Within the model, these protrusions are related to the up-buckled Si atoms, which are the only ones seen in the image. The dark center of the "flower" is then related to the bottom flat Si hexagon with predominately sp^2 hybridized Si atoms. In fact, simulated STM images, based on DFT calculations of the local electron density of states according to the Tersoff–Hamann approach [40,41] for the model in Figure 5.11, are very similar to the one observed experimentally [39].

The discussed structural parameters of epitaxial silicene underline the 2D character and the similarities to free-standing silicene. In the next two sections, we will discuss the electronic and vibrational properties of these layers which further support such assignment.

5.4.4 Electronic Properties of Epitaxial Silicene on Ag(111)

The electronic properties of any crystalline material strongly depend on the atomic arrangement and bonding configuration of its lattice, which also accounts for 2D crystalline materials such as silicene. We have seen in Section 5.2 that even with low buckling, the silicene lattice exhibits similar electronic properties as a fully planar structure, i.e., Dirac cones at the $\overline{K}$ points of the BZ. However, the atomic structure of epitaxial silicene on Ag(111) undergoes a more substantial structure modification because of the interaction with the substrate, as discussed in Section 5.4.3. Hence, we can expect that this modification has also a stronger impact on the electronic properties.

Indeed, theoretical DFT calculation show that these linear electronic dispersions at the $\overline{K}$ points are not preserved for (3×3) epitaxial silicene on silver [42–47]. It turns out that the d-orbitals of the Ag atoms hybridize with the p-orbitals of Si atoms resulting in interface states [44]. Such states were also found experimentally by angle-resolved photoemission spectroscopy (ARPES) and initially interpreted as branches of a Dirac cone [6]. Later ARPES results identified, however, these states to result from hybridized states localized at the silicene/Ag(111) interface [43,48].

ARPES maps filled electronic states of a crystal below the Fermi level and measures photoelectrons emitted from the surface (or 2D layer) at different emission angles. By rotating the crystal around high-symmetry points of the BZ, the electronic band structure can be probed.

The silicene/Ag interface state was found to exhibit a linear dispersion at the $\overline{K}$ point of the (3×3) BZ, which also coincides with the $\overline{M}$ point of the silver BZ, shown in Figure 5.12a (adapted from Ref. [43]). In the ARPES result, it is seen that such a band is absent when no Si is deposited, when only linear bands of Ag *sp* bulk states are present (Figure 5.12a). These silver *sp* bands, found along the $\overline{\Gamma}_{Ag}$-$\overline{M}_{Ag}$ direction, can be distinguished from the silicene/Ag interface state by their different slopes (Figure 5.12b). The specific interface bands appear only when epitaxial silicene is formed on top of the Ag(111) surface. Its attribution to an interface state becomes clearer when the full band diagram is considered including both the electronic bands of a substrate and an adlayer as demonstrated in Figure 5.12b. However, the Dirac cones expected for free-standing silicene are not preserved.

The absence of real Dirac cones at the epitaxial silicene $\overline{K}$ points is also supported by scanning tunneling spectroscopy (STS) measurements in a magnetic field at low temperatures [42]. Under those conditions, a Landau-level splitting is observed for graphene which is basically related to the existence of Dirac electrons and should therefore also be observed for silicene. Such measurements, however, show no such splitting for silicene/Ag(111), which can be interpreted as the absence of Dirac electrons in its epitaxial form.

Although the electronic band structure of free-standing silicene is altered for epitaxial silicene on Ag(111), the newly emerged interface state could still attract interest due to

its steep linear dispersion. Additionally, ARPES measurements in a wider range of k values also show that cone-like dispersions are also found at the edge of the first BZ of the Ag(111) surface [49]. At points between the $\overline{M}_{Ag}$ point and the $\overline{K}_{Ag}$ point on both sides, coinciding with silicene $\overline{K}$ points, paired cones were measured by ARPES and assigned to Dirac cones. These bands only appear when epitaxial silicene is present on the Ag(111) surface and stem from the interaction between these two systems.

The discussion of the electronic properties of epitaxial silicene on Ag(111) shows that the metallicity of the substrate disturbs the properties of the Si adlayer. If there is a high density of states around the Fermi level, purely silicene-related bands cannot be electronically decoupled from those of the substrate, and thus hybridization occurs. Such outcome is not new and also found for the combination of Si and transition metals, such as crystalline $CaSi_2$ [4]. According to the model of this crystal structure, it consists of pure Si layers with a buckled honeycomb structure of mixed sp^2/sp^3 Si hybridization and layers of non-interconnected Ca atoms, both alternating along the c axis. The hybridization between the Ca $3d$ and the Si $2p$ levels changes the semi-metallic character of the Si layers into a pure metallic one. Furthermore, the charge transfer between the layers is shown to stabilize these crystals. This is another example showing the tendency of Si to

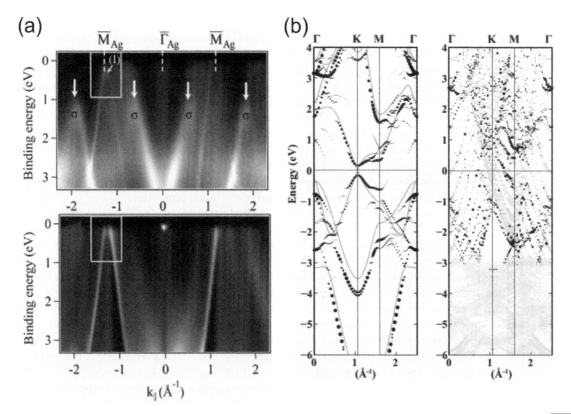

FIGURE 5.12 (a) ARPES maps measured on (3×3) epitaxial silicene/Ag(111) and the pure Ag(111) surface along the $\overline{\Gamma}_{Ag} - \overline{M}_{Ag}$ direction at a photon energy of 135 eV (adapted from Ref. [43]). (b) Simulated electronic band structure of free-standing silicene, free-standing (3×3)-reconstructed silicene, and epitaxial (3×3) silicene/Ag(111), where the light gray dispersion lines stem from silver states and dark gray dispersion lines from states related to the silicene layer (adapted from Ref. [44]).

form aromatic lattices only if stabilized, here, by transition metals. This stabilization, in turn, significantly perturbs the electronic properties of the Si structure.

In order to find an epitaxial silicene structure that facilitates the conservation of the electronic properties of free-standing silicene, a substrate with weaker interaction would be needed, i.e., with van der Waals interaction. There have been a couple of theoretical calculations, suggesting that such substrates could be a H-terminated Si surface [50,51] and pure [52] or graphene-covered SiC surface [53], but experimental confirmation is yet missing.

We will see in the next section that the vibrational properties of the epitaxial silicene layer on Ag(111) are more robust than the electronic properties and are much closer to the ones expected for free-standing silicene.

5.4.5 Vibrational Properties of Epitaxial Silicene/Ag(111)

Epitaxial silicene on Ag(111) exhibits vibrational properties which help to understand its relation to free-standing silicene and underline its 2D character. But before going into a detailed discussion of this topic, we would like to recall the main considerations for the vibrational signature of silicene.

First of all we saw in Section 5.2 that the out-of-plane transversal optical phonon mode (ZO) of free-standing silicene has a very low energy value at the Γ point of around 180 cm^{-1}, while the other two phonon modes, the in-plane transversal and longitudinal optical phonon modes, are degenerate with an energy value of 580 cm^{-1}, three times as high as the one of the ZO mode. Their energy is also higher in comparison to that of the optical L(T)O mode of bulk Si, found at around 520 cm^{-1}.

The phonon frequencies (energies) can be probed by means of optical or electron spectroscopy. Two powerful spectroscopic methods are Raman spectroscopy, probing phonons at Γ, and high-resolution electron energy loss spectroscopy (HREELS), probing phonons close to the zone edge and high-symmetry points. In order to map the phonon dispersions over the whole BZ, neutron scattering could be

applied. Neutron scattering requires large bulk samples, a nuclear reactor as a neutron source, and a triple-axis spectrometer which make it an experimentally complicated technique. Raman spectroscopy, on the other hand, is relatively simple and a well-established non-destructive method to study 2D materials. Furthermore, Raman spectrometer can easily be attached to UHV preparation or analysis chambers and enables the *in situ* investigation of a 2D material without exposing it to air. As we have seen above, elemental 2D materials with low buckling are chemically reactive and would be modified or destroyed under ambient conditions by adsorption processes. Hence, *in situ* Raman spectra can be utilized as a fingerprint of a 2D material and allows, for example, its modification to be followed. Much more than this, the vibrational properties determined by Raman spectroscopy give deep insight into the structural, chemical, and electronic properties of the material.

The Raman signature of epitaxial (3×3) silicene on Ag(111) is shown in Figure 5.13a. When compared to free-standing silicene, the expectations are not completely fulfilled. In the low-energy region, two modes are observed at 175 and 216 cm^{-1} instead of only one. At the same time, a band appears at 514 cm^{-1}, roughly 60 cm^{-1} lower than expected for the LO/TO mode. However, the similarities to free-standing silicene are much more when compared to bulk-like Si.

Whether these observed experimental bands of epitaxial silicene correspond to the phonon modes expected for a 2D layer can be better concluded from their symmetries. The symmetries can be determined by polarization-dependent Raman measurements. By choosing the polarization of the incident and/or the scattered light, one can suppress or allow bands of a certain symmetry. The band symmetry is reflected by a second-rank tensor, the so-called "Raman tensor".

The symmetry of this tensor is determined in accordance to the point group symmetry of the given crystalline system, and it is identical to the symmetry elements of this group. When the crystal symmetry is known, the symmetries of all possible vibrations can easily be determined.

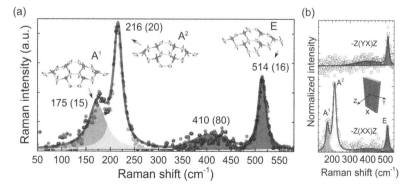

FIGURE 5.13 (a) Raman spectrum of epitaxial (3×3) silicene on Ag(111) with characteristic phonon modes at 175, 216, 410, and 514 cm^{-1}. The linewidth (full width at half maxima (FWHM)) is given in brackets. The vibrational motion patterns of the three intense modes are shown as insets with the eigenvectors of the calculated phonon modes. (b) Polarization-dependent Raman measurement for the same sample as in (a) recorded in parallel $\bar{z}(xx)z$ and crossed $\bar{z}(yx)z$ polarization configurations (adapted from Ref. [14]).

Since epitaxial (3×3) silicene belongs to the C_{6v} group, only two symmetries of the Raman-active vibrations can be observed: A-type and E-type symmetries. Moreover, the bands can differ by their degeneracy. A-symmetry modes are non-degenerate, and their Raman tensor has only diagonal non-zero elements, while E-symmetry modes are double-degenerate having both diagonal and non-diagonal matrix components. For the C_{6v} symmetry point group, we obtain the following three Raman tensors $A(z)$, E_1, and E_2:

$$A(z) = \begin{pmatrix} a & 0 & 0 \\ 0 & a & 0 \\ 0 & 0 & b \end{pmatrix}, \quad E_1 = \begin{pmatrix} 0 & 0 & c \\ 0 & 0 & c \\ c & c & 0 \end{pmatrix},$$

$$E_2 = \begin{pmatrix} d & -d & 0 \\ -d & -d & 0 \\ 0 & 0 & 0 \end{pmatrix}.$$

The selection rules for these two symmetry types demonstrate that E modes can be observed in parallel and crossed polarization configurations, while A modes are seen only in parallel configuration. Parallel configuration denotes the case when the vector of electric field of the incident (i) light is parallel to the one of the collected scattering (s) light, whereas they are perpendicular in the crossed configuration. For the 2D layer, e_i and e_s can be aligned along the (100) or (010) direction, and the intensity of a specific mode is given by

$$I \sim |e_i \cdot R \cdot e_s|^2, \qquad (5.8)$$

where R is the Raman tensor. This way we find that the low-frequency modes of epitaxial silicene have an A symmetry and the one at 514 cm^{-1} has an E symmetry as it is deduced from the polarization-dependent Raman measurements (Figure 5.13b). The A modes correspond to an out-of-plane atomic motion and are denoted as A^1 and A^2. The high-frequency mode is related to in-plane vibration and denoted as E. Independent group symmetry analysis shows that the out-of-plane vibration of free-standing silicene also has an A symmetry, while the in-plane double-degenerate mode is of E symmetry [54].

The expected frequency position of the ZO phonon for free-standing silicene around 180 cm^{-1} coincides pretty well with the A^1 band at 175 cm^{-1}. The origin of the second out-of-plane mode is found in the (3×3) reconstruction of the silicene lattice. Since this unit cell of the epitaxial silicene is bigger, it contains more atoms giving rise to more phonon modes. The total number of optical (Raman- and infrared-active) vibrations of the (3×3) structure can be calculated by $3N - 3$, where N is the number of atoms in the unit cell (here 18), giving 51 normal modes. Of course, experimental Raman spectra do not show all of these modes as their intensities might be very low. But it is obvious that epitaxial silicene should exhibit more modes than free-standing silicene.

DFT calculations for epitaxial silicene on Ag(111) show that the system has numerous vibrational modes in the range from 100 to 514 cm^{-1} [14]. Their activity can be mainly determined by the mode symmetry. Amidst all the modes, only three were found to fulfil the symmetry pattern required for a vibration to be Raman-active.

These modes are found at frequencies of 173, 204, and 514 cm^{-1}. Overall these frequencies match nicely the frequencies of the two A modes and the E mode (175, 216, and 514 cm^{-1}, respectively). The smaller deviations of the energetic position in particular of the A modes are likely related to the DFT calculations, known to under- or overestimate the calculated energetic phonon positions [55]. Besides the frequency, the symmetry gives a much stronger evidence for the matching of the calculated and the experimental phonon modes. The calculated vibrational motion patterns in Figure 5.13a (insets) show that the difference between the two out-of-plane A vibrations lies in the mutual motion of the bottom- and upper-buckled Si atoms. The calculated motion pattern of the E mode at 514 cm^{-1} has a strict in-plane character, which confirms the assignment to an in-plane LO/TO mode. Its energy is redshifted, whereas the LO/TO mode of free-standing silicene should be blueshifted with respect to bulk Si.

The reason for the redshift is closely related to what has already been discussed for the electronic properties of epitaxial silicene on Ag(111): The π-bonding for this system is greatly reduced due to the strong hybridization with the silver electronic states. This means that the total bonding strengths and thus the force constants are less enhanced in bulk Si compared to free-standing silicene. Moreover, the Si$-$Si bond length is larger than expected, which also contributes to the lowering of the phonon frequency. Although the Raman band at 514 cm^{-1} has much lower energy than the LO/TO phonon mode of free-standing silicene, its motion pattern and its symmetry clearly corroborate the assignment.

Our final remarks on the vibrational properties of epitaxial silicene concern the widths of the Raman bands, which are related to the phonon lifetime. The FWHM of each band is shown next to its position in Figure 5.13a. The usual values for the FWHM of Raman modes of a single crystal are found in the range of 2–4 cm^{-1}, which is several times smaller than those of epitaxial silicene/Ag(111). Since the linewidth broadening cannot be related to the disorder or other ways of the phonon lifetime reduction, it can be assumed that it stems from a significant phonon–electron scattering, i.e., a strong electron–phonon coupling. Indeed, when the thermal evolution of the linewidth for each band is followed, no thermal contribution to the broadening is observed up to 300°C, which demonstrates that the temperature-related phonon–phonon scattering plays a small role in the linewidth broadening of the Raman bands of epitaxial silicene/Ag(111) [14].

5.4.6 Similarities and Differences to Graphene

In the last three sections, we have seen that the properties of epitaxial silicene on Ag(111) are modified with respect to free-standing silicene but still clearly show its 2D character.

In this part, we want to make a comparison to graphene, the prototypical member of the class of 2D materials.

We have seen that epitaxial silicene is based on a honeycomb atomic lattice, similar to graphene, but with low buckling, also expected for free-standing silicene, resulting from the different atomic properties of Si, if compared to C. The electronic properties are influenced by the Ag substrate but show linear bands at the $\overline{K}$ point of the BZ of silicene and even some paired cone-like bands at the Ag BZ edge. Epitaxial silicene is metallic and thus differs fundamentally from bulk-like Si.

Very interesting is also the comparison of their vibrational properties. In Figure 5.14, the calculated phonon dispersions of graphene and free-standing silicene are depicted (adapted from Ref. [56]). Both phonon dispersions show many similarities such as the presence of a low-energy ZO phonon and the double-degenerate LO/TO mode described earlier. The most evident difference found here are the phonon energies, which are much lower for silicene accounting mainly for the bigger mass of the Si atoms. Some other differences are slopes of the dispersions and the number of anti-crossing points such as the one at the $\overline{K}$ points of both materials. The latter is related to the different symmetry group and the buckling of silicene. The symmetry group for the planar honeycomb structure of graphene is D_{6h}, and it is D_{3d} for the buckled honeycomb lattice of silicene. Obviously, the two spatially separated sublattices of the latter must reduce its symmetry. At the same time, the corrugation within the layer changes the selection rules for the vibrations. The ZO phonon has now an A symmetry and is now allowed but was forbidden in the planar lattice of graphene, where it has a B symmetry. When compared to the experimental Raman spectra, the one of epitaxial silicene fits even better to the theoretical expectations: ZO modes of A symmetry at lower frequencies and an E mode at higher frequencies softened, however, with respect to bulk-like Si, as discussed in Section 5.4.5. For graphene, the ZO mode is symmetry forbidden, and defect-related phonon modes (D mode) or a mode related to double resonance (2D mode) dominate the Raman lineshape, instead of the LO/TO mode (here known as the G mode). Non-idealities are thus more prominent than the expected modes in the case of graphene.

Despite these differences, epitaxial silicene and graphene have similarities in terms of their vibrational properties also. A softening of the LO/TO mode, as observed for epitaxial silicene, is also found for epitaxial graphene if grown on substrates significantly interacting with the graphene layer, such as Co(0001) and Ni(111) [57]. Moreover, the electronic dispersion is also modified in the same way for epitaxial silicene, and the Dirac cones at the $\overline{K}$ points are not preserved for a substrate interaction. None of these observations are made if graphene is grown on metallic substrates which show only weak interaction, such as Ru(0001) and Cu(111). This clearly demonstrates the role of the interaction and at the same time the similarities between epitaxial silicene and graphene, in spite of the evident differences.

5.5 Chemical Stability and Modification of Epitaxial Silicene

5.5.1 Oxidation of Epitaxial Silicene

The fundamental physical properties of epitaxial silicene/Ag(111), discussed above, are crucial to understand its interaction with organic or inorganic molecules or atomic species. Firstly, we look at graphene as a reference. It is well-known that an aromatic molecule such as benzene exhibits a much higher rigidity than non-aromatic molecules. Graphene, having a fully planar structure, consisting of C=C double bonds, was shown to possess the same level of stability [58,59]. Oxidation of graphene takes place only in acidic environments and at elevated temperatures [60]. Such harsh conditions are required to break the strong double bonds and saturate the DBs with O atoms or oxygen-containing functional groups.

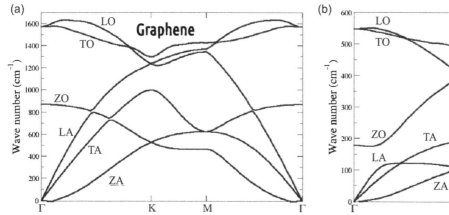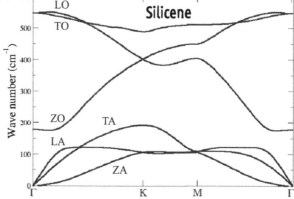

FIGURE 5.14 Phonon dispersions of free-standing graphene (a) and free-standing low-buckled silicene (b). The labels LO, TO, and ZO stand for longitudinal, transversal, and flexural optical phonons, respectively. The labels LA, TA, and ZA stand for longitudinal, transversal, and flexural acoustical phonons, respectively (adapted from Ref. [56]).

For epitaxial silicene, the π-bonds are weaker, and it should be more sensitive to oxidation. Indeed, the oxidation of (3×3) epitaxial silicene on Ag(111) occurs very fast and is completed at around 100 Langmuirs (L) and would only take a millionths of a second under ambient conditions [61].

This oxidation behavior can nicely be seen by *in situ* Raman spectroscopy. Figure 5.15a shows Raman spectra recorded on epitaxial silicene after incremental oxidation dosages of 10 L. It is found that the Raman signature of pristine epitaxial silicene vanishes after an oxygen dosage of 100 L. The gradual fading of the phonon modes indicates a typical chemisorption process, characterized by an exponential decay of the mode intensity as a marker of the amount of material oxidized, shown in Figure 5.15b. Smaller differences in the slopes of the decrease of the different modes are related to their different origins. The modes also show a slightly different frequency shift with oxygen exposure (see Figure 5.15c). The A^2 mode remains at the initial spectral position, whereas the E mode softens with increasing oxygen dosages. This is related to the fact that the out-of-plane A^2 mode is more sensitive to adsorption process, whereas the in-plane E mode reacts only weakly.

These results demonstrate that epitaxial silicene is very reactive to oxygen and thus not stable in air, just as expected. Transfer through air will therefore always require some protection of the silicene layer, either by a capping layer or by keeping it in a non-reactive atmosphere.

5.5.2 Hydrogenation of Epitaxial Silicene

The high reactivity of epitaxial silicene also has advantages over non-reactive graphene. Surface adsorption and doping are key factors to modify or adjust the electronic material properties. An example of such a functionalization of silicene is the hydrogenation. Theoretical studies show that hydrogenation of free-standing silicene should lead to a non-zero electronic band gap due to further separation of the Si sublattices [62,63]. Hydrogen binds to both sides of the silicene layer, i.e., to both sublattices.

The hydrogenation of epitaxial silicene on Ag(111) was also investigated experimentally by means of STM, Raman spectroscopy, and electron energy loss spectroscopy [64–66]. The experiments have shown that epitaxial silicene adsorbs H atoms only from the top side, while the bottom side is protected by the silver substrate. Figure 5.16a shows an STM image measured after the hydrogenation of (3×3) epitaxial silicene. It is found that the H atoms modify the lattice of the silicene layer, breaking the mirror symmetry of its unit cell by bonding to six Si atoms on one side and only one Si atom on the other side [64]. According to the results obtained by Raman spectroscopy, the bond length as well as the bond angles increase, indicating bigger separation of the two sublattices [65]. This is concluded from the evolution of the vibrational modes of epitaxial silicene during hydrogenation shown in Figure 5.16b. The E mode softens upon hydrogenation, shifting from 514 to 480 cm^{-1}, while the A^2 mode stiffens with a shift of about 40 cm^{-1}. Such behavior of the spectral bands agrees with an enlargement of the Si—Si bond length and the emergence of the additional restoring force in the out-of-plane direction, exerted by adsorbed H atoms.

Additionally, many more phonon modes appear after hydrogenation which can be assigned to different Si—H bonds. The degree of the silicene hydrogenation can be tracked by determination of the intensity ratio between the new and the initial phonon modes [65]. Despite these

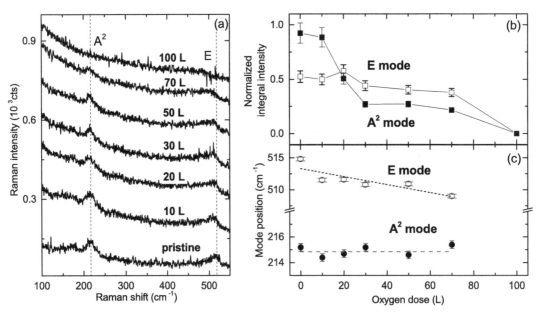

FIGURE 5.15 (a) Raman spectra of epitaxial silicene/Ag(111) recorded during a step-wise exposure to molecular oxygen, in incremental steps of 10 L. (b) Normalized integral intensities and (c) mode positions of the A^2 and E modes plotted as functions of the oxygen exposure (adapted from Ref. [61]).

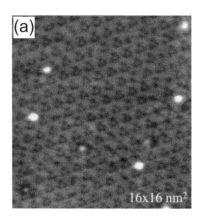
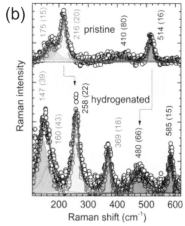

FIGURE 5.16 (a) STM topography image of epitaxial silicene/Ag(111) after hydrogenation (adapted from Ref. [64]). (b) Raman spectra of pristine and hydrogenated silicene measured *in situ*. The Raman bands are labeled by their energetic positions. The linewidth values of the modes is given in parenthesis. The single components are numerically fitted (adapted from Ref. [65]).

significant structural changes, hydrogenated silicene retains a metallic character due to the unaltered substrate–adlayer interaction. Remarkably, the hydrogenation can be reversed via annealing hydrogenated silicene at temperatures of about 230°C, where the H atoms desorb, restoring the initial (3×3) superstructure of epitaxial silicene [64]. Note, that the temperature (or energy) which is required for the de-hydrogenation of epitaxial silicene is comparably low.

The illustrated way to functionalize epitaxial silicene allows us to design the composites of silicene and more intricate chemical species applicable for efficient catalysis as well as for templating more complicated epitaxial growth. A recent study has shown that when complex organic molecules, such as phthalocyanines, are adsorbed on the surface of epitaxial silicene/ZrB$_2$, they tend to bond in a preferable lattice site, related to the boundary domain within the Si sheet [67]. Smaller aromatic molecules, such as benzene, were reported to align well with the honeycombs of silicene, passivating its surface [68]. Such chemisorption can play an important role for tuning the properties of silicene itself and its decoupling from the substrate.

5.6 Outlook and Perspectives

We have seen above that epitaxial (3×3) silicene on Ag(111) clearly demonstrates the properties of a 2D material system, but these properties differ to a certain extent from the ones expected for free-standing silicene. The reason for this difference is the significant interaction with the underlying silver substrates which alters in particular the electronic properties. On the other hand, the substrate is firstly important as a template to facilitate the formation of the epitaxial silicene layer and secondly stabilizes it. Without the substrate, the Si layer is likely to undergo a transition into its thermodynamically stable form, sp^3-bonded diamond-like Si [5].

However, other substrates might still support silicene but show a weaker interaction with the 2D Si layer. Theoretical investigations have shown that, for example, hydrogenated

Si and Ge(111) surfaces [51], H-passivated Si-terminated SiC(0001) [50] or ... solid Ar [69] might be used as such substrates. But these DFT-based calculations only show that if the silicene layer was formed on these substrates, it would be thermodynamically (meta) stable and preserve its Dirac electronic properties. Unfortunately, these calculations cannot give any hint on how or if these layers can also be synthesized experimentally on those substrates. Future work is needed to find and evaluate such substrates in order to obtain silicene and other elemental 2D systems like germanene or stanene that still preserve the envisaged new properties.

If these electronic properties are preserved, these 2D materials might also preserve the topologically non-trivial properties expected. An intriguing property of these 2D materials is that their electronic band gaps are controllable by applying a gate voltage [70,71]. This way these 2D materials can undergo a topological phase transition from a quantum spin-Hall (QSH) insulator to a quantum valley-Hall (QVH) insulator. Because for the larger spin-orbit coupling in these materials compared to graphene, such topological non-trivial properties are expected to occur at realistic temperature or even room temperature, which underlines their importance for future application in nano-electronics. Other interesting topological properties are achievable, for example, by controlling their topological edge states via a gate voltage which would allow the realization of a field effect topological transistors [72].

Also the apparent disadvantage of these materials of having a (low) buckled honeycomb lattice might turn into an advantage since it allows the control of electronic properties of these materials via chemical functionalization or external fields. This could be efficiently utilized in a transistor, where the electronic band gap of silicene depends on the electric field applied perpendicularly to the lattice plane [73,74]. DFT results show that an orthogonal electric field separates the two sublattices, which leads to a band gap opening, similar to the hydrogenation discussed above.

Among all such possible applications, where in particular one-atom-thin 2D Si layers can efficiently be exploited, an interesting option is the combination with Si-based microelectronics, entirely based on bulk Si as an active material. Using silicene instead, e.g., in a transistor, could make it even smaller. Such silicene field effect transistors (FETs) could be part of future technological application. First reports on the realization of silicene FETs have paved the way in this direction and demonstrated that these technologically challenging issues can be overcome [75].

5.7 Summary

In this chapter, we have discussed the epitaxial synthesis of silicene on Ag(111) templates with emphasis on the properties of these 2D layers. In order to understand this new Si allotrope, some challenges had to be overcome, in particular the significant interaction between silicene and the silver substrate.

Although theoretical works had investigated structure and properties of free-standing silicene layers in detail, including, for example, its low-buckled honeycomb arrangement, their adaptation for the real existing silicene layer was not simple. Only the combination with experimental studies on the structural, electronic, and vibrational properties allowed to unveil the nature of these epitaxial silicene layers on Ag(111) and demonstrated its two-dimensionality. These results also show that the properties of the epitaxial films are altered by the interaction with the substrate.

Despite this, epitaxial silicene serves as a rich playground for future investigations. It can be used as a template for possible functionalization and adjustment of its properties. Such studies can also pave the way to find lead substrate materials with an interaction that is partly or fully lifted. In this case, all the predicted application directions could be realized.

References

1. Landau, L. and Lifshitz, E., *Statistical Physics. Part 1*, volume 5, Fizmatlit, Moscow, 2004.
2. Castro Neto, A. H., Guinea, F., Peres, N. M. R., Novoselov, K. S., and Geim, A. K., The electronic properties of graphene. *Reviews of Modern Physics* **81** (2009) 109.
3. Pauling, L., *General Chemistry*, Dover Publications, New York, 3rd revised edition, 1988.
4. Takeda, K. and Shiraishi, K., Theoretical possibility of stage corrugation in Si and Ge analogs of graphite. *Physical Review B* **50** (1994) 14916.
5. Hoffmann, R., Small but strong lessons from chemistry for nanoscience. *Angewandte Chemie International Edition* **52** (2013) 93.
6. Vogt, P. et al., Silicene: Compelling experimental evidence for graphenelike two-dimensional silicon. *Physical Review Letters* **108** (2012) 155501.
7. Cahangirov, S., Topsakal, M., Akturk, E., Sahin, H., and Ciraci, S., Two- and one-dimensional honeycomb structures of silicon and Germanium. *Physical Review Letters* **102** (2009) 236804.
8. Guzman-Verri, G. G. and Lew Yan Voon, L. C., Electronic structure of silicon-based nanostructures. *Physical Review B* **76** (2007) 075131.
9. Ariel, V. and Natan, A., Electron effective mass in graphene, in *International Conference on Electromagnetics in Advanced Applications (ICEAA), 2013*, pp. 696–698, IEEE, 2013.
10. Cahangirov, S., Topsakal, M., Akturk, E., Sahin, H., and Ciraci, S., Two- and one-dimensional honeycomb structures of silicon and Germanium. *Physical Review Letters* **102** (2009) 236804.
11. Hull, R., *Properties of Crystalline Silicon*, IET, 1999, Google-Books-ID: C_TWB_0rRLgC.
12. Takagi, N. et al., Silicene on Ag(1 1 1): Geometric and electronic structures of a new honeycomb material of Si. *Progress in Surface Science* **90** (2015) 1.
13. Tubino, R., Piseri, L., and Zerbi, G., Lattice dynamics and spectroscopic properties by a valence force potential of diamondlike crystals: C, Si, Ge, and Sn. *The Journal of Chemical Physics* **56** (1972) 1022.
14. Solonenko, D. et al., 2D vibrational properties of epitaxial silicene on Ag(111) *2D Materials* **4** (2017) 015008.
15. Fleurence, A. et al., Experimental evidence for epitaxial silicene on diboride thin films. *Physical Review Letters* **108** (2012) 245501.
16. Meng, L. et al., Buckled silicene formation on Ir(111). *Nano Letters* **13** (2013) 685.
17. Lin, C.-L. et al., Structure of silicene grown on Ag(111). *Applied Physics Express* **5** (2012) 045802.
18. Feng, B. et al., Evidence of silicene in honeycomb structures of silicon on Ag(111). *Nano Letters* **12** (2012) 3507.
19. Aizawa, T., Suehara, S., and Otani, S., Silicene on zirconium carbide (111). *The Journal of Physical Chemistry C* **118** (2014) 23049.
20. Castrucci, P. et al., Raman investigation of air-stable silicene nanosheets on an inert graphite surface. *Nano Research* (2018) 1.
21. Le Lay, G., Quentel, G., Faurie, J., and Masson, A., Epitaxy of noble metals and (111) surface superstructures of silicon and germanium part I: Study at room temperature. *Thin Solid Films* **35** (1976) 273.
22. Pohl, U. W., *Epitaxy of Semiconductors: Introduction to Physical Principles*, Springer Science & Business Media, Berlin, 2013, Google-Books-ID: DShEAAAAQBAJ.
23. Frank, F. C. and Merwe, J. H. v. d., One-dimensional dislocations. I. Static theory. *Proceedings of the Royal Society of London. Series A* **198** (1949) 205.
24. Vogt, P. et al., Silicene: Compelling experimental evidence for graphenelike two-dimensional silicon. *Applied Physics Letters* **104** (2014) 021602.

25. Solonenko, D., Gordan, O. D., Lay, G. L., Zahn, D. R. T., and Vogt, P., Comprehensive Raman study of epitaxial silicene-related phases on Ag(111). *Beilstein Journal of Nanotechnology* **8** (2017) 1357.

26. Liu, Z.-L. et al., Various atomic structures of monolayer silicene fabricated on Ag(111). *New Journal of Physics* **16** (2014) 075006.

27. Chen, L. et al., Spontaneous symmetry breaking and dynamic phase transition in monolayer silicene. *Physical Review Letters* **110** (2013).

28. Chiappe, D., Grazianetti, C., Tallarida, G., Fanciulli, M., and Molle, A., Local electronic properties of corrugated silicene phases. *Advanced Materials* **24** (2012) 5088.

29. Grazianetti, C., Chiappe, D., Cinquanta, E., Fanciulli, M., and Molle, A., Nucleation and temperature-driven phase transitions of silicene superstructures on Ag(111). *Journal of Physics: Condensed Matter* **27** (2015) 255005.

30. Moras, P., Mentes, T. O., Sheverdyaeva, P. M., Locatelli, A., and Carbone, C., Coexistence of multiple silicene phases in silicon grown on Ag(111). *Journal of Physics: Condensed Matter* **26** (2014) 185001.

31. Arafune, R. et al., Structural transition of silicene on Ag(111). *Surface Science* **608** (2013) 297.

32. Collins, R. et al., Evolution of microstructure and phase in amorphous, protocrystalline, and microcrystalline silicon studied by real time spectroscopic ellipsometry. *Solar Energy Materials and Solar Cells* **78** (2003) 143.

33. Ishidate, T., Inoue, K., Tsuji, K., and Minomura, S., Raman scattering in hydrogenated amorphous silicon under high pressure. *Solid State Communications* **42** (1982) 197.

34. Fukaya, Y. et al., Structure of silicene on a Ag(111) surface studied by reflection high-energy positron diffraction. *Physical Review B* **88** (2013) 205413.

35. Resta, A. et al., Atomic structures of silicene layers grown on Ag(111): Scanning tunneling microscopy and noncontact atomic force microscopy observations *Scientific Reports* **3** (2013) 2399.

36. Liu, Z.-L. et al., The fate of the $2\sqrt{3} \times 2\sqrt{3}R(30°)$ silicene phase on Ag(111). *APL Materials* **2** (2014) 092513.

37. Sheng, S. et al., Vibrational properties of a monolayer silicene sheet studied by tip-enhanced raman spectroscopy. *Physical Review Letters* **119** (2017) 196803.

38. Acun, A., Poelsema, B., Zandvliet, H. J. W., and Gastel, R. v., The instability of silicene on Ag(111). *Applied Physics Letters* **103** (2013) 263119.

39. Vogt, P. et al., Silicene: Compelling experimental evidence for graphenelike two-dimensional silicon. *Physical Review Letters* **108** (2012) 155501.

40. Tersoff, J. and Hamann, D. R., Theory and application for the scanning tunneling microscope. *Physical Review Letters* **50** (1983) 1998.

41. Tersoff, J. and Hamann, D. R., Theory of the scanning tunneling microscope. *Physical Review B* **31** (1985) 805.

42. Lin, C.-L. et al., Substrate-induced symmetry breaking in silicene. *Physical Review Letters* **110** (2013) 076801.

43. Mahatha, S. K. et al., Silicene on Ag(111): A honeycomb lattice without Dirac bands. *Physical Review B* **89** (2014) 201416.

44. Cahangirov, S. et al., Electronic structure of silicene on Ag(111): Strong hybridization effects. *Physical Review B* **88** (2013) 035432.

45. Guo, Z.-X., Furuya, S., Iwata, J.-i., and Oshiyama, A., Absence and presence of Dirac electrons in silicene on substrates. *Physical Review B* **87** (2013).

46. Gori, P., Pulci, O., Ronci, F., Colonna, S., and Bechstedt, F., Origin of Dirac-cone-like features in silicon structures on Ag(111) and Ag(110). *Journal of Applied Physics* **114** (2013) 113710.

47. Wang, Y.-P. and Cheng, H.-P., Absence of a Dirac cone in silicene on Ag(111): First-principles density functional calculations with a modified effective band structure technique. *Physics Review B* **87** (2013) 245430.

48. Avila, J. et al., Presence of gapped silicene-derived band in the prototypical (3 × 3) silicene phase on silver (111) surfaces. *Journal of Physics: Condensed Matter* **25** (2013) 262001.

49. Feng, Y. et al., Direct evidence of interaction-induced Dirac cones in a monolayer silicene/Ag(111) system. *PNAS* **118** (2016) 14656–14661.

50. Liu, H., Gao, J., and Zhao, J., Silicene on substrates: A way to preserve or tune its electronic [roperties. *The Journal of Physical Chemistry C* **117** (2013) 10353.

51. Kokott, S., Pflugradt, P., Matthes, L., and Bechstedt, F., Nonmetallic substrates for growth of silicene: An ab initio prediction. *Journal of Physics: Condensed Matter* **26** (2014) 185002.

52. Li, P. et al., Topological Dirac states beyond π-orbitals for silicene on SiC(0001) surface. *Nano Letters* **17** (2017) 6195, arXiv: 1710.00325.

53. Matusalem, F., Koda, D. S., Bechstedt, F., Marques, M., and Teles, L. K., Deposition of topological silicene, germanene and stanene on graphene-covered SiC substrates. *Scientific Reports* **7** (2017) 15700.

54. Ribeiro-Soares, J., Almeida, R. M., Cancado, L. G., Dresselhaus, M. S., and Jorio, A., Group theory for structural analysis and lattice vibrations in phosphorene systems. *Physical Review B* **91** (2015) 205421.

55. Lazzeri, M., Attaccalite, C., Wirtz, L., and Mauri, F., Impact of the electron-electron correlation on

phonon dispersion: Failure of LDA and GGA DFT functionals in graphene and graphite. *Physical Review B* **78** (2008) 081406.

56. Gori, P., Pulci, O., Vollaro, R. d. L., and Guattari, C., Thermophysical properties of the novel 2D materials graphene and silicene: Insights from Ab-initio calculations. *Energy Procedia* **45** (2014) 512.

57. Usachov, D. Y. et al., Raman spectroscopy of lattice-matched graphene on strongly interacting metal surfaces. *ACS Nano* **11** (2017) 6336.

58. Zanetti, J. E. and Egloff, G., The thermal decomposition of benzene. *Journal of Industrial & Engineering Chemistry* **9** (1917) 350.

59. Ni, Z. H. et al., The effect of vacuum annealing on graphene. *Journal of Raman Spectroscopy* **41** (2010) 479.

60. Greer, E. N. and Topley, B., Attack of oxygen molecules upon highly crystalline graphite. *Nature* **129** (1932) 904.

61. Solonenko, D., Selyshchev, O., Zahn, D. R., and Vogt, P., Oxidation of *Epitaxial Silicene*. *Physica Status Solidi B* (2018). doi: 10.1002/pssb.201800432. submitted.

62. Osborn, T. H., Farajian, A. A., Pupysheva, O. V., Aga, R. S., and Lew Yan Voon, L. C., Ab initio simulations of silicene hydrogenation. *Chemical Physics Letters* **511** (2011) 101.

63. Zolyomi, V., Wallbank, J. R., and Fal'ko, V. I., Silicane and germanane: Tight-binding and first-principles studies. *2D Materials* **1** (2014) 011005.

64. Qiu, J. et al., Ordered and reversible hydrogenation of silicene. *Physical Review Letters* **114** (2015) 126101.

65. Solonenko, D. et al., Hydrogen-induced $sp^2 - sp^3$ rehybridization in epitaxial silicene. *Physical Review B* **96** (2017) 235423.

66. Medina, D. B., Salomon, E., Le Lay, G., and Angot, T., Hydrogenation of silicene films grown on Ag(111). *Journal of Electron Spectroscopy and Related Phenomena* **219** (2017) 57.

67. Warner, B. et al., Guided Molecular Assembly on a Locally Reactive 2D Material. *Advanced Materials* **29** (2017) 1703929–1703936.

68. Stephan, R., Hanf, M.-C., and Sonnet, P., Molecular functionalization of silicene/Ag(111) by covalent bonds: A DFT study *Physical Chemistry Chemical Physics* **17** (2015) 14495.

69. Sattar, S., Hoffmann, R., and Schwingenschlögl, U., Solid argon as a possible substrate for quasi-freestanding silicene. *New Journal of Physics* **16** (2014) 065001.

70. Ezawa, M., A topological insulator and helical zero mode in silicene under an inhomogeneous electric field. *New Journal of Physics* **14** (2012) 033003.

71. Drummond, N. D., Zolyomi, V., and Fal'ko, V. I., Electrically tunable band gap in silicene. *Physical Review B* **85** (2012) 075423.

72. Ezawa, M., Quantized conductance and field-effect topological quantum transistor in silicene nanoribbons. *Applied Physics Letters* **102** (2013) 172103.

73. Gurel, H. H., Ozcelik, V. O., and Ciraci, S., Effects of charging and perpendicular electric field on the properties of silicene and germanene. *Journal of Physics: Condensed Matter* **25** (2013) 305007.

74. Yan, J.-A., Gao, S.-P., Stein, R., and Coard, G., Tuning the electronic structure of silicene and germanene by biaxial strain and electric field. *Physical Review B* **91** (2015).

75. Tao, L. et al., Silicene field-effect transistors operating at room temperature. *Nature Nanotechnology* **10** (2015) 227.

6

Emissive Nanomaterials and Liquid Crystals

Marianne E. Prévôt, Julie P. Vanegas, Elda Hegmann and Torsten Hegmann
Kent State University

Julia Pérez-Prieto
University of Valencia

Yann Molard
Université de Rennes 1, CNRS

6.1 Introduction – Key Concepts

6.1.1 Liquid Crystals

Brief Overview

As a classical scientific oxymoron, Otto Lehmann introduced the term "liquid crystals" in 1889. Discovered by botanist Friedrich Reinitzer in 1888, and described by him as double melting, we now know that the specific cholesterol derivative, cholesteryl benzoate, studied by Reinitzer forms two liquid crystalline phases, the cholesteric (or chiral nematic, N*) and a nematic blue phase (BP*) [1]. This notation of "flowing crystals", which blurs the borders of the usual classification of the fundamental states of matter, is included in the field of "soft matter", popularized by Pierre-Gilles de Gennes [2]. Liquid crystals are mesomorphic materials. Composed by organic molecules called "mesogens", they present phases with a degree of organization intermediate between the crystalline solid phase and the isotropic liquid phase.

Two pathways, largely depending on the molecular structure of the constituent molecules, can generate the liquid crystalline state. When the emergence of the mesomorphic state depends on the molecular concentration in a solvent, the liquid crystals are termed *lyotropic*. Mesophases occurring during temperature variations are described as "pure" and called *thermotropic* mesophases. These mesophases are classified according to their degree of order and according to the morphology and the chemical structure and shape of the mesomorphic molecules. In the case of thermotropic liquid crystals, phase transitions are driven by changes in temperature (Figure 6.1). The level of organization generally decreases with increasing temperature.

Liquid Crystal Phases – Types and Properties

Another classification of liquid crystals could be done according to their shape. The shapes of the molecule forming the thermotropic liquid crystal phase(s) are called rod-like (or calamitic) when the molecule assuming free rotation around its long axis describes a cylinder, disc-like (or discotic) when, as the name suggests, the molecule has the shape of a disc, and bent-core (banana-like), when the shape resembles, well, a banana. This is, of course, not an exhaustive list, but a representation of conceivably the three most important cases (Figure 6.2) [3].

Each of these primary shapes is capable of forming a variety of supramolecular organizations in liquid crystal phases. Figure 6.3 provides a representation of some of these, formed according to the shape of the constituent molecules. In the nematic phase, the organization corresponds to long-range orientational order in one direction, layering in smectic phases and stacking in columnar phases, respectively, is observed when, in addition to this ordering, translation order in one or two directions is added. Each of these phase classes is additionally divided into subclasses characterized by their degrees of symmetry breaking.

Georges Friedel, a French mineralogist and crystallographer, in 1922, undertook the first principle classification of the mesophases formed by calamitic molecules [5]. From a study of their optical properties by polarized microscopy, he distinguishes, in descending order of temperature:

- **A liquid state**, perfectly isotropic, where the molecules do not have a specific position or a particular orientation: they move and fluctuate in a completely random way.

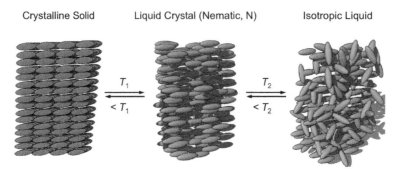

FIGURE 6.1 Graphical representation of the formation of a thermotropic (here nematic) liquid crystal phase (or a sequence of thermotropic mesophases) following a change in temperature.

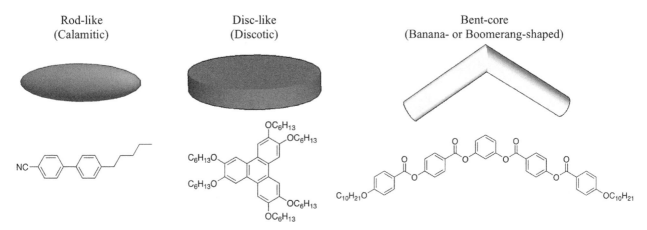

FIGURE 6.2 Principal shape and prominent molecular example of typical calamitic, discotic, and bent-core liquid crystals.

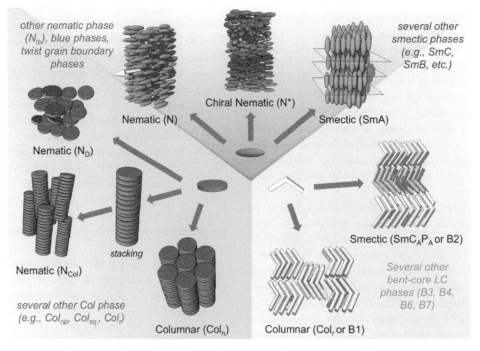

FIGURE 6.3 Examples of the most important and technologically relevant liquid crystal phases formed by rod-like, disc-like, and bent-core liquid crystal molecules. Abbreviations: N_{tb} = twist-bend nematic phase, N_D = discotic nematic phase, N_{Col} = columnar nematic phase, subscripts for the columnar phases: ob = oblique; sq = square; r = rectangular; h = hexagonal, $SmC_A P_A$ = anticlinic antiferroelectric polar smectic-C phase, for the various B phases see [4].

- **A nematic state (N)** in which there is a long-distance orientation order. The molecules are oriented on average in the same direction according to a director **n** (arrow in Figure 6.4). This phase is invariant by translations and rotations around an axis parallel to the director **n**. The centers of gravity of the molecules do not possess any order of position. The molecules can point arbitrarily on average in both directions, because these two positions are equiprobable. From a microscopic point of view, the order of the phase is characterized by the order parameter S; the average parameter calculated from the distribution function of the orientation of the molecules relative to the director **n** (arrow in Figure 6.4). The typical value of the nematic order parameter is between 0.4 and 0.7 [6].

In the cholesteric (or chiral nematic, N^*) phase the mean orientation of the molecules rotates in a direction perpendicular to the director (helical axis). The cholesteric phase is thus considered a twisted nematic phase. It is characterized by a helical pitch (that is commonly temperature-dependent), which represents the distance between the molecules possessing the same orientation after a full 360° twist.

- **A smectic state,** which features an additional modulation parallel to the long-range orientational order of the nematic phase. The centers of gravity of the molecules are no longer arranged randomly but are located in parallel and equidistant planes generating a layered structure. Within these layers, the molecules are free to diffuse. Many smectic phases are described in the literature, and their structure depends on the orientation of the molecules in relation to the layer normal: phases with tilted and non-tilted molecules. The smectic phase constituting the simplest layered arrangement is known as smectic-A (SmA), where the director **n** is perpendicular to the plane of the layers. If the director is inclined at an angle θ to the normal to the layers, the smectic-C phase (SmC) is obtained. The period of the layers is about $a \cdot cos\theta$, with "a" being the molecular length. There are also hexagonal phases characterized by the existence of a long-range order of orientation of neighboring molecules. These phases are therefore more ordered than the previous ones. For example, in the hexagonal SmB phase, the molecules normal to the layers are arranged in a hexagonal network whose orientation is conserved at a long distance, unlike the centers of gravity of the molecules whose positions are correlated only at short range.

- **A crystalline state** in which the molecules are constrained to maintain both a specific position in three-dimensional space, but also a particular orientation. Molecules can be subjected to vibrational movements due to thermal fluctuation but, on average, the order is total, and molecular energy of interactions dominant.

This classification has been enriched over the years, scientist creating and understanding the complex assembly of various new phases with increasingly complex and exotic materials. Liquid crystals today present a large polymorphism.

Optical and Electrical Properties of Liquid Crystal Phases

Shape anisotropy of liquid crystal molecules and/or their aggregates results in birefringence due to the existence of two indices of refraction according to the directions of propagation and polarization of the light and according to the molecular orientation. The refractive index is related to the permittivity of the medium that is mathematically described by a second-order tensor and represented by an ellipsoid whose half-axis lengths are the main refractive indices. Figure 6.5 represents this ellipsoid of the indices. Since the molecule exhibits rotational symmetry about its axis, all the vibrations perpendicular to this axis are equivalent. In other words, the light propagating in the direction of the molecule vibrates identically in any direction perpendicular to that axis. This means that the refractive indices

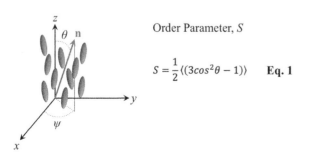

Order Parameter, S

$$S = \frac{1}{2} \langle (3cos^2\theta - 1) \rangle \quad \textbf{Eq. 1}$$

FIGURE 6.4 Definition of the order parameter, S, of the nematic liquid crystal phase.

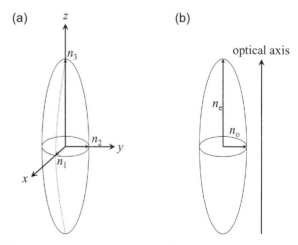

FIGURE 6.5 Ellipsoid indices: (a) for a birefringent medium and (b) in the particular case of a positive uniaxial birefringent medium.

n_1 and n_2 are identical; nematic and smectic-A phases are therefore uniaxial. Thus, linear polarized light will split into two components through the medium. These two waves are called ordinary and extraordinary, with respective refractive indices n_o and n_e that are perpendicular and parallel polarized to the plane containing the optical axis according to the diagram (b) in Figure 6.5 [7, 8].

Each component propagates in the birefringent medium at a different speed according to the index considered: n_o for the ordinary component and $n(\theta)$ for the extraordinary index resulting from the projection along the two proper axis. The birefringence is then written as:

$$\Delta n = n\left(\theta\right) - n_o \tag{6.2}$$

with

$$n\left(\theta\right) = \frac{n_e n_o}{\sqrt{n_e^2 cos^2\theta + n_o^2 sin^2\theta}} \tag{6.3}$$

where θ is the angle between the optical axis and the light wave propagation direction. In the configuration described in Figure 6.5, the ordinary index is constant, whatever the inclination due to the symmetry of revolution. After crossing the medium, one of the waves will therefore delay behind the other, which results in a phase shift $\Delta\varphi(\lambda)$ after recombination given by the relation:

$$\Delta\varphi\left(\lambda\right) = \frac{2\pi e}{\lambda}\Delta n \tag{6.4}$$

with e being the thickness of the sample.

Depending on the nature of the liquid crystal molecules, rod-like liquid crystals possess either a permanent or an induced dipole. In the presence of an applied electric field, the charged parts undergo opposing forces, proportional to their charge and to the field, which cause a molecular rotation until the dipole is aligned parallel to the field.

Defects in Nematic and Smectic – A Liquid Crystals

Considering the prevalent use of nematic liquid crystals in optical device and particularly display applications, a description of defects and types of alignment will focus exclusively on nematic liquid crystals. The same is true for the vast majority of experiments and applications of liquid crystals interfaced in some way with emissive nanomaterials – nematic liquid crystals are used in most studies. Without particular surface treatment, the nematic phase has a heterogeneous planar orientation on glass substrates: the director **n** is parallel to the substrates but points in different directions. Locally, the minimization of the elastic energy causes the molecules to orient themselves with those in their immediate vicinity. But at larger distances, the director field may adopt a configuration where the direction of the director changes abruptly. Such a structure gives rise to a so-called "Schlieren" (German for striations) texture between crossed polarizers. An example is shown in Figure 6.6.

Black brushes corresponding to the extinction direction of the nematic liquid crystal characterize such disclinations.

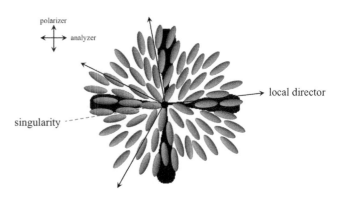

FIGURE 6.6 Four-brush disclination (strength $s = +1$) in a nematic liquid crystal *Schlieren* texture.

The points where two or four branches meet correspond to singularities of the director and are called "disclinations". In the two-dimensional case we can classify them according to their strength s, defined by:

$$s = \frac{\omega}{2\pi} \tag{6.5}$$

where ω is the angle traversed by vectors tangent to the governing field for a complete turn in the trigonometric direction around the singularity line. The points where four brushes meet correspond to disinclinations with $s = \pm 1$. When two brushes intersect, $s = \pm\frac{1}{2}$. These cases are illustrated in Figure 6.7 [8–10].

The SmA phase forms triangular patterns that resemble fans. These patterns can display different birefringence colors depending on the thickness of the sample. Two defect lines, around which the layers bend, a conjugated hyperbola and an ellipse, characterize them. They are called focal conics (Figure 6.8) [8]. As we will see in Section 6.3.1, defect lines in both nematic and smectic-A liquid crystal phases (textures) are used to assemble emissive nanomaterials.

Alignment of Liquid Crystals

Almost any surface causes the director to orient in a specific direction near the surface. The molecules attain an azimuthal orientation that is gradually transferred throughout the cell volume. There are three main types of the liquid crystal director alignment near a solid substrate or at the free surface (i.e. air), namely planar, homeotropic, and titled orientations.

The surface anchoring designates the orientation direction of the director compared to a surface or an interface. The anchoring energy corresponds to the energy needed to deviate the director of the azimuthal (φ_0) and polar (θ_0) anchoring angle defined by the polar coordinates with respect to the interface. The anchoring energies can be decoupled into azimuthal and polar energy (denoted ω_a and ω_z, respectively) for small deviations with the following expressions:

$$\omega_a = \frac{1}{2}\omega_\varphi sin^2\left(\varphi - \varphi_0\right) \tag{6.6}$$

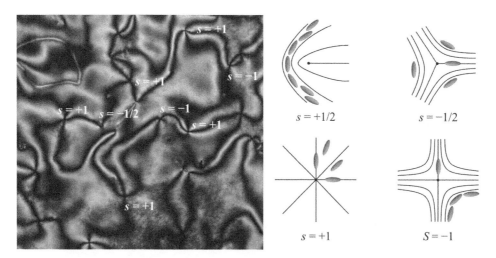

FIGURE 6.7 Types of disclinations found in Schlieren textures. Polarized optical photomicrograph of nematic Schlieren texture (left), schematic drawings of the defects (right).

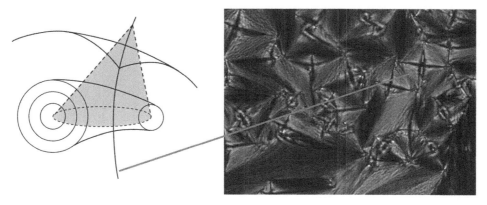

FIGURE 6.8 Focal conic texture (see polarized optical photomicrograph on the right) formed due to the formation of a cyclide and a hyperbola in the smectic-A phase.

$$\omega_z = \frac{1}{2}\omega_\theta sin^2\left(\theta - \theta_0\right) \qquad (6.7)$$

The anchoring is called homeotropic when $\theta = 90°$: i.e. the director orientation is normal to the interface. The anchoring is called planar when $\theta = 0°$: i.e. the director orientation is parallel to the interface. Homogeneous planar orientation appears when the director is oriented uniformly over the surface and φ is fixed. Heterogeneous planar orientation corresponds to a non-uniform orientation of the director over the surface and φ has different fixed values in different points of the surface. In case of tilted orientation, θ is fixed and φ is arbitrary (Figure 6.9).

A surface treatment is used to orient liquid crystals. We distinguish to main group of treatments: mechanical treatment and chemical treatment. Homogeneous planar orientation of the director is usually obtained by rubbing a priori spin-coated polymer (commonly some type of polyimide) surfaces with velvet cloth in some defined direction. Homeotropic orientation can be prepared by coating the substrate with a surfactant. Tilted orientation can be obtained with an oblique evaporation thin film of SiO_x on the substrate. There are many more types of alignment,

for example, using photoalignment layers, and these are discussed in the following review [11].

Orientational and positional order of uniaxial nematic phase contributes to the elasticity of the liquid crystal. By confining the nematic phase the director field can be distorted, its direction progressively changing from point to point. The deformations in liquid crystals can be described in terms of three basic types of deformation, which associate three elastic constants for the liquid crystal, each corresponding to a type of deformation: K_1 for splay deformations, K_2 for the twist deformation, and K_3 for the bend deformations, which cause a change in the free energy of the system (Figure 6.10). If these deformations remain weak at the molecular scale, the free elastic energy, called the Frank–Oseen free energy, takes the form:

$$F_d = \frac{1}{2}K_1(\nabla\cdot\hat{n})^2 + \frac{1}{2}K_2(\hat{n}\cdot\nabla\times\hat{n})^2 + \frac{1}{2}K_3(\hat{n}\times\nabla\times\hat{n})^2 \quad (6.8)$$

Equation 6.8 is the basis for the theoretical treatment of defects and textures of nematic LCs [12–14]. Distortions that do not correspond to minima are unstable and should therefore not occur in stable textures.

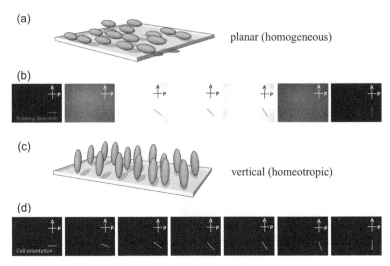

FIGURE 6.9   Main types of liquid crystal orientation near surfaces: (a) planar anchoring and (b) polarized optical photomicrographs of a thin film of a nematic liquid crystals in a cell favoring planar anchoring depending on the cells rubbing direction (indicated) with respect to the microscope's crossed polarizers, (c) homeotropic, and (d) polarized optical photomicrographs of a thin film of a nematic liquid crystals in a cell favoring homeotropic anchoring depending on the cells in-plane orientation (indicated) with respect to the microscope's crossed polarizers; A = analyzer, P = polarizer.

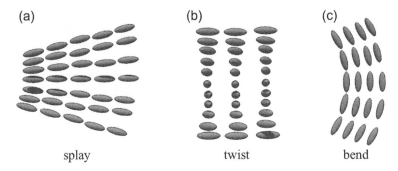

FIGURE 6.10   (a) Splay, (b) twist, and (c) bend deformation in nematic liquid crystals.

6.1.2   Luminescent Nanomaterials

Luminescence – Origin

Luminescence stems from the Latin "lumen" meaning "light". The German physicist Eilhardt Wiedemann in an attempt to design "cold" light, i.e. the energy involved does not derive from the temperature, introduced the expression in 1888 – interestingly enough at the same year liquid crystals were discovered. A luminophore refers to material characterized by the ability to emit visible light [15]. Luminophores can be either organic, inorganic, or hybrid organic/inorganic. A prefix is added to specify the nature of energy in the luminescence process; photoluminescence implies photon absorption, electroluminescence results from the application of an electric field. For example, hexanuclear molybdenum complexes involve d-d transitions, which according to the transition rules are normally prohibited. However, there is a probability that they take place. The simplified Perrin–Jablonski diagram represents the electronic states of a photoluminescent molecule and the transitions between these states. The vertical axis is in units of energy, the horizontal axis differentiates the states according to their spin multiplicity (Figure 6.11). Generally, photoluminescence phenomena obey to Stokes' law; accordingly the energy of light emission (i.e. wavelength) is typically lower (the wavelength longer) than its excitation counterpart.

Luminescence – Concepts

Spin Multiplicity (singlet/triplet states): After the excitation, the electron is found on an energy level higher than the initial state. By definition, a singlet state is the electron configuration in which all the electrons are paired with opposite spins (e.g., the singlet ground state and singlet excited states). By contrast, in the triplet state, there are two unpaired electrons with the same spin (e.g., the triplet ground state and triplet excited states). This principle is illustrated in Figure 6.12. Thus, the total spin of the molecule S is 0 in the singlet state and 1 in the triplet state. Multiplicity corresponds to the degeneracy of the electronic wave functions, i.e. to the number of electronic wave functions differing only in the orientation of their spin. Multiplicity is therefore a function of the magnitude of S according to Hund's rule: $M = 2S + 1$. The multiplicity is 1 for singlet states and 3 for triplet states.

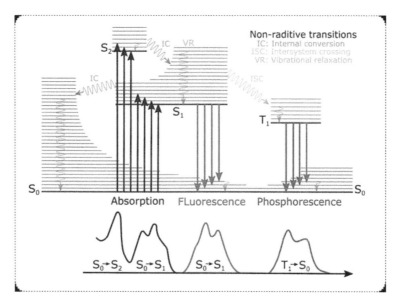

FIGURE 6.11 Simplified diagram of Perrin–Jablonski for an open-shell system. The vertical axis represents the energy, and the horizontal axis represents the states according to their multiplicity of spin.

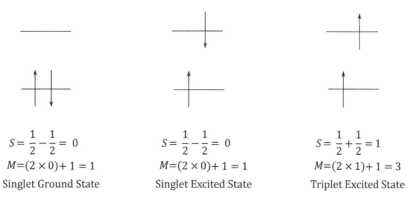

$$S = \frac{1}{2} - \frac{1}{2} = 0$$
$$M = (2 \times 0) + 1 = 1$$
Singlet Ground State

$$S = \frac{1}{2} - \frac{1}{2} = 0$$
$$M = (2 \times 0) + 1 = 1$$
Singlet Excited State

$$S = \frac{1}{2} + \frac{1}{2} = 1$$
$$M = (2 \times 1) + 1 = 3$$
Triplet Excited State

FIGURE 6.12 Singlet and triplet states.

On the Perrin–Jablonski diagram in Figure 6.11, the singlet states are denoted as S_0, S_1, S_2, with 0 corresponding to the fundamental state, 1 the first excited state, and 2 the second excited state. T_1 represents the first excited triplet state. In this diagram, two types of arrows are used: straight and wavy arrows. Light excitation of a system promotes an electron from a filled orbital to an empty one. The photoexcitation involves unpairing of electrons but they still have opposite spins; in addition, the nuclei retain their ground state configuration (vertical transition, Franck–Condon principle). Subsequently, energy is dissipated by:

1. Vibrational relaxation (VR, vertical wavy arrow): it is a non-radiative process that occurs between vibrational levels (the electron does not change from one electronic level to another).

2. Internal conversion (IC, vertical wavy arrow): it is a non-radiative process in which the excited electron goes from a vibrational level in one electronic state to another vibrational level in a lower electronic state), this requires a strong overlap between the vibrational and electronic energy levels.

3. Fluorescence (vertical arrow): this is a radiative process, which consists of transition of the electron to electronic states with the same multiplicity, usually between the first excited electronic state and the ground state.

4. Intersystem crossing (ISC, horizontal wavy arrow). This is a non-radiative process in which the electron changes spin multiplicity from an excited singlet state to a triplet excited state. This process is several orders of magnitude slower than fluorescence.

5. Phosphorescence (vertical arrow): this is a radiative process, which consists of transition of the electron to electronic states with different multiplicity. Each emission or absorption of light (energy, E) by a luminophore is related to the frequency ν of this light by the relation $E = h\nu$ (where h is the Planck constant). The $h\nu_A$, $h\nu_F$, $h\nu_P$ energies on the Perrin–Jablonski diagram

in decreasing order are those of the absorbed photon, the fluorescence-emitted photon, and the phosphorescent-emitted photon, respectively. The Stokes displacement, $\Delta\lambda$, refers to the difference (in wavelength or frequency units) between the band maxima of the absorption and emission spectra of the same electronic transition. The wavy arrows correspond to nonradiative energy losses. Thus, not all electronic transitions are radiative; they obey selection rules [16].

Radiative Transitions: Optical properties are governed by conservation laws (energy, angular momentum) by the energy gap between the quantum level (directly impacting the transmission and absorption wavelengths as shown in the simplified Perrin–Jablonski diagram in Figure 6.11) and by the selection rules that precisely allow transitions between quantum levels.

Allowed Transitions – Selection Rules: In the dipolar electric approximation, some transitions between electronic states are forbidden. First, in the case of centrosymmetric molecules, Laporte's rule imposes Δ with the angular momentum; it is said to be a reversal of symmetry between the initial state and the final state. In particular, transitions of type d-d are forbidden. A second selection rule is given by the relation $\Delta s = 0$. It stipulates that transitions between different multiplicity states are forbidden: singlet–triplet and singlet–triplet transitions are forbidden, therefore of low probabilities, while singlet–singlet and triplet–triplet transitions are allowed. The radiative transitions are generally characterized in intensity by their "oscillator strength" (i.e. the probability of an optical transition, either absorption or emission, between energy levels), which corresponds to a dimensionless number less than 1. The oscillator forces are defined in a vacuum by:

$$f_{ij} = \frac{2m\omega}{3\overline{h}e^2(4\pi\varepsilon_0)^{-1}}\,|M_{ij}|^2 \qquad (6.9)$$

where m is the mass of the electron, ω the frequency of the transition, e the charge of the electron, ε_0 the permittivity of vacuum, M_{ij} the transition matrix element, and $\hbar$ the reduced Planck constant (i.e. $\frac{h}{2\pi}$). However, the selection rules can be modified if spin–orbit coupling and vibronic transitions are taken into account.

Modification of Selection Rules: Spin–orbit coupling is the interaction between two magnetic moments: the one generated by the orbital movement of the electron around the nucleus and that intrinsic to the electron. As a result, a wave function relative to a singlet (or triplet) state always contains a small fraction of a wave function of a triplet (or singlet) state. Thus, considering the spin–orbit coupling, there is a slight but not insignificant probability that the transition between a singlet state and a triplet state takes place and vice versa. Such forbidden transitions are all the more active as the spin–orbit coupling is strong, which is the case for metals. Nevertheless, they involve very small intensities compared to allowable spin transitions.

Finally, the radiative relaxation of an excited electron can occur between levels of the same multiplicity (fluorescence), but also with spin reversal (phosphorescence) – see straight light gray and straight dark gray arrows, respectively, in the Perrin–Jablonski diagram (Figure 6.11).

Vibronic coupling transforms an orbital forbidden to an active transition. In vibronic coupling, there is temporary disappearance of the symmetry center as a result of the vibration movements. The atoms that make up the molecule can vibrate around their equilibrium position, which gives an additional energy to the molecule in addition to the electronic energy (dashed gray sine wave-like arrows in the Perrin–Jablonski diagram (Figure 6.11).

Fluorescence vs. Phosphorescence: Distinction between fluorescence and phosphorescence mainly concerns the transition kinetics and the Stokes displacement. It has been found experimentally that the transitions between states of different multiplicity are 10^3–10^5 times slower than those of the same multiplicity [17]. In the case of fluorescence, the excited species return very quickly (10^{-10}–10^{-7} s) to their original energy state. During phosphorescence, they pass through an intermediate energy state, or they remain a certain time in the excited state before returning to the initial state. The intersystem crossing may be fast enough (10^{-9}–10^{-7} s) to compete with other de-excitation pathways, namely fluorescence and internal conversion. The characteristic time, τ_L, of a phosphorescent transition extends from 10^{-6} s to several hours. Due to the de-excitation pathway implied in fluorescence and phosphorescence, less Stokes displacement is expected in the case of fluorescent emission compared with phosphorescent emission.

Luminescence Quantum Yield: Due to the competition between radiative and non-radiative transitions, the magnitude of the luminescence quantum yield or quantum luminescence efficiency needs to be defined. This quantity corresponds to the ratio between the number of emitted photons and the number of absorbed photons. In other words, this quantum yield ϕ_L is a function of the production yield of the radiative states. We have seen that the total spin of the molecule is 0 in the singlet state and 1 in the triplet state. Thus, for a singlet state, the quantum number M_s associated with this magnitude is zero. Hence there is only one possible state represented by the spin wave function $\varphi_{s=0,\,M_s=0} = \varphi_{0,0}$. For a triplet state, three states are possible, corresponding to the three spin functions $\varphi_{1,-1}$, $\varphi_{1,0}$, and $\varphi_{1,1}$. The spin statistic therefore indicates that three times more triplet states will be formed than singlet states. Thus, by considering only the theory related to the spin statistic, phosphorescent materials should reach quantum efficiencies superior to those obtained in fluorescent molecules (internal quantum efficiency consideration only), giving rise to more energy efficient devices [18].

Emission Quenching designates, in the broad sense, any process decreasing the intensity of the emitted signal and consequently the luminescence quantum efficiency of a luminophore. It is a non-radiative relaxation of excited electrons to a fundamental (ground) state. This extinction can be the result of intermolecular collisions, in which case the electronic energy is converted into kinetic energy and vibrations. It can also result from the formation of a non-luminescent complex, a change in the experimental conditions (pH, solvent, etc.) of the molecule [19].

Photobleaching occurs when a fluorophore permanently loses the ability to fluoresce due to its degradation.

Upconversion – Anti-Stokes Photoluminescence: In principle, combinations of lanthanides, uranides, and various transition metal species embedded in solids or in the form of upconverting nanoclusters can produce anti-Stokes upconversion emissions. Photon upconversion is the sequential absorption of two or more photons by a species eventually resulting in the emission of light at shorter wavelength than the excitation wavelength. Several different mechanisms have been recognized to be involved in upconversion either alone or in combination. The three basic upconversion mechanisms are energy transfer upconversion (ETU), excited-state absorption (ESA), and energy migration-mediated upconversion as summarized in Figure 6.13 [20].

Energy transfer upconversion (ETU) occurs in systems where absorption and emission do not take place within the same center. This phenomenon does not necessarily imply any charge transport. It implies a sensitizer (S) that absorbs the light and an activator (A) to which the ion energy is transferred, and which emits the output photon. This phenomenon could be resonant non-radiative transfer and phonon-assisted non-radiative transfer [21]. These different cases are illustrated on top of the Figure 6.14. Activators (A) have several excited states. Though, considering the case of the activator in an excited state, the exchange energy is a relative energy difference and not an absolute energy (see bottom in Figure 6.14). The situation could repeat itself several times at the activator, i.e. n-photon upconversion by energy transfer is possible.

Upconversion can only occur in materials with more than one metastable excited state, in which multiphonon relaxation processes are not predominant. The sublevels implying in the luminescence process should not strongly

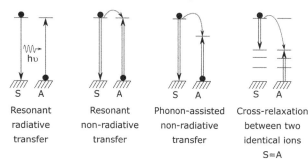

S and A are intially from ground state

| Resonant radiative transfer | Resonant non-radiative transfer | Phonon-assisted non-radiative transfer | Cross-relaxation between two identical ions S=A |

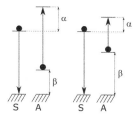

S is intially from ground state and A is in an excited state

FIGURE 6.14 Various basic energy transfer processes between two species: note that activator species (A) receiving the energy from the sensitizer (S) is initially in its ground state. Cross-relaxation is the special case where S is identical to A. Double arrows symbolizes Coulombic interaction.

participate in the metal–ligand bonding to reduce the electron–phonon interaction and consequently multiphonon relaxation processes.

Fluorescent Metal Nanomaterials

Metal nanomaterials whose diameter is smaller than 2 nm are known as metal nanoclusters. Metal nanoclusters, such as those of Ag and Au are a fairly recently developed class of emissive nanomaterials, featuring unusual physicochemical, optical, electronic and chemical properties, different from other species known to date, especially from those of their plasmonic nanoparticle counterparts.

They are characterized by highly controllable fluorescence emission, which largely depends on the nature of the metal as well as of the anchoring group, large Stokes shift, absence of intermittency in the emission, and high photostability. These very low cytotoxic nanoparticles can be prepared easily and with high size control, and their surface can be easily functionalized (Figure 6.15). Exhibiting acceptable fluorescence quantum yields (less than 10%), the luminescence maximum depends on the nature of the ligand-anchoring group, which is attributed to charge transfer between this group and the cluster surface. It has recently been reported that rigidification of the nanocluster ligand shell through host–guest interactions drastically enhances the nanocluster luminescence [22].

Fluorescent metal nanoparticles are commonly prepared via two strategies including a top-down and a bottom-up method. In the first approach, "top-down", a physical method is used in which applying an external force to a

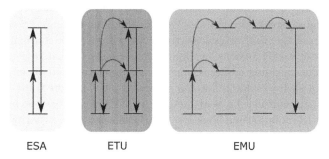

ESA   ETU   EMU

FIGURE 6.13 Mechanisms for emission of upconverting nanoparticles: ETU = energy transfer upconversion, ESA = excited-state absorption, and EMU = energy migration-mediated upconversion.

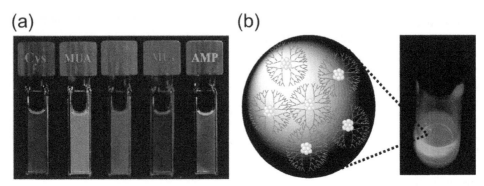

FIGURE 6.15 Photographs of fluorescent metal nanoparticles under UV light irradiation: (a) emissive gold nanoparticles decorated with cysteine (Cys), mercaptoundecanoic acid (MUA), 11-mercaptoundecanol (MU_1 and MU_2), and adenosine monophosphate (AMP).(Reprinted with permission from Ref. [23], Copyright © 2017, Wiley-VCH Verlag GmbH & Co. KGaA, Weinheim.) (b) Emissive gold nanoparticles decorated poly(amidoamine) (PAMAM). (Reprinted with permission from Ref. [24], Copyright © 2003, American Chemical Society.)

solid (volumetric material) results in its disintegration into secondary particles (dust) and, subsequently, nanoparticles. This can be achieved by chemical treatment ("etching"). Generally, these methods give rise to structures with imperfections in the surface of the material and size distributions greater than 10 nm.

Alternatively, in the "bottom-up" approach, a chemical method is employed that proceeds via the reduction of metal ions, followed by the formation of emissive metal nanoparticles after the nucleation of the formed metal atoms. The control of the aggregation step is most critical for the control of size and size distribution of the final nanoparticles. This method of synthesis commonly results in smaller nanoparticles, greater control of size and shape, and nanomaterials with few surface imperfections.

At present, a general mechanism for the origin of fluorescence of metal nanoclusters has not been established. Some mechanisms have been proposed that could explain the phenomenon, but they could indeed be different for each type depending on the size, the ligands lining the surface, and even the synthetic route [25].

Whetten and co-workers suggested that the luminescence is caused by radiative recombination between a higher excited state and the ground state (HOMO-LUMO gap) [26]. Murray et al. as well as Jin et al., however, hypothesized that the mechanism could be associated with interband recombination transitions between the filled $5d^{10}$ and the $6(sp)^1$ conduction band [27, 28]. Others propose that the luminescence is due to numerous factors such as the metal core and its intrinsic effects of quantization, the surface of the nanoparticle including ligand–metal, metal–metal interactions, or solvent and valence state effect. Numerous examples also highlight that the observed fluorescence might depend on the synthesis pathway itself. For example, the work by Jin and co-workers [29] shows that the thiol surface ligands (—SR) play a significant role in enhancing the fluorescence of gold nanoclusters. Specifically, surface ligands can influence the fluorescence in two different ways: (i) charge transfer from the ligands to the metal nanocluster core through the Au—S bonds and (ii) direct donation

of delocalized electrons of electron-rich atoms or groups of the ligands to the cluster core. In general, fluorescent metal nanoparticles show great promise in chemical and biosensing, optoelectronic applications and catalysis, as well as in biological and nanomedicine applications such as biolabeling, bioimaging, and targeted cancer therapy.

Upconversion Nanoparticles (UCNPs)

Lanthanide (rare earth)-doped UCNPs are dilute host-guest systems where trivalent lanthanide ions are dispersed as a guest in an appropriate dielectric host lattice with a dimension of less than 100 nm. The lanthanide dopants are the active centers of emission. Through a specific selection of lanthanides, UCNPs can display wavelength (color) selective upconversion ranging from the NIR over visible (green, red) to the UV via two-photon or multi-photon mechanism. Unique narrow emission bands are recorded after NIR light excitation (800 nm, 980 nm) of theses nanoparticles by using a low-power continuous-wave diode laser. These properties combined with their high stability, low cytotoxicity, good photostability, and non-photoblinking or photobleaching make UCNPs unique optical tools, among others, in biological studies. The only drawback of UCNPs is their relatively low upconversion quantum yield.

Thanks to the pioneering theoretical work of Judd [30] and Dieke [31] we currently have a good understanding of many of the spectroscopic properties of Ln^{3+} ions incorporated into single crystals. The UC photoluminescence arises from the 4f-4f orbital electronic transitions with concomitant wave functions localized within a single lanthanide ion. The shielding of 4f electrons by the outer complete 5s and 5p shells results in line-like sharp emissions (Figure 6.16).

In addition, rare-earth-doped UCNPs can be excited by NIR radiation and possess several advantages such as narrow emission peaks, large anti-Stokes shifts, excellent signal-to-noise ratios, and improved detection sensitivity due to an absence of autofluorescence, deeper NIR light penetration into biological tissue causing less photo-damage to biological samples, and excitation via low-power NIR lasers that are compact and inexpensive. They are usually produced with a

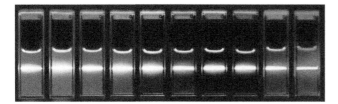

FIGURE 6.16 A general approach to fine-tuning the upconversion emission colors based upon a single host source of NaYF$_4$ nanoparticles doped with Yb^{3+}, Tm3$^+$, and Er^{3+}. (Reprinted with permission from Ref. [32], Copyright © 2008, American Chemical Society.)

hydrophobic ligand (e.g., oleic acid) by following a bottom-up procedure. Subsequent ligand exchange with a more hydrophilic ligand can lead to biocompatible NPs, while using a functional ligand can further enhance the functionality of the nanoparticle.

Rare earth doped UCNPs can be synthesized by a variety of techniques. Examples of the size and shape variations accessible are shown in Figure 6.17.

Quantum Dots (QDs)

QD semiconductors in the range of 2–10 nm in core diameter are made of tens, hundreds, even thousands of atoms of group II-B and VI-A elements such as CdSe, CdTe, and ZnO, group III-A and V-A elements such as InP and InAs, as well as group IV-A and VI-A elements such as PbS. Their size- and shape-dependent optical and electronic

properties that vastly differ from those of the bulk and those of atomic/molecular systems are attributed to the three-dimensional quantum confinement, i.e. the strong confinement of electrons and holes in the case where the radius of a particle is close to the exciton Bohr radius [38, 39]. In a semiconductor, an electron can be promoted from the valence to the conduction band, leaving a hole or "absence of an electron" in the valence band upon irradiation by photons with greater energy than the bandgap. Therefore, this hole is assumed to behave as a "particle" with a certain effective mass and positive charge. The bound state of the electron–hole pair is called "exciton" [40]. Luis Brus developed the relation between size and bandgap energy of semiconductor nanocrystals by applying the particle in a sphere model approximation to the bulk Wannier Hamiltonian [40, 41]. According to the approximation, the lowest Eigenvalue in a quantum-confined system is given by Eq. 6.10 [40]:

$$E_{g,QD} = E_{g,b} + \left(\frac{\hbar^2}{8R^2}\right)\left(\frac{1}{m_e}+\frac{1}{m_h}\right) - \left(\frac{1.8e^2}{4\pi\varepsilon_0\varepsilon R}\right) \quad (6.10)$$

where the first two energy terms (E) are the bandgap energies of the quantum dot and bulk solid, respectively. R is the radius of the quantum dot, m_e is the effective mass of the electron in the solid, e is elementary charge of the electron, $\hbar$ is again the reduced Planck's constant, the m_h is the effective mass of the hole in the solid, and ε is the dielectric constant of the solid. The middle term and the last term represent the exciton of a "particle-in-a-box-like" and Coulomb attraction of the electron–hole

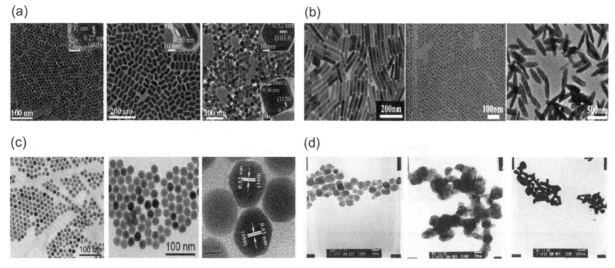

FIGURE 6.17 Transmission electron microscopy (TEM) images of representative shapes of various rare-earth-doped nanoparticles: (a) rare earth (RE) trifluoroacetate RE(CF$_3$COO) used to dope (left to right) α-NaYF$_4$ nanopolyhedra, β-NaYF$_4$ nanorods, and β-NaHoF$_4$ hexagonal nanoplates synthesized via thermal decomposition (Adapted with permission from Ref. [33], Copyright © 2006, American Chemical Society.) (b) Nano-/microparticles obtained by hydro(solvo)thermal reaction based on β-NaYF$_4$ (left to right) Yb^{3+}, Er^{3+} nanorods (Adapted with permission from Ref. [34], Copyright © 2006 American Chemical Society.); α-NaYF$_4$ nanospheres, and YF$_3$ rice-like nanoparticles. (Adapted with permission from Ref. [35], Copyright © 2006, American Chemical Society.) (c) By Ostwald ripening of (left to right) NaGdF$_4$:Yb (18 mol%) and NaGdF$_4$:Tm (1 mol%)/NaGdF4:Eu (10 mol%) core/shell nanoparticles at different magnifications. (Adapted with permission from Ref. [36], Copyright © 2010 WILEY-VCH Verlag GmbH & Co. KGaA, Weinheim.) (d) By co-precipitation EDTA/Ln^{3+} and then annealed at (left to right) 400°C for 5 h, 500°C for 1 h, and at 600°C for 15 min. (Adapted with permission from Ref. [37], Copyright © 2007, Elsevier.)

pair, respectively. As we can see from Eq. 6.10, E_g of QDs scales with diameter. As the dimension (i.e. the radius, R) decreases, the bandgap energy increases. In other words, a smaller QD emits light with a shortest wavelength (higher energy) as governed by the bandgap [42]. Hence, the photoluminescence colors of QDs can be tuned by controlling the particle size. This also means that fluorescence spectroscopy can be utilized to determine the relative size and size distribution of QDs using the emission wavelength, where the relative size distribution is determined by the full width at half maximum (FWHM) of the emission peak [43]. Figure 6.18 shows the size-dependent photoluminescence of semiconductor nanocrystals throughout the visible region of the electromagnetic spectrum.

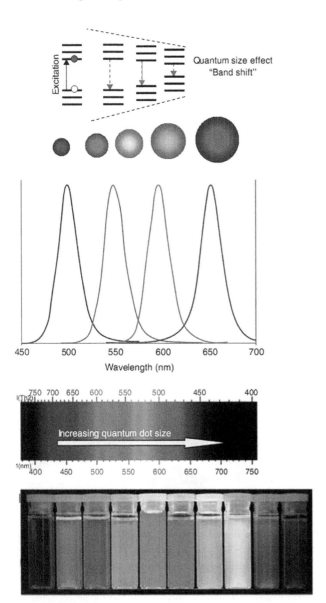

FIGURE 6.18 Size-dependent photoluminescence (PL) of semiconductor QDs. Electronic structure of QDs correlated with the QD radius and resulting due to quantum confinement. (Reproduced with permission from Ref. [44], Copyright © 2010, Wiley-VCH Verlag GmbH & Co. KGaA, Weinheim.)

Although size control in QD synthesis allows for tuning the wavelength as expected from the bandgap approximation, problems associated with the surface of QDs known as "surface traps" frequently cause emission(s) at wavelengths higher than expected. In QDs with very high surface-to-volume ratio, surface traps are due to imperfections (surface defects) and dangling bonds. Such surface trap sites have their intrinsic electronic energy states, which are usually localized within the QD's bandgap [44] as first suggested by Efros and Rosen. Defects on the surface of the QD act as temporary surface traps for electrons or holes (i.e. charge carriers), leading to non-radiative relaxation, therefore preventing radiative recombination that greatly reduces the quantum yield [45]. The resulting low fluorescence quantum yield, broad fluorescence range, and blinking due to the charge recombination pathway are among the main problems in utilizing QDs for optical applications. A review article by Carrillo-Carrión et al. elaborates on QD luminescence and important aspects affecting the photoactivation [46]. One approach to minimize the undesirable effects of surface traps and increasing quantum yields in QDs is a combination of two or more semiconductors or doping one semiconductor with a small quantity of another metallic or semiconductor element. Combination of two or more semiconductors forming heterojunctions in so-called "core–shell" semiconductor QDs can diminish the instability caused by the abundance of surface states. In these core–shell systems, a semiconductor QD core is surrounded by another semiconducting material with wider bandgap as a shell to passivate the surface states of the nanocrystal [47] (Figure 6.19).

Metal Atom Clusters

Metal atom clusters have been defined by F. A. Cotton in 1964 as a "finite group of metal atoms which are held together by bonds directly between the metal atoms" [49]. These metallic aggregates are linked to different kinds of ligands to form nanometer-sized cluster motifs. The nature and the number of metals and ligands involved govern the architecture of these motives. Thus, there are clusters with different nuclearity. Figure 6.20 shows clusters of linear, triangular, and tetrahedral geometry, obtained from two, three, and four metal atoms, respectively.

Among this wide family, octahedral nanoclusters based on a molybdenum, tungsten, or rhenium scaffold have remarkable luminescence properties. The elementary units in the chemistry of octahedral clusters are the configurations $[M_6L_{18}]^n$ and $[M_6L_{14}]^n$ represented in Figure 6.21 with M designating the metal, L the ligand, and n the cluster charge [51]. However, only the $[M_6L_{14}]^n$-based motifs can be emissive with appreciable quantum yields. According to the description by Schäfer and co-workers [52], the 14 ligands bonded to the metal core in $[M_6L_{14}]^n$, 8 are said to be *inner*, noted L^i and capping the faces of the metallic metal octahedron, and 6 are called *apical*, noted L^a and are in the terminal position.

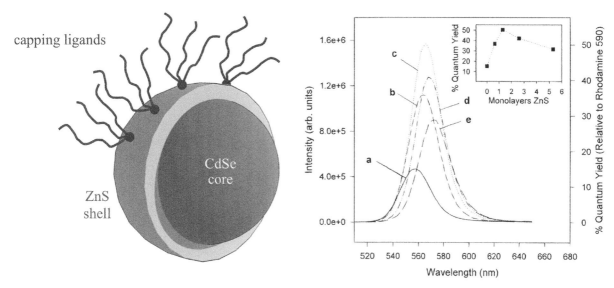

FIGURE 6.19 Representative model (ligands capping the entire QD surface: e.g., trioctylphosphine oxide = TOPO) and photoluminescence (PL) spectra for a series of ZnS overcoated QDs with $42 \pm 10\%$ Å diameter CdSe cores. The spectra are for: **a** 0, **b** 0.65, **c** 1.3, **d** 2.6, and **e** 5.3 monolayers ZnS coverage. The position of the maximum in the PL spectrum shifts, and the spectrum broadens with increasing ZnS coverage. Inset: The PL quantum yield is charted as a function of ZnS coverage. The PL intensity increases with the addition of ZnS reaching, 50% at ~1.3 monolayers, and then declines steadily at higher coverage. (Reproduced with permission from Ref. [48], Copyright © 1997, American Chemical Society.)

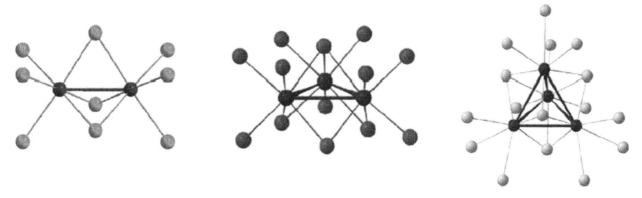

FIGURE 6.20 Schematic illustration of the geometry of a linear, triangular, and tetrahedral cluster (left to right). Metallic atoms are represented in a darker and non-metallic ligands in a lighter grayscale tone. (Reproduced with permission from Ref. [50].)

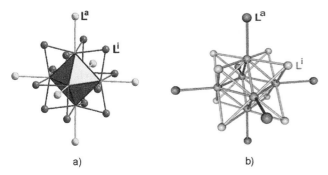

FIGURE 6.21 (a) $[M_6L_{18}]^n$ and (b) $[M_6L_{14}]^n$ configuration. (Reproduced with permission from Ref. [50].)

The $[M_6L^i_8L^a_6]^n$ structure is found in many compounds whose physicochemical properties depend on: (i) the nature of the metallic core (Mo, Re, or W), (ii) the nature of the ligands surrounding the metallic core (inner ligands could be either halogens or chalcogens, while there is a very wide range of possible apical donor ligands ranging from a cyano to halogen groups as well as organic anionic or neutral groups), and (iii) the number of valence electrons per cluster, i.e. the number of available electrons for metal–metal bonds.

As an example, the electronic emission spectra of molybdenum clusters cover a large window of wavelengths typically from 500 to 900 nm with a maximum emission around 700 nm. The emission characteristics are a mixture of metal center localized transitions and charge transfer transitions from the metal centers to the ligands (Figure 6.22).

The lifetime and luminescence quantum yields of these octahedral molybdenum clusters are listed in Table 6.1. Lifetimes on the order of a few hundred microseconds (typically 200–300 μs) are a signature of the phosphorescent character of these luminophores. By changing the coordination sphere

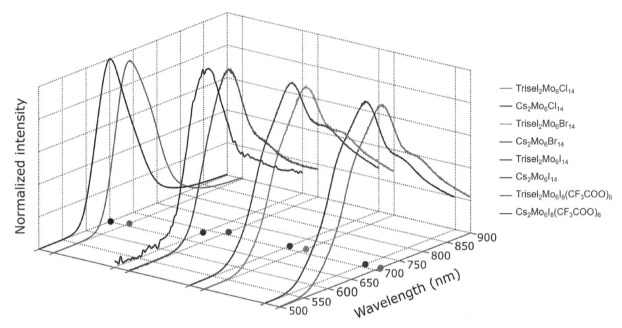

FIGURE 6.22 Photoluminescence spectra of a range of $[Mo_6L_{14}]^n$ clusters. The emission maxima are centered around 700 nm independent of the specific chemical nature or structure of the ligands (for the chemical structure of the Trisel ligand, see Table 6.2). (Reproduced with permission from Ref. [50].)

TABLE 6.1 Average Quantum Yields (φ_L) in Degassed Acetonitrile and Luminescence Lifetimes (τ_L) of Various $[Mo_6L_{14}]^n$ Clusters Both at 25°C [50]

Cluster	φ_L	$\tau_L/\mu s$
$(Bu_4N)_2[Mo_6Br_{14}]$	0.20	100
$(Bu_4N)_2[Mo_6I_{14}]$	0.15	90
$(Bu_4N)_2[Mo_6Br_8(CF_3COO)_6]$	0.30	335
$(Bu_4N)_2[Mo_6I_8(CF_3COO)_6]$	1	220
$(Bu_4N)_2[Mo_6Br_8(C_3F_7COO)_6]$	0.35	370
$(Bu_4N)_2[Mo_6I_8(C_3F_7COO)_6]$	0.59	300

of the luminescent $[Mo_6L^i{}_8]^{4+}$ cluster core, the quantum yields of the $[Mo_6L^i{}_8L^a{}_6]^n$ clusters can be varied considerably. For example, substitution of the iodine atoms in the apical position by trifluoroacetate groups results in increased quantum yields.

The excited states in these cluster systems evolve into geometry. At the excited state, there is a large deviation of the O_h symmetry. Because the geometries of the excited state are very different from those related to the fundamental state no significant changes are observed between photoluminescence spectra depending on the nature of the ligand. Since excitation occurs at a constant geometry and strong deformation of the excited O_h symmetry is hypothesized, the clusters are expected to show large Stokes shifts. Indeed, $[Mo_6L_{14}]^n$ clusters do show a very large Stokes shift of up to 300 nm [53]. Due to their favorable emission properties, these types of luminescent metal clusters are sought for biomedical applications, as solar concentrators as well as for lighting and display devices. Several of these applications require well-aligned and self-assembled superstructures of these metal clusters, which as we will see in Section 6.2.2 can be accomplished by a variety of mesogenic (liquid crystalline) ligands.

6.2 Self-Assembly of Luminescent Nanomaterials

Liquid crystal nanoscience has evolved into two major research thrusts. In one, liquid crystals are used either as templates for the syntheses of nanomaterials or they are used as organizing anisotropic fluids (hosts) to assemble and manipulate nanoscale materials. The other focuses on approaches where nanomaterials (here focusing on emissive nanomaterials) impact liquid crystal properties based on very fundamental interactions between liquid crystal host molecules and emissive nanomaterial additives. In the latter, the modulation of optical and electro-optical properties for potential application in devices such as displays or light shutters crystallized as the main focus. Many important concepts and strategies of liquid crystal nanoscience, including studies elucidating the roles of nanomaterial shape, size, and functionalization, were already summarized in a number of review articles [54, 55].

Of course, there is also significant research on a combination of emissive nanomaterials and liquid crystals, where the two materials only share an interface or where the liquid crystal modulates the emission of the luminescent nanomaterials. The reader will be familiar with QD displays, where the superior emissive properties of semiconductor nanocrystals (brightness, color purity) serve as part of the display backlight [56]. Additional polarized emission using quantum rods has been described by a number of groups. Hens and co-workers, for example, described large-scale, electro-switchable polarized emission from semiconductor nanorods aligned using polymeric nanofibers (Figure 6.23) [57].

TABLE 6.2 Properties of Selected Clustomesogens Depending on the Type of Pro-Mesogenic Unit and the Approach to Connect the Octahedral Mo Clusters to the Pro-Mesogenic Units

Cluster	Pro-Mesogenic Units	Approach	φ_L	Phase Transitions	$T/°C$	Polarized optical microscopy (POM; White Light)	POM (Irradiation $\lambda_{exc.}$ = 380−420 nm)
$(Bu_4N)_2[Mo_6Br_8Ln_6]$	HLn (n=9)	Covalent	0.1	g−SmA SmA−Iso	18 132		
$Cs_2[Mo_6Br_{14}]$		Supra-molecular	0.2	g−Col$_r$ Col$_r$−Iso	97 159		
$[Mo_6Br_{14}]^{2-}$	Trisel =	Electrostatic	0.2	g−N N−Iso	52 97		
$[Mo_6I_8(OCOC_2F_5)_6]^{2-}$	Trisel =	Electrostatic	0.7	g−N N−Iso	20 64		

Source: Reproduced with permission from Refs. [64, 66−68], Copyright © 2015 and 2016, Royal Society of Chemistry; Copyright © 2015 Wiley-VCH Verlag GmbH & Co. KGaA, Weinheim.

g, glass transition; SmA, smectic-A phase; Col$_r$, rectangular columnar phase; N, nematic phase; Iso, isotropic liquid phase.

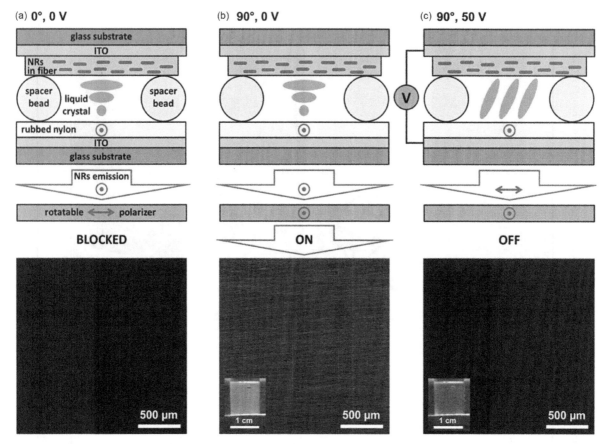

FIGURE 6.23 Schematic of liquid crystal cell with CdSe/CdS@SiO$_2$@PVP (polyvinylpyridine) nanofibers and corresponding fluorescence microscopy images depending on the polarizer orientation and applied electric field: (a) polarizer parallel to nanofibers (0°), no electric field applied, (b) polarizer rotated perpendicularly (90°), no electric field applied (ON-position), (c) polarizer maintained perpendicular (90°), 50 V electric field is applied (OFF-position); insets: fluorescence microscopy images in (b) and (c) are pictures of the LC cell after UV excitation. (Reproduced with permission from Ref. [57], Copyright © 2015, American Chemical Society.)

Others expanded this approach by adjusting the shapes of core and shell or core–shell semiconductor nanocrystals to obtain tunable emission colors as well as polarized emission. Again, stretched polymer films were employed to align the emissive nanomaterials, here specifically a copolymer poly(butyl methacrylate)-co-(isobutyl methacrylate) as shown in Figure 6.24 [58].

The use of liquid crystals as a matrix or a state of matter in intimate contact with the emissive nanomaterials is an emerging strategy for organizing inorganic nanomaterials in the form of a composite. The liquid crystalline state combines order and mobility (fluidity) on a molecular level, which is why materials with liquid crystalline properties appear as perfect candidates for the construction and organization of nanostructured materials with "built-in" self-repair or self-correction mechanisms. They de facto act as solvent to disperse nanomaterials and by their anisotropic nature ensure the organization of the material in several dimensions.

Nature is full of multiphase materials called functionalized organic–inorganic nanocomposites. The shells of crustaceans are beautiful specimens. In crabs, carbohydrate chains, known as chitin, associate with proteins to form a tubular structure. Since chitin contains chiral carbons, the structure is organized into a helix. It is within this overall structure that the process of biomineralization takes place, which constitutes the structural features of their shell. In the way of structured shell crabs, liquid crystal-nanoparticle hybrids aim to combine the intrinsic properties of clusters with those of liquid crystals to form a self-organizing unit [59].

Combining organic and inorganic properties in a single material is a very old challenge that was realized from the beginning of the industrial era with, for example, the use of inorganic pigments suspended in an organic mixture in industrial paint and the development of companies such as DuPont, Dow Corning, and 3M. For 30 years, the challenge has been to functionalize these organic–inorganic hybrid materials; where both organic and inorganic groups contribute to providing additional functions for more sophisticated materials.

6.2.1 Luminescent Nanorods

The rod-like shape of various emissive nanomaterials, similar to rod-like liquid crystals, gives rise to self-assembly into quasi-liquid crystalline motifs (phases). One of the

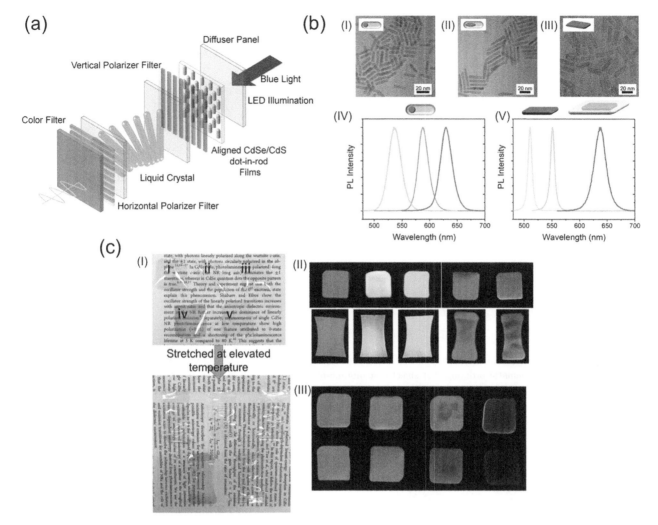

FIGURE 6.24 (a) Schematic of a liquid crystal display using anisotropic nanocrystals as backlight, which absorb unpolarized light from blue LED emitters and re-emit linearly polarized light with near unity quantum efficiency and polarization direction aligned with the vertical polarizer filter. (b) CdSe-based luminescent anisotropic nanocrystals: TEM images of (I) CdSe/CdS dot-in-rods; (II) CdSe/CdS rod-in-rods; (III) CdSe nanoplatelets; (IV) emission spectra of a colloidal solution of CdSe/CdS dot-in-rods; (V) emission spectra of bare CdSe nanoplatelets and CdSe/CdS core−shell nanoplatelets. (c) Composites consisting of luminescent nanocrystals dispersed in poly(butyl-co-isobutyl methacrylate). (I) Films under ambient light before and after stretching consisting of CdSe/CdS dot-in-rods: (i) 4.5 nm core, (ii) 3.8 nm core, (iii) 2.1 nm core, (iv) CdSe/Cd$_{1-x}$Zn$_x$S nanoplatelets, (v) CdSe nanoplatelets. (II) Same films under ambient light before and after stretching. (III) Films of varying concentration of nanocrystals under UV light. The top row of films consists of core−shell quantum dots; the bottom row of films consists of similar concentrations of CdSe/CdS rod-in-rods. (Reproduced with permission from Ref. [58], Copyright © 2016, American Chemical Society.)

earliest reports by Alivisatos and co-workers highlights that both the concentration and the aspect ratio are critical parameters for the formation of stable nematic-type liquid crystal phases [60, 61]. Just recently, Smalyukh et al. demonstrated that this concept could be broadened by the use of a polymeric depletant resulting in lyotropic liquid crystal self-assembly of upconverting nanorods (UCNRs, Ln^{3+}-doped hexagonal β-NaYF$_4$ rod-like nanocrystals; aspect ratios of 6.5 and 9.3 depending on the doped Ln^{3+} ion). The self-assembly arises from the repulsive steric and electrostatic colloidal interactions facilitated by the anisometric geometric shape of the UCNRs and depletion forces induced by the non-adsorbing polymer (in this case dextran). The resulting self-assembled

colloidal superstructures exhibit various nematic-, biphasic- as well as smectic-like ordering, including stable and long-lived metastable configurations with uniaxial or multi-axial ordering (Figure 6.25) [62].

In other examples, nanorods can be aligned, for example, in an end-to-end fashion using specifically engineered liquid crystal defect arrays. Lacaze and co-workers demonstrated the use of oriented linear arrays of smectic-A defects, the so-called smectic oily streaks, enables the orientation of gold nanorods (GNRs). For GNRs capped with mesogenic ligands oriented and end-to-end self-assembled short GNR chains were observed when a sufficiently high density of the GNRs Was Used. Strongly anisotropic light absorption and on−off switching of the GNR luminescence, both controlled by

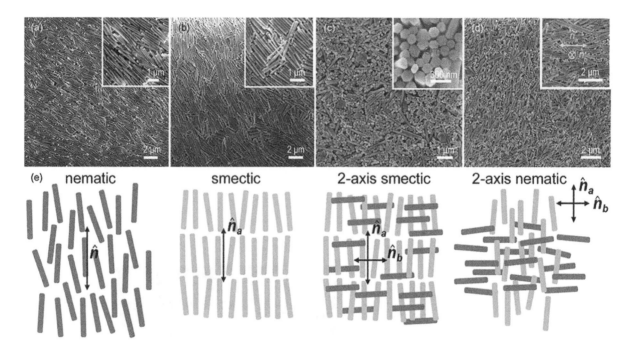

FIGURE 6.25 (a–d) Scanning electron microscopy (SEM) images and (e) graphical depictions showing the colloidal self-assembled UCNR morphologies: (a) nematic ordering, (b) smectic-membrane-like ordering, (c) smectic membranes with two-axis orientational ordering, and (d) nematic ordering (two-axis ordering of UCNRs). The initial concentrations of UCNRs are ~25, 35, 50, and 65 mg mL^{-1} (a)–(d), respectively. Insets show enlarged views with greater details of the ordering. (Reproduced with permission from Ref. [62], Copyright © 2018, Royal Society of Chemistry.)

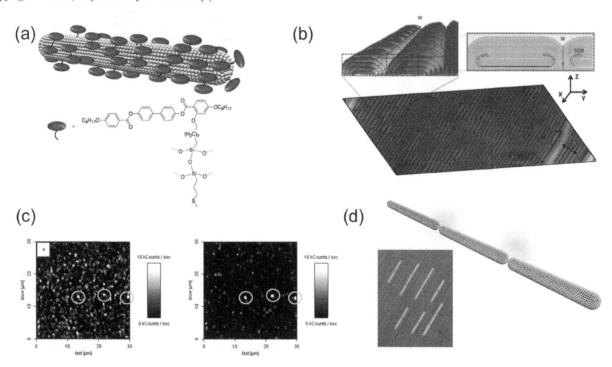

FIGURE 6.26 (a) Schematic representation of the side-on nematic liquid crystal-coated GNRs. (b) Oily streaks structure showing a few smectic layers, organized in parallel hemi-cylinders containing curvature walls (W), oriented along OX (top) and polarized optical microscope image of these oily streaks (bottom). (c) Two-photon luminescence of a sample of the GNRs in 8CB (laser excitation wavelength: 820 nm; average power: 500 μW). The same zone was raster scanned with a 600 nm beam waist (lateral extension, gray circle, shown as inset in image on the left) using an excitation polarization either along (left image) or perpendicular to (right image) the oily streaks. (d) Schematic showing the two-photon luminescence due to the highly enhanced fields created between end-to-end stacked GNRs and overlaid image of the expected orientation and organization of the GNRs in the patterned smectic-A defects. (Reproduced with permission from Ref. [63], Copyright © 2017, American Chemical Society.)

incident light polarization were observed, indicating that the orientation of the GNRs was in fact mostly parallel to the oily streaks, thus demonstrating a favorable trapping of the GNRs in the smectic dislocations. Here, strongly anisotropic field enhancement effects for GNR chains, particularly when the longitudinal mode is excited, were shown to lead to strongly enhanced luminescence (Figure 6.26 on opposite page) [63].

These are just a few examples that demonstrate a combination of the liquid crystalline state and the emissive properties of certain nanomaterials can be coupled in mutually beneficial ways. Another pathway to accomplish this is to combine the emissive properties of metal clusters directly with liquid crystal molecular motifs.

6.2.2 Clustomesogens

The relative bulkiness of transition metal clusters and the isometric shape of the most intensely investigated octahedral $M_6X_8L_6$ coordination clusters are, at a first glance, not favorable for the creation of rod- or disc-like shapes that commonly favor liquid crystal phase formation. Thus, the strategy involves the functionalization of these clusters with anisotropic organic liquid crystal promoters. As such, the class of so-called *clustomesogens* is now an integral part of liquid crystal nanoscience. As depicted in Figure 6.27, rod-shaped geometries favor self-organization in lamellar or nematic phases, while disc-shaped morphologies of such clusters generally lead to columnar phases [64, 65].

Clustomesogens were developed using covalent, supramolecular, or electrostatic interactions. The pro-mesogenic (i.e. biphenyl, cyanobiphenyl (CB), or cholesteryl) units were terminally attached via a spacer chain to the $[M_6X^i_8]^{m+}$ core.

In the covalent approach, the pronounced ionic–covalent character of the M-L^a bond allows the exchange of apical ligands by organic moieties without modifying the $[M_6X^i_8]^{m+}$ core structure. Supramolecular clustomesogens have been generated by host–guest interactions between crown ether derivatives containing LC promoters and alkali cations and electrostatic interactions between the crown ether complex and the anionic counterion. In the electrostatic approach,

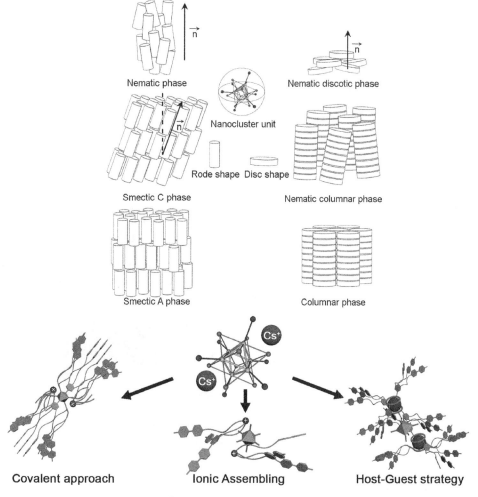

FIGURE 6.27 Covalent, ionic, and macrocyclic (i.e. crown ether) approach to clustomesogens forming a variety of liquid crystal phases, largely depending on the number (density) of pro-mesogenic ligands around the cluster core. (Adapted with permission from Ref. [65], Copyright © 2016, American Chemical Society.)

anisotropic building blocks consist in polyanionic inorganic moieties surrounded by several pro-mesogenic organic cations.

The complex supermolecular self-organization depends on several parameters. Control over the type of liquid crystal phase formed is based on the mesogenic units' density around the metal core. A low density of mesogenic units results in an overall rod-like shape, whereas at higher density a disc-like geometry is adopted (similar to morphologies formed by dendrimers). The more mesogens surround the metal core, the more the role of the organic part becomes dominant in the self-organization process. A low mesogen density gives rise to self-organization where the organic and inorganic parts play a role and where neighboring clusters are in close proximity. In this configuration, electrostatic repulsion between clusters and cores occurs, and the lamellar organization is lost in favor of a nematic phase.

As the emissive excited state of these transition metal cluster is mainly metal-centered, the coordination of the pro-mesogenic ligands around the cluster induces no significant change and no alteration in the emission properties of the cluster core: the emission maxima are constant and the high quantum yields are preserved. A list of some specific examples is provided in Table 6.2 on page 6-15.

Clustomesogens show several major characteristics: a deep red luminescence with quantum yields ranging from 0.1 to 0.7 depending on the nature of the ligands associated with the inorganic cluster, a self-organization at the nanometer scale as a result of molecular ordering events that produce supramolecular rods or discs, the presence of organic units ensuring their solubility in organic solvents, and birefringence. They are not only useful as bulk material but also dispersed as emissive dopant (or additives) in commercial liquid crystals. For example, the presence of cyanobiphenyl units within the organic matrix in some of the clustomesogens (see Table 6.2) provides a structural analogy to certain commercial liquid crystal singles such as those of Merck marketed under the name of nCB or nOCB, where n is the number of carbon atoms in the aliphatic chain (see, for example, 5CB in Figure 6.2), or mixtures such as E44 or E7. While these particular mixtures are no longer marketed for display applications, they are well known, regularly studied, and serve as model systems for many fundamental experiments.

Hence, especially the nematic clustomesogens can be used to obtain homogeneous mixtures with commercial nematic liquid crystals because two liquid crystalline compounds are miscible if they form the same phase (Figure 6.28).

An only marginal variation (decrease) in the clearing point (nematic to isotropic phase transition) on both heating and cooling with increasing concentration of the clustomesogens indicates that as dopants these clustomesogens are essentially fully miscible in the nematic host matrix. In this way, integration of clustomesogens within a host liquid crystal matrix provides access to phosphorescent hybrid materials possessing a nematic phase over a wide temperature range including ambient temperature that could also be used for electro-optic switching or display devices (Figure 6.29) [68].

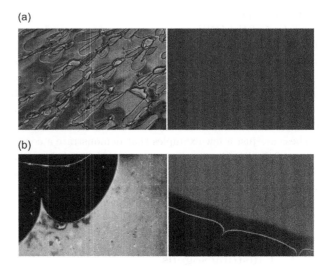

FIGURE 6.28 Photomicrographs of mixtures of clustomesogens in nematic liquid crystal hosts: (a) using the covalent approach – E44/TBA$_2$[Mo$_6$Br$_8$L$_6$] (9:1) under white light and crossed polarizers (left) and UV irradiation (right) and (b) using the ionic approach – E44/Trisel$_2$[Mo$_6$Br$_{14}$] (9:1) under white light and crossed polarizers (left) and UV irradiation (right). (Reproduced with permission from Refs. [66, 69], Copyright © 2014 Wiley-VCH Verlag GmbH & Co. KGaA, Weinheim; Copyright © 2015, Royal Society of Chemistry.)

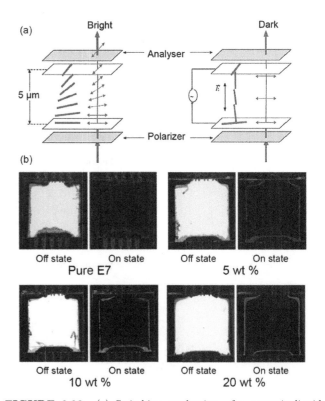

FIGURE 6.29 (a) Switching mechanism of a nematic liquid crystal in a conventional twisted nematic cell geometry. (b) Photographs showing the electrically controlled liquid crystal orientation in twisted cells containing E7/Trisel$_2$[Mo$_6$I$_8$(OCOC$_3$F$_7$)]$_6$ mixtures at various concentrations. (Reproduced with permission from Ref. [68], Copyright © 2015, Wiley-VCH Verlag GmbH & Co. KGaA, Weinheim.)

6.3 Luminescent Nanomaterials as Additives

6.3.1 Upconverting Nanomaterials in Liquid Crystals

UCNPs are starting to replace the use of small photo-responsive organic molecules such as azobenzenes or diarylethene derivatives in host–guest liquid crystal systems. UCNPs excited in the NIR do not require prolonged exposure to the high-energy light thereby decreasing the risk of photodegradation inherent to their organic counterparts [70,71].

In several studies, UCNPs were doped into LC hosts, serving as a controller of LC alignment. Li and co-workers [72] took advantage of low-energy NIR light to reversibly change the helical pitch of a photoresponsive, self-assembled cholesteric liquid crystal containing upconverting β-NaGdF$_4$:TmYb nanoparticles. The net result was a controllable as well as reversible change in the refractive index of the liquid crystal phase due to the isomerization of the azobenzene mesogens from their *trans*-isomers to their *cis*-isomers (Figure 6.30).

Another impressive example is the system developed by Yu and co-workers [73], where NaYF4:Yb, Tm upconversion nanophosphors were incorporated into an azotolane-containing cross-linked liquid crystal polymer

(CLCP) film. The resulting composite film generated fast bending upon exposure to continuous-wave near-IR light at 980 nm (Figure 6.31). This process occurs because the

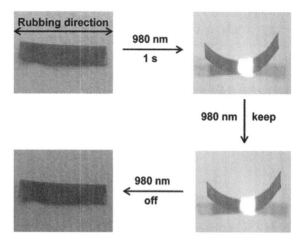

FIGURE 6.31 Photographs of the azotolane CLCP/UCNP composite film bending toward the light source and along the alignment direction of the mesogens, remaining bent in response to the continuous-wave NIR irradiation at 980 nm (power density = 15 W cm^{-2}), and becoming flat after light source was removed (film size: 8 mm × 2 mm × 20 μm). (Reproduced with permission from Ref. [73], Copyright © 2011, American Chemical Society.)

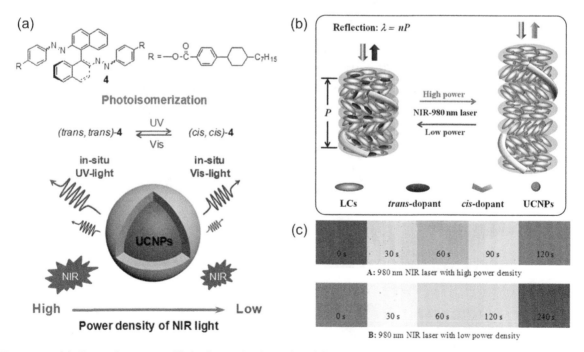

FIGURE 6.30 (a) Chemical structure of light-driven chiral switch and description of photoisomerization, triggered in a remote control process using in situ UV light generated under the NIR and in situ visible light generated under the same wavelength NIR light but with low excitation power density when the NaGdF$_4$:TmYb UCNPs were irradiated by a 980 nm NIR laser. (b) Schematic showing the reversible tuning of self-organized cholesteric superstructures using the chiral switch shown in (a) and UCNPs upon irradiation of NIR laser at different power densities. (c) Reflection of the N*-LC doped with 3 wt.% of the chiral molecular switch and 1.5 wt.% UCNPs (10-μm-thick planar cell, room temperature, polarized reflective mode microscope): **A**, upon irradiation with 980 nm NIR laser at high power density (2 W mm^{-2}) and **B**, irradiation with 980 nm NIR laser at low power density (0.15 W mm^{-2}). (Reproduced with permission from Ref. [72], Copyright © 2014, American Chemical Society.)

upconversion luminescence of the nanophosphors leads to *trans-cis* photoisomerization of the azotolane units and a concomitant alignment change of the mesogens triggered by a judicious choice of the photochromic units in CLCPs and UCNPs.

Guo and co-workers demonstrated the use of UCNPs. To create code patterns that could be used for anti-counterfeiting, a strategy was developed to modulate the luminescence intensity of UCNPs in liquid crystal network composite films featuring luminescence patterns. Variation of the luminescence intensity was realized by a change of the liquid crystal molecular orientation upon applying an electric field. A significant contrast of the luminescence intensity between the scattering state and the homeotropic

state thus permitted the fabrication of both micro- as well as macro-sized upconversion luminescence patterns (Figure 6.32) [74].

6.3.2 Quantum and Carbon Dots in Liquid Crystals

By far the most significant body of work in this burgeoning field is focusing on dispersions of QDs and carbon dots (C-dots) in nematic, chiral nematic, and ferroelectric liquid crystal phases (based on the chiral smectic-C phase, SmC*). This prominence of studies in this sub-filed arises from the predominant use of these particular phases in liquid crystal display devices similar to the aforementioned experiments on

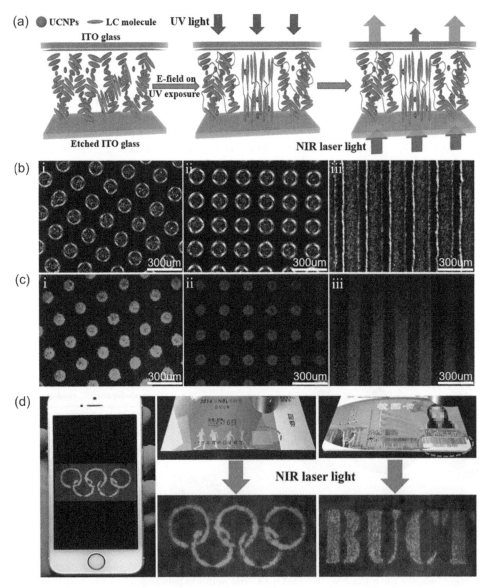

FIGURE 6.32 (a) Schematic models of upconverting luminescent patterning fabrication; (b) polarizing optical microscopy images (crossed polarizers); and (c) luminescence images of UCNP/LCN films with: (i) dot pattern, (ii) dot pattern, and (iii) fringe pattern. (d) Examples such as: patterned image identification with an Apple iPhone 5 and a 20× objective of interlocked Olympics rings; and letters of a two-colored luminescent patterns. (Reproduced with permission from Ref. [74], Copyright © 2017, Wiley-VCB Verlag GmbH & Co. KGaA, Weinheim.)

clustomesogens as additives in nematic phases. Several recent review articles summarized the state-of-the-art in this area [54, 75–77], which is why we will only highlight the most important concepts. As for other surface-functionalized nanomaterials, emissive nanomaterials generally are used to affect either the electro-optic response or the alignment (anchoring) of liquid crystal molecules or both in some specific cases [78–81]. As we will see quickly in discussing these examples, the ways nanomaterials affect the alignment of liquid crystals does determine their beneficial or detrimental contributions to several parameters affecting the response of the liquid crystal host phase to applied electric fields.

The attraction of using emissive nanomaterials stems particularly from the fact that their distribution or aggregation in a given liquid crystal medium can be traced by fluorescence confocal microscopy [82,83]. Hirst and co-workers demonstrated what is now commonly assumed, which is that nanoparticles are better soluble in the isotropic liquid than in the ordered liquid crystalline (in this particular case nematic) phase. Moreover, the fully formed N-phase produced three-dimensional QD assemblies that are situated at the defect points in the LC volume (Figure 6.33). Small-angle X-ray scattering data confirmed this model [84].

Martin Urbanski summarized specifically how the dispersibility of nanoparticles (emissive or not) affects the electro-optic response of nematic liquid crystals [85]. Nanoparticles or QDs with a high dispersibility in the nematic liquid crystal host are homogeneously dispersed in the LC phase and their influence on the alignment layers is negligible. Contrary, strong surface coverage, i.e. the N-LC expelled initially dispersed nanoparticle inclusions, causes

homeotropic alignment, which constitutes the foundation for unusual textural effects or reverses switching phenomena as described by our group. An intermediate degree of NP surface coverage of the two confining substrates leads to an increase of pretilt angle combined with apparent reduced dielectric anisotropy $\Delta\varepsilon$ and perceived lower threshold voltages V_{th} compared to the neat nematic LC host (Figure 6.34) [86].

The scenario depicted on the left in Figure 6.34 can also be used to create emissive patterned nanoparticle alignment layers. For example, Sharma and co-workers demonstrated the use of emissive carbon dots (C-dots) as patterned alignment layers for N-LCs, which might pave the way for a combination of backlighting and simultaneous alignment layer (Figure 6.35) [87]. If combined with shape anisotropy, emissive nanorods with surface chemistry inducing homeotropic or planar anchoring could provide polarized emissive alignment layers that would eliminate the various layers used in LCDs to date.

Finally, there are also numerous papers that describe the use of emissive nanomaterials for ferroelectric liquid crystal devices based on the chiral smectic-C phase (SmC*) [88]. A review article by Kumar and co-workers summarizes some of the key findings in this sub-area of research [89].

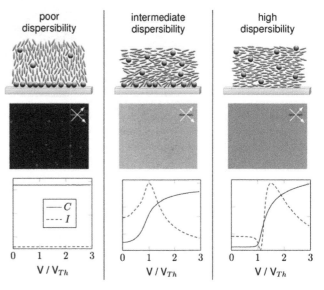

FIGURE 6.34 Illustration of three possible degrees of surface interactions of nanoparticles or quantum dots with the alignment layers: (a) no surface coverage of particles, therefore no influence on the initial alignment layers, (b) medium surface coverage that influences the initial tilt angle and anchoring strength. The determination of bulk properties via Fréedericksz transition is possible, but requires additional numerical simulations, (c) strong surface coverage can induce homeotropic alignment. For materials with positive dielectric anisotropy no Fréedericksz-like switching is possible, but electroconvection could be observed. (Reproduced with permission from Ref. [86], Copyright © 2014, Wiley-VCH Verlag GmbH & Co. KGaA, Weinheim.)

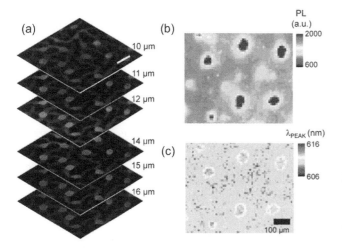

FIGURE 6.33 (a) Fluorescence confocal microscopy images of the QD-doped N-LC sample at the depths indicated. (b) and (c) Spatially resolved photoluminescence maps of the sample: (b) large-scale scan showing QD emission intensity and (c) peak emission wavelength (all figures have identical scale bar). (Reproduced with permission from Ref. [84], Copyright © 2013, Royal Society of Chemistry.)

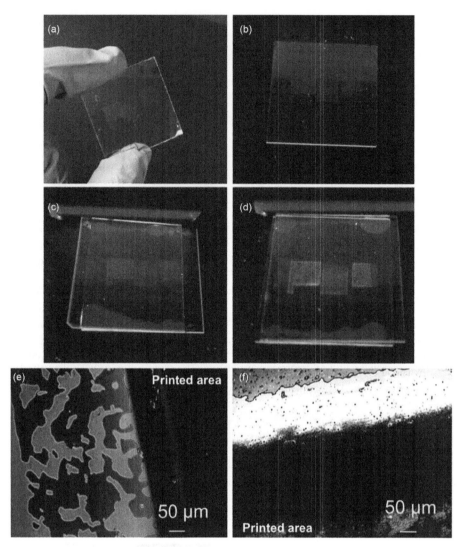

FIGURE 6.35 (a) C-dots printed on 30° SiO$_x$ after 1 h in ambient light and (b) under UV light ($\lambda_{exc.} = 365$ nm). (c) C-dots printed in single pass and (d) in three passes showing the modulation of emission intensity with increasing C-dot layers printed on the substrate. (e and f) Polarized optical photomicrographs of textures and observed alignment patterns (90° crossed polarizers). (Reproduced with permission from Ref. [87], Copyright © 2016, Taylor & Francis.)

6.4 Concluding Remarks

The combination of liquid crystals as a state of matter and a concept of light emission after excitation, each discovered in 1888, have come a long way and now form the basis of some of the most advanced liquid crystal displays on the market as well as research at the frontiers of soft condensed matter physics and chemistry. This chapter gave a brief overview of some of the key characteristics of each involved class of materials, liquid crystal phases, and a selection of emissive nanomaterials. Emissive nanomaterials can nowadays be fabricated in a wide range of shapes and sizes, and we expect research in this direction to provide further advances and new applications. Upconverting nanomaterials as well as emissive nanoclusters will play key roles in this research in the near future, and current research most likely scratches just the surface of what is possible. The wealth of data on metal nanoparticles in the various

liquid crystal phases will find new stimulus and find new insights from the use of emissive metal nanoclusters, where research in combination with liquid crystals is just beginning to emerge. Aside from their use in the omnipresent displays, we anyway all look at for too long every day, this particular materials combination will open up new areas of applications in sensors as well as other responsive devices such as security tags, anti-counterfeit measures, and smart coatings.

References

1. Mitov M. (2014) Liquid-crystal science from 1888 to 1922: Building a revolution. *ChemPhysChem* 15(7):1245–1250.
2. de Gennes P.-G. (2003) *Simple Views on Condensed Matter* (World Scientific, River Edge, NJ) 3rd Ed.

3. Tschierske C. (1998) Non-conventional liquid crystals—the importance of micro-segregation for self-organisation. *J. Mater. Chem.* 8(7): 1485–1508.

4. Reddy R.A. & Tschierske C. (2006) Bent-core liquid crystals: Polar order, superstructural chirality and spontaneous desymmetrisation in soft matter systems. *J. Mater. Chem.* 16(10):907–961.

5. Friedel G. (1922) Les états mésomorphes de la matière. *Ann. Phys.* 18:273.

6. Collings P.J. & Hird M. (1997) *Introduction to Liquid Crystals Chemistry and Physics* (Taylor & Francis, London; Bristol, PA).

7. Demus D. & Richter L. (1978) *Textures of Liquid Crystals* (Deutscher Verlag für Grundstoffindustrie, Leipzig) 1st Ed.

8. Dierking I. (2003) *Textures of Liquid Crystals* (Wiley-VCH, Weinheim).

9. Lavrentovich O.D. (2001) *Defects in Liquid Crystals: Computer Simulations, Theory, and Experiments* (Kluwer Academic Publishers, Dordrecht; Boston, MA).

10. Chandrasekhar S. (1992) *Liquid Crystals* (Cambridge University Press, Cambridge England; New York) 2nd Ed.

11. Ishihara S. (2005) How far has the molecular alignment of liquid crystals been elucidated? *J. Disp. Technol.* 1(1):30–40.

12. Liu Y.N. & Wang W. (2018) The oseen-frank limit of onsager's molecular theory for liquid crystals. *Arch. Ration. Mech. An.* 227(3):1061–1090.

13. Majumdar A. & Zarnescu A. (2010) Landau-de gennes theory of nematic liquid crystals: The oseen-frank limit and beyond. *Arch. Ration. Mech. An.* 196(1):227–280.

14. Alouges F. & Ghidaglia J.M. (1997) Minimizing oseen-frank energy for nematic liquid crystals: Algorithms and numerical results. *Ann. I. H. Poincare-Phy.* 66(4):411–447.

15. Queisser H.J. (1981) Luminescence, Review and Survey. *J. Lumin.* 24–25:3–10.

16. Leverenz H.W. (1968) *An Introduction to Luminescence of Solids* (Dover Publications, New York).

17. Pope M., Swenberg C.E., & Pope M. (1999) *Electronic Processes in Organic Crystals and Polymers* (Oxford University Press, New York) 2nd Ed.

18. Baldo M.A. et al. (1998) Highly efficient phosphorescent emission from organic electroluminescent devices. *Nature* 395(6698):151–154.

19. Valeur B. & Berberan-Santos M.N. (2013) *Molecular Fluorescence: Principles and Applications* (Wiley, Weinheim).

20. Yang W.F., Li X.Y., Chi D.Z., Zhang H.J., & Liu X.G. (2014) Lanthanide-doped upconversion materials: Emerging applications for photovoltaics and photocatalysis. *Nanotechnology* 25(48): 482001.

21. Zhou J., Liu Q., Feng W., Sun Y., & Li F.Y. (2015) Upconversion Luminescent materials: Advances and applications. *Chem. Rev.* 115(1):395–465.

22. Deng H.H. et al. (2017) Fabrication of water-soluble, green-emitting gold nanoclusters with a 65% photoluminescence quantum yield via host-guest recognition. *Chem. Mater.* 29(3):1362–1369.

23. Londono-Larrea P., Vanegas J.P., Cuaran-Acosta D., Zaballos-Garcia E., & Perez-Prieto J. (2017) Water-soluble naked gold nanoclusters are not luminescent. *Chem. Eur. J.* 23(34):8137–8141.

24. Zheng J., Petty J.T., & Dickson R.M. (2003) High quantum yield blue emission from water-soluble Au-8 nanodots. *J. Am. Chem. Soc.* 125(26):7780–7781.

25. Lu Y.Z. & Chen W. (2012) Sub-nanometre sized metal clusters: From synthetic challenges to the unique property discoveries. *Chem. Soc. Rev.* 41(9): 3594–3623.

26. Link S. et al. (2002) Visible to infrared luminescence from a 28-atom gold cluster. *J. Phys. Chem. B* 106(13):3410–3415.

27. Huang T. & Murray R.W. (2001) Visible luminescence of water-soluble monolayer-protected gold clusters. *J. Phys. Chem. B* 105(50):12498–12502.

28. Zhu M., Aikens C.M., Hollander F.J., Schatz G.C., & Jin R. (2008) Correlating the crystal structure of A thiol-protected Au-25 cluster and optical properties. *J. Am. Chem. Soc.* 130(18):5883–5885.

29. Wu Z.K. & Jin R.C. (2010) On the ligand's role in the fluorescence of gold nanoclusters. *Nano Lett.* 10(7):2568–2573.

30. Judd B.R. (1962) Optical absorption intensities of rare-earth ions. *Phys. Rev.* 127(3):750–761.

31. Dieke G.H., Crosswhite H.M., & Crosswhite H. (1968) *Spectra and Energy Levels of Rare Earth Ions in Crystals* (Interscience Publishers, New York).

32. Wang F. & Liu X.G. (2008) Upconversion multi-color fine-tuning: Visible to near-infrared emission from lanthanide-doped NaYF4 nanoparticles. *J. Am. Chem. Soc.* 130(17):5642–5643.

33. Mai H.X. et al. (2006) High-quality sodium rare-earth fluoride nanocrystals: Controlled synthesis and optical properties. *J. Am. Chem. Soc.* 128(19): 6426–6436.

34. Wang L.Y. & Li Y.D. (2006) $Na(Y_{1.5}Na_{0.5})F_6$ single-crystal nanorods as multicolor luminescent materials. *Nano Lett.* 6(8):1645–1649.

35. Wang X., Zhuang J., Peng Q., & Li Y.D. (2006) Hydrothermal synthesis of rare-earth fluoride nanocrystals. *Inorg. Chem.* 45(17):6661–6665.

36. Liu Y.S. et al. (2010) A strategy to achieve efficient dual-mode luminescence of Eu^{3+} in lanthanides doped multifunctional $NaGdF_4$ nanocrystals. *Adv. Mater.* 22(30):3266–3271.

37. Wei Y., Lu F.Q., Zhang X.R., & Chen D.P. (2007) Synthesis and characterization of efficient near-infrared upconversion Yb and Tm codoped

NaYF$_4$ nanocrystal reporter. *J. Alloy. Compd.* 427(1–2):333–340.

38. Talapin D.V., Rogach A.L., Kornowski A., Haase M., & Weller H. (2001) Highly luminescent monodisperse CdSe and CdSe/ZnS nanocrystals synthesized in a hexadecylamine-trioctylphosphine oxide-trioctylphospine mixture. *Nano Lett.* 1(4):207–211.

39. Franzl T. et al. (2007) CdSe : Te nanocrystals: Band-edge versus Te-related emission. *J. Phys. Chem. C* 111(7):2974–2979.

40. Brus L.E. (1984) Electron electron and electron-hole interactions in small semiconductor crystallites– the size dependence of the lowest excited electronic state. *J. Chem. Phys.* 80(9):4403–4409.

41. Brus L. (1986) Electronic wave-functions in semiconductor clusters–experiment and theory. *J. Phys. Chem.* 90(12):2555–2560.

42. Borchert H. et al. (2003) High resolution photoemission study of CdSe and CdSe/ZnS core-shell nanocrystals. *J. Chem. Phys.* 119(3):1800–1807.

43. Mandal A., Nakayama J., Tamai N., Biju V., & Isikawa M. (2007) Optical and dynamic properties of water-soluble highly luminescent CdTe quantum dots. *J. Phys. Chem. B* 111(44):12765–12771.

44. Mansur H.S. (2010) Quantum dots and nanocomposites. *Wires Nanomed. Nanobio.* 2(2):113–129.

45. Efros A.L. & Rosen M. (1997) Random telegraph signal in the photoluminescence intensity of a single quantum dot. *Phys. Rev. Lett.* 78(6):1110–1113.

46. Carrillo-Carrion C., Cardenas S., Simonet B.M., & Valcarcel M. (2009) Quantum dots luminescence enhancement due to illumination with UV/Vis light. *Chem. Commun.* (35):5214–5226.

47. Chaudhuri R.G. & Paria S. (2012) Core/shell nanoparticles: Classes, properties, synthesis mechanisms, characterization, and applications. *Chem. Rev.* 112(4):2373–2433.

48. Dabbousi B.O. et al. (1997) (CdSe)ZnS core-shell quantum dots: Synthesis and characterization of a size series of highly luminescent nanocrystallites. *J. Phys. Chem. B* 101(46):9463–9475.

49. Cotton F.A. (1964) Metal atom clusters in oxide systems. *Inorg. Chem.* 3(9):1217–1220.

50. Prévôt M. (2014) Démonstrateurs des potentialités applicatives des clustomésogènes. Université de Rennes 1, Rennes, France.

51. Cordier S. et al. (2005) Elaboration of hybrid nanocluster materials by solution chemistry. *Prog. Solid State Chem.* 33(2–4):81–88.

52. Schäfer H. & Schnering H.G. (1964) Metall-metall-bindungen bei niederen halogeniden, oxyden und oxydhalogeniden schwerer übergangsmetalle thermochemische und strukturelle prinzipien. *Angew. Chem.* 76(20):833–849.

53. Costuas K. et al. (2015) Combined theoretical and time-resolved photoluminescence investigations of $[(Mo_6Br_8Br_6{}^a)\text{-}Br\text{-}^i]^{2-}$ metal cluster units: evidence

of dual emission. *Phys. Chem. Chem. Phys.* 17(43):28574–28585.

54. Mirzaei J., Reznikov M., & Hegmann T. (2012) Quantum dots as liquid crystal dopants. *J. Mater. Chem.* 22(42):22350–22365.

55. Stamatoiu O., Mirzaei J., Feng X., & Hegmann T. (2012) Nanoparticles in liquid crystals and liquid crystalline nanoparticles. *Top. Curr. Chem.* 318:331–393.

56. Kim H.J., Shin M.H., Lee J.Y., Kim J.H., & Kim Y.J. (2017) Realization of 95% of the Rec. 2020 color gamut in a highly efficient LCD using a patterned quantum dot film. *Opt. Express* 25(10):10724–10734.

57. Aubert T. et al. (2015) Large-scale and electroswitchable polarized emission from semiconductor nanorods aligned in polymeric nanofibers. *ACS Photonics* 2(5):583–588.

58. Cunningham P.D. et al. (2016) Assessment of anisotropic semiconductor nanorod and nanoplatelet heterostructures with polarized emission for liquid crystal display technology. *ACS Nano* 10(6): 5769–5781.

59. Sanchez C., Arribart H., & Guille M.M.G. (2005) Biomimetism and bioinspiration as tools for the design of innovative materials and systems. *Nat. Mater.* 4(4):277–288.

60. Li L.S. & Alivisatos A.P. (2003) Semiconductor nanorod liquid crystals and their assembly on a substrate. *Adv. Mater.* 15(5):408–411.

61. Li L.S., Walda J., Manna L., & Alivisatos A.P. (2002) Semiconductor nanorod liquid crystals. *Nano Lett.* 2(6):557–560.

62. Xie Y. et al. (2018) Liquid crystal self-assembly of upconversion nanorods enriched by depletion forces for mesostructured material preparation. *Nanoscale* 10(9):4218–4227.

63. Rozic B. et al. (2017) Oriented gold nanorods and gold nanorod chains within smectic liquid crystal topological defects. *ACS Nano* 11(7):6728–6738.

64. Gandubert A. et al. (2018) Tailoring the self-assembling abilities of functional hybrid nanomaterials: from rod-like to disk-like clustomesogens based on a luminescent $\{Mo_6Br_8\}^{4+}$ inorganic cluster core. *J. Mater. Chem. C* 6(10):2556–2564.

65. Molard Y. (2016) Clustomesogens: Liquid crystalline hybrid nanomaterials containing functional metal nanoclusters. *Acc. Chem. Res.* 49(8):1514–1523.

66. Prévôt M. et al. (2015) Electroswitchable red-NIR luminescence of ionic clustomesogen containing nematic liquid crystalline devices. *J. Mater. Chem. C* 3(20):5152–5161.

67. Nayak S.K. et al. (2016) Phosphorescent columnar hybrid materials containing polyionic inorganic nanoclusters. *Chem. Commun.* 52(15): 3127–3130.

68. Prévôt M. et al. (2015) Design and integration in electro-optic devices of highly efficient and robust red-NIR phosphorescent nematic hybrid liquid

crystals containing $[Mo_6I_8(OCOC_nF_{2n+1})_6]^{2-}$ (n = 1, 2, 3) Nanoclusters. *Adv. Funct. Mater.* 25(31): 4966–4975.

69. Cortes M.A. et al. (2014) Thermotropic luminescent clustomesogen showing a nematic phase: A combination of experimental and molecular simulation studies. *Chem. Eur. J.* 20(28): 8561–8565.

70. Wu T. & Branda N.R. (2016) Using low-energy near infrared light and upconverting nanoparticles to trigger photoreactions within supramolecular assemblies. *Chem. Commun.* 52(56):8636–8644.

71. Cheng Z.Y. & Lin J. (2015) Synthesis and application of nanohybrids based on upconverting nanoparticles and polymers. *Macromol. Rapid Comm.* 36(9): 790–827.

72. Wang L. et al. (2014) Reversible near-infrared light directed reflection in a self-organized helical superstructure loaded with upconversion nanoparticles. *J. Am. Chem. Soc.* 136(12):4480–4483.

73. Wu W. et al. (2011) NIR-light-induced deformation of cross-linked liquid-crystal polymers using upconversion nanophosphors. *J. Am. Chem. Soc.* 133(40):15810–15813.

74. Ye S.M. et al. (2017) Modulated visible light upconversion for luminescence patterns in liquid crystal polymer networks loaded with upconverting nanoparticles. *Adv. Opt. Mater.* 5(4):1600956.

75. Sharma A., Urbanski M., Mori T., Kitzerow H.-S., & Hegmann T. (2016) Metallic and semiconductor nanoparticles in LCs. In: *Particles in Liquid Crystals*, Word Scientific (Ch. 14):495–533.

76. Hegmann T., Qi H., & Marx V.M. (2007) Nanoparticles in liquid crystals: Synthesis, self-assembly, defect formation and potential applications. *J. Inorg. Organomet. Polym. Mater.* 17(3):483–508.

77. Qi H. & Hegmann T. (2008) Impact of nanoscale particles and carbon nanotubes on current and future generations of liquid crystal displays. *J. Mater. Chem.* 18(28):3288–3294.

78. Qi H. & Hegmann T. (2009) Multiple alignment modes for nematic liquid crystals doped with alkylthiol-capped gold nanoparticles. *ACS Appl. Mater. Interf.* 1(8):1731–1738.

79. Qi H., Kinkead B., & Hegmann T. (2008) Unprecedented dual alignment mode and Freedericksz transition in planar nematic liquid crystal cells doped with gold nanoclusters. *Adv. Funct. Mater.* 18(2): 212–221.

80. Urbanski M., Kinkead B., Qi H., Hegmann T., & Kitzerow H.-S. (2010) Electroconvection in nematic liquid crystals via nanoparticle doping. *Nanoscale* 2(7):1118–1121.

81. Reznikov M., Sharma A., & Hegmann T. (2014) Ink-jet printed nanoparticle alignment layers: Easy design and fabrication of patterned alignment layers for nematic liquid crystals. *Part. Part. Syst. Charact.* 31(2):257–265.

82. Urbanski M., Kinkead B., Hegmann T., & Kitzerow H.-S. (2010) Director field of birefringent stripes in liquid crystal/nanoparticle dispersions. *Liq. Cryst.* 37(9):1151–1156.

83. Urbanski M. et al. (2016) Chemically and thermally stable, emissive carbon dots as viable alternatives to semiconductor quantum dots for emissive nematic liquid crystal-nanoparticle mixtures with lower threshold voltage. *Liq. Cryst.* 43(2):183–194.

84. Rodarte A.L., Pandolfi R.J., Ghosh S., & Hirst L.S. (2013) Quantum dot/liquid crystal composite materials: self-assembly driven by liquid crystal phase transition templating. *J. Mater. Chem. C* 1(35):5527–5532.

85. Urbanski M. (2015) On the impact of nanoparticle doping on the electro-optic response of nematic hosts. *Liq. Cryst. Today* 24(4):102–115.

86. Urbanski M., Mirzaei J., Hegmann T., & Kitzerow H.-S. (2014) Nanoparticle doping in nematic liquid crystals: Distinction between surface and bulk effects by numerical simulations. *ChemPhysChem* 15(7): 1395–1404.

87. Sharma A., Hofmann D., & Hegmann T. (2016) Patterned alignment of nematic liquid crystals generated by inkjet printing of gold nanoparticles and emissive carbon dots on both flexible polymer and rigid glass substrates. *Liq. Cryst.* 43(6): 828–838.

88. Shukla R.K. et al. (2015) Electro-optic and dielectric properties of a ferroelectric liquid crystal doped with chemically and thermally stable emissive carbon dots. *RSC Adv.* 5(43):34491–34496.

89. Singh G., Fisch M., & Kumar S. (2016) Emissivity and electrooptical properties of semiconducting quantum dots/rods and liquid crystal composites: A review. *Rep. Prog. Phys.* 79(5):056502.

Nanoscale Alloys and Intermetallics: Recent Progresses in Catalysis

Arnab Samanta and
Subhra Jana
S. N. Bose National Centre for Basic Sciences

7.1 Introduction

Nanoscience has now become a common term not only in scientific research work but also in our daily life. It is important to know the origin of the prefix *nano*, a Greek word meaning "dwarf", to understand nanoscience and nanotechnology. Nanoscience is not only limited to chemical and physical knowledge but also demands biological understanding, since biology and biochemistry also have very important role for nanoscientific advances.[1,2] One nanometer (nm) is simply one billionth of meter, i.e., $1 \text{ nm} = 1 \times 10^{-9}$ m. Nanoparticles (NPs) consist of a few hundred up to several thousand atoms are more appealing to a large community of researchers due to their large surface-to-volume ratio and surface atomic activity. About 160 years ago, Michael Faraday's legendary discovery illustrated that the scope and use of metals could be further expanded when they processed in the form of nanomaterials. In a finely divided state, as in the form of nanocrystals, metals exhibit intriguing electronic, optical, magnetic and catalytic properties. Narrow size distribution, stabilization and unique property compared to bulk make them more attractive, resulting in the development of several strategies for improving the synthesis and properties of NPs. The properties of NPs depend on their external structure (e.g., size, shape and uniformity) and can be applied into a number of technologically valuable applications ranging from catalysis,[3-5] plasmonics,[6] biosensing,[7] transportation,[8] optics,[9] magnetic application[10] to data storage.[11]

Alloys are basically a solid solution of two or more metals or a physical mixture of metals, forming a single solid phase, whereas intermetallic compounds are atomically ordered alloys having well-defined compositions and crystal structures that are different from their constituent elements (see Figure 7.1). It is therefore necessary to precise control over their particle size, shape and composition at the atomic and macroscopic dimensions to introduce a unique effect on their activity. Nanoscale alloys and intermetallics composed of two or more metallic elements have potential applications in the field of catalysis, biosensing, battery storage, data storage, etc. owing to their high activity and robust stability. Their properties can be tuned by varying the atomic ordering, composition and size. They also find structural and non-structural applications as corrosion resistant materials, high-temperature gas turbine hardware, heat treatment fixtures, magnetic materials and hydrogen storage materials.[12] Unlike disordered alloys, ordered intermetallics provide a uniform active sites on the same surface plane because of the better control over structure and electronic effects owing to their compositional and positional order.[13,14]

Intermetallics and alloys have come out as a class of novel materials with great opportunities towards the development of low-cost and high-performance industrial catalysts.[15-18] They are traditionally synthesized using metallurgical techniques, which required high-temperature reduction, followed by annealing for long periods of time.[19,20] However, based on these strategies, it is really challenging to get nanocrystalline intermetallics or alloys with high surface area, which is a significant criterion to be used them in the field of energy, environment and catalysis. Again, heat treatment at high temperature produces an undesired particle growth that may contribute to the decrease in metal mass activity in alloys. Apart from these issues, the high annealing temperature eliminates the stabilizing ligands from the surface, leaving aside a carbide coating due to the thermal

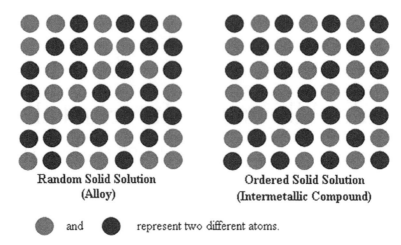

FIGURE 7.1   Schematic presentation of the crystal structure of alloys and intermetallics.

decomposition of the capped organic ligands that impede their redispersion in solution and also significantly enhance the crystallite size through sintering.[21,22]

In contrast to high-temperature synthetic routes, low-temperature solution chemistry approaches have been considered as one of the most attractive strategies to produce size, shape and composition tunable nanoscale alloys and intermetallics, including a number of new metastable phases.[23-26] Based on these routes, alloys were synthesized from the reaction of molecular precursors, e.g., metal salts, metal–organic compounds and metal–ligand complexes in a solution medium by chemical reduction of corresponding metal salts using suitable reducing agents or heated to a desired temperature in presence of stabilizing agents to regulate their growth and also to induce a repulsive force opposed to the van der Waals force, resulting in the formation of stable alloys in solution. The commonly used stabilization procedure for NPs involves electrostatic, steric and electrosteric stabilization using ligands, surfactants, soft templates, coordinating solvents, etc.[27,28] Consequently, NPs collected from their growing reaction mixture are individually distinguishable and free-standing, having single or polycrystalline domain with the desired geometry and chemical composition that imparts their characteristic properties. Furthermore, NPs can also possess a monolayer of firmly surface-bound capping materials that not only assist solubility and stability to the particles but also govern their interactions with the environment.[28] Access to these solution chemistry routes that facilitate the preparation of alloys and intermetallics unveils their potential applications due to enhanced or unique properties typically observed in the nano level.

The catalytic, electronic and optical properties of alloys are different from their corresponding monometallic counterpart due to the electronic effects[29,30] and ensemble effects[31,32] of two metals. Alloys and intermetallics have withdrawn enormous interest as a catalyst as they bring forth excellent catalytic activity in a wide range of inorganic to organic reactions and may possess enhanced activity and/or selectivity compared to single-component metal NPs. Their catalytic activity is generally regulated by their particle size, shape, morphology, crystal lattice parameters, specific surface area and modification of the type of surface defects.[33,34] The application of colloidal suspension of alloys or intermetallics in catalysis is often limited by several drawbacks. Apart from the difficulty in separation of the product and recycling of catalyst particle, their properties and activity can be highly affected and reduced once they started agglomeration and became bulk like materials. All those limitations can be overcome through the immobilization of colloidal NPs onto the solid supports that induces an outstanding stability, efficiency and enormous improvement in the selectivity of the catalyst.

In this chapter, we have discussed the synthesis of alloys and intermetallics based on the solid-phase and solution-phase synthesis techniques. Among the several low-temperature solution-phase synthesis routes, we have highlighted wet chemical reduction and thermal decomposition procedures to synthesize alloys and intermetallics. We also focus on their enhanced catalytic activity and stability as catalysts towards different inorganic to organic reactions, including hydrogenation, oxidation and electrocatalysis, following which we demonstrate an outlook on this emerging field.

7.2   The Effect of Size, Shape and Composition

Nucleation is the primary stage of any crystallization process and gives rise to the formation of nuclei, which are the building blocks for crystal growth. According to the classical nucleation theory, a thermodynamic system tends to minimize its Gibbs free energy, which in turn maximizes the entropy of the whole system. Thus, nucleation, evolution of nuclei into seeds, followed by growth of seeds are the three crucial steps for the formation of nanocrystals (NCs) and their overall growth is controlled by the competition between a decrease in bulk energy (favors growth) and an increase in surface energy (favors dissolution) and finally direct the evolution of seeds into NCs. The nucleation stage for the growth of NCs with anisotropic shapes imparts a crucial role in determining the size and shape of the evolved NCs, since the morphology and the growth rate of the seeds are controlled by an interplay between thermodynamics and kinetics.[35] Thermodynamically, NCs will grow towards the facets having the lowest energy at equilibrium, whereas the formation dynamics can affect the shape of the formed NCs and is a kinetics driven process. Any metastable shape of NCs can be arrested by tuning the reaction conditions before the reaction reaches the equilibrium stage. It should be pointed out that both size and shape of a NC play a key role in determining the total change in Gibbs free energy. With decrease in size of a NC the total surface free energy increases and the disorder-to-order transition temperature also drastically reduced.[36] This pronounced effect was found in several bimetallic systems with the help of in situ heating inside a transmission electron microscope.[37-39] Again, specific surface free energy of the individual facets has prominent influence on the disorder-to-order transition temperature and therefore it should be taken into consideration. M. Chi et al. demonstrated the structural and compositional transitions in Pt_3Co alloys during in situ heating of an individual Pt_3Co NC and investigated with the help of in situ scanning transmission electron microscopy (see Figure 7.2). The evolution of the surface area of {111}, {110} and {100} facets was evaluated as a function of annealing conditions and it was found that the {111} facet became dominant when the ordered intermetallic phase was formed.[40]

Composition plays a crucial role in determining the formation of intermetallic compounds. Apart from alloys, formation of intermetallics is limited to a few bimetallic systems that lie in a state of thermodynamic minima. Generally, metal pairs were chosen to be formed intermetallics from the corresponding bulk phase diagrams. In addition to the bulk phase diagrams, atomic radius as well as electronegativity of the metals has also been taken into consideration for the prediction of probable intermetallics formation. It is obvious that with change in stoichiometry of the metals of a specified bimetallic system, local inhomogeneity and degree of atom disorder occur to some extent. Again, if the difference in surface free energy between two metals is too high, phase segregation may take place at the surface of the NCs, resulting in a slight deviation from the desired stoichiometry.[41,42] Generally, surface segregation will be favored in a system having increased surface-to-volume ratio, and it is mostly observed during the synthesis of nanoscale alloys and intermetallic of Pt–M, where M is typically a 3d transition metal.[43]

7.3 Synthesis of Alloys and Intermetallics

The most commonly used techniques for the synthesis of NPs are top-down (physical method) and bottom-up (chemical method) approaches. In top-down method, the dimensions of bulk materials were physically reduced until they reach to a structure with at least one dimension less than 100 nm. In contrary, bottom-up synthesis involves self-assembly of individual atoms or molecules into clusters or NPs by carefully controlled chemical reaction.[44,45] The bottom-up approach is more advantageous than the top-down, since it originates nanomaterials with less defects, more homogenous chemical composition and better short- and long-range ordering. Again, two most commonly used strategies for the synthesis of intermetallic NCs are thermal-annealing and wet chemistry approaches.[42] To synthesize size-dependent and morphology-induced intermetallic NPs, great efforts

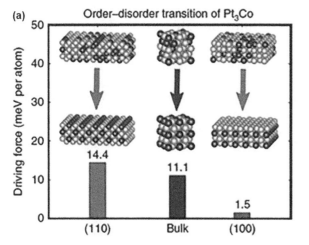
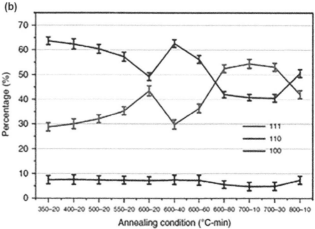

FIGURE 7.2 (a) The driving force to form ordered structures at different crystallographic facets and in the bulk determined from using first-principles density functional theory calculations, which show that the (110) facet has the highest driving force to form ordered structure. (b) The evolution of the concentrations of the surface facets, for example, –111″, –110″ and –100″, as a function of annealing condition; the concentrations were quantified by assuming a truncated cuboctahedron NP morphology. (Adapted with permission from Ref. [40].)

have been made over the past decades. Undoubtedly, with time the emergence and development of nanoscience and nanotechnology offer new opportunities in this area. In this chapter, we have discussed solid-phase and solution-phase synthesis techniques for the preparation of alloys and intermetallics.

7.3.1 Solid-Phase Synthesis

The key motivation of solid-phase synthetic approaches is to make materials with useful physical properties using several powerful and productive ways having unique advantages and disadvantages. Generally, alloys and intermetallics were synthesized using very high temperature techniques, including arc melting, radio frequency induction heating, high-energy ball milling, chemical vapor deposition, etc. In high temperature techniques, very high temperature is required to induce sufficient diffusion between the metals for making the alloys as the starting materials are solid. Recently, K.R. Ravi and coworkers have synthesized partially amorphous Mg–Zn–Ca alloy from their elemental powders using a high-energy planetary ball milling technique.[46] They synthesized an alloy with a composition of $Mg_{60}Zn_{35}Ca_5$ through elemental powder mixture of Mg, Zn and Ca metals. S. Umrao et al. have demonstrated a novel approach for preparation of $MoS_{2(1-x)}Se_{2x}$ alloys on a SiO_2/Si substrate without any surface modifications using low-pressure chemical vapor deposition with a furnace with two heating zones developed to obtain MoS_2 and $MoSe_2$.[47] J. Y. Lee and coworkers have reported the synthesis of Si–Ni alloy by arc melting followed by high-energy mechanical milling.[48] The alloy particles consist of an electrochemically active silicon phase with inactive phases, such as $NiSi_2$ and $NiSi$ distributed uniformly on the surface of the graphite. A. Onda et al. reported Ni–Sn intermetallic compounds having three different composition of Ni_3Sn, Ni_3Sn_2 and Ni_3Sn_4 by chemical vapor deposition (CVD) of $Sn(CH_3)4$ onto Ni/SiO_2.[49] PdGa intermetallic compound was prepared by melting the appropriate amounts of Pd and Ga under protective Ar atmosphere in a high frequency furnace and subsequent annealing of the obtained ingot in an evacuated quartz glass ampoule at high temperature for 170 h.[50] Based on the high-energy planetary ball milling method, amorphous $Ca_5Mg_{60+x}Zn_{35-x}$ (X = 0, 3 and 7) alloys have been synthesized through a liquid metallurgy route.[51] It

was found that with increasing reaction time Zn content in $Ca_5Mg_{60+x}Zn_{35-x}$ (X = 0, 3 and 7) alloy decreases and prolonged milling of these alloys produces the nucleation of $Mg_{102.08}Zn_{39.6}$. Similarly, monophasic $Sb_{1-x}Bi_x$ alloys were prepared by high-energy mechanical synthesis route, where Bi undergoes a complete and highly reversible alloying reaction with Sb and the later displays no electrochemical activity.[52] Zhang et al. reported the synthesis of octahedral Pt–Ni alloys on carbon support taking the advantages of both solid-state chemistry and wet chemistry synthetic approaches (Figure 7.3).[53] Pt–Ni alloy NPs are rhombic shape with straight edges and an average edge length of 5.8 ± 1.5 nm, suggesting the effectiveness of solid-state chemistry method to make uniform alloy particles.

Recently, C. L. Tracy et al. discussed hexagonal close-packed phase of the prototypical high-entropy alloy of CrMnFeCoNi through the arc melting of an equimolar mixture of the constituent metals Cr, Mn, Fe, Co and Ni under an Ar atmosphere, followed by drop casting of the resulting melt to produce an ingot of the high-entropy CrMnFeCoNi alloy.[54] The martensitic transformation of fcc to hcp structure started at 14 GPa and is due to the suppression of the local magnetic moments, which destabilized the initial fcc structure. High-entropy alloy of FeMnCoCrAl was reported by A. Marshal et al. using combinatorial sputtering and casting at room temperature.[55] Experimental and theoretical understanding indicates the formation of single phase and body-centered-cubic (bcc) solid with an Al concentration of ±6 atomic%. However, despite the several advantages of top-down approaches, there are several drawbacks that include NPs having wide size distribution with comparatively larger size and inconsistent physical and catalytic activity. This can be overcome using bottom-up approaches, since chemical methods are the most convenient ways to control the size, shape and surface properties of the NPs.

7.3.2 Solution Chemistry Synthesis

Alloy and intermetallic NPs synthesized through the bottom-up chemistry techniques give rise to the smallest nanostructures with specific size, well-defined surface composition and properties owing to the accumulation of the atoms and molecules via carefully controlled chemical reactions. It is important to note that methods that allow reactions to be carried out at lower temperature generally

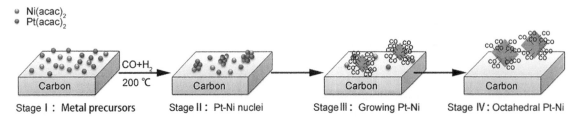

FIGURE 7.3 Schematic illustration of the formation of octahedral Pt−Ni alloys on C-support. (Adapted with permission from Ref. [53].)

produce new phases. Many approaches have been developed to make alloys and intermetallics by solution chemistry synthesis routes, such as wet chemical/photochemical reduction, thermal decomposition, co-precipitation, microemulsion, sol–gel, hydrothermal synthesis, microwave-assisted synthesis, etc. Among the solution-phase synthesis routes, we have focused mainly wet chemical reduction and thermal decomposition routes to synthesize alloys and intermetallics.

Reduction

Chemical reduction of metal salts is the simplest and the most commonly used synthetic route to produce smallest nanostructures. The synthetic approach is often similar to a "seed" model, where the beginnings are small but finally grow in complexity and completeness. There are several reports on the synthesis of alloys and intermetallics in the literature obtained by reduction route. Since the legendary discover of Michael Faraday, metal NPs and corresponding alloys have been synthesized by chemical reduction of corresponding metal salts with suitable reducing agents, such as borohydride, sodium citrates, aldehydes, alcohols, diketones, sugars, etc. Organic-phase reduction reaction has also been carried out using organic soluble reducing agents, such as superhydride, alkyl ammonium borohydride, oleylamine, and n-butyllithium, etc.

A facile approach has been demonstrated by Zhu et al. to synthesize well-defined three-dimensional Pt–M (M = Fe, Co, Cu, Ni) bimetallic alloy nanosponges in large scale by wet chemical reduction route using ethanolic solution of $NaBH_4$ in the presence of Al, which is crucial for the synthesis of the 3D porous PtM nanostructures.[56] Supported Pd–Au alloy were prepared through a colloid immobilization method using $PdCl_2$ and $HAuCl_4·3H_2O$ as metal precursors in poly(N-vinylpyrrolidone) and aqueous solution of sodium borohydride ($NaBH_4$) as reducing agent.[57] A. Paul Alivisatos and coworkers have established that trialkylborohydride molten salt (MEt_3BH, M = Na, K) itself as a highly reducing reaction medium for the synthesis of nanoscale Pt_3Y intermetallics and other early-late intermetallics, having diameter of the NPs of

5–20 nm.[58] In this procedure, there is no need of any organic ligands, as the resulting halide salt by-product prevents sintering and also facilitates dispersion of the nanoscale intermetallic onto a support. Murphy and coworkers established a synthetic route for Au–Ag alloy NPs by the reduction of $HAuCl_4$ and $AgNO_3$ using $NaBH_4$ where sodium citrate was used as a capping agent and the solution concentration was kept in such a way that there is no possibility of precipitation of AgCl during the reaction.[59] Cu–Ag alloyed NPs have also been prepared from a mixture of aqueous solution of $AgNO_3$ and $CuCl_2$ in presence of hexadecylamine and glucose as reducing agents.[60] Chandler et al. prepared bimetallic Ni–Au alloy NPs having 2 nm diameters in alkylated amine-terminated generation 5 polyamidoamine (PAMAM) dendrimers.[61] Ultra-small NiSn dendrimer-encapsulated alloy NPs were prepared in toluene via the co-complexation method under N_2 atmosphere to avoid oxidation using sodium triethylborohydride as a reducing agent.[62] Recently, Bennett et al. presented a cathodic corrosion method for the preparation of a number of alloys containing the constituent metals that are immiscible in the bulk state at room temperature (see Figure 7.4).[63] Based on this approach, they have synthesized composition controlled alloys of PtBi and PtPb with Pt compositions in the range of 60%–92%.

Au–Cu nanocubes with face-centered cubic (fcc) structure were reported using a polyol strategy, through the simultaneous reduction of Cu and gold salts in a mixture of 1,2-hexadecanediol (HDD), diphenyl ether (DPE), 1-adamantanecarboxylic acid (ACA), 1-HDA and 1-dodecanethiol (DDT).[64] Schaak and coworkers prepared AuCu and $AuCu_3$ intermetallics from PVP-stabilized Au–Cu NP aggregates obtained by borohydride reduction, followed by annealing under an inert atmosphere.[65] They also demonstrated the synthesis of intermetallic M-Sn (M = Fe, Co, Ni, Pd) NCs by sequential reduction of the metal salts in tetraethylene glycol (TEG) using $NaBH_4$, followed by heating to 170°C–205°C under Ar in the presence of PVP and/or other polymers.[66] AuPt alloy NPs and nanowires (NWs) were prepared by the same group

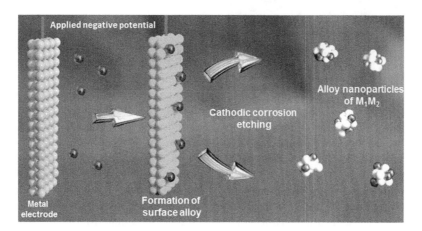

FIGURE 7.4 Schematic representation of the cathodic corrosion reaction mechanism. (Adapted with permission from Ref. [63].)

using a rapid butyllithium reduction of Au^{3+} and Pt^{4+} precursors in oleylamine at 78°C with a subsequent solution annealing for 2 h at 230°C for AuPt NPs and at 190°C for AuPt NWs, respectively.[67] They have extended the same procedure for the synthesis of intermetallic Au_3Fe, Au_3Co and Au_3Ni NCs using solution chemistry route where n-butyllithium was used as a reducing agent.[34] Yin et al. reported sub-10 nm Pt–Pd nanotetrahedrons and nanocubes with high shape selectivity via one-pot hydrothermal routes based on the concept of stabilizing particular facets with shape-selective small ions (as shown in Figure 7.5).[35] Using sodium oxalate and formaldehyde as the (111)-facet selective agent and reducing agent, respectively, single crystalline around 5 nm Pt–Pd nanotetrahedrons enclosed by four (111) facets with a shape selectivity of 70% were produced, whereas uniform single-crystalline

Pt–Pd nanocubes (8.5 ± 0.8 nm) with a shape selectivity of 88% can be obtained in presence of large amount of Br^- and tiny amount of I^- anions as the (100) facet selective agents. Pt–Pd alloy nanoflowers of 25 nm diameter with dominant (111) facets are reported through a facile cochemical reduction route in a poly(allylamine hydrochloride)-based aqueous solution.[68] Somorjai group synthesized $Rh_{0.5}Pd_{0.5}$ and $Pt_{0.5}Pd_{0.5}$ alloy NPs with diameters of ~15 nm by high-temperature polyol-based method.[69] In a similar way, they prepared several alloys of Rh_xPd_{1-x} (~15 nm), Rh_xPt_{1-x} (8–11 nm) and Pd_xPt_{1-x} (~16 nm), where $x = 0.2, 0.5, 0.8$.[70] Based on a chemical reduction technique, Sn–Ag–Cu nanoalloys were synthesized using PVP and $NaBH_4$ as surfactant and reducing agents, respectively, in diethylene glycol.[71] Leonard et al. prepared $AuCuSn_2$ ordered intermetallics with ternary composition by heating a mixture of $HAuCl_4 \cdot 3H_2O$, $Cu(C_2H_3O_2)_2$ and $SnCl_2$ (7-fold excess) in TEG and PVP, subsequently a freshly prepared dilute $NaBH_4$ solution was added and heated the solution to 120–200°C for 10 min.[25] Toshima and coworkers demonstrated the synthesis of PVP-protected Au/Pt/Ag trimetallic alloys with an average diameter of 1.5 nm by reduction of the corresponding ions with rapid injection of $NaBH_4$, whereas drop-wise addition of $NaBH_4$ into the starting solution originates particle with larger size.[72]

Thermal Decomposition

Thermal decomposition approach involves decomposition of organometallic compounds and metal-surfactant complexes in high-boiling organic solvents containing stabilizing surfactants, which in turn give rise to the monodisperse NCs of alloys and intermetallics.[73] In this procedure, nucleation started from a supersaturated solution, resulting in aggregation of atoms to form nuclei. Nuclei then grow fast, leading to the decrease in concentration of metal atoms in solution and finally grow into NCs with larger size until an equilibrium state is reached between the atoms on the surface of the NC and the atoms in the solution, without the formation of any new nuclei since concentration of atoms is below the minimum supersaturation level.[74,75] In this route, the size and morphology of NPs can be regulated by changing the ratios of the starting reagents as well as reaction temperature, reaction time and aging period.

The pioneering work by Murray et al. to achieve monodisperse cadmium chalcogenide NCs using thermal decomposition techniques has fascinated to the researchers to extend this route for the synthesis of nanoscale alloy and intermetallics.[76] Recently, Y. Yin and coworkers reported Au–Pd alloy shells on Pd NCs with controlled shapes and component by incorporating Au atoms to the surface of Pd NCs through a simple solution-phase surface alloying process (see Figure 7.6).[77] The reduction of Au(I) produces Au atoms onto the surface of Pd nanocubes and those Au atoms diffused into the lattice of Pd upon heating at 200°C, leading to the formation of the Au–Pd alloy surfaces, since Pd and Au are completely miscible in all

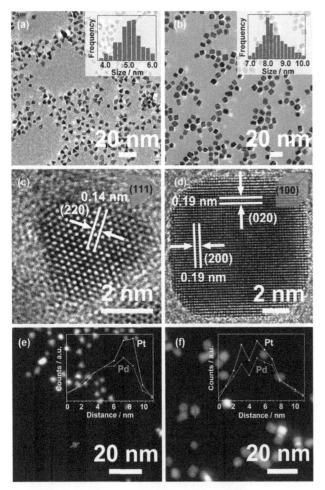

FIGURE 7.5 TEM images of Pt–Pd (a) nanotetrahedrons and (b) nanocubes, HRTEM images of a single Pt–Pd (c) nanotetrahedron and (d) nanocube, and high-angle annular dark-field (HAADF)-STEM images of Pt–Pd (e) nanotetrahedrons and (f) nanocubes. Insets in panels a and b are the size distribution histograms of the as-prepared nanotetrahedrons and nanocubes, respectively. Insets in panels e and f are the HAADF-STEM-EDS line scan profiles of a single nanotetrahedron and nanocube. (Adapted with permission from Ref. [35].)

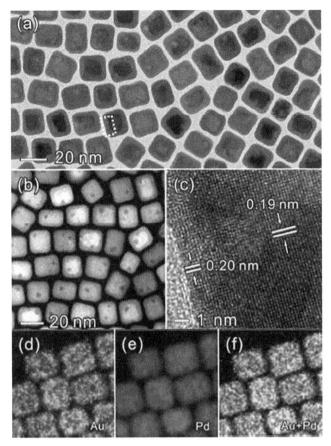

FIGURE 7.6 Electron microscopy characterizations of the as-prepared core–shell Pd@Au-Pd alloy surface nanocubes. (a) TEM image, (b) HAADF-STEM image, (c) HRTEM image and (d–f) energy dispersive X-ray Analysis (EDX) of the as-prepared core–shell Pd@Au-Pd alloy surface nanocubes. (Adapted with permission from Ref. [77].)

proportions with only a slight lattice mismatch (4.9%). Pd-M (M = Ni, Ag, Cu) alloy shells with a tunable thickness on preformed nanoscale Pd seeds was synthesized by X. Li et al. by the combination of the slow reduction of "M" ions and the subsequent diffusion of M ad-atoms into the surface lattice of Pd seeds, where Pd-M alloy shell is regulated by changing the type and amount of the metal precursor and reaction time.[78] Wang et al. reported a route for the preparation of high-indexed Pt_3Ni alloy tetrahexahedral nanoframes via by co-reducing platinum (IV) and nickel (II) precursors in 1-octadecene and oleylamine at 290°C.[79] Zhang et al. reported the synthesis of monodisperse Pt_3Ni nanoctahedra and nanocubes from their metal salts in a mixture of oleylamine and oleic acid, tungsten hexacarbonyl, $W(CO)_6$ as a shape directing material.[80] Through carbon monoxide etching of carbon-supported $PtNi_4$ tetrahexahedral NCs at high temperature transformed to Pt_3Ni alloy tetrahexahedral nanoframes, having an open, stable and high-indexed microstructure, containing a segregated Pt thin layer strained to the Pt−Ni alloy surfaces. In another report, Ding et al. presented the synthesis of tetrahexahedral and rhombic dodecahedral Pt–Ni NCs by simply changing

the ratios of oleylamine and oleic acid keeping the other synthesis parameters the same.[81] Xia and coworkers also synthesized uniform Pt–Ni octahedra using oleylamine and oleic acid as surfactants and $W(CO)_6$ as a source of CO, since CO acts as a capping agent for {111}facet in presence of Ni.[82] In an another report, truncated-octahedral Pt_3Ni NCs were synthesized from their metal precursors in diphenyl ether using a mixture of borane-tert-butylamine complex and hexadecanediol as the reducing agents, where adamantanecarboxylic acid or adamantaneacetic acid was used in determining the shape of the NCs.[83] Xu et al. synthesized 8 nm Pt–Cu nanocubes with exposed {100} crystal planes by simultaneous reduction of their corresponding metal salts using 1,2-tetradecanediol in a mixture of 1-octadecene, tetraoctylammonium bromide, oleylamine and a trace amount of 1-dodecanethiol.[84] In a similar way, composition controlled nanocubes of Pt_xCu_{1-x}[85] and Pt_3Fe[86] terminated with {111} and {100} facets, respectively, were also prepared. Cu_xPt_{100-x} alloy nanocubes with a relatively broad range of composition ratios were also reported by Zhao et al., where composition of the nanocubes was regulated by changing the ratio of metal precursors and the amounts of stabilizing/coordination agents.[87] Cable et al. demonstrated a reversible cyclic interconversion of alloys using a simple chemical process. Pt_3Sn was produced through the reaction of PtSn with K_2PtCl_6 and can be converted back to PtSn by the reaction with $SnCl_2$. Similarly, $PtSn_2$ was formed by the reaction of PtSn with $SnCl_2$, and it can be converted back to PtSn by the reaction between $PtSn_2$ and K_2PtCl_6.[88]

A facile strategy has also been developed for synthesizing hierarchical Pt–Co NWs with high-index, platinum-rich facets and ordered intermetallic structure from their metal precursors in oleylamine as solvent, cetyltrimethylammoniumchloride as the structure-directing agent, and glucose as a reducing agent.[89] Intermetallic Au_3Cu alloy truncated nanocubes with tunable sizes (15–30 nm) were achieved via phase-stabilized synthesis through heating a mixture of Cu microparticles, oleylamine, trioctylphosphine (TOP) and $AuPPh_3Cl$ at 200°C, where the long chain amine, i.e., oleylamine prevents the oxidation of Cu NPs as well as produces Au atoms from their precursor through reduction reaction.[90] Monodispersed spherical fcc $Au_{1-x}Cu_x$ alloy NPs with tunable compositions, $x = 0$ to 0.5 have also been synthesized in a mixture of oleylamine, oleic acid and 1-octadecene.[91] Somorjai and coworkers reported the composition tunable synthesis of Ni–Cu alloys ($Ni_{0.8}Cu_{0.2}$, $Ni_{0.6}Cu_{0.4}$, $Ni_{0.5}Cu_{0.5}$, $Ni_{0.4}Cu_{0.6}$, and $Ni_{0.2}Cu_{0.8}$) by thermolysis approach.[92] Cubic PtPd alloy with a dominant {100} face was synthesized through the decomposition of their metal precursors in a mixture of oleylamine and oleic acid in the presence of $Mo(CO)_6$.[93] The thermodynamically stable {100} facets of fcc PtPd alloy possibly arise due to the slow nucleation and selective growth caused by Mo^{n+} and CO. CuPd nanoalloy with bcc structure was also reported via thermal decomposition of a Pd nanoparticle@metal–organic framework composite material, and this procedure may be generalized for the synthesis of different nanoscale alloys through decomposition of metal–organic framework.[94] S. Jana et al. demonstrated the formation of Ni–Zn intermetallics from their transition metal precursors using solution chemistry route (see Figure 7.7).[95] First, Ni NPs were synthesized from nickel(II) acetylacetonate in oleylamine and then injection of diethylzinc produces solid and hollow intermetallic Ni–Zn NPs as a function of reaction time, since Zn atoms diffuse to the solvated Ni NP surfaces yields Ni–Zn intermetallics. Based on a two-step procedure, intermetallic Au–Zn alloys with varied composition were reported by the reaction between Au NPs with diethylzinc in oleylamine at 250–300°C.[96] An alloy, $Au_{22}Ir_3(PET)_{18}$, was synthesized by Bhatt et al. through the reaction between $Au_{25}(PET)_{18}$ and $Ir_9(PET)_6$, where the ligand PET is 2-phenylethanethiol.[97] Sun et al. synthesized composition and size tunable FePt alloys by the simultaneous reduction of platinum salt by a diol and thermal decomposition of iron carbonyl in a mixture of oleic acid and oleylamine, where composition of the alloys was adjusted by controlling the molar ratio of iron carbonyl to the platinum salt.[98] Similarly, $CoPt_3$ NPs of different particle sizes were achieved by the reduction of platinum salt and thermodecomposition of cobalt carbonyl in the presence of 1-adamantanecarboxylic acid in mixtures of coordinating solvents.[99] Monodisperse $Fe_{55}Pt_{45}$ NCs (4.0 ± 0.2 nm) that self-assembled into hexagonal close-packed and cubic arrays were prepared by a modified polyol process and the transformation of face-centered cubic (fcc) to face-centered tetragonal (fct) occurred at 550°C.[100]

Tri-metallic ultrathin FePtPd alloy NWs with tunable compositions and controlled length (<100 nm) and diameter (2.5 nm) were synthesized by Guo et al, through the thermal decomposition of $Fe(CO)_5$ and controlled sequential reduction of $Pt(acac)_2$ and $Pd(acac)_2$ at 240°C.[101] The same group also developed a seed mediated growth of FePt alloy over ultrathin $Fe_{36}Pt_{32}Pd_{32}$ and $Fe_{48}Pt_{37}Au_{15}$ NWs, where FePtM (M = Pd, Au) NWs act as seed particle (see Figure 7.8).[102] The resulting FePtM/FePt core/shell NWs are of 2.5 nm wide core with a tunable shell thickness of 0.3–1.3 nm, obtained through the controlled decomposition of $Fe(CO)_5$ and reduction of $Pt(acac)_2$ in the presence of FePtM NW seeds. Trimetallic PtFeCo alloy nanostructures with controllable compositions and TriStar shape were prepared via an organic-phase process from their metal salts in a mixture of 1,2-tetradecanediol, tetraoctylammonium bromide and 1-octadecene at 215–222°C.[103] Hence, the appropriate choice of metal precursors and regulation of the bonding interactions between particle surfaces and capping materials are the decisive factors in obtaining size, shape and composition tunable monodispersed alloys and intermetallics.

7.4 Catalytic Application

Alloys and intermetallics having unique catalytic, electronic and optical properties have become one of the most important materials in catalysis research to develop new catalysts with enhanced activity and selectivity because of their tunable chemical and physical properties that originate from the atomic ordering, composition and size of the constituent metals. However, their properties and activity can be highly affected and reduced once they started agglomeration and inclined to be bulk like materials, which can be restricted by loading them on solid supports or synthesize them directly over the supports.[104–106] However, direct synthesis on to the solid support has several limitations, including unwanted particle growth, difficult to achieve desired size, shape and morphology, etc. This can be addressed by synthesizing size, shape and composition controlled alloy or intermetallic NPs in solution, followed by immobilization of those preformed and well-defined NPs over solid support, resulting in the formation of heterogeneous catalysts. Moreover, the dispersion and the composition homogeneity of alloys and intermetallics are the potential knob for achieving significant catalytic efficacy during chemical reactions, since in heterogeneous catalysis, the reaction rate largely depends on the affinity between reactive species and surface sites according to the Sabatier principle.[107]

A number of fundamental studies on oxygen reduction reaction (ORR) and hydrogen evaluation reaction (HER)

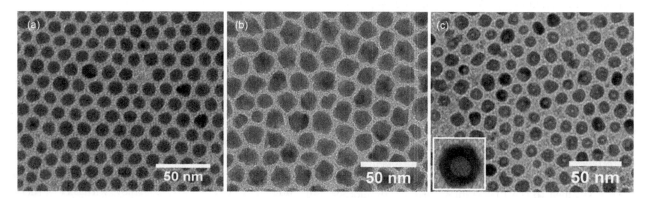

FIGURE 7.7 TEM images of (a) Ni NPs and corresponding (b) solid and (c) hollow Ni–Zn alloys. (Adapted with permission from Ref. [95])

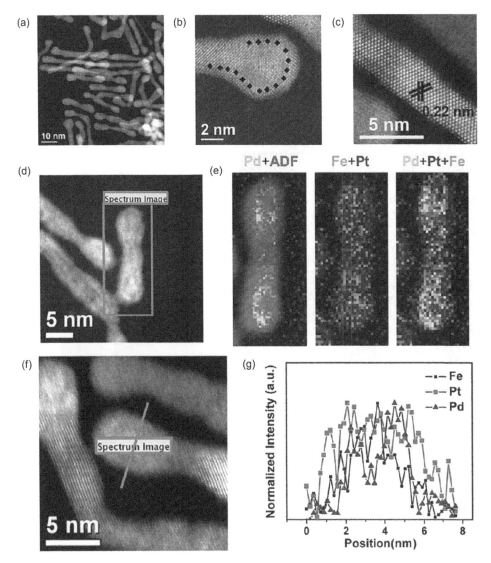

FIGURE 7.8 HAADF-STEM (a, d, f), high-resolution HAADF-STEM (b, c) and STEM-EELS (electron energy loss spectroscopy) mapping (29 × 66 pixels, spatial resolution of 3 Å) (e) images of FePtPd/FePt-0.8. (0.8 indicates FePt shell thickness) (f, g) High-resolution (1.7Å) line scan EELS analysis across one NW. (Adapted with permission from Ref. [102].)

have been demonstrated by several groups. Y. Xia and coworkers studied the catalytic activity of 9 nm Pt−Ni octahedra, after removing the surfactants through acetic acid treatment and showed that specific activity of Pt−Ni octahedra alloys is 51-fold higher in comparison to Pt/C catalyst for the ORR at 0.93 V, together with a mass activity of 3.3 A mg_{Pt}^{-1} at 0.9 V. They proposed that the enhancement of ORR activity is due to the presence of a clean and well-preserved (111) surface for the Pt−Ni octahedral.[82] Zhu et al. demonstrated enhanced catalytic efficacy of well-defined three-dimensional PtM (M = Fe, Co, Cu, Ni) alloy nanosponges for ORR relative to the commercial Pt/C catalysts because of their sponge-like structure with porous and interconnected networks, which is favorable to spread and diffuse electrolyte ions.[56] However, $Pt_{79}Fe_{21}$ exhibits highest ORR activity among all the catalysts owing to its high specific surface area and the porous structure. Shape-dependent catalytic activity of Pt_3Ni nanoalloys was

carried out by Zhang et al. for ORR, after making the thin layers of catalysts supported on glassy carbon, subsequently compared their ORR activity (see Figure 7.9).[80] They established that Pt_3Ni nanoctahedra terminated with {111} facets have fivefold higher ORR activity than that of nanocubes (terminated with {100} facets) with a similar size and both of them possess much higher efficacy compared to commercial Pt/C. Similarly, Pt−Ni nanoframes with tetrahexahedral and rhombic dodecahedral morphology are also reported as high-performance electrocatalysts for both ORR and alcohol oxidations than those of the commercial Pt/C catalyst.[81] The enhanced electrocatalytic activity of tetrahexahedral Pt−Ni nanoframes than that of rhombic dodecahedral Pt−Ni nanoframes is possibly due to their higher Ni composition and more edges and corners. The ORR activity of truncated-octahedral Pt_3Ni NCs with exposed {111} facets was more reactive than {100} facets, authenticating the general trend observed in the single-crystal surfaces.[83]

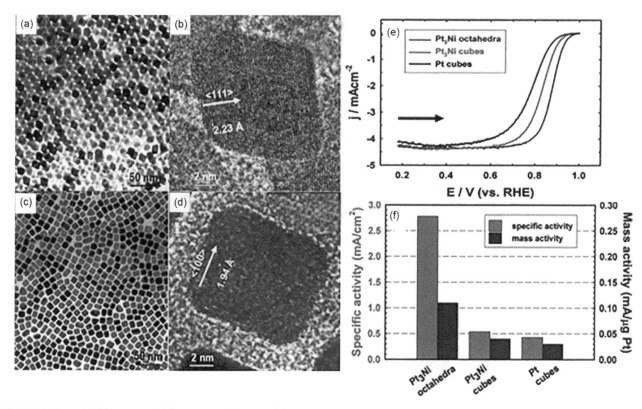

FIGURE 7.9 TEM images of (a) nanooctahedra and (c) nanocubes. High-resolution TEM images of single (b) nanooctahedra and (d) nanocube. (e) Polarization curves for ORR on Pt3Ni nanooctahedra, Pt3Ni nanocubes and Pt nanocubes supported on a rotating GC disk electrode in O_2 saturated 0.1 M $HClO_4$ solution at 295 K. (f) Comparison of the ORR activities on the three types of catalysts. Specific activity and mass activity were all measured at 0.9 V vs. RHE (reversible hydrogen electrode) at 295 K. (Adapted with permission from Ref. [80].)

Recently, Bu et al. reported the ordered Pt_3Co NWs having specific/mass activities for ORR of 39.6/33.7 times higher relative to commercial Pt/C catalyst (see Figure 7.10).[89] Based on the DFT study, they demonstrated that the active threefold hollow sites on platinum-rich high-index facets facilitate to improve the ORR activity and stability of the catalyst. The advantage of multimetallic nanostructures in tuning the stability and catalytic properties of nanocatalysts was reported in case of Au NPs coated with a Pt-bimetallic alloy, $Au/FePt_3$.[108] Here, Au core imparts stability to catalyst, whereas the topmost Pt atoms, which are electronically altered by the subsurface Fe, attribute to the higher catalytic activity of $Au/FePt_3$ towards ORR compared to $FePt_3$/C and Pt/C. Guo et al. demonstrated hell thickness and core composition-dependent electrocatalytic activity of FePtM/FePt (M = Pd, Au) NWs for ORR in 0.1 M $HClO_4$, after loaded on carbon and also showed that core/shell FePtM/FePt NWs are superior electrocatalyst than FePt alloy NWs for ORR or commercial Pt/C.[102] Among FePtM/FePt NWs, the FePtPd/FePt NWs (0.8 nm shell) has the specific activity of 3.47 mA/cm^{-2} and the mass activity of 1.68 A/mg Pt at 0.5 V (vs. Ag/AgCl), which is much higher than FePtAu/FePt (1.59 mA/cm^{-2}) and FePt (0.82 mA/cm^{-2}) or commercial Pt/C (0.24 mA/cm^{-2}). The superior ORR activity of FePtPd/FePt NWs than other catalysts may be ascribed to the electronic

effect of Pd to FePt, which further downshifts the d-band center of Pt and facilitates O_2 adsorption, activation and desorption. In a similar fashion, trimetallic TriStar-shaped PtFeCo alloy exhibited enhanced HER performance in comparison with the bimetallic PtCo and PtFe alloys.[103] The exceptional HER activity of these TriStar PtFeCo alloy has strong correlation with the chemical compositions, to which interatomic charge polarization and d electron couplings may both contribute.

A great effort has also been devoted for the preparation of nanoscale Pt-based alloys and intermetallics for the oxidation of methanol or formic acid. Recently, Lee et al. reported the enhanced electrocatalytic activity of cubic PtPd alloy than that of Pt or polycrystalline PtPd NPs for the oxidation of methanol and formic acid.[93] This enhanced electrocatalytic activity of the PtPd alloy maybe ascribed to the presence of dominant {100} facets in PtPd alloy as well as homogeneous distribution of Pt and Pd atoms. Tetragonal PtZn NPs of 3–15 nm diameter supported on carbon, synthesized by reaction of Pt NPs with Zn vapor, have shown electrocatalytic activity towards oxidation of formic acid and methanol.[109] Yan and coworkers synthesized single-crystalline Pt–Pd nanotetrahedrons and nanocubes that exhibited facet-dependent enhanced electrocatalytic activity and stability for electrooxidation of methanol compare to commercial Pt/C.[35]

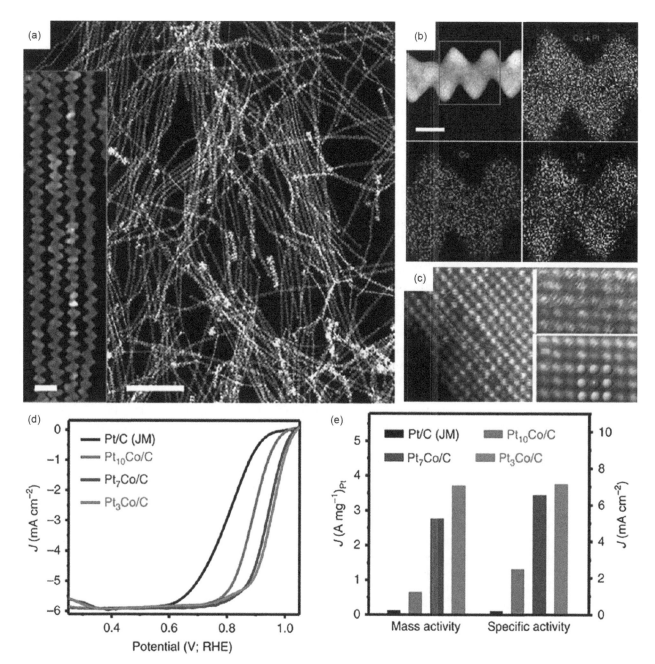

FIGURE 7.10   (a) STEM image of Pt3Co nanowires. Inset is an enlarged STEM image. (b) STEM-ADF image and EDS (energy dispersive X-ray spectroscopy) elemental mappings of the Pt3Co nanowires. (c) Enlarged HAADF-STEM and FFT-filtered HAADF-STEM images with L12-ordered intermetallic structure. The scale bars in (a), inset of a–c are 200, 20, 10 and 1 nm, respectively. (d) ORR polarization curves and (e) ORR-specific activities and mass activities of different catalysts. (Adapted with permission from Ref. [89].)

Pt–Pd nanocubes with exposed (100) facet demonstrated higher catalytic efficiency, while Pt–Pd nanotetrahedrons with (111)-facet possess better stability. The enhancement of both activity and tolerance to the poisoning effects for both the Pt–Pd alloys may be attributed to the formation of bimetallic surface atom arrangements and/or the modification of the electronic structure of surface Pt atoms due to alloying with Pd atoms. Pt–Cu nanodendrites with high alloying degree and high shape selectivity were also reported for methanol oxidation. Their enhanced electrocatalytic activity and long-term stability could be ascribed to the "synergistic effect" between Pt and Cu atoms (such as electronic effect, geometric effect, bifunctional mechanism, etc.), unique interconnected nanostructures and high CO tolerance.[110] S. Guo et al. evaluated the composition-dependent catalytic activity and stability of FePtPd alloy NWs for methanol oxidation and compared their activity with FePt NWs as well as Pd, Pt and PtPd NPs.[101] It turned out that trimetallic $Fe_{28}Pt_{38}Pd_{34}$ NWs are the most efficient methanol oxidation catalysts over Pt, PtPd and FePt catalysts. Xu et al. demonstrated superior electrocatalytic activity of Pt–Cu nanocubes for methanol oxidation compare to the spherical Pt–Cu and Pt NCs with similar sizes due to presence of {100}-terminated facets in the nanocubes.[84] In an another report, composition controlled Pt–Cu nanocubes for formic acid oxidation was also demonstrated by the same group, where $Pt_{80}Cu_{20}$ exhibits a superior electrocatalytic activity compared to the pure Pt nanocubes.[85] Recently, P. Rodriguez and coworkers synthesized $Pt_{70}Bi_{30}$ and $Pt_{90}Pb_{10}$ alloy NPs, which exhibited extraordinary activity towards formic acid oxidation in acidic media due to the level of cleanliness of the catalyst as well as composition of the metal alloys.[63] Recently, Yang and coworkers have proven the electrocatalytic activity of PtAg intermetallic NCs for the oxidation of formic acid due to the ensemble effect of Pt and Ag in PtAg intermetallic NCs.[111] Ag–Pt intermetallics demonstrate 5 times higher activity than its alloy NPs and 29 times higher than the reference Pt/C. Electrocatalytic activity of Pt_3Fe NPs towards methanol and formic acid oxidation was also reported and found that Pt_3Fe nanocubes were more reactive than that of spherical ones owing to the different crystal surfaces, further signifying the possibility of improved catalytic efficiency by controlling the shape of particles.[86] Jana and coworkers have developed a two-step procedure to fabricate $Au_{0.33}Ag_{0.66}$ alloy NPs over the surface of thiol modified halloysite clay nanotubes, resulting in environmentally benign and low-cost heterogeneous catalyst and their catalytic efficiency was examined for the reduction of nitroaromatics.[112] The enhanced catalytic activity of alloyed Cu–Ag NPs was also studied by Wu et al. for the reduction of 4-nitrophenol in comparison to pristine Ag monometallic NPs.[60] Selective hydrogenation of alkynes is an interesting reaction both from scientific as well industrial point of view to remove traces of acetylene from ethylene during the preparation of polyethylene. K. Kovnir et al. explored the concept of using intermetallic compounds with covalent bonding rather than alloys to achieve durable catalyst with pre-selected electronic and local structural properties.[50] They reported Pd–Ga intermetallic compounds as a highly selective catalyst for the semi-hydrogenation of acetylene. The covalent interaction between Pd and Ga provides in situ stability to the crystal structure as well as polarization of the Pd atoms to maximize the activation barrier for hydrogen atoms to enter into the bulk which in turn inhibit subsurface hydrogen formation and enhancing the selectivity. X. Li et al. demonstrated that Au–Pd alloy shells on Pd NCs can exhibit much enhanced activity and improved selectivity in alkyne semi-hydrogenation reactions relative to the original Pd seeds, commercial Pd/C, and even Lindlar catalysts.[77] Au–Pd alloy surfaces show turnover frequency (TOF) of 4,540 h^{-1}, which is 4 and 10 times higher compared to Pd/C (1,253 h^{-1}) and Lindlar catalyst (467 h^{-1}), respectively. Nørskov group has reported a method based on density functional theory calculations to identify promising metal combinations as a substitution to achieve alloys and intermetallics for semi-hydrogenation reactions and also pointed out that the presence of Pd atom ensembles is responsible for full hydrogenation but causes deactivation to the catalyst.[113]

Recently, Wang et al. synthesized NiCo alloy NPs using heteronuclear metal–organic frameworks as metal alloy precursors and after immobilization on the SiO_2 frameworks utilized them as a catalyst in the catalytic hydrogenation of furfuryl alcohol to tetrahydrofurfuryl alcohol under mild conditions with high yield and selectivity.[114] $NiCo/SiO_2$ catalyst was 2 and 20 times more active than only Ni and Co catalysts loaded on SiO_2, respectively, which is probably due to the synergistic effect between Co and Ni in $NiCo/SiO_2$ catalyst. Ultrathin alloy shells as heterogeneous catalysts to increase the utilization efficiency and enhance the catalytic activity of metal atoms have been evaluated for hydrogenation of chloronitrobenzene by Li et al.[78] Based on the surface atom diffusion approach, they tailored the surface electronic structure of Pd catalysts by alloying a second metal into the lattice of Pd catalysts. All the alloy surfaces (Pd–Ag, Pd–Cu and Pd–Ni) demonstrate a significantly improved selectivity for chloroanilines compared to pure Pd catalysts, but Pd–Ni surfaces possess the highest selectivity of >99%.[78] Composition-dependent activity of Cu_xPt_{100-x} nanocubes after coated on a Ti foil was investigated by Zhao et al. for the electrocatalytic CO_2 reduction.[87] They demonstrated that catalytic activity of the nanocubes increases with increasing Cu constituent and follows the order: pure $Pt > Cu_{85}Pt_{15} > Cu_{68}Pt_{32} > Cu_{32}Pt_{68} > Cu_{22}Pt_{78} >$ pure Cu. The catalytic activity of supported Pd–Au alloy catalysts was evaluated for the selective hydrosilylation of α,β-unsaturated ketones and alkynes, and the catalytic activity of this alloy was sensitive to Pd/Au atomic ratio (see Figure 7.11).[57] Interestingly, Pd–Au alloy NPs with a low Pd/Au atomic ratio was found to be highly active heterogeneous catalysts under mild reaction conditions. It should be noted that stability of alloys and intermetallics as well as the capping materials that tune the size and

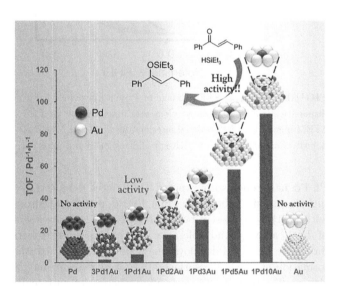

FIGURE 7.11 Relationship between Pd–Au/Pd–Pd coordination numbers ratio and the TOF based on Pd in Pd–Au/SiO_2 for alkyne hydrosilylation at 40°C. (Adapted with permission from Ref. [57].)

shape of the alloys and intermetallics, induce an effect on their catalytic activity as they consist of many unsaturated surface atom. Therefore, all these factors should be taken into consideration to design the novel catalysts, since they provide valuable insights during catalysis.

7.5 Conclusions

In this chapter, we have highlighted some of the recent progresses made in alloys and intermetallics up until now and their applications in catalysis along with stability as potential catalysts. Based on the modulated chemistry for the synthesis of alloys and intermetallics between immiscible metals, several strategies were introduced for fabricating unique nanomaterials with unusual surface properties. We have also illustrated in this chapter numerous fundamental studies in a wide range of heterogeneous catalysis from electrocatalysis to hydrogenation and oxidation reactions under mild condition using those alloys and intermetallics. It is essential to increase the proportion and intrinsic activity of active sites of a catalyst. Moreover, tuning of size, shape and composition at the atomic and nanoscale dimensions is a key aspect to introduce a major effect on their activity. Compared to disorder alloys, intermetallic compounds possess more precise control over the local geometry and electronic structure of the constituent metal atoms, it is thus necessary to standardize the composition, structure and morphology during the synthesis of nanostructures, which in turn enhance the density and intrinsic activity of active sites, leading to the upsurge in their catalytic efficacy and durability. However, their properties and activity may also be affected enormously once they started agglomeration during catalysis. This limitation could be overcome by loading the preformed alloys or intermetallics onto the solid supports, which in turn leaves an impact on catalytic reaction engineering as well as heterogeneous catalysis both from the economical and practical point of view. Although significant progresses were made by developing new-generation nanostructures with specific molecular structure together with electronic, photonic and magnetic properties, however, it is still an emerging field of research, which may bring forth significant advancement in materials science in a wide range of applications from catalysis to spectroscopy to the development of nanotechnological devices. This chapter may also be fascinated to the large number of researchers to nurture our understanding on this prospecting field.

Acknowledgment

Authors acknowledge Nano Mission Research Grant (SR/NM/NS-18/2014) and SERB Women Excellence Award (SB/WEA/08/2016) funded by Department of Science and Technology (DST), Government of India and also S. N. Bose National Centre for Basic Sciences, Kolkata, India.

References

1. Hornyak, G.L., Dutta, J., Tibbals, H.F. and Rao, A., 2008. *Introduction to Nanoscience.* CRC Press, New York.
2. Yang, Y., Jia, Y., Gao, L. et al., 2011. Fabrication of autofluorescent protein coated mesoporous silica nanoparticles for biological application. *Chemical Communications*, 47, 12167–12169.
3. Wittstock, A., Zielasek, V., Biener, J., Friend, C.M. and Bäumer, M., 2010. Nanoporous gold catalysts for selective gas-phase oxidative coupling of methanol at low temperature. *Science, 327*, 319–322.
4. Ertl, G., Knözinger, H. and Weitkamp, J., 1997. *Handbook of Heterogeneous Catalysis.* Wiley, Weinheim.
5. Jana, S., 2015. Advances in nanoscale alloys and intermetallics: Low temperature solution chemistry synthesis and application in catalysis. *Dalton Transactions, 44*, 18692–18717.
6. Cobley, C.M., Skrabalak, S.E., Campbell, D.J. and Xia, Y., 2009. Shape-controlled synthesis of silver nanoparticles for plasmonic and sensing applications. *Plasmonics, 4*, 171–179.
7. Bruchez, M., Moronne, M., Gin, P., Weiss, S. and Alivisatos, A.P., 1998. Semiconductor nanocrystals as fluorescent biological labels. *Science, 281*, 2013–2016.
8. Hamedi Shokrlu, Y., & Babadagli, T., 2013. In-situ upgrading of heavy oil/bitumen during steam injection by use of metal nanoparticles: A study on in-situ catalysis and catalyst transportation. *SPE Reservoir Evaluation & Engineering, 16*, 333–344.
9. Wang, J., Gudiksen, M.S., Duan, X., Cui, Y. and Lieber, C.M., 2001. Highly polarized photoluminescence and photodetection from single indium phosphide nanowires. *Science, 293*, 1455–1457.
10. Lu, A.H., Salabas, E.E. and Schüth, F., 2007. Magnetic nanoparticles: Synthesis, protection, functionalization, and application. *Angewandte Chemie International Edition, 46*, 1222–1244.
11. Maier, S.A., Brongersma, M.L., Kik, P.G., Meltzer, S., Requicha, A.A. and Atwater, H.A., 2001. Plasmonics—a route to nanoscale optical devices. *Advanced Materials, 13*, 1501–1505.
12. Stoloff, N.S., Liu, C.T. and Deevi, S.C., 2000. Emerging applications of intermetallics. *Intermetallics, 8*, 1313–1320.
13. Alden, L.R., Han, D.K., Matsumoto, F., Abrua, H.D. and DiSalvo, F.J., 2006. Intermetallic PtPb nanoparticles prepared by sodium naphthalide reduction of metal-organic precursors: electrocatalytic oxidation of formic acid. *Chemistry of Materials, 18*, 5591–5596.
14. Casado-Rivera, E., Volpe, D.J., Alden, L. et al., 2004. Electrocatalytic activity of ordered intermetallic

phases for fuel cell applications. *Journal of the American Chemical Society*, *126*, 4043–4049.

15. Ahmadi, T.S., Wang, Z.L., Green, T.C., Henglein, A. and El-Sayed, M.A., 1996. Shape-controlled synthesis of colloidal platinum nanoparticles. *Science*, *272*, 1924–1925.

16. Lim, B., Jiang, M., Camargo, P.H. et al., 2009. Pd-Pt bimetallic nanodendrites with high activity for oxygen reduction. *Science*, *324*, 1302–1305.

17. Jaime, M., Movshovich, R., Stewart, G.R. et al., 2000. Closing the spin gap in the Kondo insulator $Ce_3Bi_4Pt_3$ at high magnetic fields. *Nature*, *405*, 160–163.

18. Krenke, T., Duman, E., Acet, M. et al., 2005. Inverse magnetocaloric effect in ferromagnetic Ni–Mn–Sn alloys. *Nature Materials*, *4*, 450–454.

19. Novet, T. and Johnson, D.C., 1991. New synthetic approach to extended solids: Selective synthesis of iron silicides via the amorphous state. *Journal of the American Chemical Society*, *113*, 3398–3403.

20. Suryanarayana, C., 2001. Mechanical alloying and milling. *Progress in Materials Science*, *46*, 1–184.

21. Elkins, K.E., Vedantam, T.S., Liu, J.P. et al., 2003. Ultrafine FePt nanoparticles prepared by the chemical reduction method. *Nano Letters*, *3*, 1647–1649.

22. Teng, X. and Yang, H., 2003. Synthesis of face-centered tetragonal FePt nanoparticles and granular films from $Pt@Fe_2O_3$ Core– Shell Nanoparticles. *Journal of the American Chemical Society*, *125*, 14559–14563.

23. Armbrüster, M., Kovnir, K., Behrens, M. et al., 2010. Pd-Ga intermetallic compounds as highly selective semihydrogenation catalysts. *Journal of the American Chemical Society*, *132*, 14745–14747.

24. Dinega, D.P. and Bawendi, M.G., 1999. A solution-phase chemical approach to a new crystal structure of cobalt. *Angewandte Chemie International Edition*, *38*, 1788–1791.

25. Leonard, B.M., Bhuvanesh, N.S. and Schaak, R.E., 2005. Low-temperature polyol synthesis of $AuCuSn_2$ and $AuNiSn_2$: Using solution chemistry to access ternary intermetallic compounds as nanocrystals. *Journal of the American Chemical Society*, *127*, 7326–7327.

26. Vasquez, Y., Luo, Z. and Schaak, R.E., 2008. Low-temperature solution synthesis of the non- equilibrium ordered intermetallic compounds Au_3Fe, Au_3Co, and Au_3Ni as nanocrystals. *Journal of the American Chemical Society*, *130*, 11866–11867.

27. Baghbanzadeh, M., Carbone, L., Cozzoli, P.D. and Kappe, C.O., 2011. Microwave-assisted synthesis of colloidal inorganic nanocrystals. *Angewandte Chemie International Edition*, *50*, 11312–11359.

28. Cozzoli, P.D., Pellegrino, T. and Manna, L., 2006. Synthesis, properties and perspectives of hybrid nanocrystal structures. *Chemical Society Reviews*, *35*, 1195–1208.

29. Best, R.J. and Russell, W.W., 1954. Nickel, copper and some of their alloys as catalysts for ethylene hydrogenation1. *Journal of the American Chemical Society*, *76*, 838–842.

30. Schwab, G.M., 1950. Alloy catalysts in dehydrogenation. *Discussions of the Faraday Society*, *8*, 166–171.

31. Aihara, N., Torigoe, K. and Esumi, K., 1998. Preparation and characterization of gold and silver nanoparticles in layered laponite suspensions. *Langmuir*, *14*, 4945–4949.

32. Itakura, T., Torigoe, K. and Esumi, K., 1995. Preparation and characterization of ultrafine metal particles in ethanol by UV irradiation using a photoinitiator. *Langmuir*, *11*, 4129–4134.

33. Augustine, R.L. and O'Leary, S.T., 1995. Heterogeneous catalysis in organic chemistry. Part 10[1]. Effect of the catalyst support on the regiochemistry of the heck arylation reaction. *Journal of Molecular Catalysis A: Chemical*, *95*, 277–285.

34. Li, Y., Boone, E. and El-Sayed, M.A., 2002. Size effects of PVP–Pd nanoparticles on the catalytic Suzuki reactions in aqueous solution. *Langmuir*, *18*, 4921–4925.

35. Yin, A.X., Min, X.Q., Zhang, Y.W. and Yan, C.H., 2011. Shape-selective synthesis and facet-dependent enhanced electrocatalytic activity and durability of monodisperse sub-10 nm Pt–Pd tetrahedrons and cubes. *Journal of the American Chemical Society*, *133*, 3816–3819.

36. Qi, W., Li, Y., Xiong, S. and Lee, S.T., 2010. Modeling size and shape effects on the order-disorder phase-transition temperature of CoPt nanoparticles. *Small*, *6*, 1996–1999.

37. Alloyeau, D., Langlois, C., Ricolleau, C., Le Bouar, Y. and Loiseau, A., 2007. A TEM in situ experiment as a guideline for the synthesis of as-grown ordered CoPt nanoparticles. *Nanotechnology*, *18*, 375301–375306.

38. Takahashi, Y.K., Ohkubo, T., Ohnuma, M. and Hono, K., 2003. Size effect on the ordering of FePt granular films. *Journal of Applied Physics*, *93*, 7166–7168.

39. Miyazaki, T., Kitakami, O., Okamoto, S. et al., 2005. Size effect on the ordering of L 1 0 FePt nanoparticles. *Physical Review B*, *72*, 144419–144423.

40. Chi, M., Wang, C., Lei, Y. et al., 2015. Surface faceting and elemental diffusion behaviour at atomic scale for alloy nanoparticles during in situ annealing. *Nature Communications*, *6*, 8925–8933.

41. Ruban, A.V., Skriver, H.L. and Nørskov, J.K., 1999. Surface segregation energies in transition-metal alloys. *Physical Review B*, *59*, 15990–16000.

42. Yan, Y., Du, J.S., Gilroy, K.D. et al., 2017. Intermetallic nanocrystals: Syntheses and catalytic applications. *Advanced Materials*, *29*, 1605997–1605997.

43. Liao, H., Fisher, A. and Xu, Z.J., 2015. Surface segregation in bimetallic nanoparticles: A critical issue in electrocatalyst engineering. *Small, 11*, 3221–3246.

44. Schrinner, M., Ballauff, M., Talmon, Y. et al., 2009. Single nanocrystals of platinum prepared by partial dissolution of Au-Pt nanoalloys. *Science, 323*, 617–620.

45. Klabunde, K.J. and Richards, R. eds., 2001. *Nanoscale Materials in Chemistry.*Wiley, New York.

46. Ramya, M., Karthika, M., Selvakumar, R., Raj, B. and Ravi, K.R., 2017. A facile and efficient single step ball milling process for synthesis of partially amorphous Mg-Zn-Ca alloy powders for dye degradation. *Journal of Alloys and Compounds, 696*, 185–192.

47. Umrao, S., Jeon, J., Jeon, S.M., Choi, Y.J. and Lee, S., 2017. A homogeneous atomic layer $MoS_{2(1-x)}$ Se_{2x} alloy prepared by low-pressure chemical vapor deposition, and its properties. *Nanoscale, 9*, 594–603.

48. Park, M.S., Rajendran, S., Kang, Y.M. et al., 2006. Si–Ni alloy–graphite composite synthesized by arc-melting and high-energy mechanical milling for use as an anode in lithium- ion batteries. *Journal of power sources, 158*, 650–653.

49. Onda, A., Komatsu, T. and Yashima, T., 2001. Preparation and catalytic properties of single- phase Ni–Sn intermetallic compound particles by CVD of $Sn(CH_3)_4$ onto Ni/Silica. *Journal of Catalysis, 201*, 13–21.

50. Kovnir, K., Armbrüster, M., Teschner, D. et al., 2009. In situ surface characterization of the intermetallic compound PdGa–a highly selective hydrogenation catalyst. *Surface Science, 603*, 1784–1792.

51. Manne, B., Bontha, S., Ramesh, M.R., Krishna, M. and Balla, V.K., 2017. Solid state amorphization of Mg-Zn-Ca system via mechanical alloying and characterization. *Advanced Powder Technology, 28*, 223–229.

52. Murgia, F., Laurencin, D., Weldekidan, E.T. et al., 2018. Electrochemical Mg alloying properties along the $Sb_{1-x}Bi_x$ solid solution. *Electrochimica Acta, 259*, 276–283.

53. Zhang, C., Hwang, S.Y., Trout, A. and Peng, Z., 2014. Solid-state chemistry-enabled scalable production of octahedral Pt–Ni alloy electrocatalyst for oxygen reduction reaction. *Journal of the American Chemical Society, 136*, 7805–7808.

54. Tracy, C.L., Park, S., Rittman, D.R. et al., 2017. High pressure synthesis of a hexagonal close-packed phase of the high-entropy alloy CrMnFeCoNi. *Nature Communications, 8*, 15634–15639.

55. Marshal, A., Pradeep, K.G., Music, D. et al., 2017. Combinatorial synthesis of high entropy alloys: Introduction of a novel, single phase, body-centered-cubic FeMnCoCrAl solid solution. *Journal of Alloys and Compounds, 691*, 683–689.

56. Zhu, Z., Zhai, Y. and Dong, S., 2014. Facial synthesis of PtM (M = Fe, Co, Cu, Ni) bimetallic alloy nanosponges and their enhanced catalysis for oxygen reduction reaction. *ACS Applied Materials & Interfaces, 6*, 16721–16726.

57. Miura, H., Endo, K., Ogawa, R. and Shishido, T., 2017. Supported palladium–gold alloy catalysts for efficient and selective hydrosilylation under mild conditions with isolated single palladium atoms in alloy nanoparticles as the main active site. *ACS Catalysis, 7*, 1543–1553.

58. Kanady, J.S., Leidinger, P., Haas, A. et al., 2017. Synthesis of Pt_3Y and other early–late intermetallic nanoparticles by way of a molten reducing agent. *Journal of the American Chemical Society, 139*, 5672–5675.

59. Mallin, M.P. and Murphy, C.J., 2002. Solution-phase synthesis of sub-10 nm Au-Ag alloy nanoparticles. *Nano Letters, 2*, 1235–1237.

60. Wu, W., Lei, M., Yang, S. et al., 2015. A one-pot route to the synthesis of alloyed Cu/Ag bimetallic nanoparticles with different mass ratios for catalytic reduction of 4-nitrophenol. *Journal of Materials Chemistry A, 3*, 3450–3455.

61. Chandler, B.D., Long, C.G., Gilbertson, J.D. et al., 2010. Enhanced oxygen activation over supported bimetallic Au−Ni catalysts. *The Journal of Physical Chemistry C, 114*, 11498–11508.

62. Gates, A.T., Nettleton, E.G., Myers, V.S. and Crooks, R.M., 2010. Synthesis and characterization of NiSn dendrimer-encapsulated nanoparticles. *Langmuir, 26*, 12994–12999.

63. Bennett, E., Monzó, J., Humphrey, J. et al., 2016. A synthetic route for the effective preparation of metal alloy nanoparticles and their use as active electrocatalysts. *ACS Catalysis, 6*, 1533–1539.

64. Liu, Y. and Walker, A.R., 2010. Monodisperse gold–copper bimetallic nanocubes: Facile one-step synthesis with controllable size and composition. *Angewandte Chemie International Edition, 122*, 6933–6937.

65. Sra, A.K. and Schaak, R.E., 2004. Synthesis of atomically ordered AuCu and $AuCu_3$ nanocrystals from bimetallic nanoparticle precursors. *Journal of the American Chemical Society, 126*, 6667–6672.

66. Chou, N.H. and Schaak, R.E., 2007. Shape-controlled conversion of β-Sn nanocrystals into intermetallic M-Sn (M = Fe, Co, Ni, Pd) nanocrystals. *Journal of the American Chemical Society, 129*, 7339–7345.

67. Zhou, S., Jackson, G.S. and Eichhorn, B., 2007. AuPt alloy nanoparticles for CO-tolerant hydrogen activation: architectural effects in Au-Pt bimetallic nanocatalysts. *Advanced Functional Materials, 17*, 3099–3104.

68. Fu, G., Wu, K., Lin, J. et al., 2013. One-pot water-based synthesis of Pt–Pd alloy nanoflowers and their superior electrocatalytic activity for

the oxygen reduction reaction and remarkable methanol-tolerant ability in acid media. *The Journal of Physical Chemistry C, 117*, 9826–9834.

69. Tao, F., Grass, M.E., Zhang, Y. et al., 2008. Reaction-driven restructuring of Rh-Pd and Pt- Pd core-shell nanoparticles. *Science, 322*, 932–934.

70. Tao, F., Grass, M.E., Zhang, Y. et al., 2010. Evolution of structure and chemistry of bimetallic nanoparticle catalysts under reaction conditions. *Journal of the American Chemical Society, 132*, 8697–8703.

71. Roshanghias, A., Yakymovych, A., Bernardi, J. and Ipser, H., 2015. Synthesis and thermal behavior of tin-based alloy (Sn–Ag–Cu) nanoparticles. *Nanoscale, 7*, 5843–5851.

72. Zhang, H., Okumura, M. and Toshima, N., 2011. Stable dispersions of PVP-protected Au/Pt/Ag trimetallic nanoparticles as highly active colloidal catalysts for aerobic glucose oxidation. *The Journal of Physical Chemistry C, 115*, 14883–14891.

73. Park, J., An, K., Hwang, Y. et al., 2004. Ultra-large-scale syntheses of monodisperse nanocrystals. *Nature Materials, 3*, 891–895.

74. Xia, Y., Xiong, Y., Lim, B. and Skrabalak, S.E., 2009. Shape-controlled synthesis of metal nanocrystals: Simple chemistry meets complex physics? *Angewandte Chemie International Edition, 48*, 60–103.

75. LaMer, V.K. and Dinegar, R.H., 1950. Theory, production and mechanism of formation of monodispersed hydrosols. *Journal of the American Chemical Society, 72*, 4847–4854.

76. Murray, C., Norris, D.J. and Bawendi, M.G., 1993. Synthesis and characterization of nearly monodisperse CdE (E = sulfur, selenium, tellurium) semiconductor nanocrystallites. *Journal of the American Chemical Society, 115*, 8706–8715.

77. Li, X., Wang, Z., Zhang, Z. et al., 2017. Construction of Au-Pd alloy shells for enhanced catalytic performance toward alkyne semihydrogenation reactions. *Materials Horizons, 4*, 584–590.

78. Li, X., Wang, X., Liu, M. et al., 2018. Construction of Pd-M (M = Ni, Ag, Cu) alloy surfaces for catalytic applications. *Nano Research, 11*, 780–790.

79. Wang, C., Zhang, L., Yang, H. et al., 2017. High-indexed Pt$_3$Ni alloy tetrahexahedral nanoframes evolved through preferential CO etching. *Nano Letters, 17*, 2204–2210.

80. Zhang, J., Yang, H., Fang, J. and Zou, S., 2010. Synthesis and oxygen reduction activity of shape-controlled Pt$_3$Ni nanopolyhedra. *Nano Letters, 10*, 638–644.

81. Ding, J., Bu, L., Guo, S. et al., 2016. Morphology and phase controlled construction of Pt–Ni nanostructures for efficient electrocatalysis. *Nano Letters, 16*, 2762–2767.

82. Choi, S.I., Xie, S., Shao, M. et al., 2013. Synthesis and characterization of 9 nm Pt–Ni octahedra with a record high activity of 3.3 A/mg$_{Pt}$ for the oxygen reduction reaction. *Nano Letters, 13*, 3420–3425.

83. Wu, J., Zhang, J., Peng, Z. et al., 2010. Truncated octahedral Pt$_3$Ni oxygen reduction reaction electrocatalysts. *Journal of the American Chemical Society, 132*, 4984–4985.

84. Xu, D., Liu, Z., Yang, H. et al., 2009. Solution-based evolution and enhanced methanol oxidation activity of monodisperse platinum–copper nanocubes. *Angewandte Chemie International Edition, 48*, 4217–4221.

85. Xu, D., Bliznakov, S., Liu, Z., Fang, J. and Dimitrov, N., 2010. Composition-dependent electrocatalytic activity of Pt-Cu nanocube catalysts for formic acid oxidation. *Angewandte Chemie International Edition, 122*, 1304–1307.

86. Zhang, J., Yang, H., Yang, K. et al., 2010. Monodisperse Pt$_3$Fe nanocubes: Synthesis, characterization, self-assembly, and electrocatalytic activity. *Advanced Functional Materials, 20*, 3727–3733.

87. Zhao, X., Luo, B., Long, R., Wang, C. and Xiong, Y., 2015. Composition-dependent activity of Cu–Pt alloy nanocubes for electrocatalytic CO$_2$ reduction. *Journal of Materials Chemistry A, 3*, 4134–4138.

88. Cable, R.E. and Schaak, R.E., 2006. Reacting the unreactive: A toolbox of low-temperature solution-mediated reactions for the facile interconversion of nanocrystalline intermetallic compounds. *Journal of the American Chemical Society, 128*, 9588–9589.

89. Bu, L., Guo, S., Zhang, X. et al., 2016. Surface engineering of hierarchical platinum-cobalt nanowires for efficient electrocatalysis. *Nature Communications, 7*, 11850–11859.

90. Zhao, W., Yang, L., Yin, Y. and Jin, M., 2014. Thermodynamic controlled synthesis of intermetallic Au$_3$Cu alloy nanocrystals from Cu microparticles. *Journal of Materials Chemistry A, 2*, 902–906.

91. Motl, N.E., Ewusi-Annan, E., Sines, I.T., Jensen, L. and Schaak, R.E., 2010. Au-Cu alloy nanoparticles with tunable compositions and plasmonic properties: Experimental determination of composition and correlation with theory. *The Journal of Physical Chemistry C, 114*, 19263–19269.

92. Zhang, Y., Huang, W., Habas, S.E. et al., 2008. Near-monodisperse Ni-Cu bimetallic nanocrystals of variable composition: Controlled synthesis and catalytic activity for H$_2$generation. *The Journal of Physical Chemistry C, 112*, 12092–12095.

93. Lee, J.Y., Kwak, D.H., Lee, Y.W., Lee, S. and Park, K.W., 2015. Synthesis of cubic PtPd alloy nanoparticles as anode electrocatalysts for methanol and formic acid oxidation reactions. *Physical Chemistry Chemical Physics, 17*, 8642–8648.

94. Li, G., Kobayashi, H., Kusada, K. et al., 2014. An ordered bcc CuPd nanoalloy synthesised via the thermal decomposition of Pd nanoparticles covered

with a metal–organic framework under hydrogen gas. *Chemical Communications, 50*, 13750–13753.

95. Jana, S., Chang, J.W. and Rioux, R.M., 2013. Synthesis and modeling of hollow intermetallic Ni–Zn nanoparticles formed by the Kirkendall effect. *Nano Letters, 13*, 3618–3625.

96. Schaefer, Z.L., Vaughn II, D.D. and Schaak, R.E., 2010. Solution chemistry synthesis, morphology studies, and optical properties of five distinct nanocrystalline Au–Zn intermetallic compounds. *Journal of Alloys and Compounds, 490*, 98–102.

97. Bhat, S., Baksi, A., Mudedla, S.K. et al., 2017. $Au_{22}Ir_3$ $(PET)_{18}$: An unusual alloy cluster through intercluster reaction. *The Journal of Physical Chemistry Letters, 8*, 2787–2793.

98. Sun, S., Murray, C.B., Weller, D., Folks, L. and Moser, A., 2000. Monodisperse FePt nanoparticles and ferromagnetic FePt nanocrystal superlattices. *Science, 287*, 1989–1992.

99. Shevchenko, E.V., Talapin, D.V., Rogach, A.L. et al., 2002. Colloidal synthesis and self-assembly of $CoPt_3$ nanocrystals. *Journal of the American Chemical Society, 124*, 11480–11485.

100. Varanda, L.C. and Jafelicci, M., 2006. Self-assembled FePt nanocrystals with large coercivity: reduction of the fcc-to-$L1_0$ ordering temperature. *Journal of the American Chemical Society, 128*, 11062–11066.

101. Guo, S., Zhang, S., Sun, X. and Sun, S., 2011. Synthesis of ultrathin FePtPd nanowires and their use as catalysts for methanol oxidation reaction. *Journal of the American Chemical Society, 133*, 15354–15357.

102. Guo, S., Zhang, S., Su, D. and Sun, S., 2013. Seed-mediated synthesis of core/shell FePtM/FePt (M= Pd, Au) nanowires and their electrocatalysis for oxygen reduction reaction. *Journal of the American Chemical Society, 135*, 13879–13884.

103. Du, N., Wang, C., Wang, X. et al., 2016. Trimetallic TriStar nanostructures: Tuning electronic and surface structures for enhanced electrocatalytic hydrogen evolution. *Advanced Materials, 28*, 2077–2084.

104. Dai, J. and Bruening, M.L., 2002. Catalytic nanoparticles formed by reduction of metal ions in multilayered polyelectrolyte films. *Nano Letters, 2*, 497–501.

105. Jana, S., Praharaj, S., Panigrahi, S. et al., 2007. Light-induced hydrolysis of nitriles by photoproduced α-MnO_2 nanorods on polystyrene beads. *Organic Letters, 9*, 2191–2193.

106. Das, S. and Jana, S., 2015. A facile approach to fabricate halloysite/metal nanocomposites with preformed and in situ synthesized metal nanoparticles: A comparative study of their enhanced catalytic activity. *Dalton Transactions, 44*, 8906–8916.

107. Somorjai, G.A. and Li, Y., 2010. *Introduction to Surface Chemistry and Catalysis*. Wiley, Hoboken, NJ.

108. Wang, C., Van Der Vliet, D., More, K.L. et al., 2010. Multimetallic Au/$FePt_3$ nanoparticles as highly durable electrocatalyst. *Nano Letters, 11*, 919–926.

109. Miura, A., Wang, H., Leonard, B.M., Abrua, H.D. and DiSalvo, F., 2009. Synthesis of intermetallic PtZn nanoparticles by reaction of Pt nanoparticles with Zn vapor and their application as fuel cell catalysts. *Chemistry of Materials, 21*, 2661–2667.

110. Gong, M., Fu, G., Chen, Y., Tang, Y. and Lu, T., 2014. Autocatalysis and selective oxidative etching induced synthesis of platinum–copper bimetallic alloy nanodendrites electrocatalysts. *ACS Applied Materials & Interfaces, 6*, 7301–7308.

111. Pan, Y.T., Yan, Y., Shao, Y.T., Zuo, J.M. and Yang, H., 2016. Ag–Pt compositional intermetallics made from alloy nanoparticles. *Nano Letters, 16*, 6599–6603.

112. Jana, S. and Das, S., 2014. Development of novel inorganic–organic hybrid nanocomposites as a recyclable adsorbent and catalyst. *RSC Advances, 4*, 34435–34442.

113. Studt, F., Abild-Pedersen, F., Bligaard, T. et al., 2008. Identification of non-precious metal alloy catalysts for selective hydrogenation of acetylene. *Science, 320*, 1320–1322.

114. Wang, H., Li, X., Lan, X. and Wang, T., 2018. Supported ultrafine NiCo bimetallic alloy nanoparticles derived from bimetal-organic frameworks: A highly active catalyst for furfuryl alcohol hydrogenation. *ACS Catalysis, 8*, 2121–2128.

8

Nanoionics: Fundamentals and Applications

Joachim Maier
Max Planck Institute for Solid State Research

8.1 Introduction

Solid state research is, in a variety of subfields such as electronics, usually restricted to cases in which ionic species are assumed to be immobile. In many situations, such assumptions are critical or at least profiles may occur that are established during preparation, e.g. at high temperatures. A typical situation is met in nanoelectronics where even very low diffusion coefficients of ionic species can severely influence the interfacial behavior owing to the tiny distances involved. Moreover, there are various cases in which ionic mobility is key to the function considered. This is typically the case for electrochemically relevant solids, such as solid electrolytes, that are based on ion conduction or solid electrodes for which not only both ionic and electronic transports are indispensable but where also storage of ions and electrons and thus of neutral components is necessary.

If a high density of boundaries is involved and in particular if small systems are addressed, we refer to nanoionics, a scientific field of equal importance as nanoelectronics (Maier 2005a).

Before dealing with typical nanoionics problems, it is advisable to consider thermodynamics and kinetics of the so-called mixed conductor in which both ions and electrons are mobile. This material can serve as the general case as far as electric and chemical transport is concerned, but is also of direct practical applicability when component transport (enabled by coupled ion and electron transport) is necessary as it is the case for storage electrodes or permeation membranes.

8.2 Bulk Mixed Conductors

8.2.1 Thermodynamics

Electronic disorder in our model mixed conductor is characterized by the excitation of a regular electron to form an excess electron in the conduction band leaving a hole in a valence band:

$$\text{Nil} \rightleftharpoons \text{excess electron} + \text{electron hole}. \tag{8.1}$$

Equally internal ionic disorder occurs by exciting a regular ion into an interstitial site leaving a vacancy

$$\text{Nil} \rightleftharpoons \text{interstitial ion} + \text{ion vacancy}. \tag{8.2}$$

Coupling of ionic and electronic defect states can occur internally and/or externally whereby in the latter case the neighboring phase is involved. Restricting ourselves to a disorder in the cation lattice and to a charge of +1, one can formulate the exchange of the neutral (metal) component between the phases as

$$\begin{aligned}\text{neutral component} &\rightleftharpoons \text{interstitial atom} \\ &\rightleftharpoons \text{interstitial ion} + \text{excess electron}\end{aligned}. \tag{8.3}$$

In the latter equation the neutral atom is allowed to get ionized.

Not only can various other redundant formulations be given by linear combinations of reactions (8.1)–(8.3), but additional (i.e. non-redundant) ionic disorder types are also possible. Schottky disorder is such an example, while above we restricted to Frenkel disorder.

The complexity of the point defect situation can be enormous and it is completely justified to term the addressed field "defect chemistry". In the publication list, various detailed treatises of this field are given (Wagner and Schottky 1930, Kröger 1964, Maier 2004, Maier 2005b, Maier 2007b, Kofstad 1972, Schmalzried 1981).

The thermodynamic treatment of defect chemistry relies on the balance of the electrochemical potentials ($\tilde{\mu}$) for the above equations. (See Scheme 8.1 for details.)

Scheme 8.1 Bulk Defect Thermodynamics and Kinetics

$E(x) \rightleftharpoons P(x')$ (special case diffusion: $P \equiv E$, $x' \neq x$)

More general:

$\sum_j \nu_{jr} A_j \rightleftharpoons \text{Nil}$ (r: internal disorder reactions, internal association reactions, interaction reactions with neighboring phase)

Equilibrium: $\mathcal{A} \equiv \tilde{\mu}_P - \tilde{\mu}_E = 0$ ($\mathcal{A}$: affinity)

$$\mathcal{A}_r = \sum_j \nu_{jr} \tilde{\mu}_j = 0$$

$$\tilde{\mu}_j = \mu_j + z_j e \phi$$

$$\mu_j = \mu_j^\circ + kT \ln c_j^{\nu_j} + kT \ln \gamma_{j,\text{cfg}}$$
$$+ kT \ln \gamma_{j,\text{ncfg}} \quad (\gamma: \text{activity coefficient})$$

Dilute: $c_P / c_E = K(T)$ (K: mass action constant)

$$\Pi_r c_j^{\nu_{jr}} = K_r(T)$$

Close to equilibrium: $\mathcal{R} \propto \mathcal{A}$ ($\mathcal{R}$: reaction rate)

Special case diffusion: $J_j \propto -\nabla \tilde{\mu}_j$ (J: flux density)

$$\frac{\partial c_j}{\partial t} = -\nabla J_j + \nu_j \mathcal{R}_{rj}$$

Far from equilibrium: $\mathcal{R} \propto \sinh [\Delta \tilde{\mu} / 2kT]$

In the bulk the equations refer to the same electric potential (ϕ), so these balances can also be formulated for the chemical potentials. The chemical potential additively combines a concentration independent term (μ°) as a species-specific materials constant and a term that stems from carrier statistics and reflects the configurational entropy. Here we restrict to dilute conditions for which the latter contribution is given by $kT \ln c$ (c: concentration). Then all defect reactions can be treated by ideal mass action laws. The solution for the carrier concentrations can be obtained if the electroneutrality equation is added (and mass balance if necessary). (Scheme 8.2 gives the thermodynamic coupling of ions, electrons, components and defects in terms of chemical potentials.)

Scheme 8.2 Coupling of Chemical Potentials of Components, Ions and Defects

$$\mu_{M^+} + \mu_{e^-} = \mu_M$$
$$\mu_{M^+}^\circ + \mu_{e^-}^\circ \neq \mu_M^\circ$$
$$\mu_{M^+} = \mu_i = -\mu_\nu$$
$$\mu_{M^+}^\circ \neq \mu_i^\circ \neq \mu_\nu^\circ$$
$$\mu_{e^-} = \mu_n = -\mu_p$$
$$\mu_{e^-}^\circ \neq \mu_n^\circ \neq \mu_p^\circ$$

i: interstitial defect, ν: vacancy defect

The variables are temperature (enters mainly via $\mu^\circ(T)$), the impurity content (enters via electroneutrality equation) and the chemical potential of the neutral component (defined by the chemical environment, e.g. gas phase see Eq. (8.3)). The latter parameter can, for ideal gases, also be expressed in terms of the respective partial pressure in the gas phase. In more general cases, one uses the so-called component activity, defined by

$$\mu(\text{component}) = \mu^\circ(\text{component}) + kT \ln a(\text{component}). \tag{8.4}$$

Note that unlike for defects ideality of the component distribution is usually far from being a good approximation.

If the complexity of the defect chemical situation is low (Brouwer-conditions) and defect distribution ideal (Boltzmann-conditions) the solution for any defect j can be given in terms of a power law

$$c_j(P_k, C, T) = \alpha_j \left(\Pi_k P_k^{N_{kj}} \right) C^{M_j} \left(\Pi_r K_r(T)^{\gamma_{rj}} \right) \tag{8.5}$$

α_j are simple rational numbers, P_k denotes the partial pressure of the reversibly exchangeable component k, C is the doping content and comprises all frozen-in charged defects, K_r is the mass action constant of the defect chemical reaction r. The exponents N_{kj}, M_j, γ_{rj} are characteristic for the defect chemical situation under concern and are ideally simple rational numbers.

Equation (8.5) gives rise to sectionally linear van't Hoff diagrams ($\log c = f(1/T)$), Kröger–Vink diagrams ($\log c = f(\log a)$) and analogous diagrams for the doping content $\log c = f(\log C)$ as sketched in Figure 8.1.

Here two remarks may be added.

1. In the context of so-called ab initio thermodynamics that mostly is classic dilute defect chemistry including ab initio calculations of μ°, usually the Fermi level (electrochemical potential of electrons) is used as parameter instead of $\log a$. This is not incorrect, but one refers then to a rather indirect parameter being a function of the electronic carrier concentration rather than to an independent experimental control parameter.

2. In the typical electronic picture, ionic defects are considered as immobile and hence as dopants. This is correct for low enough temperatures. The full picture treats them as mobile defects at high temperatures (with their concentrations following Eq. (8.5)). On cooling, the concentration will freeze at a certain temperature below which we lose a control parameter (P-term) and the now constant concentration enters the doping concentration term C. Then the freezing temperature takes the role of a control parameter. Examples of such involved treatments were described in the following works: Sasaki and Maier (1999a, b), Maier (2003a), and Waser (1991).

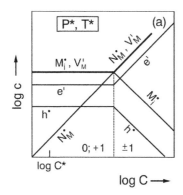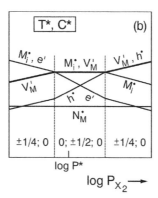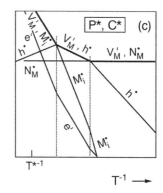

FIGURE 8.1 Typical defect chemistry for a Frenkel disordered compound M^+X^- as a function of doping content realized by substitution of M^+ by N^{2+} (a), X_2 partial pressure (b), and inverse temperature. (c) The working point P^*, T^*, C^* is indicated. (Reprinted from Maier 2004.)

8.2.2 Kinetics

Defect kinetics (cf. Scheme 8.1) can be best treated by simple chemical reaction kinetics. The formalism directly applies for the local reactions but can also be used for describing the transport process by conceiving it as a hopping process with zero standard affinity. A very general flux equation based on this approach has recently been given and can describe situations in which electric and chemical effects occur and in which severe deviations from equilibrium are allowed (Riess and Maier 2008).

$$j_k = -\frac{2kT}{z_k^2 e^2 \Delta x}\sigma_k \sinh\left(\frac{\Delta\tilde{\mu}_k}{2kT}\right). \quad (8.6)$$

$\Delta\tilde{\mu}_k \equiv$ drop of the electrochemical potential of k over the hopping distance Δx; $\tilde{\mu}_k \equiv \mu_k + z_k e\phi$ where $\mu_k \equiv$ chemical potential of k, ϕ = electrical potential; $\sigma_k \equiv$ non-equilibrium conductivity of k.

Close to equilibrium Eq. (8.6) reduces to the classic linear transport equation that combines Ohm's and Fick's law (where σ_k now adopts the equilibrium value)

$$j_k = -\frac{\sigma_k}{z_k^2 e^2}\nabla(\mu_k + z_k e\phi) \quad (8.7)$$

Eq. (8.7) reduces for Boltzmann-conditions to the Nernst–Planck equation

$$j_k = -D_k \nabla c_k - \frac{\sigma_k}{z_k e}\nabla\phi. \quad (8.8)$$

Of special importance is the chemical diffusion (Wagner 1975) that describes the kinetics of stoichiometric changes. Such variation refers to a flux of a neutral component typically enabled by an ambipolar combination of an ionic and an electronic flux. As in the bulk no excess charge can arise, electrochemical flux coupling leads to a common diffusion coefficient D^δ, which is in the case of an oxide, where valence states contribute to ionic motion, given by (Maier 1993):

$$D^\delta = \frac{kT}{4e^2}\left[2\sigma_{O^-} + 4s_{O^0}\right.$$
$$\left. + \frac{(\sigma_{O^{2-}} + 2\sigma_{O^-})(\sigma_{e^-} - \sigma_{O^-})}{\sigma}\right]\frac{d\mu_O}{dc_O} \quad (8.9)$$

For Boltzmann-conditions and with knowledge of defect chemistry $d\mu_O/dc_O$ can be further evaluated as

$$\frac{d\mu_O}{dc_O} = kT\left(\frac{\chi_{O_i''}}{c_{O_i''}} + 4\frac{\chi_{h^\cdot}}{c_{h^\cdot}}\right) = kT\left(\frac{\chi_{V_O^{\cdot\cdot}}}{c_{V_O^{\cdot\cdot}}} + 4\frac{\chi_{e'}}{c_{e'}}\right) \quad (8.10)$$

χ is a differential (un-)trapping factor.

In spite of trapping effects, D^δ can be decomposed into an effective conductivity and an effective concentration. The inverse of the bracketed term in Eq. (8.9) is proportional to a chemical resistance R^δ while the inverse of the last factor is proportional to a chemical capacitance C^δ, the product $R^\delta C^\delta$ yielding the chemical relaxation time τ^δ.

Both quantities are composed of ionic and electronic contributions. Not only is the chemical capacitance C^δ much larger than the dielectric capacitance by many orders of magnitude, it is also proportional to L (sample thickness) rather than to $1/L$. As a consequence, the chemical relaxation time is proportional to L^2/D^δ (Jamnik and Maier 2001).

Figure 8.2 sketches the concentration change on varying the outer partial pressure (fast surface reactions assumed).

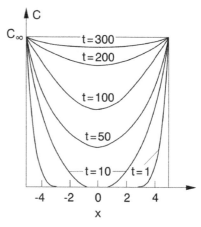

FIGURE 8.2 Concentration profile of diffusion controlled chemical incorporation from two sides in a quasi-one-dimensional situation. (Reprinted from Maier 2004.)

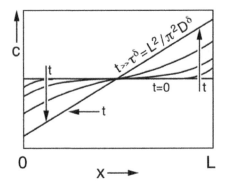

FIGURE 8.3 Concentration profiles in a bulk controlled symmetrical electric polarization with blocking electrodes in a quasi-one-dimensional situation. (Reprinted from Maier 2004.)

Figure 8.3 depicts an example of how defect concentrations change on applying an electric voltage to a sample, which is sandwiched between two usual electrodes. "Usual" means here that the electrodes allow passage for electrons but not for ions. One recognizes that a mixed conductor will unavoidably undergo internal stoichiometric polarization (Wagner 1956, Hebb 1952).

This phenomenon leads to the degradation of dielectric capacitors (Waser et al. 1990) or to low-frequency anomalies in photo-electrodes (Yang et al. 2015). In a constructive sense, it can be used to separate ionic and electronic conductivities or to fabricate resistive switches.

Solid electrolytes are characterized by high carrier concentration and mobility of ionic defects while electronic contributions are negligible. Their employment naturally implies using neighboring phases (electrodes), which allow ions to traverse, as it is the case in batteries. If both carrier types are significantly conductive, transport of the neutral component is enabled characterized by a non-zero chemical diffusion (D^δ) coefficient. Solid electrodes necessitate significant

D^δ values for the component to be stored in due time. Such phases are typically neighbored by phases, which allow only electrons (one side) or ions (other side) to traverse. High D^δ values are necessary for a high power density, while a high storage capacity requires a high δ. Here we have tacitly assumed the case of homogeneous storage. For other bulk storage modes that involve phase transformations the reader is referred to lit (Maier 2013).

8.3 Interfaces

Now we are prepared to address interfaces. For simplicity, we refer to an ideal (strain-free) abrupt contact characterized by step functions of the μ°-parameters. Interfaces of charge carrier containing systems are — except singularities (called point of zero charge) — always charged. The thermodynamic picture then is given by Figure 8.4, where the resulting electric field bends both the electronic and ionic "energy levels" (Jamnik et al. 1995, Maier 2004).

The formal treatment (cf. Scheme 8.3) has to use electrochemical potentials and to replace the electroneutrality condition by Poisson's equation (plus boundary condition). The relative concentration changes that occur in consistency with the electric field depend, for low concentrations, solely on the effective sign of the carriers. As far as the extent of the space charge zones is concerned, there are eventually two cases, the Gouy–Chapman case (Scheme 8.4) characterized by carriers that can follow the electric field (in principle it suffices that the enriched majority carrier does this) and the Mott–Schottky case (Scheme 8.5). In the latter, one majority carrier is depleted and the majority counter carrier (the dopant) spatially frozen. Then the depth of the space charge layers depends on the applied electric potential and is, due to weaker screening, typically significantly larger than the Debye length characterizing the depth of Gouy–Chapman profiles.

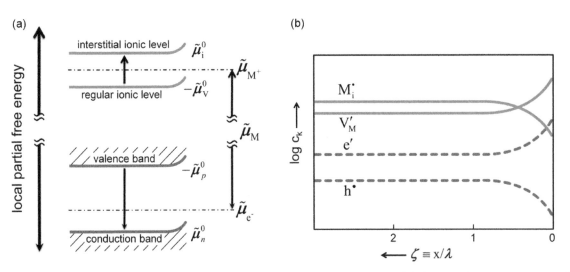

FIGURE 8.4 Ionic and electronic redistribution at a boundary of our ionic compound M^+X^- with positive counter charge in equilibrium. (a) Diagram shows the full thermodynamic picture. (a: Reprinted from Maier 2003b. Copyright (2002), with permission from Elsevier.) (b: Reprinted from Maier 2002. Copyright (2002), with permission from Elsevier.)

Scheme 8.3 Interfacial Thermodynamics: Planar System (α/β)

$$A^\alpha \rightleftharpoons A^\beta$$

Neutral ideal particles

$$\frac{c_A(\beta)}{c_A(\alpha)} = \exp-\frac{\mu^\circ_\beta - \mu^\circ_\alpha}{kT} = \exp-\frac{\Delta_{a/\beta}G^\circ}{kT}$$

Charged ideal particles

$$\left((dG)_{p,T} = \sum \tilde{\mu}_i dn_i\right)$$

$$\tilde{\mu}_i = \mu^\circ_i + kT\ln c_i + z_i e\phi(x)$$

$$\frac{c_i(\beta)}{c_i(\alpha)} = \exp-\frac{\Delta_{a/\beta}G^\circ}{kT}\exp-\frac{ze\Delta_{a/\beta}\phi}{kT}$$

$$\frac{c_i(x)}{c_i(x=\infty)} = \exp-\frac{ze\Delta\phi(x)}{kT}$$

$$\zeta_i^{1/z_i}(x) = \exp-\frac{e\Delta\phi(x)}{kT}$$

Poisson: $\phi'' \propto \sum_i z_i c_i$

Scheme 8.4 Gouy–Chapman Profile

$$\phi'' \propto \rho/F = c_+ - c_- = c_\infty(\zeta_+ - \zeta_-)$$

(Dilute: Poisson–Boltzmann)

$$\zeta_\pm = \left(\frac{1 + \vartheta_\pm \exp-\xi}{1 - \vartheta_\pm \exp-\xi}\right)^2 = \zeta_\mp^{-1}$$

$$\vartheta_\pm \equiv \frac{\zeta_{\pm 0}^{1/2} - 1}{\zeta_{\pm 0}^{1/2} + 1} \equiv -\vartheta_\mp$$

$$\zeta \equiv \frac{c}{c_\infty}; \xi \equiv \frac{x}{\lambda}$$

$$\lambda \equiv \sqrt{\frac{\varepsilon kT}{2e^2 c_\infty}}\,(\text{Debye length})$$

Effective excess parallel conductivity:

$$\sigma_m^\parallel = \sum_\pm (2\lambda/L) e u_\pm \left[2c_\infty \frac{\vartheta_\pm}{1 - \vartheta_\pm}\right]$$

Pronounced accumulation of mobile majority carrier 1:

$$\sigma_m^\parallel \simeq z_1 e u_1 (2\lambda/L)\sqrt{c_{10}c_\infty} = u_1\sqrt{2\varepsilon kT c_0}/L$$

$$\left(\zeta_1 \simeq \frac{\zeta_{10}}{(1 + \sqrt{\zeta_{10}}\xi/2)^2}\right)$$

Scheme 8.5 Mott–Schottky Profile

(Majority carrier 1 immobile, majority carrier 2 depleted, z_1, z_2: charge numbers)

$$\varsigma_1 = 1$$

$$\zeta_2 = \exp-\left|\frac{z_2}{z_1}\right|\left(\frac{x - \lambda^*}{2\lambda}\right)^2 = \exp-\left|\frac{z_2}{z_1}\right|\left(\frac{\xi - \xi^*}{2}\right)^2$$

$$\lambda^* = \lambda\sqrt{4\frac{z_1}{z_2}\ln\zeta_{20}}\,\,(\text{Mott–Schottky length})$$

Effective excess serial resistivity:

$$\Delta\rho^\perp \simeq \frac{\lambda^*/L}{|z_2|\,eu_2 c_2^*}$$

$$c_2^* \equiv c_{20}2\ln(c_{2\infty}/c_{20})$$

The kinetic treatment of charge transport has naturally to also include electric potentials, and hence the gradients of the electrochemical potential have to be taken seriously. Note that the presence of the interface introduces an anisotropy. Boundary layers that lower the conductivity are sensitively felt if electric measurements are performed perpendicular to the interface but essentially overlooked if the measurement occurs along the interfaces. The situation for enhanced boundary conductivities is exactly opposite. In a polycrystal or composite a superposition of all these effects occurs and can be handled in the case of a brick-layer system (Maier 1986).

Extensive research of the last decades has shown that interfacial charging due to ionic effects is almost ubiquitous in ionic systems and of enormous influence for ionic and electronic phenomena. In the retrospect, this is in fact not surprising as often ionic point defects are in majority and sufficiently mobile. A well-investigated example is polycrystalline $SrTiO_3$, where the electronic properties are strongly varied near grain boundaries as a consequence of the grain boundary core charge stemming from oxygen vacancy enrichment ("fellow traveler effect") (Maier 1989c, b).

Let us consider pure ionic effects first and consider several instructive examples.

So to speak the starting point of nanoionics was the finding that admixtures of insulating Al_2O_3 particles to (weak) ion conductors such as AgCl (Maier 1989b) or LiI (Liang 1973) enhances the ionic conductivity (Ag^+ or Li^+) by orders of magnitude. The reason was found to be internal cation adsorption at the alumina's surface and thus the creation of very mobile cation vacancies in the space charge zones adjacent (Maier 1995). Related effects occur at interfaces of two ionic conductors such as AgCl/AgI (Ag^+ redistribution) or CaF_2/BaF_2 (F^- redistribution) (Figure 8.5), where both space charge zones are involved ($\upsilon - i$ junction) (Maier 1995).

F^- adsorption when SiO_2 is used as adsorptive leads to an F^- vacancy enrichment. Recently such effect has been

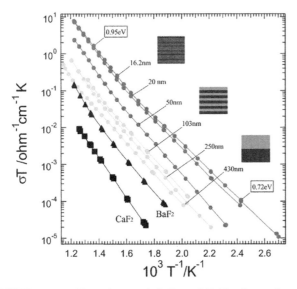

FIGURE 8.5 Heterolayers of CaF_2 and BaF_2 show substantial F^- conductivity variations as a function of individual layer thickness. Black symbols: isolated bulk phases. Light symbols: semi-infinite heterolayers. Gray symbols: heterolayers under finite boundary conditions. (Reprinted from Sata et al. 2000. Copyright (2000), with permission from Springer Nature.)

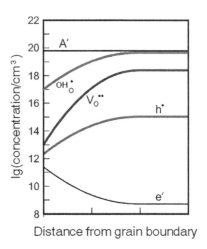

FIGURE 8.6 Behavior of protonic and oxidic point defects, as well as of electron holes and excess electrons in an acceptor-doped (ionically compensated) oxide near a positively charged boundary. (Reprinted from Maier 2014a. Copyright (2014), with permission from Springer.)

exploited for improving liquid Li-electrolytes ("soggy-sand electrolytes"). Here it is the anion of the Li-salt that is adsorbed. In this way, ion pairs are broken, the Li^+ conductivity is increased and the unfavorable anion conductivity is decreased (Pfaffenhuber et al. 2013).

An instructive heterojunction is the contact LiF/TiO_2 (Li et al. 2012). As LiF is Schottky-disordered, and TiO_2 can accommodate Li, a Li^+ transfer from LiF to TiO_2 is expected. Indeed an increased V'_{Li} conductivity was detected in LiF near the interface. The necessary Li_i defects in TiO_2 could not be directly observed, but one could observe the indirect effect on the electronic carriers in TiO_2, the conductivity of which is dominant; while pristine TiO_2 is n-type, it turns p-type at the contact owing to the field induced by Li_i enrichment. This is a nice example of the aforementioned fellow traveler effect. Very recent results point towards qualitatively similar effects at junctions of TiO_2 and halide perovskites in photoelectrochemical cells (Kim et al. 2018).

Ionic depletion layers have been investigated at grain boundaries of $SrTiO_3$ (Gregori et al. 2017). Here grain boundaries are characterized by an O^{2-} deficiency of the core leading to a depletion of the oxygen vacancy concentration in the space charge zones. At this example the full defect chemical situation could be studied, i.e. effects on $V_O^{··}$, e', $h^·$, even OH_O (OH_O corresponds to an extra H^+ on a regular O^{2-} and is of relevance in the presence of H_2O) (Figure 8.6).

According to Figure 8.6, hole conductivity decreases, the oxygen vacancy conductivity decreases, the conduction electron conductivity increases, and − if introduced − proton conductivity decreases. Not only are all these effects qualitatively in agreement with the expectations, they are also quantitatively understood.

Analogous effects are observed near dislocations (Adepalli et al. 2013, Adepalli et al. 2017). This is not surprising as low-angle grain boundaries can be thought as being composed of dislocation cores. In fact, the atomistic spacing of dislocation cores when arranged to a grain boundary network is already a genuine size effects as considered below (de Souza et al. 2005).

All the effects discussed are consequences of an introduced interfacial or dislocation core charge. Apart from the fact that the concentration variations are restricted to the immediate environment, there is a far-reaching formal similarity to the conventional (homogeneous) doping (Figure 8.7). A positive (negative) charge of the higher-dimensional defect introduced increases all the negatively (positively) charged mobile defects in the vicinity, while the response of the positively (negatively) charged ones is opposite. For this reason, the term heterogeneous doping (or higher-dimensional doping) was coined. Introducing active grain boundaries could also be called two-dimensional doping, and the introduction of charged dislocations as one-dimensional doping (as opposed to zero-dimensional doping if point defects are introduced as impurities). An interesting case of two-dimensional doping where the core charge is a priori known was discussed in Baiutti et al. (2015) (Figure 8.7). Molecular beam epitaxy allows La-O layers in La_2CuO_4 to be replaced in an atomically sharp manner by Sr-O layers. In the same way as substitutional Sr-defects (Sr^{2+} instead of Ca^{3+}) are negatively charged and give rise to an increased hole concentration; in case of two-dimensional doping the introduced layer exhibits a negative "interfacial" charge density, and holes are enriched in the space charge zones. In the same way as at low temperature, superconductivity is homogeneously introduced in the classic case, here superconductivity concentrates on the space charge zone. Interestingly, the very first layer is not superconductive that was attributed to the very

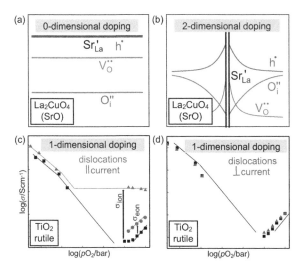

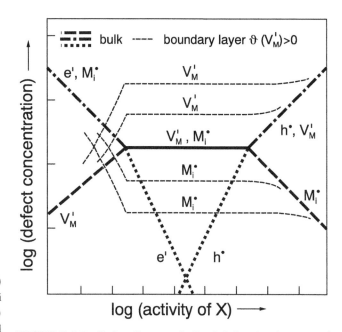

FIGURE 8.7 (a) Zero-dimensional doping in La_2CuO_4. (b) Two-dimensional doping in La_2CuO_4. (Reprinted from Baiutti et al. 2015. Copyright (2014), with permission from Springer.) (c,d) One-dimensional doping in TiO_2 in different directions. (Reprinted from Adepalli et al. 2013. Copyright (2013), with permission from Wiley-VCH.)

FIGURE 8.8 Defect diagram of a Frenkel disordered compound M^+X^- for different distances from the positively charged interface. Variation of X-activity corresponds to a storage situation. (Bold lines: bulk). (Reprinted from Maier 1989a Copyright (1989), with permission from Elsevier.)

steeply increased vacancy concentration (oxygen vacancies are doubly positive charged and enriched more steeply).

In contrast to a polycrystalline situation, a $SrTiO_3$ single crystal (Adepalli et al. 2017) in which dislocations are induced in one crystallographic direction shows conductivity enhancement only in this direction, while in the other direction bulk behavior is observed. In the case of depletion effects, bulk behavior is also observed along the dislocation lines as here bulk is always dominant. In TiO_2 such anisotropy features could be nicely confirmed (Figure 8.7) (Adepalli et al. 2013).

So far we restricted ourselves to charge transport. When stoichiometric variations are addressed (chemical diffusion) the kinetics are more complex. The reader is referred to ref. Maier (2004) for more details. Here it suffices to add two comments:

1. If chemical diffusion through space charge zones (perpendicular to the boundary) is concerned, one observes – in the steady state – a slowing down, as the ambipolar motion is determined by the sluggish component, which is typically the one that is depressed when compared to the bulk. (Interestingly, this can be violated if large driving forces are applied, see Eq. (8.6) (Riess et al. 2013).)

2. If chemical diffusion along interfaces is concerned, the description needs to be at least two-dimensional as gradients along and across occur.

Space charge effects also give rise to storage anomalies. This already follows from Figure 8.8 showing Kröger–Vink diagrams of boundary regions (Maier 1989a). An extreme situation is addressed in the case of job-sharing storage, where storage of e.g. Li $(= Li^+ + e^-)$ occurs in composites of two phases (e.g. Ru/Li_2O) of which none is able to store Li (i.e. Li^+ and e^-) (Fu et al. 2014). This is enabled by

decoupling the roles of the two carriers. The effect is of significant practical importance as the materials space for electrodes can be largely extended (Chen and Maier 2018). Furthermore, the charging process – which is a coupled ionic/electronic transport (chemical diffusion process) as described above – has the potential to be very fast as also transport is decoupled in that ions and electrons can take different pathways. The detailed thermodynamics and kinetics have been recently worked out (Chen and Maier 2017, Chen et al. 2018) allowing also supercapacitors to be integrated into the general nanoionics (defect chemical) picture (Figure 8.9).

The rise of nanoionics as an important subfield of solid state ionics is also due to the progress of nanotechnology through which a high density of defined interfaces can be produced. It is not just the high fraction of interfaces ("trivial size effects") that are to the fore but also the interference of closely neighbored interfaces generating exciting "anomalies" ("true size effects").

It is clear that the abrupt core space charge picture is a first approximation. In reality, elastic, plastic and structural effects may be expected to occur. Formally this leads to a variation of the standard chemical potential typically with the consequence of lowering space charge effects.

8.4 Size Effects

Of great fundamental interest is the size-dependent behavior of the considered functionalities. The most straightforward (non-trivial) size effect consists in the overlap of space charge zones (cf. Scheme 8.6) if the spacing of two interfaces is so

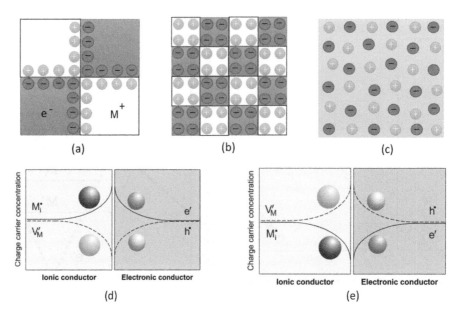

FIGURE 8.9 "Job-sharing" composites show M-excess and M-deficiency owing to space charge effects. (a) semi-infinite, (b) mesosopic, (c) homogeneous limit; (d) M-excess and (c) M-deficit. (Reprinted from Chen and Maier 2018. Copyright (2018), with permission from Springer Nature.)

narrow that in equilibrium substantial electric fields do not disappear in between (Maier 1987).

Scheme 8.6 Space Charge Overlap (Planar)

Conductivity enhancement parallel to the interface (accumulated species: species 1 in Gouy–Chapman, species 3 in Mott–Schottky case)

Gouy–Chapman:

$$\Delta\sigma_m^{\parallel} = 2u_1 \left[2kT\mathcal{E} \left(c_{10} - c_1^{\#} \right) \right]^{1/2}$$

$$\xi^{\#} \equiv 2\sqrt{\frac{c_{1\infty}}{c_1^{\#}}} \left[\mathcal{E}\left(\frac{c_{1\infty}}{c_1^{\#}}, \frac{\pi}{2} \right) - \mathcal{E}\left(\frac{c_{1\infty}}{c_1^{\#}}, \mathrm{Arcsin}\sqrt{c_1^{\#}/c_{10}} \right) \right]$$

$\mathcal{E}$: elliptical integral of first kind
$c_1^{\#}$, $\xi^{\#}$ refer to the center of symmetrical thin film

Mott–Schottky: numerical evaluation using

$$\Delta\phi\left(x\right) = -\frac{c_{2\infty}F}{2\varepsilon} \left[x^2 - Lx + \lambda^{*2} \right]$$

Owing to less step decay, for very thin films boundary values may be used

$$\Delta\rho_m^{\perp} = \frac{1}{eu_2} \left[\frac{1}{c_{20}} - \frac{1}{c_{2\infty}} \right] \simeq \frac{1}{eu_2 c_0}$$

cf. (Maier 2007a).

A well-investigated example is CaF_2/BaF_2 heterolayers (Figure 8.5) (Sata et al. 2000). Not only is for very small spacing the conductivity effect significant because of space charge overlap, also a curvature change in the conductance-size characteristic was observed. This was attributed to the fact that overlapped Mott–Schottky layers are involved (rather than overlapped Gouy–Chapman layers) (Guo and Maier 2009).

Even more illustrative is the situation in $SrTiO_3$ nanocrystals (Figure 8.10). The behavior of n-, p- and v-conductivities could be quantitatively explained as a function of oxygen partial pressure. Like in a single crystal, under such mesoscopic conditions a quasi-homogeneous situation occurs but with strongly modified defect densities (Lupetin et al. 2010). In this context it is worthy of mention that even single low-angle grain boundaries can be thought of being a superposition of dislocation cores the charge effects of which overlap (de Souza et al. 2005)

Not only do dislocations give rise to space charge effects similar to polycrystalline $SrTiO_3$, very dense dislocation networks can also show space charge overlap (Adepalli et al. 2017).

As far as electrochemical interfacial storage is concerned, a high energy density is expected if the spacing is tiny. In

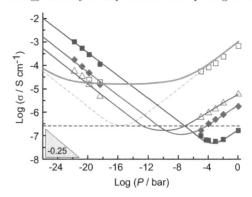

FIGURE 8.10 Nanocrystalline $SrTiO_3$ (of various grain sizes) shows substantial depression of p-type and vacancy-type but increase of n-type conductivity. Open squares refer to the bulk of the microcrystalline $SrTiO_3$ obtained by coarsening the porous nanocrystalline sample. (Reprinted from Lupetin et al. 2010. Copyright (2010), with permission from Wiley-VCH.)

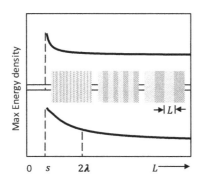 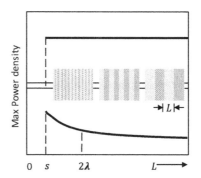

FIGURE 8.11 Size reduction in job-sharing compositions shows increase of maximum energy density and increase or constancy of power density. (Reprinted from Chen and Maier 2018. Copyright (2018), with permission from Springer Nature.)

contrast to bulk storage, this is not at the expense of power density as shown in Figure 8.11 (Chen and Maier 2018).

Even though the most straightforward one, space charge overlap is not the only mesoscopic size effect. Two more which are of fundamental thermodynamic interest shall now be discussed in the context of storage properties.

As well-known the chemical potential contains a capillary contribution (cf. Scheme 8.7) if small particles are considered (Maier 2009). In complete equilibrium the related increased capillary pressure is isotropic, as for a Wulff-shaped crystallite (and trivially for a fluid droplet), the ratio of surface tension and distance (γ/r) from the center is the same for all faces.

Core fraction: $\varphi^s \equiv V^s/V = 3s/\bar{r}$

$$\left(\begin{array}{c} s = 5\text{Å} \\ \\ \bar{r} = 50\text{Å} \end{array} \quad \text{then } \varphi^s = 30\% \right)$$

Cubic crystal: $L = 2\bar{r}$
(s: thickness of surface layer)

$$\varphi^s \simeq \frac{6s}{L} = \frac{6N^{2/3}}{N} = 6N^{-1/3}$$

(φ^s: volume fraction of surface layer)

More precisely: $\varphi^s = \frac{N - \left(N^{1/3} - 2\right)^3}{N} = 6N^{-1/3} - 12N^{-2/3} + 8N^{-1}$

Scheme 8.7 Interfacial Thermodynamics: Particles (MX)

Wulff theorem: $\gamma_j/h_j = \text{const}$
(γ_j: surface tension, h_j: distance from surface to center, of face j)
Increased chemical potential:

$$\mu_{\text{MX}} = \mu_{\text{MX}}^\infty + 2\frac{\overline{\gamma}}{\overline{r}}V_m \equiv \mu_{\text{MX}}^\infty + \alpha V_{\text{MX}}$$

$$\mu_{\text{M}} = \mu_{\text{M}}^\infty + \alpha V_{\text{M}}$$

$$\mu_{\text{defect}} = \mu_{\text{defect}}^\infty + \alpha V_{\text{defect}}$$

Sphere: $\overline{\gamma} = \gamma, \overline{r} = r$

Crystal: $\alpha \equiv \overline{\gamma}/\overline{r} = \gamma_j/h_j$ (otherwise μ_{MX} would not be uniform)

$$\overline{\gamma} = \sum_j a_j \gamma_j/a = \frac{\gamma_j}{h_j} \sum_j a_j h_j/a = \frac{\gamma_j}{h_j}\overline{r}$$

$$\overline{r} = 3V/a$$

More precisely: $\text{æ} \neq 0 \neq \omega$ (æ: edge tension, ω: corner tension)

$$\frac{1}{h_j}\left(\gamma_j + \sum_k \text{æ}_k \frac{\partial L_k}{\partial a_j} \right) = \text{const}$$

More generally: $\gamma, \text{æ}, \omega$ depend on $\overline{r}$

A simple but illustrative experiment is the investigation of the open-circuit potential of the cell

$$\text{Ag (macro)} | \text{silver electrolyte} | \text{Ag (nano)} \qquad (8.11)$$

which showed a small but significant non-zero emf that can be related to the capillary effect (Schroeder et al. 2006).

In the context of storage effects also voltage anomalies for RuO_2, $LiFePO_4$ and MoS_2 have been explained (Maier 2013). Of special interest is the extra emf for amorphous materials, which can be as high as several 500mV. In Delmer et al. (2008), it has been suggested that this value could be estimated as extra emf of a crystallite of atomistically small size (cf. also Maier 2014b). Formally speaking, this is an effect affecting μ°. This term is also varied if elastic or structural variations occur defining another length scale. Overlap phenomena due to this length scale can easily lead to phase transformations (Maier 2009).

The second effect to be discussed refers to the configurational entropy caused by the statistics of individual tiny particles rather than that of individual atoms (Maier 2009). Let us consider a concrete example namely Li storage in $FePO_4$, which exceeds the small solubility limit and leads to the formation of the Li-richer phase $LiFePO_4$. Let us assume a Li storage of 50% related to the Fe content in $FePO_4$. In a multiparticle system, various possibilities may occur to realize such a situation. In the simplest case, all particles are, in the example taken, composed of 50% $FePO_4$ and 50% $LiFePO_4$, e.g. in a core–shell arrangement. This corresponds to a very homogeneous particle arrangement but with an interface in any particle. For smaller size where the energy cost of interfaces per particle is significant, a bimodal distribution may be preferred, i.e. 50% of the particles consist of (Li saturated) $FePO_4$ and 50% of (Li deficient) $LiFePO_4$. At even smaller particles it has been shown theoretically that the miscibility gap can disappear being equivalent to a homogeneous particle situation in which any particle is composed of a (macroscopically unstable) $Li_{0.5}FePO_4$ (Abdellahi et al. 2016) phase. (Interestingly local energy minima complicate the situation additionally (Gu et al. 2011).) The high concentration of particles – if size is extremely small – may then even demand a fluctuation of the 50% value from particle to particle. Figure 8.12 refers to the distribution statistics of atoms, clusters and crystals.

If diffusion is the limiting process in the storage kinetics (consider intercalation electrodes), size reduction is a helpful recipe as the diffusional time constant is proportional to the square of the size. Simultaneously the number of particles scaling inversely with the cube of size increases substantially. In other words when the electrode particles are strongly down-sized, the problem is shifted to the network connecting the myriads of particles. Then solutions, which make the nanocircuitry most efficient, are decisive. In Zhu et al. (2017), it is shown that adaption of the circuitry to the ratio of electronic and ionic conductivity is worthwhile.

A relatively new application of nanoionics is resistive switching (Waser and Aono 2007). Here it is the mixed conduction of the materials used that enable the reversible growth of a very conductive filament which according to its length does or does not short-circuit the electrodes.

Another application of solid state ionics – and in the case of miniaturization, of nanoionics – refers to the important field of sensors and actuators.

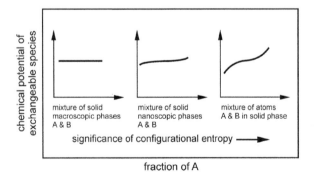

FIGURE 8.12 While in a macroscopic ensemble a certain energy gap suffices to "localize" the system in the ground state situation (left level scheme), for atomistic species (right level scheme) the gain in configurational entropy is great enough to lead to a distribution over ground state and excited state at non-zero temperatures. The nanoscopic case (center level scheme) takes an intermediate position. These configurational effects lead to smearing out of transitions that are of first order in the macroscopic case. (Reprinted from Maier 2009. Copyright (2009), with permission from Wiley-VCH.)

Certainly also in areas where ionic carriers are not directly essential for the function, nanoionics is implicitly of importance if ionic components are involved.

8.5 Summary

Nanoionics takes variations of ionic charge carrier densities at interfaces (or dislocations) seriously and becomes particularly relevant if the distances between higher-dimensional structure elements are so small that mesoscopic effects occur. The coupling with electronic effects underscores the generality of the approach. Various examples, mostly from energy research in which ionic charge carriers are explicitly important, highlight the significance of the field. Similarly relevant, but not as well investigated, are the implicit effects in the fields of solid state physics, catalysis, photoelectrochemistry and solid state chemistry.

References

Abdellahi, A., Akyildiz, O., Malik, R., Thornton, K. & Ceder, G. 2016. The thermodynamic stability of intermediate solid solutions in LiFePO4 nanoparticles. *Journal of Materials Chemistry A* 4: 5436–5447.

Adepalli, K. K., Kelsch, M., Merkle, R. & Maier, J. 2013. Influence of line defects on the electrical properties of single crystal TiO2. *Advanced Functional Materials* 23: 1798–1806.

Adepalli, K. K., Yang, J., Maier, J., Tuller, H. & Yildiz, B. 2017. Tunable oxygen diffusion and electronic conduction in SrTiO3 by dislocation-induced space charge fields. *Advanced Functional Materials* 27: 1700243(1–9).

Baiutti, F., Logvenov, G., Gregori, G., Cristiani, G., Wang, Y., Sigle, W., van Aken, P. A. & Maier, J. 2015. High-temperature superconductivity in space-charge regions of lanthanum cuprate induced by two-dimensional doping. *Nature Communications* 6: 8586(1–8).

Chen, C.-C. & Maier, J. 2018. Decoupling electron and ion storage and the path from interfacial storage to artificial electrodes. *Nature Energy* 3: 102–108.

Chen, C. C. & Maier, J. 2017. Space charge storage in composites: Thermodynamics. *Physical Chemistry Chemical Physics* 19: 6379–6396.

Chen, C. C., Navickas, E., Fleig, J. & Maier, J. 2018. Kinetics of space charge storage in composites. *Advanced Functional Materials* 28: 1705999(1–11).

de Souza, R. A., Fleig, J., Maier, J., Zhang, Z., Sigle, W. & Rühle, M. 2005. Electrical resistance of low-angle tilt grain boundaries in acceptor-doped SrTiO3 as a function of misorientation angle. *Journal of Applied Physics* 97: 053502(1–7).

Delmer, O., Balaya, P., Kienle, L. & Maier, J. 2008. Enhanced potential of amorphous electrode materials: Case study of RuO2. *Advanced Materials* 20: 501–505.

Fu, L., Chen, C.-C., Samuelis, D. & Maier, J. 2014. Thermodynamics of lithium storage at abrupt junctions:

Modeling and experimental evidence. *Physical Review Letters* 112: 208301(1–5).

Gregori, G., Merkle, R. & Maier, J. 2017. Ion conduction and redistribution at grain boundaries in oxide systems. *Progress in Materials Science* 89: 252–305.

Gu, L., Zhu, C., Li, H., Yu, Y., Li, C., Tsukimoto, S., Maier, J. & Ikuhara, Y. 2011. Direct observation of lithium staging in partially delithiated $LiFePO_4$ at atomic resolution. *Journal of the American Chemical Society* 133: 4661–4663.

Guo, X. & Maier, J. 2009. Comprehensive modeling of ion conduction of nanosized CaF_2/BaF_2 multilayer heterostructures. *Advanced Functional Materials* 19: 96–101.

Hebb, M. H. 1952. Electrical conductivity of silver sulfide. *Journal of Chemical Physics* 20: 185–190.

Jamnik, J. & Maier, J. 2001. Generalised equivalent circuits for mass and charge transport: Chemical capacitance and its implications. *Physical Chemistry Chemical Physics* 3: 1668–1678.

Jamnik, J., Maier, J. & Pejovnik, S. 1995. Interfaces in solid ionic conductors: Equilibrium and small signal picture. *Solid State Ionics* 75: 51–58.

Kim, G. Y., Senocrate, A., Moia, D., & Maier, J. 2018. Ionically generated built-in equilibrium space charge zones - a paradigm change for lead halide perovskite interfaces, *Arxiv*. http://arxiv.org/abs/1907.13618

Kofstad, P. 1972. *Nonstoichiometry, Diffusion and Electrical Conductivity in Binary Oxides*. New York: Wiley

Kröger, F. A. 1964. *Chemistry of Imperfect Crystals*. Amsterdam: North-Holland Publishing Company.

Li, C., Gu, L., Guo, X., Samuelis, D., Tang, K. & Maier, J. 2012. Charge carrier accumulation in lithium fluoride thin films due to Li-ion absorption by titania (100) subsurface. *Nano Letters* 12: 1241–1246.

Liang, C. C. 1973. Conduction characteristics of lithium iodide aluminium oxide solid electrolytes. *Journal of the Electrochemical Society* 120: 1289–1292.

Lupetin, P., Gregori, G. & Maier, J. 2010. Mesoscopic charge carriers chemistry in nanocrystalline $SrTiO_3$. *Angewandte Chemie International Edition* 49:10123–10126.

Maier, J. 1986. On the conductivity of polycrystalline materials. *Berichte der Bunsengesellschaft für Physikalische Chemie* 90: 26–33.

Maier, J. 1987. Space charge regions in solid two-phase systems and their conduction contribution - III: Defect chemistry and ionic conductivity in thin films. *Solid State Ionics* 23: 59–67.

Maier, J. 1989a. Kröger-vink diagrams of boundary regions. *Solid State Ionics* 32–33: 727–733.

Maier, J. 1989b. Space charge regions in solid two phase systems and their conduction contribution. IV. The behaviour of minority charge carriers. Part A: Concentration profiles, conductivity contribution, determination by generalized wagner-hebb-procedure. *Berichte der Bunsengesellschaft für Physikalische Chemie* 93: 1468–1473.

Maier, J. 1989c. Space charge regions in solid two phase systems and their conduction contribution. IV. The behaviour of minority charge carriers. Part B: Application to the interfaces $AgCl/\gamma\ Al_2O_3$ and SnO_2/O_2. *Berichte der Bunsengesellschaft für Physikalische Chemie* 93: 1474–1479.

Maier, J. 1993. Mass transport in the presence of internal defect reactions - concept of conservative ensembles: I, chemical diffusion in pure compounds. *Journal of the American Ceramic Society* 76: 1212–1217.

Maier, J. 1995. Ionic conduction in space charge regions. *Progress in Solid State Chemistry* 23: 171–263.

Maier, J. 2002. Nano-sized mixed conductors (Aspects of nano-ionics. Part III). *Solid State Ionics* 148: 367–374.

Maier, J. 2003a. Complex oxides: High temperature defect chemistry vs. Low temperature defect chemistry. *Physical Chemistry Chemical Physics* 5: 2164–2173.

Maier, J. 2003b. Defect chemistry and ion transport in nanostructured materials. Part II. Aspects of nanoionics. *Solid State Ionics* 157: 327–334.

Maier, J. 2004. *Physical Chemistry of Ionic Materials. Ions and Electrons in Solids*. Chichester: Wiley

Maier, J. 2005a. Nanoionics: Ion transport and electrochemical storage in confined systems. *Nature Materials* 4: 805–815.

Maier, J. 2005b. Solid state electrochemistry I: Thermodynamics and kinetics of charge carriers in solids. In *Modern Aspects of Electrochemistry*, eds. B. E. Conway, C. G. Vayenas & R. E. White, pp. 1–173. New York: Kluwer Academic/Plenum Publishers.

Maier, J. 2007a. Mass storage in space charge regions of nanosized systems (Nano-ionics. Part V). *Faraday Discussions* 134: 51–56.

Maier, J. 2007b. Solid state electrochemistry II: Devices and techniques. In *Modern Aspects of Electrochemistry*, eds. C. Vayenas, R. E. White & M. E. Gamboa-Aldeco, pp. 1–138. New York: Springer.

Maier, J. 2009. Thermodynamics of nanosystems with a special view to charge carriers. *Advanced Materials* 21: 2571–2585.

Maier, J. 2013. Thermodynamics of electrochemical lithium storage. *Angewandte Chemie International Edition* 52: 4998–5026.

Maier, J. 2014a. Nanoionics at high temperatures. In *Encyclopedia of Applied Electrochemistry*, eds. G. Kreysa, K. Ota, R. F. Savinell & T. Ishihara, pp. 1341–1346. New York: Springer Science+Business Media.

Maier, J. 2014b. Pushing nanoionics to the limits: Charge carrier chemistry in extremely small systems. *Chemistry of Materials* 26: 348–360.

Pfaffenhuber, C., Goebel, M., Popovic, J. & Maier, J. 2013. Soggy-sand electrolytes: Status and perspectives. *Physical Chemistry Chemical Physics* 15: 18318–18335.

Riess, I., Kalaev, D. & Maier, J. 2013. Currents under high driving forces. *Solid State Ionics* 251: 2–8.

Riess, I. & Maier, J. 2008. Symmetrized general hopping current equation. *Physical Review Letters* 100: 205901(1–4).

Sasaki, K. & Maier, J. 1999a. Low-temperature defect chemistry of oxides. I. General aspects and numerical calculations. *Journal of Applied Physics* 86: 5422–5433.

Sasaki, K. & Maier, J. 1999b. Low-temperature defect chemistry of oxides. II. Analytical relations. *Journal of Applied Physics* 86: 5434–5443.

Sata, N., Eberman, K., Eberl, K. & Maier, J. 2000. Mesoscopic fast ion conduction in nanometre-scale planar heterostructures. *Nature* 408: 946–949.

Schmalzried, H. 1981. *Solid State Reactions.* New York: Academic Press Inc.

Schroeder, A., Fleig, J., Maier, J. & Sitte, W. 2006. Inherent emf relaxation of electrochemical cells with nanocrystalline Ag electrodes. *Electrochimica Acta* 51: 4176–4181.

Wagner, C. 1956. Galvanische Zellen mit festen Elektrolyten mit gemischter Stromleitung. *Zeitschrift für Elektrochemie* 60: 4–7.

Wagner, C. 1975. Equations for transport in solid oxides and sulfides of transition metals. *Progress in Solid State Chemistry* 10: 3–16.

Wagner, C. & Schottky, W. 1930. Theory of controlled mixed phases. *Zeitschrift für physikalische Chemie* 11: 163–210.

Waser, R. 1991. Bulk conductivity and chemsitry of acceptor-doped strontium-titanate in the quenched state. *Journal of the American Ceramic Society* 74: 1934–1940.

Waser, R. & Aono, M. 2007. Nanoionics-based resistive switching memories. *Nature Materials* 6: 833–840.

Waser, R., Baiatu, T. & Haerdtl, K.-H. 1990. dc electrical degradation of perovskite-type titanates: I, ceramics. *Journal of the American Ceramic Society* 73: 1645–1653.

Yang, T.-Y., Gregori, G., Pellet, N., Graetzel, M. & Maier, J. 2015. The significance of ion conduction in a hybrid organic-inorganic lead-iodide-based perovskite photosensitizer. *Angewandte Chemie International Edition* 54: 7905–7910.

Zhu, C., Usiskin, R. E., Yu, Y. & Maier, J. 2017. The nanoscale circuitry of battery electrodes. *Science* 358: eaao2808(1–8).

Structure-Dynamic Approach of Nanoionics

A.L. Despotuli and
A.V. Andreeva
Russian Academy of Science

9.1   Introduction

The theory of fast ion transport (FIT) on a nanoscale – structure-dynamic approach (SDA) of nanoionics (Despotuli and Andreeva 2015) is surveyed in the context of a development of solid-state ionics (Sunandana 2015) and nanoionics (Despotuli and Nikolaichik 1993). Nanoionics is a new interdisciplinary branch of science and technology. SDA concerns methods of nanoionics.

All facts and observations are "theory-loaded". Therefore, in terms of (Popper 1959, Schuster 1995), for new scientific discipline, "a generally accepted problem-situation", i.e., an opportunity "to fit contribution into the framework of scientific knowledge", is not typical. It impels researchers "to begin afresh from the beginning" and consider "problems of extension" instead of "problems of fit".

The area of nanomaterials and nanostructures embraces a set of fundamental and applied directions. Most of them relate to nanosystems of electron conductors and/or dielectrics. There are also materials with FIT, i.e., superionic conductors (SICs), and solid electrolytes (SEs) (Despotuli and Andreeva 2010). Solid-state ionics (Sunandana 2015), supporting and supplementing the solid-state electronics, discovers a variety of applications in the modern world of high technologies. For example, ionic nanostructures play an important role in nanoelectronics and

microsystem technique that require hetero-integration of semiconductor devices, energy and power sources, sensors, actuators, etc.

Modeling and simulations of "electronic" properties and characteristics of nanostructures and nanodevices have reached a very high level. The opposite situation exists in the area of solid-state materials with ionic conductivity. For example, Materials Studio (http://accelrys.com), Atomistix ToolKit (http://quantumwise.com) and COMSOL (www. comsol.com), i.e., the leading software packages for nanoscience and technical physics, do not have any competence in modeling and simulations of space–time phenomena and effects connected with ionic transport on a nanoscale. It impedes R&D of ionic nanostructures and nanodevices.

Promising applications of ionic functional materials are wide (Despotuli and Andreeva 2018): nanoelectronics, next-generation information and communication technologies (including, e.g., neuromorphic systems for the mimicking of neuronal synapses functions), heterojunctions for energy storage/conversion, catalysts, micro/nanosystem technology, sensors and electrostatic rocket engines. On the way towards digital manufacturing technologies (www.oecd.org/mcm/documents/C-MIN-2017-5-EN.pdf), which require the information about objects during designing, processing, postprocessing and controlling stages, the absence of modern tools for a computer modeling of ion transport in nanostructures is an inadmissible situation.

Since the theoretic foresight (Lehovec 1953) and experimental discovery (Liang 1973), it has been recognized that ionic conductivity σ of interface regions (e.g., composite nanoscale materials) is usually significantly higher than in a bulk. As the next step towards nano-objects, the existence of a new branch of science and technology (nanoionics) differing by controllable and tunable ionic hopping transport on a nanoscale has been identified (Despotuli and Nikolaichik 1993).

Currently, nanoionics focuses on the area of memristive heterostructures. Appropriate nanodevices are physically reconfigurable systems with the changing of chemical composition and electrical resistance, which are defined by ionic processes on a nanoscale (Frascaroli et al. 2018, Wang et al. 2017). Nanoionic memristors combine the processing and storage capabilities induced by an external electrical stimulus like in nerve structures. R&D of memristors for both information storage and neuromorphic systems requires theoretical tools for the modeling of ionic transport in different conditions.

Various models proposed for the design and description of memristive systems are reviewed in (Panda et al. 2018). The review is addressed to the modeling community and claims, "It covers all the important models reported till now and elucidates their features and limitations" (Panda et al. 2018). According to Despotuli and Andreeva (2018), these models of nanoscopic memristors are mainly based on the relationships and equations of a macroscopic type. The reviewed models contain a set of fitting parameters and can be classified as phenomenological ones. These models operate with dynamic average variables and parameters for which distribution functions and laws of correct averaging are unknown. Even if the stochastic property is included in the model of a memristor, it is still an unfounded approximation. Therefore, the question is: What approximations for distribution functions of state variables and for laws of averaging of local variables do we need to use for ionic devices with a cross-section diameter of ~100 nm, ~10 nm and less? Or, how must the width of a tunnel barrier in memristor and the potential difference on the barrier be averaged? Answers to similar questions are important for storage/conversion energy devices, sensor, etc., where formation/relaxation of ionic space charge quickly occurs in non-stationary modes.

One more unsolved problem relates to condensed matter physics. It is universal dynamic response (UDR), i.e., the power law of conductivity that had been discovered in (Jonscher 1977),

$$\mathrm{Re}\,\sigma^*(\omega, T) \propto \omega^\alpha, \qquad (9.1)$$

where $\sigma^*(\omega, T)$ is the real part of the complex (thermally activated) conductivity, and $\alpha < \approx 1$. The law (9.1) holds in a wide frequency range. There has been no consensus until now on a standard theoretical explanation of the reasons and mechanisms of the physical averaging leading to the emergence of UDR in macroscopic solid ionic conductors.

To date (the end of 2018), about 9,500 references had been made on two works by A. K. Jonscher devoted to UDR.

In the literature, there are many ideas (interpretations) related to the law (9.1). According to Funke (1991), "Structurally disordered solid electrolytes, both crystalline and glassy, as well as ionic melts, exhibit a set of spectroscopic peculiarities for ionic conductivity that is at variance with the predictions of simple random-hopping models". The negation of simple hopping models means that the macroscopic behavior is defined by the existence of unknown complex transient states of mobile ions. However, the remarkable "universality" (9.1) refers to the independence of UDR from physical and chemical structures also as from details of ion–ion interactions.

According to Habasaki et al. (2017) "The universal properties found suggest they originate from some fundamental physics governing the motion of the ions". This statement can be understood as the existence of an unknown fundamentality in condensed matter physics. Note, the proposition by Macdonald (2000) is more applicable for the analysis of UDR in ionic conductors: "both conductive and dielectric dispersions are simultaneously important in the frequency region of interest". This proposition implies the existence of such well-known fundamentality as the Maxwell displacement currents in experimental samples. In our opinion, the law (9.1) is some result of a space–time averaging of interconnected currents, i.e., ionic hopping currents and Maxwell displacement currents on a nanoscale.

Mainstream for the realization of the idea of interconnected "conductive and dielectric dispersions" is the presentation of processes by the method of complex impedance (Z) of an appropriate equivalent electric circuit (Macdonald 1987, Uchaikin et al. 2016, Abouzari et al. 2009). In this approach, results are presented in a manner like "The Z' and Z'' versus frequency plots are well fitted to an equivalent circuit model. The circuits consist of the parallel combination of resistance (R), fractal capacitance (CPE) and capacitance (C). Furthermore, the frequency-dependent AC conductivity obeys Jonscher's universal power law" (Rhimi 2018). However, for a macroscopic sample, the physical sense of an equivalent electric circuit appears if each unit of the circuit means an elementary process. In this case, the whole circuit mimics the result of physical averaging for a set of elementary processes. Note, fractal capacitances, or, in other words, constant phase elements (CPEs), are phenomenological macroscopic objects without standard physical interpretation. CPE itself needs a definition through elementary physical processes and mechanisms just like for UDR.

Solid-state impedance spectroscopy gives roughly averaged experimental data, i.e., a large amount of information (which reflects interconnected local processes) gets lost. For clear interpretations of such data, we need theoretical approaches, which allow calculating the impedance Z through the state variables directly connected with ionic hopping transport and dielectric polarization.

A new theoretical approach addresses to the logic of elementary processes on a nanoscale and emphasizes

(Despotuli and Andreeva 2015) that solid-state ionic conductors are dynamical non-linear systems (Despotuli and Andreeva 2017). In such systems, key parameters (heights of barriers in potential landscape) depend non-locally on external influences. Dynamic behavior of an ionic space charge in parametric-dependent systems cannot be presented correctly by equivalent electric circuits where elements have constant parameters (Despotuli and Andreeva 2016, 2017). These findings are obtained within the frame of SDA by computer experiments. Calculated data are in good concordance with results of impedance spectroscopy.

SDA takes into account the main nanoscale features of crystal structures of all ionic conductors, namely, a non-uniform potential landscape. SDA does not use the derivatives on spatial coordinates in a set of differential equations, because the differentiation on space coordinates is a doubtful operation on a nanoscale.

In the subsequent chapters, the fundamentals of SDA, which describe on a nanoscale the interrelations of ion-transport and dielectric polarization processes in solid-state ionic materials, are presented. Some future prospects of non-local non-linear ionics are considered.

9.2 Nanoionics as a Branch of Nanoscience and Nanotechnology

Nanoionics (Despotuli and Nikolaichik 1993) can be defined as a new branch of science and technology with its own subject, objects and methods of researches. It is the study and application of phenomena, properties, effects and mechanisms of processes connected with FIT in all-solid-state nanoscale systems (https://en.wikipedia.org/wiki/Nanoionics). Potential applications of nanoionics are nanomaterials and electrochemical devices: SE-batteries, fuel cells, supercapacitors, sensors, memristors, etc. A practice puts forward requests for more high-speed devices. Therefore, nanoionics has to disclose the prospects of applications of materials and devices with the FIT on a nanoscale. Multidisciplinary scientific and industrial field of solid-state ionics (Sunandana 2015) considers nanoionics and nanoionic devices (https://en.wikipedia.org/wiki/Nanoionic_device) as its new division.

9.2.1 Solid-State Ionic Conductors: New Classification

Usually, solids are poor ionic conductors, specific ionic conductivity σ_i is lower than 10^{-6} S/cm (at 300 K). Only some of them are noticeable σ_i. There are families of H^+-, Li^+-, Cu^+-, Ag^+-, F^-- and O^--superionic conductors (SICs), which have $\sigma_i \sim 10^{-3}$ S/cm at $T = 300$ K or ≈ 1 S/cm at moderately high T (Sunandana 2015, https://en.wikipedia.org/wiki/Fast_ion_conductor).

Properties and characteristics of ionic conductors and nanostructures define fields of their possible applications. For example, silver/copper chalcogenides (of Ag_2S type) have a high σ_i but Ag_2S cannot be used as solid electrolytes in supercapacitors. It is due to high electronic conductivity σ_e that is a cause of leak currents in an electric double layer (EDL). However, combinations of both high σ_i and σ_e (mixed conductors) are necessary for cathode materials of solid-state galvanic cells. Therefore, evaluations of ionic conductors without data about σ_i/σ_e relations are poorly defined. The application-oriented classification of solid ionic conductors (Figure 9.1) was proposed in Despotuli and Andreeva (2005, 2009).

The lg σ_i – lg σ_e classification distinguishes for the first time a new class of ionic conductors – "advanced superionic conductors" (AdSICs). Solid electrolytes (SEs) have $\sigma_i \gg \sigma_e$ while the quantity σ_i has any value. Solids with $\sigma_i > \sim 10^{-3}$ S/cm (300 K), while σ_e has any value, are SICs. Solids with $\sigma_i > 0.1$ S/cm (300 K, the activation energy for ion transport $E_a \approx 0.1$ eV) are AdSICs. A well-known example of AdSIC-SE is $RbAg_4I_5$ with $\sigma_i > 0.25$ S/cm and $\sigma_e \sim 10^{-9}$ S/cm (300 K) in high-quality samples. According to the lg σ_i – lg σ_e classification, $RbAg_4I_5$ is AdSIC and, simultaneously, SE. Ag_2S is AdSIC but not SE, because σ_e can be large.

Crystal structures of AdSICs are close to optimum for FIT. Isostructural compounds $RbAg_4I_5$ and $CsAg_4Br_{3-x}I_{2+x}$ (Despotuli et al. 1989) have recorded high Ag^+ hopping conductivities, $\sigma_i \approx 0.3$ S/cm at 300 K and a low activation energy of ion transport $E_a \approx 0.1$ eV

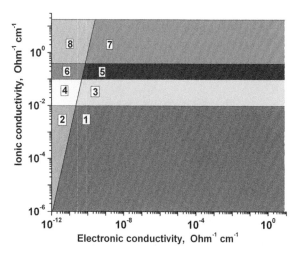

FIGURE 9.1 Classification of ionic conductors in lg σ_i – lg σ_e coordinates (S/cm). 2, 4 and 6 – known solid electrolytes (SEs), i.e., materials with $\sigma_i \gg \sigma_e$; 1, 3 and 5 – known mixed ion–electron conductors; 3 and 4 – superionic conductors (SICs), i.e., materials with $\sigma_i > 10^{-3}$ S/cm, σ_e – arbitrary value; 4 – SIC and simultaneously SE; 5 and 6 – advanced superionic conductors (AdSIC), where $\sigma_i > 10^{-1}$ S/cm (300 K), $E_a \approx 0.1$ eV, σ_e – arbitrary value; 6 – AdSIC and simultaneously SE; 7 and 8 – hypothetical AdSIC with $E_a \sim k_B T$ (300 K); 8 – hypothetical AdSIC and simultaneously SE (Despotuli and Andreeva 2009).

($\approx 4\mathrm{k}\,T_{300}$). Ionic conductivity σ_i depends on many factors. The most important factors are the heights (η) of potential barriers in preferred ion diffusion paths (FIT tunnels), i.e., paths with low values of η. Non-preferred conduction paths with higher potential barriers do not influence upon the σ_i due to exponential dependence of hopping transport on η: $\sigma_i \propto \exp(-E_a / k_B T)$, where $E_a \approx \eta_{\max}$ (the highest barrier in the FIT tunnel). The AdSIC crystal structures, present a rigid, usually close-packed sub-lattice of immobile ions. Inside the rigid sub-lattice, there are interconnected crystallographic positions forming the network of FIT tunnels.

A crystal structure of AdSIC α-AgI (Im3m space symmetry) can serve as a prototype for the elaboration of the theoretical FIT-models. The structure of α-AgI (Figure 9.2) presents a close-packed bcc sub-lattice of immobile I^--ions. Inside this rigid sub-lattice, there are tetrahedral crystallographic positions (minima of a potential landscape) available for mobile Ag^+ ions. The concentration (n^*) of these minima in the potential landscape of α-AgI is six

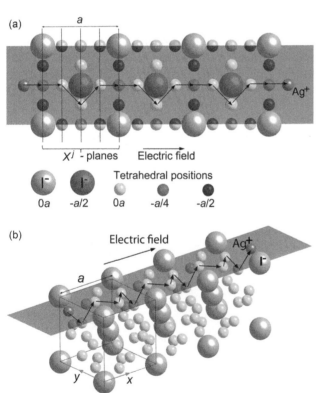

FIGURE 9.2 Crystallographic structure of AdSIC α-AgI (Im3m, a $\approx$ 0.509 nm) in two image projections (a and b). The rigid sub-lattice of AdSIC is composed by large iodine anions (large spheres). Within the I^--anion sub-lattice, there is a set of interconnected tetrahedral crystallographic positions (small light spheres), forming the <100> FIT tunnels in which Ag^+ cations (two small dark spheres) can drift in the direction of electric field. For the (a) image projection, different tints of gray color show the distance of anions I^- and tetrahedral positions available for mobile Ag^+ cations from (001) crystallographic plane (Despotuli and Andreeva 2015).

times more than the concentration of mobile Ag^+ ions. The projection of a "zig-zag" motion of Ag^+ ions in FIT tunnels (over tetrahedral crystallographic positions) coincides with the direction of external influence of current generator $G(t)$. Under action of $G(t)$, non-equilibrium concentrations of Ag^+ ions arise on the edges of nanostructure and, as a consequence, electrostatic field F is appearing. Ionic flow can be simplistically presented as a motion of cations in a layered structure. Each hop of a cation along the direction of F field leads it into the following layer (from plane X^j to plane X^{j+1} in Figure 9.2), i.e., without a "zig-zag" motion. SDA deals with such simplified layered models.

The $\sigma_i - \sigma_e$ classification also distinguishes a class of "hypothetical advanced superionic conductors" with higher ion-transport characteristics as compared with known AdSICs. Ionic solids with such characteristics have not been found in nature, but large structure fragments with "hypothetical" ion-transport characteristics have been discovered. The studies of pearceite-polybasite minerals with the general formula $(Ag, Cu)_{16}M_2S_{11}$, (M = Sb, As) (Bindi 2006) revealed superionic phases (300 K) with a trigonal (P3m1) symmetry group and activation energy $E_a \sim 0.16$ eV (the highest potential barrier in FIT tunnels). The mineral structures can be presented as regular alternation of two module layers stacked along the c axis: the first module layer (labeled A) with the general composition: $[(Ag,Cu)_6(AsSb)_2S_7]^{2-}$ and the second module layer (labeled B) with the composition: $[Ag_9CuS_4]^{2+}$. The FIT occurs in module B, which has "rings" with a diameter of an order of 8A in the mineral structures. In the rings, Ag^+ ions move with a very low $E_a < k_B T_{300}(\sim 15$ meV). This example is important because the design of nanostructures provides additional possibilities/degrees of freedom (compare to bulk structures) for achievement of record high σ_i at 300 K.

Modern methods of material engineering allow manipulating processes on molecular and atomic levels during materials synthesis. It is a way for the creation of fundamentally new architectures, new metamaterials and structures with unique characteristics and functionality. Therefore, discoveries of natural ionic conductors and creation of artificial structures with $E_a \sim k_B T_{300}$ can be expected in future.

9.2.2 Nanoionics: Term and Conception

The modern university textbook (Sunandana 2015) considers nanoionics as the section of solid-state ionics. Wikipedia (https://en.wikipedia.org/wiki/Nanoelectronics) relates nanoionics to nanoelectronics. The term and conception of nanoionics and nanoionic devices were first introduced in 1992 (Despotuli and Nikolaichik 1993): "The results obtained show that it is possible to form arrays of electrochemical devices with single elements $\sim$10 nm in size in the films". Nanoionic devices are characterized by nanoionic parameter $\lambda/L \sim 1$, where L is the size of a device structure, and λ is the characteristic size of a specific region with FIT. "Possibilities to influence on these specific regions $<\lambda>$ in a controllable manner may appear in short sized devices". The essence of physics for non-stationary non-equilibrium processes in nanoionics is the interrelated ionic currents and polarization (Maxwell displacement currents) in a crystal potential landscape, non-uniform on a nanoscale.

The analysis revealed (Despotuli and Andreeva 2005, Andreeva and Despotuli 2005) that nanoionics has two different branches of development, principally differing by material design methods. The first branch, i.e., nanosystems of class 1, deals with ionic materials created on the basis of substances with initial small σ_i at 300 K (the areas 1 and 2 in Figure 9.1). FIT in such materials arises owing to high concentrations of point defects on hetero-boundaries. The density of hetero-boundaries is large in nanomaterials. The second branch, i.e., the nanosystems of class 2, is nanoionics of AdSICs (areas 5 and 6 in Figure 9.1). Crystal structures of AdSICs are close to the optimum for FIT, therefore, distortion and violation of this structure on functional heterojunctions create a considerable reduction in conductivity σ_i, so, it is necessary as much as possible to keep the initial AdSIC structure. In AdSICs, a violation of FIT in area of non-coherent hetero-boundaries (interfaces) is considered in nanoionics as the classification criteria of AdSICs. Note, in the materials of class 1 (the areas 3 and 4 in Figure 9.1), the ion-transport characteristics remain worse in comparison with AdSICs. The nanosystems of class 2 with ion-transport characteristics at the level of the AdSIC bulk present considerable practical interest.

Nanoionics should be extended in new directions such as the interface engineering, theory of complex processes and averaging on a nanoscale, and non-linear dynamic ionics.

9.3 Structure-Dynamic Approach (SDA) of Nanoionics: Space–Time Description

Crystal potential landscape (non-uniform on a nanoscale) is a general base for solving the problems in solid-state ionics and nanoionics. In SDA, directed ionic flows (thermally activated transitions of mobile ions) in model nanostructures are due to a weak external influence, i.e., the condition of small deviations from detailed balance holds. Non-stationary ionic flows in non-uniform potential landscape induce polarization processes and origin of Maxwell displacement currents. A long-range Coulomb interaction connects all processes. It leads to the emergence of collective phenomena such as a formation and a relaxation of ionic spatial charge, an accumulation of charge on SE/EC (electronic conductor) polarizable heterojunctions, Warburg impedance, Jonsher's "universal" dynamic response (Jonscher 1977), etc. Thus, SDA gives on a deep level the space–time descriptions of local interrelated processes in model nanostructures with ionic hopping transport.

9.3.1 Main Components of SDA

SDA includes the following:

1. The layered model of a nanostructure with 1D-geometry of FIT. The model combines a set $\{X^j\}$ of parallel crystallographic planes X^j separated by potential barriers $\eta_{j,j+1}$. FIT tunnels are perpendicular to all X^j planes and parallel to a vector F_{eff} of electric field strength.

2. The method of dynamic "hidden" variables $n_i(t)$ that are concentrations of mobile ions in the minima of a potential landscape, i.e., on $\{X^j\}$ planes.

3. The physico-mathematical formalism, i.e., detailed-balance relations and a system of kinetic differential equations that operate with a set $\{n_i(t)\}$ of hidden variables.

4. The method of a uniform effective electrostatic field F_{eff} of X^j plane.

5. The new notion – "Maxwell displacement current on a potential barrier".

All computer experiments in the frame of SDA were performed in the "Wolfram Mathematica" package (www.wolfram.com/mathematica/).

9.3.2 Structural Models in SDA

Non-uniform Potential Landscape

Canonical physico-mathematical formalisms for a description of ionic transport in solids (Mehrer 2007) are based on the concept of a regular crystalline potential landscape, i.e., all heights of potential barriers η are equal. In such approaches, ion-transport characteristics (mean square displacement, diffusion coefficient, mobility, activation energy) have a clear physical meaning only at averaging on a scale that exceeds the length of ionic jump in a lattice. However, variations of η with a space coordinate x can be considerable in objects of solid-state ionics and nanoionics. For example, local areas with η variations ≈ 0.5 eV/nm exist in SEs and in SICs, various degrees of ordering occur near free surfaces and in areas of intercrystalline homo- and hetero-phase boundaries, where specific ionic dynamics can arise on a nanoscale due to a non-uniform potential landscape. Interface structures influence considerably on frequency-capacitance characteristics of EC/SIC heterojunctions (Despotuli and Andreeva 2005, 2009, 2010, 2015, Andreeva and Despotuli 2005).

Layered Nanostructures

A structural model of SDA is presented on the example of EC_b/AdSIC heterojunctions with ideally polarizable (blocking) electrode EC_b (Figure 9.3a). The electrode EC_b blocks ionic and electronic currents if the voltage of EDL has the right polarity and value. For this case, EDL can be formed by the right direction of a current generator $G(t)$.

(a)

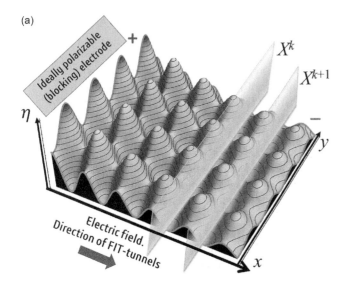

(b)

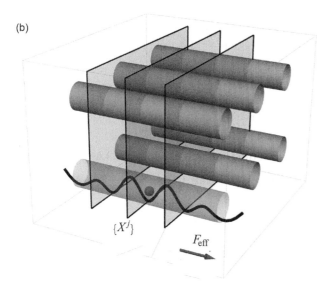

FIGURE 9.3 The structural layered models of ionic transport in SDA. (a) Non-uniform potential landscape in the area of an ideally polarizable (blocking) electrode. (b) FIT tunnels where 1D ionic flow coincides with the direction of external action (Despotuli and Andreeva 2015).

An electrochemical decomposition of AdSIC arises at achievement of the critical voltage in EDL. In working conditions under $G(t)$-action, only Maxwell displacement currents will be possible in EDL. Coherent (structure-ordered) EC_b/AdSIC heterojunctions for creation of high frequency-capacitance supercapacitors were proposed for the first time in Despotuli and Andreeva (2003), Andreeva and Despotuli (2005).

It is convenient to designate EC_b/AdSIC model hetero-junctions as $EC_b/\{X^j\}$ ($j = 1,2 \ldots M$), where $\{X^j\}$ is a set of crystallographic X^j planes, parallel to a EC_b, M is a number of planes in $\{X^j\}$. Each X^j plane (with x_j coordinates corresponds to minima in a non-uniform potential landscape. The normal to the EC_b (X^0 plane) is oriented along the conduction of FIT tunnels (Figure 9.3b) as well

as along the direction of F_{eff} field (Despotuli and Andreeva 2015).

In conditions of structural coherence on EC_b/AdSIC interface, mismatch elastic deformations in the interface area do not cause an essential worsening in high transport characteristics of AdSIC. However, even low-mismatch deformations can cause significant changes in the potential landscape of AdSIC (Despotuli and Andreeva 2003, 2005). It allows distinguishing two areas in the $EC_b/\{X^j\}$ structure: SE at the interface and AdSIC in the bulk where a crystal structure is close to optimal for FIT. An influence of a current generator on the EC_b/SE-AdSIC induces a non-equilibrium ionic space charge in a non-uniform potential landscape (SE area). In processes of relaxation, a non-equilibrium space charge shrinks near EC_b and forms an ionic capacitor plate of EDL.

In the EC_b/SE-AdSIC structural model, the 3D space of an ionic conductor is partitioned by parallel planes X^j into some set of layers, just as in lattice gas model (Liu 1980, Macdonald 1982, Macdonald and Liu 1983) of the equilibrium theory of EDL. Unlike (Liu 1980, Macdonald 1982, Macdonald and Liu 1983), the X^j planes in SDA correspond to minima of a non-uniform potential landscape. The non-uniform potential landscape $\eta(x)$ extends from EC_b ($x_0 = 0$) along the x-coordinate on several nanometers and $x = x_M$ corresponds to the right edge X^M of the $EC_b/\{X^j\}$ system. The symbol $\eta_{j,j+1}$ denotes the height of a potential barrier (in eV) between neighboring minima with the j and $j+1$ indexes. Barrier minima $\eta(x_j) = 0$ are located in points x_j ($j = 1,2 \ldots M$), and maxima are in $(x_j - \Delta/2)$, where $\Delta = x_{j+1} - x_j$, i.e., Δ is the length of ion hop. The heights of barriers decrease with increasing of the j index, i.e., with the distance from x_0. The $\eta(x_M - \Delta/2)$ equals to the barrier height $\eta_v = const$, i.e., as in the AdSIC bulk.

Along y- and z-coordinates (like along the x-coordinate), heights $\eta(x_j - \Delta/2, y_p, z_q)$ change discretely (p and q are integers from 1 to ∞) that corresponds the AdSIC periodicity. The values of $\eta(x_j - \Delta/2, y_p, z_q) = \eta(x_j - \Delta/2)$ for any p and q, i.e., layers between planes with coordinates x_j and x_{j+1}, are considered as macroscopic ones along the y- and z-directions. So, in the Cartesian xyz-system, the state of a mobile ion (cation, for definiteness) is uniquely determined by the x_j coordinate of the X^j plane. Such a layered model with 1D ionic transport catches the main structural features of SEs, while it simplifies the representation of real crystal structures in the theory.

In the initial state, a chemical potential of cations is *const* along x-coordinate when a current generator $G(t) = 0$ on the X^0 and X^M edge planes of $\{X^j\}$. So, the initial equilibrium concentrations of cations equal n_0 on all X^j ($j = 1,2 \ldots M$) planes. If a current generator specifies a current density $I(t) \neq 0$ on the X^0 and X^M edge planes, then the concentrations n_j of mobile cations change on all X^j planes (due to long-range electrostatic fields). The set of n_j concentrations is denoted by $\{n_j\}$. The weakness of external influences means that a relative change of cation concentration is $y_j(t) = (n_j - n_0)n_0 \ll 1$.

9.3.3 Method of Hidden Variables in SDA

In this section, a parallel between a method of equivalent electric circuits and classical quantum mechanics are drawn. M. Bohr and W. Heisenberg refused to consider existential (spatiotemporal) processes by means of the hidden variables, and constructed quantum mechanics only on observed quantities. The similar positivistic approach is used in the method of equivalent electric circuits for interpretation of impedance spectroscopy data. A careful fitting with equivalent circuits leads to using of elements without clear physical meaning. Really, the fitting of experimental data (where observed quantities are the current, voltage and shift of phase) to a response of the circuit means a refusal from modeling of real spatiotemporal processes. Constant phase elements (CPEs) are examples of such refusal.

In SDA, the relative change of cation concentrations $y_j(t) = (n_j - n_0)n_0$ $1 \leq j \leq M$ were called by "hidden variables" (Despotuli and Andreeva 2012), because modern experimental possibilities for defining of functions $n_j(t)$ are currently absent. The set of $\{n_j\}$ allows describing spatiotemporal processes in nanostructures with FIT.

9.3.4 Physico-Mathematical Formalism of SDA

Kinetic equations and detailed-balance relations (Despotuli and Andreeva 2012, 2015) can describe motions of classical particles performing random walks in a lattice (stochastic processes over discrete states)

$$\mathrm{d}P_i/\mathrm{d}t = \sum (P_j w_{j \to i} - P_i w_{i \to j}); \; i \neq j; \; i; \; j \in S \quad (9.2)$$

$$w_{i \to j} = w_{j \to i} = \exp\left((E_i - E_j)/k_B T\right), \quad (9.3)$$

where $w_{i \to j}\mathrm{d}t$ is a probability of transition between local states $i \leftrightarrow j$ in time interval $\mathrm{d}t$, S is a linearly ordered set, k_B is the Boltzmann constant and T is a temperature (in K). Transitions $i \leftrightarrow j$ of mobile ions (cations) are regarded as instantaneous events.

In SDA, in the linearly ordered set S, the transitions of cations are only possible between the adjacent planes $X^j \leftrightarrow X^{j+1}$ over the barriers $\eta_{j,j+1}$. A symbol P_j is a probability to find cation on X^j plane, and $(E_j - E_{j+1})$ is the energy difference for adjacent planes (minima of a potential landscape).

The principle of detailed balance is valid for the Markov stochastic processes, their statistical properties in the time $t^* + \mathrm{d}t$ ($\mathrm{d}t > 0$) only depend on the processes at t^*. For the Markov processes in $EC_b/\{X^j\}$ layered nanostructures, the condition $P_j(t) \propto n_j(t)$ is satisfied because $P_j(t) = n_j(t)/\Sigma n_j$, where n_j is time-dependent concentration of mobile cations in potential landscape minimum with a coordinate x_j ($j \in S$), the sum Σn_j is determined by the operation of a current generator $G(t)$.

After some assumptions (relating to w), the kinetic equations and detailed-balance relations were written by Despotuli and Andreeva (2012, 2015) as the conservation

law of mobile cations. Ionic currents edn_j/dt through X^j planes are

$$edn_j/dt = e(-n_j f_{j \to j+1}\rho_{j+1} - n_j f_{j \to j-1}\rho_{j-1}$$
$$+ n_{j-1}f_{j-1 \to j}\rho_j + n_{j+1}f_{j+1 \to j}\rho_j), \quad j = 1, \ldots M \quad (9.4)$$

where $f_{j \to j+1}$ is a frequency with which cations reach the potential barrier top at attempts of $j \to j+1$ transitions, and ρ_{j+1} designates a probability of vacant positions on X^{j+1} plane.

SDA assumes that a uniform effective electrostatic field F_{eff} along the x axis exists in the layer between X^j and X^{j+1} planes. The frequency $f_{j \to j+1}$ depends on the initial height of the barrier $\eta_{j,j+1}$, and on t through the field additives $(\Omega_{j+1,M})$ to the heights of barriers:

$$\Omega_{j+1,M} \equiv \frac{e^2 \Delta}{\varepsilon_0 \varepsilon_{j,j+1}} \sum_{k=j+1}^{M} \{n_k(t) - n_0\}, \quad (9.5)$$

where $\varepsilon_{j,j+1}$ is an effective dielectric susceptibility of the layer between X^j and X^{j+1} planes, ε_0 is the electric constant and e is the absolute value of an electron charge. The Eq. (9.5) take into account the existence of the charges $(-e\sum_{k=j+1}^{M}\{n_k(t) - n_0\})$ on the EC_b electrode. The value $\varepsilon_{j,j+1}$ is determined by displacements of mobile ions in minima of a potential landscape. Therefore, a significant dielectric polarization connected with mobile ions should increase with a reduction of the potential landscape depth, i.e., $\varepsilon_{j,j+1} \propto 1/\eta_{j,j+1}$, (Despotuli and Andreeva 2010). The system of Eq. (9.4) should be taken into account the boundary conditions for current density on the left and right edges of a nanostructure. Components of the Eq. (9.4) have the signs, which are defined by "a motion of cations from j minimum" (a sign minus) or "a motion of cations to j minimum" (sign plus).

The current densities in (9.4) can be defined relative to the $+x$ axis direction in new symbols: the values $en_j f_{j \to j+1}\rho_{j+1} \equiv I_{j \to j+1} > 0$ are the cation current densities over the barriers $\eta_{j,j+1}$ in the $+x$ direction (as the arrow indicates in $I_{j \to j+1}$); the values $-en_j f_{j \to j-1}\rho_{j-1} \equiv I_{j-1 \leftarrow j} < 0$ are the cation current densities over the barriers $\eta_{j-1,j}$ in the $-x$ direction (as the arrow indicates in $I_{j-1 \leftarrow j}$). Similarly, two other components in (9.4) can be written as $en_{j-1} f_{j-1 \to j}\rho_j \equiv I_{j-1 \to j} > 0$ and $-en_{j+1} f_{j+1 \to j}\rho_j \equiv I_{j \leftarrow j+1} < 0$. In these new symbols, the Eq. (9.4) are

$$en_0 dn_j/dt = -I_{j \to j+1} + I_{j-1 \leftarrow j} + I_{j-1 \to j} - I_{j \leftarrow j+1}$$
$$\equiv -I_{j,j+1} + I_{j-1,j}, \quad j = 1, \ldots M \quad (9.6)$$

where $y_j \equiv (n_j - n_0)/n_0$ value are relative changes of concentrations of mobile cations on X^j planes. The sums $I_{j \to j+1} + I_{j \leftarrow j+1} \equiv I_{j,j+1}$ (or $I_{j-1 \to j} + I_{j-1 \leftarrow j} \equiv I_{j-1,j}$) are the resultant densities of cation currents over potential barriers separating the X^j and X^{j+1} (or X^{j-1} and X^j) planes. The sign "minus" at $I_{j,j+1}$ in Eq. (9.6) means that the cation concentration on the X^j plane decreases if the resulting current vector $I_{j,j+1}$ coincides with the $+x$ direction.

9.3.5 Condition of Small External Influence

The principle of detailed balance is strictly satisfied in the equilibrium state of a system. Charge and voltage on the $EC_b/\{X^j\}$ blocking heterojunction depend on a current density of $G(t) = I_M(t)$ generator, i.e., on the integral $\int I_M(t) \, dt$. Therefore, a smallness of local current densities in $\{X^j\}$ was chosen as the criterion of a weakness of external influence.

A relaxation of an ionic space charge in the $\{X^j\}$ nanostructure is a result of $I_{j \to j+1}$ and $I_{j \leftarrow j+1}$ ionic currents (oppositely directed), which flow over all barriers in the set of $\{\eta_{j,j+1}\}$. The resulting currents $I_{j,j+1} \equiv I_{j \to j+1} + I_{j \leftarrow j+1}$ become equal to zero at $\{n_j(\infty)\}$ when the principle of detailed balance $|I_{j \to j+1}| = |I_{j \leftarrow j+1}|$ is strictly carried out. In the state $\{n_j(\infty)\}$, values of $|I_{j \to j+1}|$ modulus can be estimated (Despotuli and Andreeva 2011) by

$$|I_{j \to j+1}| \sim e n_0 \exp\left(-\eta_{j,j+1}/k_B T\right) \equiv I_{\text{ex}}. \quad (9.7)$$

The quantities I_{ex} are the exchange current densities on $\eta_{j,j+1}$ potential barriers. For example, I_{ex} are values of the order of $2.7 \cdot 10^{-1}$, $6.3 \cdot 10^{2}$ and $1.5 \cdot 10^{6}$ A m^{-2} for barriers $\eta_{j,j+1}$ with the heights 0.7, 0.5 and 0.3 eV, respectively ($n_0 = 10^{18}$m^{-2}, $T = 300$ K, and Debye frequency $\nu_D = 10^{12}$s^{-1}). During transition of the $EC_b/\{X^j\}$ system to a non-equilibrium state, the criteria of $G(t)$ smallness are

$$|I_{j,j+1}| / |I_{j \to j+1}| \ll 1 \quad (9.8)$$

$$|I_M(t)| / |I_{j \to j+1}| \ll 1 \quad (9.9)$$

The fulfillment of Eqs. (9.8) and (9.9) means the deviations of $\{X^j\}$ from equilibrium state are small.

9.3.6 Method of Uniform Effective Electrostatic Field (F_{eff}) in SDA

A real non-uniform 3D field (F_{3D}) of a charged X^j plane can be approximated with different accuracy degree. The simplest, most coarse approximation, is a uniform Gauss field F_G^j of X^j plane

$$F_G^j = \delta_j / 2\varepsilon_0 \varepsilon \quad (9.10)$$

where $\varepsilon_0 = 8.85 \cdot 10^{-12}$F/m, ε is the relative dielectric permittivity and δ_j is the surface density of an excess charge at a uniform-continuous distribution on X^j. In Despotuli and Andreeva (2015, 2016, 2018a, 2018b), it is substantiated that the F_G^j field is a suitable approximation of the F_{3D} field in SDA, and F_G^j can be used as the uniform effective field (F_{eff}) of a charged X^j plane, i.e., $F_{\text{eff}} \approx F_G^j$.

The influence of the components of F_{3D} field (induced by point charges, higher order attractive and repulsion interactions) on a heights η of potential barriers in crystallographic "FIT tunnels" was first studied by Flygare and Huggins (1973). In the substantiation of SDA, we take into account only the main component of F_{3D}, i.e., the Coulomb field $F_{\text{dis}}^{j,k,a}$ of the plane X^j with a discrete random distribution of excess point charges. In the $F_{\text{dis}}^{j,k,a}$ symbol, j is the index of X^j plane with excess charges, k is the index of transitions $X^k \to X^{k+1}$ of mobile ions, a is the index of a separate potential barrier ($1 \le a \le n_A$) between X^k and X^{k+1} planes and n_A is the number of mobile ions, which are able to take part in $X^k \to X^{k+1}$ transitions.

Vector fields $F_{\text{dis}}^{j,k,a}$ and F_G^j differ considerably. For weak fields less than $\sim 10^3$ V/cm (as in the impedance spectroscopy), the average distance r_m between excess point charges on X^j plane is larger than 300 nm. By comparison, a typical distance Δ between X^k and X^{k+1} planes is ≈ 0.15 nm (as in α-AgI), i.e., $r_m \gg \Delta$. In the macroscopic approach, an error of calculations in the approximation F_G^j is small if the inequality $r_m \ll R$ holds for point charges on X^j plane (R is the distance from an X^j plane to the point of field observation, i.e., $R \gg \Delta$). This consideration shows (with visibility of plausibility) that the F_G^j approximation of SDA contradicts the macroscopic theory. The latter requires the fulfillment of condition $r_m \ll R$. However, it is logically incorrect contradistinction, i.e., a "geometric-field paradox".

In Despotuli and Andreeva (2016), a uniform field $F_{\text{eff}}^{j,k}$ of the plane X^j (in the vicinity of the barrier $\eta_{k,k+1}$) is determined by using the corrections to field additives Ω_G^j of the F_G^j. The modified additives are $K_{|j-k|}\Omega_G^j$ (where $K_{|j-k|}$ are the correction coefficients). These additives should provide the same change-of-an-average frequency $\Delta f_{k,k+1}$ for the ionic jumps $X^k \to X^{k+1}$ as a non-uniform field $F_{\text{dis}}^{j,k,a}$ created by a discrete-random distribution of excess point charges on X^j. Thus, the uniform field $F_{\text{eff}}^{j,k}$ is an approximation of a non-uniform field $F_{\text{dis}}^{j,k,a}$ and can serves as a tool to eliminate the errors in SDA.

Calculated correction coefficients $K_{|j-k|}$ ($k = 0$ and $j = 1,2...7$) to the F_G^j field change the frequency $f_{k,k+1}$ of ion transitions just as the $F_{\text{dis}}^{j,k,a}$ field. The dependences of the correction coefficients for two characteristic lengths (λ_Q) of electrostatic screening are presented in Figure 9.4.

According to Despotuli and Andreeva (2016), the coefficients $K_{|j-k|}$ influence weakly on the dynamic properties of nanostructures with the length $L > 4$ nm (an error ~ 20 %).

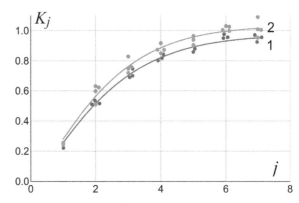

FIGURE 9.4 The dependences of the correction coefficients Kj ($\varepsilon = 50$) to the Gauss field on the index j of plane X^j for $\lambda_Q = 0.3$ nm (curve 1) and $\lambda_Q = 0.4$ nm (curve 2) (Despotuli and Andreeva 2016).

It indicates the existence of some dimensional factor that provides the relevancy of F_{G} and $F_{\mathrm{eff}}^{j,k}$ approximations on a nanoscale.

9.3.7 Novel Dimensional r_{i} Factor

Qualitative explanations of the appearance of the dimensional factor in nanoionics are given in Despotuli and Andreeva (2018a). A smallness of an average distance r_{i} between mobile ions in SICs and AdSICs predetermines the correctness of $F_{\mathrm{eff}} \approx F_{\mathrm{G}}$ approximation on a nanoscale. Figure 9.5 explains a "geometric-field paradox" in the position of charges and of the site of observation where the applicability of $F_{\mathrm{eff}} \approx F_{\mathrm{G}}$ approximations become possible.

According to Despotuli and Andreeva (2016), the change in the kinetics of ionic transport under a weak influence on a $\{X^j\}$ nanostructure is determined (in the first approximation) by the sum S of the field additives $\Omega_{k,k+1}$ to the heights of potential barriers $\eta_{k,k+1}$, which mobile ions overcome:

$$S = \sum_k \Omega_{k,k+1}. \tag{9.11}$$

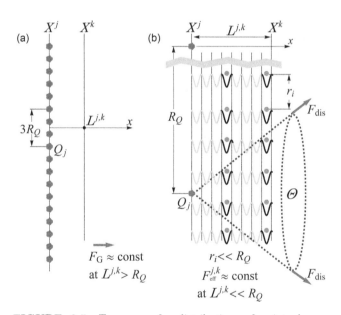

FIGURE 9.5 Two cases for distributions of point charges and positions of observation for electric field in classical electrostatics. (a) The standard condition of textbooks: the validity of uniform Gauss field $F_{\mathrm{G}} = \mathrm{const}$ $(L^{j,k})$ approximations occurs at $L^{j,k} \gg R_{\mathrm{Q}}$, where R_{Q} is the average distance between excess charges (a quasi-uniform charge distribution) on X^j plane, $L^{j,k}$ is the distance between X^j and X^k planes. (b) The condition of nanoionics: X^j plane with separate excess charges Q_j ($R_{\mathrm{Q}} > 300$ nm) is close to any of the $\{X^k\}$ planes ($L^{j,k} \sim 1$ nm) where the minima of potential landscape and mobile ions (balls of the smaller size) are. The effective field $F_{\mathrm{eff}} \approx F_{\mathrm{G}}$ approximation becomes possible at the distance of $L^{j,k} \approx 1$ nm ($L^{j,k} \ll R_{\mathrm{Q}}$), though the non-uniform field $F_{\mathrm{dis}} \propto (1/L^{j,k})^2$ exists in a solid angle Θ (Despotuli and Andreeva, 2018).

The sum (9.11) depends from the average distance r_{i} between mobile ions. The value r_{i} is usually small in AdSICs, for example, $r_{\mathrm{i}} \approx 0.6$ nm in α-AgI. Therefore, in a vicinity of each excess point charge Q_j, there is a set of mobile ions, which can make transitions between the planes X^{k-1} and X^k over potential barriers $\eta_{k,k-1} + \Omega_{k,k-1}^j$ (Figure 9.5b). Every excess charge Q_j significantly changes the field additives Ω only for those barriers of N mobile ions (the plane X^k), which are in the nearest vicinity of Q_j. Let designate this vicinity by a solid angle Θ. Vertex of angle Θ coincides with position of charge Q_j (Figure 9.5b). The sum (9.11) of the field additives Ω to the heights $\eta_{k,k+1}$ of the N barriers, confined in the angle Θ, would be $\approx N \cdot \Omega$. With an increase of the plane index k, i.e., the distance $L^{j,k} = (|k - j| \Delta) \ll R_{\mathrm{Q}}$ between the charge Q_j and plane X^k, each field additive Ω (within Θ) in (9.11) decreases as $\sim(|k - j|\Delta)^{-2}$, where Δ is the distance between X^k and X^{k+1} planes ($\Delta \approx 0.15$ nm in α-AgI). Simultaneously, with the increasing index k the number of mobile ions N in the vicinity of the angle Θ increases as $\sim(|k - j|\Delta)^2$, so for the characteristic length $L \sim r_{\mathrm{i}}$ the sum S in Eq. (9.11) becomes $S \approx N \cdot \Omega \approx \mathrm{const}$ $(L^{j,k})$ although $R_{\mathrm{Q}} \gg L^{j,k} \sim 1$ nm.

Thus, the dimensional factor of nanoionics (which provides the possibility of using approach $F_{\mathrm{eff}}^{j,k} \simeq F_{\mathrm{G}}$ at $L^{j,k} \ll R_{\mathrm{Q}}$) is an average distance r_{i} between mobile ions.

9.3.8 Average Potential Difference on a Nanoscale

Laws of averaging of local potential differences are important for nanoscience and nanotechnology, but these laws are not investigated in solid-state ionics and nanoionics. Laws of nanoscale averaging in solid-state ionic conductors relate to a multitude of phenomena and processes potentially available for observations. However, "Only the theory decides what one can observe" (Holton 2000).

Measurements of Average Potential Difference in Gedanken Experiments

Two fragments of layered SE nanostructures, which adjoin to a metal electrode, are presented schematically in Figure 9.6. If a set of layered nanostructures, i.e., metal 1/SE/metal 2 (metal 2 electrode is omitted in Figure 9.6, is under influence of a current generator $G(t)$, then time-dependent ionic currents $I_G(t)$ arise. To calculate a response of the electrode (metal 1), we need to know the distribution of ionic currents I_G among a set of layered nanostructures with different potential landscapes. However, for this purpose, it is necessary to know first the rules of spatial averaging for local potential differences in a separate $\{X^k\}$ nanostructure. The question of Gedanken experiment is: What would be an average potential difference $<V>$ between X^k and X^{k+1} in $\{X^k\}$?

Laws of spatial averaging of electrostatic potential differences in $\{X^k\}$ nanostructures have been investigated recently (Despotuli and Andreeva 2018b). In computer

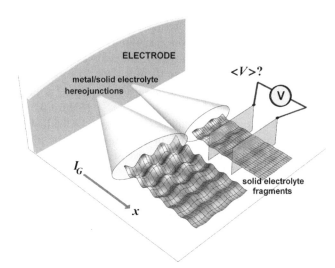

FIGURE 9.6 Two fragments of layered nanostructures metal 1/SE/metal 2 (electrode is omitted). Cation currents $I_G(t)$ are induced by a current generator $G(t)$. The scheme illustrates both a typical lateral heterogeneity and a nanoscale non-uniformity (perpendicularly to the layers) of potential landscapes. The knowledge of the average potential difference $<V>$ between layers of a nanostructure is necessary for simulation of ion-transport processes and flows of energy in parallel combinations of nanostructures (Despotuli and Andreeva 2018b).

experiments, it was used model, where mobile ions have discrete-random distributions on X^k planes in the nodes of 2D square lattice. The number of all nodes is more than the number of mobile ions. The interplanar distance Δ between neighbor parallel planes equals 0.15 nm in $\{X^j\}$ (as in AdSIC α-AgI). Under a weak electric external influence $G \neq 0$, the average distance R_Q (on the X^j plane) between excess point charges Qj can exceed 300 nm, i.e., $R_Q \gg r_i > \Delta$. Transitions $X^k \to X^{k+1}$ of ions in a non-uniform electrostatic field of a set $\{Qj\}$ of point charges may be related to some local $W^a_{k,k+1}$ and average $<W^a_{k,k+1}>$ works.

The average $<W^a_{k,k+1}>$ are calculated for two cases of geometry: random and periodic distribution of mobile ions in 2D square lattice. Data on $<W^a_{k,k+1}>$ correspond to conditions: the single charge $Q = e$ is on the X^0 plane and has the $\{0,0,0\}$ coordinates while mobile ions are on the X^k ($k = 1 \ldots 8$) planes in nodes of the 2D square lattice. The average work $<W^a_{k,k+1}>$ of the $X^k \to X^{k+1}$ ionic transitions is proportional to the sum $S_{k,k+1}$

$$<W^a_{k,k+1}> \propto \sum_{a=1}^{M} \left(\frac{1}{\sqrt{(k\Delta)^2 + y_a^2 + z_a^2}} - \frac{1}{\sqrt{(k+1)^2\Delta^2 + y_a^2 + z_a^2}} \right) \equiv S_{k,k+1},$$

$$(9.12)$$

where a random number generator specifies the y_a^2 and z_a^2 coordinates of filled nodes with mobile ions (their number equals M) in 2D lattice. Data for the case of a single charge

Q can be easily generalized for a set $\{Q\}$ of point charges (Despotuli and Andreeva 2018b).

Computer Investigation of Averaging Laws

The dependence of random variable $S_{k,k+1}$ ($\propto < W^a_{k,k+1}>$) on the k index is presented in Figure 9.7. It is a case of discrete-random distributions of mobile ions in M nodes on 2D square lattices with respect to a single charge Q. A result of single calculation of random variable $S_{k,k+1}$ for each k index is denoted in probability theory as a sample. Five series of $S_{k,k+1}$ calculations with 260 samples in each series were performed. Total number of samples equals $N = 5 \times 260 = 1,300$. Arithmetical mean of $S_{k,k+1}$ in each series is denoted as $<S_{k,k+1}>$.

The Figure 9.7 shows that the $S_{k,k+1}$ sums (averaged over 260 samples of random distributions of ions on the nodes, i.e., $<S_{k,k+1}>$) do not considerably fluctuate and change with k index. The characteristic behavior of $<S_{k,k+1}>$ was verified on stability by the variation of a method used for the generating data (random distributions vs periodical arrangement).

The dependencies of the sums $S_{k,k+1}$ on the index k were calculated also for the periodical arrangement of mobile ions in nodes of the 2D square lattice. Appropriate data are presented in Figure 9.8. The marker 1 corresponds to the $\{X^j\}$ nanostructure with the cross-section area 300 nm × 300 nm ($M \approx 250,000$ is the number of the $X^k \to X^{k+1}$ ionic transitions in the field of a single charge Q). The marker 2 corresponds to the cross-section area 30 nm × 30 nm ($M \approx 2,600$).

Thus, if $\Delta = r_i/4$, R_Q, r_i and n_0 parameters of $\{X^k\}$ nanostructures are close to the values in AdSICs then both random and periodical $S_{k,k+1}$ sums ($< W^a_{k,k+1}>$) coincide with the accuracy of ~10%. For small r_i-dimensional factor

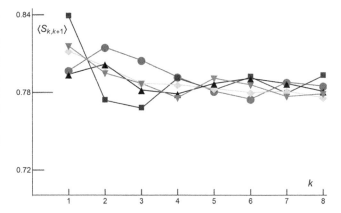

FIGURE 9.7 Arithmetical mean $< S_{k,k+1} > \propto < W^a_{k,k+1} >$ versus k index for random distributions of ions in nodes of the 2D square lattice on the X^k planes ($k = 1,2 \ldots 8$) with respect to a point charge Q on the X^0 plane. Five plot markers denote five series of calculations of $<S_{k,k+1}>$. Conditions of computer experiments: $N = 5 \times 260 = 1,300$ total number of random samples $S_{k,k+1}$ for each k index; cross-section of model nanostructure is 300 nm × 300 nm (Despotuli and Andreeva 2018b).

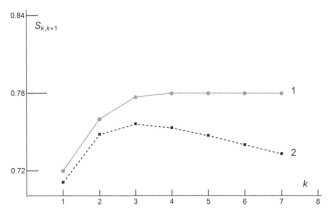

FIGURE 9.8 Sums $S_{k,k+1} \propto < W^a_{k,k+1} >$ versus k index for periodical arrangement of ions in the nodes of 2D square lattice on the X^k planes ($k = 1,2\ldots 7$) with respect to a point charge Q on the X^0 plane. Conditions of the computer experiments: 1 – model nanostructures ($M \approx 250{,}000$ is the number of the $X^k \to X^{k+1}$ ionic transitions in the field of a single charge Q) with 300 nm × 300 nm cross-section; 2 – model nanostructures ($M \approx 2{,}600$) with 30 nm × 30 nm cross-section (Despotuli and Andreeva 2018b).

(~ 0.6 nm), calculated $S_{k,k+1}$ data (Figures 9.7 and 9.8) show with the accuracy of $\sim 10\%$ that (i) $S_{k,k+1}$ sums do not depend on the k index; (ii) $S_{k,k+1}$ sums can be considered as constant if the cross-section diameter of a nanostructure exceed ~ 20–30 nm. It means that ions, which are located at distances more than ~ 20–30 nm from a charge Q, give a negligible contribution to the average work $< W^a_{k,k+1} >$ of the $X^k \to X^{k+1}$ ionic transitions.

In the context of the concept "effective uniform field" (Despotuli and Andreeva 2015, 2016), it is interesting to compare the numerical value of $< W^a_{k,k+1} > \propto < S_{k,k+1} >$ (9.12) with the similar one for a uniform F_G field, which created by the X^j plane with a uniform-continuous distribution of charge (surface density $e/1 \, \text{cm}^2$). The work in field F_G at the $X^k \to X^{k+1}$ transition of every ion on the distance Δ is denoted as $W^U_{k,k+1}$, it is

$$W^U_{k,k+1} = \left(e^2 / 2\varepsilon_0 \right) \left(\Delta / 1 \, \text{cm}^2 \right). \quad (9.13)$$

The ratio $< W^a_{k,k+1} > / W^U_{k,k+1}$ is ≈ 1.03 for $k = 1,2\ldots 8$,

i.e., in nanoionics, there is the remarkably-simple ratio for works of charge transfer in electrostatic fields:

$$< W^a_{k,k+1} > \approx W^U_{k,k+1}. \quad (9.14)$$

The simplest approximation (9.14) is very useful for the modeling of ion-transport processes and energy flows in the FIT nanostructures. Obtained data can be predicted with the high accuracy of $\sim 10\%$.

The average potential difference $< V^a_{k,k+N} >$ for N layers of the nanostructure is

$$< V^a_{k,k+N} > \approx N < W^a_{k,k+1} > /e \quad (9.15)$$

The results (9.13)–(9.15) are predetermined by the smallness of the discovered r_i-dimensional factor (≈ 0.6 nm) in

comparison with the nanostructure thickness ($|j - k| \, \Delta \sim 2 \div 3$ nm).

Probability Density Function for Local Potential Differences

Data of independent calculations of the random variables $S_{k,k+1}$ ($N = 5 \times 260 = 1{,}300$) for the ionic transitions $X^2 \to X^3$ and $X^6 \to X^7$ are demonstrated in Figure 9.9. The same data (in an averaged form) are presented in Figure 9.7. Calculated histograms for the sums $S_{2,3}$ and $S_{6,7}$ (see Figure 9.7) show the random variables $S_{k,k+1}$ obey to normal (Gaussian) distributions.

Thus, obtained results for the FIT nanostructures demonstrate that an average value of an electrostatic potential difference (mathematical expectation) in a non-uniform field F_{dis} is defined by the distribution (of local potential differences), which is close to the normal (Gaussian) one. This average potential difference equals (with the accuracy of $\sim 10\%$) the potential difference in the uniform field F_G that could be created by a uniform-continuous distribution of the same charge on the plane of $\{X^k\}$ as it is shown by the relation (9.14).

As shown in (Despotuli and Andreeva 2018b), the average potential difference $< V^a_{k,k+N} >$ (mathematical expectation) for local potential differences $V^a_{k,k+N}$ between parallel crystallographic planes X^k and X^{k+N} ($N > 1$) in a non-uniform electrostatic field F_{dis} of excess point charges on X^j plane ($k > j$) also corresponds to the normal (Gaussian) distribution. The value $< V^a_{k,k+N} >$ equals approximately (with the accuracy of $\sim 10\%$) to the potential difference between X^k and X^{k+N} planes in a uniform field F_G

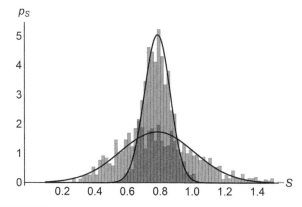

FIGURE 9.9 Probability density functions ρ_S for random sums $S_{k,k+1}$, which are proportional to the average potential difference $< V^a_{k,k+1} >$ between the nearest X^k and X^{k+1} planes. 1 – the histogram ρ_S for the sums $S_{2,3}$ and the envelope which is close to the normal distribution (mathematical expectation $\mu_{2,3} \approx 0.78$ and standard deviation $\sigma_{2,3} \approx 0.23$); conditions of computer experiments: $N = 5 \times 260 = 1{,}300$ samples of the $S_{2,3}$ sum. 2 – the histogram ρ_S for the sums $S_{6,7}$ and the envelope which is close to the normal distribution with $\mu_{6,7} \approx 0.78$ and $\sigma_{6,7} \approx 0.079$; $N = 5 \times 260 = 1{,}300$ samples of the sum $S_{6,7}$ (Despotuli and Andreeva 2018b).

created by a uniform-continuous distribution of the same value of charge on X^j plane. The knowledge of the average potential difference $< V_{k,k+N}^a >$ is necessary to describe ion-transport and energy-transformation processes in parallel–serial combinations of nanostructures.

The laws of spatial averaging of potential differences can be used for modeling of processes (with an acceptable accuracy) in ionic nanostructures with cross-section diameter $> \sim 20$ nm. Obtained simple results may serve as the initial point for the elaboration of the presented problem (physical and mathematical averaging on a nanoscale for ionic conductors) by the methods of quantum mechanics, in particular, to elucidate the "dimensional effect for relative dielectric permittivity" that is completely unexplored in nanoionics. The obtained results can be used at the creation of new theoretical system devoted to description of dynamic response in layered nanostructures with FIT.

9.3.9   New Notion: Maxwell Displacement Current on Potential Barrier

There is a paradoxical situation, the quantum theory of electronic transport in nanostructures and in disordered solids is presented in a multitude of textbooks, while the theory of FIT based on laws of classical physics exists only in debatable forms. Partly, it's probably because the reviews (Dyre et al. 2009, Sidebottom 2009, Macdonald 2010) on dynamic problems of FIT properties in disordered solids do not mention even such fundamental physical quantity as Maxwell displacement current.

The model EC_b/SE-AdSIC/EC_r system with blocking (EC_b) and reversible (EC_r) electrodes can be influenced by an impulse current generator $G(t)$. In the $+EC_b$ potential condition, total ionic space charge near EC_b electrode has a negative sign. In the approaches of the uniform effective field F_{eff}, and structural model of SDA, the electric field strength $F_{j+1,j}$ in a layer between X^j and X^{j+1} planes in the EC_b/SE-AdSIC/EC_r system can be described enough correctly by the Gauss law (9.10) (Despotuli and Andreeva 2016, 2018a, 2018b)

$$F_{j+1,j} = \frac{-en_0}{\varepsilon_0 \varepsilon_{j,j+1}} \sum_{k=j+1}^{M} y_k(t), \qquad (9.16)$$

where the sign "minus" in Eq. (9.16) shows the deficit of cations on $\{X^k\}$ planes ($y_k < 0$), i.e., at the EC_b/SE interface the vector of electric field strength $F_{j,j+1}$ directs in the $+x$ direction (from EC_b to SE).

The vector of electric induction D in the layer between X^j and X^{j+1} planes is given by the equation

$$D_{j+1,j} = \varepsilon_0 \varepsilon_{j,j+1} F_{j+1,j} \qquad (9.17)$$

The local Maxwell displacement current $I_{Dj,j+1}$ on the barrier $\eta_{j,j+1}$ can be determined through the time derivative of the x-component of the $D_{j+1,j}$ vector

$$I_{Dj,j+1} \equiv dD_{j+1,j}/dt \qquad (9.18)$$

Taking Eqs. (9.16)–(9.18) into account, the expression for $I_{Dj,j+1}(t)$ can be written as

$$I_{Dj,j+1} = en_0 \sum_{k=j+1}^{M} \frac{dy_k}{dt} \qquad (9.19)$$

Computer experiments in terms of SDA showed that the displacement current $I_{Dj,j+1}$ satisfies, with an accuracy of more than $10^{-5}\%$, the fundamental ratio of a circuit closed on a current generator:

$$I(t) = I_{j,j+1}(t) + I_{Dj,j+1}(t) \qquad (9.20)$$

i.e., in any time t, the sum of a resultant cation current $I_{j,j+1}$ and a displacement current $I_{Dj,j+1}$ on a barrier $\eta_{j,j+1}$ equals an external current $I_M(t)$ created by $G(t)$.

Equation 9.20 was also proved analytically by Despotuli and Andreeva (2015). It leads to important conclusion, namely, that Maxwell displacement currents in the hidden form are already in the physico-mathematical formalism (9.4) of SDA, i.e., the equation 9.20 is predetermined by a fundamental conservation law of mobile cations.

9.4   Some Applications of SDA

Due to physical averaging on a nanoscale, many processes and phenomena in solid-state ionics are inaccessible for modern experimental methods and are not in a focus of theory. The purpose of all computer simulations within the frame of SDA is to capture, disclose and describe dynamic phenomena, which hide in ionic conductors behind local processes of actual physical averaging.

9.4.1   Dynamic Response of Warburg Type in Complex Potential Landscape

Dynamic response of disordered solid–ionic conductors was the subject of reviews (Dyre et al. 2009, Sidebottom 2009, Macdonald 2010). Comparative computer experiments in SDA were performed using the model $EC_r/\{X^j\}(j = 1,2\ldots21)/EC_r$ nanostructures, where EC_r are the ideal-reversible electrodes attached to a current generator (G) and $\{X^j\}$ is a layered nanostructure with 1D geometry of FIT tunnels. Electrochemical reactions on EC_r electrodes occur with an infinitesimal overvoltage. Therefore, EC_r electrodes give zero contributions to impedances $Z(\omega)$ in a wide range of frequencies ω of a generator $G(\omega)$.

Computer experiments within the frame of SDA demonstrate that nanostructures with arbitrary distribution of potential barrier heights reveal the same type of behavior as disordered ionic conductors. The length of $\{X^j\}$ model nanostructures was ≈ 3 nm that corresponds to several elementary cells of SEs. Results of computer experiments are presented below in Figures 9.10–9.15. The insets in the figures show the sequence of potential barriers in the model nanostructures.

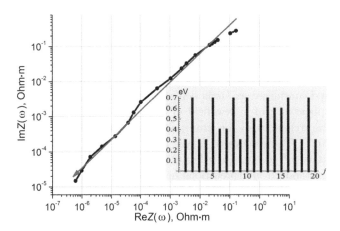

FIGURE 9.10 Calculated $\mathrm{Re}Z(\omega)$ and $\mathrm{Im}Z(\omega)$ data for the model $\{X^j\}$ nanostructure with twenty potential barriers. The range of frequencies $\omega/2\pi$ is $1 \div 10^5$ Hz; $n_0 = 10^{18}$ m^{-2}; $T = 300$ K. On the inset: the potential landscape in $\{X^j\}$ with $j = 1, 2 \ldots 21$ (Despotuli and Andreeva 2013).

The interrelation of Im and Re components of complex impedance $Z(\omega)$ for the model $\mathrm{EC_r/SE/EC_r}$ system is shown in Figure 9.10. It is Nyquist's plot, which can be approximated by the straight line with the tangent of slope ≈ 1. Such behavior of the $Z(\omega)$ impedance looks like the Warburg impedance, $W(\omega)$. In electrochemistry, impedances $W(\omega)$ appear in theory as a result of solutions of diffusion equations for systems where concentrations of particles oscillate on electrodes under small sinusoidal external influences.

9.4.2 Phase Relations Between Hidden Variables in Space–Time Processes

In the $\mathrm{EC_r/SE/EC_r}$ system (Figure 9.10), the analysis of behavior of hidden variables demonstrates that concentrations of mobile cations which overcome $\eta_{j,j+1}$ potential barriers with noticeable phase lags $\varphi_{j,j+1}$ relative to the phase of AC current generator $G(\omega)$ (as well as concentrations of cations that have failed to finish the $X^j \leftrightarrow X^{j+1}$ transitions during a half-cycle) increase with ω rise. It is capacitor behavior of barriers with large heights $\eta_{j,j+1}$. In these conditions, the distances between areas, where ionic currents $I_{j,j+1}$ dominate under $I_{Dj,j+1}$, increase, i.e., mutual capacitances between areas with low barriers $\eta_{j,j+1}$ become smaller. Such behavior can be incorrectly interpreted as the decrease of dielectric susceptibility (Dyre et al. 2009, Sidebottom 2009, Macdonald 2010) but it is a consequence of the fundamental ratio $I_M(t) = I_{Dj,j+1}(t) + I_{j,j+1}(t)$ for currents in non-uniform potential landscape.

In Figure 9.10, the highest barriers of $\{X^j\}$ nanostructure are 0.7 eV. Combinations of hidden variables allow to calculate the phase shift $\varphi_{j,j+1}$ between $I_{j,j+1}$ and voltage $V_{j+1,j}$ for all $\eta_{j,j+1}$ barriers. It appears that for the highest barriers of $\{X^j\}$ $\varphi_{j,j+1} \approx 0$ and the $I_{j,j+1}/V_{j+1,j}$ ratio

does not depend on t and ω, i.e., Ohm's law holds. The calculated specific ionic conductivity σ_i (300 K) for $\eta_{j,j+1} = 0.7$ eV is $\approx 2.1 \times 10^{-7}$ S/cm. The low potential barriers of $\{X^j\}$ influence very slightly on the $\sigma_{j,j+1}$ of highest barriers. The important conclusion is that the notion of specific ionic conductivity is inapplicable for potential barriers with small heights.

Calculations of specific ionic conductivity σ (Figure 9.11) show that Ohm's law holds also in the case of single high barrier surrounded with a number of lower barriers.

Data of Figure 9.11 were obtained for $\{X^j\}$ nanostructures with various combinations of barrier heights between the highest $\eta_{10,11}$ barrier and others lower barriers surrounding $\eta_{10,11}$.

These combinations are $\eta_{10,11} = 0.7$ eV, 0.6 eV and 0.5 eV (other barriers $\eta_{j,j+1} = 0.3$ eV); $\eta_{10,11} = 0.4$ eV (other barriers $\eta_{j,j+1} = 0.2$ or 0.3 eV); $\eta_{10,11} = 0.3$ eV (other barriers $\eta_{j,j+1} = 0.2$ eV); $\eta_{10,11} = 0.2$ eV (other barriers $\eta_{j,j+1} = 0.1$ eV). Calculated specific conductivities $\sigma_{10,11}$ for 300 K are noted by asterisks in Figure 9.11. Extrapolation gives for $\eta_{10,11} = 0.1$ eV the value $\sigma_{10,11} \approx 0.3$ S/cm that corresponds to literary data on ionic conductivity σ_i of AdSICs at 300 K (see Section 9.2).

Data in Figures 9.10–9.11 explain the independence of σ_i in AdSIC RbAg$_4$I$_5$ on ω in the wide frequency range 0.001–10^{10} Hz (Bruckner at al. 1984). Ionic transport through the highest potential barriers follows Ohm's law because the frequency dependence appears when Maxwell displacement currents $I_{Dj,j+1}(\omega)$ through the highest barriers of RbAg$_4$I$_5$ (the heights of ≈ 0.1 eV) become comparable with currents of ionic conductivity $I_{j,j+1}(\omega)$. This condition is carried out for barriers with 0.1 eV heights if frequencies exceed 10^{10} Hz.

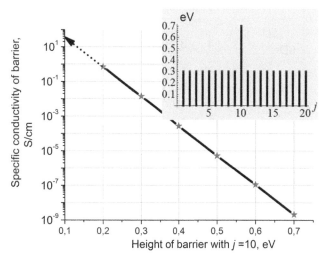

FIGURE 9.11 Specific ionic conductivity $\sigma_{10,11} \equiv I_{10,11}(t)/V_{11,10}(t)$ of the single high barrier $\eta_{10,11}$ (the values are marked with asterisks) surrounded with a number of lower barriers $\eta_{j,j+1}(j \neq 10)$ in frequency range $7 \div 700{,}000$ Hz (300 K). On an inset – a profile of the potential landscape of nanostructure with single high potential barrier $\eta_{10,11} = 0.7$ eV (Despotuli and Andreeva 2013).

9.4.3 Modeling of Processes at the Charging of Blocking Heterojunction

The behavior of capacitance $C(t)$, Maxwell displacement currents $I_{Dj,j+1}(t)$ and ionic currents $I_{j,j+1}(t)$ are presented in Figure 9.12 for the charging of the model $EC_b/\{X^j\}$ (j = 1,2...28) ideally polarizable heterojunction. The current generator operated in a galvanostatic mode, i.e., $G(t) = const$ at $t > 0$ and $G(t) = 0$ at $t \le 0$. The condition of small steric effects (an accumulated charge is small) was satisfied.

The calculated data show (Figure 9.12) that the condition of a closed circuit (20) holds for six highest barriers $\eta_{j,j+1}$ of the $EC_b/\{X^j\}$ heterojunction. The Maxwell displacement currents $I_{Dj,j+1}(t)$ dominate over the ionic conductivity currents $I_{j,j+1}(t)$ at small t. Capacitance $C(t)$, determined by the ratio of the accumulated charge to the voltage on the $EC_b/\{X^j\}$, reaches its maximum due to appearance of the steric effect at large t. The value of $C(t)$ grows with t because the width of ionic space charge becomes more narrow due to complementary processes of relaxation. Obvious, in impulse supercapacitors, function $C(t)$ must reach its maximum at small t. It is possible if the heights of potential barriers $\eta_{j,j+1}$ are small in the interface region of a device. Therefore, nanoionic devices must be fabricated by interface engineering methods (Despotuli and Andreeva 2003, Andreeva and Despotuli 2005).

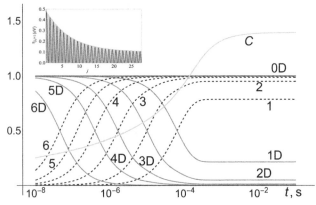

FIGURE 9.12 Time dependence of the capacitance C (in F/m^2) of ionic space charge, normalized Maxwell's displacement currents $I_{Dj,j+1}/I(t)$ and normalized cation currents $I_{j,j+1}/I(t)$ in the region of the $EC_b/\{X^j\}$(j = 1,2 ... 28) blocking heterojunction charging in the galvanostatic mode ($n_0 = 10^{18}$ m^{-2}, $T = 300$ K). The marker numbers from 1 to 6 correspond to the ionic current $I_{j,j+1}/I(t)$ over barrier $\eta_{j,j+1}$ (j = 1,2, ... 6), the markers 1D–6D corresponds to normalized Maxwell displacement current $I_{Dj,j+1}/I(t)$ through barrier $\eta_{j,j+1}$ (j = 1,2, ... 6), the marker 0D corresponds to normalized Maxwell displacement current between the EC_b and nearest to EC_b minimum potential landscape (X^1 plane with coordinate x_1). On inset – the potential landscape in the region of blocking heterojunction (Despotuli and Andreeva 2015).

9.4.4 Modeling of Universal Dynamic Response (UDR)

Since the review by Jonsher (1977), the frequency-dependent dynamic response is in a focus of researchers and considered as fundamental phenomena. This led to the appearance of a new field of investigations. Its subject is the dynamics of ion transport in solids (Habasaki et al. 2017). The discovered power law for the complex conductivity $\sigma^*(\omega)$, i.e., Re $\sigma^*(\omega) \propto \omega^n$ ($n < \approx 1$), is usually used as standard interpretations in materials science and solid-state ionics (or e.g., Rhimi 2018).

Dynamic responses of $EC_r/\{X^j\}/EC_r$ model systems were studied in computer experiments (Despotuli and Andreeva 2015, 2016). The model $\{X^j\}$ nanostructure represents centrosymmetrical potential landscape with the barrier heights changing from 0.2 to 0.8 eV, as shown in Figure 9.13 by light gray color. The formation of ionic space charge in the $\{X^j\}$ nanostructure with the L width occurs under influence of the AC current generator (amplitude of current density is $I_0 = $ const). Corresponding data are presented in Figures 9.13 and 9.14. The distributions of Maxwell displacement currents (the ratio of $I_{Dj,j+1}(t)$ to $I_{j,j+1}(t)$) in the model nanostructure are presented by dark gray color in Figure 9.13a–c for three frequencies of $G(\omega)$.

Figure 9.14 shows that real component of complex conductivity $\sigma^*(\omega)$ of the $EC_r/\{X^j\}/EC_r$ system varies with the frequency ω approximately by the power law

$$\text{Re}\,\sigma*(\omega) \equiv (L/V_0)\,I_0\,\cos\varphi \propto \omega^n, \qquad (9.21)$$

where L is the width of $EC_r/\{X^j\}/EC_r$ system; V_0 is the amplitude of AC voltage $V(t) = V_0 e^{i(\omega t+\varphi)}$ on $EC_r/\{X^j\}/EC_r$; I_0 is the amplitude of AC current density $I(t) = I_0 e^{i\omega t}$; φ is the phase shift between current and voltage; $n < \approx 1$ in a wide frequency range. Thus, the behavior Re $\sigma^*(\omega)$ is consistent with experimental data on Jonsher's "universal" dynamic response. This result may be considered as a validation of SDA.

The analysis of power law (9.21) in terms of hidden variables shows that the cosine of φ weakly depends on ω in the cases when the local $I_{Dj,j+1}(t)$ and $I_{j,j+1}(t)$ currents are of the same order in a set of potential barriers of a non-uniform potential landscape. The power law holds in a wide frequency range if the difference of potential barrier heights is rather large, while the distribution of barriers in height is rather smooth. Figures 9.13 and 9.14 show that some deviation from the $V_0 \propto 1/\omega$ dependence is defined by an increase in the width of a space charge region with the ω increasing. The influence of spatial disorder of barriers on the behavior of Re$\sigma^*(\omega)$ is weak. This leads to about constant angle of phase shift between current and voltage on the sample. The obtained result explains the methodology of impedance spectroscopy, which uses the term "element with a constant phase shift" (CPE). These "elements" (with unknown physical content) are included formally in equivalent electrical circuits for the fitting of their dynamic response to experimental data of ω behavior of SE samples

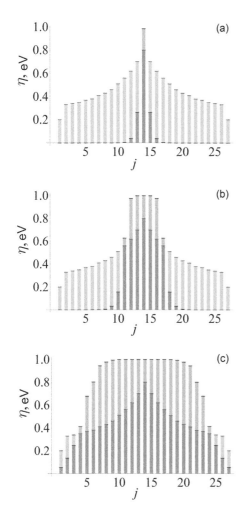

FIGURE 9.13 The ratio of amplitudes of the Maxwell displacement currents $I_{Dj,j+1}$ on potential barriers to amplitude I_0 of current of generator (dark gray color) for the $\mathrm{EC_r}/\{X^j\}/\mathrm{EC_r}$ model nanostructure with a non-uniform potential landscape, presented by light gray color. (a) Current generator with frequency $\omega/2\pi = 0.01$ Hz, $\cos\varphi = 0.383$; (b) $\omega/2\pi = 10$ Hz, $\cos\varphi = 0.156$; (c) $\omega/2\pi = 50$ kHz, $\cos\varphi = 0.129$ (Despotuli and Andreeva 2015).

Thus, the application of SDA for simulation of coupled ion-transport and dielectric polarization processes (collective phenomena) in the region of a non-uniform potential landscape clearly points out the factors and conditions, which define the origin of Jonsher's power law in model nanostructures.

9.5 A Step Towards Non-local Non-linear Dynamical Ionics

The system of ordinary differential equations (9.4) describes kinetics of space-temporal processes, i.e., ionic transitions $\dots X^{j-1} \leftrightarrow X^j \leftrightarrow X^{j+1} \dots$ over potential barriers $\dots \eta_{j-1,j},\ \eta_{j,j+1}\dots$ and Maxwell displacement currents in layered nanostructures $\{X^j\}$. The equations (9.4) can be written in general form as

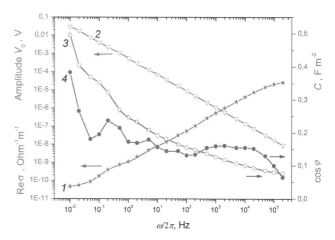

FIGURE 9.14 Frequency dependences, which are defined by UDR: 1 – Log $\mathrm{Re}\sigma^*$, where σ^* is complex conductivity; 2 – Log V_0, where V_0 is the amplitude of AC voltage; 3 – capacitance C of ionic space charge, 4 – $\cos\varphi$, where φ is the phase shift between current density and voltage on the edges of the model nanostructure, presented in Figure 9.13 (Despotuli and Andreeva 2015).

$$\frac{\mathrm{d}y_j(t)}{\mathrm{d}t} \equiv \dot{y}_j = \Phi_j(y_1, \dots y_M), \quad j = 1, \dots M;$$

$$y_j(t) = \frac{n_j(t) - n_0}{n_0}, \qquad (9.22)$$

where $y_j(t)$ is a relative change of ionic charge concentration on X^j plane, Φ_j is non-linear function $(j = 1, \dots M)$, n_0 is a surface equilibrium concentration of the mobile ions on any X^j planes in the conditions $G(t) \equiv 0$, $n_j(t)$ is a non-equilibrium surface concentration of mobile ions on X^j plane in conditions $G(t) \neq 0$. The weakness of external influence means that $y_j(t) \ll 1$, and arising field additives Ω to heights of barriers η meet conditions $\Omega_{j,j+1} \ll k_\mathrm{B}T \ll \eta_{j,j+1}$.

In $\{X^j\}$ nanostructures, a set of variables $\{y_j(t)\} \equiv |Y(t)\rangle$ influence locally and globally (non-locally) on kinetics of processes. The local effect: in (9.4), each component of $n_j f_{j\to j+1} \rho_{j+1}$ type depends on a pair of concentrations $n_j(t)$ and $n_{j+1}(t)$ and determines resulting ionic flows over a potential barrier $\eta_{j,j+1} + \Omega_{j,j+1}(t)$, where $\Omega_{j,j+1}(t)$ is a field additive to the barrier height $\eta_{j,j+1}$. The global (non-local) effect: an excess charge on any X^j plane gives own contribution to the all field additives Ω, i.e., takes part in the formation of the set $\{\Omega_{j,j+1}(t, y_1, \dots, y_M)\}$. Therefore, continuous modifications of a potential landscape $\{\eta_{j,j+1} + \Omega_{j,j+1}(t, y_1, \dots, y_M)\}$ take place (the evolution of a potential landscape). In physics, non-local continuum field theories (e.g., Eringen 2002) are applied to account the long range interactions between particles.

9.5.1 Non-commutativity of External Actions in SDA

The system (9.22) connects the set $\{y_j(t)\} \equiv |Y(t)\rangle$ with the set of the time derivatives $\{\dot{y}_j(t)\} \equiv |\dot{Y}(t)\rangle$, and with $G(t)$ function (an influence of a current generator). Quantities

$|Y(t)\rangle$ and $|\dot{Y}(t)\rangle$ with initial and boundary conditions define a trajectory (evolution) of a $\{X^j\}$ system in $2M$-dimensional phase space. All $M \sim 10...30$ components of $|Y(t)\rangle$ are interconnected by long-range Coulomb interaction. A total number of such internal relations is large, $\sim M (M-1) \sim M^2$. It is well known (Casti 1979), a large "structural connectivity" inevitably leads to the non-linear behavior of dynamic models. A theoretical analysis of such behavior is a difficult problem even for a small number of $\{y_j(t)\} \equiv |Y(t)\rangle$ components. A theoretical analysis becomes ineffective for large M. Hence, numerical methods and computer simulations should play extremely important role in dynamic non-linear ionics. An inherent non-linearity of SDA becomes clear in the computer experiments.

Let a current generator $G(t)$ creates two successive impulses of different polarity (G^- and G^+ operators) in a $\{X^j\}$ nanostructure where a non-uniform potential landscape is symmetric relatively to the center of masses. Let a Δt duration of a G-impulse satisfies the condition $\Delta t \ll \tau$ (τ is a characteristic relaxation time of a $\{X^j\}$ nanostructure). The value τ is determined by a height of the highest barrier of $\{X^j\}$. The redistribution of excess charges between minima of a potential landscape occurs at any G-impulse. The impulse G^+ modifies initial barrier heights $\ldots \eta_{k-1,k}, \eta_{k,k+1}\ldots$ of $\{X^j\}$. A consecutive impulse of G-generator with the opposite polarity G^- makes an impact on the nanostructure with another initial potential landscape, i.e., $\ldots \eta_{k-1,k} + \Omega_{k-1,k}, \eta_{k,k+1} + \Omega_{k,k+1}\ldots$

An initial equilibrium distribution of excess charges in $\{X^j\}$ ($\Delta t = 0$) is denoted here as $|Y_0\rangle = |0,\ldots,0\rangle$. The action of G^+ operator on $|Y_0\rangle$ is denoted as

$$|Y^+\rangle = G^+ |Y_0\rangle. \qquad (9.23)$$

The operator G^+ converts a set $|Y_0\rangle$ to a response $|Y^+\rangle$ of the nanostructure ($|Y_0\rangle \to |Y^+\rangle$). A set $|Y^+\rangle$ describes a non-equilibrium state of $\{X^j\}$. The action of the inverse operator G^-, i.e., $G^- = -|G^+|$ on $|Y_0\rangle$, we denote as a response $|Y^-\rangle$

$$|Y^-\rangle = G^- |Y_0\rangle \qquad (9.24)$$

A consecutive application of G^+ and G^- (or G^- and G^+) operators gives

$$|Y^{+,-}\rangle = G^- |Y^+\rangle = G^-G^+ |Y_0\rangle \qquad (9.25)$$
$$|Y^{-,+}\rangle = G^+ |Y^-\rangle = G^+G^- |Y_0\rangle, \qquad (9.26)$$

In Figure 9.15a,b, calculated data for the actions (9.25) and (9.26) are presented for the $\{X^j\}$ symmetric nanostructure. Calculations were performed within the SDA framework with model parameters $T = 300$ K and $n_0 = 10^{18}$ m^{-2} (Despotuli and Andreeva 2017). The distribution of barrier heights in $\{X^j\}$, i.e., $\ldots \eta_{k-1,k}, \eta_{k,k+1}\ldots$, is presented in the inset of Figure 9.15 a: 0.23, 0.30, 0.31, 0.32, 0.33, 0.34, 0.35, 0.37, 0.38, 0.41, 0.43, 0.46, 0.51, 0.56, 0.62, 0.73, 0.80, 0.73, 0.62, 0.56, 0.51, 0.46, 0.41, 0.38, 0.37, 0.35,

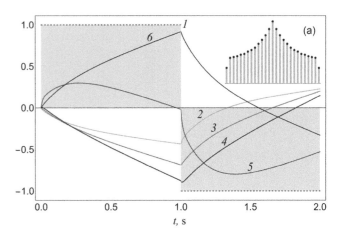
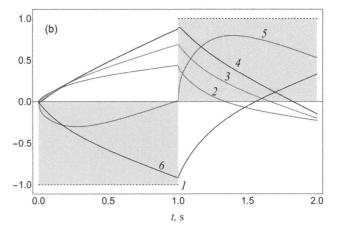

FIGURE 9.15 The influence of two sequences of current impulses ($G^+ \to G^-$ and $G^- \to G^+$) on the $\{X^j\}$ model nanostructure (in the inset) with an initial state $|Y_0\rangle = |0,\ldots,0\rangle$. Impulses create different non-equilibrium states ($|Y^{+,-}\rangle$ and $|Y^{-,+}\rangle$) with new hidden variables. (a) The sequence of current impulses $G^+ \to G^-$ (label 1, dotted line) gives the transition $|Y_0\rangle = |0,\ldots,0\rangle \to |Y^{+,-}\rangle$: $2 - y_1(t), 3 - y_2(t), 4 - y_3(t), 5 - y_4(t)$ and $6 - y_5(t)$. (b) The sequence of impulses $G^- \to G^+$ (label 1) gives the transition $|Y_0\rangle = |0,\ldots,0\rangle \to |Y^{-,+}\rangle$: $2 - y_1, 3 - y_2, 4 - y_3, 5 - y_4$ and $6 - y_5$ (Despotuli and Andreeva 2017).

0.34, 0.33, 0.32, 0.31, 0.30, 0.23 eV. The Figure 9.15 shows that consecutive current impulses $G^+ \to G^-$ or $G^- \to G^+$ create non-equilibrium states $|Y^{+,-}\rangle$ or $|Y^{-,+}\rangle$ at $\Delta t = 2$ s. The behavior of non-equilibrium state variables are shown only for five relative concentrations, i.e., for $y_1(t), y_2(t), y_3(t), y_4(t)$ and $y_5(t)$ components of the vector $\{y_j(t)\} \equiv |Y(t)\rangle$, $1 \le j \le M = 28$. Obtained data indicate that non-equilibrium states $|Y^{+,-}\rangle$ and $|Y^{-,+}\rangle$ are quite different. Namely, the corresponding components in both $|Y^{+,-}\rangle$ and $|Y^{-,+}\rangle$ sets have opposite signs:

$$|Y^{+,-}\rangle \ne |Y^{-,+}\rangle, \ |Y^{+,-}\rangle = -|Y^{-,+}\rangle. \qquad (9.27)$$

It means that the operators G^+ and G^- ($G^- = -|G^+|$) do not commutate in the condition $\Delta t \ll \tau$ when only a small part of the $\{X^j\}$ energy transfers into heat.

Thus, a non-uniform potential landscape predetermines the specificity of dynamic responses in systems with ionic conductivity. The sequence of cause–consequence relations in nanostructures is: an external action $G(t) \to$ an appearance of time-dependent distribution of excess charges on crystallographic planes $\{X^j\}$ in nanostructure $\to$ an appearance of non-uniform electric fields induced by excess charges $\to$ changes in heights of potential barriers $\to$ a change in kinetics of ionic processes $\to$ a time-dependent non-linear response of nanostructures. It means that the SDA equations define emergent properties, i.e., properties, which are not reducible to the properties of system parts.

Hence, objects of nanoionics are dynamic systems with key parameters (heights of potential barriers) depending on the external action $G(t)$. Dynamic responses of such objects cannot be described completely within a framework of linear approximations, for example, by the method of equivalent electric circuits. Appropriate theories should belong to the area of *non-local non-linear dynamical ionics*.

9.6 Further Prospects

The beginning era of nanoionic devices demands a non-local non-linear theory of dynamic response. Nanoionics tries to describe, for example, diffusion and reactions, in terms which have sense only on a nanoscale, e.g., in terms of a non-uniform potential landscape. Therefore, search for fundamental properties (which can be included in a future theory of ionic transport on a nanoscale) are very important. The theoretical system – structure-dynamic approach (SDA) of nanoionics is a step on this way.

Experimental methods of ionics have spatial and temporal restrictions on the detecting of processes; it leads to loss of the information. The combinations of the hidden variables in SDA allow not only to understand at a deeper level the ion-transport processes but give much more information in comparison with experimental data, e.g., on the frequency behavior of the impedance $Z(\omega)$. It is in accordance with the principle: "the scale is less but its manifestation is more global" (S.P. Kurdyumov).

Complex processes in ionic conductors cannot be correctly presented in terms of equivalent electric circuits with elements (resistors R and capacitors) having constant parameters. Indeed, an ionic current $I_{k,k+1}(t)$ over a potential barrier $\eta_{k,k+1}$ cannot be compared with a current in the resistor $R = const$ because the field additive Ω to the barrier height $\eta_{k,k+1}$ depends on a set of time-dependent hidden variables $\{n_i(t)\}$. Hence, for nanosystems with ionic conductivity, the search for the "best suited" equivalent electric circuits has no deep methodological basis.

The laws of spatial averaging of potential differences in ionic conductors discovered under the substantiation of SDA can be used for modeling of processes (with an acceptable accuracy) in nanostructures/nanodevices with a cross-section diameter more than $\sim$20 nm. However, the cross-section diameter of model devices can be reduced below 10 nm by using the corrected phenomenological coefficients.

In this way, obtained results can be used in nanoelectronics, for example, for modeling of processes in sub-10 nm memristors.

The results of the executed researches can serve as the initial point for the elaboration of the presented problem (physical and mathematical averaging on a nanoscale for ionic conductors) by methods of quantum mechanics, in particular, to elucidate the "dimensional effect for relative dielectric permittivity" (Despotuli and Andreeva 2018) that is completely unexplored in nanoionics.

Ideas, methods, approaches and results of non-linear dynamics (the branch of the modern theory of oscillations and waves) have actively been used in many disciplines for several decades. However, non-linear dynamics has made a weak impact on the development of solid-state ionics. The SDA demonstrates that ionic nanosystems are inherently non-linear objects. Therefore, the propagation of ideas and approaches of non-linear dynamics on the areas of solid-state ionics and nanoionics opens new possibilities for science and technology of ionic materials, structures and devices. The initiation of research programs on "dynamic non-local non-linear ionics" has to be undertaken. In such future programs, projects devoted to the advancement, development and verification of theoretical models, which take into account local Maxwell displacement currents, have to be presented. One major issue of these projects is an interrelation between Maxwell displacement current on a potential barrier and magnetic induction.

The authors expect that in future, dynamic non-local non-linear ionics (taking into account the non-uniformity of a potential landscape on a nanoscale and operating with hidden variables), new processes and phenomena will be discovered. Currently, these unstable ionic processes and phenomena (masked by the actual space–time averaging) are out from attention and can be interpreted in modern researches as the presence of noise in experimental data. Really, "Only the theory decides what one can observe" (Holton 2000).

The problem of high mobility of ions in ordered nanostructures is fundamental for various membranes and heterosystems of live organisms. Therefore, results of nanoionics on a dynamic response to electric influences have to be demanded in new multidisciplinary areas BioElectronic Medicine (https://www.src.org/library/publication/p095388/p095388.pdf)/Semiconductor Synthetic Biology (https://www.src.org/library/publication/p095387/p095387.pdf).

This study was conducted in the Institute of Microelectronics Technology and High Purity Materials of Russian Academy of Science (IMT RAS) within the framework of the State Task No. 075 - 00475-19-00.

References

Abouzari, S.M.R., Berkemeier, F., Schmitz, G. et al. 2009. On the physical interpretation of constant phase elements. *Solid State Ionics* 180: 922–927.

Andreeva, A.V. and Despotuli, A.L. 2005. Interface design in nanosystems of advanced superionic conductors. *Ionics* 11: 152–160.

Bindi, L., Evain, M., Pradel, A. et al. 2006. Fast ion conduction character and ionic phase-transitions in disordered crystals: The complex case of the minerals of the pearceite-polybasite group. *Phys. Chem. Miner.* 33: 677–690.

Bruckner, H.J., Roemer, H. and Unruh, H.G. 1984. Conductivity of the solid electrolyte RbAg4I5 in the microwave region. *Solid State Commun.* 49: 149–151.

Casti, J.L. 1979. *Connectivity, Complexity and Catastrophe in Large-Scale Systems.* Chichester: John Wiley & Sons.

Despotuli, A.L., Zagorodnev, V.N., Lichkova, N.V. and Minenkova N.A. 1989. New high-conductivity solid electrolytes $CsAg_4B_{3-x}I_{2+x}$ ($0.25 \leq x \leq 1$). *Solid State Phys.* (in Russian) 31: 242–244.

Despotuli, A.L. and Nikolaichik, V.I. 1993. A step towards nanoionics. *Solid State Ionics* 60: 275–278.

Despotuli, A.L. and Andreeva, A.V. 2003. Creation of new types of thin-film solid electrolyte supercapacitors for microsystems technology and micro (nano) electronics. *Microsyst. Techn.* (in Russian) 11: 2–10.

Despotuli, A.L., Andreeva, A.V. and Rambabu, B. 2005. Nanoionics of advanced superionic conductors. *Ionics* 11: 306–314.

Despotuli, A.L. and Andreeva, A.V. 2009. A short review on deep-sub-voltage nanoelectronics and related technologies. *Int. J. Nanosci.* 8: 389–402.

Despotuli, A.L. and Andreeva, A.V. 2010. Nanoionics: new materials and supercapacitors. *Nanotechnol. Russ.* 5: 506–520.

Despotuli, A.L. and Andreeva, A.V. 2011. Advanced nanostructures for advanced supercapacitors. *Acta Phys. Pol. A* 120: 260–265.

Despotuli, A.L. and Andreeva, A.V. 2011. "Advanced carbon nanostructures" for "advanced supercapacitors": What does it mean? *Nanosci. Nanotechnol. Lett.* 3: 119–124.

Despotuli, A.L. and Andreeva, A.V. 2012. Model, method and formalism of new approach to ion transport processes description for the solid electrolyte/electronic conductor blocking hetero-junctions. *Nano Microsyst. Tech.* (Russian) 9: 16–21.

Despotuli, A.L. and Andreeva, A.V. 2013. Maxwell displacement current in nanoionics and intrinsic ion transport properties of model nanostructures. *Nano Microsyst. Tech.* (Russian) 8: 2–9.

Despotuli, A.L. and Andreeva, A.V. 2015. Maxwell displacement current and nature of Jonsher's dynamic response in nanoionics. *Ionics* 21: 459–469.

Despotuli, A.L. and Andreeva, A.V. 2016. Method of uniform effective field in structure-dynamic approach of nanoionics. *Ionics* 22: 1291–1298.

Despotuli, A.L. and Andreeva, A.V. 2017. Dimensional factors and non-linear processes in structure-dynamic approach of nanoionics, *Nano Microsyst. Tech.* (in Russian) 19: 338–352.

Despotuli, A.L. and Andreeva, A.V. 2018. Dimensional factor and reciprocity theorem in structure-dynamic approach of nanoionics. *Ionics* 24: 237–241.

Despotuli, A.L. and Andreeva, A.V. 2018. Spatial averaging of electrostatic potential differences in layered nanostructures with ionic hopping conductivity. *J. Eleciroanal. Chem.* 829: 1–6.

Dyre, J.C., Maass, P., Roling, B. and Sidebottom, D.L. 2009. Fundamental questions relating to ion conduction in disordered solids. *Rep. Prog. Phys.* 72: 046501.

Eringen A.C. 2002. *Nonlocal Continuum Field Theories.* New York: Springer.

Frascaroli, J., Brivio, S., Covi, E. and Spiga, S. 2018. Evidence of soft bound behaviour in analogue memristive devices for neuromorphic computing. *Sci. Report.* 8: 7178.

Flygare, W.H. and Huggins, R.A. 1973 Theory of ionic transport in crystallographic tunnels. *J. Phys. Chem. Solids* 34: 1199–1204.

Funke, K. 1991. Is there a "universal" explanation for the "universal" dynamic response? *Berichte der Bunsengesellschaft für physikalische Chemie* 95: 955–964.

Habasaki, J., Leon, C. and Ngai, K.L. 2017. *Dynamics of Glassy, Crystalline and Liquid Ionic Conductors.* Cham: Springer.

Holton, G. 2000. Werner Heisenberg and Albert Einstein. *Phys. Today* 53: 38–42.

Jonscher, A.K. 1977. The "universal" dielectric response. *Nature* 267: 673–679.

Lehovec, K. 1953. Space-charge layer and distribution of lattice defects at the surface of ionic crystals. *J. Chem. Phys.* 21: 1123–1128.

Liang, C.C. 1973. Conduction characteristics of the lithium iodide-aluminum oxide solid electrolytes. *J. Electrochem. Soc.* 120: 1289–1292.

Liu, S.H. 1980. Lattice gas model for the metal-electrolyte interface. *Surf. Sci.* 101: 49–56.

Macdonald, J.R. 1982. Layered lattice gas model for the metal-electrode interface. *Surf. Sci.* 116: 135–147.

Macdonald, J.R. and Liu, S.H. 1983. An iterated three-layer model of the double layer with permanent dipoles. *Surf. Sci.* 125: 653–678.

Macdonald, J.R. 1987. Impedance spectroscopy and its use in analyzing the steady-state ac response of solid and liquid electrolytes, *J. Electroanal. Chem.* 223: 25–50.

Macdonald, J.R. 2000. Comparison of the universal dynamic response power-law fitting model for conducting systems with superior alternative models. *Solid State Ionics* 133: 79–97.

Macdonald, J.R. 2010. Addendum to "Fundamental questions relating to ion conduction in disordered solids". *J. Appl. Phys.* 107: 101101.

Mehrer, H. 2007. *Diffusion in Solids.* Berlin: Springer

Panda, D., Sahu, P.P. and Tseng, T.Y. 2018. A collective study on modeling and simulation of resistive random access memory. *Nanoscale Res. Lett.* 13: 8.

Popper, K.R. 1959. *The Logic of Scientific Discovery.* New York: Taylor & Francis.

Rhimi, T., Leroy, G., Duponchel, B. et al. 2018. Electrical conductivity and dielectric analysis of NaH_2PO_4 compound. *Ionics* 24: 3507–3514.

Sunandana, C.S. 2015. *Introduction to Solid State Ionics: Phenomenology and Applications.* Boca Raton, FL: Taylor & Francis.

Schuster, J.A. 1995. *An Introduction to the History and Social Studies of Science.* Sydney: University of Sydney.

Sidebottom, D.L. 2009. Colloquium: Understanding ion motion in disordered solids from impedance spectroscopy scaling. *Rev. Mod. Phys.* 81: 999–1014.

Uchaikin, V.V., Sibatov, R.T. and Ambrozevich, A.S., 2016. On impedance spectroscopy of supercapacitors, *Russ. Phys. J.* 59: 845–855.

Wang, Z., Wang, L., Nagai, M. et al. 2017. Nanoionics-enabled memristive devices: Strategies and materials for neuromorphic applications. *Adv. Electron. Mater.* 3: 1600510.

10

Energetic Processing of Molecular and Metallic Nanoparticles by Ion Impact

A. Domaracka, A. Mika,
P. Rousseau, and B.A. Huber
Normandie Université, ENSICAEN,
UNICAEN, CEA, CNRS, CIMAP

10.1 General Aspects of Ion-Nanoparticle Collisions

Nanoparticles (NPs), dust and aerosols are ubiquitous species in planetary atmospheres in the Solar System (e.g. Earth, Titan or Pluto) (Tielens 2013), as well as in other exoplanets (Beaulieu et al. 2012). Depending on their size, shape, chemical composition and distribution they may strongly change their physical and chemical properties and may, in particular, affect the climate and the energy balance of the planets via their interaction with the radiation field. NPs act as cloud condensation germs, and therefore, modify as well cloud properties (Twomey 1991, Zhang et al. 2015). Furthermore, organic aerosols do have astrobiological implications, as they can be a source for prebiotic material in the early Earth's or Titan's atmosphere. Observations of the Cassini-Huygens Mission, showing the presence of complex organic aerosols (haze) in Titan's upper atmosphere (Waite et al. 2007, López-Puertas et al. 2013), indicate that the present understanding of the complex organic chemistry and of nanoparticle formation from atmospheric gases in planetary atmospheres is still rather limited.

In recent decades, strong efforts have been undertaken in developing and tailoring nanoparticles for specific purposes. Such applications concern fields like catalysis, bio-sensors and imaging, medical applications as well as new energy-resources and environmental studies. A specific and important class of nanoparticles contains carbon atoms like for example clusters of fullerenes or polycyclic aromatic hydrocarbon (PAHs) molecules. These molecules are an ubiquitous component of the interstellar medium and play an important role for its physical and chemical characteristics (Joblin and Tielens 2011). They are a key link between small hydrocarbon species and much larger carbonaceous grains (Schlemmer et al. 2015). Their formation is discussed in terms of bottom-up processes described by covalent molecular growth or related to top-down processes linked to the fragmentation or sputtering of larger systems forming small hydrocarbon radicals or carbon chains (Tielens 2008). An important part of the atmospheric chemistry and the energetic processing of nanoparticles is driven by external energy sources such as photons from solar UV light, from electrons and ions being present in planet's magnetospheres, from solar wind ions or cosmic rays. Whereas the interaction with photons has been widely studied, corresponding data for ion collisions are still very limited.

Another class of nanoparticles which recently have found an increased interest in relation with the cancer treatment based on the irradiation with ion beams (Schlathölter et al. 2016, Lacombe 2014), are clusters containing metal atoms. It has been shown that the survival fraction of living cells is strongly reduced ($\sim 30\%$) when metallic NPs are present during the irradiation (Porcel et al. 2014). In order to obtain a better targeting of the tumor cells, liganded metal systems are used, like glutathione-protected gold particles (Soleilhac et al. 2017).

In contrast to collisions with photons, where after the absorption of one photon a specific well-defined energy is transferred to the NP, the internal energy of the NP after the ion collisions is not well defined as this strongly depends on the impact parameter, which is the smallest distance occurring during a straight-line trajectory. This disadvantage can partly be compensated by specific experimental techniques (coincidence methods), by working with ions in different charge states and by using different collision processes.

In the following we will discuss some typical values of energy transfers occurring in keV-MeV ion collisions with NPs of different sizes and structures as well as the processes of multi-electron capture and multi-electron emission from NPs. Furthermore, we will shortly describe penetrating collisions and sputtering phenomena.

10.1.1 Energy Transfer in Ion Collisions with NPs

In order to determine the internal energy of complex systems after ion collisions, specific experimental methods or analysis techniques have to be applied. In general, the energy distribution is wide, but is well defined for inducing specific reaction processes. For the case of small biomolecules as adenine (nucleobase: $C_5H_5N_5$) an experimental set-up based on a triple coincidence measurement has been developed for fragmentation studies of the adenine dication as a function of the excitation energy (Brédy et al. 2009). A dication is formed in a single collision of Cl^+ ions (3 keV) with a neutral molecule and the excitation energy distribution is obtained for each fragmentation channel by measuring the kinetic energy loss for the projectile in coincidence with electrons and fragments. This method is called *collision induced dissociation under energy control (CIDEC)* and is based on the formation of a negative scattered projectile as a result of double electron capture from the target molecule. The results which are shown in Figure 10.1 for various dissociation channels, indicate that the internal excitation energy of the produced dication varies with the fragmentation channel and lies in the range from 0 to 20 eV.

Another method is based on the comparison of ion mass spectra obtained in ion/molecule collisions with those measured after photoionization of the same molecule by applying the PEPICO technique (PhotoElectron-PhotoIon-Coincidence) and by applying quantum chemistry calculations (Maclot et al. 2016). As in the latter case individual photo-fragments are measured in coincidence with the energy selected photoelectrons, from the photon energy and the measured kinetic energy of the photoelectron the excitation energy can be deduced. Finally, the comparison of both mass spectra (produced by photons and ions) yields the excitation energy in the ion collision case. In Figure 10.2 an example is given for the fragmentation of the nucleoside thymidine ionized in collisions with 48 keV O^{6+} projectiles as well as by photons in the energy range from 8 to 14 eV. The result is displayed in Figure 10.2 as a function of the excitation energy defined as difference between the energy left in the target and the ionization potential. The energy distribution increases smoothly up to a maximum at around

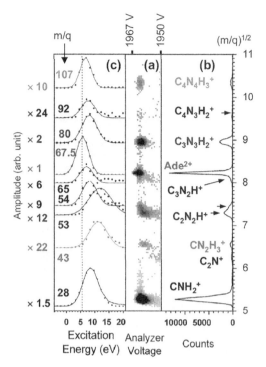

FIGURE 10.1 (c) Excitation energy distributions for different fragments produced in the dissociation of the adenine dication induced by 3 keV Cl^+ ions; (a) map between the projectile energy loss (a voltage of 1967 V corresponds to zero energy loss) and the recoil ion time-of-flight spectrum; (b) time-of-flight spectrum integrated along the energy loss. (Reprinted from R. Brédy et al., *J. Chem. Phys.* **130** (2009) 114305, with the permission of AIP Publishing.)

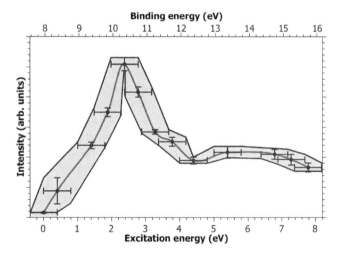

FIGURE 10.2 Distribution of the excitation energy due to ionizing collisions of O^{6+} ions with the nucleoside Thymidine. (The R^2 coefficient for the fit is 0,86.). (Reprinted with permission from S. Maclot et al., *Phys. Rev. Lett.* **117** (2016) 073201. Copyright (2016) by the American Physical Society.)

2 to 3 eV and then it extends up to 8 eV and likely also beyond.

In another study, the fragmentation of small C_n clusters ($n = 5, 7, 9$), produced in charge transfer collisions of fast ($v = 2.6$ a.u.) singly charged C_n^+ clusters with He, was studied and the branching ratios for *all* possible fragmentation channels were measured (Martinet et al. 2004). The comparison with Microcanonical Metropolis Monte Carlo simulations (Aguirre et al. 2017) based on quantum chemistry calculations allowed determining the energy distribution of the excited clusters just after the collision. Internal energies up to 40 eV have been evaluated.

In the case of sodium clusters, containing on the average $\sim$200 atoms, the minimum energy transfer was determined for multi-ionisation processes (Chandezon et al. 2001a, Chandezon et al. 2001b). The method was based on the evaluation of the smallest size of multiply charged clusters (appearance size) measured in the ion mass spectra with the aid of the model of an evaporative ensemble (Klots 1988). It turned out that the internal energy depends strongly on the charge state of the projectile. The higher the charge is, the larger is the distance at which electrons might be captured, and hence the lower the energy transfer to the multiply charged cluster. In Figure 10.3 the transferred energies are shown for collisions where five electrons were captured as a function of the projectile charge z. For proton collisions ($z = 1$) penetrating collisions are required to ionize the cluster five times. This induces a large energy transfer of $\sim$25 eV. For the highest projectile charge (Xe^{28+}), the energy transfer is as low as 1–2 eV. The transition between far and close collisions occurs evidently for $z = 5$. The experimental results (full squares) are well reproduced by theoretical calculations (open circles), based on solving the semi-classical Vlasov equation (Plagne and Guet 1999, Daligault et al. 2002).

10.1.2 Electronic and Nuclear Collisions

Ion collisions can be subdivided into elastic collisions with the nuclei of the target (nuclear stopping) as well as into processes due to the friction of the ion in the electronic cloud (electronic stopping) (Ziegler et al. 1985). The relative importance of both processes depends in addition to the target properties on the projectile mass and in particular on its velocity. In Figure 10.4 the mass/charge spectra are shown for collisions of He with the Pyrene molecule ($C_{16}H_{10}$) at center-of-mass energies of 11 keV (upper part) and 110 eV (lower part), respectively (Chen et al. 2014).

In the high energy case (upper panel), electronic stopping dominates and the electronic excitation which is induced locally along the ion trajectories is rapidly redistributed across all internal degrees of freedom. A large fraction of the collisions will lead to statistical fragmentation into the lowest dissociation energy channels at about 5 eV (loss of H or C_2H_2).

In the lower panel (110 eV case), the most prominent fragment peak is due to the loss of a single carbon atom (CH_x, $x = 0,1,\ldots$), which is a clear signature for non-statistical fragmentation. The reason is that nuclear stopping is now the dominant energy loss mechanism.

The velocity dependence of both mechanisms is shown in Figure 10.5, where the energy loss of high energy ions (Ar^+) in collisions with C_{60} fullerenes is investigated within a fully microscopic approach, called non-adiabatic quantum molecular dynamics (Kunert and Schmidt 2001). In the case of the Ar^+ projectile, the maximum of the nuclear energy loss occurs at very low velocities (at ion energies of about 1 keV) with a sharp peak reaching values of more than 400 eV. The impact parameter was chosen to be 0.2 a.u., i.e. the projectile penetrates the fullerene molecule having a radius of $\sim$7 a.u. Towards higher collision energies electronic stopping becomes dominant and reaches a value of

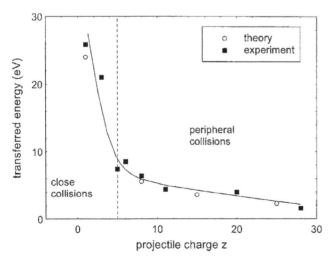

FIGURE 10.3 Minimum energy transfer for the production of 5*times charged Na_{200} clusters with projectiles in different charge states. (Reproduced from F. Chandezon et al., *Phys. Scr.* **T92** (2001) 168. Copyright 2001 IOP Publishing.)

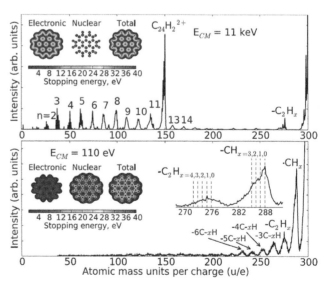

FIGURE 10.4 Upper panel: Mass to charge spectrum for He^+ + $C_{24}H_{12}$ collisions at 11 keV center-of-mass energy. Lower panel: Mass to charge spectrum for $C_{24}H_{12}^+$ + He collisions at 110 eV center-of-mass energy. The insets show results from the stopping calculations. (Reprinted from T. Chen et al., *J. Chem. Phys.* **140** (2014) 224306, with the permission of AIP Publishing.)

(a)

(b)

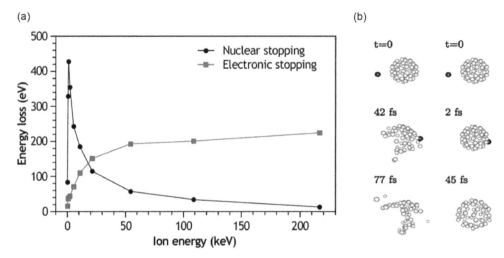

FIGURE 10.5 (a) Nuclear and electronic energy loss for Ar^+ collisions with C_{60} fullerenes (impact parameter 0.2 a.u.). (b) Structure of the fullerene molecule after the collision for two ion energies. Left series: 0.4 keV, where nuclear stopping is dominant; right series: 200 keV, where electronic stopping is favored. (Reprinted with permission from T. Kunert and R. Schmidt, *Phys. Rev. Lett.* 86 (2001) 5258. Copyright (2001) by the American Physical Society.)

$\sim$200 eV. As shown in the right part of Figure 10.5, both mechanisms lead (right after the collision) to different structures in the 50–100 fs time range. Whereas for nuclear collisions the form changes rapidly and several atoms are kicked out, electronic excitation does not drastically change the molecular form. The excitation energy which primarily is deposited in the electronic system is transferred to the nuclei by electron-phonon coupling (Grimvall and Wohlfarth 1981) on a longer time scale. Thus, the system will decay by statistical evaporation of C_2 dimers. The theoretical results are well confirmed by experimental studies (Opitz et al. 2000).

10.1.3 Multi-electron Processes: Electron Capture and Electron Emission

For collision velocities below 1 a.u., the electron capture process, transferring one or several electrons from the target molecule to a multiply charged projectile, is the dominant process. In particular for very large NPs (in the nm-range) the number of electrons can overcome the charge state of the projectile. The capture process can be described in the framework of the classical over- the-barrier model (COBM). Here it is assumed that electron transfer is possible when the potential seen by the electron moving from the target to the projectile equals the Stark-shifted binding energy of the active electron at the target (Ryufuku et al. 1980, Barany et al. 1985, Niehaus 1986, Zettergren et al. 2002). The cross section for such a process is large, when the projectile charge is high and the electron binding energy is low. For an atomic target like Ar, the ionization potential increases strongly with the degree of ionization from $\sim$15.8 to 4,426 eV for charge states $q = 0$ to $q = 17$, in particular with strong steps at electronic shell closures. As shown in Figure 10.6, this reduces strongly the cross section for multi-electron capture by roughly a factor of about 10^6 comparing single capture with the capture of 18 electrons. On the other hand,

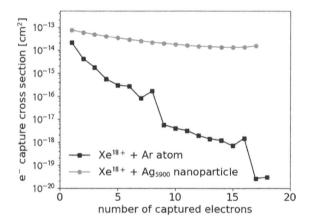

FIGURE 10.6 Cross sections for multi-electron capture calculated with the COBM-model (Zettergren et al. 2002) for non-penetrating collisions. Full dots: Xe^{18+} + Ag_{5900}; full squares: Xe^{18+} + Ar.

for metallic NPs the ionization potential IP increases only slowly with the charge state according to the relation:

$$IP = W_b + (q + 0.4) / \left(r_{WS} n^{1/3}\right)$$

where W_b is the bulk work function, q the charge and r_{WS} the Wigner Seitz radius and n the particle number (Bréchignac et al. 1989, de Heer 1993). For 6 nm Ag clusters containing approximately 5,900 Ag atoms this means that IP increases from 4.6 eV only to $\sim$13 eV for $q = 0$ to 17. Thus, multi-electron capture processes are strongly favored in collisions of multiply charged projectiles with large metallic NPs. In Figure 10.6 we consider only non-penetrating collisions to avoid contributions from nuclear collisions, i.e. did not take into account collisions with impact parameters smaller than 3 nm.

The static classical over the barrier model is found to have some drawbacks for electron capture from large metal clusters (Plagne and Guet 1999) as dynamical aspects are not well represented. These are related to the time the electrons need to pass from the cluster to the target, which is important for large impact parameter collisions. Here, the COBM model gives larger final cluster charges compared to a theoretical approach based on the solution of the self-consistent Vlasov equation (Plagne and Guet 1999).

Furthermore, in the COBM model it is assumed that the captured electron contributes fully to the shielding of the original projectile charge. However, this assumption is not valid for high projectile charge states, as in this case many electrons are captured into high-lying excited states forming multiply excited states which can decay on short time scales, i.e. during the collision time, by Auger processes which re-ionize the projectile. Thus, a projectile in a given charge state can attract more electrons from the target than its own initial charge. This has been demonstrated for the collision system $Xe^{25+} + C_{60}$, where cross sections for the following initial processes were measured:

$$Xe^{25+} + C_{60} \rightarrow Xe^{(25-s)+} + C_{60}^{r+} + (r-s)\,e^-.$$

The experimental device allowed to measure the charge state of the projectile after the collision ($Xe^{(25-s)+}$) (Martin et al. 1999, Martin et al. 2000) and to determine the number of electrons (s) stabilized at the projectile in coincidence with the number of emitted electrons (s). Thus, a total balance of the active electrons (r) could be established. In Figure 10.7 it is shown that with low probability more than $r = 60$ electrons can be taken off the molecular target within a single collision, out of which only a small number ($<s> \sim 15$) is stabilized at the projectile. The mechanism is based on multiple electron capture processes and fast electron emission by Auger processes during the collision. This keeps the charge state of the projectile during the collision high which thus can play the role of an electron "pump" into the continuum. It should be mentioned that the internal

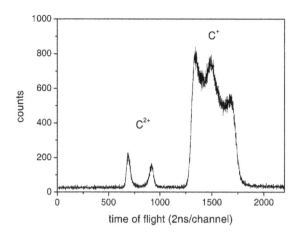

FIGURE 10.8 Time-of-flight mass spectrum showing the forward and backward peaks of high energetic C^+ and C^{2+} fragments. (Reprinted with permission from S. Tomita et al., *Phys. Rev. A* **65** (2002) 053201. Copyright (2002) by the American Physical Society.)

energy of the fullerene ion is mainly due to the high charge leading to a strong Coulomb explosion. The related kinetic energies, which are manifested in the recoil mass spectrum as forward and backward peaks of small atomic fragments (see Figure 10.8), are found to be of the order of 100–200 eV and to total internal energies of the ionized system of ~6 keV (Tomita et al. 2002a).

The examples described so far showed that, on the one hand, in collisions of ions with NPs very low energies can be transferred in distant collisions, inducing soft ionization, dissociation and fragmentation processes. On the other hand, close or penetrating collisions may lead to energy transfers of several keV and to Coulomb explosion phenomena. In photon collisions, similar phenomena can be observed. Thus, in single photon experiments the energy transfer is limited to the photon energy inducing ionization or fragmentation. However, multiphoton experiments with fs lasers, as shown for 800 nm irradiation of lead clusters containing about 400 lead atoms, can produce highly charged atomic fragments with charges up to 26 and kinetic energies in the keV range (up to 15 keV) (Lebeault et al. 2002).

The yield of secondary electrons generated by the irradiation of gold nanoparticles (Au_{32}) by fast charged projectiles has been calculated (Verkhovtsev et al. 2015), showing a strong increase of the number of emitted electrons due to collective electron excitations in the nanoparticle, in particular compared to a water target. The underlying mechanisms are related to (i) plasmon excitations in the whole nanoparticle and (ii) to the localized excitation of the d electrons (giant resonance) in individual atoms. Figure 10.8 shows the number of electrons emitted per unit length and per unit energy for both mechanisms. This work clearly shows that the decay of collective electron excitations in a nanoparticle embedded in a biological medium represents an important mechanism for the production of low energy electrons (Figure 10.9).

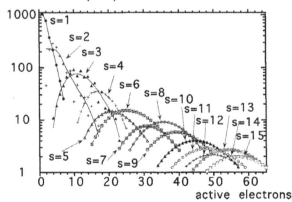

FIGURE 10.7 Cross sections for the stabilization of s electrons as a function of the number of r active electrons. (Reprinted with permission from S. Martin et al., *Phys. Rev. A* **59** (1999) R1734. Copyright (1999) by the American Physical Society.)

Summarizing box on ion-nanoparticle collisions:

- Ions can interact with the electronic system of matter, leading to friction and the *electronic energy loss* as well as by elastic nuclear collisions giving rise to *nuclear stopping processes*.

- Depending on the nature of the ion-induced process the *energy transfer* in low-energy collisions can vary between 1 eV and several keV.

- In particular for highly charged projectiles the *number of active electrons* can reach high values (approaching 100) in a single collision leading to the Coulomb explosion of the nanoparticle.

- In high-Z metal nanoparticles (Au_n) *collective excitations* (plasmon and giant resonances) increase strongly the number of ejected electrons.

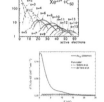

10.2 Ion Interaction with Biomolecular Clusters

Ionizing irradiation of biological matter leads to physical and chemical phenomena which are important from a fundamental point of view. In particular, corresponding studies provide a deeper understanding of cancer treatment therapies (e.g. hadron-therapy). Such detailed knowledge is required to investigate possible undesirable side effects and to possibly optimize the therapy strategies.

In general, to explore the molecular mechanisms underlying radiation damage, intensive investigations concerning ionization and fragmentation of biomolecular systems in the gas phase are required. It is the advantage of gas-phase studies that they allow isolating a system and measure its intrinsic properties. In particular, coincidence mass spectrometry analysis can be used to obtain information about the fragmentation dynamics. Therefore, such an approach has been mostly applied to amino acids, DNA building blocks, and peptides (e.g. Afrosimov et al. 2015, Alvarado et al. 2008, de Vries et al. 2003, Bari et al. 2010, Bernigaud et al. 2009, Brédy et al. 2009, Maclot et al. 2011, Milosavljević et al. 2017, Moretto-Capelle and Lepadellec 2006, Schlathölter et al. 2006; for a review article see Rousseau and Huber 2017, and references therein). However, this approach neglects any influence of the chemical environment (presence of other biomolecules and surrounding water molecules). In order to investigate the environmental effects (presence of a solvent) biomolecules have been embedded in a molecular cluster.

10.2.1 Environmental Effects

Protective Role of a Cluster Environment

Figure 10.10 shows an inclusive mass spectrum of cationic products after interaction of 300 keV Xe^{20+} ions with neutral glycine clusters (Domaracka et al. 2011, Maclot et al. 2011). A series of singly charged clusters (containing up to four molecules) is observed. To investigate the effect of the chemical environment on biomolecular fragmentation, the integrated peak intensities for the case of the isolated molecule are compared with those obtained when the molecule is embedded in a cluster. Figure 10.11 shows that the distribution of the molecular fragments is very different for both cases.

In general, for isolated amino acid molecules, ionized intact molecules are observed with very low intensities and the main dissociation channel implies the cleavage of the

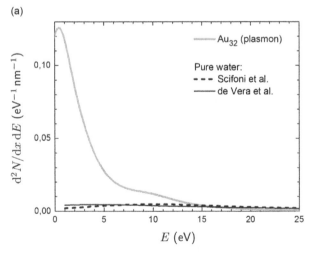

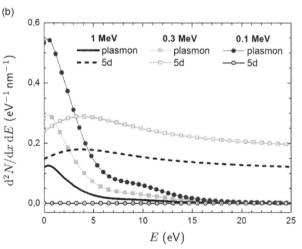

FIGURE 10.9 (a) Number of electrons per unit length and per unit energy emitted via the plasmon excitation mechanism from the Au_{32} cluster irradiated by a 1 MeV proton (thick full curve, showing a maximum at very low energies). The other curves represent the number of electrons generated from n equivalent volume of pure water. Thin and dashed curves represent the results obtained within the dielectric formalism by (Scifoni et al. 2010) and (de Vera et al. 2013), respectively. (b) Number of electrons per unit length and per unit energy produced via the plasmon and the 5d excitation mechanisms in the Au_{32} cluster irradiated by a proton of different kinetic energies. Solid line and full symbols illustrate the plasmon contribution to the electron production yield. Dashed line and open symbols show the contribution of the 5d giant resonance. (Reprinted with permission from A. V. Verkhovtsev, A. V. Korol, A. V. Solov'yov, *J. Phys. Chem. C* **119** (2015) 11000. Copyright (2015) American Chemical Society.)

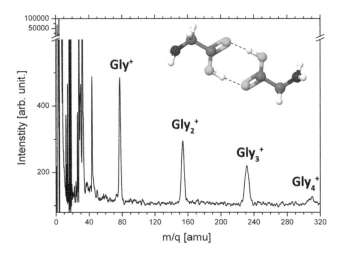

FIGURE 10.10 Mass spectrum of cationic products obtained after interaction of 300 keV Xe^{20+} ions with neutral D2-glycine clusters. (Reproduced from S. Maclot et al., *ChemPhysChem* **12** (2011) 930. Copyright (2011), with permission from WILEY-VCH Verlag GmbH & Co. KGaA, Weinheim.)

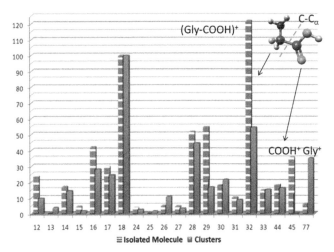

FIGURE 10.11 Histogram of peak intensities obtained after 300 keV Xe^{20+} ion collisions with isolated D2-glycine molecules (dashed bars) and their neutral molecular clusters (full bars). The peak intensities of the two spectra are normalized to the water peak. (Reproduced from S. Maclot et al., *ChemPhysChem* **12** (2011) 930. Copyright 2011, with permission from WILEY-VCH Verlag GmbH & Co. KGaA, Weinheim.)

C-C$_\alpha$ bond. The amine group preferably carries the charge leading to the formation of the (Gly-COOH)$^+$ cation; in case of the isolated D2-glycine this corresponds to the molecular mass of 32 amu (see Figure 10.11), which can further dissociate by HD loss. In the case of double electron capture the second charge is localized on the carboxyl group leading to the observation of the COOH$^+$ fragment (45 amu) (Maclot et al. 2011). Comparing the isolated and embedded molecule fragmentation, we observe that (Gly- COOH)$^+$ product is strongly reduced and the complementary fragment COOH$^+$ disappears totally for the cluster collision. Thus, when the molecules are embedded in a cluster, a protective effect of the environment is observed. The fragmentation pattern

changes: the weakest C-C$_\alpha$ bond becomes protected and stable intact amino acid molecular cations are observed as fragments of the molecular clusters (Gly)$_n^+$. The molecular cluster acts as a "buffer" for the excess energy, capable of rapidly redistributing the excess energy and charge over all cluster constituents, leading to the rupture of hydrogen bonds between the molecules (typical binding energies are of the order of hundreds of meV) cooling the total system at the same time.

New Fragmentation Channels Due to Clustering

Figure 10.12 shows the mass spectra of isolated uracil molecules (U) and of pure and mixed uracil-water clusters in the region below the monomer peak (Markush et al. 2016). Qualitatively, the fragmentation mass spectra are rather similar. However, the presence of a chemical environment leads to the formation of broader peaks which is due to the evaporation of neutral molecules from the cluster and hence to a wider distribution of kinetic fragment energies. As presented above in the case of glycine clusters, again a protective effect of the environment is found. A strong reduction of the uracil dissociation is observed in the case of mixed water-uracil clusters.

Furthermore, the molecular environment can also lead to the opening of new fragmentation pathways which is not observed in the case of isolated molecules. For uracil, a new fragment is observed at $m/z = 96$ which correspond to the loss of OH [UH-OH]$^+$. This dissociation pathway has also been observed for other nucleobase clusters (Schlathölter et al. 2006) and is related to the interplay between intra- and

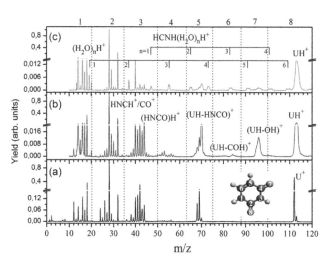

FIGURE 10.12 Mass spectrum obtained after interacion of 36 keV C^{4+} ions with isolated uracil molecules (a), pure uracil clusters (b) and nano-hydrated uracil clusters (c) in the m/z region up to the monomer. The maximum intensities of the three spectra have been normalised to 1. The regions labelled 1–8 can be roughly assigned to fragments containing 1–8 "heavy" atoms, C, N, O, respectively. In the inset, the uracil molecule and corresponding numbering of atoms is shown. (Reproduced from P. Markush et al., *Phys. Chem. Chem. Phys.* **18** (2016) 16721–16729, with permission from the PCCP Owner Societies.)

intermolecular interactions within the cluster. Uracil dimers are hold together by two O··H bonds in planar configuration. The electronic structure of the molecule is strongly modified due to hydrogen bonding, leading to the weakening of C-H and C=O bonds and giving rise to the fragment mass m/z = 96 (see Markush et al. 2016). Another new product is observed at m/z = 70 which is attributed to the loss of HNCO from the protonated molecule [UH- HNCO]$^+$. An increase of the complementary protonated molecular fragment is also observed at m/z = 44 [HNCO]H$^+$.

Intra Cluster Reactivity – Proton Transfer

Figure 10.13 presents the mass spectrum of pure uracil clusters obtained after interaction with 36 keV C^{4+} ions for clusters containing up to 20 molecules (Markush et al.

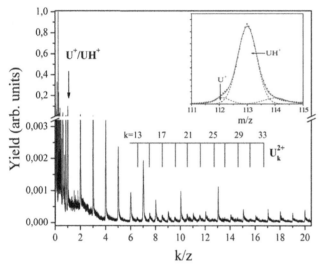

FIGURE 10.13 Mass spectrum of uracil clusters after interaction with 36 keV C^4 projectiles. The zoom-in shows the contributions of the radical cation and the protonated molecule to the monomer peak. The small peak at mass 114 corresponds to the isotopic contribution. (Reproduced from P. Markush et al., *Phys. Chem. Chem. Phys.* **18** (2016) 16721–16729, with permission from the PCCP Owner Societies.)

2016). It is interesting to note that the presence of the cluster environment leads to chemical reactivity within the clusters: namely proton transfer. The protonated uracil molecule (UH$^+$) dominates over the non-protonated one (U$^+$) (see zoom in Figure 10.13). For larger clusters the peak widths become large and an estimation of U$_k^+$ and UH$_k^+$ becomes difficult (most likely they are mixtures of both contributions).

Poully et al. (2015) have shown that the protonation probability depends on the type of biomolecules. Figure 10.14 presents ratios of the ionized parent molecule and protonated monomers of alanine and lactic acid clusters after interaction with 30 keV O^{3+} ions. Protonation dominates in both cases. However, in the case of the alanine, the radical cation (M$^+$) is barely noticeable and corresponds only to $(1.6 \pm 0.2)\%$ of the protonated species, whereas for lactic acid the proportion is much higher (12.3 ± 0.7) %. The difference in the protonation probability is consistent with the gas-phase proton affinities: lactic acid has a lower proton affinity (817,4 kJ/mol) (Berruyer-Penaud et al. 2004) than alanine (902 kJ/mol) (Bouchoux et al. 2011).

It has been shown that also the nature of the ion projectile influences the protonation of lactic acid clusters (Poully et al. 2015). Two ion beams have been used: 30 keV O^{3+} and 6 keV He$^+$. Higher proportion of the radical cation was observed for collisions with oxygen ions $(12.3 \pm 0.7)\%$ compared to a helium beam $(8.1 \pm 1.1)\%$, suggesting that protonation is favored at smaller impact parameter collisions. The same conclusion was obtained by analyzing the mass spectra as a function of the multiplicity m (number of detected fragments after the interaction). A slight decrease for $m = 1$ to 3 is observed for O^{3+} from 13.1 ± 1.0 to 11.9 ± 05. Therefore, the radical cation formation is favored in single ionization processes.

10.2.2 Mixed Biomolecule-Water Clusters: Formation of Hydrated Fragments

Hydration plays an important role in the chemistry of biological systems. Therefore, a first experimental approach consists of considering nano-hydrated clusters of molecules

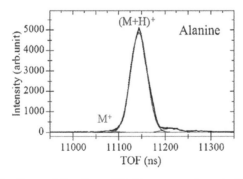
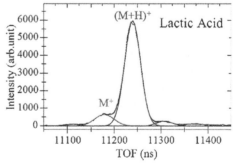

FIGURE 10.14 Zoom of the time of flight mass spectrum in the region of the monomer for alanine and lactic acid clusters after collisions with 30 keV O^{3+} ions. The contribution of radical cations (M$^+$) and protonated molecules (M+H)$^+$ is shown. Peaks on the right side of protonated species correspond to isotopic contributions of the molecule. (Reproduced from J. C. Poully et al., *ChemPhysChem* **16** (2015) 2389. Copyright (2015), with permission from WILEY-VCH Verlag GmbH & Co. KGaA, Weinheim.)

in order to mimic the presence of a simple but realistic biological environment. The formation of nano-hydrated clusters of DNA/RNA bases is of particular relevance as the presence of water molecules directly influences their conformations, stability and recognition properties (Liu et al. 2006, Liu et al. 2008, Sukhodub 1987, Schneider and Berman 1995, Prell et al. 2008, Wyttenbach and Bowers 2009).

The fragmentation of nano-hydrated uracil (Markush et al. 2016) and bromouracil (Castrovilli et al. 2017) clusters has been investigated with 36 Kev C^{4+} projectiles. The mass region of hydrated uracil clusters is shown in Figure 10.15. The zoom-in shows the region between the monomer and dimer. The series of protonated water clusters $(H_2O)_nH^+$ is observed produced by evaporation from large clusters. A second series of peaks is assigned to mixed water-uracil systems, $U(H_2O)_nH^+$. Finally, the most interesting result concerns the observation of a series of hydrated fragments $HCNH(H_2O)_nH^+$ or $CO(H_2O)_nH^+$ with $n = 0$–10 (the series $n = 0$–5 was shown in Figure 10.12). For the isolated uracil molecule the fragment $HCNH^+$ dominates (see Figure 10.12). However, the presence of an environment can change the energetics; therefore, the CO fragment could be possible as well.

Similar results have been obtained for hydrated 5-bromouracil clusters. In addition to $HCNH(H_2O)_nH^+/CO(H_2O)_nH^+$ ions, several series of protonated and hydrated fragments ($[BrC_3ONH_2]$, $[BrC_4N_2H_2O]$, $[BrC_2N]$, $[BrC_2NH]$, $[BrC_2O]$ and $[C_4N_2O_2H_2]$) were detected (Castrovilli et al. 2017). According to theoretical studies of Wang et al. (2009), the uracil molecule has four likely hydrophilic hydration sites (N1-C2, C2-N3, N3-C4 and C4-C5) where water molecules can participate in the formation of two H-bonds with two adjacent H/O atoms of uracil. In the case of bromouracil (Hu et al. 2004) the regions C4-C5 and C5-C6 are not accessible due to the size of the Br atom (see Figure 10.12 for notation).

The hydrated fragments probably originate from a nonstatistical process due to a localized energy deposition leading to ultrafast fragmentation. On the one hand, the cleavage of covalent bonds may occur on very short time scales, preventing the energy redistribution within the hydrated system and thus avoiding the evaporation of the water molecule and leading to the intermolecular hydrogen bond breaking. On the second hand, the presence of water molecules in the cluster may lead to a weakening of the bonds in the uracil or bromouracil molecules via a perturbation of the charge localization.

Summarizing box on the influence of a cluster environment:

- **The cluster acts as a buffer:**
  (i) **energy redistribution** within the clusters leads to the cleavage of the weak H-bonds, the molecular dissociation is reduced
  (ii) **charge redistribution** within the cluster leads to the closure of specific fragmentation channels
  (e.g. COOH$^+$ channel in amino acids clusters)
- **Opening of new fragmentation channels** occurs which is typical for the condensed matter
- Biomolecular clusters are **protonated**
- **Hydrated species** are observed

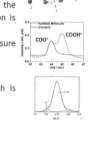

10.3 Ion-Induced Covalent Bond Formation and Molecular Growth

As discussed before, the energy which is transferred in an ion collision to the target molecule or nanoparticle can lead to the fragmentation of the molecular system, i.e. to a reduction of the molecular size. This process can follow an electronic excitation, the energy of which is transferred after some time (ps to μs) to the heavy particle system and induces the so-called *statistical* fragmentation. When the ion hits a nucleus directly with sufficient energy, the target atom will be emitted in a so-called *non-statistical* fragmentation process. If this process occurs on a very short time scale (fs), the fragments, which might be highly reactive, are still in the vicinity of the target and may react with other target constituents. Thus, in the case of a molecular cluster, reactive fragments may form new, larger molecules in the cluster, which are, as it is shown further below, covalently bound.

In the following, the energetics of different knockout processes will be discussed for ion collisions with different Polycyclic Aromatic Hydrocarbon molecules (PAHs). Furthermore, illustrating examples will be given for molecular growth processes studied in several systems.

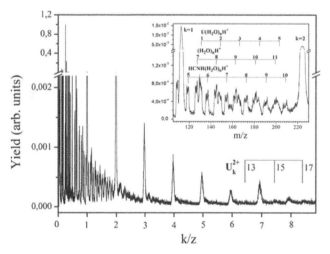

FIGURE 10.15 Mass/charge spectrum of nano-hydrated uracil clusters after interaction with 36 keV C^{4+} ions. k indicates the number of U molecules. The zoom-in presents the mass spectrum region between the monomer and dimer. The series of water-uracil clusters, protonated water clusters and hydrated HCNH/CO fragments are observed. (Reproduced from P. Markush et al., *Phys. Chem. Chem. Phys.* **18** (2016) 16721–16729, with permission from the PCCP Owner Societies.)

10.3.1   Atomic Knockout from PAH Molecules

The lowest dissociation energies for PAH molecules are of the order of 5–7 eV (Holm et al. 2011, Stockett et al. 2014a). These fragmentation channels are linked to the emission of an H atom or a C_2H_2 molecule. The dissociation energies for losses of a single C atom are substantially higher - typically of the order of 11–17 eV, depending on the position in the molecule. The values do not vary strongly for different PAH sizes (Holm et al. 2011). Indeed, when PAHs are interacting with photons (Ekern et al. 1998), electrons (Wacks and Dibeler 1959) or high-energy ions (Holm et al. 2010), the H- and C_2H_2-loss channels are dominant, as to be expected for *statistical* fragmentation processes. On the other hand, single C-loss (or CH_x loss) is typically not important in statistical fragmentation processes and is therefore a clear signature of *non-statistical knockout* (Stockett et al. 2014a). In Figure 10.16 the mass loss spectra for PAH^+ + He collisions at a center-of-mass energy of 110 eV is shown for different PAH molecules (anthracene to coronene). The shaded area identifies the loss of CH_x fragments due to knockout. The relative intensity of the CH_x-loss peak increases with the mass and size of the PAH^+ parent ion as more nuclei are contained in the cluster. The C_2H_y-loss peak, on the other hand, decreases with increasing PAH size, as emission of C_2H_y molecules in statistical fragmentation becomes less probable due to the increased number of degrees of freedom.

In Figure 10.17, we show the results from Density Functional Theory (DFT) calculations at the B3LYP/6-311++G(2d,p) level, together with the mass spectrum for He/Cor^+ collisions. The energies are given relative to that of the intact coronene cation (Cor^+) in eV. The DFT calculations indicate that these fragments are highly reactive. For the unrelaxed structures I*, II* and III* the energy differences were estimated for removing a C-atom from three different sites in the coronene molecule: (a) an inner C-atom (attached to 3 other C-atoms (I*)); (b) an outer C-atom either attached to three other C-atoms (II*) or (c) attached to 2 C-atoms and one H-atom (III*). These structures can relax to structures I, II and III, respectively. These are thermodynamically stable and may give contributions to the single carbon loss peak. When the internal energy of the systems I, II and III is sufficiently high they may undergo further statistical fragmentation.

Another type of study concerned the energies which are required to allow for knockout processes. These are described in ref. (Stockett et al. 2014b) for collisions of 100 eV He atoms with PAH and PAHN ions (anthracene $C_{14}H_{10}$, acridine $C_{13}H_9N$ and phenazine $C_{12}H_8N_2$). Molecular dynamics simulations were performed using the DL POLY Classic package (Smith and Forester 1996), where the

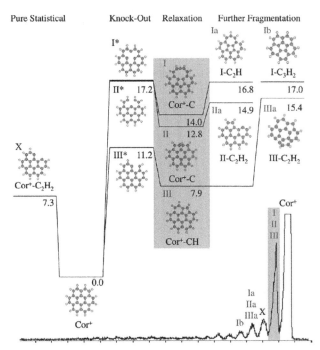

FIGURE 10.17 Likely structures of fragments produced by carbon knockout from coronene ($C_{24}H_{12}$, Cor^+). The numbers indicate DFT energies (in eV) for the separated systems in relation to the ground state of Cor^+. The labels I, II, III, etc. also indicate the fragment peaks to which they contribute. The gray area highlights structures which may contribute to the CH_x-loss peak. Structure (X) may result from C_2H_2-loss in a purely statistical process. (Reprinted with permission from M. H. Stockett et al., *Phys. Rev. A* **89** (2014) 032701. Copyright (2014) by the American Physical Society.)

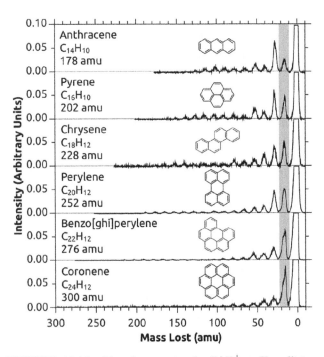

FIGURE 10.16 Mass loss spectra for PAH^+ + He collisions at 110 eV center-of-mass energy for different PAH molecules. (Reprinted with permission from M. H. Stockett et al., *Phys. Rev. A* **89** (2014) 032701. Copyright (2014) by the American Physical Society.)

molecules are modeled using the Tersoff-bond-order potential (Tersoff 1988) with parameters for the C-C, C-H, H-H and C-N interactions taken from refs. (Tersoff 1989, de Brito Mota et al. 1999, Matsunaga et al. 2000, Gatchell et al. 2015a,b). The energies transferred from the atomic projectile to the individual atoms in the molecule (Chen et al. 2014, Larsen et al. 1999, Tomita et al. 2002b) are computed using the Ziegler-Biersack-Littmark (ZBL) potential (Ziegler et al. 1985). The trajectories of all atoms are followed for up to 5 ps with a time step of 10^{-17} s.

The simulations show that the energy transfer required for a knockout process depends on the incident angle with respect to the molecular plane, as described by (Postma et al. 2014) with slightly lower values for face-on collisions. The values given in Table 10.1 apply to the knockout of different atoms (H, C or CH and N) from the PAH and PAHN $C_{14}H_{10}$, $C_{13}H_9N$, and $C_{12}H_8N_2$ molecules. The knockout of C and CH units show nearly the same typical value of 33 eV, as the H-atom is also lost when the corresponding outer C-atom is emitted. The knockout of an N-atom is found to be much lower (21 eV) in comparison with those for C atoms, a fact, which reflects that bonds between a nitrogen and a carbon atom in general are weaker than bonds between two C atoms (Jans 1967).

10.3.2 Ion-Induced Growth in PAH Clusters

Experimentally it has been shown recently, that new covalently bound molecules can be created under certain experimental conditions, inside of molecular clusters or nanoparticles (Delaunay et al. 2015). These findings have been confirmed and underlined by molecular dynamics and DFT calculations (Gatchell et al. 2015a, Gatchell et al. 2015b). As discussed in the precedent chapter, reactive species have to be produced by ion collisions in the cluster, at time scales ($\sim$1–10 fs) well below the typical dissociation time of the excited cluster ($\sim$ps), i.e. by knockout collisions. Furthermore, the cross sections for such processes have to be large. This is favored by increasing the size of the nanoparticle and by choosing the projectile mass and its velocity in a way that the collision occurs close to the maximum of the nuclear stopping power (Ziegler et al. 1985). In the experiment, several projectile ions (H^+, He^+, N^{3+}, O^{6+} and Ar^{2+}) were collided at different kinetic energies with the PAH molecule pyrene. The signature of molecular growth should be the appearance of peaks where CH_x units are added to

the clusters. The resulting experimental mass spectra are shown in Figure 10.18 (a) displaying the monomer to dimer region.

For 10 keV proton beams only very little evidence of growth processes is found. A peak of CH_x loss is observed as well as a broad shoulder extending to larger masses which is due to delayed evaporation of the dimer or trimer system (statistical processes occurring in the first extraction region of the TOF system). The signature of molecular growth becomes clearer for heavier projectiles and is strongest with 12 keV Ar^{2+} projectile ions. Here a series of peaks is observed which are separated by 12–14 mass units, i.e. CH_x units.

In a classical molecular dynamics simulation the pyrene molecules and clusters were modeled using the Adaptive Intermolecular Reactive Empirical Bond Order (AIREBO) potential (Stuart et al. 2000, Brenner et al. 2002) as defined in the LAMMPS Molecular Dynamics Simulator software suite (Plimpton 1995). The AIREBO potential is a reactive many-body potential for modeling dynamical bond forming and breaking in hydrocarbon molecules. As described in the preceding section, the interaction between the projectile ions and the pyrene molecules are modelled using the Ziegler-Biersack-Littmark (ZBL) potential (Ziegler et al. 1985), a screened coulombic potential, which describes the elastic Rutherford scattering of colliding atoms. As the ZBL potential does not take into account inelastic electronic scattering, a temperature of 2,000 K is assumed in order to compensate for this defect. The results of these simulations which describe the entire collision process from the ion impact (nuclear scattering) to the formation of new molecular species are shown in the right column of Figure 10.18 for collisions between clusters containing nine pyrene molecules and atoms at the same energies as used in the experiment. The essential features of the measured spectra are well reproduced. A movie describing the growth process in pyrene in more detail can be seen in (Rousseau 2015). Density functional tight binding calculations yield the same growth products as the classical simulations and give access to the structures of the growth products as shown for some examples in Figure 10.19.

10.3.3 Growth in Pure and Mixed Clusters of C_{60} Fullerenes: Dust Formation

One way to induce growth processes in clusters of C_{60} fullerenes is to strongly heat them in nearly central collisions with heavy ions and to form in that way a type of nanoplasma where fullerenes are strongly dissociated. A corresponding experiment has been performed with Xe^{20+} projectiles at an energy of 400 keV (Zettergren et al. 2010) colliding with clusters containing more than 15 C_{60} molecules.

As shown in Figure 10.20, large even-size carbon clusters are formed in the size range C_{96}–C_{180} when large cluster sizes (ρ_{high}) are used. An analysis of their kinetic energies indicates, that that C_{70}–C_{94} mainly are formed by

TABLE 10.1 Minimum and Typical Energies Required for the Knockout of H, C (CH), or N Atoms from Anthracene $C_{14}H_{10}$, Acridine $C_{13}H_9N$, and Phenacine $C_{12}H_8N_2$, According to the Performed MD Simulations.

	H-atom (eV)	C (CH) (eV)	N-atom (eV)
Minimum threshold value	5	16	13
Typical threshold value	10	33	24

Source: According to (Stockett et al. 2014b).

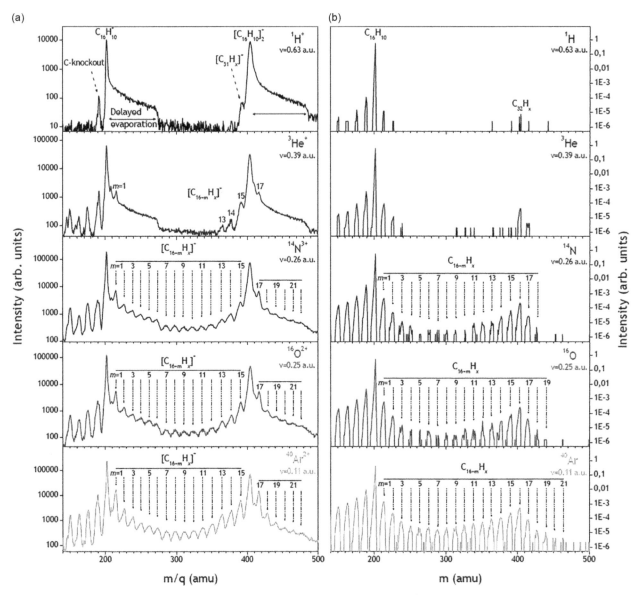

FIGURE 10.18 Comparison of experimental and simulated mass spectra. (a) experimental mass spectra for different ions colliding with pyrene clusters. (b) spectra of only covalently bound molecules from MD simulations with the same projectiles and kinetic energies as used in the experiments. (Reprinted with permission from R. Delaunay et al., *J. Phys. Chem. Lett.* (2015) 1536. Copyright (2015) American Chemical Society.)

coalescence reactions between small carbon molecules and C_{60}, whereas C_n with $n > 96$ are due to self-assembly of small molecules and to shrinking of hot giant fullerenes.

Growth experiments based on the knockout of individual C-atoms have also been performed in $He^{2+}/(C_{60})_n$ collisions at 22.5 keV energy (Zettergren et al. 2013, Wang et al. 2014, Zettergren et al. 2014). However, as in this case the probability for multiple knockout is rather low and limited to the emission of C and C_2 fragments, only C_{59} and C_{58} residues are formed. This leads after reactions with C_{60} molecules to the formation of dumb-bell shaped C_{118}^+ and C_{119}^+ clusters. Molecular Dynamics simulations (Zettergren et al. 2013) show that collisions between C_{58}^+/C_{59}^+ ions and C_{60} molecules with kinetic energies as low as about 1 eV can lead to the formation of covalently bound C_{118}^+/C_{119}^+ systems

on the picosecond timescale. A similar bonding between two intact C_{60} cages would require kinetic energies of 60–70 eV and is therefore highly unlikely (see Figure 10.21).

When replacing the He^{2+} projectile by a singly charged Ar^+ ion colliding with clusters of C_{60} fullerenes at 3 keV collision energy, the knockout probability strongly enhances and large covalent carbon nanoparticles containing several hundreds of atoms are formed (Delaunay et al. 2018). The covalent bonding can be demonstrated by the measured appearance sizes of doubly and triply charged clusters. Whereas for van der Waals clusters of C_{60} fullerenes these values are 5 and ~11 for $q = 2$ and 3, respectively, the values are smaller for the newly formed systems (~2 and 7) indicating a stronger bonding in the latter case (see Figure 10.22).

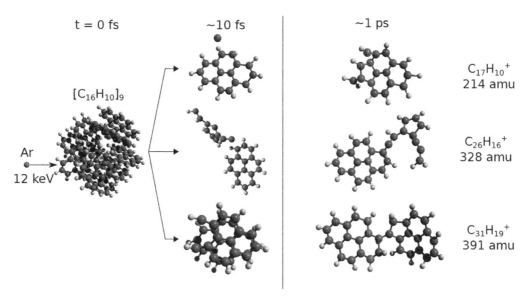

FIGURE 10.19 Three growth processes that take place in DFTB simulations of pyrene clusters following collisions with 12 keV Ar atoms. (Reprinted with permission from R. Delaunay et al., *J. Phys. Chem. Lett.* (2015) 1536. Copyright (2015) American Chemical Society.)

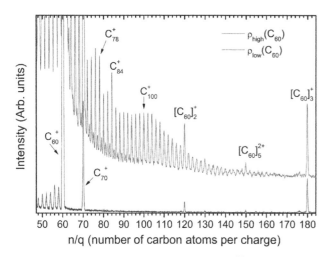

FIGURE 10.20 Intensities resulting from $Xe^{20+}+(C_{60})_m$ collisions for small and large cluster sizes (ρ_{high}, ρ_{low}). Fullerene dimers, trimers, and pentamers are indicated by $(C_{60})_2^+$, $(C_{60})_3^+$, and $(C_{60})_5^{2+}$, and other carbon clusters with C_n. (Reprinted from H. Zettergren et al., *J. Chem. Phys.* **133** (2010) 104301, with the permission of AIP Publishing.)

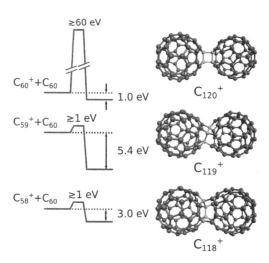

FIGURE 10.21 The effective energy barriers for covalent bond formations are $\sim$1, $\sim$1, and $>\sim$ 60 eV for C_{58}^+, C_{59}^+ and C_{60}^+ when interacting with a C_{60} molecule. The adiabatic dissociation energies are calculated at the B3LYP/6-31G(d) level of theory to be 1, 5.4, and 3 eV for the C_{120}^+, C_{119}^+, and C_{118}^+ ions in their ground states. (Reprinted with permission from H. Zettergren et al., *Phys. Rev. Lett.* **110** (2013) 185501. Copyright (2013) by the American Physical Society.)

Furthermore, the cross sections for molecular growth processes, forming covalent systems which contain more than 60 carbon atoms, were determined to be about $5 * 10^{-14}$ cm^2. This represents more than 70% of the geometrical cross sections and clearly shows the high efficiency of the underlying processes. Figure 10.23 shows possible structures of the newly formed nanoparticles which have been obtained by classical molecular dynamics calculations for the collision system Ar (3 keV) + $(C_{60})_{24}$. They contain both aromatic and aliphatic structures which are considered as dust components in space. In the extreme case, a van der Waals system containing originally 1,440 C- atoms (namely 24 C_{60} fullerene molecules) is transformed into a covalently bound dust nanoparticle containing in the extreme case 1,294 carbon atoms in a singly collision.

Very recently, ionizing collisions of He^{2+} (22.5 keV) and Ar$^+$ (3 keV) ions with mixed C_{60}/Coronene clusters have been studied by means of mass spectrometry (Domaracka et al. 2018). As shown in Figure 10.24, in the case of He^{2+} collisions, coronene, C_{60} and mixed Cor/C_{60} clusters are formed, mainly loosely bound with the exception

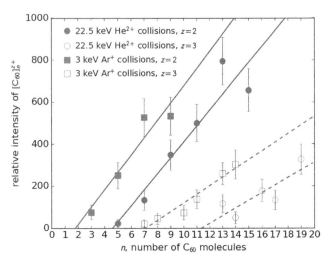

FIGURE 10.22 Relative intensities of doubly (open and full points) and triply (open and full squares) charged reaction products. Circles: van der Waals clusters produced in 22.5 keV He^{2+} collisions. Squares: covalently bound clusters formed in 3 keV Ar^{+} collisions. (Reprinted from R. Delaunay et al., *Carbon*, **129** (2018) 766 Copyright (2018), with permission from Elsevier.)

of C_{118} and C_{119} clusters. The peaks are accompanied by evaporation series, also delayed on the μs time scale. In the case of Ar^{+} collisions, as already described before, strong molecular growth due to elastic nuclear collisions occurs. A large variety of new covalently bound mixed clusters is observed, either pure coronene or fullerene systems or mixed systems containing both species are formed. A refined analysis of the region at the mixed dimer $C_{60}C_{24}H_{12}$ (mass 1,020 amu) (see Figure 10.25) shows that the system which is formed with the highest probability has lost one H-atom, $C_{60}C_{24}H_{11}$, with the mass of 1,019 amu. Also the loss of 2 H-atoms is important. This finding is in very good agreement with earlier studies of laser vaporization experiments performed in the group of H. W. Kroto (Dunk et al. 2013).

Summary box on bond formation and molecular growth:

- **Knockout** of atoms in a molecular nanoparticle can lead to a **non-statistical fragmentation process** on very short time scales (fs) and the formation of **highly reactive species** (fragments and residues).

- Due to the high density in the cluster these reactive species react with intact molecules and **form new covalently bound molecular systems.** Experiment and theory are in very good agreement.

- Ion collisions with loosely bound clusters of PAHs (pyrene to coronene) and C_{60} fullerenes show a strong molecular growth forming dust particles containing up to about 1300 C atoms.

- In the case of mixed clusters (C_{60} + coronene) a huge variety of complex systems is formed by ion collisions where nuclear stopping is dominant.

10.4 Ion-Induced Processes in Metallic Nanoparticles

10.4.1 Fission and Stability of Multiply Charged Metal Clusters

The stability of multiply charged metallic clusters can be described within the classical liquid droplet model which is the basic approach for fission processes in nuclear and cluster physics. It has its roots in the work of Lord Rayleigh, who discussed the deformation of charged liquid incompressible droplets and the interplay between surface and Coulomb forces (Lord Rayleigh 1882). Whereas short-range cohesive forces create a surface tension which tries to keep the object in a spherical form, the long-range electrostatic forces will drive the system apart into two or more fragments. Thus, the competition between both forces leads to a fission barrier which separates the bound cluster state from different fission configurations. These mechanisms are described in detail, also including quantal (shell) corrections, in the review by Näher et al. (1997).

The stability of multiply charged metal clusters is quantified by the fissility parameter X, which is defined by the

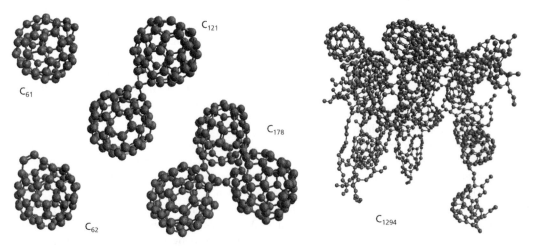

FIGURE 10.23 Some selected structures of growth products obtained from MD simulations of 3 keV Ar collisions with $(C_{60})_{24}$ clusters. Note then for the structure to the far right a different reduced scale has been used. (Reprinted from R. Delaunay et al., *Carbon*, **129** (2018) 766 Copyright (2018), with permission from Elsevier.)

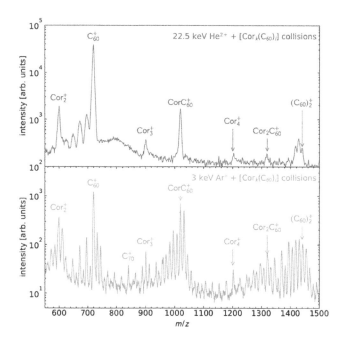

FIGURE 10.24 Mass spectrum obtained in collisions of 22.5 keV He^{2+} (upper panel) and 3 keV Ar^+ projectiles (lower panel) with $(Cor)_n(C_{60})_m$ clusters. (Reproduced from A. Domaracka et al., *Phys. Chem. Chem. Phys.* **20** (2018) 15052–15060, with permission from the PCCP Owner Societies.)

ratio of the Coulomb energy E_C and twice the surface energy E_S:

$$X = E_C/2E_S.$$

In the case of $X = 1$, both forces become equal and the fission barrier disappears. Alternatively, X can be written as

$$X = \left(q^2/n\right) / \left(q^2/n\right)_{crit}, \quad \text{where } \left(q^2/n\right)_{crit} = 16\pi r_{ws}^3 \sigma/e^2,$$

where q is the cluster charge, n the number of atoms contained in the cluster, r_{ws} the Wigner Seitz radius and

σ the surface tension. For $X = 1$ one obtains the critical charge $q_{crit}(n)$ for a given cluster size and the critical size $n_{crit}(q)$ for a given charge state. These values determine the stability limits for cluster temperatures of 0 K. However, in the experiment finite temperatures $T > 0$ K are present due to the initial cluster preparation and the applied ionization process. Thus, a metastable decay can be activated by overpassing low fission barriers. In addition, other processes like the emission of neutral particles (evaporation) may come into play, which are in competition with fission when the activation energy equals the height of the fission barrier. In this way so-called experimental appearance sizes n_{app} exist, which are larger than the corresponding values n_{crit} and which depend on various parameters like internal energy of the cluster, activation energies of evaporation processes, experimental mass resolution and hence detectability and other experimental details. We will discuss some cases further below.

In contrast to electron impact and ns-laser photoionization, ion collisions have the ability to form clusters in very high charge states with relatively low energy transfer (see Figure 10.3) and allow therefore to study cluster stability in a very wide range of charge states. This has been demonstrated in early studies of sodium clusters (Chandezon et al. 2001a, Chandezon et al. 2001b, Plagne and Guet 1999, Daligault et al. 2002) with an average cluster diameter of 1.5 nm which were produced in a cluster aggregation source equipped with an oven device. More recently, the stability of larger nanoparticles has been studied based on a magnetron cluster source (Mika 2017, Mika 2019a, Mika 2019b). For Bi clusters, containing 10 up to ∼300 Bi atoms (diameter up to 3 nm), the time-of-flight mass/charge spectrum, obtained in collisions of 120 keV Ar^{8+} projectiles, is shown in Figure 10.26.

Figure 10.26a shows the total spectrum as a function of the n/q values. The spectrum consists of two parts: A high-intensity distribution of small singly charged fragments with sizes of up to $n \sim 20$ and a low-intensity distribution at

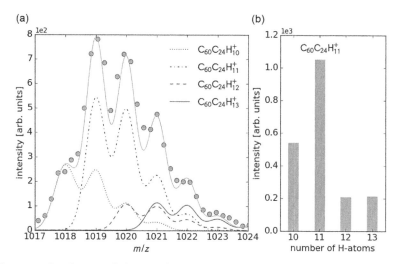

FIGURE 10.25 (a) The mass distribution of the mixed cluster dimer $C_{60}C_{24}H_x$ with corresponding isotopic contributions. (b) Intensity of each contributing cluster corrected for isotopic contributions. (Reproduced from A. Domaracka et al., *Phys. Chem. Chem. Phys.* **20** (2018) 15052–15060, with permission from the PCCP Owner Societies.)

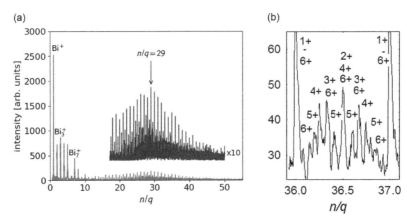

FIGURE 10.26 (a) Time-of-flight mass/charge spectrum produced in collisions of Ar^{8+} (120 keV) ions with neutral Bi clusters; (b) Zoom-in into the n/q range between 36 and 37. The numbers give the charge state of different contributions. (Taken from Mika 2017.)

larger n/q values representing multiply charged species. As shown in Figure 10.26b, clusters in charge states up to at least six are observed which are partly overlapping. Thus, the signal at $n/q = 36.5$ contains contributions from two, four and six times charged clusters, namely Bi_{73}^{2+}, Bi_{146}^{4+} and Bi_{219}^{6+}, respectively. By deconvolution of the mass resolved n/q spectra one can construct mass spectra for individual charge states as shown in Figure 10.27 for $q = 2$ to 6. The distribution for triply charged clusters starts at n-values of about 46, although the initial cluster distribution contains sizes as low as $n = 20$; that for $q = 4$ starts at about 90. These are the appearance sizes n_{app} which correspond to systems where the height of the fission barrier equals the activation energy for the evaporation of a Bi atom (~ 2 eV). When the internal energy is high the system decays for $n > n_{app}$ by the emission of a neutral particle staying in its original charge state, for example in $q = 4$. However, when n passes below n_{app}, a singly charged particle is emitted increasing in this mass range the intensity of the distribution

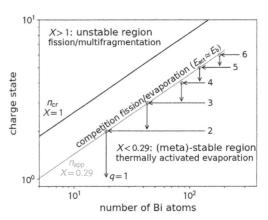

FIGURE 10.28 Stability diagram for multiply charged Bi clusters, showing n_{crit} (full line), n_{app} (full dots and dashed line) as well as decay processes. (Taken from Mika 2017, Mika 2019a). (Reprinted with permission from A. Mika et al., *Phys. Rev. A* **100** 99 (2019) 012707. Copyright (2019) by the American Physical Society.)

TABLE 10.2 Appearance Sizes for Bi_n^{q+} Clusters Produced in Collisions with 120 keV Ar^{8+} Ions

Charge state q	2	3	4	5	6
n_{app}	19 ± 3	43 ± 3	87 ± 4	121 ± 4	181 ± 4

Source: Mika 2017, Mika 2019a

in the charge state $q-1 = 3$, as to be seen in Figure 10.27. The appearance sizes for $q = 2$ to 6 are shown in Figure 10.28 as full dots and listed in Table 10.2. The dashed line which connects these points corresponds to an X-value of about 0.29, yielding appearance sizes which are much larger than the critical sizes defined by $X = 1$ (full line in Figure 10.28).

As mentioned above the measured values of n_{app} depend on various parameters, in particular on the internal energy of the decaying cluster. In Table 10.3 we show the strong increase of n_{app} for 4- times charged sodium clusters, ionized in collisions with O^{5+} projectiles, with different initial cluster temperatures prepared by passing a heat bath device before the ionization process.

More recently, appearance sizes of other types of clusters have been measured after electron impact ionization of He

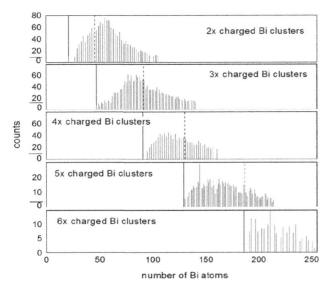

FIGURE 10.27 Mass spectra for Bi clusters in charge states 2 to 6 produced in collisions with Ar^{8+} ions. (Taken from Mika 2017.)

TABLE 10.3 Appearance Sizes for Na_n^{4+} Clusters for Collisions of O^{5+} Ions with Neutral Na_{200} Clusters with Different Initial Temperatures

T (K)	100	223	272	326	373
N_{app}	82 ± 6	99 ± 5	104 ± 6	118 ± 2	117 ± 2

Source: Ntamack et al. 2003

nanodroplets doped with coronene (Mahmoodi-Darian et al. 2018). At these very low temperatures the authors observed the appearance size for doubly charged coronene clusters to be three, i.e. Cor_3^{2+}, at variance to Cor_{15}^{2+} which had been observed in the ionization of coronene clusters by He^+ collisions at higher temperatures (Johansson et al. 2011). It should be mentioned that for collisions with highly charged ions (Xe^{25+}) the energy transfer in ionizing collisions is expected to be much lower. Therefore, the appearance size for doubly charged Coronene clusters is expected also in that case to be smaller than 15, similar to the case of sodium clusters (Table 10.3, Ntamack et al. 2003).

Returning to Bi clusters, also a much lower appearance size was reported for doubly charged systems Bi_n^{2+}, namely n_{app} ($q = 2$) $= 5$ (Schulze et al. 1988). The studies were based on electron impact ionization of small Bi clusters ($n < 40$) showing an increase of the appearance size to $n_{app}(q = 2) = 9$, when the initial cluster distribution is slightly increased. In the ion collision case, discussed above, a much wider cluster distribution was used ($n < 300$) and n_{app} ($q = 2$) was found to be 19. The difference might be explained by different isomers or binding conditions and by the fact that for very small clusters a metallic character is not yet established in the semimetal Bi, which might than be considered as a semiconductor. Indeed, structure calculations of small Bi clusters show a change in the structure at small cluster sizes, where forms that are characteristic for semiconductor systems have been calculated (Kelting et al. 2012).

The size distribution of small singly charged fission fragments is shown in Figure 10.29 for different projectile ions used to ionize neutral Bi clusters. Whereas for high projectile charges, where the average X-value of the products is expected to be high, the emission of the monomer is the dominant asymmetric fission process; for low projectile charges with lower average fissilities it is the trimer which becomes most important. This is due to different binding energies, shell and entropy effects in the formation of small size fragments as also found for other cluster systems (Näher et al. 1997).

10.4.2 Sputtering of Large Silver Nanoparticles

When considering ion collisions with metallic nanoparticles of larger sizes, as for example silver clusters containing up to 10,000 atoms or having diameters of 5–10 nm, other processes come into play and gain importance. This has two reasons: (i) On the one hand, due to the large geometrical cross section of the target, penetrating collisions become important and can overcome the role of

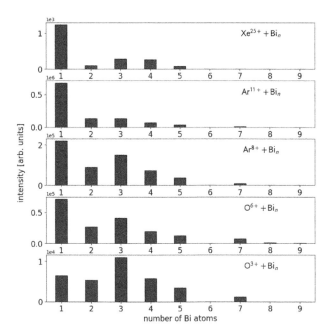

FIGURE 10.29 Size distributions for small singly charged fission fragments, produced in collisions of multiply charged ions with neutral Bi_n clusters (Taken from Mika 2017).

peripheral multi-electron capture collisions. For example for the collision system $Ar^{8+} + Ag_n$ (6 nm) the cross section for multi-electron capture in peripheral collisions (impact parameter is larger than the particle radius) can be calculated within the classical over-the barrier-model (Ryufuku et al. 1980, Barany et al. 1985, Niehaus 1986, Zettergren et al. 2002) to be $\sim 2.5 * 10^{-13}$ cm^2, which corresponds approximatively to the geometrical cross section. Thus, strong effects from penetrating collisions leading to sputtering phenomena can be expected. (ii) On the other hand, 6nm Ag nanoparticles can sustain large excess charges and are characterized by a critical charge of ~ 80. Therefore, most of the products which are formed in lower charge states, will be (meta)-stable with respect to fission. Only if the energy transfer is very high, long evaporation chains may lead to fission processes as shown in ns-laser experiments with sodium clusters (Martin et al. 1992).

In Figure 10.30 we show a TOF mass/charge spectrum obtained in collisions of Xe^{17+} projectiles at 255 keV with free Ag nanoparticles containing on the average 6,000 atoms (Mika 2017, Mika 2019b). This represents a small piece of matter with its own crystalline structure as shown in Figure 10.31 in the case of a 5 nm nanoparticle, taken with an AFM microscope. As in the previous section, dealing with fission processes, the mass/charge spectrum consists of two distributions: a series of small singly charged Ag clusters, extending to n-values of about 20, and a distribution, not resolved with respect to n/q, in the n/q region from 20 to several 1,000. (It has to be mentioned that for large systems the detection efficiency decreases). The latter distribution with lower intensity is attributed to highly charged systems formed by electron capture, whereas the intense distribution of small fragments is attributed to sputtering

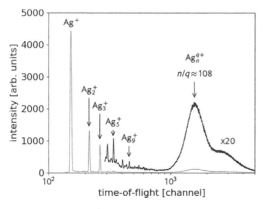

FIGURE 10.30 Mass/charge time-of-flight spectrum obtained in collisions of 255 keV Xe^{17+} ions with Ag nanoparticles of a diameter of 6 nm. (Taken from Mika 2017, 2019b). (Reprinted with permission from A. Mika et al., *Phys. Rev. B* 100 (2019) 075439. Copyright (2019) by the American Physical Society.)

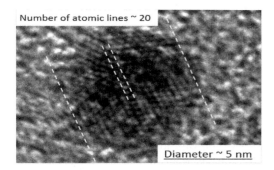

FIGURE 10.31 AFM image of a silver nano-particles with a diameter of 5 nm. (Taken from Mika 2017.)

events. This distribution is dominated by the monomer and decreases in intensity towards larger fragment sizes.

In order to identify the underlying mechanism, one might compare the spectrum produced by ion collisions (266 keV Xe^{17+}) with free nanoparticles (Mika 2017, 2019b) with that obtained after ion collisions (Xe^+) with a polycrystalline Ag surface (Harbich and Félix 2002). For this purpose peak intensities have been integrated and are shown in Figure 10.32, normalized to the monomer peak. Both spectra are very similar showing the shell closures at $n = 3$ and 9, as well as the even-odd oscillations being due to the higher stability and ionization potentials of systems containing an even number of electrons, hence favoring the presence of singly charged clusters with an odd number of atoms. This agreement supports the importance of sputtering processes. Also the yield of the small fragments (including the intensity of the monomer, dimer and trimer) produced in collisions of various projectiles in different charge states at different kinetic energies with free Ag nanoparticles compares well (see Figure 10.33) with the semiempirical formula established for sputter yields from ion/surface collisions (Matsunami et al. 1984).

From a theoretical point of view, very recently the group of H.M. Urbassek et al. has studied the sputtering of nanoparticles with the aid of MD simulations (Sandoval and Urbassek 2015). They studied the sputtering of large

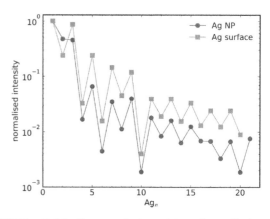

FIGURE 10.32 Integrated intensities of small fragments produced in 255 keV $Xe^{17+}/$ Ag_n nanoparticle (dots from Mika 2017, 2019b) and 20 keV Xe^+/Ag surface collisions (squares from Harbich and Félix 2002. (Reprinted with permission from A. Mika et al., *Phys. Rev. B* 100 (2019) 075439. Copyright (2019) by the American Physical Society.)

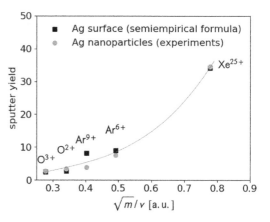

FIGURE 10.33 Dots: Ion yields (monomer, dimer and trimer) measured for different projectiles: (30 keV O^{2+}, 45 keV O^{3+}, 135 keV Ar^{9+}, 90 keV Ar^{6+}, 375 keV Xe^{25+}, taken from (Mika 2017). Squares: calculated according to Matsunami et al 1984). The experimental data are normalized for Xe^{25+}. (Reprinted with permission from A. Mika et al., *Phys. Rev. B* 100 (2019) 075439. Copyright (2019) by the American Physical Society.)

Au NPs by 200-keV neutral Xe impact. The spherical NP contained 463 878 Au atoms and had a diameter of 24.8 nm. In this case the average penetration depth of the projectile in a solid Au target is estimated to be 19.9 nm below the surface. Thus, the deposited-energy profile fills out well the present NP. Recent calculations (Järvi and Nordlund 2012, Nietiadi et al. 2014) predict that under these conditions the dependence of the sputter yield on the NP radius passes through a maximum, where the sputter yield may exceed that one for solid targets by a factor of $\sim$2.

A snapshot of a single collision event is shown in more detail in Figure 10.34. The particle ejection starts on a timescale of $\sim$1 ps, close to the impact point. The emission is found to increase until 10 ps and finally ends after several tens of ps. The ejection is found to occur from isolated spots on the NP surface, both in forward and backward direction.

FIGURE 10.34 Snapshot of a collision event shown 11 ps after particle impact. Only the particles with a kinetic energy above 0.4 eV are displayed. (Reproduced from Sandoval and Urbassek; licensee Springer. 2015 This article is distributed under the terms of the Creative Commons Attribution 4.0 International License.)

For the particular case of 200-keV Xe bombardment of Au particles, it is found that collision spikes lead to an average sputter yield of 397 atoms compared to only 116 atoms for a bulk Au target. Around 31% of the impact energy remains in the nanoparticle whereas the rest is transported away by the transmitted projectile and the ejecta. Fluctuations in the energy deposition are very strong and so are the resulting emission profiles. While sometimes only little energy is deposited close to the impact point, in other events, energy is deposited both close to the ion impact and at the ion exit point on the NP. In most events, a spike is clearly observed in the NP, which is characterized by a high density of particles moving with high energies.

The most abundant sputter products are found to be the monomer, followed by dimers and trimers and larger fragments with decreasing intensity (Järvi and Nordlund 2012). These findings agree with the experimental findings described before, although in the calculation no charge is involved and the electronic excitation is taken into account only indirectly by adding a temperature.

Ion-Shaping of Matrix-Embedded Metallic NPs

Ion beam shaping of nanoparticles embedded in a dielectric matrix has been introduced as a new technique that allows changing the morphology of the particle and its orientation in space. Nanocrystals (NC) represent a particularly interesting case due to their nonlinear optical properties and the possibility that their properties can be tuned by the surface plasmon resonance which depends on the NC size as well as their chemical surrounding. Therefore, such systems are highly relevant for various applications in nano-electronics.

In a first example (Rizza et al. 2007), it is demonstrated that chemically synthesized Au NCs with initial diameter of 15.9 ± 2.1 nm), which are embedded in a silica matrix and irradiated with 4 MeV Au^{2+} ions, can be reduced in size by the formation of halos of satellites around the original cluster. A complete dissolution and a nearly monodisperse distribution centered at a diameter of 2 nm $\pm$ 0.4 nm can be obtained by increasing the ion fluence to $8 * 10^{16}$ ions/cm^2 (see Figure 10.35). This approach is useful for a reduction of the mean size of embedded nanoparticles.

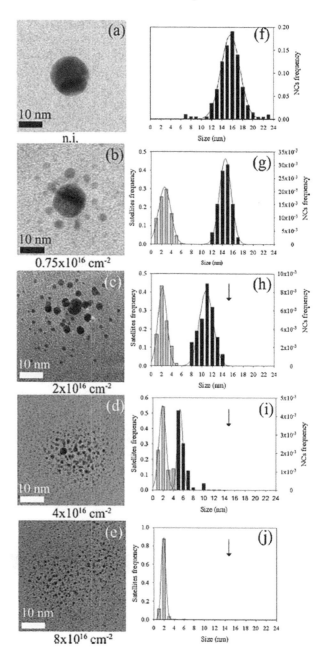

FIGURE 10.35 (a–e) Bright field TEM images of the time sequence of NC evolution under 4 MeV Au irradiation at 300 K. The samples were irradiated at increasing fluences up $8 * 10^{16}$ cm^{-2}. (f–j) The corresponding size distributions of NC and satellites. (Reprinted from G. Rizza et al., *J. Appl. Phys.* **101** (2007) 014321, with the permission of AIP Publishing.)

In another similar experiment, ion-beam-shaping mechanisms were studied for nearly monodispersed metallic nanoparticles (Au, Ag, and Au0.5Ag0.5) which were embedded in a silica matrix (Rizza et al. 2012). It is shown that several morphologies can be obtained, depending on the NP size (3–100 nm), as there are: (i) spherical shapes, (ii) facetted nanoparticles, (iii) nanorods, and (iv) nanowires. Examples are shown in Figure 10.36 for irradiations with 4 MeV Au and 74 MeV Kr ions at different fluences.

Figure 10.36 demonstrates a clear correlation between the NP size and its morphology. Four different deformation regimes can be identified: (i) Formation of satellites (<10 nm) which remain in their spherical shape; (ii) NPs in the range 10–30 nm transform continuously into nanorods and nanowires; (iii) Larger sizes (30–70 nm) show an increasing refractoriness to deformation, take longer times to be ion shaped and reveal facetted configurations. (iv) Finally, NPs being larger than 70–80 nm) are not deformed even at the highest irradiation fluence.

The analysis of the deformation rate shows that it varies strongly with the initial NP size. In Figure 10.37 the evolution of the maximum deformation length L_{max}, normalized to the NP diameter D_0, is shown as a function of the cluster size (obtained at a fluence of $5 \times 10^{14} \mathrm{cm}^{-2}$). It is observed that up to about 30 nm of the particle diameter, the normalized maximum deformation length L_{max} rapidly increases with the NP size, however, above this value it decreases towards its asymptotic value of 1.

In parallel, the thermal-spike model was implemented for three-dimensional anisotropic and composite media to study the evolution of the temperature profile within an irradiated NP (Rizza et al. 2012, Dufour et al. 2012) and to quantify

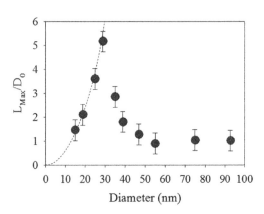

FIGURE 10.37 Evolution of the normalized maximum deformation length (L_{max}/D_0), obtained at a fluence of 5×10^{14} cm^{-2}, as a function of the initial NP size. (Reprinted with permission from G. Rizza et al., *Phys. Rev. B* **86**, 035450, 2012. Copyright (2012) by the American Physical Society.)

the vaporized/molten fraction as a function of the initial NP size. By a comparison of the model results with the experimental findings, a clear correlation was found between the deformation path and the fraction of the nanoparticle that is molten (vaporized).

The following ion-shaping mechanisms have been identified (Rizza et al. 2012): (i) Very small NPs remain spherical in form upon irradiation; (ii) Completely molten NPs transform into nano-rods and nanowires (for higher fluences); (iii) Facetted NPs are formed by partially molten NPs; (iv) Very large NPs do not melt nor deform. In particular, it was shown that the ion-shaping mechanism is effective only if the ion deposes its energy within the metallic NP.

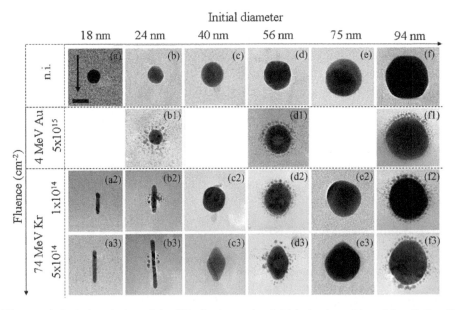

FIGURE 10.36 The morphological evolution of Au NPs for increasing initial size (x axis) and irradiation fluence (y axis). The length of the scale bar is 20 nm. The ion-beam direction is indicated by the arrow. Some of the samples [(b1), (d1), and (f1)] have been preirradiated with 4-MeV Au ions at 5×10^{15} cm^{-2} to create a halo of satellites around the central NPs (a2) to (f3). Afterward, all the samples have been irradiated up to 5×10^{15} cm^{-2} with 74-MeV Kr ions. (Reprinted with permission from G. Rizza et al., *Phys. Rev B* **86**, 035450, 2012. Copyright (2012) by the American Physical Society.)

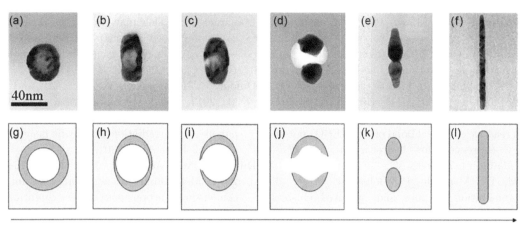

Fluence (a.u.)

FIGURE 10.38 (a–f) TEM micrographs and (g–l) sketches showing how a HNP can be transformed into a nanowire aligned along the ion beam. (Reproduced from P. E. Coulon et al., *Sci. Rep.* **6** (2016) 21116. This article is distributed under the terms of the Creative Commons Attribution 4.0 International License.)

A more recent work concerns the ion-shaping of hollow gold nanoparticles (HNPs) embedded in a silica matrix into vertically aligned prolate morphologies (Coulon et al. 2016). The HNPs had an outer diameter between 20 and 60 nm and the shell thickness varied between 3 and 14 nm. Ion irradiation has been performed with 74 MeV Kr ions at fluences up to $3.8 * 10^{14}$ cm^{-2}.

Under these conditions HNPs can be transformed into nano-wires (NWs) as shown in Figure 10.38. Simulations based on the three-dimensional thermal-spike model show that the particles undergo a partial melting where the amount of molten phase is maximal when the impact is off- center. When hitting only one hemisphere of the hollow nanoparticle, the molten phase is spatially confined. This favors the squeezing of the shell and the deformation of the HNP. Using this approach, the authors were able to qualitatively account for the different experimental observations as a function of the HNP's outer diameter, shell thickness and cavity dimension. Finally, these functional materials can be seen as building blocks for the fabrication of nanodevices with real three-dimensional architecture.

Summary box on ion interactions with metal nanoparticles:

- In collisions with multiply charged ions, 3 nm Bi nanoparticles are multi-ionized and are found to be stable above their appearance sizes which agree with predictions of the liquid drop model.

- For smaller sizes they decay by asymmetric fission emitting predominantly singly charged monomer and trimer ions.

- For silver nanoparticles with diameters of 6 nm, where cross sections for peripheral electron capture and the geometrical cross section become comparable, the emitted small fragments (dominantly monomers) are due to sputtering and the observed mass spectra compare well that of ion/surface collisions.

- Collisions of high-energy ions with matrix-embedded Au nanoparticles are a powerful tool to shape the morphology and to change the form of nanoparticles. In this way small monodisperse nanoparticles can be formed as well as nanorods and nanowires.

References

Afrosimov, V. V., Basalaev, A. A., and Vasyutinskii, O. S. 2015. Fragmentation of uracil after electron capture by doubly charged ions. *Eur. Phys. J. D* 69: 3.

Aguirre, N. F., Díaz-Tendero, S., Hervieux, P.-A., Alcamí, M., and Martín, F. 2017. M$_3$C: A computational approach to describe statistical fragmentation of excited molecules and clusters. *J. Chem. Theory Comput.* 13: 992–1009.

Alvarado, F., Bernard, J., Chen L. et al. 2008. Precise determination of 2-deoxy-D-ribose internal energies after keV proton collisions. *ChemPhysChem* 9: 1254–1258.

Barany, A., Astner, G., Cederquist, H. et al. 1985. Absolute cross sections for multi-electron processes in low energy Ar^{q+}−Ar collisions: Comparison with theory. *Nucl. Instrum. Methods Phys. Res. B* 9: 397–399.

Bari, S., Hoekstra, R., and Schlathölter, T. 2010. Peptide fragmentation by keV ion-induced dissociation. *Phys. Chem. Chem. Phys.* 12: 3376–3383.

Beaulieu, J. Dieters, S., and Tinetti, G. 2012. *Molecules in the Atmospheres of Extrasolar Planets*. Astronomical Society of the Pacific Conference Series 450 (2012): ISBN: 978-1-58381-782-7.

Bernigaud, V., Manil, B., Maunoury, L. et al. 2009. Ionisation and fragmentation of Tetraphenyl Iron (III) Porphyrin Chloride by slow multiply charged ion impact. *Eur. Phys. J.* D51: 125–130.

Berruyer-Penaud, F., Bouchoux, G., Payen, O., and Sablier, M. 2004. Structure, reactivity and thermochemical properties of protonated lactic acid. *J. Mass Spectrom.* 39: 613–620.

Bouchoux, G., Huang, S., and Inda, B. S. 2011. Acid–base thermochemistry of gaseous aliphatic α-aminoacids. *Phys. Chem. Chem. Phys.* 13: 651–668.

Bréchignac, C., Cahuzac, P., Carlier, F., and Leygnier, J. 1989. Photoionization of mass-selected K$_n^+$ ions: A test for the ionization scaling law. *Phys Rev. Lett.* 63: 1368–1371.

Brédy, R., Bernard, J., Chen, L., Montagne, G., Li, B., and Martin, S. 2009. Fragmentation of adenine under energy control. *J. Chem. Phys.* 130: 114305.

Brenner, D. W., Shenderova, O. A., Harrison, J. A., Stuart, S. J., Ni, B., and Sinnott, S. B. 2002. A second-generation reactive empirical bond order (REBO) potential energy expression for hydrocarbons. *J. Phys. Condens. Matter* 14: 783–802.

Castrovilli, M. C., Markush, P., Bolognesi, P. et al. 2017. Fragmentation of pure and hydrated clusters of 5Br-uracil by low energy carbon ions: Observation of hydrated fragments. *Phys. Chem. Chem. Phys.* 19: 19807–19814.

Chandezon, F., Lebius, H., Tomita, S., Guet, C., Pesnelle, A., and Huber, B. A. 2001a. Energy transfer in multi-ionizing ion/cluster collisions. *Phys. Scr.* T92:168–170.

Chandezon, F., Tomita, S., Cormier, D. et al. 2001b. Rayleigh instabilities in multiply charged sodium clusters. *Phys. Rev. Lett.* 87: 153402.

Chen, T., Gatchell, M., Stockett, M. H. et al. 2014. Absolute fragmentation cross sections in atom- molecule collisions: Scaling laws for non-statistical fragmentation of polycyclic aromatic hydrocarbon molecules. *J. Chem. Phys.* 140: 224306.

Coulon, P. E., Amici, J., Clochard, M. C. et al. 2016. Ion-shaping of embedded gold hollow nanoshells into vertically aligned prolate morphologies. *Sci. Rep.* 6: 21116.

Daligault, J., Chandezon, F., Guet, C., Huber, B. A., and Tomita, S. 2002 Energy transfer in collisions of metal clusters with multiply charged ions. *Phys. Rev. A* 66: 033205.

de Brito Mota, F., Justo, J. F., and Fazzio, A. 1999. Hydrogen role on the properties of amorphous silicon nitride. *J. Appl. Phys.* 86: 1843–1847.

de Heer, W. 1993. The physics of simple metal clusters: Experimental aspects and simple models. *Rev. Mod. Phys.* 65: 611–676.

de Vries, J., Hoekstra, R., Morgenstern, R. et al. 2003. Charge driven fragmentation of nucleobases. *Phys. Rev. Lett.* 91: 053401.

Delaunay, R., Gatchell, M., Maclot, S. et al. 2015. Molecular growth inside of polycyclic aromatic hydrocarbon clusters induced by ion collisions. *J. Phys. Chem. Lett.* 6: 1536–1542.

Delaunay, R., Gatchell, M., Mika, A. et al. 2018. Shock-driven formation of covalently bound carbon nanoparticles from ion collisions with clusters of C_{60} fullerenes. *Carbon* 129: 766–774.

de Vera, P., Garcia-Molina, R., Abril, I., Solov'yov, A. V. 2013. Semiempirical odel for the ion impact ionization of complex biological media. *Phys. Rev. Lett.* 110: 184104.

Domaracka, A., Capron, M., Maclot, S. et al. 2011. Ion interaction with biomolecular systems and the effect of the environment. *J. Phys. Conf. Ser.* 373: 012005.

Domaracka, A., Delaunay, R., Mika, A. et al. 2018. Ion collision-induced chemistry in pure and mixed loosely bound clusters of coronene and C_{60} molecules. *Phys. Chem. Chem. Phys.* 20: 15052–15060.

Dufour, C., Khomenkov, V., Rizza, G. and Toulemonde, M. 2012. Ion-matter interaction: The three- dimensional version of the thermal spike model. Application to nanoparticle irradiation with swift heavy ions. *J. Phys. D: Appl. Phys.* 45: 065302.

Dunk, P. W., Adjizianc, J.-J., Kaiser, N. K. et al. 2013. Metallofullerene and fullerene formation from condensing carbon gas under conditions of stellar outflows and implication to stardust. *PNAS* 110: 18081–18086.

Ekern, S. P., Marshall, A. G., Szczepanski, J., and Vala, M. 1998. Photodissociation of gas-phase polycyclic aromatic hydrocarbon cations. *J. Phys. Chem. A* 102: 3498–3504

Gatchell, M., Delaunay, R., Maclot, S. et al. 2015a. Ion-induced reactivity in pyrene clusters. *J. Phys. Conf. Ser.* 583: 012011.

Gatchell, M., Delaunay, R., Stockett, M. H. et al. 2015b. Molecular dynamics studies of impulse driven reactions in molecules and molecular clusters. *J Phys: Conf. Ser.* 635: 032043.

Grimvall, G. and Wohlfarth, E. 1981. *The Electron–Phonon Interaction in Metals,* vol 16: Selected Topics in Solid State Physics. North Holland Publishing Company, Amsterdam, New York, Oxford.

Harbich, W. and Félix, C. 2002. Mass selected cluster deposition in strongly or weakly interacting media. *C. R. Phys.* 3: 289–300.

Holm, A. I. S., Johansson, H. A. B., Cederquist, H., and Zettergren, H. 2011. Dissociation and multiple ionization energies for five polycyclic aromatic hydrocarbon molecules. *J. Chem. Phys.* 134: 044301.

Holm, A. I. S., Zettergren, H., Johansson, H. A. B. et al. 2010. Ions colliding with cold polycyclic aromatic hydrocarbon clusters. *Phys. Rev. Lett.* 105: 213401.

Hu, X., Li, H., Ding, J., and Han, S. 2004. Mutagenic mechanism of the A-T to G-C transition induced by 5-bromouracil: An ab initio study. *Biochemistry* 43: 6361–6369.

Johansson, H. A. B., Zettergren, H., Holm, A. I. S. et al. 2011. Ionization and fragmentation of polycyclic aromatic hydrocarbon clusters in collisions with keV ions *Phys. Rev. A* 84: 043201.

Jans, G. J. 1967. *Thermodynamic Properties of Organic Compounds.* Academic Press, New York.

Järvi, T. T., and Nordlund, K. 2012. Sputtering of free-standing metal nanocrystals. *Nucl. Instrum. Meth. B* 272: 66–69.

Joblin, C., and Tielens, A. G. G. M. 2011 *PAHs and the Universe: A Symposium to Celebrate the 25th Anniversary of the PAH Hypothesis.* EDP Sciences – Collection, EAS Publication Series, EAS, EDP Sciences 2011.

Kelting, R., Baldes, A., Schwarz, U. et al. 2012. Structures of small bismuth cluster cations. *J. Chem. Phys.* 136:154309.

Klots, C. E. 1988. Evaporation from small particles. *J. Phys. Chem.* 92: 5864–5868.

Kunert, T., and Schmidt, R. 2001. Excitation and fragmentation mechanisms in ion-fullerene collisions. *Phys. Rev. Lett.* 86: 5258–5261.

Lacombe S. 2014.European ITN-Project ARGENT (Advanced Radiotherapy, Generated by Exploiting Nanoprocesses and Technologies), FP7 (2014-2017) http://itn-argent.eu/

Larsen, M. C., Hvelplund, P., Larsson, M. O., and Shen H. 1999. Fragmentation of fast positive and negative C_{60} ions in collisions with rare gas atoms. *Eur. Phys. J. D* 5: 283–289.

Lebeault, M. A., Viallon, J., Chevaleyre, J. et al. 2002. Resonant coupling of small size-controlled lead clusters with an intense laser field. *Eur. Phys. J. D* 20: 233–242.

Liu, B., Brondsted-Nielsen, S., Hvelplund, P. et al. 2006. Collision-induced dissociation of hydrated adenosine monophosphate nucleotide ions: Protection of the ion in water nanoclusters. *Phys. Rev. Lett.* 97: 133401.

Liu, B., Haag, N., Johansson, H. et al. 2008. Electron capture induced dissociation of nucleotide anions in water nanodroplets. *J. Chem. Phys.* 128: 075102.

López-Puertas, M., Dinelli, B. M., Adriani, A. et al. 2013. Large abundances of polycyclic aromatic hydrocarbons in Titan upper atmosphere. *Astophys. J.* 770: 132.

Lord Rayleigh, F. R. S. 1882. XX. On the equilibrium of liquid conducting masses charged with electricity. *Phil. Mag.* 14: 184–186.

Maclot, S., Capron M., Maisonny, R. et al. 2011. Ion-induced fragmentation of amino acids: Effect of the environment. *ChemPhysChem* 12: 930–936.

Maclot, S., Delaunay, R., Piekarski, D. G. et al. 2016. Determination of energy-transfer distributions in ionizing ion-molecule collisions. *Phys. Rev. Lett.* 117: 073201.

Mahmoodi-Darian, M., Raggl, S., Renzler, M. et al. 2018. Doubly charged coronene clusters-Much smaller than previously observed. *J. Chem. Phys.* 148: 174303.

Markush, P., Bolognesi, P., Cartoni, A. et al. 2016. The role of the environment in the ion induced fragmentation of uracil. *Phys. Chem. Chem. Phys.* 18: 16721–16729.

Martin, S., Bernard, J., Chen, L., Denis, A., and Désesquelles, J. 2000. Number of electrons active in slow collisions of O^{8+} and Ar^{8+} with C_{60}. *Eur. Phys. D* 12: 27–32.

Martin, S., Chen, L., Denis, A., and Désesquelles J. 1999. Very high multiplicity distributions of electrons emitted in multicapture collision between Xe^{25+} and C_{60}. *Phys. Rev A* 59: R1734–R1737.

Martin, T. P., Näher, U., Göhlich, H. and Lange, T. 1992. Fission of highly charged sodium clusters. *Chem. Phys. Lett.* 196: 113–117.

Martinet, G., Díaz-Tendero, S., Chabot, M. et al. 2004. Fragmentation of highly excited small neutral carbon clusters. *Phys. Rev. Lett.* 93: 063401.

Matsunaga, K., Fisher, C., and Matsubara, H. 2000. Tersoff potential parameters for simulating cubic boron carbonitrides. *Jap. J. Appl. Phys.* 39: L48–L51.

Matsunami, N., Yamamura, Y., Itikawa, Y. et al. 1984. Energy dependence of the ion-induced sputtering yields of monatomic solids. *At. Data Nucl. Data Tables* 31: 1–80.

Mika, A. 2017. PhD thesis: Interactions of slow multiply charged ions with large, free radiosensitizing metallic nanoparticles. Normandie Université (Caen, France).

Mika, A., Rouseau, P., Domaracka, A. and Huber, B. A. 2019a. Fragmentation of Bi clusters by multiply charged ions, *Phys. Rev. A* 99: 012707.

Mika, A., Rousseau, P., Domaracka, A. and Huber, B. A. 2019b. Interaction of multiply charged ions with large free silver nanoparticles: Multielectron capture, fragmentation, and sputtering phenomena, *Phys. Rev. B* 100: 075439.

Moretto-Capelle, P. and Lepadellec, A. 2006. Electron spectroscopy in proton collisions with dry gas- phase uracil base. *Phys. Rev. A* 74: 062705.

Näher, U., Bjørnholm, S., Frauendorf, S., Garcias, F., and Guet, C. 1997. Fission of metal clusters. *Phys. Rep.* 285: 245–320.

Nietiadi, M. L., Sandoval, L., Urbassek, H. M., and Möller W. 2014. Sputtering of Si nanospheres. *Phys Rev. B* 90: 045417.

Milosavljević, A. R., Rousseau, P., Domaracka, A., Huber, B. A., and Giuliani, A. 2017. Multiple electron capture from isolated protein poly-anions in collision with slow highly charged ions. *Phys. Chem. Chem. Phys.* 19: 19691–19698.

Ntamack, G. E., Chandezon, F., Manil, B. et al. 2003. Stability and fragmentation processes of highly charged sodium clusters. *Eur. Phys. J. D* 24: 153–156.

Niehaus, A. 1986. A classical model for multiple-electron capture in slow collisions of highly charged ions with atoms. *J. Phys. B* 19: 2925.

Opitz, J., Lebius, H., Tomita, S. et al. 2000. Electronic excitation in $H^{+}-C_{60}$ collisions: Evaporation and ionization. *Phys. Rev. A* 62: 022705.

Prell, J. S., O'Brien, J. T., Holm, A. I. S., Leib, R. D., Donald, W. A., and Williams, E. R. 2008. Electron capture dissociation by a hydrated gaseous peptide: Effects of water on fragmentation and molecular survival. *J. Am. Chem. Soc.* 130: 12680–12689.

Plagne, L. and Guet, C. 1999. Highly ionized but weakly excited metal clusters in collisions with multicharged ions. *Phys. Rev. A* 59: 4461–4469.

Plimpton, S. 1995. Fast parallel algorithms for short–range molecular dynamics. *J. Comput, Phys.* 117: 1–19.

Porcel, E., Tillement, E., Lux, F. et al. 2014. Gadolinium-based nanoparticles to improve the hadrontherapy performances. *Nanomed. Nanotechnol. Biol. Med.* 10:1601–1608.

Postma, J., Hoekstra, R., Tielens, A. G. G. M., and Schlathölter, T. 2014. Molecular dynamics study on slow ion interactions with the polycyclic aromatic hydrocarbon molecule anthracene. *Astrophys. J.* 783: 61.

Poully, J. C., Vizcaino, V., Schwob, L. et al. 2015. Formation and fragmentation of protonated molecules after ionization of amino acid and lactic acid clusters by collision with ions in the gas phase. *ChemPhysChem* 16: 2389–2396.

Rizza, G., Cheverry, H., Gacoin, T., Lamasson, A., and Henry, S. 2007. Ion beam irradiation of embedded nanoparticles: Toward an in situ control of size and spatial distribution. *J. Appl. Phys.* 101: 014.

Rizza, G., Coulon, P. E., Khomenkov, V. et al. 2012. Rational description of the ion-beam shaping mechanism. *Phys. Rev. B* 86: 035450.

Rousseau, P. 2015. CNRS-Institut de Physique: Des collisions avec des ions pour la croissance de molécules au sein d'un agrégat. www.cnrs.fr/inp/spip.php?article 3833

Rousseau, P. and Huber, B. A. 2017. Ion collisions with biomolecules and biomolecular clusters. In *Nanoscale Insights into Ion-Beam Cancer Therapy*, ed. A. V. Solov'yov, pp. 121–158. Springer, Cham

Ryufuku, H., Sasaki, K., and Watanabe, T. 1980. Oscillatory behavior of charge transfer cross sections as a function of the charge of projectiles in low-energy collisions. *Phys. Rev. A* 21: 745–750.

Sandoval, L. and Urbassek, H. M. 2015. Collision-spike sputtering of Au nanoparticles. *Nanoscale Res. Lett.* 10: 314.

Schlathölter, T., Alvarado, F., Bari, S. et al. 2006. Ion-induced biomolecular radiation damage: From isolated nucleobases to nucleobase clusters. *Chem. Phys. Chem.* 7: 2339–2345.

Schlathölter, T., Eustache, P, Porcel, E. et al. 2016. Improving proton therapy by metal-containing nanoparticles: Nanoscale insights. *Int. J. Nanomed.* 2016: 1549–1556.

Schlemmer, S., Giesen, T., and Mutschke, H., (2015) *Laboratory Astrochemistry: From Molecules through Nanoparticles to Grains*, Wiley-VCH, ISBN: 978-3-527-40889-4

Schneider, B., and Berman, H. M. 1995. Hydration of the DNA bases is local. *Biophys. J.* 69:2661.

Schulze, W., Winter, B. and Goldenfeld, I. 1988. Stability of multiply positively charged homonuclear clusters. *Phys. Rev. B* 38: 12937–12941.

Scifoni, E., Surdutovich, E., and Solov'yov, A. V. 2010. Spectra of secondary electrons generated in water by energetic ions. *Phys. Rev. E* 81: 021903.

Soleilhac, A., Bertorelle, F., Comby-Zerbino, C. et al. 2017. Size characterization of glutathione-protected gold nanoclusters in the solid, liquid and gas phases. *J. Phys. Chem. C* 121: 27733–27740.

Smith, W., and Forester, T. 1996. DL'POLY'2.0: A general-purpose parallel molecular dynamics simulation package. *J. Mol. Graphics* 14: 136–141.

Stockett, M. H., Adoui, L., Alexander, J. D. et al. 2014a. Nonstatistical fragmentation of large molecules. *Phys. Rev. A* 89: 032701.

Stockett, M. H., Gatchell, M., Alexander, J. D. et al. 2014b. Fragmentation of anthracene $C_{14}H_{10}$, acridine $C_{13}H_9N$ and phenazine $C_{12}H_8N_2$ ions in collisions with atoms. *Phys. Chem. Chem. Phys.* 16: 21980–21987.

Stuart, S. J., Tutein, A. B., and Harrison, J. A. 2000. A reactive potential for hydrocarbons with intermolecular interactions. *J. Chem. Phys.* 112: 6472–6486.

Sukhodub, L. 1987. Interactions and hydration of nucleic acid bases in a vacuum. Experimental study. *Chem. Rev.* 87: 589–606.

Tielens, A. G. G. M. 2008. Interstellar polycyclic aromatic hydrocarbon molecules. *Ann. Rev. Astron. Astrophys.* 46:289–337.

Tielens, A. G. G. M. 2013. The molecular universe. *Rev. Mod. Phys.* 85: 1021–1081.

Tomita, S., Hvelplund, P., Nielsen, S. B., and Muramoto, T. 2002a. C_{59}-ion formation in high-energy collisions between cold C_{60}^- and noble gases. *Phys. Rev. A* 65, 043201.

Tomita, S., Lebius, H., Brenac, A., Chandezon, F., and Huber, B. A. 2002b. Kinetic-energy release and fragment distribution of exploding, highly charged C_{60} molecules. *Phys. Rev. A* 65: 053201.

Tersoff, J. 1988. New empirical approach for the structure and energy of covalent systems. *Phys. Rev. B* 37: 6991–7000.

Tersoff, J. 1989. Modeling solid-state chemistry: Interatomic potentials for multicomponent systems. *Phys. Rev. B* 39: 5566–5568 (Erratum 1990. *Phys. Rev. B* 41: 3248).

Twomey, S. 1991. Aerosols, clouds, and radiation. *Atmos. Environ.* 25 A: 2435–2442.

Verkhovtsev, A. V., Korol, A. V., and Solov'yov, A. V. 2015. Electron production by sensitizing gold nanoparticles irradiated by fast ions. *J. Phys. Chem. C* 119: 11000–11013.

Wacks, M. E., and Dibeler, V. H. 1959. Electron impact studies of aromatic hydrocarbons. I. Benzene, naphthalene, anthracene, and phenanthrene. *J. Chem. Phys.* 31, 1557.

Waite Jr., J. H., Young, D. T., Cravens T. E. et al. 2007. The process of tholin formation in titan's upper atmosphere. *Science* 316: 870–875.

Wang, Y., Zettergren, H., Rousseau, P. et al. 2014. Formation dynamics of fullerene dimers C_{118}^+, C_{119}^+, C_{120}^+. *Phys. Rev. A* 89: 062708.

Wang, F.-F., Zhao, D.-X., and Yang, Z.-Z. 2009. Theoretical studies of uracil-$(H_2O)_n$ ($n = 1–7$) clusters by ab initio and ABEEM$\sigma\pi$/MM fluctuating charge model. *Chem. Phys.* 360: 141–149.

Wyttenbach, T., and Bowers, M. 2009. Hydration of biomolecules. *Chem. Phys. Lett.* 480: 1–16.

Zettergren, H., Johansson, H. A. B., Schmidt, H. T. et al. 2010. Magic and hot giant fullerenes formed inside ion irradiated weakly bound C_{60} clusters. *J. Chem. Phys.* 133: 104301.

Zettergren, H., Rousseau, P., Wang, Y. et al. 2013. Formations of dumbbell C_{118} and C_{119} inside clusters of C_{60} molecules by collision with α particles. *Phys. Rev. Lett.* 110: 185501.

Zettergren, H., Rousseau, P., Wang, Y. et al. 2014. Bond formation in C_{59}^{+} – C_{60} collisions. *J. Phys. Conf. Ser.* 488: 012028.

Zettergren, H., Schmidt, H. T., Cederquist, H. et al. 2002. Static over-the-barrier model for electron transfer between metallic spherical objects. *Phys. Rev. A* 66: 032710.

Zhang, X, West, R. A., Irwin P. G. J., Nixon, C. A., and Yung, Y. L. 2015. Aerosol influence on energy balance of the middle atmosphere of Jupiter. *Nat. Com.* 6: 10231.

Ziegler, J. F., Biersack, J. P., and Littmark, U. 1985. *The Stopping and Range of Ions in Matter*, vol. 1; Pergamon, New York.

11

Nanoscale Fluid Dynamics

Ravi Radhakrishnan,
N. Ramakrishnan,
David M. Eckmann, and
Portonovo S. Ayyaswamy
University of Pennsylvania

11.1 Introduction

Nanoscale fluid dynamics (NFD) as defined in this chapter is the study of the motion of nanosized particles (NPs) that are suspended in an external liquid medium. Such a suspension is often referred to as a nanofluid. The liquid medium itself may be Newtonian or non-Newtonian, static or flowing under the influence of an external pressure gradient, unbounded or confined in a tube-like vessel. In addition, there may be temperature gradients in the medium which may cause heat transport in addition to the mass transfer. The nanosize is typically in the range of 1–100 nanometer (nm) and would correspond with, for example, the diameter of a spherical NP. NPs with diameters in the range of 2–10 nm are classified as quantum dots, and these will not be discussed in this chapter. The NPs may be of various shapes, while the spherical NP is among the most commonly studied in the literature. The shapes considered in this chapter will be restricted to those of spheres, spheroids, and ellipsoids.

Based on experimental observations, it is now well known that under identical external conditions, transport properties such as diffusivity, viscosity, thermal conductivity, and electrical conductivity of nanofluids are significantly different from those of suspensions containing larger size particles. However, at the time of this writing, how the NP dispersion in the host medium influences these properties is still being intensely debated (see [1–4]). Clearly, for a given sum total of particle volumes in a suspension, the cumulative interfacial surface area of the particles that is exposed to the fluid will be larger with smaller size particles. Macroscopic properties and emergent behavior of the nanofluid dependent on surface area will be impacted by this feature, and this is one reason for the comparatively enhanced transport noted with nanofluids. Apart from this, there are other important reasons such as the ones related to the dynamics of the NP random motion in a static or a flowing suspension (Brownian interactions and diffusivities), the nature of the proximity-dependent interaction of an NP with a confining boundary, and so on. These special features will be discussed in this chapter.

At present, a very large amount of research on a worldwide basis is being undertaken to ascertain and provide the reasons for the observed behavior of nanofluids and NFD. An important motivating factor for this large interest is the immediate impact on the associated technologies. A nanofluid with an enhanced thermal conductivity and hence a high heat transfer coefficient will serve to very efficiently cool a tiny computer chip, thus enabling very high processing power for the system as a whole. In a completely different context, drug (for example, an antibiotic) laden, optimally "functionalized", sized, and shaped NPs may successfully negotiate their way through a micronscale blood vessel and deliver the drug to the intended target such as an endothelial cell surface on an inflamed tissue. The implications are profound. The targeted drug delivery in this example would very much depend on the diffusivity of the NPs in a non-Newtonian fluid (blood) flow containing red blood cells and other constituents. The major aim of this chapter is to discuss the fluid dynamics aspects associated with NP suspensions whether static or flowing.

11.2 Foundations

11.2.1 Conservation Equations

The study of NFD as described in this chapter is largely based on concepts from non-equilibrium statistical mechanics combined with those from continuum fluid mechanics and transport that govern NP behavior in an external viscous fluid medium. In a fluid, the molecules are in continual random thermal motion consistent with its temperature. The dynamics at this molecular level can be described based on transitions between microstates. A microstate defines the complete set of positions and momenta of all the particles/molecules of the system. For molecular systems, the microstate of the system with a given set of positions and momenta at a given time t depends only on the microstate at the immediately preceding time step. This memory-less feature is referred to as a Markov process, and all Markov processes obey the master equation [5]. The probability to access a microstate defined by a given value of the microstate variables y is denoted by $P(y,t)$, which, for a general dynamical process at non-equilibrium, is time-dependent. Every Markov process is governed by a set of probability balance equations, collectively referred to as the master equation given by

$$\frac{\partial P(y,t)}{\partial t} = \int dy'\{w(y|y')P(y',t) - w(y'|y)P(y,t)\}. \quad (11.1)$$

Here, y and y' denote different microstates, and $w(y|y')$ is the transition probability (which is a rate of transition in units of a frequency) from state y' to state y. Macroscopic conservation equations can be derived from the master equation by taking the appropriate moment:

$$\frac{\partial}{\partial t}\langle y\rangle = \int y\frac{\partial P(y,t)}{\partial t}dy = \int\int dy\,dy'(y'-y)w(y'|y)P(y,t). \quad (11.2)$$

Indeed, a reduced form of the master equation is the Boltzmann equation [6], where the microstates defined in terms of the positions and momenta of all particles (assumed to be hard spheres) are reduced to a one-particle (particle j) distribution by integrating over the remaining $n-1$ particles and where the operator for the total derivative d/dt is expressed as the operator for the partial derivative $\partial/\partial t$ plus the convection term $\mathbf{u}\cdot\frac{\partial}{\partial r}$, where $\mathbf{u}$ is the velocity. The moments of the Boltzmann equation were derived by Enskog for a general function ψ_j [6]. Substituting ψ_j as m_j, the mass of particle j, yields the continuity equation, as $m_j v_j$, the momentum of particle j, yields the Navier–Stokes equation, and as $\frac{1}{2}m_j v_j^2$, the kinetic energy of the particle, yields the energy equation, which together represent conservation equations that are the pillars of continuum hydrodynamics.

11.2.2 Thermal and Brownian Effects

One of the main attributes of NFD that differentiates it from traditional hydrodynamics is that the fluid mechanics and thermal effects have to be treated with equal importance. It is worth noting that while the thermal effects and fluctuations are described within the scope of the master equation (Eq. (11.1)), by taking the moment to derive the conservation law (Eq. (11.2)), oftentimes the thermal effects are averaged out to produce only a mean-field equation. Indeed, the continuity, Navier–Stokes, and energy equations cannot accommodate thermal fluctuations that are inherent in Brownian motion even though such effects are fully accommodated at the level of the parent master equation. Therefore, NFD must be approached differently than traditional hydrodynamics.

One approach is to start with the mean-field conservation equation such as the Boltzmann equation and add the thermal fluctuations as a random forcing term, which results in the Boltzmann–Langevin equation derived by Zwanzig and Bixon [7]. This will amount to the same governing equation if such random fluctuating terms are added as random stress terms to the Navier–Stokes equations. This latter approach is referred to as the fluctuating hydrodynamics (FHD) approach first proposed by Landau and Lifshitz [8]. In the FHD formulation, the fluid domain satisfies

$$\nabla\cdot\mathbf{u} = 0, \quad (11.3)$$

$$\rho\frac{D\mathbf{u}}{Dt} = \nabla\cdot\boldsymbol{\sigma}, \quad (11.4)$$

where $\mathbf{u}$ and ρ are the velocity and density of the fluid, respectively, and $\boldsymbol{\sigma}$ is the stress tensor given by

$$\boldsymbol{\sigma} = -p\mathbf{J} + \mu[\nabla\mathbf{u} + (\nabla\mathbf{u})^T] + \mathbf{S}. \quad (11.5)$$

Here, p is the pressure, $\mathbf{J}$ is the identity tensor, and μ is the dynamic viscosity. The random stress tensor $\mathbf{S}$ is assumed to be a Gaussian white noise that satisfies

$$\langle S_{ij}(\boldsymbol{x},t)\rangle = 0, \quad (11.6)$$

$$\langle S_{ik}(\boldsymbol{x},t)S_{lm}(\boldsymbol{x}',t')\rangle = 2k_BT\mu(\delta_{il}\delta_{km} + \delta_{im}\delta_{kl})$$
$$\times\delta(\boldsymbol{x}-\boldsymbol{x}')\delta(t-t'), \quad (11.7)$$

where $\langle\cdot\rangle$ denotes an ensemble average, k_B is the Boltzmann constant, T is the absolute temperature, and δ_{ij} is the Kronecker delta. The Dirac delta functions $\delta(\boldsymbol{x}-\boldsymbol{x}')$ and $\delta(t-t')$ denote that the components of the random stress tensor are spatially and temporally uncorrelated. The mean and variance of the random stress tensor of the fluid are chosen to be consistent with the fluctuation–dissipation theorem [9]. By including this stochastic stress tensor due to the thermal fluctuations in the governing equations, the macroscopic hydrodynamic theory is generalized to include the relevant physics of the mesoscopic scales ranging from tens of nanometers to a few microns.

An alternative approach to NFD (and one that is different from FHD) is to start with a form of the master equation referred to as the Fokker–Planck equation. Formally, the Fokker–Planck equation is derived from the master equation by expanding $w(y'|y)P(y,t)$ as a Taylor series in powers of

$r = y' - y$. The infinite series is referred to as the Kramers–Moyal expansion, while the series truncated up to the second derivative term is known as the Fokker–Planck or the diffusion equation, which is given by [5]:

$$\frac{\partial P(y,t)}{\partial t} = -\frac{\partial}{\partial y}\{a_1(y)P\} + \frac{\partial^2}{\partial y^2}\{a_2(y)P\}. \quad (11.8)$$

Here, $a_n(y) = \int r^n w dr$. The solution to the Fokker–Planck equation yields the probability distribution of particles which contains the information on Brownian effects. At equilibrium (i.e., when all the time dependence vanishes), the solution can be required to conform to the solutions from equilibrium statistical mechanics. This approach leads to a class of identities for transport coefficients, including the famous Stokes–Einstein diffusivity for particles undergoing Brownian motion to be discussed later in this chapter. Moreover, there is a one-to-one correspondence between the Fokker–Planck equation and a stochastic differential equation (SDE) that encodes for a trajectory of a Brownian particle. The generalized Fokker–Planck equation is written in terms of a generalized order parameter S, given by

$$\frac{\partial P(S,t)}{\partial t} = \frac{D}{k_B T}\frac{\partial}{\partial S}\left\{P\frac{\partial F(S)}{\partial S}\right\} + D\frac{\partial^2 P}{\partial S^2}, \quad (11.9)$$

where $F(S)$ is the free energy density (also referred to as the Landau free energy) along S [10], D is the diffusion coefficient along S which is also related to the a_ns of the original Fokker–Planck equation, i.e., $a_2 = 2D$. The quantity $k_B T$ which has the units of energy is called the Boltzmann factor and serves as a scale factor for normalizing energy values in NFD. Corresponding to every generalized Fokker–Planck equation (Eq. (11.9)), there exists an SDE given by

$$\frac{\partial S}{\partial t} = -\frac{D}{k_B T}\frac{\partial F(S)}{\partial S} + \sqrt{2D}\xi(t), \quad (11.10)$$

where $\xi(t)$ represents a unit-normalized white noise process. The SDE encodes for the Brownian dynamics (BD) of the particle in the limit of zero inertia. The corresponding equation when inertia of the particle is added is often referred to as the Langevin equation. In summary, Brownian or thermal effects are described within the hydrodynamics framework using either the FHD approach or the BD/Langevin equation approach.

11.2.3 Multiphase NFD: Stochastic Dynamics of NP

Thus far, our discussion of NFD has been general and applicable mostly to single-phase flows such as those of pure fluids or dispersion of a miscible dye in a single phase. However, the simplest non-trivial example with technological relevance in NFD is when an NP is suspended in a fluid creating a moving interface as the particle experiences Brownian motion. We will discuss the general frameworks for describing its dynamics as well as the equilibrium properties of such a system.

An NP experiencing random motion in a fluid is influenced by hydrodynamic interactions. The fluid around the particle is dragged in the direction of motion of the particle. On the other hand, the motion of the particle is resisted by viscous forces arising due to its motion relative to the surrounding fluid. In this context, it is helpful to recall the results for the motion of a sphere in steady Stokes flow ($Re \leq O(1)$, with Re based on the radius of the particle). Since the spherical shape NP is among the most commonly studied one in the literature related to NP motion, the Stokes law for drag force, $f_D = 6\pi\mu U a$, where f_D is the drag force on a sphere in steady Stokes flow, a is the radius of sphere, μ is the dynamic viscosity of the fluid, and U is the translational speed of the sphere in the direction of its motion, is frequently invoked. The quantity $\zeta^{(t)} = 6\pi\mu a$, separately, is called the Stokes dissipative friction force coefficient for a spherical NP or simply the friction force coefficient. Similarly, a rotational friction coefficient, $\zeta^{(r)} = 8\pi\mu a^3$, defined for a rotating sphere (see [11]) is used in the context of describing NP rotation. However, there is a basic difference between a particle in steady Stokes flow and an NP in Brownian motion. With an NP, the momentum of the fluid surrounding the particle at any instant is related to its history. This can be understood in light of the linear response theory which is the foundation of non-equilibrium thermodynamics. A system at equilibrium evolving under a Hamiltonian $\mathcal{H}$ experiences a perturbation $\Delta\mathcal{H} = fA$, where f is the field variable (such as an external force) and A is the extensive variable (such as the displacement) that is conjugated to the field. The perturbation throws the system in a non-equilibrium state, and when the field is switched off, the system relaxes back to equilibrium in accordance with the regression process as described by Onsager [12–14]:

$$\Delta\bar{A}(t) = (f/k_B T)\langle\Delta A(0)\Delta A(t)\rangle, \quad (11.11)$$

where $\Delta A(t) = A(t) - \langle A\rangle$. The above identity holds under linear response, when $\Delta\mathcal{H}$ is small or, equivalently, when $\Delta A(t, \lambda f) = \lambda\Delta A(t, f)$. The most general form to relate the response A to the field f under the linear response is given by

$$\Delta\bar{A}(t) = \int_{-\infty}^{\infty} \chi(t - t')f(t')dt'. \quad (11.12)$$

Here, we have further assumed that physical processes are stationary in the sense that they do depend on not the absolute time but only the time elapsed, i.e., $\chi(t, t') = \chi(t - t')$. One can use the linear-response relationship to derive an equation for the dynamics of NP interacting with a thermal reservoir of a fluid (also called a thermal bath). The dynamics of the particle (in one dimension along the x coordinate for simplicity of illustration) is given by $m\frac{d^2 x}{dt^2} = -\frac{dV(x)}{dx} + f$, where $V(x)$ is the potential energy function and f is any external driving force including random Brownian forces from the solvent degrees of freedom. The thermal bath will experience forces f_b in the absence of the particle, and when the particle is introduced, the perturbation will change the bath forces to f. This change $f - f_b$ can be can be described under linear response as

$$\Delta \bar{f}(t) = f - f_b = \int_{-\infty}^{\infty} dt' \chi_b(t - t') x(t'). \qquad (11.13)$$

Using this relationship, and by performing integration by parts, the dynamics of the particle can be written as

$$m \frac{dU}{dt} = -\frac{dV(x)}{dx} + f_r - \int_{-\infty}^{t} dt' \zeta_b(t - t') U(t'). \qquad (11.14)$$

Here f_r is the random force from the bath that is memoryless, $U = dx/dt$, and $\chi_b = -d\zeta_b/dt$. This form of the equation for the dynamics of the NP is referred to as the generalized Langevin equation (GLE), and it encodes for the memory/history forces. We note that while the parent equation (i.e., the master equation) is Markovian, the memory emerges as we coarse-grain the timescales to represent the fluid–particle interaction and is a consequence of the second law of thermodynamics. The strength of the random force that drives the fluctuations in the velocity of a NP is fundamentally related to the coefficient representing the dissipation or friction present in the surrounding viscous fluid. This is the fluctuation–dissipation theorem [15]. The friction coefficient, ζ, associated with NP motion is time dependent (see Eq. (11.14)) and is no longer given by the constant Stokes value. In any description of NP motion, therefore, the mean and the variance of the thermal fluctuations have to be chosen to be consistent with the fluctuation–dissipation theorem. In order to achieve thermal equilibrium, the correlations between the state variables should be such that there is an energy balance between the thermal forcing and the dissipation of the system as required by the fluctuation–dissipation theorem [15,16].

11.2.4 Equilibrium and Transport Properties

According to equilibrium statistical mechanics, in a uniform temperature fluid, the molecular velocities will be Maxwellian, and the energy components related to the various degrees of freedom will satisfy the equipartition principle. Indeed, the solutions to the Fokker–Planck equation written for the velocity variable at steady state yields the Maxwell–Boltzmann distribution consistent with the picture from equilibrium statistical mechanics. If an NP is introduced into the fluid medium, it will experience molecular collisions and the associated fluctuating impulses. As a net result, the fluctuating NP will translate in the fluid in a random manner while experiencing rotation. If the bulk fluid is driven by an external pressure gradient, the random translational and rotational motions will still be significant at very low Reynolds numbers, Re (say, Re based on the vessel diameter). In NFD, most of the quantities associated with the fluid and the NP are evaluated by ensemble averaging as noted and defined earlier. In a numerical simulation, this ensemble average is obtained by averaging over

successive configurations that are generated in the process of simulation. Customarily implicit to this averaging is the ergodic assumption that an ensemble average of a property of a system over many replicas is the same as an average taken over a long enough time of one particular replica of the system that is being actually numerically simulated. If the NP and the surrounding fluid are in thermal equilibrium, just as for the fluid molecules, the velocity components of the NP will also be Maxwellian and the NP, energy components related to the various degrees of freedom will also satisfy the equipartition principle. Thus, the equilibrium probability density function (PDF) of any one Cartesian component of the velocity of the NP, U_i, will follow the Maxwell–Boltzmann (MB) distribution,

$$P(U_i) = \sqrt{\left(\frac{m}{2\pi k_B T}\right)} \exp\left\{-\frac{mU_i^2}{2k_B T}\right\}, \qquad (11.15)$$

where m is the NP mass and the equilibrium statistics of the three components U_i along the three coordinate directions are independent of each other; note we denote the velocity of the fluid using v and that of the NP using U. In thermal equilibrium, the mean (or the average) value of U_i is

$$\langle U_i \rangle = 0, \qquad (11.16)$$

and the mean squared value is

$$\langle U_i^2 \rangle = \frac{k_B T}{m}. \qquad (11.17)$$

From the equipartition theorem, at thermal equilibrium, the translational and rotational temperatures of the NP are given by

$$T^{(t)} = \frac{m \langle \mathbf{U}^2 \rangle}{3k_B}; \quad T^{(r)} = \frac{\mathbf{I} \langle \boldsymbol{\omega}^2 \rangle}{3k_B}, \qquad (11.18)$$

where $\mathbf{U}$ and $\boldsymbol{\omega}$ are the translational and angular velocities of the NP and $\mathbf{I}$ is its moment of inertia.

Another important application of the Onsager regression relationship (Eq. (11.11)) is the emergence of a class of relationships that relate transport properties to correlation functions that are known as the Green–Kubo relationships [14,17]. These relationships are also a consequence of the fluctuation–dissipation theorem. In NFD, the velocity autocorrelation function (VACF) and the angular velocity auto correlation function (AVACF) play important roles. Most importantly, they enable the calculation of the diffusion of the particle in the medium. The VACF correlates the velocity of an NP with itself and reveals the effect of various forces on the translational motion of the NP. The VACF is a generalization of its mean squared value (or its variance) and is the primary characteristic of the "memory" that this random variable possesses [17]. An NP rotating with a time-dependent angular velocity will experience a torque, and rotational relaxations are affected by the long-time persistence of the angular velocity. AVACF will establish this relationship between angular velocity and relaxation [18]. The VACF and AVACF, normalized by $3k_B T/m$ and $3k_B T/\mathbf{I}$, respectively, are expressed by the following:

$$\frac{\langle \mathbf{U}(t)\mathbf{U}(0)\rangle}{3k_B T/m}, \quad \frac{\langle \boldsymbol{\omega}(t)\boldsymbol{\omega}(0)\rangle}{3k_B T/\mathbf{I}}. \tag{11.19}$$

The VACF and the AVACF display characteristic scaling dependencies on time reflective of the hydrodynamic interaction of the NP with the fluid. This will become clear in subsequent sections where we discuss the VACFs of NP in unbounded and bounded fluid media.

We characterize the net spatial movement of a randomly moving NP by its mean square displacement (MSD) over time. Linear and angular mean square displacements serve as the measures of diffusion of the NP in the medium. Letting $\mathbf{r}(t)$ denote the distance travelled by a NP in time t starting from a fixed origin, we write this and its square by

$$\mathbf{r}(t) = \int_0^t \mathbf{U}(\tau)d\tau, \quad \langle r^2(t)\rangle = \int_0^t \int_0^t \mathbf{U}(\tau)\mathbf{U}(\tau')d\tau d\tau'. \tag{11.20}$$

Equivalently, for N number of NPs,

$$\langle r^2(t)\rangle = \frac{1}{N}\sum_{i=1}^N \langle |r_i(t) - r_i(0)|^2\rangle. \tag{11.21}$$

In Eq. (11.20), with $\tau' = \tau + \xi$, integrating over τ, and taking the ensemble average,

$$\langle r^2(t)\rangle = 2\int_0^t (t-\xi)\langle \mathbf{U}(\xi)\mathbf{U}(0)\rangle d\xi \tag{11.22}$$

$$= 2t\int_0^t \langle \mathbf{U}(\xi)\mathbf{U}(0)\rangle d\xi - 2\int_0^t \xi\langle \mathbf{U}(\xi)\mathbf{U}(0)\rangle d\xi. \tag{11.23}$$

In Eq. (11.23), for NP motion in a non-confined uniform fluid and for large t, the VACF may $\to 0$ rendering the first integral to be finite and the second one to yield a constant. In that limit, the MSD is given by the Einstein equation [19],

$$\langle r^2(t)\rangle \to 2dD^{(t)}t + C, \tag{11.24}$$

where d is the dimensionality of the system, $D^{(t)}$ is the diffusion coefficient or diffusivity, and C is yet another constant. The Stokes–Einstein relationship is expressed by

$$D^{(t)} = \frac{k_B T}{\zeta^{(t)}}. \tag{11.25}$$

From the Einstein equation, the diffusion coefficient may be expressed in terms of the VACF as

$$D^{(t)} = \frac{1}{d}\int_0^\infty \langle \mathbf{U}(t)\mathbf{U}(0)\rangle dt, \tag{11.26}$$

where d is the dimensionality of the system. Equation (11.26) is an example of the Green–Kubo relation. The Einstein equation and the Green–Kubo relation are derived by using the Onsager regression relationship (Eq. (11.11)) to relate the macroscopic concentration variable $C(r,t)$ described by Fick's law of diffusion in mass transfer theory to the microstate variable $P(r,t)$ [14]. They must be used

only for calculations in homogeneous fluids [20]. Analogous to translation, we have the rotational diffusivity $D^{(r)}$ in terms of the rotational MSD,

$$\langle r_m^2(t)\rangle \to 6D^{(r)}t. \tag{11.27}$$

This would hold for NP rotation in a uniform fluid at large t. The corresponding Stokes–Einstein–Debye relationship is expressed by

$$D^{(r)} = \frac{k_B T}{\zeta^{(r)}}. \tag{11.28}$$

We conclude this section by noting that the different timescales for Brownian motion of a spherical NP are (i) hydrodynamic timescale, $\tau_\nu = a^2/\nu$ (the timescale for momentum to diffuse over a distance equal to the radius of the NP); (ii) Brownian timescale, $\tau_b = m/\zeta^{(t)}$ (Brownian relaxation time over which velocity correlations decay in the Langevin equation); and (iii) Diffusive timescale, $\tau_d = a^2\zeta^{(t)}/k_B T$ (Brownian diffusion time over which the NP diffuses over a distance equal to its radius).

11.3 Computational Methods and Implementation

11.3.1 The FHD Method

Over the past decades, numerical simulations of the FHD approach have been carried out employing the finite volume method [21,22], lattice Boltzmann method (LBM) [23–29] and stochastic immersed boundary method [30]. A coarse-graining methodology has been developed to bridge molecular dynamics and FHD simulations [31,32]. Serrano and Español [33] and Serrano et al. [34] have solved the FHD Navier–Stokes equations without a particle using the finite volume Lagrangian discretization in a moving Voronoi grid. They have ensured that their discretized governing equations cast in the GENERIC (General Equation for Non-Equilibrium Reversible/Irreversible Coupling) formalism [35,36] satisfy the fluctuation–dissipation theorem. The GENERIC formalism proposed by Grmela and Öttinger [35] and Öttinger and Grmela [36] ensures the correct treatment of thermal fluctuations and FHD. Patankar has simulated the thermal motion of two-dimensional particles in a stationary medium with the finite element method (FEM) [26]. Sharma and Patankar [21] have employed a finite volume method based on distributed-Lagrangian multiplier (DLM) to simulate the thermal motion of particles. The computational domain is periodic in all directions, and the thermal fluctuations are included in the fluid equations using the random stress tensor. They have validated the numerical results by comparison with analytical expressions. Nie and Lin [29] have employed the fluctuating LBM to simulate Brownian motion of particles and have validated their numerically obtained VACF by comparison with theoretical predictions. It is shown that the temperatures characterizing translational motion of the particle in three coordinate directions agree with each other after a lapse of time, but the predicted particle temperature is 15% lower than the effective temperature of fluid fluctuations. This is in accord with the earlier findings of Ladd [25], who first proposed the use of fluctuating LBM. Adhikari et al. [27] have established agreement between fluctuation and dissipation by introducing ghost noise to the fluctuating LBM in the formulation (see Dünweg and Ladd [28] for further discussions of [27]).

In this section, we illustrate the direct numerical simulations (DNSs) based on the arbitrary Lagrangian–Eulerian (ALE) FEM [37–40] to accurately resolve the fluid–particle interfacial motions. Both translational and rotational motions of an NP in (i) a stationary fluid medium and (ii) Poiseuille flow are investigated. An unstructured finite element mesh, generated by the Delaunay–Voronoi method [41], has enabled a significantly higher number of mesh points in the regions of interest (i.e., close to the particle and wall surfaces compared to the regions farther away). This feature also keeps the overall mesh size computationally reasonable even with an NP moving in a very large domain [37,38,42–44]. Thermal fluctuations are included in the equations of linearized hydrodynamics by adding stochastic components to the stress tensor as white noise in space and time as prescribed by the FHD method [11,45]. As noted in Español et al. [45], "even though the original equations of FHD are written in terms of stochastic partial differential equations, at a very fundamental level the inclusion of thermal fluctuations requires always the notion of a 'mesoscopic cell' in order to define the fluctuating quantities". In Español et al. [45], it is shown that FHD equations discretized in terms of finite element shape functions based on the Delaunay triangulation satisfy the fluctuation–dissipation theorem. The numerical schemes for the implementation of thermal fluctuations in the Landau–Lifshitz Navier–Stokes equations are expected to perform very delicate tasks [46,47], and obtaining accurate numerical results is a challenging endeavor.

11.3.2 Numerical Implementation of the FHD Method

We describe the implementation for a general ellipsoidal NP immersed in an incompressible, quiescent, or flowing Newtonian fluid contained in a cylindrical tube Σ, as shown in Figure 11.1. This section is derived from the methodology described in [48], where complete details may be found. The inlet and outlet boundaries are denoted by Σ_i and Σ_o, respectively; Σ_w is the wall boundary; and the particle surface is denoted by Γ_p. The dimensions of the particle are denoted by a, b, and c, and the length and diameter of the tube are L and D, respectively, as shown in Figure 11.1a, b. The position of the particle (i.e., its center of mass) is expressed either in terms of r, the radial distance from the tube axis, or h, the radial distance as measured from the wall boundary. The angular orientation of the particle is measured in terms of the inclination angle θ which denotes an in-plane tilt (in the x–z plane). At $\theta = 0°$, a is the dimension of the NP along the tube axis, while b and c are those along the radial directions, and this is illustrated in Figure 11.1b. We describe the shape of the NC with respect to its dimensions at $\theta = 0°$ and define the aspect ratio as $\varepsilon = a/c$. Hence in our notations, $a < b = c$ for an oblate spheroid ($\varepsilon < 1.0$) and $a > b = c$ for a prolate spheroid ($\varepsilon > 1.0$).

In view of the asymmetric shape and the orientation of the spheroid, yet one more measure of length becomes relevant in our problem. With reference to Figure 11.1a, it may be noted that ζ_0 is the maximum value from among the projections of a, b, and c on a plane perpendicular to the cylinder axis. For example, $\zeta_0 = b/2$ when $\theta = 0°$, and $\zeta_0 = a/2$ when $\theta = 90°$. For notational simplicity, we define the nondimensional separation between the NC and the wall, in terms of h and ζ_0, as $\tilde{h} = (h - \zeta_0)/\zeta_0$.

The equations for the fluid domain were presented in Section 11.2.2. The translational and rotational motions of a rigid particle suspended in the fluid satisfy

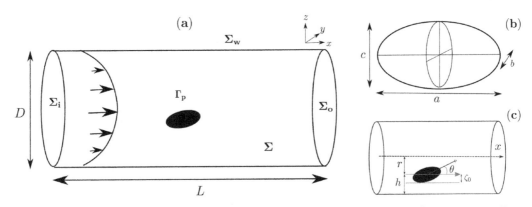

FIGURE 11.1 Schematic representation of (a) a spheroid bounded by a circular tube of length L and diameter D with a Poiseuille flow along the x direction, (b) the dimensions of a spheroid denoted by a, b, and c. Panel (c) shows the various length scales in the system: (i) the proximity of the particle from the wall boundary described either in terms of its radial distance r or its separation from the wall $h = D/2 - r$ and (ii) ζ_0 which denotes the maximum radial size of the particle – the value of ζ_0 is a function of θ.

$$m\frac{d\boldsymbol{U}}{dt} = \boldsymbol{G} - \int_{\Gamma_p} \underline{\boldsymbol{\sigma}} \cdot \hat{\boldsymbol{n}} \, ds, \qquad (11.29)$$

$$\underline{\mathbf{I}}\frac{d\boldsymbol{\Omega}}{dt} + \boldsymbol{\Omega} \times (\underline{\mathbf{I}}\boldsymbol{\Omega}) = -\underline{\mathbf{R}}^T \int_{\Gamma_p} (\boldsymbol{x} - \boldsymbol{X}) \times (\underline{\boldsymbol{\sigma}} \cdot \hat{\boldsymbol{n}}) \, ds, \qquad (11.30)$$

where $\boldsymbol{U} = (U_x, U_y, U_z)^T$ and $\boldsymbol{X} = (X_x, X_y, X_z)^T$ are the translational velocities and the position of the center of mass of the NC, respectively, in the Cartesian frame (x, y, z). $\boldsymbol{\Omega} = (\Omega_1, \Omega_2, \Omega_3)^T$ is rotational velocity of the particle in the body fitted frame of reference given by $(1, 2, 3)$. The mass and the moments of inertia of the particle are given by m and $\underline{\mathbf{I}}$, respectively, and $\boldsymbol{G}$ represents a body force such as gravity. $\hat{\boldsymbol{n}}$ is the outward drawn unit normal to the particle surface. Here, the moment of inertia $\underline{\mathbf{I}}$ is also defined with respect to the body frame attached to the particle. $\underline{\mathbf{R}}$ is the rotational matrix that transforms the body frame quantities to the inertial frame (x, y, z). The rotational matrix is defined in terms of the quaternions [49]: $\boldsymbol{q} = (q_0, q_1, q_2, q_3)^T$ with $\|q\|^2 = q_0^2 + q_1^2 + q_2^2 + q_3^2 = 1$,

$$\underline{\mathbf{R}} = \begin{pmatrix} 2(q_0^2 + q_1^2) - 1 & 2(q_1 q_2 - q_0 q_3) & 2(q_1 q_3 + q_0 q_2) \\ 2(q_1 q_2 + q_0 q_3) & 2(q_0^2 + q_2^2) - 1 & 2(q_2 q_3 - q_0 q_1) \\ 2(q_1 q_3 + q_0 q_2) & 2(q_2 q_3 + q_0 q_1) & 2(q_0^2 + q_3^2) - 1 \end{pmatrix}. \qquad (11.31)$$

The position $\boldsymbol{X}$ and the quaternions $\boldsymbol{q}$ of the particle evolve in time according to

$$\frac{d\boldsymbol{X}}{dt} = \boldsymbol{U} \qquad (11.32)$$

and

$$\frac{d\boldsymbol{q}}{dt} = \frac{1}{2} \begin{pmatrix} 0 & -\Omega_1 & -\Omega_2 & -\Omega_3 \\ \Omega_1 & 0 & \Omega_3 & -\Omega_2 \\ \Omega_2 & -\Omega_3 & 0 & \Omega_1 \\ \Omega_3 & \Omega_2 & -\Omega_1 & 0 \end{pmatrix} \boldsymbol{q}. \qquad (11.33)$$

The initial conditions are given by

$$\boldsymbol{U}(t = 0) = 0, \quad \boldsymbol{\Omega}(t = 0) = 0, \quad \boldsymbol{u}(t = 0) = 0 \text{ in } \Sigma, \qquad (11.34)$$

and the boundary conditions are given by

$$\boldsymbol{u} = \boldsymbol{u_{in}} \quad \text{on } \Sigma_i \quad \text{(inlet)}, \qquad (11.35)$$

$$\underline{\boldsymbol{\sigma}} \cdot \hat{\boldsymbol{n}} = 0 \quad \text{on } \Sigma_o \quad \text{(outlet)}, \qquad (11.36)$$

$$\boldsymbol{u} = 0 \quad \text{on } \Sigma_w \quad \text{(wall boundary)}, \qquad (11.37)$$

$$\boldsymbol{u} = \boldsymbol{U} + \underline{\mathbf{R}}\boldsymbol{\Omega} \times (\boldsymbol{x} - \boldsymbol{X}) \quad \text{on } \Gamma_p \quad \text{(particle surface)}. \qquad (11.38)$$

The above formulation is numerically solved, and the details are provided in the next subsection.

11.3.3 The Weak Formulation

Let $\mathcal{V}$ be the function space given by

$$\mathcal{V} = \left\{ \begin{array}{l} \boldsymbol{V} = (\boldsymbol{U}, \boldsymbol{\Omega}, \boldsymbol{u}, p) | (\boldsymbol{U}, \boldsymbol{\Omega}) \in \mathcal{R}^3, \, \boldsymbol{u} \in \mathcal{H}^1, \, p \in \mathcal{L}^2 \\ \\ \boldsymbol{u} = 0 \text{ on } \Sigma_w, \, \boldsymbol{u} = \boldsymbol{U} + \underline{\mathbf{R}}\boldsymbol{\Omega} \times (\boldsymbol{x} - \boldsymbol{X}) \text{ on } \Gamma_p \\ \\ \boldsymbol{u} = \boldsymbol{u}_{in} \text{ on } \Sigma_i, \, p = 0 \text{ and } \underline{\boldsymbol{\sigma}} \cdot \hat{\boldsymbol{n}} = 0 \text{ on } \Sigma_o \end{array} \right\}, \qquad (11.39)$$

where $\mathcal{H}^1$ is the Hilbert space for the fluid velocity field. The test function space $\mathcal{V}_0$ is the same as $\mathcal{V}$, except that $\boldsymbol{u} = 0$ on Σ_i and Σ_o, and hence

$$\tilde{\boldsymbol{V}} = (\tilde{\boldsymbol{U}}, \tilde{\boldsymbol{\Omega}}, \tilde{\boldsymbol{u}}, \tilde{p}) \in \mathcal{V}_0. \qquad (11.40)$$

Multiplying Eq. (11.4) by the test function for the fluid velocity $\tilde{\boldsymbol{u}}$ and integrating over the fluid domain at time t yields

$$\int_\Sigma \rho^{(f)} \frac{D\boldsymbol{u}}{Dt} \cdot \tilde{\boldsymbol{u}} \, dv - \int_\Sigma (\nabla \cdot \underline{\boldsymbol{\sigma}}) \cdot \tilde{\boldsymbol{u}} \, dv = 0. \qquad (11.41)$$

Upon integration by parts, the second term may be expressed as

$$\int_\Sigma (\nabla \cdot \underline{\boldsymbol{\sigma}}) \cdot \boldsymbol{u} \, dv = -\int_\Sigma \underline{\boldsymbol{\sigma}} : \nabla \boldsymbol{u} \, dv + \int_{\Gamma_p} (\underline{\boldsymbol{\sigma}} \cdot \hat{\boldsymbol{n}}) \cdot \tilde{\boldsymbol{u}} \, ds, \qquad (11.42)$$

and the last term of Eq. (11.42) may be rewritten using Eqs. (11.29) and (11.30) as

$$\int_{\Gamma_p} (\underline{\boldsymbol{\sigma}} \cdot \hat{\boldsymbol{n}}) \cdot \tilde{\boldsymbol{u}} \, ds = \int_{\Gamma_p} (\underline{\boldsymbol{\sigma}} \cdot \hat{\boldsymbol{n}}) \cdot (\tilde{\boldsymbol{U}} + (\underline{\mathbf{R}}\tilde{\boldsymbol{\Omega}}) \times (\boldsymbol{x} - \boldsymbol{X})) \, ds$$

$$= \tilde{\boldsymbol{U}} \cdot \int_{\Gamma_p} \underline{\boldsymbol{\sigma}} \cdot \hat{\boldsymbol{n}} \, ds + (\underline{\mathbf{R}}\tilde{\boldsymbol{\Omega}}) \cdot \int_{\Gamma_p} (\boldsymbol{x} - \boldsymbol{X}) \times (\underline{\boldsymbol{\sigma}} \cdot \hat{\boldsymbol{n}}) \, ds$$

$$= -\tilde{\boldsymbol{U}} \cdot \left(m\frac{d\boldsymbol{U}}{dt} - \boldsymbol{G}\right) - (\underline{\mathbf{R}}\tilde{\boldsymbol{\Omega}}) \cdot \left(\underline{\mathbf{R}}\left[\underline{\mathbf{I}}\frac{d\boldsymbol{\Omega}}{dt} + \boldsymbol{\Omega} \times \underline{\mathbf{I}}\boldsymbol{\Omega}\right]\right)$$

$$= -\tilde{\boldsymbol{U}} \cdot \left(m\frac{d\boldsymbol{U}}{dt} - \boldsymbol{G}\right) - \tilde{\boldsymbol{\Omega}} \cdot \left(\underline{\mathbf{I}}\frac{d\boldsymbol{\Omega}}{dt} + \boldsymbol{\Omega} \times \underline{\mathbf{I}}\boldsymbol{\Omega}\right). \qquad (11.43)$$

From Eqs. (11.5) and (11.41)–(11.43), we get the weak formulation for the combined fluid–particle momentum equations:

$$\int_\Sigma \rho^{(f)} \frac{D\boldsymbol{u}}{Dt} \cdot \tilde{\boldsymbol{u}} \, dv - \int_\Sigma p\nabla \cdot \tilde{\boldsymbol{u}} \, dv$$

$$+ \int_\Sigma (\mu(\nabla \boldsymbol{u} + (\nabla \boldsymbol{u})^T) + \underline{\mathbf{S}}):$$

$$\nabla \tilde{\boldsymbol{u}} \, dv + \tilde{\boldsymbol{U}} \cdot \left(m\frac{d\boldsymbol{U}}{dt} - \boldsymbol{G}\right)$$

$$+ \tilde{\boldsymbol{\Omega}} \cdot \left(\underline{\mathbf{I}}\frac{d\boldsymbol{\Omega}}{dt} + \boldsymbol{\Omega} \times (\underline{\mathbf{I}}\boldsymbol{\Omega})\right) = 0, \qquad (11.44)$$

with

$$\int_\Sigma \tilde{p}(\nabla \cdot \boldsymbol{u}) \, dv = 0. \qquad (11.45)$$

11.3.4 ALE Mesh Movement

An ALE technique is used to handle the movement of the particle in the fluid domain, see [38]. The material derivative of $u(x,t)$ in an ALE formulation is given as

$$\frac{Du}{Dt} = \frac{\delta u}{\delta t} + [(u - u_m) \cdot \nabla]u, \tag{11.46}$$

where

$$\frac{\delta u}{\delta t} = \frac{\partial}{\partial t} u(x(\phi,t),t)|_{\phi \text{ is fixed}} \quad \text{and} \quad \frac{d}{dt} x(\phi,t) = u_m, \tag{11.47}$$

are the time derivatives of the velocity and the mesh velocity, respectively, with the former being defined in a fixed referential frame ϕ.

The mesh velocity u_m in Eq. (11.47) is set to follow the motion of the particles and the motion of the confined fluid and is computed using Laplace's equation in the fluid domain:

$$\nabla \cdot (\epsilon_e \nabla u_m) = 0 \qquad \text{in } \Sigma, \tag{11.48}$$

subject to boundary conditions:

$$u_m = U + \underline{R}\Omega \times (x - X) \qquad \text{on } \Gamma_p, \tag{11.49}$$

$$u_m = 0 \qquad \text{on } \Sigma_w + \Sigma_i + \Sigma_o. \tag{11.50}$$

Here, ϵ_e controls the deformation of the mesh, and we choose it to be $\epsilon_e = 1/V_e$, where V_e is the volume of the tetrahedral element. Similarly, the acceleration a_m of the mesh vertices is chosen to satisfy

$$\nabla \cdot (\epsilon_e \nabla a_m) = 0 \qquad \text{in } \Sigma, \tag{11.51}$$

with boundary conditions:

$$a_m = \frac{dU}{dt} + \underline{R}\frac{d\Omega}{dt}(x - X)$$
$$+ \underline{R}\Omega \times (\underline{R}\Omega \times (x - X)) \text{ on } \Gamma_p, \tag{11.52}$$

$$a_m = 0 \quad \text{on } \Sigma_w + \Sigma_i + \Sigma_o. \tag{11.53}$$

The linear weak formulations for the mesh velocity and acceleration are solved using the biconjugate gradient stabilized method. The positions of the mesh vertices are updated using the second-order forward Euler scheme:

$$x_m^{n+1} = x_m^n + u_m^n(x^n)\Delta t + \frac{1}{2}a_m^n(x^n)\Delta t^2. \tag{11.54}$$

11.3.5 Temporal Discretization

We use an adaptive second-order backward finite difference method to discretize the time derivatives in Eq. (11.44) which are given by

$$\frac{Du}{Dt} \approx C_1 \frac{u^{n+1}(x) - u^n(x')}{\Delta t_n} + C_2 \frac{\delta u^n(x')}{\delta t}$$
$$+ \quad [(u^{n+1}(x) - u_m^{n+1}(x)) \cdot \nabla]u^{n+1}(x), \tag{11.55}$$

$$\frac{dU}{dt} \approx C_1 \frac{U^{n+1} - U^n}{\Delta t_n} + C_2 \frac{\delta U^n}{\delta t}, \tag{11.56}$$

$$\frac{d\Omega}{dt} \approx C_1 \frac{\Omega^{n+1} - \Omega^n}{\Delta t_n} + C_2 \frac{\delta\Omega^n}{\delta t}, \tag{11.57}$$

where $C_1 = \frac{\Delta t_n}{2\Delta t_n + \Delta t_{n-1}}$ and $C_2 = \frac{\Delta t_n + \Delta t_{n-1}}{2\Delta t_n + \Delta t_{n-1}}$, with $\Delta t_n = t_{n+1} - t_n$ being the time step for integration.

However, we use a second-order finite difference scheme to discretize the position and the orientation (represented by quaternions) of the particle as

$$X^{n+1} = X^n + \Delta t_n U^n + \frac{(\Delta t_n)^2}{2}\frac{dU^n}{dt}, \tag{11.58}$$

$$q^{n+1} = q^n + \Delta t_n \frac{dq^n}{dt} + \frac{(\Delta t_n)^2}{2}\frac{d^2 q^n}{dt^2}. \tag{11.59}$$

The derivatives of q^n are computed using Eq. (11.33).

Using Eqs. (11.55)–(11.57), the weak formulation of the governing equations (see Eq. (11.44)) may now be expressed as

$$\int_\Sigma \rho^{(f)} \left(\frac{C_1}{\Delta t_n}u^{n+1}(x) + ((u^{n+1}(x) - u_m^{n+1}(x)) \cdot \nabla)u^{n+1}(x) \right)$$

$$\cdot \tilde{u}dv - \int_\Sigma p^{n+1}(x)\nabla \cdot \tilde{u}\,dv + \int_\Sigma \left(\mu(\nabla u^{n+1}(x) \right.$$

$$+ (\nabla u^{n+1}(x))^T) + S^{n+1}(x)) : \nabla \tilde{u}\,dv + \frac{C_1}{\Delta t_n}m\tilde{U}U^{n+1}$$

$$+ \tilde{\Omega} \cdot \left(\frac{C_1}{\Delta t_n}\underline{I}\Omega^{n+1} + \Omega^{n+1} \times (\underline{I}\Omega^{n+1}) \right)$$

$$= \int_\Sigma \rho^{(f)} \left(\frac{C_1}{\Delta t_n}u^n(x') - C_2\frac{\delta u^n(x')}{\delta t} \right)$$

$$\cdot \tilde{u}\,dv + \left(\frac{C_1}{\Delta t_n}mU^n - C_2 m\frac{dU^n}{dt} + G \right)\tilde{U}^n$$

$$+ \tilde{\Omega} \cdot \underline{I}\left(\frac{C_1}{\Delta t_n}\Omega^n - C_2\frac{d\Omega^n}{dt} \right) \tag{11.60}$$

and

$$\int_\Sigma \tilde{p}(\nabla \cdot u^{n+1}(x))\,dv = 0. \tag{11.61}$$

The location of the grid in the new domain x and its correspondence to the old domain x' follows Eq. (11.54). Since the nodes on the particle surface are also updated by Eq. (11.54), these node positions may move away from the body surface, and hence we need to reset the surface nodes at each time step.

11.3.6 Spatial Discretization

1. *Surface/boundary mesh:* The boundaries of the computational domain are discretized as described in [38]. Briefly, as shown in Figure 11.2c, we start by approximating the surface of a unit sphere by

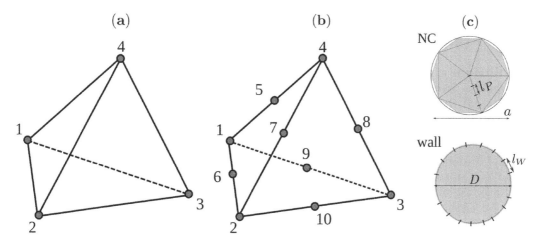

FIGURE 11.2 The four-node and ten-node tetrahedrons used in the finite element representation of the computational domain are shown in panels (a) and (b), respectively. The top panel in (c) shows an icosahedron used in the discretization of a spherical particle of diameter a which is later mapped to a spheroid—here l_P denotes the mesh length on the particle surface. The lower panel in (c) shows the cross section of a cylindrical tube of diameter D. The mesh size on the particle surface is denoted by l_W.

an icosahedron and further subdivide the faces of the icosahedron into a triangular mesh with a predefined characteristic length l_P. The triangular mesh on the icosahedron is stereographically projected to construct the boundary mesh for a spheroidal particle with specified values of a, b, c, and θ. Similarly, the cylindrical wall boundary is discretized into a triangular mesh with a characteristic length l_W. In the following, we will describe the finite element mesh parameters used in our calculations in terms of l_P and l_W.

2. *Volume mesh:* The fluid domain is discretized by tetrahedral finite elements generated using Delaunay–Voronoi methods. The discrete solution for the fluid velocity is approximated by piecewise quadratic functions and is assumed to be continuous over the domain. We use ten-node tetrahedral elements (Figure 11.2b) to locally interpolate the velocity. On the other hand, the pressure and the stress are piecewise linear and continuous and are interpolated using four-node tetrahedral elements (Figure 11.2a). The four-node and ten-node elements used to interpolate the stress and the velocity are known to satisfy the Ladyzhenskaya–Babuska–Brezzi conditions for stability [38].

For a given finite element mesh, the combined fluid–solid weak formulation (Eq. (11.60)) reduces to a nonlinear system of algebraic equations, which is solved by a Newton–Raphson algorithm. Similarly, the mesh velocity (Eq. (11.48)) and mesh acceleration (Eq. (11.51)) can also be reduced to linear systems of algebraic equations. These coupled systems are solved by a multigrid preconditioned conjugate gradient method.

11.3.7 Random Stress Tensor on a Mesh

We now describe the procedure to numerically generate the random stresses associated with the unstructured tetrahedral mesh. The random stress at each node on the computational domain depends on the volumes of the tetrahedrons associated with it.

The components of the random stress tensor $\underline{\mathbf{S}}^{(i)}$ in the ith tetrahedral element, with volume $V_e^{(i)}$, is approximated from Eq. (11.7) as

$$\langle S_{xx}\rangle^{(i)} = \langle S_{yy}\rangle^{(i)} = \langle S_{zz}\rangle^{(i)} = 0, \qquad (11.62)$$

$$\langle S_{xy}\rangle^{(i)} = \langle S_{yz}\rangle^{(i)} = \langle S_{zx}\rangle^{(i)} = 0, \qquad (11.63)$$

$$\langle S_{xx}^2\rangle^{(i)} = \langle S_{yy}^2\rangle^{(i)} = \langle S_{zz}^2\rangle^{(i)} = \frac{4k_BT\mu}{V_e^{(i)}\Delta t}, \qquad (11.64)$$

$$\langle S_{xy}^2\rangle^{(i)} = \langle S_{yz}^2\rangle^{(i)} = \langle S_{zx}^2\rangle^{(i)} = \frac{2k_BT\mu}{V_e^{(i)}\Delta t}, \qquad (11.65)$$

where Δt is the time step for the numerical simulation. The total stress on a node is then computed as

$$\underline{\mathbf{S}} = \mathcal{C}\sum_{i=1}^{N_e}\underline{\mathbf{S}}^{(i)}, \qquad (11.66)$$

with $\mathcal{C} = 1$ when the node is inside the computational domain and $\mathcal{C} = \sqrt{2}$ when the node is on a boundary surface. N_e is the number of tetrahedrons associated with this node. At a boundary node, since we consider the spheroidal particles to be solid, the tetrahedral volume $V_e^{(i)}$ underestimates the total volume defined by the Dirac delta function $\delta(\boldsymbol{x}-\boldsymbol{x}')$, given in the right-hand side of Eq. (11.7). Ignoring the effect of the particle curvature on the estimate for $V_e^{(i)}$, we approximate the effective volume as $\delta(\boldsymbol{x}-\boldsymbol{x}') = 2/V_e^{(i)}$. Using this estimate in Eqs. (11.64) and (11.65) and summing over all tetrahedral elements linked to a given node leads to the general equation given in Eq. (11.66).

11.4 Illustrative Examples and Select Applications

11.4.1 Equilibrium and Transport Properties from an FHD Numerical Study

The results shown in Figure 11.3 are obtained from five different realizations of an FHD simulation, with each realization consisting of $N = 20,000$ time steps. The error bars have been plotted from standard deviations of the temperatures obtained with the different realizations, based on which the statistical error is established to be less than 5%. It is noted that due to the incompressibility assumption invoked in the model equations, the particle mass m is augmented by an added mass $m_0/2$, $M = m + m_0/2$, where m_0 is the mass of the displaced fluid and M is the virtual or added mass of the particle [50–54]; the added mass correction is, in general, a function of distance from the wall, see [55] for example. We note that the added mass terms described above pertain only to spherical NP and the expressions for the added mass terms for ellipsoidal NPs are described in [48].

The close agreement with the Maxwell–Boltzmann distribution displayed in Figure 11.3 demonstrates that in the numerical simulation, the correct temperature T_0 is

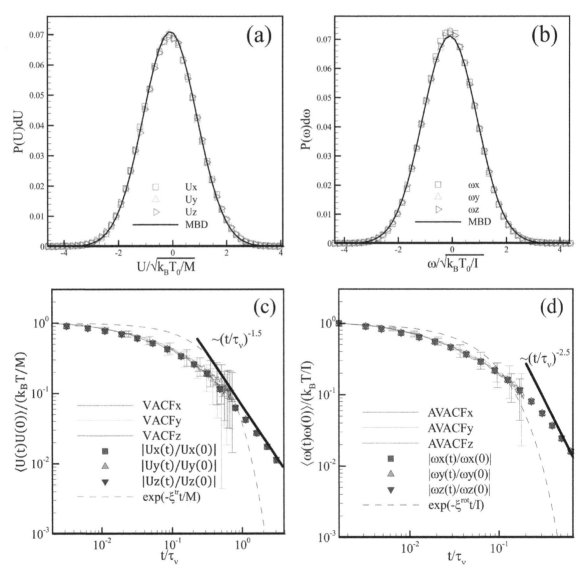

FIGURE 11.3 The evolution of translational and rotational dynamics of a spherical particle ($a = 250$ nm) in the inertial regime, when immersed in a Newtonian fluid (with properties of water) and confined at the center of a cylindrical vessel ($D = 20$ μm), simulated using the FHD approach at temperature $T_0 = 310K$. (a, b) The probability distributions of the Cartesian components of the translational and rotational velocities showing the adherence to the corresponding Maxwell Boltzmann distributions. (c, d) The short-time ($t \sim \tau_\nu = a^2/\nu$, where $\nu = \mu/\rho$ and ρ is the fluid density) evolution of the VACF and AVACF in the inertial regime showing the correct asymptotic transition from the exponential behavior ($\xi^{(tr)} = 6\pi\mu a$ and $\xi^{(rot)} = 8\pi\mu a^3$) for $t \to 0$ to algebraic behavior for $t > \tau_\nu$.

being realized for both translational and rotational degrees of freedom, panels (a,b), while simultaneously adhering to the correct temporal correlations, panels (c,d). Based on the numerical simulation, for the parameters considered, the translational VACF follows an exponential decay in the range for $t \leq 0.343\tau_\nu$ and an algebraic tail for $t \geq 1.202\tau_\nu$. The simulation also predicts the transitional range. Similarly, the rotational VACF follows an exponential decay, for $t \leq 0.115\tau_\nu$ and an algebraic tail for $t \geq 0.495\tau_\nu$. The error bars have been plotted from standard deviations of the decay at particular time instants obtained with 45 different realizations. For the system studied, Zwanzig and Bixon [51] have shown that for constant friction coefficient $\zeta^{(t)}$, the VACF of the particle in a simple fluid obeys

$$\langle \mathbf{U}(t)\mathbf{U}(0)\rangle = \frac{3k_B T}{M}\, e^{-\zeta^{(t)}t/M};$$
$$\langle \boldsymbol{\omega}(t)\boldsymbol{\omega}(0)\rangle = \frac{3k_B T}{\mathbf{I}}\, e^{-\zeta^{(r)}t/\mathbf{I}}, \qquad (11.67)$$

which denote exponential decays, while for the time-dependent friction coefficient, the decay of the VACF at long times obeys a power-law [9]:

$$\langle \mathbf{U}(t)\mathbf{U}(0)\rangle \simeq \left(\frac{k_B T \rho^{(f)1/2}}{4\pi^{3/2}\mu^{3/2}}\right) t^{-3/2};$$
$$\langle \boldsymbol{\omega}(t)\boldsymbol{\omega}(0)\rangle \simeq \left(\frac{3k_B T \rho^{(f)3/2}}{32\pi^{3/2}\mu^{5/2}}\right) t^{-5/2}, \qquad (11.68)$$

The results computed using FHD model are in good agreement with the predictions of Eqs. (11.67) and (11.68) for short and long times, respectively, and also predict the transition between the two. For the parameters considered, the translational VACF follows an exponential decay in the range for $t \leq 0.343\tau_\nu$ and an algebraic tail for $t \geq 1.202\tau_\nu$. Similarly, the rotational VACF follows an exponential decay, for $t \leq 0.115\tau_\nu$ and an algebraic tail for $t \geq 0.495\tau_\nu$ as has been determined from different realizations.

NP Diffusivity

Diffusivity in Unbounded Medium: The two panels in Figure 11.4 show both the short- and long-time translational and rotational MSDs of a neutrally buoyant NP ($a = 250$ nm) initially placed at the center of a large cylindrical vessel of diameter and length $D = 10\ \mu$m and $L = 10\ \mu$m, respectively. The fluid medium is static.

In the regime where the particle's motion is dominated by its own inertia (ballistic), $0.346\tau_\nu \leq t \leq 0.63\tau_\nu$ (translation), and $0.174\tau_\nu \leq t \leq 0.316\tau_\nu$ (rotation), the translational and rotational motions of the particle follow $(3k_B T/M)t^2$ and $(3k_B T/\mathbf{I})t^2$, respectively. In the diffusive regime, $t \gg \tau_b$, and when $t \geq 7\tau_\nu$ (translation) and $t \geq 1.2\tau_\nu$ (rotation), the translational and rotational MSDs increase linearly in time to follow $6D_\infty^{(t)}t$ and $6D_\infty^{(r)}t$, respectively, where $D_\infty^{(t)} = k_B T/\zeta^{(t)}$ and $D_\infty^{(r)} = k_B T/\zeta^{(r)}$ ($\zeta^{(r)} = 8\pi\mu a^3$) are the translational and rotational self-diffusion coefficients. The MSDs in an intermediate regime between the ballistic and the diffusive are related to hydrodynamic memory effects, and these are also displayed in the figures. It is also observed from Figure 11.4 that in the diffusive regime, the translational and rotational MSDs of the particle follow the Stokes–Einstein [56,57] and Stokes–Einstein–Debye [58] relations, respectively. Huang et al. [54] have provided experimental results for the Brownian motion of an NP in a liquid, and the numerical results here are in good agreement with them. The particle (translational and rotational) diffusivities calculated from the VACF are consistent with those predicted from the Stokes–Einstein and Stokes–Einstein–Debye relationships.

Wall Confinement Effects

In NFD of NPs in an unbounded fluid domain in the inertial regime, the transition from exponential behavior of the VACF to algebraic behavior is attributed to influences of

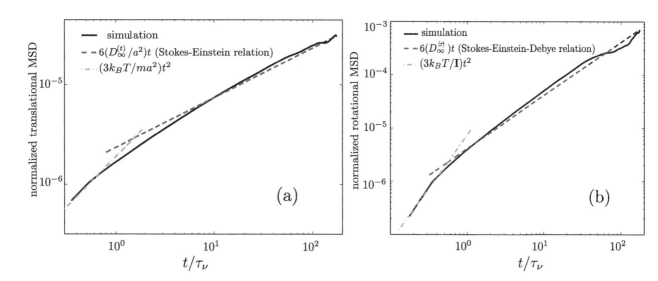

FIGURE 11.4 The normalized translational (a) and rotational (b) MSDs of a neutrally buoyant Brownian particle ($a = 250$ nm) in a stationary fluid medium computed using FHD simulations. The time on the horizontal axis has been scaled by τ_ν, the characteristic relaxation time for the particle.

the fluid inertia. However, in many technological applications, boundary effects and confining potentials are important, which make the evaluation of particle motion and associated transport much more complicated. The timescales of these motions overlap with the inertial timescale of the fluid, and as such, the temporal correlations in the inertial regime are strongly influenced by the confining boundaries and potentials. How the presence of a bounding planar wall alters the algebraic scaling of the VACF in the inertial regime has been analyzed by Gotoh and Kaneda [59], Pagonabarraga et al. [60,61], Felderhof [62], and Franosch and Jeney [63]. Collectively, these studies have investigated the dynamics of the motion of a spherical particle near the wall by confining the particle to different distances h from the wall. In the absence of an external potential, the parallel motion in the bulk regime ($h/a \to \infty$) showing the $t^{-1.5}$ scaling of the VACF transitions to a $t^{-2.5}$ scaling with positive amplitude in the near-wall regime (($h/a > 1$), while for the perpendicular motion, a $t^{-3.5}$ scaling for the intermediate times is followed by a long-time tail that exhibits a $t^{-2.5}$ scaling with negative amplitude. In the presence of a confining potential, the VACF is characterized by a $t^{-3.5}$ scaling in the bulk regime and a $t^{-4.5}$ scaling in the near-wall regime ($\frac{h}{a} > 1$). For a particle in the lubrication regime (where $(h-a)/a \to 0$), a detailed and consistent study of the hydrodynamic interactions and important aspects of the cylindrical wall effects in the lubrication and other hydrodynamic regimes are available in [64]. Also, for the majority of scaling relationships for VACF under various spatial and adhesion regimes that have been obtained through the asymptotic analysis of linear hydrodynamics, their validity through DNSs have been established in [64].

Figure 11.5 illustrates typical configurations explored in [64]: A single particle of typical radius $a = 250$ nm (and varied between 250 nm and 1 μm due to the relevance of this size range in pharmacological/rheological experiments) is suspended in a stationary fluid at a distance h, from the cylindrical wall of the vessel. The vessel dimensions are chosen to nominally be diameter $D = 5$ μm (and varied from 2.5 μm to 50 μm) and length $L = 10$ μm (and varied from 5 μm to 25 μm) in order to mimic the representative dimensions of blood capillaries and lymph nodules. To explore the effect of the proximity to the wall on NP motion, three typical simulation setups have been used: particle is located at the center of the vessel corresponding to $(h-a)/a = 9$ (> 1) or the bulk or core regime (Figure 11.5a), particle is placed near the wall but outside the lubrication layer with $(h-a)/a = 1$ or the near-wall regime (Figure 11.5b), and particle is inside the lubrication layer at a distance $(h-a)/a = 0.2$ (< 1) or the lubrication regime (Figure 11.5c). It should be noted that in the near-wall regime, the equations may result in a stiff system; this feature is evident from the analytical expression for the parallel component of the VACF for a particle located close to a wall, in which the series expansion of the admittance tensor contains a coefficient with a pre-factor that scales

as $(h/a)^2$ [62]. In this regime, the numerical stability issues have been addressed by employing a finer mesh; i.e., the mesh density of the particle surface is doubled in comparison to the nominal value. In [64], in order to exclude errors dependent on computational platform, the same calculations were performed on servers with different hardware configurations, and the results were found to be mutually consistent.

The results for the long-time behavior of the velocity (or the VACF) in the inertial regime for a particle at different locations relative to the boundary are analyzed together with the effect of the wall curvature and are displayed in Figure 11.6 for the translational velocity components. The limit $(D/d) \to \infty$ denotes the limit of the infinite planar wall for the parameter range examined in the study. An interesting effect observed in this study on cylindrical wall effects is that, apart from the short-time exponential decay and the intermediate-time algebraic decay, at much longer times, $t > C_1^{tr}\tau_D$ or $C_1^{rot}\tau_D$, with $\tau_D = D^2/\nu$, a second exponential decay ($\exp\left(-C_2^{tr}t/\tau_D\right)$ or $\exp\left(-C_2^{rot}t/\tau_D\right)$ with prefactors C_2^{tr} and C_2^{rot}) occurs. In the insets of Figure 11.6a for different D/d, the timescales at which the second exponential decay appears, where the particle velocity deviates from the algebraic scaling at least by 10%, are compared. This characteristic time is found to be only a function of D/d or τ_D/τ_ν. Detailed results for a particle located in the near-wall regime $(h-a)/a = 1$ have been presented in [64]. The parallel and perpendicular components are illustrated in Figure 11.6c, d for translational velocities. The results for the translational motion show that, for $D/d < 20$, the velocity decays exponentially without a clear intermediate algebraic scaling. For larger diameters, after the initial Stokes-exponential decay, algebraic correlations are observed, where the parallel motion displays a $t^{-2.5}$ scaling and the perpendicular motion first displays a $t^{-3.5}$ scaling behavior at intermediate times ($t \sim h^2/\nu$) followed by a $t^{-2.5}$ scaling with a negative sign (anticorrelation) due to the wall reflection of the diffused vortex. Eventually, the algebraic decay transitions to a final exponential decay due to the wall confinement. Figure 11.6d also illustrates that the presence of a curved wall causes an anticorrelation to occur at later times compared to those for a particle near a planar wall [62]. Similar trends are observed for the angular velocity relaxation where it first shows an initial exponential decay characterized by the instantaneous Stokes drag followed by an algebraic decay ($t^{-2.5}$ scaling for rotation about the parallel axis and $t^{-3.5}$ for perpendicular axis [60]) and a long-time second exponential decay. In Figure 11.6e,f, the time evolution of the velocity for a particle in the lubrication layer or $(h-a)/a < 1$ is depicted. The general characteristics of the velocity temporal response are similar to those for the near-wall case. However, the enhanced Stokes drag for the lubrication layer leads to a more distinct separation between the two exponential decays such that the intermediate algebraic decay is manifest even for smaller vessel diameters. We again observe the anticorrelation in the smaller vessel occurs at later times, indicating that the vessel

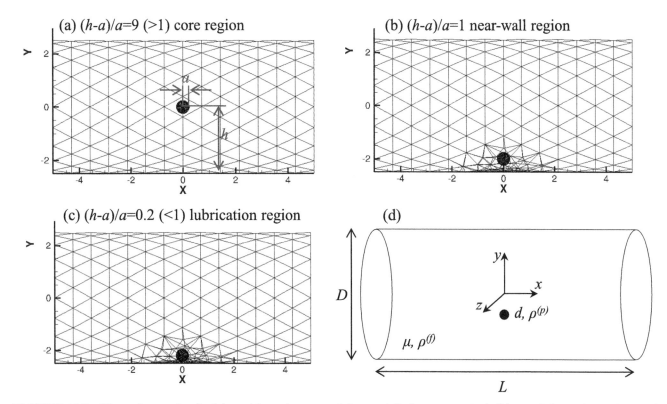

FIGURE 11.5 The surface meshes for (a) particle at the center of the vessel (bulk or core region), (b) particle located near the vessel wall (near-wall region), and (c) particle in the lubrication layer (lubrication region). (d) Schematic view of an NP in a cylindrical vessel.

curvature constrains the evolution of the particle motion. Eventually, the algebraic decay changes to a final exponential correlation due to the presence of strong confinement of the vessel wall. The angular velocity relaxation about the parallel axis exhibits the same general trend as in the near-wall case where the anticorrelation is noted to occur at later timescales for smaller tube diameters. In the lubrication regime, this effect is significant for rotation about the perpendicular axis such that anticorrelation is observed. The diffusion coefficient of the NP at different distances from the confining wall can be obtained from the VACF by solving the corresponding integral in the Green–Kubo relationship, as demonstrated in [64].

11.4.2 Construction of Thermostats

As stated earlier, an NP suspended in a fluid undergoes random motion due to the thermal fluctuations in the fluid. The fluid may be static or flowing under an external pressure gradient. In determining the translational and rotational motions of the NP in an incompressible Newtonian fluid (static or flowing), the performance of three formulations in adhering to thermal equipartition while simultaneously preserving the hydrodynamic interactions and correlations are examined: (i) the FHD method, (ii) the generalized Langevin dynamics (GLD) method, and (iii) the hybrid method (HM). Each formulation has its own strengths and shortcomings. Also, each formulation requires evaluation by a suitable numerical procedure. As a general rule, for

stochastic numerical simulations, independent of the specific procedure, a large number of realizations are required to acquire satisfactory statistics of the dynamical properties. This is usually computationally challenging and intensive. Apart from the three mentioned above, there is a computationally simpler method called the Deterministic method that provides sufficiently accurate long-time behavior of the VACF and AVACF, but this method will not be able to track the actual stochastic NP trajectory. These various methodologies will be briefly discussed in the following.

The FHD essentially consists of adding stochastic stresses (random stress) to the stress tensor in the momentum equation of the fluid [11]. The stochastic stress tensor depends on the temperature and the transport coefficients of the fluid medium [9,33]. FHD numerical simulations have been carried out by employing the finite volume method [21,22,33, 34], LBM [23–29], FEM [45,50,65], and stochastic immersed boundary method [30]. DNSs of FHD approach have been carried out by employing the finite volume method [21,46, 47,66], LBM [23–29], smoothed-particle method [67,68], and stochastic Eulerian–Lagrangian method [30,69]. A comparison of these methods are described in Uma et al. [50]. In the following, we extend the work of Uma et al. [50] by employing the ALE FEM to account for the fluid–particle interaction. We note that a key feature and strength of the ALE method is that, due to the adaptive mesh approach, it can resolve multiple hydrodynamic regimes such as bulk, near-wall, and lubrication amidst arbitrarily shaped boundaries. In the GLE, the effects of thermal fluctuations are

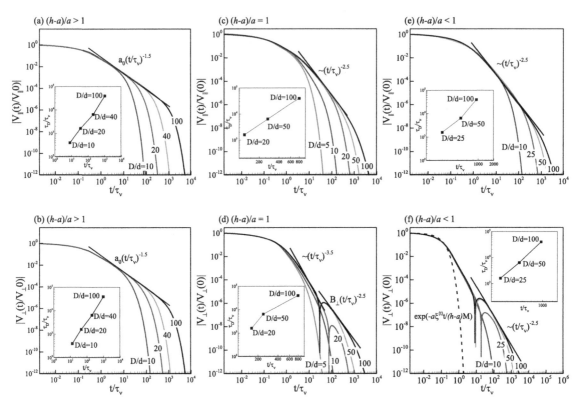

FIGURE 11.6 The effect of confinement and curvature of the cylindrical wall on the translational velocity of a 500 nm-diameter particle in a stationary fluid (calculated by the Deterministic method), located in the center of the vessel with (a) parallel $(h-a)/a > 1$ (axial) and (b) perpendicular $(h-a)/a > 1$ (radial) directions, near the wall in a direction (c) parallel $(h-a)/a = 1$ and (d) perpendicular $(h-a)/a = 1$ to the wall, and in the lubrication zone (e) parallel $(h-a)/a < 1$ and (f) perpendicular $(h-a)/a < 1$. $B_\perp = (h^2/a^2 - 5/9)/4\sqrt{\pi}$. The inset in each panel shows the corresponding comparison between τ_D and the time at which the second exponential decay $C_1^{tr}\tau_D$ appears. The coefficient C_1^{tr} is approximately 0.025, 0.017, and 0.019 for bulk, near-wall, and lubrication regimes, respectively. These coefficients have been determined by plotting the axes in a semi-log scale and then fitting the data.

incorporated as random forces and torques in the particle equation of motion [68,70–75]. The properties of these forces depend on the grand resistance tensor. The tensor in turn depends on the fluid properties, particle shape, and its instantaneous location such as its proximity to a wall or a boundary. In the GLE approach, a robust thermostat can be implemented by suitably tuning the noise spectrum of the random forces and torques by adding memory, but the coupling of the thermostat to the fluid equations of motion alters the true hydrodynamic behavior as quantified by the nature of the VACF and the value of the diffusion coefficient computed using MSD vs. time [75]. Both the Markovian (white noise) and non-Markovian (Ornstein–Uhlenbeck (O–U) noise and Mittag-Leffler (ML) noise) processes may be considered. For the non-Markovian approach, an appropriate choice of colored noise is required to satisfy the power-law decay in the VACF at long times. The non-Markovian ML noise simultaneously satisfies the equipartition theorem and the hydrodynamic correlations for a range of memory correlation times. The O–U process, however, may not provide the appropriate hydrodynamic correlations.

For NP motion in an incompressible fluid, the FHD resolves the hydrodynamics correctly but does not impose

the correct equipartition of energy based on the NP mass because of the added mass of the displaced fluid. In contrast, the Langevin approach with an appropriate memory is able to show the correct equipartition of energy but not the correct short- and long-time hydrodynamic correlations. A third approach referred to as the hybrid approach [76], which is described below in Section 11.4.3, shows for the first time that it is possible to simultaneously satisfy the equipartition theorem and the (short- and long-time) hydrodynamic correlations. In effect, this results in a thermostat that also simultaneously preserves the true hydrodynamic correlations. Thus, the hybrid approach enables a thermostat for the NP which maintains a set temperature and the correct thermal distributions (i.e., preserve the canonical ensemble), while simultaneously preserving the hydrodynamic effects (i.e., velocity autocorrelation and diffusion coefficient).

All the three methods described above are computationally expensive (large computational overhead), with the GLE perhaps the least expensive. A fourth procedure called the Deterministic method enables a computationally inexpensive calculation to study the long-time behavior of the VACF and the AVACF of an NP in a quiescent medium. The rationale for this method is derived both from the fluctuation–dissipation relation, which states that the temporal correlation in the thermal stresses is equivalent to the correlation in the hydrodynamic memory of a stationary fluid [77], and the Onsager regression hypothesis, which states that the regression of microscopic thermal fluctuations at equilibrium follows the macroscopic law of relaxation of small non-equilibrium disturbances [12,13]. Related to this, earlier studies [55,64,78,79] have shown that the averaged time correlation in the velocity of a Brownian particle in a stationary medium, is equivalent to that for a driven particle computed in the absence of thermal fluctuations. It must be noted that the trajectories identified by Deterministic simulations are not reflective of that for a fluctuating particle.

11.4.3 The Hybrid Method

While the added mass correction has a physical basis (i.e., for the incompressible fluid, the displaced mass of the fluid is essentially translated along with the particle), the FHD approach has its limitations for defining free energy landscapes of reactions occurring on the NP surfaces when such reactions are coupled to NP transport. In nanoscale mass transport, such limitations arise, e.g., during adhesion of nanocarriers to cells mediated by receptor–ligand, in the presence of convective hemodynamic flow. Since the FHD simulations solve the stochastic Navier–Stokes equation, which includes the inertial timescale, it captures the memory effects due to fluid inertia. Therefore, from the perspective of realizing a thermostat that maintains the particle at the correct temperature T_0 based on m (rather than M), one can view the NP motion as being described by a GLE which incorporates a memory function given by

$$m\frac{dU}{dt} = -\int_{-\infty}^{t} \zeta^{(t)}(t-t')U(t')\ dt' + R(t) \qquad (11.69)$$

for the Brownian particle, where $\zeta^{(t)}(t)$ is the frictional force memory kernel. Here, the random force $R(t)$ is zero-centered and obeys the fluctuation–dissipation theorem (and hence can serve as a suitable thermostat to maintain a constant temperature T_0) [15,57], namely,

$$\langle R(t)R(t')\rangle = k_B T_0 \zeta^{(t)}(|t-t'|). \qquad (11.70)$$

Therefore, a time-correlated noise is added to the particle equations of motion, Eqs. (11.29) and (11.30), similar to that in the GLE (Eq. (11.69)), resulting in

$$m\frac{dU}{dt} = -\int_{\partial\Sigma_p} [-p\boldsymbol{J} + 2\ \mu\ D[\boldsymbol{u}] + \boldsymbol{S}(\boldsymbol{x},t)] \cdot \hat{\boldsymbol{n}}\ ds$$
$$+ \int_{-\infty}^{t_-} \boldsymbol{\xi}(t')\ e^{-|t-t'|/\tau_1}\ dt', \qquad (11.71)$$

$$\frac{d(\boldsymbol{I\omega})}{dt} = -\int_{\partial\Sigma_p} (\boldsymbol{x}-\boldsymbol{X}) \times [(-p\boldsymbol{J} + 2\ \mu\ D[\boldsymbol{u}]$$
$$+\ \boldsymbol{S}(\boldsymbol{x},t)) \cdot \hat{\boldsymbol{n}}]\ ds + \int_{-\infty}^{t_-} \boldsymbol{\mu}(t')\ e^{-|t-t'|/\tau_2}\ dt', \qquad (11.72)$$

where the random force ξ and torque μ are given by

$$\xi(t') = \int_{\partial\Sigma_p} S(x',t') \cdot \hat{n}\ ds, \qquad (11.73)$$

$$\mu(t') = \int_{\partial\Sigma_p} (\boldsymbol{x}'-\boldsymbol{X}) \times (S(x',t') \cdot \hat{n})\ ds. \qquad (11.74)$$

The time integral in Eqs. (11.71) and (11.72) excludes the frictional force and torque at the time instant t since it has already been accounted for in the hydrodynamic force and torque terms, respectively. The characteristic timescales for memory for translational motion, i.e., τ_1, and rotational motion, i.e., τ_2, account for the history of the previous fluctuations, which directly impact not only the particle motion but also the fluid stresses due to the two-phase coupling of the weak formulation. The variables τ_1 and τ_2 are tuned in order to satisfy equipartition based on m (rather than on M) and $\boldsymbol{I}$. The methodology to compute these parameters is shown in [80], see also [81]. It is clear that in the limit of the characteristic memory times $\tau_1, \tau_2 \to 0$ (i.e., in the absence of memory), Eqs. (11.71) and (11.72) reduce to the Eqs. (11.29) and (11.30), respectively, which correspond to the Markovian FHD. The success and validity of this hybrid approach (i.e., by combining the Markovian FHD for the fluid along with the non-Markovian O–U noise in the equations of motion of the NP), in simulating the translational and rotational Brownian motion of an NP in an incompressible Newtonian stationary fluid medium is demonstrated in [80]; the results show that the hybrid approach can serve as an accurate thermostat without the need for the added mass correction while simultaneously capturing the correct hydrodynamic correlations [80]. This denotes a major advance since it enables the accommodations of asymmetric carrier shapes in confined geometries with attendant wall effects.

11.4.4 Coarse-Graining Inspired by Memory Function

In the description of the dynamics of nanosized Brownian particles in an unbounded fluid domain in the inertial regime, the transition from exponential behavior of the VACF to algebraic behavior is attributed to influences of the fluid inertia; this is explicit in the asymptotic analysis of the linearized hydrodynamic equations for a Newtonian fluid by Hauge and Martin-Löf [9] and Zwanzig and Bixon [51]. However, in many technological applications, boundary effects and confining potentials are important, which make the evaluation of particle motion and associated transport much more complicated. The timescales of these motions overlap with the inertial timescale of the fluid, and as such, the temporal correlations in the inertial regime are strongly influenced by the confining boundaries and potentials. Theoretical analyses on this front are described in Gotoh and Kaneda [59], Pagonabarraga et al. [60,61], Felderhof [62], and Franosch and Jeney [63]. Collectively, these studies have investigated the dynamics of the motion of a spherical particle near the wall by confining the particle to different distances h from the wall. In the absence of an external potential, the parallel motion in the bulk regime $(h/a \to \infty)$ showing the $t^{-1.5}$ scaling of the VACF transitions to a $t^{-2.5}$ scaling with positive amplitude in the near-wall regime $(h/a > 1)$, while for the perpendicular motion, a $t^{-3.5}$ scaling for the intermediate times is followed by a long-time tail that exhibits a $t^{-2.5}$ scaling with negative amplitude. In the presence of a confining potential, the VACF is characterized by a $t^{-3.5}$ scaling in the bulk regime and a $t^{-4.5}$ scaling in the near-wall regime. For a particle in the lubrication regime (where $(h-a)/a \to 0$), only the steady translational and rotational motions have been investigated: e.g., for motion perpendicular to the wall, the spherical particle experiences a translational friction coefficient that is enhanced by a factor of $a/(h-a)$ [82]; for rotation around the axis perpendicular to the wall, the torque coefficient is augmented by a factor of $1.2 - 3(\pi^2/6 - 1)(h/a - 1)$ [83,84]. A detailed and consistent study of the hydrodynamic interactions in the lubrication layer and important aspects of the wall effects in the lubrication and other hydrodynamic regimes is available in [64]. Also, for the majority of scaling relationships for VACF under various spatial and adhesion regimes that have been obtained through the asymptotic analysis of linear hydrodynamics, their validity through DNSs has been established as discussed earlier in Section 11.4.1.

Incorporation of the Effects of Boundaries on Hydrodynamic Interactions

The equation of stochastic motion for each component of the velocity of an NP immersed in a fluid in an unbounded domain (in the limit of the linearized Navier–Stokes equation) takes the form of a GLE and is given by

$$M\frac{dU}{dt} = -6\pi\mu a U(x,t) + 3a^2\sqrt{\pi\rho\mu}$$
$$\times \int_{-\infty}^{t} |t - t'|^{-\frac{3}{2}} U(x,t')\mathrm{d}t' - kx(t) + R(t).$$

$$(11.75)$$

Comparing Eqs. (11.69) and (11.75) yields $\zeta^{(t)}(t) = 12\pi\mu a\delta(t) - 3a^2\sqrt{\pi\rho\mu}t^{-\frac{3}{2}}$ and $R(t) = R_w(t) + R_c(t)$, with a white noise correlation $\langle R_w(t)R_w(t')\rangle = 12\pi\mu a k_B T\delta(t - t')$, a colored noise correlation $\langle R_c(t)R_c(t')\rangle = -3a^2\sqrt{\pi\rho\mu}k_B T|t - t'|^{-\frac{3}{2}}$, and $\langle R_w(t)R_c(t')\rangle = \langle R_w(t)\rangle\langle R_c(t')\rangle = 0$. Equation (11.75) representing a particle in an unbounded fluid domain can be extended to incorporate the effect of the boundaries on hydrodynamic interactions. In a recent study, Yu et al. [55] formulated a composite GLE that explicitly encodes the transition from a bulk domain to a near-wall domain as the particle approaches a confining boundary. This is accomplished by using a pertinent bridging function that transitions the GLE from that for the bulk regime at early times (when the momentum diffusion from the particle is yet to reach the boundary, i.e., $t \lesssim h^2/\nu$) to that for near-wall regimes for later times (when the reflected

momentum wave from the boundary begins to impact the temporal velocity correlations of particle motion, i.e., $t \gtrsim h^2/\nu$). For perpendicular motion, the composite GLE for that case is given by

$$M\frac{dU_\perp}{dt} = -6\pi\mu a\beta U_\perp(x,t) - A_1(t)\int_{-\infty}^{t} |t - t'|^{-\frac{3}{2}}U_\perp(x,t')\mathrm{d}t'$$
$$- A_2(t)\int_{-\infty}^{t} |t - t'|^{-\frac{5}{2}}U_\perp(x,t')\mathrm{d}t' - kx(t) + R(t),$$

$$(11.76)$$

where $\beta = \left(1 - \frac{9a}{8h}\right)^{-1}$, $M = (3m/2)\left(1 - \frac{a^3}{8h^3}\right)^{-1}$, and $A_1(t) = -3a^2\sqrt{\pi\rho\mu}\left(e^{-\frac{t}{\tau_w}}\right)$, and $A_2(t) = \frac{9}{8}am\sqrt{\frac{\rho}{\pi\mu}}\beta^2\left(1 - e^{-\frac{t}{\tau_w}}\right)$ complements $A_1(t)$; here, $\tau_w = h^2/\nu$.

The random forces and force correlations consistent with the fluctuation–dissipation theorem are given by

$$R(t) = R_w(t) + e^{-\frac{t}{\tau_w}}R_{c1}(t) + \left(1 - e^{-\frac{t}{\tau_w}}\right)R_{c2}(t) \text{ with}$$
$$\langle R_w(t)R_w(t')\rangle = 12\pi\mu a\beta k_B T\delta(t - t'),$$

$$\langle R_{c1}(t)R_{c1}(t')\rangle = -3a^2\sqrt{\pi\rho\mu}k_B T|t - t'|^{-\frac{3}{2}},$$

$$\langle R_{c2}(t)R_{c2}(t')\rangle = \frac{9}{8}am\sqrt{\frac{\rho}{\pi\mu}}\beta^2 k_B T|t - t'|^{-\frac{5}{2}},$$

$$\langle R_w(t)R_{c1}(t')\rangle = \langle R_w(t)R_{c2}(t')\rangle = \langle R_{c1}(t)R_{c2}(t')\rangle = 0.$$

Deterministic Method for VACF and AVACF Calculations

For stochastic simulations, a large number of realizations are required to reach satisfactory statistics of the dynamical properties. Since $\zeta^{(t)}(t)$ is hydrodynamic in origin, the scaled relaxation of $U(t)$ can also be obtained in the absence of the random force $R(t)$. This procedure is known as the Deterministic method. The Deterministic method is based on the Onsager regression hypothesis, which states that the regression of microscopic thermal fluctuations at equilibrium follows the macroscopic law of relaxation of small non-equilibrium disturbances [12,13]. Mathematically , this yields $\langle\Delta U(t)\rangle/\langle\Delta U(0)\rangle = \frac{\langle\Delta U(t)\Delta U(0)\rangle}{\langle\Delta U(0)^2\rangle}$, where $\Delta U(t) = U(t) - \langle U(t)\rangle$. This provides the framework for our employing the Deterministic method for evaluating VACFs and AVACFs. Here, we also perform Deterministic simulations in which the particle is driven initially by a weak impulse giving $U(0)$ in the absence of the $R(t)$. The correlation between a macroscopically driven $U(0)$ and the subsequent $U(t)$ would be equivalent to the calculated VACF (also denoted as $C_v(t)$) obtained from the stochastic simulations, according to Onsager's regression hypothesis [85].

In Figure 11.7a, for a non-neutrally buoyant Brownian particle at various distances from the wall, the Deterministic numerical solutions to the composite GLE have been compared with the analytical solutions of the linearized Navier–Stokes equation for the same particle–wall system in a quiescent fluid. This shows excellent agreement between

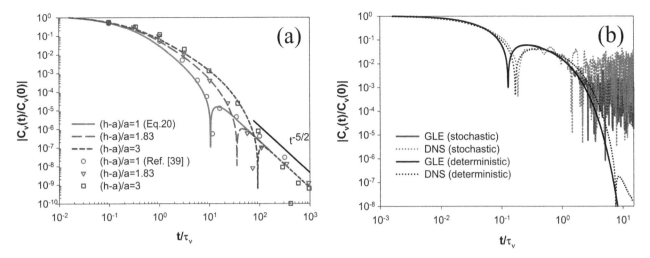

FIGURE 11.7 (a) Normalized VACF of a Brownian particle buoyant non-neutrally near an infinite plane wall in an incompressible, quiescent fluid medium for different separations from the wall. The symbols are the corresponding results from [63], and the lines are the predictions from the composite GLE simulations reported by [55]; here $\rho_p/\rho = 2.25$ with ρ_p being the density of the particle. (b) Normalized VACF of a neutrally buoyant Brownian particle in the lubrication regime with $h/a = 1.14$ in the presence of a harmonic spring (representing strong adhesion with $k = 1$ N/m) and comparison with DNS and FHD simulations.

theory and simulations. In Figure 11.7b, the Deterministic (i.e., without thermal noise) and the stochastic (i.e., with thermal noise) results for the composite GLE are compared to those from the FHD method for a particle in the lubrication regime and bound via a harmonic potential, again showing excellent agreement between the coarse-grained GLE approach [55] and the full-scale FHD/DNS simulations [64].

In Vitoshkin *et al.* [64], the dynamics and correlations in the presence of flow, wall-confinement, and adhesion interactions have been resolved by carrying out the FHD method in three different hydrodynamic regimes, namely bulk $((h-a)/a \gg 1)$, near-wall $((h-a)/a \sim 1)$, and lubrication $((h-a)/a \ll 1)$ regimes. The formulation and results of that study provide a systematic approach for studying the temporal hydrodynamic correlations in the presence of a curved vessel wall, particle size, particle–fluid density variations, adhesive interactions (confining potential), and low Reynolds number flows, especially focusing on the lubrication regime. The significance of the DNS approach using FHD in terms of capturing the VACF of NP motion under a variety of confinement and flow conditions can be recognized in relation to the GLE formalism for NP motion. Indeed, the composite GLE framework [55] correctly captures the different hydrodynamic correlations at different timescales, when boundaries, confining potentials, and flow fields are introduced. Therefore, the VACF determined by DNS provides a direct input to the GLE by defining the memory function. As demonstrated here, the GLE then represents a reduced dimensional and computationally efficient framework for encoding NP dynamics. We note that parallel approaches to bridge the molecular and hydrodynamic scales in pure fluids by combining molecular dynamics simulations and linearized FHD equations have been described by Chu et al. [31,86–88].

GLEs for Bridging Colloidal and Molecular Scale Dissipation

In nanoscale mass transport, the adhesive dynamics of NPs are governed by the binding interactions, which are more often than not mediated by receptor–ligand complexes. The example of a harmonic spring in Figure 11.7b represents such an interaction at a coarse-grained level. The confining potential can have a range of values for the reactive compliance (or alternatively the stiffness) depending on the receptor–ligand pair, the use of spacers or tethers, the properties of the NP, or the adhesive surface (e.g., flexible versus rigid); within the harmonic approximation, in which the adhesive potential is approximated as $V(x) = \frac{1}{2}kx^2$, the values for k typically fall in the range of $10^{-6} - 10$ N/m. Such potentials influence NP dynamics in an obvious way through influencing the energy landscape of adhesion, i.e., through $V(x)$. However, in the inertial timescale of $t \sim a^2/\nu$, which has been the main focus of this chapter, the internal dynamics of receptor–ligand complexes can also influence NP adhesion, independent of the energy landscape, $V(x)$. Single-molecule laser spectroscopy studies [89] have shown that fluctuations within a single protein exhibit wide relaxation spectrum characterizing a long-time memory, implying that the protein dynamics should also be described suitably by a GLE. Here, we present such an approach, where the colloidal hydrodynamics of an NP attached to a confining surface through a receptor–ligand bond is coupled to the internal frictional relaxation (e.g., dissipation associated with solvent relaxation dynamics) of the receptor–ligand complex itself, see Figure 11.8a. If we consider the dimensions of the receptor–ligand complex (in the scale of $\sim 10^1$ nm), which are much smaller than those of the NP ($a \sim 10^2$ nm), the particle dynamics itself occurs in the lubrication regime (i.e., with $(h-a)/a \ll 1$). Therefore, in

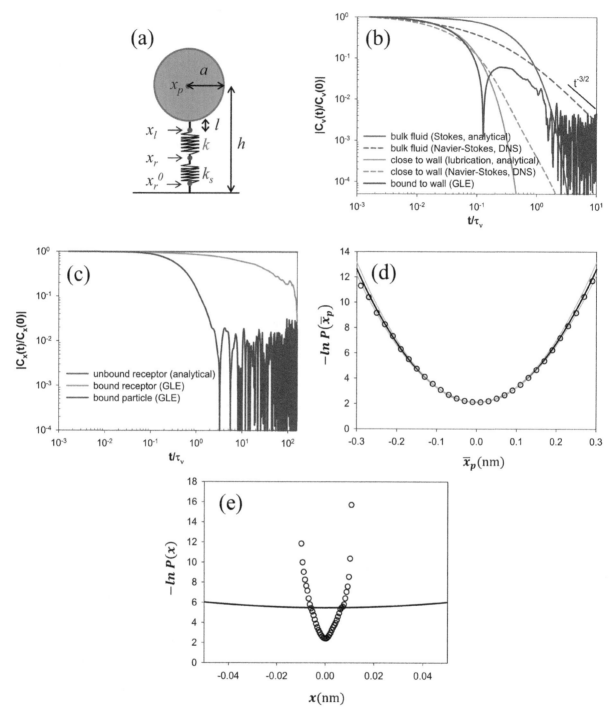

FIGURE 11.8 (a) Schematic of the coarse-grained model considered. The ligand-functionalized NP of radius a is attached to the wall through ligand–receptor binding interaction (first spring with the force constant k), including the relaxation of receptor internal dynamics subject to the conformational potential (second spring with the force constant k_s). The particle's center of mass is located at h from the wall. The circular dots from top to bottom: particle center-of-mass position x_p, ligand tip position x_l, receptor tip position x_r, and receptor tip position at equilibrium x_r^0. (b) Normalized VACFs for the NP for different conditions as marked in the legend: from top, in the bulk fluid predicted from the Stokes equation with $C_v(t) = (k_B T/M) \times \exp\{-6\pi\mu a t/M\}$, in the bulk fluid obtained from the Navier-Stokes equation, close to the wall predicted from the lubrication theory with $C_v(t) = (k_B T/M) \times \exp\{-6\pi\mu a^2 t/(M(h-a))\}$, close to the wall obtained from the Navier–Stokes equation, and attached to the vessel wall calculated from the current GLE simulations. (c) Normalized position autocorrelation functions for the receptor tip and the NP obtained from the current GLE simulations compared with the analytical solution for the free protein reported in [89,90]. Simulated probability distributions of the relative positions of (top) the NP center of mass and (bottom) the receptor tip. The solid curves are the equilibrium Boltzmann distributions for the harmonically bound NP position, with the shaded region denoting the 5% error about the MBD, panel (d), and the free receptor tip position, panel (e).

order to precisely encode the position-dependent, enhanced hydrodynamic drag when the particle binds to the surface in such a regime and simultaneously account for the dynamical relaxation of the ligand–receptor pair, we formulate and simultaneously solve two coupled GLEs:

$$M\frac{d^2\overline{x_p}(t)}{dt^2} = -\int_{-\infty}^{t}\zeta_p^{(t)}(t-t')\frac{d\overline{x_p}}{dt'}dt'$$
$$+ k(x(t) - \overline{x_p}(t)) + R_p(t) \qquad (11.77)$$

$$0 = -\int_{-\infty}^{t}\zeta_r^{(t)}(t-t')\frac{dx}{dt'}dt' - (k+k_s)x(t)$$
$$+ k\overline{x_p}(t) + R_r(t). \qquad (11.78)$$

Equation (11.77) corresponds to the NP equation of motion subject to a strong lubrication force and adhesion, while Eq. (11.78) describes the receptor dynamics. $\overline{x_p} = x_p - a - l - x_r^0$ is the rescaled NP center-of-mass position, l the ligand length, and x_r^0 the equilibrium position of the receptor tip; $x = x_r - x_r^0$ is the instantaneous receptor tip position relative to its equilibrium value. The hydrodynamic memory function in Eq. (11.77), $\zeta_p^{(t)} = 2\zeta_{p,0}^{(t)}\delta(t-t')$ with $\zeta_{p,0}^{(t)} = 6\pi\mu a\{a/(h(t)-a)\}$, denotes the enhanced resistance of the NP in the lubrication layer. $\zeta_r^{(t)}(t-t') = \zeta_r^0(\lambda_r)|t-t'|^{\lambda_r}$ in Eq. (11.78) is the memory function for internal dynamics of receptor with ζ_r^0 being the fiction coefficient and λ_r being the power-law index. Following the fluctuation–dissipation theorem [15], the random forces of the particle (R_p) and receptor (R_r) are related to the frictional terms via $\langle R_p(t)R_p(t')\rangle = 2k_BT\zeta_{p,0}^{(t)}\delta(t-t')$ and $\langle R_r(t)R_r(t')\rangle = k_BT\zeta_r^{(t)}(t-t')$.

In Figure 11.8b–d, we depict our results from the stochastic simulations of Eqs. (11.77) and (11.78) at $T = 310$ K for an NP ($a = 250$ nm) adhered to a planar wall (with equilibrium separation $x_r^0 = 19$ nm, $l = 15$ nm, and $k = 1$ N/m), see [55]. The conformational force constant $k_s = 1.88$ N/m, the friction coefficient $\zeta_r^0 = 1.32$ Ns$^{0.5}$m^{-1}, and the power-law index $\lambda_r = 0.5$ for the receptor protein are taken from the measurements of [90].

In Figure 11.8b,c, we present the normalized VACFs of the NP $C_v(t) = \langle U(t)U(0)\rangle$ for different conditions and the normalized position autocorrelation functions of the NP ($C_x(t) = \langle x_p(t)x_p(0)\rangle$) and of the receptor tip ($C_x(t) = \langle x(t)x(0)\rangle$) as functions of the scaled time, t/τ_ν. It is evident that the nature of the VACF for a bound NP is completely distinct from that for an unbound NP either in the bulk fluid or close to the vessel wall. For a free NP in the bulk, the analytical solution of the Stokes equation predicts an exponentially decaying $C_v(t)$, and the DNS of the Navier–Stokes equation [64] yields a $C_v(t)$ that exhibits a $t^{-3/2}$ long-time tail. When the particle moves to the proximity of the wall but is not bound to the wall, the lubrication force enhances the hydrodynamic resistance felt by the particle and leads to a faster decay of $C_v(t)$, where the steady-state lubrication theory predicts an exponential decay with an augmented drag coefficient, and the long-time tail of the DNS solution is altered by the wall-induced

vortex reflection. Once the particle is bound to the wall due to ligand–receptor interactions, the strong adhesion leads to oscillations in $C_v(t)$ that prevail in the hydrodynamic correlations. In the results for $C_x(t)$, the long-time decay for the bound NP is consistent with that observed in $C_v(t)$. In terms of the receptor dynamics, the much faster-decaying $C_x(t)$ for the bound receptor tip compared with the slowly-decaying correlation for a free receptor indicates that the dynamical relaxation of the protein is strongly coupled with the instantaneous motion of the NP. In Figure 11.8d,e, the probability distributions of the NP center-of-mass position and the receptor tip position are presented. We find, not surprisingly, that the presence of ligand–receptor relaxation makes the NP distribution close to the Boltzmann distribution for a harmonically bound particle. However, the NP binding makes conformational distribution of the receptor highly constrained, as evidenced by the much narrower receptor tip distribution compared to the unbound receptor. These observations suggest that the fluctuations of the antibody–antigen complex indeed play a significant role in the NP binding and relaxation dynamics.

11.5 Conclusions

Multiscale models of NP dynamics that incorporate multiple scales (e.g., the macroscopic regime, the lubrication regime, and the adhesion regime), and which couple directly with the associated transport dynamics, provide the algorithmic methodology for rapid and accurate prescription for optimal NP design. However, due to competing hydrodynamic and adhesion forces at disparate length scale, traditional multiscale frameworks are not easily amenable for addressing the associated challenges. In this chapter, we have primarily described and focused on foundations, numerical implementation, dynamics of Brownian motion of an NP, hydrodynamic interactions in the presence of flow fields, and hydrodynamic confinement. FHD, GLEs, hybrid scheme, and the Deterministic method have all been discussed in detail. Illustrative computations and results for VACF and AVACF at various regimes have been presented. In particular, the determination of the autocorrelation functions accurately for long times (compared to the inertial relaxation time) using the Deterministic approach provides a direct route for computing transport properties using the appropriate Green–Kubo formulas [91]. Several works in the literature have reported calculations of transport properties for models of simple and complex fluids [92,93]. Encouraged by these studies which focused on the molecular level, we propose that the Green–Kubo methodologies can be used to study the properties of suspensions of NPs and nanocomposites at the mesoscopic scale, when combined with computational methods that resolve hydrodynamic interactions and thermal effects simultaneously, including the examples described in this chapter.

Acknowledgments

We thank Uma Balakrishnan, Helena Vitoshkin, Hsiu-Yu Yu, Zahera Jabeen, Yaohong Wang, Karthik Mukundakrishnan, Ryan P. Bradley, Neeraj Agrawal, Tirumani Swaminathan, Arijit Sarkar, Samaneh Farokhirad, and Jin Liu for insightful discussions. This work was supported in part by the National Institutes of Health (NIH) Grant Nos. R01 EB006818 and U01 EB016027 and the National Science Foundation (NSF) Grant No. NSF CBET 1236514. We acknowledge shared computational resources provided by Extreme Science and Engineering Discovery Environment (XSEDE) (Grant No. MCB060006) and NSF DMR 1120901.

References

1. R. Prasher, P. Bhattacharya, and P. Patrick E. Brownian-motion-based convective-conductive model for the effective thermal conductivity of nanofluids. *J. Heat Transfer*, 128(6):588–595, 2006. doi: 10.1115/1.2188509.

2. P. Keblinski, R. Prasher, and J. Eapen. Thermal conductance of nanofluids: is the controversy over? *J. Nanopart. Res.*, 10(7):1089–1097, 2008. ISSN 1572-896X. doi: 10.1007/s11051-007-9352-1.

3. J. Buongiorno et al. A benchmark study on the thermal conductivity of nanofluids. *J. Appl. Phys.*, 106(9): 094312, 2009. doi: 10.1063/1.3245330.

4. E. E. (Stathis) Michaelides. *Fundamentals of Nanoparticle Flow and Heat Transfer*, pp. 1–45. Springer International Publishing, Cham, 2014. ISBN 978-3-319-05621-0. doi: 10.1007/978-3-319-05621-0_1.

5. N.G. Van Kampen. *Stochastic Processes in Physics and Chemistry* (3rd edition). Elsevier, Amsterdam, 2007. doi: https://doi.org/10.1016/B978-0-444-52965-7.50023-4.

6. S. Chapman, D. Burnett, and T. G. Cowling. *The Mathematical Theory of Non-uniform Gases* (3rd edition). Cambridge University Press, Cambridge, 1970. ISBN 9780521408448.

7. M. Bixon and R. Zwanzig. Boltzmann–Langevin equation and hydrodynamic fluctuations. *Phys. Rev.*, 187(1):267–272, 1969.

8. L. D. Landau and E. M. Lifshitz. *Fluid Mechanics, Second Edition: Volume 6 (Course of Theoretical Physics)* (2nd edition). Course of Theoretical Physics, Vol. 6. Butterworth-Heinemann, Oxford, 1987.

9. E. H. Hauge and A. Martin-Löf. Fluctuating hydrodynamics and Brownian motion. *J. Stat. Phys.*, 7(3): 259–281, 1973. doi: 10.1007/BF01030307.

10. P. M. Chaikin and T. C. Lubensky. *Principles of Condensed Matter Physics*. Cambridge University Press, Cambridge, 2000.

11. L. D. Landau and E. M. Lifshitz. *Fluid Mechanics*. Pergamon Press, London, 1959.

12. L. Onsager. Reciprocal relations in irreversible processes. I. *Phys. Rev.*, 37:405–426, 1931a.

13. L. Onsager. Reciprocal relations in irreversible processes. II. *Phys. Rev.*, 38:2265–2279, 1931b.

14. D. Chandler. *Introduction to Modern Statistical Mechanics*. Oxford University Press, Oxford, 1987a.

15. R. Kubo. The fluctuation-dissipation theorem. *Rep. Prog. Phys.*, 29(1):255–284, 1966a. doi: 10.1088/0034-4885/29/1/306.

16. R. Kubo, M. Toda, and N. Hashitsume. *Nonequilibrium Statistical Mechanics* (2nd edition), volume II. Springer-Verlag, Berlin, 1991.

17. V. Balakrishnan. *Elements of Nonequilibrium Statistical Mechanics*. CRC Press, Boca Raton, FL2008.

18. B. J. Berne. Hydrodynamic theory of the angular velocity autocorrelation function. *J. Chem. Phys.*, 56:2164–2168, 03 1972.

19. A. Einstein. Theoretical remarks on Brownian motion. *Zeit. f. Elektrochemie*, 13:41–42, 1907.

20. P. Liu, E. Harder, and B. J. Berne. On the calculation of diffusion coefficients in confined fluids and interfaces with an application to the liquid-vapor interface of water. *J. Phys. Chem. B*, 108(21): 6595–6602, 2004. doi: 10.1021/jp0375057.

21. N. Sharma and N. A. Patankar. Direct numerical simulation of the Brownian motion of particles by using fluctuating hydrodynamic equations. *J. Comput. Phys.*, 201(2):466–486, 2004. doi: 10.1016/j.jcp.2004.06.002.

22. A. Donev, E. Vanden-Eijnden, A. L. Garcia, and J. B. Bell. On the accuracy of explicit finite-volume schemes for fluctuating hydrodynamics. *Commun. Appl. Math. Comput. Sci.*, 5(2):149–197, 2010a.

23. A. J. C. Ladd. Short-time motion of colloidal particles: Numerical simulation via a fluctuating Lattice-Boltzmann equation. *Phys. Rev. Lett.*, 70(9): 1339–1342, Mar 1993. doi: 10.1103/PhysRevLett.70.1339.

24. A. J. C. Ladd. Numerical simulations of particulate suspensions via a discretized Boltzmann equation. Part 1. Theoretical foundation. *J. Fluid Mech.*, 271:285–309, 1994a. doi: 10.1017/S0022112094001771.

25. A. J. C. Ladd. Numerical simulations of particulate suspensions via a discretized Boltzmann equation. Part 2. Numerical results. *J. Fluid Mech.*, 271:311–339, 1994b. doi: 10.1017/S0022112094001783.

26. N. A. Patankar. Direct numerical simulation of moving charged, flexible bodies with thermal fluctuations. In *Technical Proceedings of the 2002 International Conference on Computational Nanoscience and Nanotechnology*, vol. 2, pp. 93–96. Nano Science and Technology Institute, 2002.

27. R. Adhikari, K. Stratford, M. E. Cates, and A. J. Wagner. Fluctuating lattice–Boltzmann. *EPL (Europhys. Lett.)*, 71(3):473–479, 2005.

28. B. Dünweg and A. J. C. Ladd. Lattice–Boltzmann simulations of soft matter systems. *Adv. Polym. Sci.*, 221:89–166, 2008. doi: 10.1007/978-3-540-87706-6_2.

29. D. Nie and J. Lin. A fluctuating lattice-Boltzmann model for direct numerical simulation of particle Brownian motion. *Particuology*, 7(6):501–506, 2009. ISSN 1674-2001. doi: 0.1016/j.partic.2009.06.012.

30. P. J. Atzberger, P. R. Kramer, and C. S. Peskin. A stochastic immersed boundary method for fluid-structure dynamics at microscopic length scales. *J. Comput. Phys.*, 224(2):1255–1292, 2007. ISSN 0021-9991. doi: 10.1016/j.jcp.2006.11.015.

31. N. K. Voulgarakis and J.-W. Chu. Bridging fluctuating hydrodynamics and molecular dynamics simulations of fluids. *J. Chem. Phys.*, 130(13):134111, 2009. doi: 10.1063/1.3106717.

32. N. K. Voulgarakis, S. Satish, and J. W. Chu. Modeling the nanoscale viscoelasticity of fluids by bridging non-Markovian fluctuating hydrodynamics and molecular dynamics simulations. *J. Chem. Phys.*, 131(23):234115, 2009a. doi: 10.1063/1.3273210.

33. M. Serrano and P. Español. Thermodynamically consistent mesoscopic fluid particle model. *Phys. Rev. E*, 64(4):046115, 2001. doi: 10.1103/PhysRevE.64.046115.

34. M. Serrano, D.F. Gianni, P. Español, E.G. Flekkøy, and P.V. Coveney. Mesoscopic dynamics of Voronoi fluid particles. *J. Phys. A Math. Gen.*, 35(7):1605, 2002. doi: 10.1088/0305-4470/35/7/310.

35. M. Grmela and H.C. Öttinger. Dynamics and thermodynamics of complex fluids. I. Development of a general formalism. *Phys. Rev. E*, 56(6):6620–6632, 1997. doi: 10.1103/PhysRevE.56.6620.

36. H.C. Öttinger and M. Grmela. Dynamics and thermodynamics of complex fluids. II. Illustrations of a general formalism. *Phys. Rev. E*, 56(6):6633–6655, 1997. doi: 10.1103/PhysRevE.56.6633.

37. H.H. Hu. Direct simulation of flows of solid-liquid mixtures. *Int. J. Multiphase Flow*, 22(2):335–352, 1996. ISSN 0301-9322. doi: 10.1016/0301-9322(95)00068-2.

38. H. H. Hu, N. A. Patankar, and M. Y. Zhu. Direct numerical simulations of fluid-solid systems using the arbitrary Lagrangian-Eulerian technique. *J. Comput. Phys.*, 169(2):427–462, 2001. ISSN 0021-9991. doi: 10.1006/jcph.2000.6592.

39. L. Zhang, A. Gerstenberger, X. Wang, and W. K. Liu. Immersed finite element method. *Comput. Methods Appl. Mech. Eng.*, 193(21-22):2051–2067, 2004. ISSN 0045-7825. doi: 10.1016/j.cma.2003.12.044.

40. X. S. Wang, L. T. Zhang, and W. K. Liu. On computational issues of immersed finite element methods. *J. Comput. Phys.*, 228(7):2535–2551, 2009. ISSN 0021-9991. doi: 10.1016/j.jcp.2008.12.012.

41. P. L. George. *Automatic Mesh Generation: Application to Finite Element Methods*. Wiley, New York, 1991.

42. T. N. Swaminathan, K. Mukundakrishnan, and H. H. Hu. Sedimentation of an ellipsoid inside an infinitely long tube at low and intermediate Reynolds numbers. *J. Fluid Mech.*, 551:357–385, 2006a. doi: 10.1017/S0022112005008402.

43. T. N. Swaminathan, H. H. Hu, and A. A. Patel. Numerical analysis of the hemodynamics and embolus capture of a Greenfield vena cava filter. *J. Biomech. Eng.*, 128(3):360–370, 2006b. doi: 10.1115/1.2187034.

44. K. Mukundakrishnan, H. H. Hu, and P. S. Ayyaswamy. The dynamics of two spherical particles in a confined rotating flow: pedalling motion. *J. Fluid Mech.*, 599:169–204, 2008. doi: 10.1017/S0022112007000092.

45. P. Español, J.G. Anero1, and I. Zúñiga. Microscopic derivation of discrete hydrodynamics. *J. Chem. Phys.*, 131:244117, 2009. doi: 10.1063/1.3274222.

46. J.B. Bell, A.L. Garcia, and S.A. Williams. Numerical methods for the stochastic Landau-Lifshitz Navier-Stokes equations. *Phys. Rev. E*, 76(1):016708, Jul 2007. doi: 10.1103/PhysRevE.76.016708.

47. S.A. Williams, J.B. Bell, and A.L. Garcia. Algorithm refinement for fluctuating hydrodynamics. *Multiscale Model. Simul.*, 6:1256–1280, 2008. doi: 10.1137/070696180.

48. N. Ramakrishnan, Y. Wang, D. M. Eckmann, P. S. Ayyaswamy, and R. Radhakrishnan. Motion of a nano-spheroid in a cylindrical vessel flow: Brownian and hydrodynamic interactions. *J. Fluid Mech.*, 821:117–152, June 2017.

49. M. P. Allen and D. J. Tildesley. *Computer Simulation of Liquids (Oxford Science Publications)* (reprint edition). Oxford University Press, Oxford, 1989.

50. B. Uma, T. N. Swaminathan, R. Radhakrishnan, D. M. Eckmann, and P. S. Ayyaswamy. Nanoparticle Brownian motion and hydrodynamic interactions in the presence of flow fields. *Phys. Fluids*, 23(7): 073602, 2011a.

51. R. Zwanzig and M. Bixon. Hydrodynamic theory of the velocity correlation function. *Phys. Rev. A*, 2(5): 2005–2012, 1970. doi: 10.1103/PhysRevA.2.2005.

52. R. Zwanzig and M. Bixon. Compressibility effects in the hydrodynamic theory of Brownian motion. *J. Fluid Mech.*, 69:21–25, 1975. doi: 10.1017/S0022112075001280.

53. T. Li, S. Kheifets, D. Medellin, and M.G. Raizen. Measurement of the Instantaneous Velocity of a Brownian Particle. *Science*, 328(5986):1673–1675, 2010. doi: 10.1126/science.1189403.

54. R. Huang, I. Chavez, K.M. Taute, B. Lukic, S. Jeney, M.G. Raizen, and E. Florin. Direct observation of the full transition from ballistic to diffusive Brownian motion in a liquid. *Nat. Phys.*, 7:576–580, 2011. doi: 10.1038/nphys1953.

55. H.-Y. Yu, D. M. Eckmann, P. S. Ayyaswamy, and R. Radhakrishnan. Composite generalized Langevin equation for Brownian motion in different hydrodynamic and adhesion regimes. *Phys. Rev. E*, 91(5):052303, May 2015.

56. A. Einstein. On the molecular-kinetic theory of the movement by heat of particles suspended in liquids at rest. *Ann. Phys.*, 17(8):549–560, 1905. ISSN 1521-3889. doi: 10.1002/andp.19053220806.

57. R. Zwanzig. *Nonequilibrium Statistical Mechanics.* Oxford University Press, Oxford, 2001.

58. D. M. Heyes, M. J. Nuevo, J. J. Morales, and A. C. Branka. Translational and rotational diffusion of model nanocolloidal dispersions studied by molecular dynamics simulations. *J. Phys. Condens. Matter*, 10(45):10159–10178, 1998. doi: 10.1088/0953-8984/10/45/005.

59. T. Gotoh and Y. Kaneda. Effect of an infinite plane wall on the motion of a spherical Brownian particle. *J. Chem. Phys.*, 76:3193–3197, 1982. doi: 10.1063/1.443364.

60. I. Pagonabarraga, M. H. J. Hagen, C. P. Lowe, and D. Frenkel. Algebraic decay of velocity fluctuations near a wall. *Phys. Rev. E*, 58:7288–7295, 1998a. doi: 10.1103/PhysRevE.58.7288.

61. M. H. J. Hagen, I. Pagonabarraga, C. P. Lowe, and D. Frenkel. Algebraic decay of velocity fluctuations in a confined fluid. *Phys. Rev. Lett.*, 78:3785–3788, 1997. doi: 10.1103/PhysRevLett.78.3785.

62. B. U. Felderhof. Effect of the wall on the velocity autocorrelation function and long-time tail of Brownian motion. *J. Phys. Chem. B*, 109:21406–21412, 2005. doi: 10.1063/1.2084948.

63. T. Franosch and S. Jeney. Persistent correlation of constrained colloidal motion. *Phys. Rev. E*, 79:031402, 2009. doi: 10.1103/PhysRevE.79.031402.

64. H. Vitoshkin, H.-Y. Yu, D. M. Eckmann, P. S. Ayyaswamy, and R. Radhakrishnan. Nanoparticle stochastic motion in the inertial regime and hydrodynamic interactions close to a cylindrical wall. *Phys. Rev. Fluids*, 1(5):054104, September 2016.

65. P. Español and I. Zúñiga. On the definition of discrete hydrodynamic variables. *J. Chem. Phys.*, 131:164106, 2009. doi: 10.1063/1.3247586.

66. A. Donev, E. Vanden-Eijnden, A. Garcia, and J. Bell. On the accuracy of finite-volume schemes for fluctuating hydrodynamics. *Comm. App. Math. And Comp. Sci.*, 5(2):149–197, 2010b.

67. T. Iwashita, Y. Nakayama, and R. Yamamoto. Velocity autocorrelation function of fluctuating particles in incompressible fluids. *Prog. Theor. Phys.*, 178: 86–91, 2009. doi: 10.1143/PTPS.178.86.

68. T. Iwashita, Y. Nakayama, and R. Yamamoto. A numerical model for Brownian particles fluctuating in incompressible fluids. *J. Phys. Soc. Jpn.*, 77(7): 074007, 2008a. doi: 10.1143/JPSJ.77.074007.

69. P. J. Atzberger. Stochastic Eulerian Lagrangian methods for fluid-structure interactions with thermal fluctuations. *J. Comput. Phys.*, 230(8): 2821–2837, 2011.

70. D. L. Ermak and J. A. McCammon. Brownian dynamics with hydrodynamic interactions. *J. Chem. Phys.*, 69(4):1352–1360, 1978. doi: 10.1063/1.436761.

71. J. F. Brady and G. Bossis. Stokesian dynamics. *Ann. Rev. Fluid Mech.*, 20(1):111–157, 1988. doi: 10.1146/annurev.fl.20.010188.000551.

72. D. R. Foss and J. F. Brady. Structure, diffusion and rheology of Brownian suspensions by Stokesian dynamics simulation. *J. Fluid Mech.*, 407:167–200, 2000. doi: 10.1017/S0022112099007557.

73. A. J. Banchio and J. F. Brady. Accelerated Stokesian dynamics: Brownian motion. *J. Chem. Phys.*, 118 (22):10323–10332, 2003. doi: 10.1063/1.1571819.

74. T. Iwashita and R. Yamamoto. Short-time motion of Brownian particles in a shear flow. *Phys. Rev. E*, 79(3):031401, Mar 2009. doi: 10.1103/PhysRevE.79.031401.

75. B. Uma, T. N. Swaminathan, P. S. Ayyaswamy, D. M. Eckmann, and R. Radhakrishnan. Generalized Langevin dynamics of a nanoparticle using a finite element approach: Thermostating with correlated noise. *J. Chem. Phys.*, 135(11):114104, 2011b.

76. B. Uma, D. M. Eckmann, P. S. Ayyaswamy, and R. Radhakrishnan. A hybrid formalism combining fluctuating hydrodynamics and generalized Langevin dynamics for the simulation of nanoparticle thermal motion in an incompressible fluid medium. *Mol. Phys.*, 110(11-12):1057–1067, June 2012.

77. R. Kubo. The fluctuation-dissipation theorem. *Rep. Prog. Phys.*, 29(1):255–284, January 1966b.

78. I. Pagonabarraga, M. H. J. Hagen, C. P. Lowe, and D. Frenkel. Algebraic decay of velocity fluctuations near a wall. *Phys. Rev. E*, 58(6):7288–7295, December 1998b.

79. T. Iwashita, Y. Nakayama, and R. Yamamoto. A Numerical Model for Brownian Particles Fluctuating in Incompressible Fluids. *J. Phys. Soc. Jpn.*, cond-mat.soft(7):074007–074007, July 2008b.

80. B. Uma, P. S. Ayyaswamy, R. Radhakrishnan, and D. M. Eckmann. Fluctuating Hydrodynamics Approach for the Simulation of Nanoparticle Brownian Motion in a Newtonian Fluid. *Int. J. Micro-Nano Scale Transp.*, 3(1):13–20, 2013.

81. B. Uma, T. N. Swaminathan, P. S. Ayyaswamy, D. M. Eckmann, and R. Radhakrishnan. Generalized Langevin dynamics of a nanoparticle using a finite element approach: Thermostating with correlated noise. *J. Chem. Phys.*, 135:114104, 2011c. doi: 10.1063/1.3635776.

82. G. L. Leal. *Advanced Transport Phenomena.* Cambridge University Press, New York, 2007.

83. Jeffery G. B. On the steady rotation of a solid of revolution in a viscous fluid. *Proceedings of the London Mathematical Society*, s2_14(1):327–338, 1915. doi: 10.1112/plms/s2_14.1.327.

84. A.J. Goldman, R.G. Cox, and H. Brenner. Slow viscous motion of a sphere parallel to a plane wall: Motion through a quiescent fluid. *Chem. Eng. Sci.*, 22(4):637–651, 1967. ISSN 0009-2509. doi: https://doi.org/10.1016/0009-2509(67)80047-2.

85. D. Chandler. *Introduction to Modern Statistical Mechanics*. Oxford University Press, New York, 1987b.

86. B. Z. Shang, N. K. Voulgarakis, and J.-W. Chu. Fluctuating hydrodynamics for multiscale modeling and simulation: Energy and heat transfer in molecular fluids. *J. Chem. Phys.*, 137(4):044117, 2012. doi: 10.1063/1.4738763.

87. N. K. Voulgarakis, S. Satish, and J.-W. Chu. Modeling the nanoscale viscoelasticity of fluids by bridging non-Markovian fluctuating hydrodynamics and molecular dynamics simulations. *J. Chem. Phys.*, 131(23):234115, 2009b. doi: 10.1063/1.3273210.

88. N. K. Voulgarakis, S. Satish, and J.-W. Chu. Modelling the viscoelasticity and thermal fluctuations of fluids at the nanoscale. *Mol. Simul.*, 36(7-8):552–559, 2010. doi: 10.1080/08927022.2010.486832.

89. S. Kou and X. S. Xie. Generalized Langevin equation with fractional Gaussian noise: Subdiffusion within a single protein molecule. *Phys. Rev. Lett.*, 93(18):180603, October 2004.

90. W. Min, G. Luo, B. J. Cherayil, S. C. Kou, and X. S. Xie. Observation of a power-law memory kernel for fluctuations within a single protein molecule. *Phys. Rev. Lett.*, 94(19):198302, May 2005.

91. D. J. Searles and D. J. Evans. The fluctuation theorem and Green–Kubo relations. *J. Chem. Phys.*, 112(22):9727–9735, 2000. doi: 10.1063/1.481610.

92. J. Ratanapisit, D.J. Isbister, and J. Ely. Transport properties of fluids: Symplectic integrators and their usefulness. *Fluid Phase Equilib.*, 183184:351–361, 2001. doi: 10.1016/S0378-3812(01)00447-2.

93. A. S. Henry and G. Chen. Spectral phonon transport properties of silicon based on molecular dynamics simulations and lattice dynamics. *J. Comput. Theor. Nanosci.*, 5(2):141–152, 2008. doi: doi:10.1166/jctn.2008.2454.

Transport in Nanoporous Materials

Saad Alafnan
King Fahd University of Petroleum and Minerals (KFUPM)

I. Yucel Akkutlu
Texas A&M University (TAMU)

12.1 Introduction

Fluid transport in nanoporous materials has been an active area of research for a variety of applications such as nuclear, pharmaceutical, and biomedical processes. Mechanics of fluid flow in nanopores has the added complexities due to the amplified fluid–pore wall interactions. As the size of the flow paths become smaller, these molecular level interactions may lead to deviation from classical flow into different modes of solid-surface-driven diffusion. Various flow studies using simple (Lennard-Jones) fluids such as methane have shown the presence of adsorptive–diffusive transport. In this chapter, we intend to cover these studies. In addition, nonporous naturally occurring materials such as geological porous media will be covered. Rocks could have interesting multiscale pore structures, where a nanopore network storing hydrocarbon fluids could be hydraulically in communication with a large-scale feature such as micro-fracture. The fluid transport in such multiscale pore networks finds application in natural gas production and subsurface carbon dioxide storage.

Navier–Stokes equations govern inertial and viscous forces acting on the fluids:

$$x\text{-direction}: \rho \left(\frac{\partial v_x}{\partial t} + v_x \frac{\partial v_x}{\partial x} + v_y \frac{\partial v_x}{\partial y} + v_z \frac{\partial v_x}{\partial z} \right)$$
$$= -\frac{\partial P}{\partial x} + \rho g_x + \mu \left(\frac{\partial^2 v_x}{\partial x^2} + \frac{\partial^2 v_x}{\partial y^2} + \frac{\partial^2 v_x}{\partial z^2} \right),$$

$$y\text{-direction}: \rho \left(\frac{\partial v_y}{\partial t} + v_x \frac{\partial v_y}{\partial x} + v_y \frac{\partial v_y}{\partial y} + v_z \frac{\partial v_y}{\partial z} \right)$$
$$= -\frac{\partial P}{\partial y} + \rho g_y + \mu \left(\frac{\partial^2 v_y}{\partial x^2} + \frac{\partial^2 v_y}{\partial y^2} + \frac{\partial^2 v_y}{\partial z^2} \right),$$

$$z\text{-direction}: \rho \left(\frac{\partial v_z}{\partial t} + v_x \frac{\partial v_z}{\partial x} + v_y \frac{\partial v_z}{\partial y} + v_z \frac{\partial v_z}{\partial z} \right)$$
$$= -\frac{\partial P}{\partial z} + \rho g_z + \mu \left(\frac{\partial^2 v_z}{\partial x^2} + \frac{\partial^2 v_z}{\partial y^2} + \frac{\partial^2 v_z}{\partial z^2} \right).$$

$$(12.1)$$

The general form of Navier–Stokes equations is usually reduced and then linearized so they can be solved analytically or numerically. An example of what would be the flow in cylindrical geometry derived by Hagen and Poiseuille from the original Navier–Stokes equations where inertial forces are assumed very small compared to viscous effect and the gravitational force is negligible:

$$v = -\frac{R^2}{8\mu} \left(\frac{\partial P}{\partial x} \right). \qquad (12.2)$$

In porous media, a similar relationship is used for the flow with a coefficient called permeability k governing the tendency of the porous media to pass the fluid. The relationship is known as Darcy's law:

$$v = -\frac{k}{\mu} \left(\frac{\partial P}{\partial x} \right). \qquad (12.3)$$

Darcy's law can help relate the pressure forces to the flow rate. For example, it has been utilized in the simulation of the fluid transport from the reservoir to the wellbore. However, Darcy's law has its own limitations. When the length scale of the area available for the flow drops to the nanoscale, Navier–Stokes equations and their linearized forms including Darcy's law may break down. In other words, they become insufficient to describe the transport in nanoporous materials.

12.2 Flow Regimes in Nanoporous Materials

At nanoscale level, the molecule–wall interactions become more frequent inducing a molecular self-diffusion known as Knudsen diffusion. Moreover, fluid molecules in nanoporous materials are subject to a higher degree of attraction forces from the surfaces making them vulnerable to rapid adsorption and desorption. The adsorbed layer itself can be mobile under some conditions and hence act as an additional mode of transport (Kou et al. 2016; Wasaki and Akkutlu 2015a).

Viscous flow, Knudsen diffusion, and surface diffusion are three transport mechanisms that coexist at the nanoscale. Describing the transport in the nanoporous materials requires building a flow model beyond Navier–Stokes equations to reflect the multiphysics nature of the transport (Malek and Coppens 2003; Riewchotisakul and Akkutlu 2015; Kou et al. 2016; Wang et al. 2017).

1. Viscous flow

Under a pressure gradient, viscous flux is observed. This mode of transport has applications in different fields of study. Viscous flow is modeled in fluid mechanics either by Navier–Stokes equations or their linearized forms such as Hagen–Poiseuille equations. In porous media, Darcy's law relates flow rate to the permeability, viscosity, and pressure gradient. Similar to other physical properties, the viscosity which relates the change in the shear stress to the shear velocity of the fluids is a function of temperature and pressure (i.e. $\mu = f(P, T)$). It can be measured experimentally or empirically.

Riewchotisakul and Akkutlu (2015) performed a molecular dynamics simulation of methane in a single nanopore. The viscosity of the methane in the bulk phase was found to behave independently of the degree of confinement. On the other hand, the viscosity of mixtures is sensitive to the length scale. Bui and Akkutlu (2017) studied a mixture of CH_4, C_2H_6, C_3H_8, C_4H_{10}, and C_5H_{12} in a single nanopore system under continuous depletion of pressure using Monte Carlo and molecular dynamics simulations. They showed that as the pressure decreases, lighter molecules such as methane and ethane escape the nanopore more easily compared to the heavier ones. That would result in compositional differences between molecules inside the pore and at the bulk phase. The measured viscosity inside the nanopore would be higher than that predicted using empirical correlations by a factor ranging from 2 to 5. Hence, an adequate description of fluid viscosity in nanoporous materials requires modeling the compositional change as a result of confinement at a given temperature and pressure.

2. Free molecular (Knudsen) diffusion

In the kinetic theory of gases, the mean free path l is defined as the average distance a molecule would travel before colliding with another molecule. The mean free path is proportional to temperature T and inversely proportional to the pressure P and the square of molecular diameter d^2. The mean free path is related to the aforementioned factors through Boltzmann constant K_B:

$$l = \frac{K_B T}{\sqrt{2}\pi d^2 P}. \tag{12.4}$$

Molecules flowing in nanoporous media would have a mean free path comparable to the length scale of that porous media. Hence, the probability of wall–molecule collisions is higher than intermolecular collisions inducing a transport mechanism known as free molecular (Knudsen) diffusion (Figure 12.1).

Knudsen number Kn is defined as the ratio of the mean free path l to the size of nanopore L. It serves as an indicator of the relative importance of Knudsen diffusion over pressure forces:

$$Kn = \frac{l}{L} = \frac{K_B T}{\sqrt{2}\pi d^2 P L}. \tag{12.5}$$

In general,

- $Kn < 0.01$ for continuum flow
- $0.1 > Kn > 0.01$ for slip flow
- $10 > \mathrm{Kn} > 1$ for transition flow
- $Kn > 10$ for free molecular diffusion.

Knudsen diffusion can be related to the mass or molar gradient. Knudsen velocity is expressed mathematically as follows:

$$v_{Kn} = D_K \nabla n, \tag{12.6}$$

where J_{Kn} is the Knudsen velocity, D_K is the diffusivity coefficient, and n is the number of

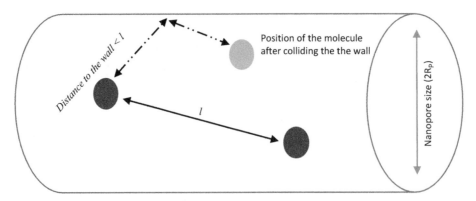

FIGURE 12.1   Molecule–wall interactions in a nanopore of size R_P where $l > R_P$ resulting in free molecular diffusion.

moles. For gases, the number of moles can be expressed in terms of pressure using the real gas law:

$$v_{Kn} = \frac{1}{P} D_K \frac{\partial P}{\partial x}. \tag{12.7}$$

In the presence of a multicomponent system, Knudsen diffusion is associated with a diffusion driven by the concentration gradient (i.e. *molecular diffusion*). An effective diffusivity coefficient of component A in the presence of another component is defined as

$$\frac{1}{D_{Ae}} = \frac{(1 - \sigma y_A)}{D_{AB}} + \frac{1}{D_{KA}}, \tag{12.8}$$

where y_A is the mole fraction of component A, D_{AB} is the molecular diffusivity coefficient of A in the presence of B, and D_{KA} is the Knudsen diffusivity coefficient of component A.

3. Surface diffusions

The interactions between fluid particles and the surfaces of the nanoporous materials can cause adsorption. If the adsorbed layer is subject to mass gradient, molecules can diffuse by the surfaces. That would be an additional transport mechanism known as surface diffusion. This phenomenon has been observed and reported in the literature; see, for example, Oura et al. (2003), Riewchotisakul and Akkutlu (2015), and Feng and Akkutlu (2015). The adsorbed layer concentration C_μ can be linked to the maximum adsorption capacity $C_{\mu s}$ and Langmuir pressure P_L of a given surface using Langmuir isotherm:

$$C_\mu = C_{\mu s} \frac{P}{P + P_L}. \tag{12.9}$$

In porous media, a pressure gradient exists across the direction of the flow. This variation of the pressure would impose concentration gradient along the adsorbed layer. The general equation governing the surface diffusion is given as follows:

$$J_s = D_s \frac{\partial C_\mu}{\partial x}, \tag{12.10}$$

where J_s is the surface diffusion flux. Substituting the Langmuir isotherm and expressing flux in velocity for a single nanopore using the density of the adsorbed layer ρ_s,

$$v_s = \frac{D_s C_{\mu s} P_L}{\rho_s (P + P_L)^2} \frac{\partial P}{\partial x}. \tag{12.11}$$

The surface diffusion involves multiple mechanisms which subjects estimating the diffusivity coefficients to high degree of uncertainty (Oura et al. 2003). For example, the measured surface diffusivity coefficients for shales range between $10^{-12} \text{m}^2/\text{s}$ and $1 \text{ m}^2/\text{s}$ (Etminan et al. 2014). Surface diffusion mechanisms include movement of single molecules to adjacent sites and cluster diffusion where grouped molecules move together through glide and shear diffusion (Kou et al. 2017) (Figure 12.2).

Molecular dynamics can be a handy tool for the assessment of diffusivity coefficient. For instance, the surface diffusion of methane molecules in organic nanoporous materials of shale formations was studied by Riewchotisakul and Akkutlu (2015) using MD simulation of methane molecules in graphite nanopores at some range of pressure and temperature. They found the velocity of the adsorbed molecules linearly related to pressure gradient:

$$v_s = C_{sv} \frac{\partial P}{\partial x}, \tag{12.12}$$

where C_{sv} is the cluster diffusivity coefficient of methane in nanoporous graphite and is equal to $5 \times 10^{-4} \text{ nm}^2/\text{psi ps}$.

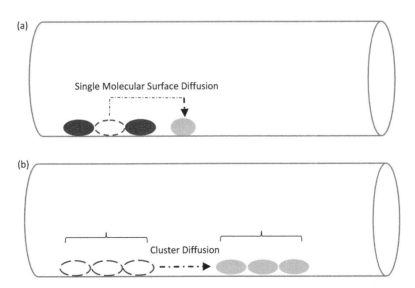

FIGURE 12.2 Schematic of the two mechanisms of surface diffusion. (a) Surface diffusion of single molecules. (b) Cluster diffusion.

4. Multiphysics transport in a single capillary

The aforementioned transport mechanisms can be combined in one equation to describe the total transport velocity. For methane transport in a single nanopore of graphite, the total velocity v_t is given as (Kou et al. 2017)

$$v_t = \left[\frac{R_a^2}{8\mu} + \frac{D}{P} + \frac{\rho_{ads}}{\rho_{bulk}} C_{sv,\,smooth} \frac{\left(R_{tube}^2 - R_a^2\right)}{R_a^2} \right] \frac{dP}{dx}.$$

$$(12.13)$$

The first term in the right-hand side represents viscous flow. The second and third terms represent the Knudsen and the cluster diffusions, respectively. In some condition, one or more mechanisms may prevail and dictate the overall transport.

12.3 Macroscopic Transport Model

Permeability, which is a macroscopic parameter of porous materials, is an intrinsic property that is only dependent on the pore sizes, their interconnectivity, and distribution. For nanoporous materials, the permeability is found to be pressure dependent and deviates from its intrinsic value (Alafnan and Akkutlu 2017). The single capillary equation can give insight to fluid mechanics microscopically. However, a macroscopic quantity such as a modified version of permeability would be more useful to facilitate the continuum modeling of nanoporous materials.

The concept of digital porous media or pore network modeling (PNM), which is simply a network of interconnected capillaries at nodes, can be used to upscale that single tube equation (Blunt 2001). The pore network model was first introduced by Fatt in 1956 to relate microscopic properties to meaningful macroscopic coefficients. In this approach, pores are used as means of fluid storage, and they are interconnected by capillaries which serve as means of fluid transport. Pores and capillaries could have different sizes and shapes to digitally replicate certain cases of porous media (Joekar-Niasar and Hassanizadeh 2012). The mass balance is written at each pore (Alafnan and Akkutlu 2017):

$$\sum_1^n \rho_{fi} A_i v_i + \frac{d(\rho_f V_P)}{dt} = 0.$$

$$(12.14)$$

The first term represents the fluid fluxes from n number of connected capillaries, and the second term is the accumulation of the mass of a fluid with density p_f inside a pore

of volume V_P. The velocity of fluid v_i in a capillary i is given by Hagen–Poiseuille equation or a modified one such as equation 12.13 for multiphysics transport.

12.4 Transport of Natural Gas in the Nanoporous Media of Shale Formations

An example of naturally occurring nanoporous media is the shale matrix where large quantities of hydrocarbons are stored in the remains of organic matter. This organic matter is rich in nanopores. Under a pressure gradient, hydrocarbons are produced from the organic nanopores to the wellbore in a multimechanism transport process. The transport phenomena in these nanopores have been studied intensively in the literature (Javadpour 2009; Ambrose et al. 2012; Wang and Reed 2009; Kang et al. 2011; Sakhaee-Pour and Bryant 2012; Wasaki and Akkutlu 2015; Kou et al. 2017; Alafnan and Akkutlu 2017). Equation 12.13 is developed to describe the viscous-diffusive transport of natural gas in a single nanopore. If we use Eq. 12.13 in 12.14 for v_i at each pore, the final equation would be as follows:

$$aP_m^2 + bP_m + cP_S^2 + dP_N^2 + eP_{NE}^2 + fP_{NW}^2 + gP_{SE}^2 + hP_{SW}^2$$
$$+ iP_{UN}^2 + jP_{DS}^2 + kP_{UNE}^2 + lP_{UNW}^2 + \ldots + c1 = 0,$$

$$(12.15)$$

where each of a, b, c, e … represents a group of pressure dependent parameters such as gas density, viscosity, and diffusivity coefficients. The pressure variables P_N, P_S, P_{NE} … are the pressures of the adjacent nodes (pores) (Table 12.1).

The system of equations can be represented in the matrix and then solved at each time step using Newton–Raphson method as

$$\boldsymbol{P}_{t+1} = \boldsymbol{P}_t + \boldsymbol{J}^{-1} f(\boldsymbol{P}_t),$$

$$(12.16)$$

where $\boldsymbol{P}_{t+1}$ is the pressure matrix at the new time step to be solved from the previous time step $\boldsymbol{P}_t$. $\boldsymbol{J}$ is the Jacobian matrix which is a matrix of the first-order partial derivatives of $\boldsymbol{f}(\boldsymbol{P}_t)$. Using the information of the pressure at the nodes, the total flux rate q is computed. This approach can be useful in obtaining the upscaled transport coefficient (i.e. *apparent permeability*).

Figure 12.3 shows a 3-D pore network model created with the sizes of pores and throats randomly sampled from predefined distributions of typical shale formation. Inlet and outlet pressure values are specified to impose a transport

TABLE 12.1 Description of the Pressure Variables in the 3-D Interconnectivity of PNM

Pressure Variable	Description	Pressure Variable	Description
P_N	P of the pore to the north in the same layer	P_{US}	P of the pore to the south in the layer above
P_S	P of the pore to the south in same layer	P_{USE}	P of the pore to the south east in the layer above
P_{NE}	P of the pore to the north east in the same layer	P_{USW}	P of the pore to the south west in the layer above
P_{NW}	P of the pore to the north west in the same layer	P_{DN}	P of the pore to the north in the layer below
P_{SW}	P of the pore to the south west in the same layer	P_{DNE}	P of the pore to the north east in the layer below
P_{SE}	P of the pore to the south east in the same layer	P_{DNW}	P of the pore to the north west in the layer below
P_{UN}	P of the pore to the north in the layer above	P_{DS}	P of the pore to the south in the layer below
P_{UNE}	P of the pore to the north east in the layer above	P_{DSE}	P of the pore to the south east in the layer below
P_{UNW}	P of the pore to the north west in the layer above	P_{DSW}	P of the pore to the south west in the layer below

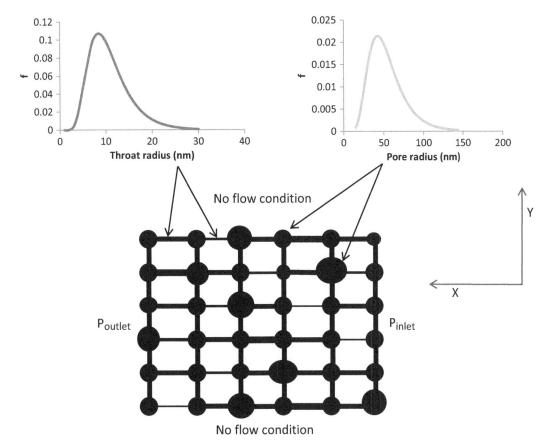

FIGURE 12.3   Schematic of the pore network model representing the organic matter of shale formations. The design parameters are given in Table 12.2.

TABLE 12.2   Design Parameters for the Nanoporous Organic Matter of Shale Formations

Pore Network Properties	
R_{throats}, nm (average)	8
R_{pores}, nm (average)	40
c_{om}, psi^{-1}	6×10^{-6}
Adsorption Parameters and Gas Properties	
P_L, psi	1,500
V_s, ft^3/ton^1	100
D_k, nm^2/s	$f_1(r_p, T)$
D_s, nm^2/psi-s	5×10^{-12}
Viscosity, poise-s	$f_2(P, T)$
Z	$f_3(P, T)$
Boundary and Initial Conditions	
P_i, psi	250–5,000
$\Delta P_{\text{outlet}-\text{inlet}}$, psi	10
Temperature, R	745

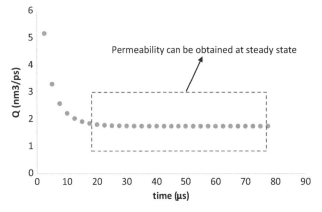

FIGURE 12.4   The rate of outflow of natural gas from the PNM. Steady state is reached after nine time steps for $P_{\text{avg}} = 250$ psi (i.e. the average pressure is maintained by fixing the inlet pressure to 245 psi and the outlet to 255 psi). Similar graphs are generated for each P_{avg} to identify the steady-state time at which the permeability can be obtained.

driving force. The idea is to mimic Darcy's experiments using a digital copy of nanoporous media.

At the steady state where the rate of outflow of natural gas is constant, the apparent permeability at the average pressure is obtained (i.e. k_{app} at $P_{\text{avg}} = \frac{P_{\text{inlet}} + P_{\text{outlet}}}{2}$):

$$k_{\text{app}} = \frac{\mu Q}{A \frac{\partial P}{\partial x}}. \tag{12.17}$$

The same procedure is repeated at different average pressures to establish the relationship between apparent permeability and the pressure as shown in Figures 12.4 and 12.5.

The previous approach can be applied to transport phenomena in other naturally or artificially occurring nanoporous media to obtain useful parameters such as mass transfer coefficients, chemical reaction equilibrium constants, and other macroscopic quantities.

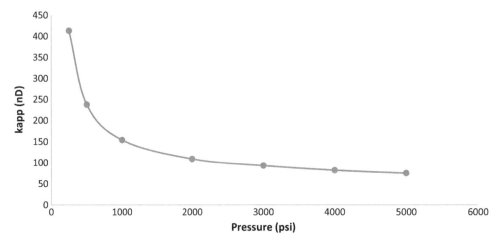

FIGURE 12.5   The permeability of the nanoporous organic rocks is sensitive to the pressure. The term apparent permeability k_{app} is used to reflect its dependency on the pressure.

References

Alafnan, S.F.K. and I.Y. Akkutlu. Matrix-fracture interactions during flow in organic nanoporous materials under loading. *Transp Porous Med* (2017).

Ambrose, R.J., Hartman, R.C., Diaz-Campos, M., Akkutlu, I.Y., Sondergeld, C.H. Shale gas in-place calculations part I—new pore-scale considerations. *SPE J.* 17(1): 219–229 (2012)

Blunt, M.J. Flow in porous media–pore-network models and multiphase flow. *Curr. Opin. Colloid Interface Sci.* **6**: 197–207 (2001). doi:10.1016/S1359-0294(01)00084-X

Bui, K., Akkutlu, I.Y. *Hydrocarbons Recovery from Model-Kerogen Nanopores.* Society of Petroleum Engineers (2017). doi:10.2118/185162-PA

Etminan, S.R., Javadpour, F., Maini, B.B. et al. Measurement of gas storage processes in shale and of the molecular diffusion coefficient in kerogen. *Int. J. Coal Geol.* **123**: 10–19 (2014).

Feng, F. and Akkutlu, I.Y. (2015) Flow of Hydrocarbons in Nanocapillary: A Non-Equilibrium Molecular Dynamics Study. SPE-177005, paper presented at the SPE Asia Pacific Unconventional Resources and Exhibition held in Brisbane, Australia, November 9–11.

Javadpour, F. (2009) *Nanopores and Apparent Permeability of Gas Flow in Mudrocks (Shales and Siltstone).* Petroleum Society of Canada. doi:10.2118/09-08-16-DA

Joekar-Niasar, V., Hassanizadeh, S.M. Analysis of fundamentals of two-phase flow in porous media using dynamic pore-network models: A review [electronic resource]. *Crit. Rev. Environ. Sci. Technol.* **42**(18), 1895–1976 (2012)

Kang, S., Fathi, E., Ambrose, R.J., Akkutlu, I.Y., Sigal, R.F. Carbon dioxide storage capacity of organic-rich shales. *SPE J.* **16**(4), 842–855 (2011)

Kou, R., Alafnan, S., & Akkutlu, I.Y. (2016) Multi-scale analysis of gas transport mechanisms in Kerogen. *Transp Porous Med*, 1–27. doi:10.1007/s11242-016-0787-7

Malek, K., Coppens, M.O. Knudsen self- and Fickian diffusion in rough nanoporous media. *J. Chem. Phys.* **119**(5): 2801–2811 (2003)

Oura, K., Lifshits, V.G., Saranin, A.A., Zotov, A.V., Katayama, M. *Surface Science: An Introduction*, pp. 325–340. Springer, Berlin (2003). ISBN 3-540-00545-5

Riewchotisakul, S., Akkutlu, I.Y. Adsorption-enhanced transport of hydrocarbons in nanometer-scale organic pores. *SPE-175107*, Paper Presented During the SPE Annual Technical Conference and Exhibition in Houston, Texas, September 28–30 (2015)

Riewchotisakul, S. and Akkutlu, I.Y. 2016. Adsorption enhanced transport of hydrocarbons in organic nanopores. *SPE Journal*, 21 (6): 1960–1969.

Sakhaee-Pour, A., Bryant, S.L. Gas permeability of shale. *SPE Reserv. Eval. Eng.* **15**(4): 401–409 (2012)

Wang, F.P., Reed, R.M. Pore networks and fluid flow in gas shales. *Paper SPE 124253* Presented at the SPE Annual Technical Conference and Exhibition, New Orleans, 4–7 October (2009)

Wang, J., Yuan, Q., Dong, M., Cai, J., Yu, L. Experimental investigation of gas mass transport and diffusion coefficients in porous media with nanopores. *Int. J. Heat Mass Transfer* **115**, **Part B**: 566–579 (2017)

Wasaki, A., Akkutlu, I.Y. Dynamics of fracture-matrix coupling during shale gas production: Pore compressibility and molecular transport effects. *SPE 170830 Society of Petroleum Engineers* (2015b). doi:10.2118/175033-MS

Wasaki, A., Akkutlu, I.Y. Permeability of organic-rich shale. *SPE J.* **20**(6): 1384–1396 (2015a)

Beyond Phenomena: Functionalization of Nanofluidics Based on Nano-in-Nano Integration Technology

Yan Xu
Osaka Prefecture University
Japan Science and Technology Agency (JST)

13.1 Introduction

Nanofluidics is the science and technology which deal with the behaviors and applications of fluids confined in and/or around geometries with nanoscale characteristic dimensions. Some issues associated with nanoscale fluidics have been occasionally dealt with by researchers in membrane science, colloid science, and chemical engineering for many decades. Nonetheless, nanofluidics has never attracted as much attention as it does now, owing to advances in nanofabrication, which give a boost to the recent growth of nanofluidics (Xu 2018; Haywood et al. 2015; Mawatari et al. 2014; Sparreboom et al. 2009). The increased availability of nanofabrication methods over the past two decades has allowed the lithographic fabrication of well-defined nanofluidic geometries such as nanochannels (or nanofluidic channels) to form novel nanofluidic devices (Figure 13.1) (Chen & Zhang 2018; Duan et al. 2013; Mijatovic et al. 2005). Such devices, usually solid-state, planar, and transparent, are compatible with some available tools used in chemistry, physics, biology, and engineering. Such devices therefore provide accessible experimental platforms allowing the study of nanofluidics by researchers with different backgrounds, and in turn new tools and methodologies are increasingly being developed.

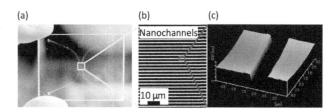

FIGURE 13.1 (a) A picture of a standard nanofluidic device having lithographically fabricated nanochannels. (b) A scanning electron microscopy (SEM) image of the nanochannels. (c) An atomic force microscopy (AFM) image of a single nanochannel.

As a result, exploration and application of nanoscale fluid behaviors and related phenomena in a controllable, reproducible, and predictable way gradually become possible (Napoli et al. 2010; Xu, Jang et al. 2012; Prakash & Conlisk 2016; Lin et al. 2018; Harms et al. 2015a). In this chapter, the topics are mainly discussed in the context of nanofluidics confined in such lithographically nanofabricated devices with well-defined nanofluidic environments (herein referred to as "nanofluidic devices").

Nanofluidics is usually considered to be a realm evolved from microfluidics but is obviously not just an extension of microfluidics. New physical phenomena and mechanisms, which are not observed at microscales or in bulk, start to emerge and dominate at nanoscales, opening up a virgin research territory to explore new scientific insights and applications of fluids. The representative phenomena and effects that have been unveiled include nonlinear transport (such as concentration polarization (Kim et al. 2007; Kim et al. 2010), ion-current rectification (Vlassiouk & Siwy 2007; Perry et al. 2010)), and altered liquid properties of water (such as lower dielectric constant (Hibara et al. 2002; Xu & Xu 2015), higher viscosity (Tas et al. 2004; Hibara et al. 2002; Xu & Xu 2015), and higher proton mobility (Liu et al. 2005; Tsukahara et al. 2007), than in the bulk), mostly stemming from ultra-high surface-area-to-volume ratios, surface charge, and electric double layer overlap exhibited in nanochannels (Figure 13.2). The discussions of these phenomena, as shown in Figure 13.2, mainly focus on a range of length scales between 1 nm and 100 nm, which are the typical scales of nanofluidics (herein referred to as "classical nanofluidics") (Sparreboom et al. 2009; Bocquet & Tabeling 2014). In many occasions, the discussions even extend to scales over 100 nm to several 100 nm, which falls in a range of scales sometimes called "extended nanofluidics" (10–1,000 nm; Figure 13.2) (Tsukahara et al. 2010). From a perspective of application potential, the length scales of the so-called extended nanofluidics embrace a much wider range of nanometric objects, for example, nanoparticles, DNAs, viruses, exosomes, lipid vesicles, and polymers (Figure 13.2), which are of great interest in chemistry, biology, materials science, physics, energy, bioengineering, drug discovery, and

clinical medicine. In addition, even in the cases of handling such nanometric objects in extended nanofluidic environments with features larger than 100 nm (i.e., beyond the range of classical nanofluidics), the length scales of interspaces/fluids among the nanometric objects or between the channel walls and the nanometric objects often lie in the range of typical scales of classical nanofluidics (i.e., 1–100 nm). In such cases, physics and principles of classical nanofluidics can be of course applied or further explored. From these views, therefore, "nanofluidics" discussed in this chapter covers both regimes of "classical nanofluidics" and "extended nanofluidics" (Figure 13.2).

In the past decade, nanofluidic transport phenomena and effects have been intensively investigated, and a few striking applications (for example, nanofluidic diodes or transistors (Kalman et al. 2008; Guan et al. 2011; Zheng et al. 2017) and resistive-pulse sensing (Piruska et al. 2010; Zhou et al. 2011; Harms et al. 2015b)) have been demonstrated. Nevertheless, nanofluidics is still in its young stage, especially in comparison with its booming "brother" microfluidics. Because nanochannels are ultra-small closed-space systems, most standard tools widely used in microfluidics as well as in nanosciences for open planar systems are difficult to directly apply to the field of nanofluidics. As a result, the functionalization of nanofluidic channels is still one of most challenging issues toward the development of nanofluidics. For open planar systems, many physical, chemical, and mechanical methods have been developed to functionalize the systems at zero, one, two, or three dimensions in the nanometer

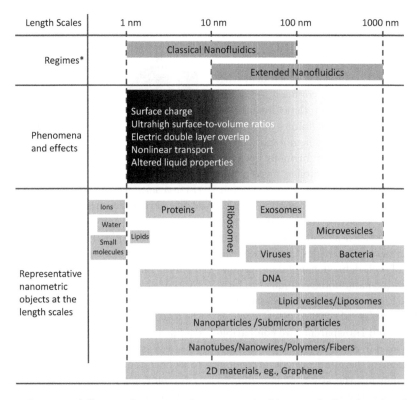

FIGURE 13.2   Regimes, phenomena/effects, and representative nanometric objects at the length scales of nanofluidics. (Reproduced with permission (Xu 2018). Copyright 2017, Wiley-VCH.)

range (Park et al. 2012; Gates et al. 2005; Ostrikov 2005; Tseng et al. 2003; Whitesides 2002; Haynes & Van Duyne 2001). Unfortunately, for small closed-space systems like nanofluidic channels, most of these methods face obstacles in either principles or practices, due to the physical barrier resulting from the "closed" feature of these systems. The lack of methodologies and technologies for the functionalization of nanofluidics not only hinders the involvement of much more researchers in many other fields to work with nanofluidics, but also impedes the transition of nanofluidics research from the current physics-centered stage to the next application-oriented stage.

Recently, a methodology called "nano-in-nano integration" has been developed to conquer the challenge (Xu, Matsumoto et al. 2015; Xu & Matsumoto 2015). The nano-in-nano integration technology based on the methodology has been established by taking advantage of two key techniques previously developed. One is a technique capable of non-damage bonding of nanofluidic devices under mild conditions (for example, conditions at low temperature (Xu, Wang et al. 2012; Xu, Matsumoto et al. 2015) or even room temperature (Xu et al. 2013)). Another is a technique capable of the placement control with nanometer-scale accuracy permitting multiple-step nanofabrication (Xu, Matsumoto et al. 2015). As a powerful technology of functionalization of nanofluidics, the nano-in-nano integration technology allows the integration of a variety of functional (for example, fluidic, electrical, optical, thermal, magnetic, chemical, and biological) components in tiny nanochannels. In this chapter, these two key techniques supporting the nano-in-nano integration are first described, and then the details of the nano-in-nano integration technology are addressed, followed by a review of related techniques and some represented applications using the nano-in-nano integration.

13.2 Low-Temperature Bonding of Nanofluidic Devices

Nanochannels (or nanofluidic channels) are the core components of most nanofluidic devices (Figure 13.1). The methods for the fabrication of nanochannels vary depending on the substrate materials. Silicon was used to fabricate nanochannels in the early time via the direct transfer of nanofabrication technologies well established in the field of microelectronics (Han & Craighead 1999; Haneveld et al. 2003; Perry & Kandlikar 2006). In the past years, the fabrication of nanochannels in two-dimensional nanoscales with high resolution, reproducibility, and flexibility on glass (fused-silica glass, in general) substrates has been developed. Afterward, glass became a major substrate material for the fabrication of nanofluidic devices, because of its excellent properties which are favorable for chemical and biological studies and applications. These excellent properties include superior optical transparency, thermal stability, chemical/biological inertness, mechanical

robustness, and hydrophilic nature which are favorable for liquid introduction.

One of the main impediments to the widespread use of nanofluidic devices for various applications, however, is the lack of feasible approaches to non-damage bonding of glass nanofluidic devices under mild conditions. In standard practice, microfluidic/nanofluidic chip devices are fabricated by bonding of upper and lower substrates containing micro-/nanofluidic channels and other structures. Bonding is the last and most critical step in manufacture of chip devices. Practically, most failures of chip manufacture result from the bonding step, especially in the case of glass chip manufacture. The bonding between glass substrates has been achieved using anodic bonding performed around 300°C (Fonslow & Bowser 2005; Queste et al. 2010), wet bonding based on surface pretreatment using hydrofluoric acid solution that is a wet etchant of glass (Xu et al. 2007; Chen et al. 2006), and fusion bonding through direct contact performed in vacuum at extremely high temperature, for example, around 1,000°C for fused-silica glass (Mellors et al. 2008; Xu et al. 2010). Because an additional interlayer between glass substrates is required in anodic bonding and the wet-etching effect results in possible damage to nanostructures in the wet-bonding process, both anodic bonding and wet bonding are considered to be unsuitable for application in the bonding of glass chips with fine nanostructures. Fusion bonding has been widely used in the bonding of glass nanofluidic chips (Tsukahara et al. 2008). Nevertheless, various functional materials usually cannot tolerate high temperatures (~1,000°C) or vacuum required in fusion bonding. This makes it nearly impossible to integrate electrical, optical, thermal, chemical, and biological units into the glass nanofluidic chip, consequently impeding practical applications of nanofluidic chips. Therefore, the glass bonding has become a technical bottleneck to broaden the applications of nanofluidic chips. It is greatly desired that the bonding of glass nanofluidic chips could be realized at low temperature or even much ideally at room temperature in ambient air without using harsh chemical solutions or vacuum.

In semiconductor industry, surface plasma activation methods such as reactive ion etching (RIE), inductively coupled plasma (ICP), and UV radiation plasma carried out in different atmospheres (argon, oxygen, hydrogen, and nitrogen) are conducted to enhance the surface attraction and bonding process of semiconductor wafers (Si, Ge, and GaAs) (Galchev et al. 2011; Visser et al. 2002; Zucker et al. 1993). Among them, a two-step surface activation process called sequential plasma surface activation (Suga et al. 2004; Howlader et al. 2006; He et al. 2018) is a famous one. In the process, wafer or substrate surfaces are treated with an oxygen RIE plasma, followed by a nitrogen microwave (MW) radical activation, as shown in Figure 13.3. The O_2 RIE plasma removes contaminants and reactivates native oxides on surfaces. Subsequent processing with N_2 radicals is considered to further generate chemically reactive surfaces. Strong bonding of semiconductor wafers (Si/Si, Si/Ge, and

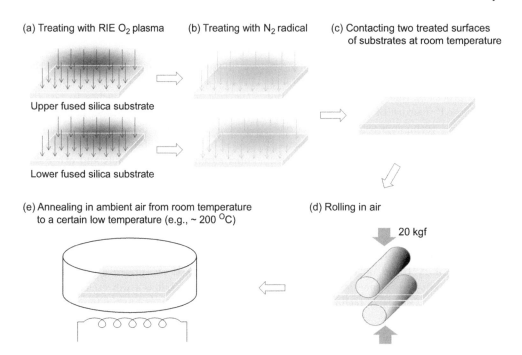

(a) Treating with RIE O$_2$ plasma   (b) Treating with N$_2$ radical   (c) Contacting two treated surfaces of substrates at room temperature

Upper fused silica substrate

Lower fused silica substrate

(e) Annealing in ambient air from room temperature to a certain low temperature (e.g., ~ 200 $^\circ$C)   (d) Rolling in air

20 kgf

FIGURE 13.3 A schematic drawing of the process of low-temperature glass bonding using a two-step plasma surface activation process. (Reproduced with permission (Xu, Wang et al. 2012). Copyright 2012, Springer.)

SiO$_2$/Ge) at low temperature or even room temperature has been accomplished owing to the combined effect of oxygen plasma and N$_2$ radicals (Wang et al. 2008; Howlader et al. 2009). Based on these preliminary studies, the methods are considered to have potential to be applied for the bonding of glass substrates and glass nanofluidic chip devices at low temperature (Howlader et al. 2011).

Direct bonding of fused-silica glass nanofluidic chip at low temperature in ambient air using the sequential plasma surface activation process was performed and evaluated (Xu, Wang et al. 2012). The bonding processes are shown in Figure 13.3. To get void-free and strong bonding interface at low temperature, the parameters of two-step plasma activation were optimized carefully. Generally, a short term of O$_2$ RIE plasma treatment less than 10 s is enough to remove surface contaminants (Wang et al. 2008; Howlader et al. 2009). Empirically speaking, the higher the surface hydrophilicity, the easier the wafer bonding is. A longer term of O$_2$ RIE plasma treatment could further improve the surface hydrophilicity, but it may increase the surface roughness and introduce nanodefects on fabricated nanostructures. In order to achieve a non-destructive surface with high surface hydrophilicity, subsequent N$_2$ radicals without bombardment energy were chosen for activation instead of a long-term O$_2$ RIE plasma treatment. The treatment of O$_2$ RIE plasma for 15 s followed by N$_2$ radicals for 30 s was elucidated to be optimal, because the process could remove surface contaminants sufficiently, reduce the surface roughness (Ra < 0.25 nm), and retain the relatively high hydrophilicity (water contact angle around 30°) of surfaces. The effect of plasma discharge power was investigated, and as a result, the optimized discharge powers were fixed at 50 W and 2,500 W for O$_2$ RIE plasma and N$_2$ radicals,

respectively. In addition, further study revealed that the process did not damage the nanostructures even in the case of the nanochannels being extremely shallow (Xu, Wang et al. 2012).

The bonding energy was evaluated by using a crack-opening test (Maszara et al. 1988), which is a standard test measuring the bonding energy. As shown in Figure 13.4, the bonding energy of fused-silica substrates bonded at different low temperatures increases with the increase of temperature from the room temperature (25°C) to 230°C. Although stronger bonding was also achieved at 300°C, 230°C would be an optimal temperature to achieve effective

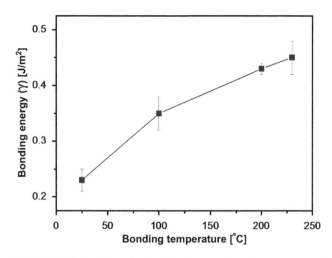

FIGURE 13.4 The relationship between bonding energy and bonding temperature in low-temperature bonding of fused-silica glass substrates. (Reproduced with permission (Xu, Wang et al. 2012). Copyright 2012, Springer.)

bonding, taking account of the bonding energy, voids, and the bonding temperature. The bonding strength at 230°C was further measured by a tensile pulling test (Vallin et al. 2005), which is another standard test measuring bonding strength. The test revealed that the bonding strength was at least 11 MPa, which is strong enough for withstanding some subsequent mechanical processes.

The optimized low-temperature bonding process was applied to bonding of standard nanofluidic devices (Figure 13.5a) with two side microfluidic channels bridged with 50 parallel nanofluidic channels (200 nm in depth, 698 nm in width, 2.4 mm in length, and spaced by 4.12 μm). The bonding was evaluated by introducing a solution of sulforhodamine B (50 μM) driven by air pressure from one side microfluidic channel to nanofluidic channels (Figure 13.5b). The real-time fluorescence imaging revealed that the liquid was successfully introduced into the nanofluidic channels. No fluorescence was observed in bonding areas between nanofluidic channels during long-term continuous operation at 450 kPa (Figure 13.5c), suggesting that no leakage occurred even at the high pressure, which can satisfy most nanofluidic chip operations.

Therefore, the bonding technique using the two-step plasma surface activation process enabled the bonding of nanofluidic devices at low temperature (optimized to around 200°C) under non-vacuum condition, with high bonding strength, non-destructive features, and non-leakage under high-pressure nanofluidic conditions. In contrast to the conventional glass fusion bonding, the low-temperature bonding allows the integration of a wide range of functional components during the manufacture of nanofluidic chips.

13.3 Room-Temperature Bonding of Nanofluidic Devices

As described in Section 13.2, the experiments of the low-temperature bonding suggest that the sequential surface activation comprising treatments with different plasma can be an effective strategy to reduce the bonding temperature of glass substrates. Hence, it was hypothesized that the equivalent or improved effect of surface activation could be obtained using a one-step plasma surface treatment composed of multiple gas sources. Several multiple gas source compositions were investigated, and an O_2/CF_4 gas mixture plasma treatment was found to possess high capability to activate surfaces of semiconductor wafers such as silicon (Wang & Suga 2012). This surface activation process was applied to bonding of glass nanofluidic devices (Xu et al. 2013). The one-step surface activation process using an optimized O_2/CF_4 gas mixture plasma treatment enabled the further decrease of the temperature for fused-silica glass bonding to room temperature (~25°C), and thereby the bonding of fused-silica glass nanofluidic chips was first achieved at room temperature (Xu et al. 2013).

The bonding energy of the bonded glass substrates was investigated with varying the CF_4 flow rate (Figure 13.6). In the case of fluorine atomic concentration at 0% which refers to using conventional O_2 plasma treatment, the bonding energy of the bonded glass substrates was (0.85 ± 0.20) J/m^2. When a CF_4 gas flow with a very small flow rate was introduced, the bonding energy was effectively enhanced. The maximum bonding energy was achieved when the atomic concentration of fluorine was 0.2%. In this case, the bonding energy was (1.12 ± 0.10) J/m^2, a sufficiently high bonding energy $(>1.0 \ J/m^2)$ meeting the requirement for withstanding various subsequent harsh mechanical processes such as grinding and polishing (Ma et al. 2007). In addition, a further increase in the atomic concentration of fluorine did not result in further improvement in bonding energy. Conversely, it led to decreases in bonding energy, as shown in Figure 13.6. Therefore, 0.2% is considered to be an optimal atomic concentration of fluorine in O_2/CF_4 plasma gas for glass bonding at room temperature under the experimental conditions. Moreover, the process is very friendly to fine nanostructures, according to a morphology evaluation,

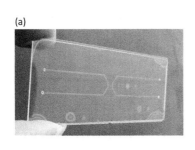

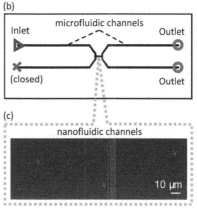

FIGURE 13.5 Evaluation of (a) a low-temperature (230°C) bonded nanofluidic chip under nanofluidic conditions by (b) continuously introducing a fluorescent dye solution (sulforhodamine B, 50 μM) into the central nanofluidic channels from the inlet on the left-side microfluidic channel at 450 kPa. (c) A fluorescence image of the nanofluidic channels after 100-min introduction of the fluorescent dye solution shows no leakage occurred. (Reproduced with permission (Xu, Wang et al. 2012). Copyright 2012, Springer.)

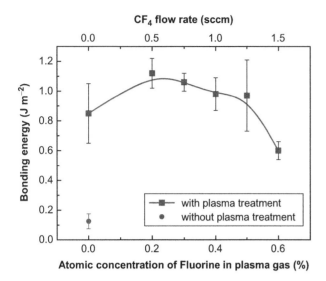

FIGURE 13.6 The relationship between bonding energy and atomic concentration of fluorine in plasma gas (or CF_4 flow rate) in room-temperature (25°C) bonding of fused silica glass substrates. (Reproduced with permission (Xu et al. 2013). Copyright 2013, The Royal Society of Chemistry.)

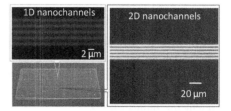

FIGURE 13.7 Evaluation of a nanofluidic chip with 1D and 2D nanochannels bonded at room temperature (25°C) under nanofluidic conditions by continuously introducing a fluorescent dye solution (sulforhodamine B, 50 μM) at 2.5 MPa. Fluorescence images show no leakage occurred in both 1D and 2D nanochannel areas during the liquid introduction. (Reproduced with permission (Xu et al. 2013). Copyright 2013, The Royal Society of Chemistry.)

which revealed the O_2/CF_4 plasma treatment did not bring about detectable damage on the nanostructures (Xu et al. 2013). Hence, the process is considered to be appropriate to the bonding of glass nanofluidic chips.

The surface characterization for hydrophilicity was further investigated by the water contact angle measurements. The contact angle on glass surface was smaller than 3° after it was treated by O_2 plasma. On the other hand, the contact angle of glass surfaces treated by O_2/CF_4 plasma (in the case of the atomic concentration of fluorine = 0.2%) was about 33°. The result implies that the glass surfaces treated by O_2/CF_4 plasma seem less hydrophilic than those treated by O_2 plasma, resulting in fewer water molecules adsorbed on surfaces. It is speculated that the CF_4 gas in plasma treatment could optimize the amount of water molecules across the bonding interface which is considered to be important in a bonding process (Xu, Wang et al. 2012), providing enhanced bonding even at room temperature. Nevertheless, the mechanism should not only be complicated but also be diverse, and further investigation from the perspectives of surface chemistry and physics is necessary for future studies and would greatly help to elucidate the mechanism.

Figure 13.7 shows a picture and evaluation results of a glass nanofluidic device bonded under the room-temperature bonding under the optimal conditions. The chip device contains two sets of micro-/nanofluidic channel hybrid. While the nanofluidic channels in the upper set are typical 1D nanochannels (10.4 μm in width, 400 nm in depth, 600 μm in length, and spaced by 10.0 μm) with only depth in nanometer scale, the nanofluidic channels in the lower set are typical 2D nanochannels (320 nm in width, 400 nm in depth, 600 μm in length, and spaced by 400 nm) with both width and depth in nanometer scale (Figure 13.7). 1D and 2D nanochannels have been widely

used in nanofluidics. Figure 13.7 demonstrates fluorescence images of 1D nanochannels and 2D nanochannels, respectively, when a solution of sulforhodamine B (50 μM) was introduced at a high pressure of 2.5 MPa. The pressure is not only much higher than the pressures used in common nanofluidic applications (usually several hundred kilopascals (kPa) at most) but also higher than the high pressure level required in some special applications such as high-pressure nanochromatography (pressures over one megapascal (MPa) are generally necessary). Fluorescence was observed in both the 1D and the 2D nanochannels, indicating that the introduction of the liquid to the nanochannels was achieved. No fluorescence was observed in the non-channel areas, revealing that no detectable leakage occurred during the continuous operation at high pressure. It is worth noting that fluorescence was also not observed in narrow bonded areas such as the spaces between neighboring 2D nanochannels, which were specially designed to be as narrow as only 400 nm in width. These results reveal that the bonding was quite strong even at a nanometer-scale level. Therefore, the room-temperature bonding has the capability to endure even high-pressure nanofluidic conditions.

To be brief, a strong, fine nanostructure-friendly, and high-pressure-resistant bonding was achieved under a very mild room-temperature condition using the one-step surface activation process based on O_2/CF_4 plasma treatment. The use of the room-temperature bonding allows for overcoming the technical bottleneck resulting from the bonding process in the nanofluidic field and makes it possible to integrate a variety of electrical, optical, thermal, chemical, and biological units into the glass nanofluidic chips before bonding.

13.4 Nano-in-Nano Integration Technology

13.4.1 Concept of Nano-in-Nano Integration

As the space size decreases to the micro-/nanometer scale, the surface-area-to-volume ratio increases significantly. As a result, the properties of the internal surface of a small closed space dominate a variety of phenomena and processes in the space. Hence, fabrication, modification, and control of the local internal surface are indispensable strategies for the functionalization of small closed-space systems. Surface patterning is a basic way to fabricate, modify, and control the properties of the local surface of a material. In the field of microfluidics, photo-patterning is a predominant method to pattern the internal surface of a closed microfluidic channel (Priest 2010; Jang et al. 2010), in addition to few other patterning methods (Delamarche et al. 2005). On the one hand, the method takes advantage of the nature of light (commonly, UV light); light can pass through the transparent physical barrier of the substrate layer and thereby enables the patterning of various photo-reactive materials on the internal surface of the "closed" microfluidic channel, using a photo-mask. On the other hand, the patterning capability of the method is restricted by the nature of light, namely, diffraction limit, as well as the resolution limit of the photo-mask and other instrumental limitations; consequently, the feature sizes and the placement precision of the pattern are generally in the micrometer range and practically difficult to be further downscaled to the nanometer range. Obviously, nanopatterning of the internal surfaces of nanofluidic channels exceeds the capability of the photo-patterning. Therefore, in order to functionalize nanofluidics, methods capable of nanopatterning in nanofluidic channels have been required.

Unlike the *patterning-after-bonding* strategy of the photo-patterning in microfluidics, a different general methodology (Figure 13.8) was proposed to achieve nanopatterning in nanofluidic channels (Xu, Matsumoto et al. 2015). The method comprises a primary *nano-in-nano patterning before bonding* by utilizing top-down approaches and a secondary *molecule patterning after bonding* by utilizing bottom-up approaches. The method does not use UV light, and therefore the capability of the patterning is not restricted by the diffraction limit of light and the resolution limit of the photo-mask. The method enables the site-specific fabrication of a variety of nanopatterns with arbitrary shapes and sizes in single nanofluidic channels, nanofluidic channel-networks, and arrayed nanofluidic channels. Not only metallic but also molecular nanopatterns and nanoarrays can be formed in nanofluidic channels by utilizing the method.

In brief, the method (Figure 13.8) is described as follows. First, gold is pre-nanopatterned in open nanofluidic channels utilizing top-down processes of electron beam lithography, deposition, and lift-off which are guided by a high-precision placement control technique. Then, the gold-nanopatterned open nanofluidic channels are bonded without damage utilizing a gold-pattern-friendly bonding technique to form closed nanofluidic channels. Next, a solution of functional molecules bearing thiol or disulfide groups is introduced into the nanofluidic channels. Consequently, a functional molecular nanopattern specifically forms on the surface of the gold nanopattern in the nanofluidic channels by bottom-up molecular self-assembly, owing to well-known thiol–gold or disulfide–gold interactions (Love et al. 2005).

13.4.2 Placement Control with Nanometer-Scale Accuracy Permitting Multiple-Step Nanofabrication

The method of the nano-in-nano integration includes two key techniques to surmount two hurdles in the establishment of the novel patterning method as described in the following.

One hurdle was how to precisely pattern gold in open nanofluidic channels. As 3D hierarchical nanostructures formed by dissimilar materials, open glass nanofluidic channels with gold nanopatterns cannot be fabricated by a one-step process utilizing current available nanofabrication approaches. Therefore, a two-step process consisting of first nanofabrication of nanofluidic channels followed by second process of gold nanodeposition on exact locations in the open nanofluidic channels in a controlled manner (i.e., *nano-in-nano patterning*) is unavoidable but greatly challenging.

This hurdle was surmounted by a placement control technique with superhigh precision (Xu, Matsumoto et al. 2015). The technique was based on accurate positioning of the fabrication placement assisted by a pair of gold reference marks pre-fabricated on the substrate. The gold reference marks could be searched manually and recognized by a backscatter detector of an electron beam (EB) system when being scanned with the EB, according to a backscattered electron yield different from that of the glass substrate. The details of positioning using the reference marks during the electron beam lithography process can refer to the related paper (Xu, Matsumoto et al. 2015). By using the reference marks, a placement control with superhigh precision of several tens of nanometers in both X-axis and Y-axis directions was achieved, as indicated in Figure 13.9. For example, the super-high-precision placement control enabled the site-specific fabrication of a very challenging nanopattern whose

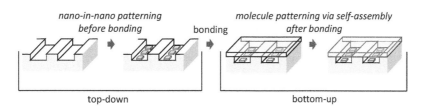

FIGURE 13.8 A schematic drawing of nano-in-nano integration. (Reproduced with permission (Xu, Matsumoto et al. 2015). Copyright 2015, The Royal Society of Chemistry.)

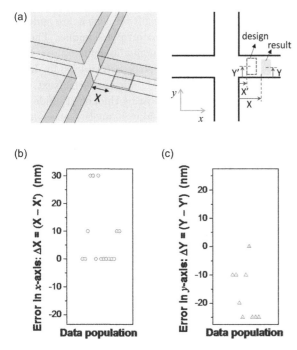

FIGURE 13.9 Evaluation of precision of the placement control permitting multiple-step nanofabrication. (a) The precision in terms of placement errors in both X-axis ($\Delta X = X - X'$) and Y-axis ($\Delta Y = Y - Y'$) directions was characterized by comparison of the actual (result) placement measured by SEM (X and Y, respectively) and the design placement (X' and Y', respectively) of a rectangular gold nanopattern (300 nm $\times$ 470 nm) fabricated in an open cross-shaped nanofluidic channel-network (520 nm wide, 240 nm deep) with the placement control technique. (b) The placement error in X-axis direction: (8 ± 16) nm in average, 30 nm in max., and 0 nm in min. (Data are the mean $\pm$ SD, $n \geq 9$.) (c) The placement error in Y-axis direction: (-16 ± 9) nm in average, 0 nm in max., and -25 nm in min. (Data are the mean $\pm$ SD, $n \geq 9$.) (Reproduced with permission (Xu, Matsumoto et al. 2015). Copyright 2015, The Royal Society of Chemistry.)

one side (Figure 13.10c) or both upper and lower sides (Figure 13.10d,h) are exactly located at edges of the bottom of the nanofluidic channels (800 nm wide, 300 nm deep), without any protrusion of gold to the outside of the nanofluidic channels. Moreover, the fabricated gold reference marks were quite robust even after experiencing several times of electron beam lithography, dry etching, and various harsh liquid processes of pattern development, metal lift-off, and substrate cleaning. The gold reference marks could work efficiently for nanofabrication with more than two steps, practically. For example, Figure 13.10j shows a scanning electron microscope (SEM) image of a gold-bottomed nanowell array in arrayed nanofluidic channels. The structure is a double nano-in-nano structure, which was fabricated by a three-step nanofabrication (Figure 13.10j).

Another hurdle was how to bond gold-patterned nanofluidic channels without destroying the pre-patterned gold. As aforementioned, the conventional fusion bonding was required to be performed in vacuum at over 1,000°C,

which is near the melting point of gold at atmospheric pressure. The ultra-high temperature would inevitably lead to destroying the gold patterns. This hurdle can be certainly surmounted by using the low-temperature/room-temperature bonding techniques (Xu, Wang et al. 2012; Xu et al. 2013) described in Sections 13.2 and 13.3. The hurdle can be also overcome by a gold-pattern-friendly bonding at 600°C (Xu, Matsumoto et al. 2015), which is much below the melting point of gold and hence does not damage the gold patterns and their functions. In comparison with the low-temperature/room-temperature bonding, although gold-pattern-friendly bonding requires a relatively higher bonding temperature, its process is much simpler and lower in cost. Further details of the gold-pattern-friendly bonding can referred to in the related paper (Xu, Matsumoto et al. 2015).

13.4.3 Nano-in-Nano Patterns and Structures

The nano-in-nano patterning supported by the above-described two key techniques enabled site-specific fabrication of gold nanopatterns in glass nanofluidic channels (i.e., nano-in-nano patterns and structures), with actual dimensions approximately same as targeted dimensions (Xu, Matsumoto et al. 2015). Figure 13.10 shows the SEM or AFM images of nanodots (290 nm; Figure 13.10a), nanowires (66 nm wide; Figure 13.10b), nanostripes (570 nm wide; Figure 13.10c), and nanowells (700 nm square, 200 nm deep; Figure 13.10i) with gold bottoms (600 nm square; Figure 13.10j) fabricated in nanofluidic channels (800 nm wide, 300 nm deep), representing quasi-0D, quasi-1D, 2D, and 3D nanopatterns, respectively. These reproducible nanostructures had feature sizes ranging from tens to hundreds of nanometers. These nanostructures are typical structures widely used in a variety of systems in nanoscience and nanotechnology but were fabricated for the first time in nanofluidic channels. Multiple patterns with arbitrary shapes in a single nanofluidic channel were also achieved. For example, while Figure 10d shows the SEM image of a series of gold patterns with gradient of length (890, 520, 330, 160 nm) in a single nanofluidic channel (800 nm wide, 300 nm deep), Figure 13.10e demonstrates the SEM image of diamond-, square-, triangle-, and heart-shaped gold nanopatterns in a single nanofluidic channel (680 nm wide, 300 nm deep). Moreover, the ability of site specificity was further demonstrated by the fabrication of a nanometer-level-positioning-required structure in a cross-shaped nanofluidic channel (520 nm wide, 240 nm deep), as shown in Figure 13.10f. The rectangular gold nanopattern (300 nm $\times$ 470 nm) positioned at a specified distance (for example, 50, 100, 200, 300, and 400 nm) with respect to the boundary of a branch of the nanofluidic channel-network was precisely fabricated as desired. In addition, these nanopatterns could be fabricated as nanoarrays in arrayed nanofluidic channels, as shown in Figure 13.10g–j, for example.

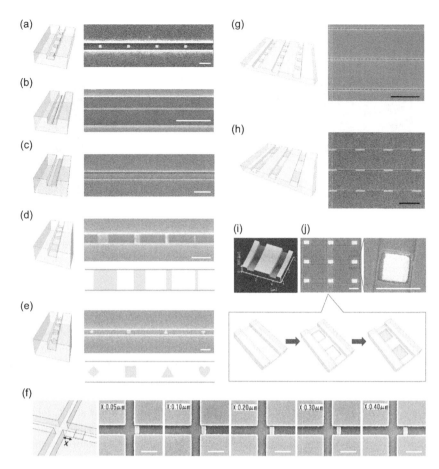

FIGURE 13.10 (a–h, j) SEM images with corresponding schematic drawings and (i) an AFM image of a variety of nano-in-nano patterns and structures. White scale bar is 1 μm, and black scale bar is 10 μm. (Reproduced with permission (Xu, Matsumoto et al. 2015). Copyright 2015, The Royal Society of Chemistry.)

These versatile nano-in-nano patterns and structures are very promising to work as fluidic, electrical, optical, thermal, magnetic, chemical, and biological functional units in nanofluidic chips after further design, characterization, optimization, and integration. Therefore, the nano-in-nano integration technology opens up a way to *in situ* integrate a wide range of functional material components (such as fluidic, chemical, biological, electrical, optical, thermal, and magnetic ones) in nanofluidic channels, which has been greatly challenging in the field of nanofluidics. Some demonstrations of functional component integration of nanofluidics based on the nano-in-nano integration technology are addressed in the following sections.

13.5 Functional Integration of Molecular Nanoarrays in Nanofluidic Channels

A self-assembled molecular nanoarray in arrayed nanofluidic channels (Figure 13.11) was fabricated in a nanofluidic chip by utilizing the nano-in-nano integration technology (Xu, Matsumoto et al. 2015). The nanofluidic chip contains a standard micro-/nanofluidic channel hybrid (Figure 13.11a).

In the chip, two side microfluidic channels (3 μm in depth and 500 μm in width) are bridged by 100 parallel arrayed nanofluidic channels (300 nm deep, 800 nm wide, 400 μm long, spaced by 2 μm), with a gold nanoarray (500 nm square, spaced by 2 μm) in each nanofluidic channel. The structure provides 160 × 100 (= 16,000) nanodot-shaped gold array on an area of 400 μm × 280 μm (= 0.112 mm²) in the chip, as partially shown in Figure 13.11b. Each nanofluidic channel has an ultra-small volume of approximately 96 fL (femtoliter, fL = 10^{-15} L). Thus, the arrayed nanofluidic channels are appropriate to handle 9.6 pL (picoliter, pL = 10^{-12} L) liquid, which is the same as the volume of a single mammalian cell. All liquids were introduced to the arrayed nanofluidic channels through the microfluidic channels by air pressure which was regulated using a 0.1-kPa-resolved pressure control system. An ethanol solution of HOOC-$(CH_2)_{10}$-S-S-$(CH_2)_{10}$-CONH-fluorescein was introduced to form the molecular nanoarray by molecular self-assembly in the nanofluidic channels. HOOC-$(CH_2)_{10}$-S-S-$(CH_2)_{10}$-CONH-fluorescein is a fluorescein disulfide, which can form self-assembled monolayers (SAM) on gold via disulfide–gold interactions. It is widely used to visualize the defects and ordering of SAM structure, owing to its easily detectable fluorescence. A high-density nanoarray (160 × 100) of the

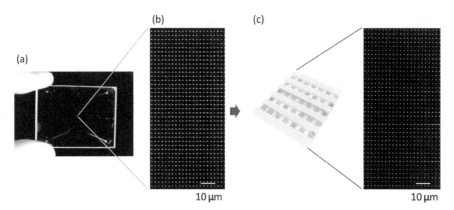

FIGURE 13.11 (a) A self-assembled fluorescent molecular nanoarray fabricated in a nanofluidic device with gold nanoarrays in arrayed nanofluidic channels. (b) A bright-field image of the gold nanoarrays. (c) Fluorescent images of the molecular nanoarray in the arrayed nanofluidic channels. (Reproduced with permission (Xu, Matsumoto et al. 2015). Copyright 2015, The Royal Society of Chemistry.)

fluorescein disulfide SAM was confirmed on the entire 400 μm × 280 μm tiny area of the arrayed nanofluidic channels by a fluorescence microscope. Figure 13.10c shows a fluorescence image of a part of the molecular nanoarray. The nanodot-shaped fluorescence array (Figure 13.10c) with a high resolution indicates the disulfide molecules were successfully self-assembled on the gold nanoarray in the nanofluidic channels. The fluorescence of the nanoarray was detected after continuous flow of absolute ethanol under air pressure of 430 kPa for 18 h. These results reveal that the high-density fluorescent molecular nanoarray was stable even under high-pressure fluidic conditions for a long time.

Therefore, the method makes it possible to accurately immobilize chemical and biological molecules in nanofluidic channels *via* self-assembly on gold. Such high-density fluorescent molecular nanoarrays in arrayed femtoliter nanofluidic channels would be very useful for single-cell omics studies, high-throughput single-molecule detection, accurate operation of liquids of ultra-small volumes, and single-molecule manipulation in the future.

13.6 Functional Integration of Active Nanovalves in Nanofluidic Channels

The ability to control a fluid in whether the macroscopic or the microscopic world affects fields as diverse as chemistry, mechanics, physics, materials science, energy, biology, drug discovery, and clinical medicine. Many sophisticated methodologies have been developed to control small amounts of fluids (Beebe et al. 2000; Burns 1998; Pompano et al. 2011), but the active regulation of fluids at subpicoliter (i.e, $<10^{-12}$ L) scales remains challenging. Nanofluidics is a potential solution to handle fluids in quantities as small as femtoliters to attoliters (aL = 10^{-18} L). Nonetheless, major challenges need to be overcome, in particular the establishment of methodology for the creation of fluidic components (i.e., valves) within tiny nanofluidic channels.

By taking advantage of the nano-in-nano integration technology, the fabrication of soft-matter-based, self-actuated, active nanovalves inside femtoliter-scale nanofluidic channels was achieved (Xu et al. 2016). The approach involves a well-tailored thermoresponsive polymer, which is synthesized on purpose to possess suitable well-defined characteristics as a trigger to allow strict, local nanofluidic operations. The approach enables the active regulation of femtoliter-scale fluids inside nanofluidic channels in response to a temperature change, which does not require the integration of additional acting-components and can be conveniently regulated externally.

The thermoresponsive polymer enabling the concept is a newly synthesized short-chain (weight-average molecular weight ≈ 10,000 and degrees of polymerization ≈ 100) poly(*N*-isopropylacrylamide) (PNIPAM) terminated with a trithiocarbonate (TTC) group at one end (referred to as TTC-*t*-PNIPAM; Figure 13.12a). PNIPAM is a famous temperature-responsive polymer, exhibiting a reversible phase transition at its lower critical solution temperature (LCST or cloud point (CP)) in an aqueous solution. When heated above LCST, the hydrated extended PNIPAM chains instantaneously convert into dehydrated compact aggregates. Due to this property, PNIPAM has been widely used in cell sheet tissue engineering (Matsuda et al. 2007) and drug delivery systems (Qin et al. 2006; Klaikherd et al. 2009) in a variety of forms, such as polymer bushes. Patterned PNIPAM brushes have been commonly fabricated by electron beam irradiation of *N*-isopropylacrylamide (NIPAM) monomers on cell culture dishes or substrates (Matsuda et al. 2007). Nevertheless, the method is not suitable for patterning of PNIPAM brushes in a nanofluidic channel, because an electron beam cannot pass through the solid-state substrate layer of the closed channel. In contrast, the use of the nano-in-nano technology allows the patterning of nanoscale PNIPAM brushes in nanofluidic channels *via* self-assembly on gold. In this study, to achieve the self-assembly, PNIPAM terminated with a TTC group was prepared by using a reversible addition fragmentation chain transfer

(a)

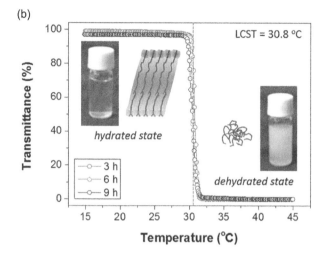

(b)

FIGURE 13.12 (a) Molecular structure of TTC-*t*-PNIPAM and (b) thermoresponsive phase transition of aqueous solutions of TTC-*t*-PNIPAM synthesized with varying the polymerization time (3, 6, and 9 h). (Reproduced with permission (Xu et al. 2016). Copyright 2016, Wiley-VCH.)

(a)

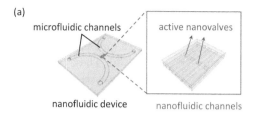

(b) (c)

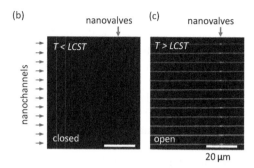

FIGURE 13.13 (a) A schematic drawing of the active nanovalves fabricated in a standard nanofluidic device. (b) and (c) Their thermoresponsive valving performance in nanofluidic channels. (Reproduced with permission (Xu et al. 2016). Copyright 2016, Wiley-VCH.)

(RAFT) polymerization technique. Light transmittance (%) versus temperature curves (Figure 13.12b) revealed that water solutions of the synthesized TTC-*t*-PNIPAM exhibited a common sharp phase transition at 30.8°C (LCST or CP), regardless of the polymerization time. Similar to a thiol group, which is the most notable functional group for linkage to gold surfaces, the TTC group has also been used to directly assemble polymer brushes on gold surfaces (Duwez et al. 2006).

A nanofluidic chip fabricated on fused-silica substrates was used to evaluate the performance of TTC-*t*-PNIPAM brushes in the active regulation of nanofluidic flows (Figure 13.13a). The chip comprises a standard micro-/nanofluidic channel hybrid (Figure 13.11a). In the parallel 2D nanofluidic channels (760 nm wide, 240 nm deep, and 300 μm long, spaced by 5.4 μm), rectangular gold nanopatterns (650 nm wide, 1580 nm long, and 70 nm thick) for subsequent self-assembly of TTC-*t*-PNIPAM brushes were accurately fabricated at places of 165 μm from the left ends of the channels (Figure 13.13a), by using the nano-in-nano integration technology. The nano-in-nano structures thereby allow for defining parallel ultra-small liquid spaces, i.e., 54.7 fL in volume for each, in the chip during fluidic operations.

TTC-*t*-PNIPAM brushes were fabricated on the gold nanopatterns within the arrayed nanofluidic channels through a continuously nanofluidic introduction of the ethanol solution of TTC-*t*-PNIPAM (2.00 mM) above

LCST, followed by a stationary incubation below LCST (i.e., at room temperature) overnight, as well as a subsequent rinse with continuously nanofluidic introduction of absolute ethanol above LCST. No clogging occurred during the entire liquid introduction, suggesting that the molecular weight (or the chain length) of TTC-*t*-PNIPAM was suitable for the strict nanofluidic operations. After draining the liquid from the whole channels and further drying the channels by introduction of nitrogen, the chip was ready for use.

The evaluation of active nanofluidic valving performance of the nanoscale polymer brushes was performed by observing the nanofluidic flow behaviors below and above LCST (30.8°C) of TTC-*t*-PNIPAM. The change of the temperature in the nanofluidic channels was induced by heating the chip with the thermoplate. Pure water was first introduced from the left microfluidic channel into the arrayed nanofluidic channels by a small air pressure at 10 kPa while heating the system above LCST. The microscopic observation of the area of the right microfluidic channel revealed that water flowed through the arrayed nanofluidic channels and continuously reached the right microfluidic channel during the introduction. Such fluid behavior suggests that the nanofluidic channels were in the "open" state when heated above LCST of TTC-*t*-PNIPAM. After the right microfluidic channel was fully filled with water, the system was cooled to ambient air temperature. In order to confirm the "closed" state of the nanofluidic channels below LCST, water was changed with a solution of sulforhodamine B (50 μM), which is an easily-detected fluorescent dye (excitation (Ex)/emission (Em) = 565/586 nm in water) with good thermal stability. Air pressure was continuously applied in approximately 10 kPa/min steps from 10 to 200 kPa, which is a quite high pressure enabling liquid introduction for many current nanofluidic applications, but no obvious fluorescence was observed in the area of nanofluidic channels at all pressures using a fluorescence microscope (Figure 13.13b). These results imply that the hydrated TTC-*t*-PNIPAM brushes enabled blocking of the nanofluidic channels and allowed for the resistance to high pressures, and hence the nanofluidic channels were in the "closed" state when cooled below LCST of TTC-*t*-PNIPAM.

Then the chip was heated again while continuously applying a pressure of 200 kPa. In contrast, strong fluorescence was observed in all nanofluidic channels immediately after heating for approximately 10–15 s (Figure 13.13c), revealing that the solution of sulforhodamine B flowed through the arrayed nanofluidic channels. It is considered that this quick valving performance responding to the temperature change should be owing to not only the sharp phase transition temperature of TTC-*t*-PNIPAM but also the very short heat transfer distance in the tiny nanofluidic channels. Therefore, the nanofluidic channels returned to the "open" state again and maintained the "open state" while heating. All these results revealed that the nanoscale TTC-*t*-PNIPAM brushes locally self-assembled in the nanofluidic channels possessed the active valving ability to regulate femtoliter-scale fluids with quick response to external temperature changes.

The ability to accurately assemble functional soft-matter structures within tiny nanofluidic channels, while not bringing about the most common, critical challenging problem in nanofluidics, i.e., undesired channel clogging, has the potential to initiate the development of nanofluidic systems with versatile capabilities, which are urgently needed in the nascent field of nanofluidics. This approach provided the first solution to the active regulation of local flows in a nanofluidic system and could be extended to build well-controlled, functional nanofluidic systems, allowing complex fluidic processes to be performed at the nanometer scale. The soft-matter-regulated nanovalves within nanofluidic channels described here improved the ability to handle small amounts of fluids downscaled to femtoliters. This ability could be further improved if the heating and cooling of the polymer brushes can be regulated in a pinpoint manner with nanoscale precision and fast conversion within the nanofluidic channels. Such future improvement could permit the operation of fluids at much smaller volume scales, for example, the formation of attoliter-volume droplets, which are promising for building continuous processes enabling the free manipulation of single molecules in the liquid phase.

13.7 Functional Integration of Electrodes in Nanofluidic Channels

As described in the section *Introduction*, in nanofluidics, new physical phenomena and mechanisms, which are not observed at microscales or in bulk, start to emerge and dominate at nanoscales. These phenomena are considered to be closely associated with the electrical double layer (EDL) on the solid/fluid (i.e., fused-silica glass/water) interface in a nanofluidic channel, where the EDL is an organized charge structure near a charged surface in contact with a fluid. The understanding of the science behind these phenomena is paramount in the road to the development of potential epoch-making applications by fully taking advantage of them. It has been suggested that the applications may include, but are not limited to, ultra-high-efficiency energy converters, ultra-high-sensitive analytical devices, and ultra-fast chemical reactors (Xu & Matsumoto 2015). So far, however, the understanding of these phenomena is still poor and seriously impeded by the lack of tools for the *in situ* investigation of nanospaces of nanofluidic channels, due to various physical barriers and challenges resulting from their very small and closed features. A glass nanofluidic device (Figure 13.14a), involving a pair of nanoscale probe electrodes (Figure 13.14b) which are accurately embedded in two ends of a single 2D nanofluidic channel (i.e., with both width and depth in the nanometer range), was designed and fabricated by using nano-in-nano integration technology. With the device, the first *in situ* electrokinetic probing of water confined in the femtoliter nanospace of the 2D nanofluidic channel under pressure-driven flow conditions (Figure 13.14d) was provided (Xu & Xu 2015).

The *in situ* measurement of streaming currents in a single 2D nanofluidic channel was performed on the fabricated nanofluidic device integrated with functional electrodes. Electrokinetic measurements focusing on streaming currents are useful for gaining an insight into the electrical state of the solid/fluid interface. Streaming currents are electric currents induced in channel or pore geometries with charged walls when counterions in the EDL are displaced through the geometries by pressure gradients. Streaming currents have been widely studied on channels and capillaries at the microscale (Yang et al. 2004; Mansouri et al. 2008; Martins et al. 2013) but are still largely unexplored on channels at the nanoscale, due to the lack of methods for their direct measurement in nanofluidic channels. So far, only a few groups have attempted to derive the state of streaming currents in nanofluidic channels by measuring total streaming currents flowing through micro/nano hybrid systems (Morikawa et al. 2010; van der Heyden et al. 2005). For example, measurements of streaming currents between either millimeter-scale reservoirs bridged by a single 1D (i.e., only the depth in the nanometer range) nanofluidic channel (van der Heyden et al. 2005), or microfluidic channels bridged by arrayed nanofluidic channels (Morikawa et al. 2010), have been reported. In addition to the use of chip devices with simple structures, both methods share a common advantage of the ease of setup, because bulk electrodes for the measurements could be easily inserted into the large reservoirs or the inlets/outlets of the microfluidic channels with the use of hands. Nevertheless, the advantage meanwhile restricts them from obtaining the *in situ*

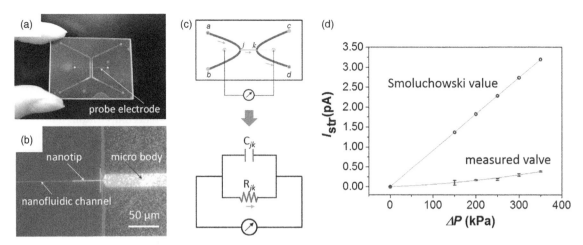

FIGURE 13.14 (a) A nanofluidic device with (b) a pair of nanoscale probe electrodes which are accurately embedded in two ends of a single 2D nanofluidic channel for the *in situ* measurement of streaming currents in the single nanofluidic channel. (c) A simplified fluidic circuit for the measurement and its corresponding equivalent electrical circuit, where C_{jk} and R_{jk} represent the capacitance between the probe electrodes and the solution resistance in the nanochannel, respectively. (d) The measurements reveal a discrepancy between *in situ* nanoscopic experimental results (measured value) and macroscopic laws of electrokinetic phenomena (Smoluchowski value), suggesting unusual liquid properties confined in the nanochannel. (Reproduced with permission (Xu & Xu 2015). Copyright 2015, Wiley-VCH.)

streaming currents in the nanofluidic channels, because the streaming currents originating from the large reservoirs and the long microfluidic channels may also be measured at the same time. In contrast, the methodology employed in this study involved the use of the nanofluidic device with a pair of integrated probe electrodes for the *in situ* measurement of streaming currents in a single 2D nanofluidic channel. Owing to the *in situ* integrated nano-in-nano electrodes, the electrical state in the nanofluidic channel space between two nano tips of the probe electrodes can be directly detected according to an equivalent electrical circuit shown in Figure 13.14c.

In situ streaming current measurements were performed with the nanofluidic chip (Figure 13.14a). All data determined at each ΔP during the pressure cycles were plotted in terms of mean I_{str} as a function of ΔP (Figure 13.14d). The error bars indicating the standard deviations of I_{str} at each ΔP are quite small, revealing a good repeatability of the measurements during the pressure cycles. The measured values were further compared with the theoretical streaming currents (i.e., Smoluchowski values), which were theoretically calculated using bulk parameters according to the Helmholtz–Smoluchowski (HS) equation (Delgado et al. 2007). The HS equation is the elementary theory describing streaming current phenomena. As shown in Figure 13.14d, the measured values are quite lower than the Smoluchowski values at each pressures. In comparison with the conventional *ex situ* measurements on a nanochannel with the equivalent radius (r) at the same scale (Morikawa et al. 2010), where the measured values for pure water were approximately two times (varying from 1.4 to 2.4 times for different cases) lower than the Smoluchowski values, the *in situ* measurements in this study exhibited much larger differences (10 times lower in average) between the measured values and the Smoluchowski values. The experimental

values obtained via the *in situ* measurements could better support the existence of unique liquid phenomena, such as decreased dielectric constant (several times) and increased viscosity (e.g., four times in a nanochannel with r at the same scale) of water confined in fused-silica glass nanofluidic channels, as previously observed with other methods (Hibara et al. 2002; Tas et al. 2004).

In short, the measurement results well support the existence of unique liquid phenomena such as lower dielectric constant and higher viscosity of water in glass nanofluidic channels and suggest that the HS equation may no longer be valid for the case. Therefore, the device and the method of this study can offer a useful nanoscale tool allowing *in situ* understanding of new physical effects of liquids confined in glass nanofluidic channels.

13.8 Regeneration of Nanofluidic Chips Integrated with Functional Components

Besides the high cost of the substrate (e.g., 30–40 US$ for a fused-silica glass of 40 mm long, 30 mm wide and 0.7 mm thick), glass nanofluidic chips are fabricated using expensive nanofabrication technologies, such as electron beam lithography incorporated with plasma dry etching or focused ion beam lithography through delicate, sophisticated, and time-consuming processes performed in clean rooms, and hence are not affordable for only single use. The fabrication cost increases significantly especially in the case of nanofluidic devices integrated with nano-in-nano functional components, because much more complicated fabrication processes are required compared to the simple nanofluidic devices with only bare nanochannels. Therefore, regeneration methods allowing the reuse of nanofluidic chips, especially those with nano-in-nano functional components, are greatly desired to efficiently and economically promote and accelerate the research and applications on nanofluidics.

Recently, a general methodology for the regeneration of glass nanofluidic chips using a multiple-step thermochemical decomposition process at high temperatures (Figure 13.15) has been developed (Xu, Wu et al. 2015). The thermochemical decomposition occurs during a well-designed sequential thermal treatment at controlled temperatures and pressures in a vacuum furnace (Figure 13.15a), as described in Figure 13.15b. The method takes advantage of exceptional properties of the fused-silica glass in strong heat resistance (strain point: 1,070°C; annealing point: about 1,140°C; softening point: about 1,665°C) and extremely low thermal expansion (coefficient of thermal expansion near zero). These exceptional properties allow numerous thermal

treatments of glass nanofluidic chips at high temperatures (for example, at 1,000°C) without changing the nanochannel structures. Thereby, the sequential thermal treatment at high temperatures enables the revival of "dead" nanofluidic chips by totally removing adsorbed dust, clogged particles, and reacted/interacted organic matter in nanochannels via the thermochemical decomposition. Furthermore, the method not only allows the regeneration of total glass nanofluidic chips but also is appropriate for the regeneration of nanofluidic chips with nano-in-nano functional components composed of dissimilar materials such as metallic nanoarrays by performing the thermal treatment at suitable high temperatures, for example, as shown in Figure 13.16. In addition, the method is simple, as no complicated operations are required because all steps of the thermal treatment can be carried out using a commercial, automatic,

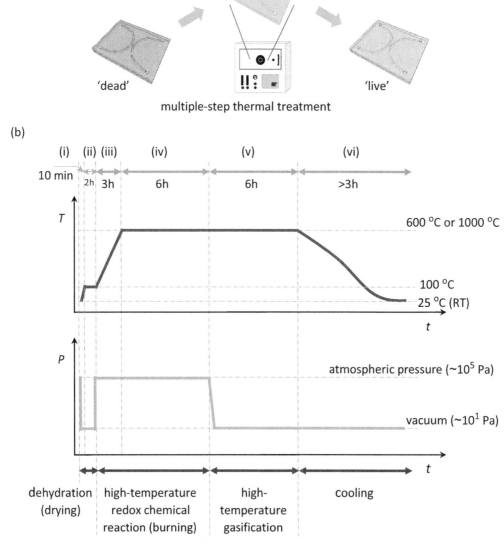

FIGURE 13.15 A schematic drawing of (a) the regeneration of glass nanofluidic devices through (b) a multiple-step sequential thermochemical decomposition process at high temperatures. (Reproduced with permission (Xu, Wu et al. 2015). Copyright 2015, The Royal Society of Chemistry.)

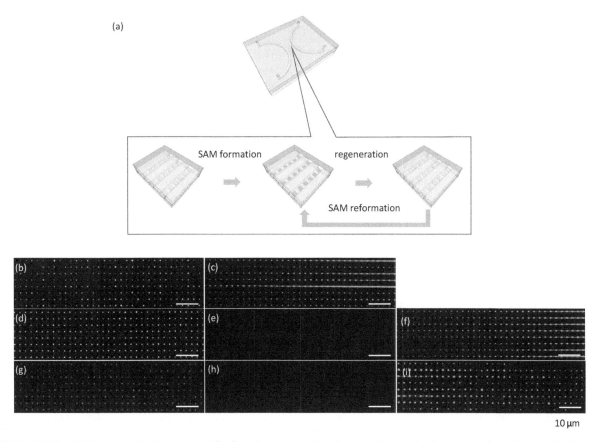

FIGURE 13.16 (a) A schematic drawing and (b–i) evaluation results of regeneration and reformation of a self-assembled fluorescent molecular nanoarray on gold nanoarrays in arrayed nanofluidic channels. (b) Bright-field and (c) fluorescence images of the fluorescent molecular nanoarray before the regeneration. (d) Bright-field and (e) fluorescence images of the gold nanoarrays after the first regeneration; (f) a fluorescence image of the reformed fluorescent molecular nanoarray after the first regeneration. (g) Bright-field and (h) fluorescence images of the gold nanoarrays after the third regeneration; (i) a fluorescence image of the reformed fluorescent molecular nanoarray after the third regeneration. (Reproduced with permission (Xu, Wu et al. 2015). Copyright 2015, The Royal Society of Chemistry.)

program-controlled vacuum furnace. Therefore, the method is very useful for the efficient and economic promotion of both fundamental research and practical applications on nascent nanofluidics.

13.9 Conclusion

The lack of the methodologies and technologies for the functionalization of nanofluidic channels is one of the most critical issues which have greatly impeded the progress of the nascent field of nanofluidics. As reviewed in this chapter, the nano-in-nano integration technology based on a novel methodology and a series of key techniques opens up a versatile and precise way for the functionalization of nanofluidics, bringing a variety of novel and potential applications to explore new chemistry, biology, and materials science through nanofluidics. The nano-in-nano integration technology would contribute to increase the involvement of much more researchers in many other fields to work with nanofluidics and thereby promote the transition of nanofluidics research from the current physics-centered stage to the next application-oriented stage.

Acknowledgments

This work was supported in part by JSPS KAKENHI (Grant No. JP18H01848), MEXT KAKENHI (Grant Nos. JP17H05468, JP19H04678), JST PRESTO (Grant No. JPMJPR18H5), and the National Natural Science Foundation of China (NSFC) (Grant No. 21628501).

References

Beebe, D.J. et al., 2000. Functional hydrogel structures for autonomous flow control inside microfluidic channels. *Nature*, 404(6778), pp. 588–590.

Bocquet, L. & Tabeling, P., 2014. Physics and technological aspects of nanofluidics. *Lab on a Chip*, 14(17), p. 3143.

Burns, M.A., 1998. An integrated nanoliter DNA analysis device. *Science*, 282(5388), pp. 484–487.

Chen, L. et al., 2006. Bonding of glass-based microfluidic chips at low- or room-temperature in routine laboratory. *Sensors and Actuators B: Chemical*, 119(1), pp. 335–344.

Chen, X. & Zhang, L., 2018. Review in manufacturing methods of nanochannels of bio-nanofluidic chips. *Sensors and Actuators B: Chemical*, 254, pp. 648–659.

Delamarche, E., Juncker, D. & Schmid, H., 2005. Microfluidics for processing surfaces and miniaturizing biological assays. *Advanced Materials*, 17(24), pp. 2911–2933.

Delgado, A.V. et al., 2007. Measurement and interpretation of electrokinetic phenomena. *Journal of Colloid and Interface Science*, 309(2), pp. 194–224.

Duan, C., Wang, W. & Xie, Q., 2013. Review article: Fabrication of nanofluidic devices. *Biomicrofluidics*, 7(2), p. 026501.

Duwez, A.-S. et al., 2006. Dithioesters and trithiocarbonates as anchoring groups for the "Grafting-To" approach. *Macromolecules*, 39(8), pp. 2729–2731.

Fonslow, B.R. & Bowser, M.T., 2005. Free-flow electrophoresis on an anodic bonded glass microchip. *Analytical Chemistry*, 77(17), pp. 5706–5710.

Galchev, T.V, Welch, W.C. & Najafi, K., 2011. A new low-temperature high-aspect-ratio MEMS process using plasma activated wafer bonding. *Journal of Micromechanics and Microengineering*, 21(4), p. 045020.

Gates, B.D. et al., 2005. New approaches to nanofabrication: Molding, printing, and other techniques. *Chemical Reviews*, 105(4), pp. 1171–1196.

Guan, W. et al., 2011. Field-effect reconfigurable nanofluidic ionic diodes. *Nature Communications*, 2, p. 506.

Han, J. & Craighead, H.G., 1999. Entropic trapping and sieving of long DNA molecules in a nanofluidic channel. *Journal of Vacuum Science & Technology A*, 17, p. 2142.

Haneveld, J. et al., 2003. Wet anisotropic etching for fluidic 1D nanochannels. *Journal of Micromechanics and Microengineering*, 13(4), pp. S62–S66.

Harms, Z.D. et al., 2015a. Conductivity-based detection techniques in nanofluidic devices. *The Analyst*, 140(14), pp. 4779–4791.

Harms, Z.D. et al., 2015b. Single-particle electrophoresis in nanochannels. *Analytical Chemistry*, 87(1), pp. 699–705.

Haynes, C.L. & Van Duyne, R.P., 2001. Nanosphere lithography: A versatile nanofabrication tool for studies of size-dependent nanoparticle optics. *The Journal of Physical Chemistry B*, 105(24), pp. 5599–5611.

Haywood, D.G. et al., 2015. Fundamental studies of nanofluidics: Nanopores, nanochannels, and nanopipets. *Analytical Chemistry*, 87(1), pp. 172–187.

He, R., Yamauchi, A. & Suga, T., 2018. Sequential plasma activation methods for hydrophilic direct bonding at sub-200°C. *Japanese Journal of Applied Physics*, 57(2S1), p. 02BD03.

van der Heyden, F.H.J., Stein, D. & Dekker, C., 2005. Streaming currents in a single nanofluidic channel. *Physical Review Letters*, 95(11), p. 116104.

Hibara, A. et al., 2002. Nanochannels on a fused-silica microchip and liquid properties investigation by time-resolved fluorescence measurements. *Analytical chemistry*, 74(24), pp. 6170–6176.

Howlader, M.M.R. et al., 2011. Nanobonding technology toward electronic, fluidic, and photonic systems integration. *IEEE Journal of Selected Topics in Quantum Electronics*, 17(3), pp. 689–703.

Howlader, M.M.R. et al., 2009. Role of heating on plasma-activated silicon wafers bonding. *Journal of The Electrochemical Society*, 156(11), p. H846.

Howlader, M.R. et al., 2006. Sequential plasma activated process for silicon direct bonding. In *ECS Transactions*, 3, pp. 191–202.

Jang, K. et al., 2010. An efficient surface modification using 2-methacryloyloxyethyl phosphorylcholine to control cell attachment via photochemical reaction in a microchannel. *Lab on a Chip*, 10(15), p. 1937.

Kalman, E.B., Vlassiouk, I. & Siwy, Z.S., 2008. Nanofluidic bipolar transistors. *Advanced Materials*, 20(2), pp. 293–297.

Kim, S.J. et al., 2007. Concentration polarization and nonlinear electrokinetic flow near a nanofluidic channel. *Physical Review Letters*, 99(4), p. 044501.

Kim, S.J., Song, Y.-A. & Han, J., 2010. Nanofluidic concentration devices for biomolecules utilizing ion concentration polarization: Theory, fabrication, and applications. *Chemical Society Reviews*, 39(3), p. 912.

Klaikherd, A., Nagamani, C. & Thayumanavan, S., 2009. Multi-stimuli sensitive amphiphilic block copolymer assemblies. *Journal of the American Chemical Society*, 131(13), pp. 4830–4838.

Lin, L., Chen, Q. & Sun, J., 2018. Micro/nanofluidics-enabled single-cell biochemical analysis. *TrAC Trends in Analytical Chemistry*, 99, pp. 66–74.

Liu, S. et al., 2005. From nanochannel-induced proton conduction enhancement to a nanochannel-based fuel cell. *Nano letters*, 5(7), pp. 1389–1393.

Love, J.C. et al., 2005. Self-assembled monolayers of thiolates on metals as a form of nanotechnology. *Chemical Reviews*, 105(4), pp. 1103–1170.

Ma, X. et al., 2007. Void-free low-temperature silicon direct-bonding technique using plasma activation. *Journal of Vacuum Science & Technology B: Microelectronics and Nanometer Structures*, 25(1), p. 229.

Mansouri, A., Kostiuk, L.W. & Bhattacharjee, S., 2008. Streaming current measurements in a glass microchannel array. *The Journal of Physical Chemistry C*, 112(42), pp. 16192–16195.

Martins, D.C. et al., 2013. Streaming currents in microfluidics with integrated polarizable electrodes. *Microfluidics and Nanofluidics*, 15(3), pp. 361–376.

Maszara, W.P. et al., 1988. Bonding of silicon wafers for silicon-on-insulator. *Journal of Applied Physics*, 64(10), pp. 4943–4950.

Matsuda, N. et al., 2007. Tissue engineering based on cell sheet technology. *Advanced Materials*, 19(20), pp. 3089–3099.

Mawatari, K. et al., 2014. Extended-nanofluidics: Fundamental technologies, unique liquid properties, and application in chemical and bio analysis methods and devices. *Analytical Chemistry*, 86(9), pp. 4068–4077.

Mellors, J.S. et al., 2008. Fully integrated glass microfluidic device for performing high-efficiency capillary electrophoresis and electrospray ionization mass spectrometry. *Analytical Chemistry*, 80(18), pp. 6881–6887.

Mijatovic, D., Eijkel, J.C.T. & van den Berg, A., 2005. Technologies for nanofluidic systems: Top-down vs. bottom-up—a review. *Lab on a Chip*, 5(5), p. 492.

Morikawa, K. et al., 2010. Streaming potential/current measurement system for investigation of liquids confined in extended-nanospace. *Lab on a Chip*, 10(7), p. 871.

Napoli, M., Eijkel, J.C.T. & Pennathur, S., 2010. Nanofluidic technology for biomolecule applications: A critical review. *Lab on a Chip*, 10(8), p. 957.

Ostrikov, K., 2005. *Colloquium*: Reactive plasmas as a versatile nanofabrication tool. *Reviews of Modern Physics*, 77(2), pp. 489–511.

Park, K.-C. et al., 2012. Nanotextured silica surfaces with robust superhydrophobicity and omnidirectional broadband supertransmissivity. *ACS Nano*, 6(5), pp. 3789–3799.

Perry, J.L. & Kandlikar, S.G., 2006. Review of fabrication of nanochannels for single phase liquid flow. *Microfluidics and Nanofluidics*, 2(3), pp. 185–193.

Perry, J.M. et al., 2010. Ion transport in nanofluidic funnels. *ACS Nano*, 4(7), pp. 3897–3902.

Piruska, A. et al., 2010. Nanofluidics in chemical analysis. *Chemical Society Reviews*, 39(3), pp. 1060–1072.

Pompano, R.R. et al., 2011. Microfluidics using spatially defined arrays of droplets in one, two, and three dimensions. *Annual Review of Analytical Chemistry*, 4(1), pp. 59–81.

Prakash, S. & Conlisk, A.T., 2016. Field effect nanofluidics. *Lab on a Chip*, 16(20), pp. 3855–3865.

Priest, C., 2010. Surface patterning of bonded microfluidic channels. *Biomicrofluidics*, 4(3), p. 032206.

Qin, S. et al., 2006. Temperature-controlled assembly and release from polymer vesicles of poly(ethylene oxide)-block-poly(N-isopropylacrylamide). *Advanced Materials*, 18(21), pp. 2905–2909.

Queste, S. et al., 2010. Manufacture of microfluidic glass chips by deep plasma etching, femtosecond laser ablation, and anodic bonding. *Microsystem Technologies*, 16(8–9), pp. 1485–1493.

Sparreboom, W., van den Berg, A. & Eijkel, J.C.T., 2009. Principles and applications of nanofluidic transport. *Nature Nanotechnology*, 4(11), pp. 713–720.

Suga, T., Kim, T.H. & Howlader, M.M.R., 2004. Combined process for wafer direct bonding by means of the surface activation method. In *2004 Proceedings. 54th Electronic Components and Technology Conference (IEEE Cat. No.04CH37546)*. IEEE, Piscataway, NJ, pp. 484–490.

Tas, N.R. et al., 2004. Capillary filling speed of water in nanochannels. *Applied Physics Letters*, 85(15), pp. 3274–3276.

Tseng, A.A. et al., 2003. Electron beam lithography in nanoscale fabrication: Recent development. *IEEE Transactions on Electronics Packaging Manufacturing*, 26(2), pp. 141–149.

Tsukahara, T. et al., 2008. Development of a pressure-driven nanofluidic control system and its application to an enzymatic reaction. *Analytical and Bioanalytical Chemistry*, 391(8), pp. 2745–2752.

Tsukahara, T. et al., 2010. Integrated extended-nano chemical systems on a chip. *Chemical Society Reviews*, 39(3), p. 1000.

Tsukahara, T. et al., 2007. NMR study of water molecules confined in extended nanospaces. *Angewandte Chemie International Edition*, 46(7), pp. 1180–1183.

Vallin, Ö., Jonsson, K. & Lindberg, U., 2005. Adhesion quantification methods for wafer bonding. *Materials Science and Engineering: R: Reports*, 50(4–5), pp. 109–165.

Visser, M.M. et al., 2002. Strength and leak testing of plasma activated bonded interfaces. *Sensors and Actuators A: Physical*, 97–98, pp. 434–440.

Vlassiouk, I. & Siwy, Z.S., 2007. Nanofluidic diode. *Nano Letters*, 7(3), pp. 552–556.

Wang, C., Higurashi, E. & Suga, T., 2008. Void-free room-temperature silicon wafer direct bonding using sequential plasma activation. *Japanese Journal of Applied Physics*, 47(4), pp. 2526–2530.

Wang, C. & Suga, T., 2012. Investigation of fluorine containing plasma activation for room-temperature bonding of Si-based materials. *Microelectronics Reliability*, 52(2), pp. 347–351.

Whitesides, G.M., 2002. Self-assembly at all scales. *Science*, 295(5564), pp. 2418–2421.

Xu, Y. et al., 2010. A microfluidic hydrogel capable of cell preservation without perfusion culture under cell-based assay conditions. *Advanced Materials*, 22(28), pp. 3017–3021.

Xu, Y. et al., 2013. Bonding of glass nanofluidic chips at room temperature by a one-step surface activation using an O_2/CF_4 plasma treatment. *Lab on a Chip*, 13(6), p. 1048.

Xu, Y., Wang, C. et al., 2012. Low-temperature direct bonding of glass nanofluidic chips using a two-step plasma surface activation process. *Analytical and Bioanalytical Chemistry*, 402(3), pp. 1011–1018.

Xu, Y., Jang, K. et al., 2012. Microchip-based cellular biochemical systems for practical applications and fundamental research: From microfluidics to nanofluidics. *Analytical and Bioanalytical Chemistry*, 402(1), pp. 99–107.

Xu, Y. et al., 2007. Microfluidic flow control on charged phospholipidpolymer interface. *Lab Chip*, 7(2), pp. 199–206.

Xu, Y., 2018. Nanofluidics: A new arena for materials science. *Advanced Materials*, 30(3), p. 1702419.

Xu, Y., Wu, Q. et al., 2015. Regeneration of glass nanofluidic chips through a multiple-step sequential thermochemical decomposition process at high temperatures. *Lab Chip*, 15(19), pp. 3856–3861.

Xu, Y., Matsumoto, N. et al., 2015. Site-specific nanopatterning of functional metallic and molecular arbitrary features in nanofluidic channels. *Lab on a Chip*, 15(9), pp. 1989–1993.

Xu, Y. & Matsumoto, N., 2015. Flexible and in situ fabrication of nanochannels with high aspect ratios and nanopillar arrays in fused silica substrates utilizing focused ion beam. *RSC Advances*, 5(62), pp. 50638–50643.

Xu, Y., Shinomiya, M. & Harada, A., 2016. Soft matter-regulated active nanovalves locally self-assembled in femtoliter nanofluidic channels. *Advanced Materials*, 28(11), pp. 2209–2216.

Xu, Y. & Xu, B., 2015. An integrated glass nanofluidic device enabling in-situ electrokinetic probing of water confined in a single nanochannel under pressure-driven flow conditions. *Small*, 11(46), pp. 6165–6171.

Yang, J., Masliyah, J.H. & Kwok, D.Y., 2004. Streaming potential and electroosmotic flow in heterogeneous circular microchannels with nonuniform zeta potentials: Requirements of flow rate and current continuities. *Langmuir*, 20(10), pp. 3863–3871.

Zheng, Y.-B. et al., 2017. A temperature, pH and sugar triple-stimuli-responsive nanofluidic diode. *Nanoscale*, 9(1), pp. 433–439.

Zhou, K., Perry, J.M. & Jacobson, S.C., 2011. Transport and sensing in nanofluidic devices. *Annual Review of Analytical Chemistry*, 4(1), pp. 321–341.

Zucker, O. et al., 1993. Application of oxygen plasma processing to silicon direct bonding. *Sensors and Actuators A: Physical*, 36(3), pp. 227–231.

Classical Density Functional Theory and Nanofluidics: Adsorption and the Interface Binding Potential

P. Yatsyshin,
M.-A. Durán-Olivencia
and S. Kalliadasis
Imperial College London

14.1 Introduction

Interest in wetting has been rapidly growing over the past few decades across different applied and theoretical fields of study. On the applied front, it plays a crucial role in many technological processes, from the vapor-liquid-solid growth of nanowires, to labs-on-chip and to applications in superhydrophobicity and nanofluidics [1–6]. It is also rich in fundamental questions. The physical mechanisms responsible for the height of liquid–gas interface during adsorption are directly linked to the small-scale workings of the atomic world. In particular, changes between the so-called Wenzel and Cassie–Baxter states of drops sitting on rough surfaces, as well as the effects of contact angle hysteresis during droplet spreading still remain in debate [7,8]. At the same time, the competition between the fluid–fluid and fluid–substrate interactions and interface fluctuations often leads to the appearance of exciting interfacial phase transitions and associated critical phenomena [9–19].

The fact that intermolecular interactions at small scales are non-local, as well as the typical presence of a wide range of length scales in problems of wetting, spurs on theoreticians to develop appropriate microscopic approaches. In the present contribution to the *21st Century Nanoscience Handbook*, we focus on the application of a self-consistent statistical–mechanical framework, known as classical density functional theory (DFT), to problems of wetting, in particular adsorption of nanofluidic systems such

as nanodrops/nanobubbles. The cornerstone of DFT is to represent the grand free energy of the fluid, possibly in contact with a substrate, as a functional of the one-body fluid density. This functional can be shown to generate a hierarchy of N-body density correlation functions, which in principle can be used to compute any quantity of interest through proper averaging [20]. The one-body fluid density is of particular interest in wetting, because it allows us to compute such important properties as interface contact angels, surface tensions, liquid and gas adsorptions, and so on. Given a model free-energy functional, the one-body fluid density can be obtained numerically by unconstrained minimization. Moreover, the uniform or bulk limit of the free-energy functional yields the binodal of the fluid. In other words, the free-energy functional offers the means to introduce the spatial dependence of the fluid density into the thermodynamic equation of state. Hence, DFT retains systematically all microscopic details but at a computational cost much lower than that of molecular simulations. On the other hand, macroscopic approaches with microscale effects are not rigorous as intermolecular forces are added to the macroscopic models in an *ad hoc* manner (typically as an additional body force-pressure). DFT thus offers an attractive alternative to all other approaches.

In the present Chapter, we consider a prototypical problem of droplet adsorption on a flat surface and discuss the application of classical DFT to the computation of drop shapes and contact angles. We highlight a connection to the well-known Derjaguin's disjoining pressure approach and demonstrate how DFT can be used rationally and systematically to determine an approximate interface binding potential, whose derivative gives the disjoining pressure.

14.2 A Simple DFT Functional

We will employ a minimalistic free energy functional, which nevertheless captures qualitatively the physics governing the equilibrium of a liquid–gas interface. Details about the approximation can be found in, e.g., our recent work [21]. More accurate approximations do exist, such as weighted density approximation and fundamental measure theory, and can capture layering effects and even freezing, e.g. [20,22,23]. We consider all interactions to be modeled by the Lennard-Jones (LJ) potential with the depth of the well ε and range σ:

$$\varphi_{\sigma,\varepsilon}^{\mathrm{LJ}}(r) = 4\varepsilon \left[\left(\frac{\sigma}{r} \right)^{12} - \left(\frac{\sigma}{r} \right)^{6} \right]. \qquad (14.1)$$

In wetting, the fluid is typically brought in contact with the substrate which exerts a potential $V_{\mathrm{ext}}(\mathbf{r})$ on the fluid. In the case of a planar LJ wall, positioned at $y = 0$, the potential $V_{\mathrm{ext}}(y)$ can be obtained by integrating $\varphi_{\sigma_{\mathrm{w}},\varepsilon_{\mathrm{w}}}^{\mathrm{LJ}}(r)$ with the wall–fluid LJ parameters ε_{w} and σ_{w} over a half-space:

$$V_{\mathrm{ext}}(y) = 4\pi\rho_{\mathrm{w}}\varepsilon_{\mathrm{w}}\sigma_{\mathrm{s}}^3 \left[-\frac{1}{6}\left(\frac{\sigma_{\mathrm{w}}}{H_0 + y}\right)^3 + \frac{1}{45}\left(\frac{\sigma_{\mathrm{w}}}{H_0 + y}\right)^9 \right], \qquad (14.2)$$

where ρ_{w} is the average density of the wall material and H_0 is a near-wall cut-off, introduced to avoid a non-physical divergence of $V_0(y)$ at fluid–substrate contact. Additionally, H_0 can be used to calibrate the effects of the fluid–substrate repulsions, which lead to the near-wall layering of the fluid density [19]. These layering affects can change the values of contact angles and wetting temperatures but do not qualitatively affect the liquid-vapor coexistence occurring far from the substrate. More details on the effect of $H_0 > 0$ are given in Appendix 1 of Ref. [24].

At the given temperature T and chemical potential μ, the equilibrium fluid density profile $\rho(\mathbf{r})$ minimizes the grand free energy functional [20,25–28]:

$$\Omega[\rho(\mathbf{r})] = F[\rho(\mathbf{r})] - \int d\mathbf{r}\, \rho(\mathbf{r}) \left(\mu - V_{\mathrm{ext}}(\mathbf{r}) \right), \qquad (14.3)$$

where $F[\rho(\mathbf{r})]$ is the "intrinsic" free energy functional. We employ the following approximation for the latter:

$$F[\rho(\mathbf{r})] = \int d\mathbf{r}\, [f_{\mathrm{id}}(\rho(\mathbf{r})) + \rho(\mathbf{r})\psi(\rho(\mathbf{r}))]$$
$$+ \frac{1}{2}\int d\mathbf{r}\int d\mathbf{r}'\, \rho(\mathbf{r})\rho(\mathbf{r}')\varphi_{\mathrm{attr}}(|\mathbf{r}-\mathbf{r}'|), \qquad (14.4)$$

where $f_{\mathrm{id}}(\rho) = k_{\mathrm{B}}T\rho\left(\ln\left(\lambda^3\rho\right) - 1\right)$ is the ideal gas free energy density, λ is the thermal wavelength, and $\psi(\rho)$ is

the Carnahan–Starling free energy density of a hard sphere fluid with molecular radii σ:

$$\psi(\rho) = k_{\mathrm{B}}T\, \frac{\eta(4 - 3\eta)}{(1 - \eta)^2}, \qquad \eta = \pi\sigma^3\rho/6. \qquad (14.5)$$

In Eq. (14.4), $\varphi_{\mathrm{attr}}(|\mathbf{r} - \mathbf{r}'|)$ describes intermolecular attractions, which are treated as a perturbation over the excluded volume interactions. This approach closely follows the Barker–Henderson thermodynamic perturbation theory [29]:

$$\varphi_{\mathrm{attr}}(r) = \begin{cases} 0, & r \le \sigma, \\ \varphi_{\sigma,\varepsilon}^{\mathrm{LJ}}(r), & r > \sigma. \end{cases} \qquad (14.6)$$

It can be shown that in the limit of a uniform fluid with $\rho(\mathbf{r}) = \rho$, the approximation for $F[\rho(\mathbf{r})]$ in Eq. (14.4) is equivalent to the so-called random phase approximation for the bulk pair correlation function [20,25,26]. The first integral term in Eq. (14.4) is the approximate free-energy functional of a hard sphere fluid in a local density approximation [25]. As mentioned earlier, this local treatment of intermolecular repulsions neglects the near-wall oscillations of the fluid density profile. Nevertheless, since such fluid layering does not qualitatively affect the physics of the liquid–gas interface, the functional in Eqs. (14.3)–(14.6) is sufficient for a qualitative study of liquid adsorption at temperatures above the freezing transition.

Taking the functional derivatives, we arrive at the Euler-Lagrange equation for the fluid density profile $\rho(\mathbf{r})$, which extremizes the grand potential (14.3):

$$k_{\mathrm{B}}T \ln \rho(\mathbf{r}) + \psi(\rho(\mathbf{r})) + \rho(\mathbf{r})\psi_\rho'(\rho(\mathbf{r}))$$
$$+ \int \rho(\mathbf{r}')\varphi_{\mathrm{attr}}(|\mathbf{r}-\mathbf{r}'|)\, d\mathbf{r}' + V_{\mathrm{ext}}(\mathbf{r}) - \mu = 0, \qquad (14.7)$$

where $\psi_\rho'(\rho)$ is the derivative of (14.5) with respect to ρ. We solve the above equation numerically. More details on the approximations involved in (14.3)–(14.6), as well as full details of the numerical scheme we use to minimize the grand free-energy functional in Eq. (14.3) can be found in the Appendix and in Refs [19,22,24]. In what follows, it is convenient to work in a system of units where σ and ε in Eq. (14.6) are set as the units of length and energy, respectively. In the bulk limit, this leads to the fluid critical temperature $T_{\mathrm{c}} \approx 1.006$. We further set $\rho_{\mathrm{w}} = 1$ and $H_0 = 5$ in (14.2) and fix the values of the LJ parameters for the wall material: $\varepsilon_{\mathrm{w}} = 0.4$ and $\sigma_{\mathrm{w}} = 2$. We choose these specific values of parameters to get a relatively high wetting temperature, thus smoothing the liquid–gas interfaces. A smoother interface requires less denser mesh and is easier to deal with in computations.

14.3 Macroscopic Drops and Planar Films

First, consider a bulk fluid of constant ρ_{b}. In this case, integration in the free-energy functional (14.3) becomes a multiplication by the fluid volume V, and the pressure P of the fluid is found from the thermodynamic relation $\Omega = -PV$. At the same time, the chemical potential can be obtained from the Euler–Lagrange equation (14.7):

$$P(T, \rho_b) = \rho_b k_B T \frac{1 + \eta + \eta^2 - \eta^3}{(1 - \eta)^3} - \frac{16\pi}{9} \rho_b^2 \sigma^3 \varepsilon, \quad (14.8)$$

$$\mu(T, \rho_b) = k_B T \ln \rho_b + \psi(\rho_b) + \rho_b \psi'_\rho(\rho_b) - \frac{32\pi}{9} \rho_b \sigma^3 \varepsilon. \quad (14.9)$$

Bulk liquid–gas coexistence, or binodal, is given by the equality of pressures and chemical potentials in the gas and liquid phases:

$$P(T, \rho_l) = P(T, \rho_g) = P_{sat}(T),$$
$$\mu(T, \rho_l) = \mu(T, \rho_g) = \mu_{sat}(T), \quad (14.10)$$

where P_{sat} and μ_{sat} are the pressure and chemical potential at coexistence (also called saturation), ρ_g and ρ_l are the densities of saturated gas and liquid. The boundaries of stability of gas and liquid are given by the spinodals:

$$\partial P / \partial \rho \big|_{\rho_g, \rho_l} = 0. \quad (14.11)$$

The liquid–gas critical point at T_c and ρ_c satisfies the following equation:

$$\partial P / \partial \rho_c = \partial^2 P / \partial \rho_c^2 = 0. \quad (14.12)$$

When a fluid at bulk coexistence is brought in contact with a planar substrate, a macroscopic drop may become adsorbed at the fluid–substrate interface. The contact angle of such drop is given by the Young equation, which expresses the balance of the different surface tensions:

$$\gamma_{lg} \cos \Theta_Y(T) = \gamma_{wl} - \gamma_{wg}, \quad (14.13)$$

where γ_{lg}, γ_{wl} and γ_{wg} are the liquid–gas, wall–liquid, and wall–gas surface tensions, respectively. The value of temperature at which Θ_Y vanishes is known as the wetting temperature of the substrate, T_w. Above T_w, the wall is completely wet by liquid, and below it, the wall is partially wet or dry. Using DFT, we can obtain the density profiles of the coexisting liquid and gas in contact with the wall, and the corresponding surface tensions are given by the following expression:

$$\gamma = \frac{1}{\mathcal{A}} (\Omega[\rho(y)] + PV), \quad (14.14)$$

where $\mathcal{A}$ is the interface area. Figure 14.1 shows the bulk phase diagram in the $\rho - T$ plane, the intersection of curves $\gamma_{wl}(T) + \gamma_{wg}(T)$ and $\gamma_{lg}(T)$, which can be used to compute $T_w = 0.915$, and the curve $\Theta_Y(T)$. At the wetting temperature, $\Theta_Y(T)$ vanishes as $O(\sqrt{(T_w - T)})$. The agreement between the contact angles obtained from the Young equation and DFT has been established in our previous studies [30,31].

The Young contact angle describes macroscopic drops at bulk coexistence. For nanodrops, the apparent contact angle

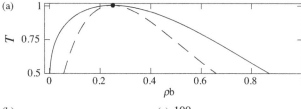

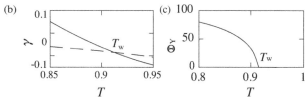

FIGURE 14.1 Bulk and planar wetting phase diagrams of an LJ fluid. (a) Bulk binodal (solid curve) and spinodal (dashed curve), coming together at the critical point (black circle). (b) The sum $\gamma_{wl}(T) + \gamma_{wg}(T)$ (solid curve) and $\gamma_{lg}(T)$ (dashed curve), computed from DFT, using Eq. (14.14). (c) The dependence of the Young contact angle on temperature, computed using Eq. (14.13).

cannot be easily expressed as a balance of surface tensions. In fact at the nanoscale, the system size becomes an independent thermodynamic field which can induce deviations from Young. A manifestation of this is that the contact angle of small drops, adsorbed below coexistence on substrate impurities, or of metastable drops, is known to be size-dependent. It is certainly of interest to go beyond the simple expression (14.13) that uses familiar macroscopic terms, such as interface and surface tension, in order to describe the angles of nanodrops in terms of intermolecular effects and the characteristic dimensions of the system. In order to do so, we first need to appreciate the important role played by the surface binding potential (especially in the vicinity of the substrate; it vanishes far from it), which arises from the LJ interactions between the fluid and the substrate.

Consider now adsorption on an ideally planar substrate. The solutions of the DFT Eq. (14.7) have the same symmetry as $V_{ext}(y)$, i.e., the wall adsorbs films of liquid or gas, depending on the bulk density. When the bulk is gas, the wall can adsorb thermodynamically stable liquid films, exhibiting approach to complete wetting [19]. We can quantify liquid adsorption in terms of the height of the adsorbed film l:

$$l(T, \mu) = \frac{\int_0^\infty (\rho(x) - \rho_b) dx}{\Delta \rho(T, \mu)}, \quad (14.15)$$

where ρ_b is the bulk fluid density and $\Delta \rho$ is the difference between the average density of the adsorbent (which is close to ρ_l) and the bulk fluid density. Additionally, we can obtain the "excess free energy density" ω^{ex} of the adsorbate by subtracting the bulk energy from the grand free-energy functional (14.3) and rewriting it in the following form:

$$\Omega[\rho(x); T, \mu] + PV = \mathcal{A} \int_0^\infty \omega^{ex}(x; \rho(x), T, \mu) dx. \quad (14.16)$$

In Figure 14.2a, we represent an isotherm $l(\mu)$, corresponding to the adsorption of liquid film at $T = 0.88 < T_{\rm w}$. It is given as a function of the "disjoining" chemical potential $\Delta\mu = \mu - \mu_{\rm sat}$. We notice the turning points of the isotherm, which provide the spinodals for the surface phases of liquid and gas films, respectively. In Figure 14.2b, we plot two representative density profiles of thin and thick adsorbed films. We further notice that both have plateaus near the densities of coexisting gas and liquid (designated by horizontal lines in the figure). In fact, thin and thick adsorbed films correspond to distinct thermodynamic surface phases, which can coexist during the so-called wall *pre-wetting transition*. If the bulk liquid–gas coexistence is reached via a thermodynamic route, along the thin–thick film coexistence, then coexistence must be approached at the wall wetting temperature, where the thick coexisting film becomes infinitely thick [32]. Evidently, since the hard sphere part of the free-energy functional in Eq. (14.4) is local with respect to the fluid density, the computed density profiles in Figure 14.1b do not exhibit any near-wall layering, which, strictly speaking, is present in the actual fluid. Nevertheless, as already mentioned, above the LJ bulk triple point, such layering does not affect the physics of phase transitions associated with the liquid–gas interface. Figure 14.2c depicts the excess free-energy density profiles, corresponding to the density profiles in (b). We can see that the wall–fluid and liquid–gas interfaces of the density profiles correspond to oscillations in $\omega^{\rm ex}(x)$ about zero. At the same time, inside the near-bulk phases of gas and adsorbed liquid, $\omega(x)$ is almost constant, and $\omega^{\rm ex}(x) \approx 0$.

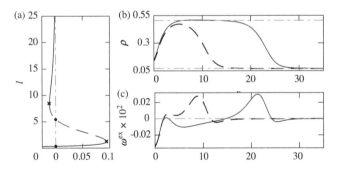

FIGURE 14.2 Representative picture of adsorption on a planar wall. (a) Isotherm of wetting by liquid film. Thermodynamically stable and unstable branches are plotted with solid and dashed curves, respectively. The limits of film metastability (spinodals) are designated by symbols. Vertical dash-dotted line demarcates gas–liquid coexistence at $\Delta\mu = 0$, and its intersection with the isotherm is designated by filled circles. (b) Representative density profiles of adsorbed liquid film at $l = 8.5$ (dashed) and $l = 21.2$ (solid). The width of the liquid–gas interface is proportional to the bulk correlation length. Horizontal dash-dotted lines designate gas and liquid densities at bulk coexistence. (c) Excess free-energy density profiles, corresponding to the profiles in (b). Horizontal dash-dotted line is drawn at the bulk value (here, zero).

14.4 Derjaguin's Approach to Wetting. Interface Binding Potential

In the previous section, our model led to film adsorption as a result of the correctly chosen thermodynamic point (i.e., T and μ) in the DFT equation (14.7). We did not have to assume the existence of the liquid–gas interface, as it appeared naturally during wetting. We also saw that the height of the adsorbed film can be easily obtained from the density profile. The problem of computing the height of an adsorbed planar film can be addressed differently, by using a conceptually simpler Hamaker-like approach, which can be traced back to the works of Derjaguin (see, e.g., Ref. [33] for discussion). It proceeds by assuming that a sharp liquid–gas interface of vanishing width is placed at a distance l from the substrate. Then, the height of such interface must minimize the following free-energy functional [34]:

$$\Omega_l\left[l; T, \mu\right] = \int_{-\infty}^{+\infty} \left[\gamma_{\rm lg}(T)\sqrt{1 + \left(\frac{dl}{dx}\right)^2} + W\left(l\left(x\right); T, \Delta\mu\right)\right]dx,$$
(14.17)

where the first term is the free energy (per unit area) of the liquid–gas interface and the second term is the thermodynamic work of adsorbate formation:

$$W\left(l\left(x\right); T, \mu\right) = \Delta\mu\Delta\rho(T, \mu)l\left(x\right)$$
$$+ \gamma_{\rm wl}(T) + \gamma_{\rm lg}(T) + g\left(l\left(x\right); T\right),$$
(14.18)

where $g(l; T)$ is the interface binding potential, which accounts for the interatomic interactions between the liquid–gas interface and the substrate. Using the Hamaker approach, it is easy to find that for LJ forces and a flat substrate, $g(l; T) = \mathcal{O}\left(l^{-2}\right)$, as $l \to \infty$. Now the equilibrium film height can be obtained by minimizing the free energy (14.17). One way to do this is by solving the following Euler–Lagrange equation:

$$\frac{d^2l}{dx^2} = \frac{1}{\gamma_{\rm lg}}\left(1 + \left(\frac{dl}{dx}\right)^2\right)^{\frac{3}{2}} \dot{W}\left(l; T, \mu\right),$$
(14.19)

where $\dot{W}\left(l; T, \mu\right)$ stands for the derivative of $W(l; T, \mu)$ with respect to l.

It is noteworthy that in contrast to the DFT Euler–Lagrange equation (14.7), Eq. (14.19) is local, which makes it significantly easier to solve. The models underlying Eqs. (14.17) and (14.19) are widely used in fluid mechanics. The interface binding potential is equal to the antiderivative of the disjoining pressure, which enters the hydrodynamic description given by the Navier–Stokes equation, and its long-wavelength approximations, as an additional body force-pressure [35,36]. In macroscopic applications, the precise form of $g(l)$ is not very important, and many authors simply use the asymptote $\mathcal{O}\left(l^{-2}\right)$. However, as the system sizes get smaller, the error in neglecting the short-range behavior of $g(l)$ grows [19]. Additionally, at the nanoscale, the interface can hardly be thought of as vanishingly sharp, as is evident from Figures 14.2b, c. The aim is then to track down the key ingredient of the model (14.17) – the binding potential – from DFT and its behavior for sufficiently small l.

Over the years, several approaches to computing the binding potential were developed. Assuming a sharp liquid–gas interface of the adsorbent in a DFT grand free energy similar to (14.3), Dietrich and co-workers obtained the analytic form of the higher-order corrections to $g(l)$ for long-ranged intermolecular forces [37]. Although such analytic approximations for $g(l)$ are valuable for fundamental theoretical studies, the availability of highly accurate DFT functionals capable of achieving quantitative accuracy for many systems (see, e.g., Refs [38,39]) spurs on theoreticians to develop numerical methods for computing $g(l)$ from a given DFT functional. This is also valuable for applied studies. For example, once $g(l)$ is obtained from DFT, the disjoining pressure can be computed and used in fluid mechanical computations of droplet spreading.

There are a number of recent works addressing the computation of the binding potentials using DFT and molecular dynamics simulations. Notably, MacDowell and co-workers obtained the binding potential using molecular dynamics simulations [40]. Archer and co-workers used a numerical approach for obtaining $g(l)$ from a DFT functional [41] by applying constrained functional minimization. Imposing the conservation of adsorption, the authors obtained density profiles of unstable liquid films whose energies give $g(l)$. However DFT is a grand-canonical framework, and imposing constraints, strictly speaking, corresponds to enforcing a certain canonical ensemble, which in turn breaks the internal consistency of the DFT approach. We outline here an alternative way to compute $g(l)$, using the isotherm of adsorption and the set of density profiles associated with it.

14.5 Obtaining the Binding Potential from DFT

Since the Derjaguin picture operates with a macroscopic quantity $l(x)$, there will inevitably remain a certain conceptual difficulty in applying a fully microscopic approach to computing $g(l)$. In particular, one needs to agree on how to define a *thermodynamically unstable* adsorbed liquid film of height l. Clearly, such a definition cannot be unique, which ultimately leads to different possibilities for constructing microscopic numerical approximations of $g(l)$. In the present work, we will make use of the adsorption isotherm, computed at fixed T, such as the one depicted in Figure 14.2. Every point (μ_0, Γ_0) on the isotherm corresponds to a density profile $\rho_{\mu_0}(x)$ of adsorbed liquid film, which extremizes the grand free-energy functional $\Omega[\rho; T, \mu_0]$ in Eq. (14.3). We note that $\rho_{\mu_0}(x)$ does not extremize $\Omega[\rho; T, \mu]$ at any other values of $\mu \neq \mu_0$. It follows that ignoring the dependence on μ within the set $\{\rho(x)\}$

of the density profiles associated with the isotherm, we can view it as a set of thermodynamically unstable adsorbed liquid films of different heights. If we can compute the free energies of the profiles $\{\rho(x)\}$, as well as their respective film heights, we will thus obtain the numerical approximation to $g(l)$ as a data table.

Setting the chemical potential to the bulk coexistence value μ_{sat}, we treat the set of the density profiles, giving rise to the isotherm at temperature T, as the profiles of thermodynamically unstable adsorbed films. The DFT functional allows us to find the work of adsorbate formation entering Eq. (14.18), given the density profile $\rho_{\mu_0}(x)$, which minimizes the DFT grand free energy at some $\mu_0 \neq \mu_{\text{sat}}$:

$$W = \int_0^\infty \Big[\omega(x; \rho_{\mu_0}(x), T, \mu_{\text{sat}}) - \lim_{x \to \infty} \omega(x; \rho_{\mu_0}(x), T, \mu_{\text{sat}}) \Big] dx,$$

(14.20)

where $\omega(x; \rho(x), T, \mu)$ denotes the grand free-energy density:

$$\Omega[\rho(x), T, \mu] = A \int_0^\infty \omega(x; \rho(x), T, \mu). \quad (14.21)$$

As mentioned above, the main trick to get to unstable profiles is to treat $\rho(x)$ and μ in the equation above as independent and to fix $\mu = \mu_{\text{sat}}$. For profiles of adsorbed liquid films that extremize the DFT grand free energy at coexistence, the grand free-energy density $\omega(x; \rho_{\text{sat}}(x), T, \mu_{\text{sat}})$ tends to $-P$, as $x \to \infty$. For LJ forces, it is easy to work out that $\omega(x; \rho_{\text{sat}}(x), T, \mu_{\text{sat}}) = O(1/x^3)$ for large x. If the profile $\rho_{\mu_0}(x)$ is plugged in place of $\rho(x)$ in $\Omega[\rho(x), T, \mu_{\text{sat}}]$, this will shift the bulk limit of the corresponding free-energy density $\omega(x; \rho_{\mu_0}(x), T, \mu_{\text{sat}})$:

$$\lim_{x \to \infty} \omega(x; \rho_{\mu_0}(x), T, \mu_{\text{sat}}) = -P + (\mu_0 - \mu_{\text{sat}})\rho_{\mu_0}^{\text{b}}, \quad (14.22)$$

where $\rho_{\mu_0}^{\text{b}}$ is the bulk limit of the profile $\rho_{\mu_0}(x)$. The height of the unstable liquid film whose density profile is $\rho_{\mu_0}(x)$ at bulk coexistence can be obtained as done in Eq. (14.15) but with the difference between coexistence densities in the denominator:

$$l = \frac{\int_0^\infty (\rho_{\mu_0}(x) - \rho_{\mu_0}^{\text{b}}) dx}{\rho_{\text{l}} - \rho_{\text{g}}}. \quad (14.23)$$

Thus, by computing an isotherm at temperature T with the corresponding set of density profiles $\{\rho(x)\}$ and using Eqs. (14.20)–(14.23), we can tabulate $W(l; T, \mu)$ from Eq. (14.18). In Figure 14.3, we represent the work of film formation at $T = 0.88$ and $\Delta\mu = 0$, computed from the isotherm in Figure 14.2a. The symbols and branch line styles are the same as those on the isotherm. The unstable concave branch of $W(l; T, \mu_{\text{sat}})$ (dashed) is bound by the isotherms' spinodals. We can further immediately make out the stable and metastable branches, which are both convex. We notice that due to phase coexistence at pre-wetting, one of the convex branches (the stable one) must have uniformly smaller free energy than the metastable one. The unstable branch of

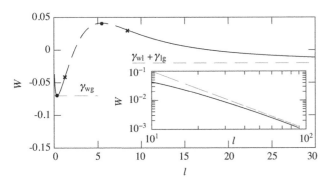

FIGURE 14.3 Thermodynamic work of film formation in Eq. (14.18), computed at μ_{sat} using the DFT isotherm at $T = 0.88$ from Figure 14.2. Filled circles and symbols, respectively, correspond to those in Figure 14.2a, designating the extrema and inflection points of $g(l)$. The dashed branch between the inflection points corresponds to unstable films. It connects the branches of stable thin and metastable thick films. Dashed horizontal lines are drawn at the values of surface tensions, as indicated in the figure. These provide the local minima of $g(l)$. Notice that the second minimum at $\gamma_{\mathrm{wl}} + \gamma_{\mathrm{lg}}$ is reached as $l \to \infty$. The inset shows the asymptotic behaviour of $g(l)$ on a log-log plot, which is in agreement with the macroscopic theory for LJ intermolecular forces. Here the dashed gray line is a guide to the eye drawn at $l \propto x^{-2}$.

$W(l; T, \mu_{\mathrm{sat}})$ ends at the inflection points corresponding to the spinodals of the isotherm. The curve $W(l; T, \mu_{\mathrm{sat}})$ has three extrema: a local minimum at γ_{wg}, followed by the maximum on the unstable branch, which is further followed by the subsequent (metastable) local minimum at $\gamma_{\mathrm{wl}} + \gamma_{\mathrm{lg}}$. The latter minimum is achieved at $x \to \infty$ and corresponds to a completely wet wall. Since $T = 0.88 < T_{\mathrm{w}}$, the infinitely thick film must be metastable, which is highlighted by the relatively higher respective branch of $W(l; T, \mu_{\mathrm{sat}})$.

Our computed $W(l; T, \mu_{\mathrm{sat}})$ seems to beautifully capture the physics of phase coexistence, which is also reflected by the hysteresis S-loop of the adsorption isotherm (the so-called *van der Waals loop*). Thus, the local minimum and maximum of $W(l; T, \mu_{\mathrm{sat}})$ near the origin reflect the fact that we have two solutions to the DFT Euler–Lagrange equation, which can be found at the intersections of the isotherm $l(\mu)$ with the saturation line $\Delta\mu = 0$ [gray dash-dotted line in Figure 14.2a]. In fact, if we compute $W(l; T, \mu)$ at any other μ, we will *only* find local extrema if a straight line drawn at the respective $\Delta\mu$ on top of the adsorption isotherm intersects with $l(\mu)$. We point out that the values of the surface tensions designated in Figure 14.3 with horizontal lines were obtained independently of computing $W(l; T, \mu_{\mathrm{sat}})$, from Eq. (14.13). These values were used here to verify the correctness of the proposed procedure for obtaining $W(l; T, \mu)$. Additionally, we checked if the asymptote of $W(l; T, \mu)$ follows the theoretical prediction for LJ forces of $O(1/x^2)$. As can be seen from the inset, our computed $W(l; T, \mu_{\mathrm{sat}})$ recovers this theoretical asymptote. We thus conclude that the proposed scheme seems to yield a sensible result for $W(l; T, \mu_{\mathrm{sat}})$.

14.6   Direct Comparison Between the DFT and Derjaguin Approaches

In order to perform a direct comparison between the models (14.3) and (14.17), we need to be able to obtain the three-phase contact line using both models under the same conditions, given by T and μ. As far as model (14.17) goes, the procedure outlined in the previous section allows us to compute $g(l)$, which can be used to solve the Euler–Lagrange equation (14.19) and obtain the contact line $l(x)$. We solve equation (14.19) as an initial value problem on a non-equidistant grid, given by our tabulation of $g(l)$. Our starting point is the value of l at the minimum of $g(l)$, and we use a small number, e.g., 10^{-3} to initialize dl/dx. This physically corresponds to a film of vanishing slope. Notice that when $\dot{W} \equiv 1$, the solutions to Eq. (14.19) are circles of curvature γ_{lg}. If $\dot{W} \equiv 0$, the equation defines a straight line – the three-phase line at saturation. Physically this means that at $\Delta\mu = 0$, the solutions must yield a flat liquid–gas interface, inclined to the wall at the angle Θ_{Y}, and at $\Delta\mu \neq 0$, they must trace near-circular droplets or bubbles.

Obtaining the three-phase contact line using the DFT grand free-energy functional (14.3) is more complicated. As we have seen, the density profiles which minimize $\Omega[\rho(\mathbf{r}); T, \mu]$ must have the same symmetry as the external adatom potential $V(\mathbf{r})$. Thus, for an ideally planar wall with the potential given in Eq. (14.2), we should expect only profiles of the kind shown in Figure 14.2 b,c, i.e., with the liquid–gas interface parallel to the wall surface. To get to the solutions representing a three-phase contact line, we can restrict the amount of matter available to the system, forcing it to nucleate a new phase on the wall surface. This would correspond to a canonical statistical–mechanical ensemble and is a valid approach. The shortcoming of such an approach is that computationally we would have to take extra care to distinguish between the effects on the shape of the contact line, associated with the finite size of the system, and those associated with the boundaries of the computational box. We would also have to augment the Euler–Lagrange equation (14.7) with the constraint, representing the fact that the average number of particles is kept constant. This average number of particles must further be properly related to the bulk thermodynamic point, T and μ, to make sure that such quasi-canonical treatment is in full agreement with the grand-canonical functional (14.3).

Here we propose an alternative approach, where we alter the wall potential in a way which is insignificant for the purposes of our investigation but which would break the planar symmetry of the system. The planar symmetry breaking can be achieved, e.g., by slightly lowering the well depth of the fluid–substrate potential locally within a small vicinity of $x = 0$. Since the translational symmetry of $V(x)$ is broken, $\rho(x)$ which extremize $\Omega[\rho(x); T, \mu]$ will be non-planar, i.e., they should exhibit a three-phase contact line. At the same time, inserting such a small impurity into the

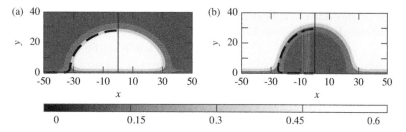

FIGURE 14.4 Comparison between DFT (contour plots) and Derjaguin model (dashed curves) at $T = 0.85 < T_{\mathrm{w}}$ ($\Theta_{\mathrm{Y}} = 65°$) showing a nanodrop (a) at $\Delta\mu = -1.8 \times 10^{-2}$ and a nanobubble (b) at $\Delta\mu = 0.9 \times 10^{-2}$, both having the same height of approximately 30σ. The data is represented by the contour plot, where lighter colors correspond to higher density (see gray bar). The superimposed solutions of equation (14.19) were translated along the x-axis to match the top points of the drop and bubble.

substrate will not change the physics of the system in any significant way, because the actual values of the wall–fluid potential are altered only *locally*. In effect, we are inserting a nucleation seed and obtaining the profiles of the three-phase contact line by allowing the system to create a supercritical nucleation cluster. Thus, we changed ε_{w} to $0.95\varepsilon_{\mathrm{w}}$ in 1σ-vicinity of $x = 0$.

When the fluid above the wall is in the gas phase (i.e., $\Delta\mu < 0$), a local increase of the wall–fluid interactions leads to the nucleation of a nanodrop. Conversely, we get nanobubble nucleation when $\Delta\mu > 0$ by locally lowering the wall–fluid potential. Two representative examples of the density profiles we obtained in this way are provided in Figure 14.4, where we selected the nanodrop and the nanobubble to be of the same height. Approaching bulk coexistence (i.e., $|\Delta\mu| \to 0$) results in the growth of the adsorbed drop/bubble. As its radius grows, the contact angle approaches the Young value Θ_{Y}. Since our system is undersaturated, the configurations we observe are unstable in an open system. On the bulk phase diagram in Figure 14.1a, our system is between the spinodal and binodal, inside the metastable region. Thus, the nanodrops and nanobubbles we obtain are indeed the critical clusters for heterogeneous nucleation on the impurity, inserted in the wall. It also follows that the binding potential we computed earlier is the energy barrier for this nucleation event.

It should be noted that to achieve the same height of the nanobubble and the nanodrop, we had to approach bulk coexistence a lot closer in the case of the nanobubble ($|\Delta\mu = 0.9|$) than in the case of the nanodrop ($|\Delta\mu = 1.8|$). When considered at the same values of $\Delta\mu = 0.9$, the nanobubbles have significantly smaller heights than the nanodrops. This happens because the liquid surrounding the nanobubbles interacts with the substrate much stronger than the gas surrounding the drops. The classical Laplace theory dictates that both the nanodrops and the nanobubbles should have the same size and spherical shape, given by the Laplace radius $R = \gamma_{lg}/\Delta\mu\Delta\rho$. The fact that the nanobubbles appear flatter is solely due to the interface binding potential. This is also apparent, because in the limit $\Delta\mu \to 0$, both configurations must tend to the same flat liquid–gas interface with the inclination angle Θ_{Y}.

In Figure 14.4, we directly superimpose the solutions of the Derjaguin model by translating them along the x-axis to align the top points with those of the DFT nanodrop and nanobubble. As can be seen, the comparison is excellent: the Derjaguin contact line $l(x)$ agrees well with the level sets of the DFT contour plot. It appears that the local Derjaguin approach fully captures the shape of the adsorbed configurations, including the heights of the nanodrops and the nanobubbles and their lateral extents. Notice that the Derjaguin contact line "wraps around" the nanobubble in Figure 14.4b. Obviously, the ordinary differential equation (14.19) cannot yield a bifurcating solution, so the contact line can follow only one "side" of the interface. We used the isotherm of adsorbing liquid film, so our binding potential corresponds to the liquid phase being on the "inside" of the interface. The direct comparison represented in Figure 14.4, as well as the physics exhibited by $g(l)$ and summarised in Figure 14.3, serve as compelling arguments in favor of our method for computing $g(l)$ using DFT.

14.7 Conclusion

In this Chapter, we have discussed the application of classical DFT to problems in adsorption and wetting. Today the cookbook of DFT approximations is very thick, offering sophisticated and accurate functionals which capture the thermodynamic behavior of many types of real fluids, from simple liquids such as noble gases to solutions of colloids and macromolecules. Classical DFT is a microscopic approach, with the typical system sizes in the nanoscale, i.e., of the order of several molecular diameters. In this respect, it is certainly of interest to be able to upscale the information captured by DFT for use in mesoscale (order of hundreds of molecular diameters) and even in the macroscale modelling of static and dynamic wetting phenomena. One way to implement this is via parameter passing, by computing the interface binding potential, which enters well-known continuum-mechanical descriptions of fluid behavior at small scales. We have offered a scheme which takes as input simple one-dimensional DFT computations of adsorbed flat films and returns the binding potential between the liquid–gas interface and the adsorbing wall. Our computed binding potential satisfies the theoretical predictions for asymptotic behavior and also qualitatively captures the physics of wetting.

To further assess the reliability of the computed binding potential, we performed a direct comparison between the droplets and bubbles obtained from unconstrained minimization of the DFT functional in two dimensions and the local Derjaguin model of wetting, which requires the binding potential as input. Along the way, we developed a simple method to capture non-planar solutions of the DFT Euler–Lagrange equation on planar geometry. These solutions are of theoretical value on their own because they provide direct access to critical nucleation clusters of gas and liquid. The direct comparison revealed remarkable agreement between the local Derjaguin model and the fully microscopic nonlocal DFT in terms of the density profiles. Although the level of agreement demonstrated by Figure 14.4 seems remarkable, we point out that the DFT employed here is local in its treatment of the excluded volume interactions, which means that the near-wall structure of the adsorbed droplets is not correctly captured within the distances of 2–3σ from the wall, where the density profiles should exhibit layering. Using a more sophisticated hard-sphere DFT functional such as that in the weighted density approximation and fundamental measure theory [22,23] may yield further insights into approximating the binding potential at small distances. Nevertheless, on the fundamental level, one must appreciate the fact that $g(l)$ is a non-local function of $l(x)$. For this reason, any local treatments of wetting phenomena in terms of partial differential equations are necessarily restricted in accuracy.

Finally, it is certainly of interest to map out the full phase space in the problem of adsorbed nanodrops and nanobubbles and understand to what extent their stability is affected by the spinodals of the planar films. It is also of interest to obtain the binding potential, using the (unstable) isotherm of gas adsorption. Given the present results, we expect that the Derjaguin model employing such isotherms should yield contact lines which "wrap around" the liquid phase, i.e., symmetric to what can be seen in Figure 14.4b. With regard to gas adsorption, one should bear in mind that the binding potential obtained from the gas adsorption isotherm cannot have two minima with an attractive wall that favors liquid, because thick gas films on such wall are thermodynamically unstable. For such a gas, $g(l)$ can only have a single minimum, at small value of l, corresponding to thin metastable gas films. This thin gas film can be seen in Figure 14.4a between the adsorbed drop and the substrate. Another interesting potential direction for future research is to consider layering transitions, where liquid adsorption isotherms exhibit multiple S-loops. These transitions happen at low T, close to the bulk triple point of liquid–gas–solid coexistence in the fluid, and must lead to several local minima in $g(l)$. It is anticipated that the shape of the three-phase contact line in this case may be stair-like near the substrate.

Acknowledgments

We acknowledge financial support from the Engineering and Physical Sciences Research Council of the UK through Grant Nos. EP/L027186 and EP/L020564 and from the European Research Council through Advanced Grant No. 247031.

Appendix: Aspects of Numerical Implementation of the DFT Euler–Lagrange Equation

Here we discuss some challenges associated with the numerical solution of Eq. (14.7) in one and two dimensions. First, we note that the non-local term in (14.7) can be simplified considerably by integrating over the directions of constant density. Given our chosen system of units, where $\sigma = 1$ and $\varepsilon = 1$, the expression (14.6) in the planar problem of film adsorption takes the following form:

$$\varphi_{\text{attr}}(y) = \begin{cases} -\dfrac{6\pi}{5}, & \text{if } |y| \leq 1, \\[2mm] 4\pi\left(\dfrac{1}{5y^{10}} - \dfrac{1}{2y^4}\right), & \text{if } |y| > 1. \end{cases} \tag{14.24}$$

The problem of droplet adsorption is numerically two-dimensional:

$$\varphi_{\text{attr}}(x,y) = \begin{cases} 2\displaystyle\int_{\sqrt{1-r^2}}^{\infty} z \; \varphi_{1,1}^{6-12}\left(\sqrt{r^2+z^2}\right), & \text{if } r \leq 1, \\[4mm] \dfrac{3\pi}{2}\left[-\left(\dfrac{1}{r}\right)^5 + \dfrac{21}{32}\left(\dfrac{1}{r}\right)^{11}\right], & \text{if } r > 1, \end{cases} \tag{14.25}$$

where $r = \sqrt{x^2+y^2}$. In what follows, we highlight the main challenges of numerical implementation of Eq. (14.7).

There are two main challenges associated with the non-local Euler–Lagrange equation of DFT. The first one is to discretize the non-local convolution-like term, and the second one is to solve the resulting system of non-linear algebraic equations. Thus, existing methods in the literature can be categorized according to the choice of quadrature and the non-linear solver. Many authors argue in favor of employing a fast Fourier transform, with subsequent Piccard-like iterations to get to the final solution, see, e.g., the review in Ref. [42]. But it is widely accepted amongst the computational community that Newton-like solvers may be more efficient and numerically stable [43]. In our recent studies, we highlighted the advantages of evaluating the quadratures in real space by using pseudospectral collocation methods and the Gauss quadrature [16,22]. If properly implemented, such approach is capable of achieving exponential convergence rates with the size of the computational grid, while Fourier transforms can have only algebraic convergence.

A faster convergence rate allows us to use significantly fewer grid points while retaining high accuracy. This is especially important in two-dimensional problems.

In a spectral method, we discretize $\rho(\mathbf{r})$ on a non-uniform grid of collocation points, constructing a global interpolating function. It can be shown that when the collocation points which cover the physical domain are obtained from the set of so-called Chebyshev points via a conformal map, the numerical convergence of the quadrature is exponential [44]. The challenge of practical implementation is to choose the conformal map in such a way that the set of collocation points is dense near the adsorbing walls (where the density is expected to vary sharply), and at the same time, there are enough grid points far from the walls so that not only the liquid–gas interface but also the asymptote of the density decay to the bulk value is captured [23]. According to Eq. (14.20), capturing the correct asymptote of the density profile is crucial in obtaining the right asymptote of the binding potential as shown in the inset of Figure 14.3.

In practice, we tried to position the last point at a distance of about $10^3\sigma$ from the adsorbing wall. For planar adsorption, we used grids of about 130 points to capture all the asymptotes and get correct binding potential. When dealing with adsorption of drops and bubbles, we used about 100 grid points along the x-axis and about 50 grid points along the y-axis. We exploited the fact that the expected solutions must possess reflective symmetry about the origin for accuracy control. After the integral is discretized, we solve the resulting non-linear system using Newton's method. Although the method requires some type of matrix inversion, the number of iterations is typically quite small. This should be contrasted with Piccard-like schemes, which do not require matrix inversions but take hundreds of iterations. We found that both in one and two dimensions, Newton's algorithm outperforms Piccard, but this can change in three-dimensional problems, due to much higher system sizes. In practice, we are capable of achieving a relative tolerance of 10^{-7} within two to three Newton steps. The computation is further accelerated by repackaging quadratures as matrix-vector products as described in Ref. [22]. In practice, the computation of a two-dimensional density profile, such as one of those shown in Figure 14.4, takes several seconds on an ordinary desktop machine.

When dealing with systems that can undergo a phase transition, there are additional problems associated with choosing the initial guess in the region of metastability. For example, from Figure 14.2a, it can be seen that for μ falling inside the S-loop, there must be three solutions, corresponding to stable, unstable, and metastable adsorbed liquid films. Obviously, all three configurations must solve Eq. (14.7) at the same μ and T, requiring one to be particularly careful with the initial guess. The problem of optimally choosing the initial guess may be solved systematically by employing a pseudo arc-length continuation technique [22,45]. It proceeds by treating the control parameter (μ in our case) in the discretized system of equations as an unknown and closing the system with a geometric constraint of curve continuity in the phase space of the discretized solutions. By starting at a very small value of μ, far from any phase transitions, where the system is mostly gas-like, we are able to trace the solutions through the region of metastability. At every step along μ, the continuation algorithm described in Ref. [22] serves to update the value of μ in such a way as to provide an optimal initial guess for the subsequent step.

References

1. T.M. Squires and S. Quake, Microfluidics: Fluid physics at the nanoliter scale, *Rev. Mod. Phys.* **77**, 977 (2005).

2. M. Rauscher and S. Dietrich, Wetting Phenomena in Nanofluidics, *Annu. Rev. Mater. Res.* **38**, 143 (2008).

3. Z. Gou and W. Liu, Biomimic From the Superhydrophobic Plant Leaves in Nature: Binary Structure and Unitary Structure, *Plant Sci.* **172**, 1103 (2007).

4. A. Calvo, B. Yameen, F.J. Williams, G.J.A.A. Soler-Illia and O. Azzaroni, Mesoporous Films and Polymer Brushes Helping Each Other to Modulate Ionic Transport in Nanoconfined Environments. An Interesting Example of Synergism in Functional Hybrid Assemblies, *J. Am. Chem. Soc.* **131**, 10866 (2009).

5. K.W. Schwarz and J. Tersoff, From Droplets to Nanowires: Dynamics of Vapor-Liquid-Solid Growth, *Phys. Rev. Lett.* **102** (2009).

6. R.E. Algra, M.A. Verheijen, L.F. Feiner, G.G.W. Immink, W.J.P. van Enckevort, E. Vlieg and E.P.A.M. Bakkers, The Role of Surface Energies and Chemical Potential during Nanowire Growth, *Nano Lett.* **11**, 1259 (2011).

7. D. Lohse and X. Zhang, Surface Nanobubbles and Nanodroplets, *Rev. Mod. Phys.* **87**, 981 (2015).

8. N. Savva and S. Kalliadasis, Dynamics of Moving Contact Lines: A Comparison Between Slip and Precursor Film Models, *Europhys. Lett.* **94**, 64004 (2011).

9. A.O. Parry, C. Rascón and A.J. Wood, Universality for 2D Wedge Wetting, *Phys. Rev. Lett.* **83**, 5535 (1999).

10. A.O. Parry, C. Rascón and A.J. Wood, Critical Effects at 3D Wedge Wetting, *Phys. Rev. Lett.* **85**, 345 (2000).

11. C. Rascón and A.O. Parry, Geometry-Dominated Fluid Adsorption on Sculpted Solid Substrates, *Nature* **407**, 986 (2000).

12. A.O. Parry, C. Rascón, N.B. Wilding and R. Evans, Condensation in a Capped Capillary is a

Continuous Critical Phenomenon, *Phys. Rev. Lett.* **98**, 226101 (2007).

13. K. Binder, Modelling of Wetting in Restricted Geometries, *Annu. Rev. Mater. Res.* **38**, 123 (2008).

14. A. Checco, B.M. Ocko, M. Tasinkevych and S. Dietrich, Stability of Thin Wetting Films on Chemically Nanostructured Surfaces, *Phys. Rev. Lett.* **109**, 166101 (2012).

15. A.O. Parry, C. Rascón, E.A.G. Jamie and D.G.A.L. Aarts, Capillary Emptying and Short-Range Wetting, *Phys. Rev. Lett.* **108**, 246101 (2012).

16. P. Yatsyshin, N. Savva and S. Kalliadasis, Geometry-Induced Phase Transition in Fluids: Capillary Prewetting, *Phys. Rev. E* **87**, 020402(R) (2013).

17. C. Rascón, A.O. Parry, R. Nurnberg, A. Pozzato, M. Tormen, L. Bruschi and G. Mistura, The Order of Condensation in Capillary Grooves, *J. Phys.: Condens. Matter* **25**, 192101 (2013).

18. P. Yatsyshin, N. Savva and S. Kalliadasis, Wetting of Prototypical One- and Two-Dimensional Systems: Thermodynamics and Density Functional Theory, *J. Chem. Phys.* **142**, 034708 (2015).

19. P. Yatsyshin, N. Savva and S. Kalliadasis, Density Functional Study of Condensation in Capped Capillaries, *J. Phys.: Condens. Matter* **27**, 275104 (2015).

20. J.F. Lutsko, Recent Developments in Classical Density Functional Theory, *Adv. Chem. Phys.* **144**, 1 (2010).

21. P. Yatsyshin, A.O. Parry, C. Rascón and S. Kalliadasis, Classical Density Functional Study of Wetting Transitions on Nanopatterned Surfaces, *J. Phys.: Condens. Matter* **29**, 094001 (2017).

22. P. Yatsyshin, N. Savva and S. Kalliadasis, Spectral Methods for the Equations of Classical Density-Functional Theory: Relaxation Dynamics of Microscopic Films, *J. Chem. Phys.* **136**, 124113 (2012).

23. A. Nold, B.D. Goddard, P. Yatsyshin, N. Savva and S. Kalliadasis, Pseudospectral Methods for Density Functional Theory in Bounded and Unbounded Domains, *J. Comput. Phys.* **334**, 639 (2017).

24. P. Yatsyshin and S. Kalliadasis, Mean-field Phenomenology of Wetting in Nanogrooves, *Mol. Phys.* **114**, 2688 (2016).

25. R. Evans, The Nature of the Liquid-Vapour Interface and Other Topics in the Statistical Mechanics of Non-Uniform, Classical Fluids, *Adv. Phys.* **28**, 143 (1979).

26. R. Evans, Density functionals in the theory of nonuniform fluids, In *Fundamentals of Inhomogeneous Fluids* edited by D. Henderson (New York: Dekker), p. 85.

27. J. Wu, Density Functional Theory for Chemical Engineering: From Capillarity to Soft Materials, *AIChE J.* **52**, 1169 (2006).

28. J. Landers, J.Yu. Gor and A.V. Neimark, Density Functional Theory Methods for Characterization of Porous Materials, *Colloid Surface A* **437**, 3 (2013).

29. J.A. Barker and D. Henderson, Perturbation Theory and Equation of State for Fluids. II. A Successful Theory of Liquids, *J. Chem. Phys.* **47**, 4714 (1967).

30. A. Pereira and S. Kalliadasis, Equilibrium Gas-Liquid-Solid Contact Angle From Density-Functional Theory, *J. Fluid Mech.* **692**, 53 (2012).

31. A. Nold, D.N. Sibley, B.D. Goddard and S. Kalliadasis, Nanoscale Fluid Structure of Liquid-solid-vapour Contact Lines for a Wide Range of Contact Angles, *Math. Model. Nat. Phenom.* **10**, 111 (2015).

32. P. Yatsyshin, A.O. Parry and S. Kalliadasis, Complete Prewetting, *J. Phys.: Condens. Matter* **28**, 275001 (2016).

33. D. Henderson, Disjoining Pressure of Planar Adsorbed Films, *Eur. Phys. J. Spec. Top.* **197**, 115 (2011).

34. G.A. Darbellay and J.M. Yeomans, An Interface Potential Approach to Capillary Condensation in a Rectangular Groove, *J. Phys. A-Math. Gen.* **25**, 4275 (1992).

35. A. Oron, S.H. Davis and S.G. Bankoff, Long-scale Evolution of Thin Liquid Films, *Rev. Mod. Phys.* **69**, 931 (1997).

36. M.C. Dallaston, M.A. Fontelos, D. Tseluiko and S. Kalliadasis, Discrete Self-similarity in Interfacial Hydrodynamics and the Formation of Iterated Structures, *Phys. Rev. Lett.* **120**, 034505 (2018).

37. S. Dietrich and M. Napiorkowski, Analytic Results for Wetting Transitions in the Presence of Van der Waals Tails, *Phys. Rev. A* **43**, 1861 (1991).

38. M. Zeng, J. Mi and C. Zhong, Wetting behavior of spherical nanoparticles at a vapor-liquid interface: a density functional theory study, *Phys. Chem. Chem. Phys.* **13**, 3932 (2011).

39. Y.X. Yu, A Novel Weighted Density Functional Theory for Adsorption, Fluid-Solid Interfacial Tension, and Disjoining Properties of Simple Liquid Films on Planar Solid Surfaces, *J. Chem. Phys.* **131**, 024704-1–024704-11 (2009),

40. L.G. MacDowell, Computer Simulation of Interface Potentials: Towards a First Principle Description of Complex Interfaces?, *Eur. Phys. J. Spec. Top.* **197** (1), 131–145 (2011).

41. A.P. Hughes, U. Thiele and A.J. Archer, Influence of the Fluid Structure on the Binding Potential: Comparing Liquid Drop Profiles from Density Functional Theory with Results from Mesoscopic Theory, *J. Chem. Phys.* **146** (6), 064705 (2017).

42. R. Roth, Fundamental Measure Theory for Hard-Sphere Mixtures: a Review, *J. Phys. Condens. Matter* **22**, 063102 (2010).

43. L.J.D. Frink, A.G. Salinger, M.P. Sears, J.D. Weinhold and A.L. Frischknecht, Numerical Challenges in the Application of Density Functional Theory to Biology and Nanotechnology, *J.Phys.: Condens. Matter* **14**, 12167 (2002).

44. N. Hale and L.N. Trefethen, New Quadrature Formulas from Conformal Maps, *SIAM J. Numer. Anal.*, **46** (2), 930 (2008).

45. A.G. Salinger and L.J.D. Frink, Rapid Analysis of Phase Behavior with Density Functional Theory. I. Novel Numerical Methods, *J. Chem. Phys.* **118** (16), 7457 (2003).

15

Water Flow in Graphene Nanochannels

Seungha Shin
The University of Tennessee – Knoxville

15.1 Introduction

Water transport through nanoscale confinements is of great interest in various research areas (fluid and thermal transport, manufacturing, biology, materials science, etc.), as it can be applied to innovative, practical applications. Through a large surface area–volume ratio, nanosized (<100 nm) channels can enhance heat dissipation for thermal management (Hanks et al., 2018), energy conversion/storage (Park et al., 2014; Ghasemi et al., 2014), water filtration and desalination (Lee et al., 2011; Yang et al., 2013; Lee et al., 2015), and biological and chemical separations (Han et al., 2000). In addition, the small scale of flow rate and its delicate control enable realization of novel drug delivery and lab-on-a-chip devices (Kovarik et al., 2009; Naguib et al., 2004; Yang et al., 2017).

Water molecules (H_2O) inside nanoscale channels are under a strong influence of interaction with channel surface, and this influence depends on channel material in addition to size; thus, a selection of channel material is important in water-flow nanochannel applications. Here, graphene and carbon nanotubes (CNTs), which have been intensively studied for the last two decades, are considered as channel materials. Graphene and CNTs, as allotropes of carbon, consist of sp^2-bonded carbon atoms in a hexagonal honeycomb lattice and possess exceptional electrical and thermal transport and mechanical strength and other unique physical properties (Novoselov, 2011; Nomura et al., 2007; Frank et al., 2007; Falkovsky, 2008; Balandin, 2011).

Graphene, as a two-dimensional (2D) material with a single or a few layers of carbon atoms can be an ultimately thin membrane. Because a flux across a membrane scales inversely with the membrane's thickness, atomically thin and mechanically strong graphene membranes offer the promise of greatly increased water permeability for water filtration and desalination (Cohen-Tanugi et al., 2012). Moreover, its high thermal conductivity ensures an effective heat dissipation in thermal management applications. CNTs are hollow, cylindrical structures, essentially sheets of graphene rolled into cylinders, and basically share superior properties of graphene. As innate nanochannels with atomic smoothness, CNTs are ideal materials for studying such water flows in nanochannels, and their properties can be further controlled by the angle at which they are rolled (their "chirality"), diameter, and wall thickness (in single-walled or multiwalled CNTs) (Knowles, 2009).

In nanochannels, atomic-level properties (e.g., structure and interaction) are more prominent because of the molecular dimension, and unique properties of graphene and CNTs will also affect water behaviors, and thus, properties quite different from those seen for the bulk counterpart are expected and have been reported. These nanoconfinement characteristics underlie the functionality of many nanochannel applications and motivate further investigations along with attractive properties of graphene and CNTs. In this chapter, thermophysical properties of confined water and hydrodynamic and thermal transport of water flow through graphene or CNT nanochannels are discussed, highlighting their distinct features different from macroscale properties. Recent research and possible applications of water-flow graphene/CNT nanochannels are also introduced.

15.2 Thermophysical Properties of Water in Nanochannels

Thermophysical properties of water (boiling point, freezing point, atomic structure, etc.) are very well known; however, when water is in graphene or CNT nanochannels, dominant confinement effects and the resulting unique atomic structures lead to thermodynamic properties very different from those of bulk water.

The water structure and ice formation in nanochannels have been investigated by experimental measurements and computational simulations. Measurements to identify atomic-scale water structure employ electron microscopy (Naguib et al., 2004), X-ray diffraction (Maniwa et al., 2002; Maniwa et al., 2005), nuclear magnetic resonance (NMR) (Ghosh et al., 2004), neutron diffraction (Takaiwa et al., 2008), and vibrational spectroscopy (Byl et al., 2006), and molecular dynamics (MD) simulation, where physical motion of molecules are simulated by classical equations of motion (Frenkel et al., 2002), is widely used in computational modeling (Koga et al., 2001). Figure 15.1a,b shows micrographic images of a nanochannel and water-filled channel obtained via transmission electron microscopy (TEM), and this system can be modeled using MD as in Figure 15.1c. In experiments and simulations on nanochannels, size (diameter) of channels, thickness of channel wall (e.g., in single-, double-, and multiwalled nanotubes), pressure, and temperature are controlled to understand thermophysical behaviors.

The previous studies have shown that the confined water freezes into square, pentagonal, hexagonal, and heptagonal ice nanotubes, and unexpectedly it does so either continuously (unlike any bulk substances, including bulk water) or discontinuously (despite the fact that it is essentially in one dimension), depending on the diameter of CNTs or the applied pressure (Takaiwa et al., 2008). The formation of different ice structures is attributed to the energy minimization process, which maximizes the number of hydrogen bonds under cylindrical confinement.

In MD simulations, thermodynamic states of water within CNT channels are examined, varying a temperature and CNT diameter. Takaiwa et al. (2008) proposed a phase diagram of water in single-walled CNT via thermodynamic study and claimed that the global maximum in the melting curve appears at 1.1 nm of diameter, where water freezes in a square-ice nanotube even at room temperature, as shown in Figure 15.2a. Experimental infrared spectroscopic measurements in a recent study reported different water structures: (i) 1D structure of confined water in less than 0.8 nm of diameter nanotubes, (ii) its ice-like structure with a diameter range of 1.1–1.2 nm, and (iii) the presence of a layer with "free" OH bonds facing the nanochannel wall for water in a larger-diameter nanotube. The observation of ice-like water corroborates the theoretical approach developed to explain the spontaneous filling of hydrophobic CNTs (Pascal et al., 2011), a counter-intuitive phenomenon.

Similar to freezing, water boiling in nanochannels deviates from bulk water due to the confinement, and a recent computational study using classical MD simulations reported the boiling behaviors in CNT nanochannels (Chaban et al., 2011). In the research, CNT diameter greatly affects the liquid-/vapor-phase transition temperature (smaller the channel dimension, higher the boiling temperature), and vapor pressures of several hundred atmospheres can be achieved easily by a small temperature growth above the phase transition (see Figure 15.2b).

As in CNTs, peculiar water structures are also observed in graphene nanocapillaries, where water is locked between two graphene sheets. High-resolution electron microscopy imaging and MD simulations showed that the nanoconfined water at room temperature forms "square ice"—a phase having symmetry qualitatively different from the conventional tetrahedral geometry of hydrogen bonding between water molecules (See Figure 15.3a–c). Square ice has a high packing density with a lattice constant of 2.83 Å and can assemble in bi- and trilayer crystallites (Algara-Siller et al., 2015). Due to the atomic structure and interfacial interaction, thermodynamic properties of water confined in graphene are also expected to deviate from the bulk counterpart.

15.3 Hydrodynamic Transport in Nanochannels

15.3.1 Macroscale Transport

Pressure difference is a driving force of fluid motion, and friction and drag during the fluid flow limit a velocity of fluid (u_f). In a hydrodynamic transport analysis, the calculation of the pressure drop for a given flow rate (power

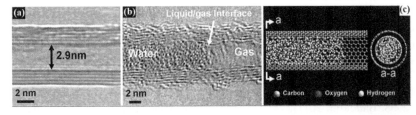

FIGURE 15.1 (a) Micrograph image of an empty nanotube with an inner diameter of 2.9 nm obtained from transmission electron microscopy (TEM). (b) Partially filled water in the 4-nm CNT channel. (c) Molecular simulation of water in a (30, 30) nanotube with a diameter of 4.07 nm, illustrating how water is arranged inside the nanotube of the same diameter as in (b) (Naguib et al., 2004).

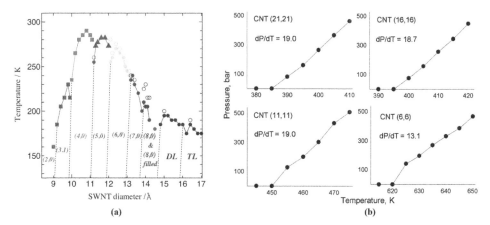

FIGURE 15.2 (a) Calculated phase change temperature of water in single-walled CNTs at atmospheric pressure with respect to the CNT diameter. Squares, triangles, and filled circles denote the temperature above which an ice phase becomes unstable and breaks into clusters upon heating, the temperature for continuous freezing and melting, and the freezing point at which liquid water freezes abruptly. Open circles indicate the existence of hysteresis, i.e., the highest temperature at which ice does not melt upon heating (Takaiwa et al., 2008). (b) Vapor pressure of the confined water with respect to temperature within CNTs of different sizes (Chaban and Prezhdo, 2011).

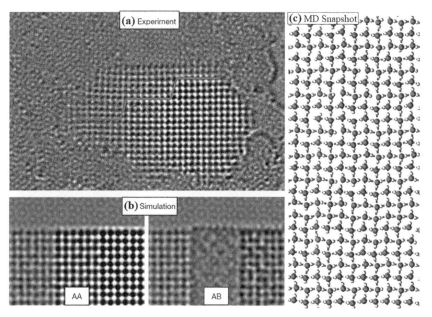

FIGURE 15.3 (a) Isolated crystallite with a varying number of layers within encapsulating graphene layers. Changes in the contrast averaged over the corresponding parts of the image occur in quantized steps. (b) Simulated TEM images for monolayer, bilayer, and trilayer ice with AA and AB stacking, respectively. The AA stacking agrees well with the experimental images, whereas the AB stacked ice results in the qualitatively different appearance. (c) Typical snapshot of MD-simulated water (dark spheres for oxygen atoms and white ones for hydrogen) in a graphene nanocapillary (Algara-Siller et al., 2015).

requirement = pressure drop × volume flow rate) is of main interest. Pressure drop (Δp) is obtained by integrating the shear stress on fluid, and the shear stress (τ_s) in a Newtonian fluid is proportional to velocity gradient ($\partial u_f/\partial z$), i.e.,

$$\tau_s = \mu \frac{\partial u_f}{\partial z}. \quad (15.1)$$

Here, the z direction is normal to the fluid–channel interface, and the proportionality constant μ is the dynamic viscosity of fluid (N-s/m^2). Fluid velocity distribution should be identified in the analysis, and processing the resulting velocity, shear stress and pressure drop are calculated.

In a macroscale analysis, a simulated system is regarded as continuum, and the behavior of a liquid can be described in terms of infinitesimal volume elements that are small compared to the flow domain but have well-defined thermophysical properties. Applying Newton's second law to a system of volumetric elements gives rise to the Cauchy and Navier–Stokes equations, which can be used to derive the Poiseuille and other continuum-level flow relations (Landau et al., 1986).

For an internal flow, the friction factor, instead of calculating detailed velocity profile, is often introduced to facilitate the friction calculation and defined as

$$f = -\frac{(dp/dx)\, D_h}{\rho u_m^2/2}, \qquad (15.2)$$

where x is the flow direction, ρ is the fluid density, u_m is the mean fluid velocity, and D_h is the hydraulic diameter, defined as the ratio of four times flow cross-sectional area (A_c) to wetted perimeter (P) or $4A_c/P$. In macroscale, the friction factor is a function of dimensionless Reynolds number (Re, ratio of inertia to viscous force; i.e., $\rho u_m D_h/\mu$) and surface roughness, and Moody's chart (Moody, 1944) is used for quick calculations of friction and pressure drop.

15.3.2 Nanoscale Considerations in Hydrodynamic Transport

An important step toward understanding liquid flow in nanoscale systems is to predict the transition from continuum to subcontinuum transport as the flow area decreases. In a system where the size of a liquid molecule is comparable to the size of the flow domain, however, the notions of a representative volumetric element and continuum-based relations are invalid. Within such "subcontinuum" systems, the movement of individual molecules must be considered when predicting mass and momentum transport (Verweij et al., 2007).

In addition, as the channel size decreases, a larger area of channel is under the influence of the surface, and in a very small dimension of channel, surface–fluid molecule interactions dominate over fluid–fluid interactions. The significance of surface interaction, or interfacial effect, in surface–fluid systems can be characterized by the Knudsen number (Kn),

which is defined as the ratio of the effective mean free path of fluid molecules (λ) to the channel dimension ($Kn = \lambda/D_h$) (Kaviany, 2014). For liquid water, the mean free path λ is approximated as the average intermolecular distance between two water molecules (or molecular diameter); i.e., $\lambda \sim (V_m/N_A)^{1/3} \sim 0.3$ nm, where V_m is the molar volume and N_A is Avogadro's number (Holt et al., 2006; Probstein, 2005; Gad-el-Hak, 2001). The smallest diameter of water-containing CNTs, which have been examined so far (Chaban and Prezhdo, 2011), is ~ 0.9 nm (c.f. the smallest CNT diameter ~ 0.3 nm (Zhao et al., 2004)), and Kn of water-flow nanochannel ranges between 0.003 and 0.3 ($0.003 < Kn < 0.3$). This range is neither fully in the free molecular nor fully in the continuum regime (Kandlikar et al., 2005; Gad-el-Hak, 2005), requiring both nanoscale and continuum approaches for the analysis of hydrodynamic transport.

Nanoscale features in mass, momentum, and energy distributions are usually neglected in macroscale analysis, but they should be included in the nanoscale consideration due to more significant influences on system properties. Specifically, density profile shows high oscillations in liquid layers adjacent to micro- and nanochannel walls, and as a consequence, there is an inhomogeneity of the liquid flow, which is not included in the classical hydrodynamic approach. As in Figure 15.4 from MD simulations (Marable et al., 2017), the oscillation has a different shape (maximum position and amplitude) depending on intermolecular interaction strength (larger the interatomic potential parameter ε_{c-o}, stronger the interaction between water and surface). Since, in nanochannels, dimension of a density-oscillating

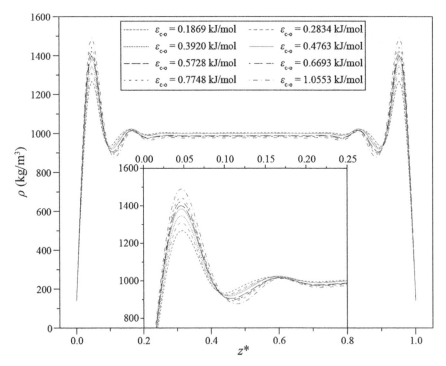

FIGURE 15.4 Water density profiles (ρ) within a nanochannel of two parallel graphene layers for increasing values of solid–fluid interaction strengths (ε_{c-o}). The distance between the two walls (H) is 5 nm, and z^* is the normalized z coordinate ($z^* = z/H$). The inset is a magnified image of density at the interfacial region (Marable et al., 2017).

region near a channel surface is comparable to the channel size, this interfacial density distribution should be properly addressed in nanoscale while ignored in macroscale analysis.

In addition to the interfacial density profile, velocity slip on the fluid–channel interface is more distinct and important in nanoscale analysis. In macroscale analysis, interfacial fluid is assumed to have the same velocity as the channel wall; thus, the velocity of the fluid relative to the channel surface is regarded as zero, and this is termed "no-slip boundary condition" ($\Delta u_{f/s} = u_f - u_s = 0$, where u_f and u_s are the velocities of interfacial fluid and solid surface). However, a finite strength of intermolecular interaction between fluid and solid surface molecules allows interfacial fluid molecules to have a net velocity over the surface, i.e., the surface velocity slippage occurs. This slip condition is used for the fluid boundary with surface in a hydrodynamic analysis, and a velocity (or hydrodynamic) slip length (L_s) characterizes the slippage. L_s for a stationary channel is calculated using the interface velocity and velocity gradient as

$$L_s = \frac{u_{f,z=0}}{(\partial u_f/\partial z)_{z=0}}. \tag{15.3}$$

Figure 15.5 summarizes three boundary conditions for a Poiseuille flow of a stationary channel: (i) no slip, (ii) partial slip, and (iii) perfect slip.

With a weaker surface–fluid molecule interaction, less molecules stick on the surface, resulting in a larger slip as shown in Figure 15.6. Not only the slippage but also the surface wettability of liquid is directly influenced by the surface–fluid interaction strength (relative to fluid intermolecular interaction); a stronger surface–fluid interaction leads to a better wettability or smaller wetting contact angle. In particular, for water molecules, hydrophobic surfaces with large water contact angles originate from weak surface–water contact angles. Since water molecules are known to interact weakly with pristine graphene and CNTs, graphene and CNTs are hydrophobic (having a large wetting contact angle) and also lead to a large slippage.

The nanoscale behaviors mentioned above, i.e., subcontinuum, density oscillation, and surface slippage, need to be addressed in the analysis of nanoscale channel flow. Moreover, the effects of the nanoscale behaviors are more significant in channels made from hydrophobic graphene/CNTs. Thus, although continuum assumption and classical hydrodynamics have been used to explain liquid transport at the nanoscale (Park and Jung, 2014;

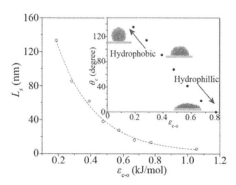

FIGURE 15.6 Velocity slip lengths (L_s) for varying well-fluid interaction strengths (ε_{c-o}) in MD simulations of a Poiseuille flow between parallel graphene plates. Inset shows the wetting contact angle (θ_c) increases and the graphene becomes more hydrophobic as ε_{c-o} decreases (Marable et al., 2017).

Bocquet et al., 2010; Travis et al., 2000; Thomas et al., 2008), hydrodynamic transport in graphene/CNT channels cannot be accurately described using conventional continuum fluid mechanics with its associated linear constitutive relations and no-slip boundary conditions (Thomas et al., 2009; Podolska et al., 2013). Velocity profiles and viscosities predicted by classical Navier–Stokes are violated at certain critical channel widths and diameters (Podolska and Zhmakin, 2013). In the case of water, this critical width or diameter of the channel corresponds to approximately 1.4 nm for the velocity profile and 5.1 nm for the viscosity (Sofos et al., 2009); i.e., below this dimension of water-flow channel, nanoscale analysis considering atomistic details should be employed. For enhanced modeling, MD method (simulations), which directly addresses behaviors of all confined atoms, has been used for theoretical and numerical studies (Squires et al., 2005; Eijkel et al., 2005; Whitesides, 2006; Phillips et al., 2005; Bizzarri et al., 2002; Åman et al., 2003; Senapati et al., 2003; Heslot et al., 1989; Varshney et al., 1994; Thomas and McGaughey, 2009).

15.3.3 Water Flow in Nanochannels

The most distinct feature of water flow in graphene/CNT is the exceptionally high flow rate. Recent experiments (Majumder et al., 2005; Holt et al., 2006; Whitby et al., 2008) and MD simulations (Suk et al., 2008; Joseph et al., 2008; Corry, 2008; Thomas et al., 2010; Nicholls et al., 2012) have shown that water is transported through CNTs at

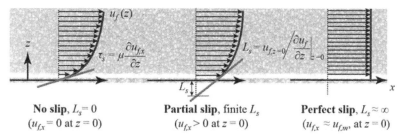

FIGURE 15.5 Velocity slip length calculation and three slip cases (no-slip, partial-slip, and perfect-slip conditions).

unexpectedly high flow rate. The flow rates of pressure-driven water through membranes of 1.6 and 7 nm diameter CNTs (Majumder et al., 2005; Holt et al., 2006) are two to five orders-of-magnitude greater than those predicted by the continuum-based no-slip Hagen–Poiseuille relation.

A popular approach to explain this enhancement is to introduce a slip length into the mathematical model; that is, the no-slip boundary conditions are replaced by the slip conditions. The surface slip results in a flow rate larger than that without slip. For a Poiseuille flow, flow enhancement (ξ) by the slip effect is calculated as (Kannam et al., 2012)

$$\xi = \frac{Q_S}{Q_N} = \left(1 + \frac{6L_s}{\delta}\right) \text{ for planar flow, and} \quad (15.4a)$$

$$\xi = \left(1 + \frac{4L_s}{R}\right) \text{ for circular flow,} \quad (15.4b)$$

where Q_S is the flow rate with the slip, Q_N is the non-slip flow rate, L_s is the slip length, and δ and R represent the channel size (height and radius, respectively). Thus, a larger slip induces a larger flow rate, and the enhancement is more significant in a smaller-scale channel. However, the flow enhancement values, observed in experiments and numerical simulations, are much larger than those obtained from this calculation; thus, additional mechanisms have been suggested to explain the extraordinary flow enhancement.

In addition to the slippage, depletion layers or density oscillations near water/channel (Alexeyev et al., 1996; Neto et al., 2005; Cottin-Bizonne et al., 2005; Joseph and Aluru, 2008) affect interfacial-layer (near the channel wall) hydrodynamics, resulting in a lower viscosity (Podolska and Zhmakin, 2013). More fundamentally, as discussed above, a different water structure induced by nanoscale confinement is also considered as a possible reason for the exceptional flow enhancement. Specifically, the existence of low-dimensional ice at room temperature has been proposed to explain the fast water permeation through hydrophobic nanocapillaries, including CNTs and graphene-based membranes (Algara-Siller et al., 2015).

15.3.4 Applications of Water Transport in Graphene/CNT Nanochannel

The reported high flow rate in nanochannel has a great effect on molecular sieving, chemical detection, and drug delivery fields, where such high flow rates would significantly increase device efficiency, accuracy, and throughput (Noy et al., 2007).

Water purification and desalination are promising approaches to supply new fresh water in the context of a rapidly growing global water gap. Filtration and separation using membranes, among various approaches, has attracted attention due to its higher energy efficiency (Elimelech et al., 2011; Spiegler et al., 2001; Addams et al., 2009). In this approach, excellent selectivity, fast permeation, and robust membrane structure to withstand the applied pressure are required for more efficient processes (Huang et al., 2013). Graphene is an excellent candidate because of its monoatomic thickness, 2D structure, high mechanical strength, chemical inertness, and impermeability to all gases (Lee et al., 2008; Novoselov et al., 2012; Bunch et al., 2008).

Theoretical studies have demonstrated that a monoatomically thick porous graphene membrane exhibits superior separation performance for gases (Jiang et al., 2009), water (Cohen-Tanugi and Grossman, 2012), and ions (Suk et al., 2010; Sint et al., 2009). In addition, adjustment of pore size and chemical functionalization in nanopore graphene can further improve the desalination performance (Cohen-Tanugi and Grossman, 2012) (Figure 15.7a). Experimental studies on nanostrand-channeled graphene oxide ultrafiltration membranes with a network of nanochannels

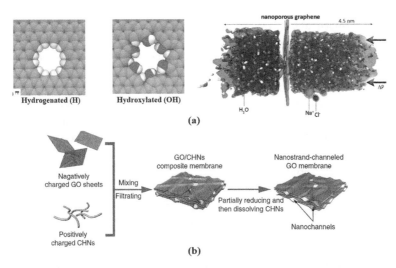

FIGURE 15.7 For aqueous separation applications, (a) MD simulation study of hydrogenated (H) and hydroxylated (OH) graphene pores for calculation of water permeability and salt rejection (Cohen-Tanugi and Grossman, 2012) and (b) experimental study of water flow in graphene-oxide (GO) membrane (Huang et al., 2013).

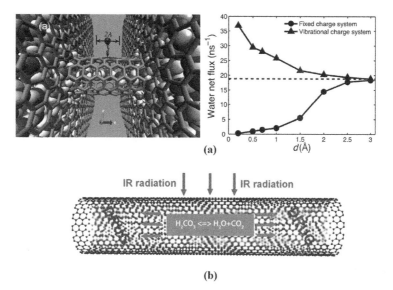

FIGURE 15.8 Nanochannel flow control by using (a) fixed and vibrational charges (Kou et al., 2015) and (b) infrared radiation heating and chemical reaction (Chaban and Prezhdo, 2011).

with a narrow size distribution (3–5 nm, Figure 15.7b) reveal a tenfold enhancement of permeance without sacrificing the rejection rate compared with that of graphene oxide membranes (and more than 100 times higher than that of commercial ultrafiltration membranes with similar rejection) (Huang et al., 2013). The flow enhancement is attributed to the porous structure and significantly reduced channel length.

Above the flow enhancement, advanced control of hydrodynamic flow should be developed for more extensive applications of nanochannels. One novel approach to control the water flow in a nanochannel is to employ the electrostatic force. While a stationary charge outside a nanochannel impedes water permeation across the nanochannel (electrogating) (Joseph et al., 2003; Li et al., 2007), a vibrational charge outside the nanochannel can promote water flux (electropropelling). This difference in water transport is a result of the vibrational-charge-induced disruption of the hydrogen bond inside the nanochannel (Figure 15.8a) (Kou et al., 2015). Pressure difference in nanochannels can be created by a chemical reaction or phase change induced by the local heating using laser (Figure 15.8b), and the water flow can be controlled by adjusting this pressure (Chaban and Prezhdo, 2011).

15.4 Thermal Transport through Nanochannel Convection

15.4.1 Macroscale Convection Heat Transfer

When water flows through a solid channel at a different temperature, heat transfer occurs between water and the channel, and this heat transfer is regarded as internal convection (Bergman et al., 2011). A local convection heat flux ($q_{s/f}$, W/m²) is proportional to the temperature

difference between the mean fluid temperature ($T_{m,f}$, K) and channel surface temperature (T_s, K), and this relation is known as Newton's cooling law:

$$q_{s/f} = h\left(T_s - T_{m,f}\right), \tag{15.5}$$

where the proportionality constant in this linear relation, h, is the convection heat transfer coefficient (W/m²-K) and characterizes the convection thermal transport.

In convection heat transfer, thermal energy transport occurs not only through interactions between neighboring atoms/molecules induced by thermal vibration, but also through the flow of fluid molecules carrying thermal energy. The former mechanism is termed "diffusion" or "conduction", and the latter is "advection". Since the fluid flow velocity affects temperature distribution as well as advection heat transfer, the analysis of convective heat transfer cannot be decoupled with hydrodynamic analysis.

As in the macroscale hydrodynamics consideration, macroscale convective heat transfer analysis simplifies the fluid–surface interfacial conditions. The velocity and temperature of interfacial fluid are assumed to be identical to those of the channel surface; i.e., no velocity slip and no temperature jump (no thermal slip) conditions. With these no-slip approximations, the advection can be neglected at the interface as the relative fluid velocity is zero; then, heat transfer between surface and water will be equivalent to thermal diffusion (i.e., conduction) in fluid near the interface, and according to the Fourier law, heat flux ($q_{s/f}$) from surface to fluid is

$$q_{s/f} = -k_f\left(\frac{\partial T_f}{\partial z}\right)_{z=0}, \tag{15.6}$$

where k_f is the fluid's thermal conductivity and z is the coordinate normal to channel surface (interface at $z = 0$). Using Eqs. (15.5) and (15.6), the convection heat transfer coefficient h can be obtained by using the fluid temperature distribution and conductivity as

$$h = \frac{k_f (\partial T_f / \partial z)_{z=0}}{T_{m,f} - T_s}. \tag{15.7}$$

To calculate temperature distribution for the macroscale convection heat transfer, the momentum (Navier–Stokes) and energy equations are employed in modeling with no-slip boundary conditions. In general, convection heat transfer results have been reported in terms of the dimensionless Nusselt number (Nu) defined as

$$Nu = \frac{hL_c}{k_f}, \tag{15.8}$$

where L_c is the characteristic length of a system, and hydraulic diameter (D_h) is used as L_c for internal convection. Nu is a function of the Reynolds (Re) and Prandtl (Pr, ratio of momentum and thermal diffusivities) numbers, and its correlation is dependent on the channel shape, surface roughness, and other flow characteristics. Nu correlations for various convection cases have been suggested from various macroscale studies.

15.4.2 Thermal Transport in Micro- or Nanoscale Channels

Micro- or nanoscale channels possess very small Re ($\ll 1$) due to their small hydraulic diameter, leading to laminar flow. With very ordered stream and less mixing, convective heat transfer by laminar flow is less effective than that by turbulent flow. However, despite the small heat transfer in a single channel, an array of micro- or nanochannels ensures a larger area of fluid–channel interaction with a given flow (or volume or mass rate); thus, the incorporation of micro- or nanochannels as a cooling method to efficiently remove excess heat is one avenue suggested for thermal management in these small-scale devices (Tuckerman et al., 1981; Pop, 2010; Wu et al., 2003; Peng et al., 1995).

As macroscale simplifications, such as continuum, no interfacial velocity slip, and no fluid density oscillation, should be adjusted in nanoscale hydrodynamic modeling, the approximation of interfacial fluid temperature to the channel surface temperature cannot be justified in nanoscale thermal transport. Due to the mismatch in thermal vibration and finite strength of interfacial interaction, thermal boundary resistance, also called the Kapitza resistance ($R_{t,s/f}$, m^2-K/W), exists between water and surface, which causes the interfacial temperature jump; i.e., $\Delta T_{s/f} = T_s - T_{f,z=0} \neq 0$. The temperature jump is proportional to heat flux as well as the thermal boundary resistance ($\Delta T_{s/f} = q_{s/f} R_{t,s/f}$). Similar to the velocity slip length, the thermal slip (or Kapitza) length (L_k) can also be defined using temperature distribution as

$$L_k = -\frac{T_s - T_{f,z=0}}{(\partial T_f / \partial z)_{z=0}}. \tag{15.9}$$

In macroscale channels, this interfacial temperature jump is negligible compared with water temperature distribution.

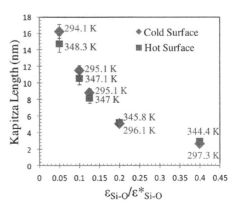

FIGURE 15.9 Thermal slip (Kapitza) length values for different normalized interaction strength between the Si substrate and water, $\varepsilon_{\text{Si-O}}/\varepsilon^*_{\text{Si-O}}$ ($\varepsilon^*_{\text{Si-O}} = 0.12088$ eV) cases obtained at hot and cold surfaces. Surface temperatures are also given on the plot (Barisik and Beskok, 2014).

While the boundary resistance is independent of system dimension, bulk thermal resistance of water increases as a system dimension increases; thus, the interfacial temperature jump becomes comparable to water temperature distribution within micro- or nanoscale channels and cannot be ignored in heat transfer analysis. In particular, in nanoscale thermal transport, interfacial effects are dominant, and many have studied the interfacial thermal transport.

Thermal transport across the interface of water and surface is strongly dependent on interaction strength between the surface and water molecules. As the interfacial interaction increases, not only do the wetting contact angle and velocity slip length decrease (more hydrophilic and less slippage), but the interfacial thermal resistance and Kapitza length also decrease (enhanced thermal transport) (Barisik et al., 2014) (Figure 15.9).

15.4.3 Water Convection in Graphene/CNT Channels

Graphitic materials, including 2D graphene and CNTs, have been identified as materials to improve microchannel thermal performance due to their excellent thermal conductivity, high surface-area-to-volume ratio, and ability to most effectively reduce the overall thermal resistance, as compared to that of more common silicon and aluminum substrates, when used as the primary substrate material of a microchannel heat sink (Shkarah et al., 2013; Ghosh et al., 2008; Duarte et al., 2015). In addition, the temperature affects water properties, such as viscosity and density, and heating can be used for hydrodynamic control (Chaban and Prezhdo, 2011), and thus, beside heat transfer applications, thermal transport by water flow in graphene/CNT nanochannels should be elucidated for various hydrodynamic applications of water-flow nanochannels.

Despite its significance and potential applications, the study of thermal transport in water-flow graphene or CNT nanochannels has been relatively unexplored compared with

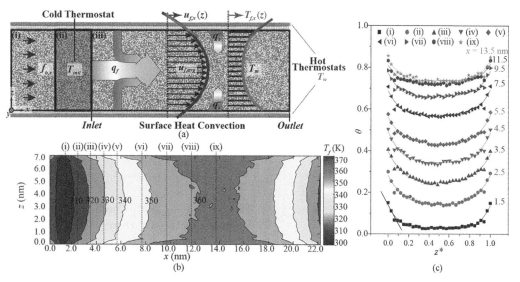

FIGURE 15.10 (a) Schematic of MD simulation for water convection heat transfer in graphene nanochannel. The domain is separated into three regions: (i) forcing region, (ii) temperature rescaling region, and (iii) data collection region. (b) Isotherm diagram of temperature development (isothermal lines are inverted to better guide the eye.). (c) Development of non-dimensional temperature ($\theta = (T - T_{init})/(T_w - T_{init})$, where T_w is the wall temperature and T_{init} is the initial water temperature in the rescaling region) profiles across the channel height (z-axis) at various x-directional points in the direction of the flow (Marable et al., 2017).

the hydrodynamic transport, and especially experimental measurement of convective thermal transport in nanochannels is rarely reported due to challenges in characterizing nanoscale thermal properties.

The study of submicrometer environments is necessary to fully understand momentum and heat transfer deviation from traditional theory (Kim et al., 2010). Since graphene has a hydrophobic and strongly bonded 2D lattice, graphene interlayer interactions are much weaker than intralayer interactions; therefore, more emphasis is placed on graphene–fluid interactions to dictate overall system performance, and several studies have been conducted to understand and enhance water–graphene interfacial thermal transport (Cao et al., 2018; Pham et al., 2016). The convective behaviors of liquid flow have been studied mostly using theoretical analysis and molecular simulations. Previous reports indicate that temperature discontinuity (or temperature jump) at the wall due to interfacial thermal resistance decreases interfacial fluid temperature gradients and ultimately yields convective behaviors deviated from those at the macroscale (Ghasemi et al., 2011; Barrat et al., 2003; Thekkethala et al., 2015; Ge et al., 2015; Shkarah et al., 2013).

In a recent study, water convection in a graphene nanochannel was thoroughly investigated via MD simulations (Marable et al., 2017). In MD, a force to maintain a flow velocity and thermostats to create a temperature difference between water and graphene channel are applied, and velocity, temperature distribution, and heat flow rate are calculated by processing position, velocity, and energy profiles of water molecules (Figure 15.10a–c). Using the resulting velocity and temperature distribution and heat flow within a channel, hydrodynamic and thermal

slip lengths and Nu can be calculated. In the simulations, nanochannel size, graphene–water interaction strength, velocity, and graphene channel temperature are controlled by adjusting atomic configuration, water–carbon interaction parameter, applied force, and thermostat temperatures. Through these parametric changes, effects of microscopic mechanisms (velocity and thermal slippage) on nanochannel convection were examined. Nusselt number of macroscale, fully developed laminar flow ($Nu_{D,fd}$) is theoretically calculated and found to be 7.541; however, various nanoscale effects lead to the deviation from this number. The recent study (Marable et al., 2017) demonstrates that the fully developed internal Nu, which represents the internal convection within a graphene nanochannel, is strongly correlated with a dimensionless thermal slip length normalized by twice the channel height ($L_k/2H$) as shown in Figure 15.11. Using similar simulation approaches, nanochannel convection in CNTs can be further studied to identify effects of curved surface. Moreover, experimental investigations are urgently demanded to verify the computational and theoretical results on nanochannel heat transfer.

15.5 Summary and Prospect

Nanochannels will be more considered to innovate thermal management, filtration/separation process, energy conversion, and biomedical devices. As discussed, nanoscale confinements cause water behaviors to deviate from those in macroscale systems, and an understanding and analysis of these peculiar properties are required for nanochannel applications. Thermophysical, hydrodynamic, and thermal transport properties of water in nanochannels are significantly dependent on interactions between water and channel

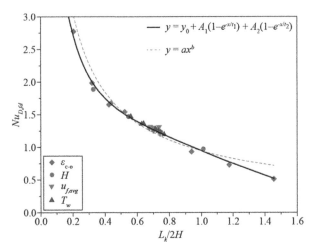

FIGURE 15.11 Fully developed Nusselt number ($Nu_{D,fd}$) of internal convection in water-flow graphene nanochannels. In various cases employing different interaction strengths (ε_{c-o}), channel height (H), average flow velocity ($u_{f,avg}$), and channel wall temperature (T_w), $Nu_{D,fd}$ and thermal slip length (L_k) were calculated, and this analysis shows possible equations of correlation between $Nu_{D,fd}$ and the normalized thermal slip length values ($L_k/2H$).

surface; thus, water properties have been studied under a variety of channel conditions, including channel materials. In particular, this chapter focuses on nanochannels made from graphene and CNTs, because their superior features in mechanical strength, electrical and thermal conductivity, chemical selectivity, and property controllability make them more attractive as nanochannel materials.

Water confined in graphene/CNT nanochannels possesses different atomic structures depending on channel size, resulting in different boiling and freezing temperatures. The unique atomic structure of water also affects hydrodynamic transport. Hydrophobicity of graphene and CNTs results in a large slip length, which leads to a larger flow rate than macroscale prediction. Convection heat transfer in nanochannel also presents deviation from macroscale laminar flow and strong dependence on surface interaction and channel size. Despite extensive studies on water flow in graphene/CNT nanochannels, not all water behaviors in nanochannels are clearly explained, and further investigations should be conducted. Advances in technologies will enable unprecedented experiments and simulations on nanoscale characterizations of water-flow graphene/CNT nanochannels, which can enhance our understanding on nanoscale phenomena and design of nanochannel systems.

References

Addams, L., Boccaletti, G., Kerlin, M., and Stuchtey, M. 2009. *Charting our Water Future: Economic Frameworks to Inform Decision-Making*. McKinsey & Company, New York.

Alexeyev, A. A. and Vinogradova, O. I. 1996. Flow of a liquid in a nonuniformly hydrophobized capillary. *Colloids and Surfaces A: Physicochemical and Engineering Aspects*, 108: 173–179.

Algara-Siller, G., Lehtinen, O., Wang, F. C. et al. 2015. Square ice in graphene nanocapillaries. *Nature*, 519: 443–445.

Åman, K., Lindahl, E., Edholm, O., Håkansson, P. and Westlund, P.-O. 2003. Structure and dynamics of interfacial water in an L_α phase lipid bilayer from molecular dynamics simulations. *Biophysical Journal* 84: 102–115.

Balandin, A. A. 2011. Thermal properties of graphene and nanostructured carbon materials. *Nature Materials* 10: 569–581.

Barisik, M. and Beskok, A. 2014. Temperature dependence of thermal resistance at the water/silicon interface. *International Journal of Thermal Sciences* 77: 47–54.

Barrat, J. L. and Chiaruttini, F. 2003. Kapitza resistance at the liquid-solid interface. *Molecular Physics* 101: 1605–1610.

Bergman, T. L., Lavine, A. S., Incropera, F. P., and DeWitt, D. P. 2011. *Fundamentals of Heat and Mass Transfer*. John Wiley & Sons, New York.

Bizzarri, A. R. and Cannistraro, S. 2002. Molecular dynamics of water at the protein–solvent interface. *The Journal of Physical Chemistry B* 106: 6617–6633.

Bocquet, L. and Charlaix, E. 2010. Nanofluidics, from bulk to interfaces. *Chemical Society Reviews* 39: 1073–1095.

Bunch, J. S., Verbridge, S. S., Alden, J. S. et al. 2008. Impermeable atomic membranes from graphene sheets. *Nano Letters* 8: 2458–2462.

Byl, O., Liu, J. C., Wang, Y. et al. 2006. Unusual hydrogen bonding in water-filled carbon nanotubes. *Journal of the American Chemical Society* 128: 12090–12097.

Cao, B. Y., Zou, J. H., Hu, G. J. and Cao, G. X. 2018. Enhanced thermal transport across multilayer graphene and water by interlayer functionalization. *Applied Physics Letters* 112: 041603.

Chaban, V. V. and Prezhdo, O. V. 2011. Water boiling inside carbon nanotubes: Toward efficient drug release. *ACS Nano* 5: 5647–5655.

Cohen-Tanugi, D. and Grossman, J. C. 2012. Water desalination across nanoporous graphene. *Nano Letters* 12: 3602–3608.

Corry, B. 2008. Designing carbon nanotube membranes for efficient water desalination. *The Journal of Physical Chemistry B* 112: 1427-1434.

Cottin-Bizonne, C., Cross, B., Steinberger, A. and Charlaix, E. 2005. Boundary slip on smooth hydrophobic surfaces: Intrinsic effects and possible artifacts. *Physical Review Letters* 94: 056102.

Duarte, J. M. C., Contreras, I. M. A. and Cely, C. R. C. 2015. An optimal high thermal conductive graphite microchannel for electronic devices cooling. *Revista Facultad de Ingeniería* 77: 143–152.

Eijkel, J. C. T. and Berg, A. V. D. 2005. Nanofluidics: what is it and what can we expect from it? *Microfluidics and Nanofluidics* 1: 249–267.

Elimelech, M. and Phillip, W. A. 2011. The future of seawater desalination: Energy, technology, and the environment. *Science* 333: 712–717.

Falkovsky, L. A. 2008. Optical properties of graphene. *Journal of Physics: Conference Series* 129: 012004.

Frank, I., Tanenbaum, D. M., Van der Zande, A. M. and McEuen, P. L. 2007. Mechanical properties of suspended graphene sheets. *Journal of Vacuum Science & Technology B: Microelectronics and Nanometer Structures Processing, Measurement, and Phenomena* 25: 2558–2561.

Frenkel, D. and Smit, B. 2002. *Understanding Molecular Simulation: From Algorithms to Applications*. Elsevier (formerly published by Academic Press), San Diego, CA.

Gad-El-Hak, M. 2001. *The MEMS Handbook*. CRC Press, Boca Raton, FL.

Gad-El-Hak, M. 2005. Differences between liquid and gas transport at the microscale. *Bulletin of the Polish Academy of Sciences: Technical Sciences* 53: 301–316.

Ge, S., Gu, Y. W. and Chen, M. 2015. A molecular dynamics simulation on the convective heat transfer in nanochannels. *Molecular Physics* 113: 703–710.

Ghasemi, H., Ni, G., Marconnet, A. M. et al. 2014. Solar steam generation by heat localization. *Nature Communications* 5: 4449.

Ghasemi, H. and Ward, C. A. 2011. Mechanism of sessile water droplet evaporation: Kapitza resistance at the solid-liquid interface. *Journal of Physical Chemistry C* 115: 21311–21319.

Ghosh, S., Calizo, I., Teweldebrhan, D. et al. 2008. Extremely high thermal conductivity of graphene: Prospects for thermal management applications in nanoelectronic circuits. *Applied Physics Letters* 92: 151911.

Ghosh, S., Ramanathan, K. V. and Sood, A. K. 2004. Water at nanoscale confined in single-walled carbon nanotubes studied by NMR. *Europhysics Letters* 65: 678–684.

Han, J. and Craighead, H. G. 2000. Separation of long DNA molecules in a microfabricated entropic trap array. *Science* 288: 1026–1029.

Hanks, D. F., Lu, Z., Sircar, J. et al. 2018. Nanoporous membrane device for ultra high heat flux thermal management. *Microsystems and Nanoengineering* 4: 1.

Heslot, F., Fraysse, N. and Cazabat, A. M. 1989. Molecular layering in the spreading of wetting liquid-drops. *Nature* 338: 640–642.

Holt, J. K., Park, H. G., Wang, Y. M. et al. 2006. Fast mass transport through sub-2-nanometer carbon nanotubes. *Science* 312: 1034–1037.

Huang, H. B., Song, Z. G., Wei, N. et al. 2013. Ultrafast viscous water flow through nanostrand-channelled graphene oxide membranes. *Nature Communications* 4: 2979.

Jiang, D. E., Cooper, V. R. and Dai, S. 2009. Porous graphene as the ultimate membrane for gas separation. *Nano Letters* 9: 4019–4024.

Joseph, S. and Aluru, N. 2008. Why are carbon nanotubes fast transporters of water? *Nano Letters* 8: 452–458.

Joseph, S., Mashl, R. J., Jakobsson, E. and Aluru, N. R. 2003. Electrolytic transport in modified carbon nanotubes. *Nano Letters* 3: 1399–1403.

Kandlikar, S., Garimella, S., Li, D., Colin, S. and King, M. R. 2005. *Heat Transfer and Fluid Flow in Minichannels and Microchannels*. Elsevier, Amsterdam.

Kumar Kannam, S., Todd, B. D., Hansen, J. S. and Daivis, P. J. 2012. Slip length of water on graphene: Limitations of non-equilibrium molecular dynamics simulations. *Journal of Chemical Physics* 136: 024705.

Kaviany, M. 2014. *Heat Transfer Physics*. Cambridge University Press, Cambridge.

Kim, B. H., Beskok, A. and Cagin, T. 2010. Viscous heating in nanoscale shear driven liquid flows. *Microfluidics and Nanofluidics* 9: 31–40.

Knowles, H. J. 2009. Carbon nanotubes and nanofluidic transport. *Advanced Materials* 21: 3542–3550.

Koga, K., Gao, G. T., Tanaka, H. and Zeng, X. C. 2001. Formation of ordered ice nanotubes inside carbon nanotubes. *Nature* 412: 802–805.

Kou, J., Yao, J., Lu, H. et al. 2015. Electromanipulating water flow in nanochannels. *Angewandte Chemie International Edition* 54: 2351–2355.

Kovarik, M. L. and Jacobson, S. C. 2009. Nanofluidics in lab-on-a-chip devices. *Analytical Chemistry* 81: 7133–7140.

Landau, L. D, and Lifshits, E. M. 2003. *Theoretical Physics* (Hydrodynamics vol VI). : Fizmatlit, Moscow, 736

Lee, B., Baek, Y., Lee, M. et al. 2015. A carbon nanotube wall membrane for water treatment. *Nature Communications* 6: 7109.

Lee, C., Wei, X. D., Kysar, J. W. and Hone, J. 2008. Measurement of the elastic properties and intrinsic strength of monolayer graphene. *Science* 321: 385–388.

Lee, K. P., Arnot, T. C. and Mattia, D. 2011. A review of reverse osmosis membrane materials for desalination—Development to date and future potential. *Journal of Membrane Science* 370: 1–22.

Li, J. Y., Gong, X. J., Lu, H. J. et al. 2007. Electrostatic gating of a nanometer water channel. *Proceedings of the National Academy of Sciences* 104: 3687–3692.

Majumder, M., Chopra, N., Andrews, R. and Hinds, B. J. 2005. Nanoscale hydrodynamics: Enhanced flow in carbon nanotubes. *Nature* 438: 44.

Maniwa, Y., Kataura, H., Abe, M. et al. 2002. Phase transition in confined water inside carbon nanotubes. *Journal of the Physical Society of Japan* 71: 2863–2866.

Maniwa, Y., Kataura, H., Abe, M. et al. 2005. Ordered water inside carbon nanotubes: Formation of pentagonal to octagonal ice-nanotubes. *Chemical Physics Letters* 401: 534–538.

Marable, D. C., Shin, S. and Yousefzadi Nobakht, A. 2017. Investigation into the microscopic mechanisms influencing convective heat transfer of water flow in graphene nanochannels. *International Journal of Heat and Mass Transfer* 109: 28–39.

Moody, L. F. 1944. Friction factors for pipe flow. *Transactions of the ASME* 66: 671–684.

Naguib, N., Ye, H. H., Gogotsi, Y. et al. 2004. Observation of water confined in nanometer channels of closed carbon nanotubes. *Nano Letters* 4: 2237–2243.

Neto, C., Evans, D. R., Bonaccurso, E., Butt, H.-J. and Craig, V. S. 2005. Boundary slip in Newtonian liquids: A review of experimental studies. *Reports on Progress in Physics* 68: 2859–2897.

Nicholls, W. D., Borg, M. K., Lockerby, D. A. and Reese, J. M. 2012. Water transport through (7, 7) carbon nanotubes of different lengths using molecular dynamics. *Microfluidics and Nanofluidics* 12: 257–264.

Nomura, K. and MacDonald, A. H. 2007. Quantum transport of massless Dirac fermions. *Physical Review Letters* 98: 076602.

Novoselov, K. S. 2011. Graphene: Materials in the Flatland (Nobel Lecture). *Angewandte Chemie-International Edition* 50: 6986–7002.

Novoselov, K. S., Fal'ko, V. I., Colombo, L. et al. 2012. A roadmap for graphene. *Nature* 490: 192–200.

Noy, A., Park, H. G., Fornasiero, F. et al. 2007. Nanofluidics in carbon nanotubes. *Nano Today* 2: 22–29.

Park, H. G. and Jung, Y. 2014. Carbon nanofluidics of rapid water transport for energy applications. *Chemical Society Reviews* 43: 565–576.

Pascal, T. A., Goddard, W. A. and Jung, Y. 2011. Entropy and the driving force for the filling of carbon nanotubes with water. *Proceedings of the National Academy of Sciences* 108: 11794–11798.

Peng, X. F., Wang, B. X., Peterson, G. P. and Ma, H. B. 1995. Experimental investigation of heat-transfer in flat plates with rectangular microchannels. *International Journal of Heat and Mass Transfer* 38: 127–137.

Pham, A. T., Barisik, M. and Kim, B. 2016. Interfacial thermal resistance between the graphene-coated copper and liquid water. *International Journal of Heat and Mass Transfer* 97: 422–431.

Phillips, J. C., Braun, R., Wang, W. et al. 2005. Scalable molecular dynamics with NAMD. *Journal of Computational Chemistry* 26: 1781–1802.

Podolska, N. I. and Zhmakin, A. I. 2013. Water flow in micro- and nanochannels. Molecular dynamics simulations. *Journal of Physics: Conference Series* 461: 012034.

Pop, E. 2010. Energy dissipation and transport in nanoscale devices. *Nano Research* 3: 147-169.

Probstein, R. F. 2005. *Physicochemical Hydrodynamics: An Introduction*. John Wiley & Sons, New York.

Senapati, S. and Berkowitz, M. L. 2003. Water structure and dynamics in phosphate fluorosurfactant based reverse micelle: A computer simulation study. *Journal of Chemical Physics* 118: 1937–1944.

Shkarah, A. J., Bin Sulaiman, M. Y., Ayob, M. R. B. and Togun, H. 2013. A 3D numerical study of heat transfer in a single-phase micro-channel heat sink using graphene, aluminum and silicon as substrates. *International Communications in Heat and Mass Transfer* 48: 108–115.

Sint, K., Wang, B. Y. and Kral, P. 2009. Selective ion passage through functionalized graphene nanopores. *Journal of the American Chemical Society* 130: 16448–16449.

Sofos, F., Karakasidis, T. E. and Liakopoulos, A. 2009. Non-equilibrium molecular dynamics investigation of parameters affecting planar nanochannel flows. *Contemporary Engineering Sciences* 2: 283–298.

Spiegler, K. & El-Sayed, Y. 2001. The energetics of desalination processes. *Desalination* 134: 109–128.

Squires, T. M. and Quake, S. R. 2005. Microfluidics: Fluid physics at the nanoliter scale. *Reviews of Modern Physics* 77: 977–1026.

Suk, M. E. and Aluru, N. R. 2010. Water transport through ultrathin graphene. *Journal of Physical Chemistry Letters* 1: 1590–1594.

Suk, M. E., Raghunathan, A. V. and Aluru, N. R. 2008. Fast reverse osmosis using boron nitride and carbon nanotubes. *Applied Physics Letters* 92: 133120.

Takaiwa, D., Hatano, I., Koga, K. and Tanaka, H. 2008. Phase diagram of water in carbon nanotubes. *Proceedings of the National Academy of Sciences* 105: 39-43.

Thekkethala, J. F. and Sathian, S. P. 2015. The effect of graphene layers on interfacial thermal resistance in composite nanochannels with flow. *Microfluidics and Nanofluidics* 18: 637–648.

Thomas, J. A. and McGaughey, A. J. H. 2008. Reassessing fast water transport through carbon nanotubes. *Nano Letters* 8: 2788–2793.

Thomas, J. A. and McGaughey, A. J. H. 2009. Water flow in carbon nanotubes: Transition to subcontinuum transport. *Physical Review Letters* 102: 184502.

Thomas, J. A., McGaughey, A. J. H. and Kuter-Arnebeck, O. 2010. Pressure-driven water flow through carbon nanotubes: Insights from molecular dynamics simulation. *International Journal of Thermal Sciences* 49: 281–289.

Travis, K. P. and Gubbins, K. E. 2000. Poiseuille flow of Lennard-Jones fluids in narrow slit pores. *Journal of Chemical Physics* 112: 1984–1994.

Tuckerman, D. B. and Pease, R. F. W. 1981. High-performance heat sinking for VLSI. *Electron Device Letters* 2: 126–129.

Varshney, A., Brooks, F. P. and Wright, W. V. 1994. Computing smooth molecular-surfaces. *IEEE Computer Graphics and Applications* 14: 19–25.

Verweij, H., Schillo, M. C. and Li, J. 2007. Fast mass transport through carbon nanotube membranes. *Small* 3: 1996–2004.

Whitby, M., Cagnon, L., Thanou, M. and Quirke, N. 2008. Enhanced fluid flow through nanoscale carbon pipes. *Nano Letters* 8: 2632–2637.

Whitesides, G. M. 2006. The origins and the future of microfluidics. *Nature* 442: 368–373.

Wu, H. Y. and Cheng, P. 2003. Friction factors in smooth trapezoidal silicon microchannels with different aspect ratios. *International Journal of Heat and Mass Transfer* 46: 2519–2525.

Yang, H. Y., Han, Z. J., Yu, S. F. et al. 2013. Carbon nanotube membranes with ultrahigh specific adsorption capacity for water desalination and purification. *Nature Communications* 4: 2220.

Yang, L., Guo, Y. J. and Diao, D. F. 2017. Structure and dynamics of water confined in a graphene nanochannel under gigapascal high pressure: Dependence of friction on pressure and confinement. *Physical Chemistry Chemical Physics* 19: 14048–14054.

Zhao, X., Liu, Y., Inoue, S. et al. 2004. Smallest carbon nanotube is 3 Å in diameter. *Physical Review Letters* 92: 125502.

Transport of Water in Graphene Nanochannels

Chinh Thanh Nguyen,
Alper Tunga Celebi,
and Ali Beskok
Southern Methodist University

16.1 Introduction

Graphene is a two-dimensional (2D) carbon-based nanomaterial consisting of a single layer of carbon atoms in honeycomb shaped lattice. Graphene has gained worldwide attention due to its outstanding properties such as high mechanical strength [1], high thermal and electrical conductivity [2,3], chemical adjustability [4,5], and optical transparency [6]. For nanofluidic applications such as biotechnology (drug delivery, DNA sequencing, cancer detection, etc.), biosensing (detection of pathogens, measurement of clinical parameters, monitoring of environmental pollutants, etc.), nanolubrication, nanopumping, and nanofiltration, graphene has special importance as the channel material or semipermeable porous membrane that allows enhanced fluid flow rates and high selectivity. These valuable aspects are mainly associated with two unique properties of graphene [7]. First, graphene is hydrophobic. The interaction strength between water atoms is stronger than that of the water/carbon atoms, making water atoms recede from walls. As a result, water contact angle on a monolayer of graphene is in the range of 95°–100°, indicating a hydrophobic characteristic [8,9]. Second, due to four covalent bonds formed by each carbon atom in the graphene layer, the surface density of graphene is much higher than that of many other materials used in nanofluidic applications. These strong covalent bonds create smooth surfaces, resulting in a low surface potential corrugation. Consequently, liquids on graphene surfaces show weak interfacial friction and large slip, leading to high flow rates [10].

In order to design and develop devices for practical nanofluidic applications, a fundamental understanding on the behaviors of liquids confined in nanoscale channels

should be achieved. Therefore, this chapter gives insights on the transport of water in graphene nanochannels based on the previous computational and experimental studies. The computational material covered in this chapter is based on our molecular dynamics (MD) studies taken from the following publications where simulation details can be found [11–14]. Sections 16.2 and 16.3 introduce transport of water in uncharged and charged planar graphene nanochannels, respectively. Section 16.4 discusses transport of water through nanoporous graphene membranes and potential applications in water purification. Conclusions are given in Section 16.5.

16.2 Transport of Water in Uncharged Graphene Nanochannels

Liquid flows at nanoscale can be driven by external means such as electric field, pressure, shear, and temperature gradient [11]. This section focuses on force-driven water flow simulations in graphene nanochannels to investigate the variation in liquid and transport properties including density distributions, velocity profiles, slip lengths, and flow rates. Figure 16.1 illustrates a three-dimensional simulation domain consisting of deionized water molecules confined in parallel graphitic walls made of three layers. The flow was induced by externally applied constant forces on water molecules in x-direction.

Density is one of the significant parameters that can affect water transport properties in graphene nanochannels. Due to the wall force field effect, water in nanoconfinements shows a near-wall layering region and a bulk region at the channel center [15–17]. Understanding density

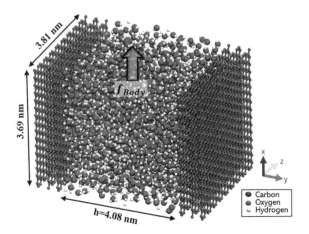

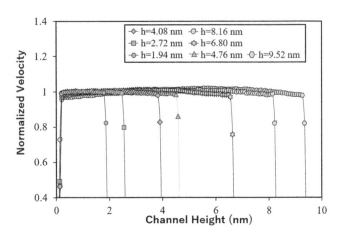

FIGURE 16.1 MD simulation domain for water flows in planar graphene nanochannels.

FIGURE 16.3 Velocity profiles of water confined in uncharged graphene nanochannels at various channel heights.

layering is important because it can lead to spatially varying density-dependent transport properties [7, 18]. In graphene nanochannels, water exhibits weak density layering owing to the hydrophobic nature of graphene. Figure 16.2 presents water density distributions in graphene nanochannels as a function of channel height. Three distinct density peaks are observed in all cases, and the highest peak is located approximately 3.15 Å away from the solid walls. We found that the locations and magnitude of the density peaks are independent of the channel height. The density layering disappears and converges to a bulk value in a couple of molecular diameters away from the wall [19]. It should also be noted that the average density in the layering region is typically lower than its thermodynamic value [7]. As channels become larger than 70 nm, the effect of density layering vanishes, and the desired density at a specified thermodynamic state was obtained. In order to establish a fixed thermodynamic state at various channel heights, the bulk density constant is kept at 997 kg/m³, and temperature is maintained at 300 K using thermostats.

Figure 16.3 shows velocity profiles normalized by the corresponding channel-average velocities for different channel heights. Velocity profiles are crucial to characterize the flow in nanochannels in non-equilibrium MD simulations. Typically, a curve-fitting to parabolic velocity profiles in the streamwise direction is applied to determine the transport properties such as viscosity and slip length. However, velocity profiles of water in uncharged graphene nanochannels exhibit plug-flow behavior with a large velocity slip at solid–liquid interfaces mainly due to the low friction nature of graphene. This plug-like flow behavior is independent of the channel height as can be observed in the figure. For such cases, one can assume that the slip velocity is equal to the average channel velocity. This assumption becomes particularly useful in the quantification of the slip lengths in hydrophobic nanochannels using plug-like velocity profiles [12, 20].

Liquids in nanoconfinements show no slip, stick-slip or velocity slip at solid–liquid interfaces depending on the solid–liquid interaction strength. At weak interfacial strengths, a finite velocity slip (u_s) is observed at interface and is typically quantified by the concept of "slip length". Slip length is the extrapolated distance relative to the wall where the tangential velocity component vanishes. Velocity slip can be defined with a Navier-type slip condition as follows:

$$u_s = u_l - u_w = \beta \frac{du}{dy}, \qquad (16.1)$$

where u_l and u_w refer to liquid and wall velocities, respectively, while β is the slip length. Since water flows in hydrophobic graphene channels show flat velocity profiles across the channel, we use a "plug-flow methodology" to determine the slip length. This method employs conservation of linear momentum in the flow direction which relates the wall shear (τ_w) to the total body force (f) applied on liquid molecules as $\tau_w = \frac{fh}{2}$ [12]. Then, this relation is combined with Navier-type slip equation given in Eq. (16.1) and the constitutive equation $\left(\tau_w = \mu \frac{du}{dy}\right)$ for incompressible Newtonian liquids to obtain

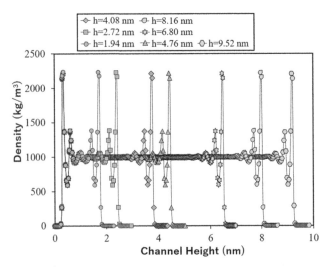

FIGURE 16.2 Density distributions of water confined in uncharged graphene nanochannels at various channel heights.

$$\beta = \frac{2\mu u_{\mathrm{s}}}{fh} = \frac{2\mu \overline{u}}{fh}, \qquad (16.2)$$

where μ is the apparent viscosity of the liquid, h is channel height, and $\overline{u}$ is the average velocity across the channel assuming insignificant differences in velocities at the channel center and channel wall due to the plug-like velocity profile. It is noted that the viscosity utilized here is considered as the thermodynamic viscosity of water (853 μPa.s) at $T = 300$ K and $\rho_{\mathrm{bulk}} = 997$ kg/m^3. In Table 16.1, we show the variation of slip length in graphene nanochannels for different channel heights. It is seen that the slip length is independent of the channel heights for heights larger 2 nm. The predicted slip length of water in those channels was found approximately 64 nm. However, for channels smaller than 2 nm, the slip length becomes larger. This is attributed to the inaccurate definitions of local thermodynamic equilibrium and bulk liquid properties such as density, velocity, and viscosity while the flow shows a discrete liquid transport. For such scales, constitutive equations for shear stress are incompatible, and the calculation of the slip length using constitutive equations and thermodynamic viscosity becomes irrelevant. In Figure 16.4, we provide an extensive comparison to our results with the previously reported MD values [21–30]. It was seen that our results are in good agreement with several reported values, whereas there are also several mismatches. These differences are mainly due to the intrinsic algorithmic details such as the use of different water models and interaction parameters, neglecting nonlinear flow contribution as well as not properly fixing the thermodynamic state of the simulations. Furthermore, simulation results can be compared with few experimental studies in literature. In a very recent report by Xie et al., slip lengths of water inside graphene nanochannels were found in the range of 1–200 nm for different channel heights, where these results followed a log normal distribution [31]. Statistical median shows approximately 16 nm considering actual graphene coverage quality. Maali et al. used atomic force microscopy (AFM) to measure the slip length of water on a graphitic surface and found slip length of approximately 8 nm [32]. These experimentally reported slip length values are smaller than those reported using simulations. This could be attributed to the unavoidable roughness of experimentally prepared graphene surfaces whereas the graphene surfaces in computational studies are perfectly smooth.

Large slip at water–graphene interface significantly increases the actual flow rate in graphene nanochannels. When quantifying flow rate in nanochannels, slip effect

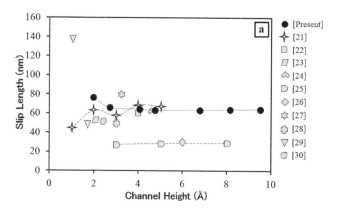

FIGURE 16.4 Variation of slip length of water in graphene nanochannels as a function of channel height.

must be taken into consideration in momentum equation. For one-dimensional (1D), fully developed, incompressible, force-driven Newtonian fluid flows between planar channels, the Navier–Stokes equation is reduced to

$$\frac{d^2u}{dy^2} = -\frac{f}{\mu}, \qquad (16.3)$$

where $u(y)$ is the streamwise velocity field. Considering a Navier-type slip boundary condition at the liquid–solid interfaces ($y = 0$ and $y = h$) as given in Eq. (16.1), the velocity profile of an incompressible liquid between two parallel plates with a channel height of h becomes

$$u(y) = \frac{fh^2}{2\mu}\left(\left(\frac{y}{h}\right)^2 - \left(\frac{y}{h}\right) - \left(\frac{\beta}{h}\right)\right). \qquad (16.4)$$

Integration of the velocity profiles across the channel height with constant width (W) gives the volumetric flow rate with slip correction (Q_{slip}) as follows:

$$Q_{\mathrm{slip}} = \frac{fWh^3}{12\mu}\left(1 + 6\left(\frac{\beta}{h}\right)\right). \qquad (16.5)$$

The above equation implies that it is possible to observe slip-enhanced force-driven flows by choosing hydrophobic channels with large slip lengths. Water slip length of 64 nm in 4 nm graphene nanochannels in our results indicates a flow enhancement factor of 97 using Eq. (16.5).

16.3 Surface Charge-Controlled Transport of Water in Graphene Nanochannels

Motivated by high in-plane electrical conductivity and good structural integrity, graphene surfaces can be used as gated electrodes in nanofluidic systems. For such systems, electrically charged surfaces can greatly alter the liquid properties such as density, velocity, viscosity, and slip length. This could be used to regulate flows for various applications such as nanovalves [12] and water purification [33]. We performed non-equilibrium molecular dynamics (NEMD)

TABLE 16.1 Slip Length Values of Water in Graphene Nanochannels at Various Sizes

Graphene Channels	
h (nm)	β (nm)
9.52	64.2
8.16	63.8
6.80	63.0
4.76	63.1
4.08	64.1
2.72	65.8
1.94	76.4

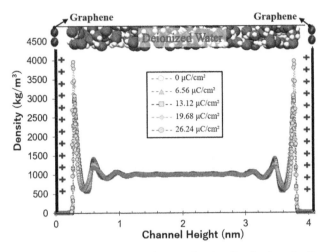

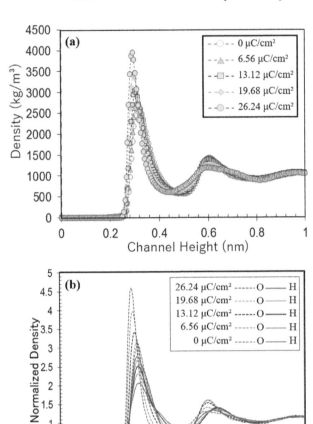

FIGURE 16.5 Variation of water density distributions in the charged graphene nanochannels at various surface charge densities.

FIGURE 16.6 (a) Interfacial density distribution of water in charged channels at various surface charge densities and (b) hydrogen and oxygen densities normalized by their average values at various surface charge densities.

simulations for force-driven flows between two charged planar graphene surfaces with different surface charge densities [12,13]. Force-driven water flows were induced by applying a constant force to each water molecule, and the solid surfaces were charged by introducing simple point charges to each carbon atom at the innermost graphene layer. Considering five different surface charge densities in the range of 0–26.24 $\mu C/cm^2$, graphene electrodes were assigned either equal positive charges or equal opposite charges. Opposite charges induce an electric field from positively charged wall toward negatively charged one, while no electric field is developed across identically charged walls. Figure 16.5 shows the density profiles for different surface charge densities. Water density in the near-wall region increases with the increase of surface charge, which is mainly due to the increased electrostatic interaction with charge. Furthermore, we observed that the density profiles show an asymmetry when the surfaces are oppositely charged, while no asymmetry develops between identically charged channels. This asymmetric behavior in density profiles becomes more distinguishable at large surface charge densities, ultimately resulting in asymmetric velocity profiles as well as different slip lengths on each wall [12]. The interfacial density behavior of water is further elucidated by showing density profiles within 1-nm distance from the wall. It is seen that the magnitude of the first density peak increases with increasing surface charge, whereas the second density peaks are reduced as shown in Figure 16.6a. In the absence of surface charge, the location of the first density peak is approximately a molecular diameter away from the wall. It slightly shifts closer to the wall as the surface charge increases because of increased electrostatic interactions. Figure 16.6b presents oxygen and hydrogen densities normalized by their average values near a positive wall, showing that oxygen atoms are facing the graphene surfaces. This is due to the fact that the water molecules arrange their dipole moments depending on the sign and magnitude of the surface charge, where negative oxygens point toward positively charged surface while positive hydrogens point

away. In other words, the magnitude and sign of surface charge critically affects the molecular orientations of water. Figure 16.7a shows velocity profiles (normalized with their channel-average values) as a function of the surface charge. Plug-like velocity profile is reserved for a zero-charge (pristine) surface. As electrical charge on the surface increases, parabolic velocity profiles are formed accompanied with changes in the slip lengths as shown in Figure 16.7b. For parabolic profiles, slip length is determined by polynomial curve-fitting [12,20] while slip length for flat velocity profiles is calculated using plug-flow methodology as explained in Section 16.2. Velocity profiles are symmetric with respect to the channel center for graphene channel with identical charges. For the cases of opposite charges, there is an asymmetry in velocity profiles at large surface charges depending on the asymmetry in the density [12]. While slip length for the uncharged graphene nanochannel is found as large as 64 nm, the increase of charge results in a drastic decrease in the slip length, which is attributed to the enhanced electrostatic interaction between the wall and liquid molecules. In addition, we elucidated the variations of the viscosity and

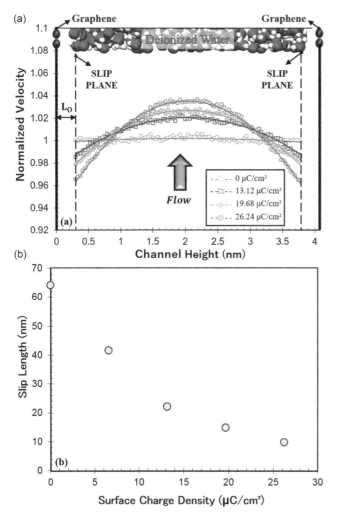

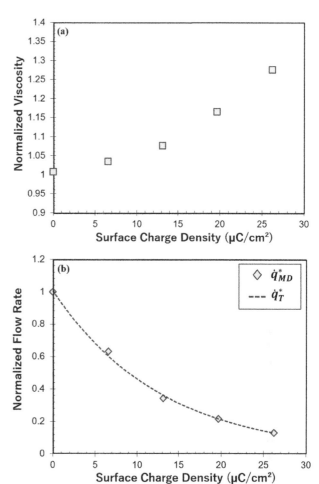

FIGURE 16.7 (a) Velocity profiles normalized by the channel average velocities at various surface charge densities and (b) variation of slip length as a function of the surface charge.

FIGURE 16.8 (a) Variation of normalized viscosity as a function of the surface charge density and (b) normalized flow rates for different charges on graphene surfaces. $\dot{q}^*_{MD}$ is the MD-predicted flow rates normalized by those of electrically neutral case, while $\dot{q}^*_T$ is the theoretical flow rate normalized by that of electrically neutral case.

flow rates with applied surface charge. Interestingly, results show a nonlinear increase in water viscosity as a function of the surface charge density, which is illustrated by the variation of normalized viscosity with that of water in uncharged graphene nanochannels at the given thermodynamic state in Figure 16.8a. Both results for graphene assigned identical or equal opposite charges show a drastic increase in viscosity with increased surface charge density [12,13]. The increase in water bulk viscosity with surface charges is also in good agreement with results reported by Qiao and Aluru [34]. This was attributed to the enhanced hydrogen bonding population between water molecules facilitated by surface charge. It inhibits the motion of water molecules and leads to increase in water viscosity. Figure 16.8b shows MD-predicted and theoretical volumetric flow rates (normalized by the flow rate in the case of uncharged graphene nanochannels) of water inside positively charged graphene channels as a function of surface charge density. MD predictions are in good agreement with the analytical flow rate calculations. In addition, we observed that volumetric flow rate shows a nonlinear decrease with increased charge, which

is consistent with the decrease in slip length and increase in viscosity. Tunable surface-charge-dependent properties of water densities, velocity profiles, slip lengths, viscosities, and flow rates enable active controls of water flows in graphene nanochannels.

16.4 Transport of Water through Nanoporous Graphene Membranes and the Applications in Water Purification

Nanoporous graphene membranes also attracted considerable attention due to their high potential in water purification and molecular species selection applications. They can impede the transport of salt ions or other molecular species depending on the pore sizes. For such systems, intermolecular interactions between these atomic species and the pore edges prevent them from passing through the pore, while water molecules can flow through the pores. It should also be noted that the hydrated diameter of most ions in water is larger than the effective size of a water molecule (Na^+ 7.2 Å, K^+ 6.6 Å, Ca^{2+} 8.2 Å, Mg^{2+} 8.6 Å and Cl^- 6.6 Å); thus, the main salt rejection mechanisms are size exclusion and steric exclusion of the hydration shell [35]. From the hydrodynamics perspective, the flow rate across the membrane is inversely proportional to the membrane thickness. Therefore, one-atomic-layer-thick nanoporous graphene membranes can induce significant reduction in the power consumption at a desired flowrate with high ion rejection rates. For applications, it is critical to understand water transport mechanisms through nanoporous graphene membranes.

We recently simulated pressure-driven saltwater flow through pristine nanoporous graphene membranes mimicking reverse osmosis (RO) water desalination processes [14]. Flows through the nanoporous membranes were created by moving two specular reflection boundaries located at the upstream and downstream of the simulation domain with equal speed as illustrated in Figure 16.9. This approach preserves the MD simulation volume and specifies a flow rate. The induced pressures in the feed and permeate reservoirs were calculated using Irving–Kirkwood relation, and pressure drop was defined as the difference of the two pressures. Figure 16.10 shows the variation of average normal stress (i.e., pressure) in the streamwise direction, which exhibits fluctuations near the nanoporous graphene membranes. We have previously shown anisotropic normal stress distribution near the walls [36–38]. Only sufficiently away from the solid surfaces, the three components of the normal stresses become equal, and the typical definition of "pressure" with isotropic normal stresses are observed. Previously, pressure-driven flows were created

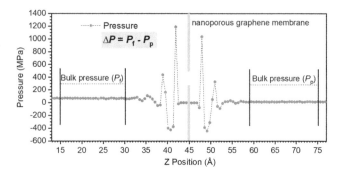

FIGURE 16.10 Typical pressure distribution along the z-direction when the two reflection boundaries are moving. Bulk pressures in the feed and permeate sides are the average of local pressures in the defined bulk regions.

either by applying constant forces on rigid pistons made from single-layer graphene sheets [39,40] or by applying constant forces on water molecules [41,42]. The former method can lead to uncertainties due to the effects of van der Waals (vdW) interactions between the pistons and water, whereas the latter method does not exactly mimic pressure-driven flows in actual membrane systems in which pressure gradient is not constant or unidirectional. We also attempted to create better statistical averages for salt rejection rates by using various initial conditions of salt distributions for every simulation of the pressure-driven flows. Three different pore sizes were selected based on the hydraulic diameters of 9.90 Å, 11.57 Å, and 14.40 Å due to the hexagonal structure of graphene. As shown in Figure 16.11a, the distribution of water molecules is found in axial direction with high peaks near the membranes, which is similar to that reported by Suk and Aluru [41]. No accumulation of salt ions (Na^+ and Cl^-) near the nanoporous graphene membrane is observed as shown in Figure 16.11b. We also realized that pressure drop necessary to obtain a specific flow rate is higher for smaller pore diameters and that pressure drop linearly increases with increasing the flow rate. The well-known linear relationship between pressure drop and flow rate was also reported in other studies [41,43,44]. The pressure drop required for a specified flow rate depends on the viscosity of water and the pore diameter but not pore length due to the one-atom-thick graphene membranes. By applying the Buckingham Pi theorem (BPT), the functional relationship between pressure drop, flow rate, pore diameter, and viscosity was established as follows:

$$\Phi = C_o \frac{\Delta P D_h^3}{\mu}, \tag{16.6}$$

where ΔP, Φ, D_h, and μ are the notations for pressure drop, volumetric flow rate, hydraulic diameter, and dynamic viscosity, respectively; C_o is a constant. Interestingly, Equation (16.6) has the same form with that obtained from the Stokes equations for an incompressible pressure-driven flow through a circular pore on an infinitely thin plate where $C_o = 1/24$ [45,46]. The viscosity of 0.6 M NaCl solution is estimated based on our previous study which shows

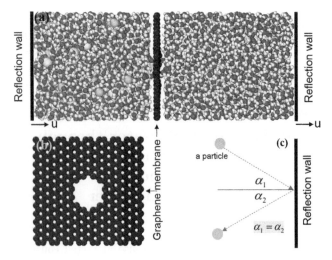

FIGURE 16.9 (a) Schematics of the simulation domain. The size of sodium ions and chloride ions are exaggerated for better visualization. (b) Typical structure of a graphene membrane with a pore in the middle. (c) Definition of a specular reflection wall.

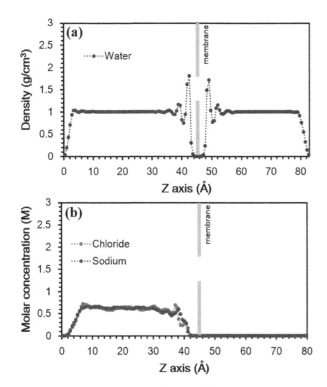

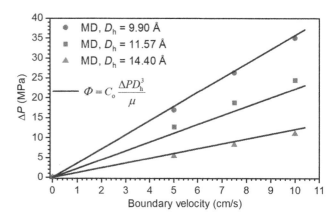

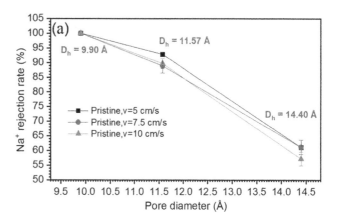

FIGURE 16.12 Pressure difference versus boundary velocity for different pore diameters for pristine nanoporous graphene membranes. Both the data obtained from MD simulations and from our predictions using Eq. (16.6) are presented. In Eq. (16.6), $C_o = 1/38$ is used for all three different cases of pore diameter.

FIGURE 16.11 Density distribution of water along the streamwise (z) direction (a) and ionic concentration distribution of sodium and chloride ions in the z-direction (b) for a system with pristine graphene membrane.

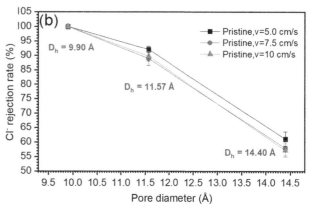

FIGURE 16.13 (a) Sodium ion rejection rates for different pore diameters and boundary velocities (or equivalently flow rate). (b) Chloride ion rejection rates for different pore diameters and boundary velocities (or equivalently flow rate).

that adding ions into water increases the water viscosity [11]. Utilizing the estimated viscosity and the obtained data of pressure and flow rate, we found that C_o is in the range of $1/38.62$–$1/38.17$. The range of C_o is approximately one and a half times smaller than that obtained from continuum analysis. It implies that the pressure drop needed with nanoporous graphene membranes is higher than that predicted by continuum analysis for a specified flow rate. This is mainly attributed to the dominance of vdW forces exerted by the nanoporous membranes on the salt solution, making it more difficult for liquid particles to flow through the nanopores. Figure 16.12 illustrates a good agreement between the MD results and our proposed analytical predictions. Importantly, we observed that a rejection rate of 100% for both sodium and chloride ions can be attained with 9.90 Å pore diameter nanoporous graphene membranes. This reported pore diameter associated with a rejection rate of 100% is even larger than 5.5 Å as identified by Cohen-Tanugi and Grossman [39] or 7.5 Å as claimed by Konatham et al. [47]. Our results confirmed that the obtained volumetric flow rate of deionized water can be as high as 3.733×10^{-12} l/h-pore with the required pressure drop of 35.02 ± 1.53 MPa, which is 52% lower to that shown by Cohen-Tanugi and Grossman [43]. This interesting discrepancy is due to the differences in the methodologies of simulating pressure-driven flows and the use of different force fields. In addition, our results showed that increasing the pore diameter to a value larger than 9.9 Å decreases rejection rate as shown in Figure 16.13. In an

experimental study, Surwade et al. found that nanoporous single-layer graphene membranes provide nearly 100% salt (KCl) rejection rate and a high water permeability for pore diameters ranging from 0.5 to 1.0 nm [48]. This result is in good agreement with our predictions from the MD simulations.

16.5   Conclusions

In this chapter, we systematically summarized our studies on water transport in planar graphene nanochannels and through nanoporous graphene membranes. The potential of using planar graphene nanochannels in nanofluidic applications was highlighted by elucidating liquid and flow properties including density, velocity, viscosity, slip length, and flow rates. Force-driven water flows in uncharged channels show plug-like velocity profiles with large slip lengths at solid–liquid interfaces. The slip lengths are found to be independent from the channel heights for heights larger than 2 nm. High slip length results in enhanced flow rate of water in nanoscale channels. It was also shown that the transport of water in graphene channels can be actively controlled by adding charges on the graphene surfaces. With increased surface charge, water molecules accumulate near the walls due to enhanced electrostatic interactions, and velocity profiles transform into a parabolic shape rather than the plug-like form. In addition, the viscosity of deionized water increases while slip length decreases with increased surface charge. The combined effect of slip reduction and viscosity enhancement decreases the flow rate in graphene nanochannels. Furthermore, the transport of ionized water through nanoporous graphene membranes was discussed in terms of density distribution, pressure distribution, permeability, and selectivity. The results show that nanoporous graphene membranes can be used in water desalination systems due to their high water permeability and salt rejection rates.

References

1. Lee, C., Wei, X., Kysar, J. W., & Hone, J. (2008). Measurement of the elastic properties and intrinsic strength of monolayer graphene. *Science*, 321(5887), 385–388.

2. Balandin, A. A., Ghosh, S., Bao, W., Calizo, I., Teweldebrhan, D., Miao, F., & Lau, C. N. (2008). Superior thermal conductivity of single-layer graphene. *Nano Letters*, 8(3), 902–907.

3. Chen, J. H., Jang, C., Xiao, S., Ishigami, M., & Fuhrer, M. S. (2008). Intrinsic and extrinsic performance limits of graphene devices on SiO$_2$. *Nature Nanotechnology*, 3(4), 206.

4. Schedin, F., Geim, A. K., Morozov, S. V., Hill, E. W., Blake, P., Katsnelson, M. I., & Novoselov, K. S. (2007). Detection of individual gas molecules adsorbed on graphene. *Nature Materials*, 6(9), 652.

5. Elias, D. C., Nair, R. R., Mohiuddin, T. M. G., Morozov, S. V., Blake, P., Halsall, M. P., ... & Novoselov, K. S. (2009). Control of graphene's properties by reversible hydrogenation: Evidence for graphane. *Science*, 323(5914), 610–613.

6. Nair, R. R., Blake, P., Grigorenko, A. N., Novoselov, K. S., Booth, T. J., Stauber, T., ... & Geim, A. K.

(2008). Fine structure constant defines visual transparency of graphene. *Science*, 320(5881), 1308–1308.

7. Kannam, S. K., Todd, B. D., Hansen, J. S., & Daivis, P. J. (2011). Slip flow in graphene nanochannels. *The Journal of Chemical Physics*, 135(14), 016313.

8. Taherian, F., Marcon, V., van der Vegt, N. F., & Leroy, F. (2013). What is the contact angle of water on graphene? *Langmuir*, 29(5), 1457–1465.

9. Shih, C. J., Wang, Q. H., Lin, S., Park, K. C., Jin, Z., Strano, M. S., & Blankschtein, D. (2012). Breakdown in the wetting transparency of graphene. *Physical Review Letters*, 109(17), 176101.

10. Thompson, P. A., & Troian, S. M. (1997). A general boundary condition for liquid flow at solid surfaces. *Nature*, 389(6649), 360.

11. Celebi, A. T., & Beskok, A. (2018). Molecular and continuum transport perspectives on electroosmotic slip flows. *The Journal of Physical Chemistry C*, 122(17), 9699–9709.

12. Celebi, A. T., Barisik, M., & Beskok, A. (2017). Electric-field-controlled transport of water in graphene nano-channels. *The Journal of Chemical Physics*, 147(16), 164311.

13. Celebi, A. T., Barisik, M., & Beskok, A. (2018). Surface charge-dependent transport of water in graphene nano-channels. *Microfluidics and Nanofluidics*, 22(1), 7.

14. Nguyen, C. T., & Beskok, A. (2018). Saltwater transport through pristine and positively charged graphene membranes. *The Journal of Chemical Physics*, 149(2), 024704.

15. Chernov, A. A., & Mikheev, L. V. (1988). Wetting of solid surfaces by a structured simple liquid: Effect of fluctuations. *Physical Review Letters*, 60(24), 2488.

16. Mamontov, E., Burnham, C. J., Chen, S. H., Moravsky, A. P., Loong, C. K., De Souza, N. R., & Kolesnikov, A. I. (2006). Dynamics of water confined in single- and double-wall carbon nanotubes. *The Journal of Chemical Physics*, 124(19), 194703.

17. Gruener, S., Wallacher, D., Greulich, S., Busch, M., & Huber, P. (2016). Hydraulic transport across hydrophilic and hydrophobic nanopores: Flow experiments with water and n-hexane. *Physical Review E*, 93(1), 013102.

18. Koplik, J., & Banavar, J. R. (1995). Continuum deductions from molecular hydrodynamics. *Annual Review of Fluid Mechanics*, 27(1), 257–292.

19. Ghorbanian, J., & Beskok, A. (2016). Scale effects in nano-channel liquid flows. *Microfluidics and Nanofluidics*, 20(8), 121.

20. Ghorbanian, J., Celebi, A. T., & Beskok, A. (2016). A phenomenological continuum model for force-driven nano-channel liquid flows. *The Journal of Chemical Physics*, 145(18), 184109.

21. Sam, A., Hartkamp, R., Kannam, S. K., & Sathian, S. P. (2018). Prediction of fluid slip in cylindrical

nanopores using equilibrium molecular simulations. *Nanotechnology*, *29*(48), 485404.

22. Kumar Kannam, S., Todd, B. D., Hansen, J. S., & Daivis, P. J. (2012). Slip length of water on graphene: Limitations of non-equilibrium molecular dynamics simulations. *The Journal of Chemical Physics*, *136*(2), 024705.

23. Xiong, W., Liu, J. Z., Ma, M., Xu, Z., Sheridan, J., & Zheng, Q. (2011). Strain engineering water transport in graphene nanochannels. *Physical Review E*, *84*(5), 056329.

24. Koumoutsakos, P., Jaffe, R. L., Werder, T., & Walther, J. H. (2003). On the validity of the no-slip condition in nanofluidics. *Nanotech*, *1*, 148–151.

25. Ramos-Alvarado, B., Kumar, S., & Peterson, G. P. (2016). Hydrodynamic slip length as a surface property. *Physical Review E*, *93*(2), 023101.

26. Thomas, J. A., & McGaughey, A. J. (2008). Reassessing fast water transport through carbon nanotubes. *Nano Letters*, *8*(9), 2788–2793.

27. Falk, K., Sedlmeier, F., Joly, L., Netz, R. R., & Bocquet, L. (2010). Molecular origin of fast water transport in carbon nanotube membranes: Superlubricity versus curvature dependent friction. *Nano Letters*, *10*(10), 4067–4073.

28. Huang, D. M., Cottin-Bizonne, C., Ybert, C., & Bocquet, L. (2008). Aqueous electrolytes near hydrophobic surfaces: Dynamic effects of ion specificity and hydrodynamic slip. *Langmuir*, *24*(4), 1442–1450.

29. Wei, N., Peng, X., & Xu, Z. (2014). Understanding water permeation in graphene oxide membranes. *ACS Applied Materials & Interfaces*, *6*(8), 5877–5883.

30. Wagemann, E., Oyarzua, E., Walther, J. H., & Zambrano, H. A. (2017). Slip divergence of water flow in graphene nanochannels: The role of chirality. *Physical Chemistry Chemical Physics*, *19*(13), 8646–8652.

31. Xie, Q., Alibakhshi, M. A., Jiao, S., Xu, Z., Hempel, M., Kong, J., ... & Duan, C. (2018). Fast water transport in graphene nanofluidic channels. *Nature Nanotechnology*, *13*, 238–245.

32. Maali, A., Cohen-Bouhacina, T., & Kellay, H. (2008). Measurement of the slip length of water flow on graphite surface. *Applied Physics Letters*, *92*(5), 053101.

33. Ho, T. A., & Striolo, A. (2015). Promising performance indicators for water desalination and aqueous capacitors obtained by engineering the electric double layer in nano-structured carbon electrodes. *The Journal of Physical Chemistry C*, 119(6), 3331–3337.

34. Qiao, R., & Aluru, N. R. (2005). Atomistic simulation of KCl transport in charged silicon nanochannels: Interfacial effects. *Colloids and Surfaces A: Physicochemical and Engineering Aspects*, 267(1–3), 103–109.

35. Thomas, M., Corry, B., & Hilder, T. A. (2014). What have we learnt about the mechanisms of rapid water transport, ion rejection and selectivity in nanopores from molecular simulation? *Small*, *10*(8), 1453–1465.

36. Barisik, M., & Beskok, A. (2011). Equilibrium molecular dynamics studies on nanoscale-confined fluids. *Microfluidics and Nanofluidics*, *11*(3), 269–282.

37. Nguyen, C. T., & Kim, B. (2016). Stress and surface tension analyses of water on graphene-coated copper surfaces. *International Journal of Precision Engineering and Manufacturing*, *17*(4), 503–510.

38. Nguyen, C. T., Barisik, M., & Kim, B. (2018). Wetting of chemically heterogeneous striped surfaces: Molecular dynamics simulations. *AIP Advances*, *8*(6), 065003.

39. Cohen-Tanugi, D., & Grossman, J. C. (2012). Water desalination across nanoporous graphene. *Nano Letters*, *12*(7), 3602–3608.

40. Heiranian, M., Farimani, A. B., & Aluru, N. R. (2015). Water desalination with a single-layer MoS_2 nanopore. *Nature Communications*, *6*, 8616.

41. Suk, M. E., & Aluru, N. R. (2013). Molecular and continuum hydrodynamics in graphene nanopores. *RSC Advances*, *3*(24), 9365–9372.

42. Wang, Y., He, Z., Gupta, K. M., Shi, Q., & Lu, R. (2017). Molecular dynamics study on water desalination through functionalized nanoporous graphene. *Carbon*, *116*, 120–127.

43. Cohen-Tanugi, D., & Grossman, J. C. (2014). Water permeability of nanoporous graphene at realistic pressures for reverse osmosis desalination. *The Journal of Chemical Physics*, *141*(7), 074704.

44. Muscatello, J., Jaeger, F., Matar, O. K., & Müller, E. A. (2016). Optimizing water transport through graphene-based membranes: Insights from nonequilibrium molecular dynamics. *ACS Applied Materials & Interfaces*, *8*(19), 12330–12336.

45. Sampson, R. A. (1891). On Stokes's current function. *Philosophical Transactions of the Royal Society of London. A*, *182*, 449–518.

46. Jensen, K. H., Valente, A. X., & Stone, H. A. (2014). Flow rate through microfilters: Influence of the pore size distribution, hydrodynamic interactions, wall slip, and inertia. *Physics of Fluids*, *26*(5), 052004.

47. Konatham, D., Yu, J., Ho, T. A., & Striolo, A. (2013). Simulation insights for graphene-based water desalination membranes. *Langmuir*, *29*(38), 11884–11897.

48. Surwade, S. P., Smirnov, S. N., Vlassiouk, I. V., Unocic, R. R., Veith, G. M., Dai, S., & Mahurin, S. M. (2015). Water desalination using nanoporous single-layer graphene. *Nature Nanotechnology*, *10*(5), 459–464.

17

Nanoscale Magnetism

Roopali Kukreja
University of California Davis

Hendrik Ohldag
University of California Santa Cruz
Stanford University

17.1 Introduction

Magnetism in materials is one of the fundamental physical phenomena that have been employed in applications like compasses for centuries. Ferromagnetic materials have been used in various applications throughout the history of mankind. Aristotle is attributed to having the first scientific discussion on magnetism with Thales of Miletus around 585 B.C. [1]. Around the same time, in ancient India, Sushruta, a surgeon, used magnet for surgical purposes for the first time. [2]. In ancient China, the earliest literary reference to magnetism lies in a book of the 4th century BC named after its author, *The Master of Demon Valley* [3]. By the 12th century, medieval Chinese navigators were using magnetic materials in compass needles for navigation. However, the fundamental phenomena underlying magnetic order were only recently understood after the advent of quantum mechanics in the 20th century, and it revolves around the property of electron called 'spin'. In the early 20th century, Weiss used the concept of 'molecular field' which aligns the spins present in a ferromagnetic material and is responsible for long-range magnetic ordering [4], although the origin of this field and its huge magnitude could not be explained. It was with the development of symmetrization postulate classifying electrons as Fermions, which manifested as exchange interaction between electrons provided the field necessary to achieve ferromagnetic ordering. It should be noted that the exchange interaction is a purely quantum mechanical concept originating from the fact that the electrons are Fermions following Pauli's Exclusion Principle [5]. This 'exchange field' defines the spin system in a material and is responsible for the spin alignment, either ferromagnetic or antiferromagnetic. On the other hand, on a macroscopic scale, magnetism was believed to be fully understood after the publication of Maxwell's laws and theory of electromagnetism, and magnetism research therefore attracted mainly the interest of engineers rather than physicists. This changed dramatically with the advent of vacuum technology, which allowed controlled thin-film growth on the monolayer scale resulting in well-defined growth of magnetic thin films, surfaces and interfaces. Consequently, nanoscale magnetism became highly relevant for magnetic storage technologies in the 1960s. With the discovery of many interesting interfacial phenomenon such as giant magnetoresistance (GMR), spin injection, spin transfer torque (STT), spin pumping etc., fundamental magnetism research has undergone a magnificent revival in past 50 years.

Hard disk drives based on tiny magnetic bits became the core of large-scale data storage devices because of their immense recording capacity, price per unit of storage, write latency and product lifetime. The parallel or anti-parallel alignment of ferromagnet (FM), read as '1' or '0' bit by the read head of the drive, made it possible to read and write information [6]. The building blocks of a hard drive are the magnetic bits where tiny FMs are aligned from North to South or South to North. These tiny FMs generate stray field which is read out by the 'read head' of the hard disk drive as '1' or '0' bit. By reading these bits, the information stored in hard drive is retrieved. This information can also be written by setting or aligning these bits using magnetic field [6]. In the past decade, different ways for manipulating the magnetization, such as using current [7,8] or optical laser [9], have gained much interest as these are easier to confine and manipulate in comparison to the magnetic field. Also, with the ever increasing areal density of these bits, understanding the spin dynamics and spin transport in these nanoscale devices has become extremely important. In this chapter, we lay out the details of such nanoscale devices including STT-based devices as well highlight active areas of research in this field.

Figure 17.1 shows different methods for manipulating magnetism on the nanoscale. This can be achieved by

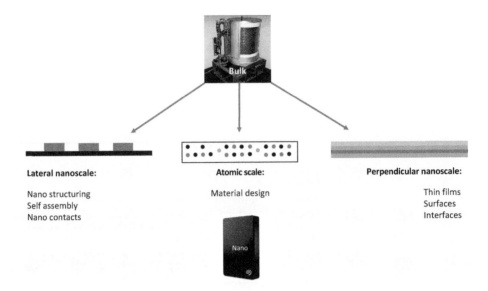

FIGURE 17.1   Methods to manipulate magnetism at nanoscale include thin film growth, heterostructuring and confinement.

confining and defining structures laterally via assembly of nanoparticles or lithography-based patterning, or one can also achieve magnetism in the perpendicular direction by thin film growth and heterostructuring. For completeness, we also mention the possibility of studying magnetism on the nanoscale by directly controlling and manipulating the atomic arrangement in alloys, oxides and other complex compounds. All these methods provide a disruptive approach for engineering state-of-the-art nanoscale devices. While classical engineering required shaping and connecting already existing materials in new and efficient ways, modern day nanoengineering starts with the design of the material on the nanoscale. To limit the scope of this chapter, we will not go into details of thin film growth and leave this to other publications focusing on material science issues.

This chapter will begin by presenting the current state of the field of nanoscale magnetism with focus on magnetic systems confined by interfaces and physical boundaries in artificial nanostructures or layered materials. By creating well-defined interfaces between different magnetic or even non-magnetic materials, it is possible to introduce new magnetic anisotropies or additional terms to the magnetic energy. One example is the presence of unidirectional anisotropies caused by the interaction between FMs and anti-ferromagnets (AFMs) leading to an effect called exchange bias. Another prominent example is the Dzyaloshinskii–Moriya interaction (DMI) that manifests itself at interfaces between ferromagnetic materials and materials with a high spin-orbit coupling (SOC), leading to a non-collinear arrangement of the magnetization in the ground state and fascinating new magnetic spin structures like skyrmions. Following the discussion on static interactions, we will address spin transport across magnetic layers. Spin injection, spin-orbit interaction and other emerging phenomena at magnetic interfaces and nanostructures have particular relevance in defining spin transport, which will also be discussed in this chapter. They act as additional contributions to the magnetic free energy

and have to be considered in the Landau–Lifshitz–Gilbert equation, which is used to describe the temporal evolution of magnetic systems.

17.2   Spatially Confined Magnetism – Nanoparticles

The most straightforward way of creating nanoscale magnetic systems is by either directly producing magnetic nanoparticles or growing magnetic films and using lithography-based approaches to pattern the films on the nanoscale into the desired shape. Both approaches are very popular, and there are myriads of examples in the literature on the experimental realization. Due to the sheer amount of research, we simply cannot present a comprehensive review of this field, but we provide a brief overview here. The motivation for developing magnetic nanoparticles with desired properties are as follows:

1. Magnetic nanoparticles are inherently mobile, which means that they can be manipulated and assembled to form a desired arrangement. Also, they can be functionalized so that they will attach themselves to preferred locations. This can have multiple applications in life science and drug delivery.

2. When a magnetic sample shrinks, its properties are less determined by the bulk of the material but more determined by its surface and the interface with the support. Also, external fields become more relevant. This can lead to completely unexpected behavior and new, previously unsuspected properties.

Magnetic nanoparticles are either produced chemically from a solution of a magnetic compound or physically by sublimation from a solid source at high temperature. They can

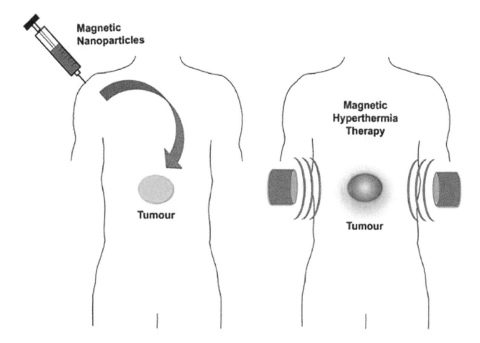

FIGURE 17.2 Magnetic nanoparticles can be injected into the human body and functionalized so that they preferably attach to tumor tissue. The particles will then be excited from outside the body with an electromagnetic field; the excitation will produce heat that can locally destroy the tumor without affecting healthy tissue. (University of York)

range in size from a few nanometers to hundreds of nanometers depending on the application. They may be synthesized from a single material (e.g. Fe, Co, Ni), oxide (e.g. Fe_3O_4, CoO) or complex compound (e.g. $LaSrMnO_3$). The fact that they are individual mobile particles offers the possibility to assemble them in any structure or to dispatch them over a distance to take over a defined function. But even a single particle can be fascinating. For example, in 1956, Meiklejohn and Bean [10] set out to study the magnetic properties of Co nanoparticles with an average size of 20 nm. When they measured the magnetic properties of these particles they were surprised to see that the hysteresis loops were not symmetric with respect to zero applied field, meaning the magnetization $\mathbf{M}(\mathbf{H})$ after application of a positive field was not simply the inverse of $\mathbf{M}(-\mathbf{H})$. This was highly unusual, because until then hysteresis loops with only even symmetries (uniaxial, cubic, etc.) were observed. The reason for this behavior was the formation of CoO layer on the outside of Co particles, resulting in a so-called core–shell nanoparticle, with a Co core and a CoO shell. Such core–shell nanoparticles are very common nowadays, and by controlling the chemical composition and thickness of each layer, it is possible to fine-tune the magnetic properties. In the example of Meiklejohn and Bean, the outer CoO layer provided an antiferromagnetic shell for the ferromagnetic core. Due to its compensated magnetic structure, the AFM is not sensitive to external magnetic fields, and any excess magnetization in the AFM* can induce a preferred

direction of the magnetization or unidirectional anisotropy. This microscopically induced unidirectional anisotropy will lead to a macroscopic shift of the hysteresis loop. This is a clear example of how simply shrinking the size of a magnetic object can lead to completely new properties, due to the dominant role played by surfaces and interfaces. We will come back to this particular system in the next section.

For now, we will focus on two applications that provide a pedagogical illustration of the efficient use of nanoparticles in magnetism. Because of their mobility, nanoparticles are well-suited in life sciences and medicine for the treatment of diseases like cancer. The materials used for magnetic nanoparticles in medicine are often biocompatible (e.g. Fe), and using appropriate ligands, they can be modified such that after they are released into the body, they will attach to tumor tissue as shown in Figure 17.2. Once in place, the particle can then be activated from the outside using radiowaves or microwaves. If the patient is placed in a magnetic field, the resonance frequency can be tuned to be far away from any resonance frequency of organic molecules. The magnetization of the particle will follow the external microwave field as in a conventional ferromagnetic resonance experiment, with two possible consequences. Depending on the actual location of the particle (inside or outside a cell) and the frequency used for the treatment, the particle itself may start to oscillate and destroy the cancerous cell through physical pressure, or if the particle is immobilized, the oscillating magnetic moment will produce enough heat locally to destroy the cell (hyperthermia). In both situations, the cancerous cells are destroyed with minimal effect on surrounding healthy tissue. However, in both cases, it is

---

* Excess magnetization exists at grain boundaries, surfaces and interfaces where the bulk symmetry of the antiferromagnet is broken.

necessary to precisely control the magnetic properties of these particles.

The second example of magnetic nanoparticles application is self-assembly. By growing thin films of materials using atomic deposition techniques, one can essentially grow films of nanoparticles as shown on the left-hand side of Figure 17.3, where the particles arrange themselves in a honeycomb lattice. The particles will form a regular array, where the variation in particle size and spacing is small compared to their distance. Even more importantly, each of the magnetic particles is magnetically independent, meaning if one applies a local magnetic field, it is possible to reverse the magnetization or information stored in each particle without affecting the neighboring particle. This method can provide a unique way to achieve even higher storage densities than possible today. For example, current storage densities are typically 1 Tbit/in^2, which could only increase slightly in future using conventional approaches by potentially limiting data retention times. The reason is that magnetic bits become thermally unstable as the size decreases, once the energy stored in the magnetic bit becomes comparable to the thermal energy at room temperature. The shape anisotropy of a nanoparticle adds to the magnetic energy, which is why it is possible to obtain smaller yet stable bits using nanoparticles. The example shown in the figure uses 10 nm particles corresponding to a storage density of 10 Tbit/in^2, and it is conceivable to decrease the particle size to 5 nm or increase the storage density to 40 Tbit/in^2.

17.3 Magnetic Interfaces

17.3.1 Chemical versus Magnetic Interfaces

Interfaces between different magnetic materials may exhibit properties that cannot be found in either of the two materials that form the interface. One of the reasons behind this observation is the fact that the magnetic interactions can reach across a few lattice sites. In the previous section, we briefly discussed the observation of exchange bias in Co/CoO core–shell nanoparticles, which we will now further elaborate. In 1956, Meiklejohn and Bean [10] discovered that the Co/CoO core–shell nanoparticles that the hysteresis loops of these particles are shifted horizontally along the field axis. This means that ideally a particle will always exhibit the same magnetization direction if the field is fully removed and the horizontal loop shift is larger than the coercive field of the FM. This effect is referred to as *exchange bias* or unidirectional anisotropy and is very useful for the fabrication of magnetic reference layers, since an AFM/FM exchange coupled bilayer can be engineered such that it is not sensitive to external fields. Despite the fact that the effect had been discovered almost 60 years ago, a fundamental understanding has been elusive until recent time. The use of X-ray spectroscopy and microscopy in this area has been extremely successful and has provided detailed insight into the exchange bias phenomenon. [11,12].

In order to fully understand what happens upon deposition of a ferromagnetic metal onto an antiferromagnetic oxide surface, let us briefly look at the chemistry of such an interface. One may already intuitively expect that an interface between a metal and an oxide is not sharp, because chemical forces that are not negligible at $T > 0$ might drive oxygen atoms to diffuse from the oxide into the metal layer. The chemical reaction at the interface can be visualized by employing the elemental and chemical sensitivity of X-ray absorption spectroscopy (XAS). For example, Figure 17.4 shows the Co L$_3$ and Ni L$_2$ XAS spectra were obtained

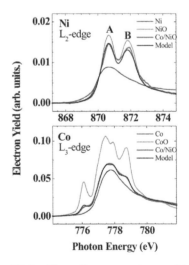

FIGURE 17.4 Absorption spectra obtained from (1 nm)Co/(1 nm)NiO in Co/NiO. These are compared to the reference metal and oxide spectra. Also shown is a fit to the Co/NiO spectra, assuming that 1 monolayer of Co is oxidized and 1 monolayer of NiO is reduced at the interface. (Reprinted figure with permission from [13]. Copyright (2001) by the American Physical Society.)

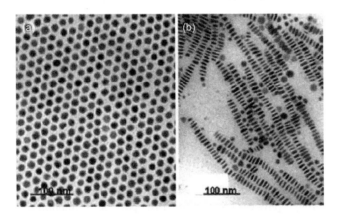

FIGURE 17.3 Magnetic nanoparticles can arrange themselves in a way so that their magnetization can be individually reversed without affecting neighboring bits or in tracks similar to the arrangement on a hard drive disk. (University of Nebraska)

from a (1 nm)Co/(1 nm)NiO sample. Also shown are reference spectra for the metal and for the bare oxide surface. The spectra obtained from the Co/NiO sample is shown as well. The XAS intensity for the Ni L_2 edge appears to be lower compared to the oxide sample, indicating the increasing metallic character of Ni atoms in the film. On the other hand, the spectrum obtained at the Co L_3 resonance exhibits a multiplet character and increased peak intensity. Both observations are an indicator of more oxidic character of the original metallic Co film.

Altogether the lineshapes observed for the samples suggest that their XAS spectra can be represented as a superposition of the pure metal and the oxide spectra. To corroborate this assumption, the last spectrum in the graph shows a linear combination of metal and oxide spectra to best fit the unknown spectra. An excellent agreement is achieved demonstrating that, indeed at the interface, a certain amount of Co is oxidized to CoO and NiO is reduced at the same time to Ni. The thickness of the interfacial layer consisting of $CoNiO_x$ can be derived from the weighting factors used for the fit. Considering an electron escape length of 2.5 nm, one estimates that about a monolayer of Ni is formed on top of the NiO and a monolayer of CoO at the bottom of Co. Therefore the thickness of the interfacial $CoNiO_x$ layer is estimated to be about two monolayers [13,14]. Several different groups have further investigated the relative alignment between the AFM spin axis and the FM moment [15,16] using X-ray photoemission electron microscopy (X-PEEM) and have reported either parallel or perpendicular alignment depending on the chemical state of the interface. In general, one finds that a stronger parallel uniaxial correlation between the two layers implies a higher amount of oxygen transfer at the interface.

Another fascinating observation upon deposition of a ferromagnetic material like Co on antiferromagnetic NiO is that the easy axis in the AFM changes its direction. The origin of the reorientation of the AFM is more challenging to investigate, in particular for the case of weak interfacial coupling between Co and NiO resulting in a perpendicular orientation. The magnetization of the Co layer in the coupled system wants to be aligned parallel to the surface plane, because the ferromagnetic film has a strong shape anisotropy which wants to minimize any stray field emerging from the sample. The antiferromagnetic axis in NiO naturally has a component out of plane, but it is usually found that it will rotate into the plane in the coupled system [16] as well. In the case of parallel exchange coupling between the two layers, this is easy to understand; however, the so-called *spin-flop* (perpendicular) coupling could be realized without reorientation of the AFM spin axis.

Figure 17.5 shows Ni X-ray magnetic linear dichroism (XMLD), Co X-ray magnetic circular dichroism (XMCD) and O X-ray linear dichroism (XLD) images of a bare NiO(001) surface and of a Co/NiO bilayer. Ni XMLD images show AFM domains in NiO, while Co XMCD images show ferromagnetic domains in Co. The third type of dichroism at the oxygen K-edge is observed because the AFM domains

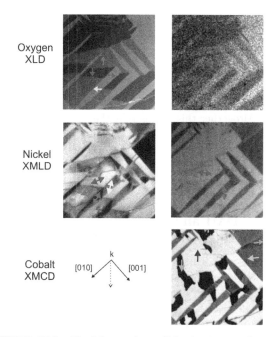

FIGURE 17.5 The left row shows dichroism images obtained of the bare surface, while the right row shows images obtained from the same spot after deposition of a thin ferromagnetic Co film. The orientation of the antiferromagnetic spin axis and the direction of the ferromagnetic moments in each domain are indicated in the Ni L-edge XMLD and Co L-edge XMCD images. Upon Co deposition, (100) domain walls vanish in the antiferromagnetic and in the O K-edge XLD images. (Reprinted figure with permission from [17]. Copyright (2009) by the American Physical Society.)

are correlated with crystallographic twin domains with different orientation of the c-axis. The first row reveals the correlation between spin and twin domains. Upon deposition of a thin Co film (2.5 nm) and formation of a thin interface mixed oxide layer, only two S-domains remain at the NiO surface due to the reorientation of the antiferromagnetic axis. For example, we find that the domain wall between some antiferromagnetic domains vanishes while the domain wall between other domain remains. If the uncompensated Ni moments caused by the chemical reduction at the interface are not the reason for the reorientation, then maybe the oxidized Co sites play a role. For this purpose, we take a close look at the contrast between the corresponding T-domain walls observed with O XLD. The overall contrast in this image is now strongly reduced due to the Co layer on top which reduces the electron yield arising from the NiO layer.

A detailed analysis of the contrast across the domain walls reveals a qualitatively similar development of the O XLD contrast compared to the Ni XMLD contrast. While the contrast across the domain walls, which vanish in the AFM picture does not completely vanish in the O LD, it is still much stronger reduced (by about a factor of 3) than the contrast across the domain wall that does not vanish in the AFM after deposition of the ferromagnetic Co layer. This observation indicates that the crystallographic T-domains undergo a structural reorientation in a similar manner as the antiferromagnetic domains undergo a spin reorientation. Such behavior can be readily explained with an interfacial lattice distortion of NiO toward a CoO-like structure. The (001) surface of CoO would only exhibit two T-domains as observed here due to the tetragonal distortion of the CoO lattice in contrast to the rhombohedral distortion of the NiO lattice. Altogether this means that the observed rotation of the antiferromagnetic the spin axis from < 121 > to < 110 > is caused by the change in magnetic anisotropy from a situation that is typical for NiO toward a CoO-like symmetry with an easy magnetic axis of < 110 > [17].

17.3.2 DMI and the Importance for New Magnetic Topologies

In recent years, chiral magnetic structures have emerged with interesting discoveries of magnetic skyrmions, fast-moving chiral domain walls, topological Hall effect and magnetoelectric effects, which have provided a unique way to confine and control magnetism at nanoscale. Chirality in magnetism defines a rotational sense for the magnetization, which can rotate clockwise or counterclockwise along an axis of the magnetic object. These chiral states have been shown to host many interesting properties that rely on this local rotation or twisting. Specifically, skyrmions, a particular class of chiral magnetization structure, are attractive candidates for information technology as they are inherently nanosized and respond more efficiently to spin-polarized currents compared to conventional domain walls. This energy-efficient aspect of chiral magnets, combined with their miniature dimensions, makes them technologically appealing as they can be integrated into current designs for high-density storage. The key ingredient for stabilizing these chiral magnetic configurations is the DMI, which occurs due to the SOC in the systems with broken inversion symmetry. In 2007, the interfacial DMI due to the broken inversion symmetry at the surface was experimentally observed which opened the route to DMI stabilized skyrmions at interfaces such as the nanoskyrmion lattice of Fe/Ir (111).

DMI is a fundamental ingredient responsible for twisting the magnetization of two coupled spins $\vec{S}_1$ and $\vec{S}_2$, as the overall energy can be reduced between the two spins by canting their relative orientation as dictated by the DMI vector. Moriya proposed that the origin of this anisotropic magnetic exchange interaction can be derived by considering SOC within a superexchange model. Later, Smith and Fert derived that DMI between $\vec{S}_1$ and $\vec{S}_2$ can also stem from an indirect exchange mechanism mediated by conduction electrons. Here, similar to the usual Ruderman–Kittel–Kasuya–Yosida (RKKY) interaction, conduction electrons locally couple to the atomic spins become spin polarized and thus mediate the interaction. In addition, the electrons spin-orbit scatter at the non-magnetic host atoms (Figure 17.6). This process leads to a long-range DMI between the atomic spins, where not only the amplitude but also the orientation of the DM vector oscillates as a function of separation. The

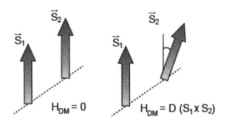

FIGURE 17.6 DMI is an asymmetric interaction and is responsible for twisting the magnetization of two coupled spins S1 and S2, to reduce the overall energy of the system.

magnetization of the coupled pair of quantum spins with spin operators $\vec{S}_1$ and $\vec{S}_2$ for DMI can then be quantified by the following Hamiltonian: $H = -D\vec{S}_1 x \vec{S}_2$.

As the orientation of the DM vector in Eq. (17.1) dictates a sense of rotation of $\vec{S}_2$ compared with $\vec{S}_1$, the distance-dependent oscillation of D induced by the indirect exchange mechanism inherently contains the possibility to tune the chirality by the separation of the pair. This tells us that energy can be gained by introducing an angle between $\vec{S}_1$ and $\vec{S}_2$. However, the deviation from collinearity has to be in the right direction, while the opposite direction is energetically costly; thus one particular sense of spin rotation is favored. Hence, DMI is an asymmetric interaction.

17.3.3 Skyrmions

Magnetic skyrmion is a particle-like distortion to an otherwise uniform magnet as shown in Figure 17.7. Tony Skyrme theorized that these particles are topologically protected, meaning that they are defined by a topological integer, N_{sk}, that will not change by deforming the field around it. Cubic and non-centrosymmetric magnets are one of the possible skyrmion-hosting materials. In these materials, non-collinear spin configurations, which include screw spin and skyrmion states, can occur. The skyrmion structure can vary depending on the different values of vorticity m and helicity γ. Skyrmion structures have been studied by neutron scattering, Lorentz transmission electron microscopy (LTEM) and spin-resolved scanning tunneling microscopy. Neutron scattering experiments on MnSi and Fe1–xCoxSi have identified an 'A' phase which includes a two-dimensional skyrmion crystal (SkX) phase [18].

Skyrmions can be generated in different magnetic material systems via different possible mechanisms including (i) long-ranged magnetic dipolar interactions, (ii) the relativistic DMI, (iii) frustrated exchange interactions and (iv) four-spin exchange interactions. Mechanism (i) applies to magnetic thin films with perpendicular easy-axis anisotropy, where magnetic charge creates a competition between in-plane magnetization preferred by the dipolar interaction and out-of-plane magnetization preferred by the anisotropy. The competition creates periodic stripes where the magnetization rotates perpendicular to the thin film. The applied magnetic field perpendicular to the thin film transforms the periodic stripes to a periodic array of skyrmions or magnetic

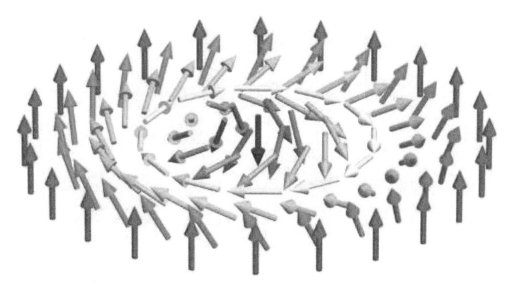

FIGURE 17.7 A magnetic skyrmion is a topological magnetic structure. It can be generated by unwinding a hedgehog onto a plane.

bubbles. In this case, the size of skyrmions ranges from 100 nm to 1 μm. Mechanism (ii) applies to non-centrosymmetric magnets such as MnSi and FeGe. The DMI determines the size of the skyrmions, which typically range from 5 to 100 nm. The skyrmions in (i) and (ii) are typically larger in size and have many internal degrees of freedom and hence are highly mobile. For (iii) and (iv), the skyrmion size is around 1 nm, the same order as the lattice constant. The stabilization mechanism effects which values of γ and m minimize the associated energy cost. In case (i), to suppress the energy cost, m = 1 and $\gamma = \pm\pi/2$ are required [1].

The emergent electromagnetic field (EEMF) describes the unique topological properties of skyrmions. For example, the topological Hall effect (THE) is induced by the emergent magnetic field of the skyrmions on conduction electrons. THE measurements have been done to show that the SkX phase region is more stable in thin films.

A key advantage of skyrmions over other magnetic nanostructures is their unique dynamics when combined with conduction electrons or electric polarization. An ultralow current density of the order 10^6 A m^{-2} can drive skyrmion crystal motion; this current is five to six orders of magnitude smaller than that needed for ferromagnetic domain wall motion. The small critical current density needed for skyrmions allows for the manipulation of information with low power consumption per bit in the slow-speed regime [18].

17.4 Spin Transport in Nanostructures and Across Interfaces

17.4.1 Giant Magnetoresistance (GMR)

GMR was independently discovered by Albert Fert [19] and Peter Grünberg [33] in 1988, for which they were jointly awarded 2007 Nobel Prize in Physics. The basic idea of GMR effect is that the resistance of two ferromagnetic layers

decoupled by a spacer layer (typically Cu or Au) is dependent upon their relative magnetic alignment. This structure is called a spin valve device. If both the layers are aligned parallel to each other, then the resistance is lower compared to the case when they are aligned antiparallel to each other. The difference in resistance of up to 110% has been observed between two configurations [32,33]. A few years later after the discovery of GMR, every hard disk drive was using this effect to 'read' orientation of magnetic bits which are used to store data. GMR effect is also used in biosensors, microelectromechanical systems and magnetoresistive random-access memory (MRAM) applications.

Figure 17.8 taken from Ref. [21] shows GMR-based read head of a hard drive. One of the two magnetic layers in the read head of hard drive is a magnetically fixed layer. The other FM is 'free' to rotate its magnetization parallel or antiparallel with respect to the 'fixed' layer. Whenever the free layer encounters a change in the magnetization direction, i.e. between two adjacent bits on the disk, it changes its direction accordingly. The bits with opposite magnetization alignment have stray fields associated with them which aligns the free magnetic layer on the read head. In this way, by applying constant current to the read head, one can use the measured voltage (and thus resistance) at read head to determine the logical '0' and '1' depending upon the parallel or antiparallel alignment of the two ferromagnetic layers.

17.4.2 Spin Transfer Torque

The understanding of methods to manipulate magnetism was transformed in 1996, when Slonczewski [7] and Berger [8] independently predicted that the free layer in spin valve structure can be switched by current instead of the field. If high-enough current density is passed through a spin valve stack as shown in Figure 17.9, the fixed FM acts as a spin polarizer/filter, and the current is spin polarized; i.e. it has a net magnetic moment. This spin-polarized current

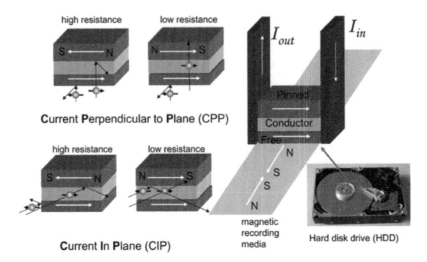

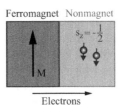

FIGURE 17.8   Giant magnetoresistance is currently used in read heads of hard disk drives.

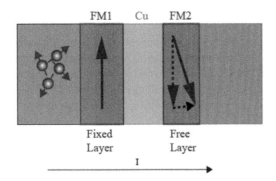

FIGURE 17.9   Spin Transfer Torque: Using current to manipulate the magnetization of the free layer.

FIGURE 17.10   Spin Injection: Spin-polarized current injected from a FM into NM.

applies a torque on the free layer when it passes through the free layer. It transfers its angular momentum to the free layer and switches the magnetization of the free layer. This phenomenon is called STT, where current is used to manipulate magnetic alignment of the free layer. As current can be constrained and manipulated much easily in comparison to fields, STT is a promising candidate for future data storage and non-volatile memory applications. In 2012, Everspin Technologies successfully produced a 64 MB STT-RAM, based on STT effect [22].

The key mechanism underlying GMR and STT effect is spin injection from ferromagnetic layer to non-magnetic layer (which then traversed through the second ferromagnetic layer). As mentioned above, the current which goes through the fixed ferromagnetic layer in the Figure 17.10 gets spin polarized. This spin-polarized current results in various kinds of interesting phenomenon, spin accumulation at the interface where spins are injected from a FM to nonmagnet (NM) called as spin injection, switching of the other ferromagnetic layer (STT, described above) and can also excite spin waves or vortex oscillations. In the following, we briefly describe the spin transport and spin accumulation behavior underlying these effects.

17.4.3   Spin Transport Across a FM-NM Interface

Mott's two-current model forms the basis of understanding the transport in ferromagnetic material well below the Curie temperature [23]. This model was further developed by Fert and Campbell [24]. Mott postulated that itinerant s electrons carry the electric current, and the electrical resistance is due to the scattering processes where the s electrons jump from the s-band to the d-band. Additionally, Fermi's golden rule states that the scattering probability of the conduction electrons is proportional to density of final states, i.e. the density of empty d states above the Fermi level. Resistivity of the transition metals with a partially filled d shell is thus produced by the dominant effect of itinerant sp electrons scattering on the localized d hole states. This scattering is spin selective where the two spin states (spin up and spin down) carry the electric current in parallel without much mutual interactions, i.e. no spin flips. We can formulate the two-current model by denoting the spin relaxation time for the separate up and down channels by $\tau_\uparrow$ and $\tau_\downarrow$. Since conduction may occur in two separate channels, the conductivities of two channels add in series according to Matthiessen's rule. We refer the readers to Refs [5] for detailed discussion on Mott's spin current model.

It is important to note that in Mott's model, the rate $1/\tau$ of electronic transition is simply due to coulomb scattering of charge. At low temperature, it arises from lattice defects or impurities, and at higher temperatures, it is dominated by atomic displacements due to thermal motion (phonons). And since the coulomb interaction does not act on spin, spin flips are forbidden, and the two spin channels are independent of each other. Hence, the itinerant sp electrons scatter into available d-states of the same spin.

This spin selective scattering process is one of the reasons for spin filtering when a current traverses through a ferromagnetic layer. In a FM like Co and Ni, we can see from the density of states that, as shown in Figure 17.11, the minority channel has higher number of holes available compared to the majority channel. As the scattering strength of itinerant sp electrons is proportional to number of d-holes available at the Fermi level, the minority electrons undergo higher scattering in the Co (or Ni) layer, resulting in a preferential loss of minority spin component. Thus the spin-dependent transport results in a spin filtering effect [5].

Another source of spin filtering effect is spin-dependent reflection at the interface [25]. When current is injected at the interface, both transmission and reflection of the current occurs. The reflection and transmission probability is spin dependent for a FM with more minority spins being reflected. This leads to different transverse spin components reflected and transmitted and thus to a discontinuity in the transverse spin current injected into the FM. Hence, the interface plays an important role in spin filtering the majority and minority channels.

Both the spin selective scattering within the FM, referred to as bulk contribution, and the spin-dependent reflection and transmission at the interface, referred to as the interface contribution, cause the spin filtering effect by a FM. Though most of the studies assume a transparent interface and only include the bulk contribution due to its simplicity, it has been shown that the interface contribution is indeed very important and is bigger than the bulk contribution [5]. This

stresses the role played by the interface in spin injection and spin transport.

When this spin-filtered current enters a non-magnetic layer, spin accumulation at the interface happens due to the mismatch in conductivities (for spin-up and spin-down channels) of ferromagnetic and non-magnetic layers, and it exponentially decays away from the interface. This spin accumulation results in a transient induced magnetic moment in the non-magnetic layer.

Let us take the conductivity of the FM as σ_F and of NM as σ_N. In the FM, as we saw earlier, the conductivities of two (spin-up and spin-down) channels would be different due to different scattering strengths. Thus, for the FM, we define, $\sigma_F^\uparrow = \alpha_F \sigma_F$ and $\sigma_F^\downarrow = (1 - \alpha_F)\sigma_F$, where the dimensionless factor $0 < \alpha_F < 1$ accounts for the asymmetry of conduction in the majority and minority channels. This can also be used for a NM where $\alpha = 0.5$, as there is no asymmetry in spin signal.

This asymmetry in conductivity inside the FM leads to asymmetric currents. We then have $j^\uparrow = \beta_F j$ and $j^\downarrow = (1 - \beta_F)j$. As $j = j^\uparrow + j^\downarrow$, the asymmetry parameter β for the current lies in the range $0 < \beta < 1$. For a NM, the value of β_N would be 0.5 as both up and down spin channels are equally present. The spin-polarized current from a FM can then be given by

$$P = \frac{j^\uparrow - j^\downarrow}{j^\uparrow + j^\downarrow} = 2\beta - 1. \tag{17.1}$$

As the spin-polarized current from the FM ($\beta_F \neq 0.5$) enters N, we will have $\beta_N \neq 0.5$ and $P \neq 0$ in the NM close to the interface. It will take time or distance from the interface for the current to equilibrate to the value of $\beta_N = 0.5$, as the transitions between spin channels are rare in non-magnetic metals. This causes a drop in voltage at the interface, called the spin accumulation voltage, V_{AS} (the derivation of which can be found elsewhere [5]), which is given by

$$eV_{AS} = \left[\alpha_F - \frac{1}{2}\right][\mu^\uparrow(0) - \mu^\downarrow(0)]. \tag{17.2}$$

This is the spin accumulation voltage which is essentially the jump in averaged chemical potential that develops at the F–N interface due to conductivity mismatch. The actual difference in the chemical potential for the two spin states is referred to as the spin voltage ($\mu^\uparrow - \mu^\downarrow$) and reflects the difference in number of electrons in the two spin channels. The spin voltage decreases away from the interface, as in the bulk of the NM, the majority and minority populations are equal. The decay of the spin voltage away from the interface is governed by the spin diffusion equation which is similar to the Fick's diffusion law [5], where spin diffusion length is given by $\Lambda = \sqrt{D\tau_{se}}$, where D is the diffusion constant and τ_{se} is the spin diffusion time or spin relaxation time or spin flip time.

For the FM, the equilibration involves transitions between opposite spin states. One of the major processes is the inelastic scattering by exciting spin waves (magnons) providing a transfer for energy and angular momentum.

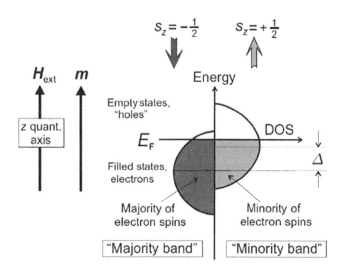

FIGURE 17.11 Stoner model, the localized 3d electrons from majority and minority bands in a ferromagnet, which are separated by exchange splitting from [42].

In Co, the equilibration length $\Lambda = 5$ nm, and the spin relaxation time is $\tau_{se} = 0.4$ ps [5]. For the NMs, spin equilibration process is dominated by spin dephasing process. Spin dephasing occurs due to the presence of spin orbit coupling when electrons scatter on the phonons or the atomic impurities and defects, i.e. coloumbic scattering process. This leads to electrons experiencing different effective magnetic paths based on the difference in the diffusive path taken, resulting in a spin precession around this magnetic field. This spin precession leads to spin randomization with equal average up and down probabilities along any chosen quantization axis. During this process, the angular momentum is directly transferred to the lattice. In Cu, this equilibration length is $\Lambda = 350$ nm [26], and the spin relaxation time is $\tau_{se} = 4.7$ ps [5].

Spin injection was first measured by Johnson and Silsbee across a FM/NM interface in a FM/NM/FM device in 1985 [27]. Since then, it has become a very exciting field of research although its quantitative understanding is still complicated by a lack of knowledge of interfacial effects on the degree and sign of the spin polarization of the injected electrons.

17.4.4 Spin Hall Effects

Before we discuss spin Hall effect, let us briefly discuss Hall effect. The Hall effect occurs due to the Lorentz force of an applied magnetic field which acts on moving carriers resulting in the transverse deflection of electrons. For ferromagnetic materials, SOC generates an asymmetric deflection of charge carriers with respect to spin direction. Thus, a charge current generates a polarized charge current, resulting in anomalous Hall effect (AHE). Similar to Hall effect, the AHE can be detected electrically in a FM via a transverse voltage due to difference in population of spin-up and spin-down electrons.

Spin Hall effect (SHE) was first measured in 2004 and is observed when an unpolarized charge current generates a transverse pure spin current. In non-magnetic materials, this effect manifests as edge spin accumulation with opposite polarization on the opposite edges of the material. The reciprocal effect of SHE is called inverse spin Hall effect (ISHE) which happens when a pure spin current generates a transverse charge current. In both cases (SHE and ISHE), the material must possess SOC. The main difference between AHE, SHE and ISHE is that while AHE relates charge degrees of freedom via relativistic spin-orbit interaction, the SHE and ISHE correlate the charge degrees of freedom (conserved quantity) to spin degree of freedom (non-conserved quantity). Note that in case of AHE, a charge current generates a polarized transverse charge current. In the SHE, an unpolarized charge current generates a pure transverse spin current.

The experimental observations of SHE and ISHE have been performed using optical methods. For SHE, experimental observations confirmed that there are two mechanisms possible: intrinsic (p GaAs by Wunderlich et al.

[34,35]) and extrinsic (n GaAs by Kato et al. [36,37]). For ISHE, the experiment was performed by a two-color optical excitation technique with perpendicular polarizations [38]. The spin current produced by the laser excitations is transferred due to the ISHE into a transverse electrical current, resulting in a spatially dependent charge accumulation which was detected by the optical transmission signal of a probe laser beam. Optical measurements were also performed on timescales shorter than the scattering time and provided a direct demonstration of the intrinsic SHE signal.

SHE induced torque (damping-like and field-like) can induce ferromagnetic resonance (FMR) oscillations, STT and drive domain walls. If these current-induced torques arise only from the absorption of the spin current generated by the SHE in the NM, the analysis of field-symmetric and field-antisymmetric contributions of the detected DC output voltage at FMR allows for a quantitative determination of the strength and symmetry of the SHE-induced torques, as well as the spin Hall angle of the NM.

Additionally, in multilayer systems where the interface breaks the structural inversion symmetry, another microscopic effect accompanies SHE–STT, and it is called spin galvanic phenomenon. The term spin galvanic effect (SGE) is derived from the analogy to the galvanic (voltaic) cell. Instead of a chemical reaction, however, it is the spin polarization that generates an electrical current (voltage) in the SGE. Inversely, an electrical current generates the spin polarization in the ISGE. SHE and ISGE go hand in hand and are known as companion phenomena, both allowing for electrical alignments of spins in the same structure. ISGE can generate relativistic spin-orbit torque. Current-induced torques generated by the companion spin Hall and spin galvanic effect can also be extended to AFMs as they can have the microscopically staggered nature of the induced effective fields and can thus couple strongly to the Neel order [39]. Since external electric fields couple weakly to antiferromagnetic moments, the electrically generated staggered fields are rather unique in providing efficient means for the manipulation of AFMs. Three microscopic mechanisms have been postulated by which AHE, SHE and ISHE occur: intrinsic, skew-scattering and side jump [28]. These mechanisms are the results of coherent band mixing effects induced by the external electric field and the disorder potential. Thus the models to describe them are much more complex than a simpler single-band diagonal transport model. As they belong to a class of coherent interference transport phenomena, they cannot be directly explained using traditional semiclassical Boltzmann theory. The key recent development that led to a better understanding of the mechanism was linking directly the semiclassical and microscopic theory of spin-dependent Hall transport. This link between the semiclassically defined processes and their fully equivalent multiband microscopic theories was established by fully generalizing the Boltzmann transport theory to take interband coherence effects into account.

Two contributions to anomalous or spin Hall conductivity are identified as τ_1 and τ_0. The contribution proportional to τ_1 is the skew-scattering mechanism. The contribution to τ_0 comes from both intrinsic and side jump processes. The first term (intrinsic) arises from the evolution of spin-orbit coupled quasiparticles as they are accelerated by an external electric field in the absence of disorder. The second term arises from scattering events from impurities that do not include the skew-scattering contribution. In order to extend the theory to describe SHE and ISHE, the coupling of these spin-current generating mechanisms to spin-charge drift diffusion transport equations is required. These drift equations used are based on a particular experiment, as depending on the measurements, the spin accumulation induced by the SHE may vary, e.g. in non-local transport measurements versus FMR-based mechanisms. Thus for SHE the spin Hall conductivity and its consequences have to be ultimately coupled to the spin accumulation that it induces which can depend on the method of measurements. In following, we provide a brief description of the three postulated microscopic mechanism [28]:

Intrinsic mechanism: The intrinsic contribution of the spin Hall conductivity is dependent only on the band structure of the perfect crystal. It arises from the non-equilibrium electron dynamics of the Bloch electrons as they are accelerated in an electric field and undergo spin precession due to the induced momentum-dependent magnetic field. Akin to the Lorentz force at work in the ordinary Hall effect (OHE), electrons moving in a crystal structure experience an internal spin orbit force from a momentum-dependent magnetic field. In FMs, this leads to an intrinsic AHE. In NMs, this leads to an intrinsic SHE. The contribution of intrinsic SHE has been evaluated with several mathematical techniques including Bloch state formalism, Kubo formula for spin Hall conductivity in an ideal lattice, semiclassical theory of the wave packet dynamics and *ab initio* theory. It has been shown that the largest contribution to the spin Hall conductivity arises, similar to AHE, whenever the bands connected via SOC are near each other at the Fermi energy. The calculated spin Hall conductivities are predicted to be large in these transition metals, and in particular, a sign change is predicted going from Pt to Ta which has been observed in experiments. The intrinsic mechanism is dominant in materials with strong SOC, and for these materials, calculations based on the intrinsic SHE can provide a good estimate for overall SHE/spin Hall angles.

Skew-scattering mechanism: Asymmetric scattering rate present in the collision term of Boltzmann transport equation. This mechanism dominates in nearly perfect crystals and is proportional to the Bloch state transport. In the presence of SOC, a transition which is right handed with respect to the magnetization direction has a different probability than the corresponding left-handed transition. A disorder potential is formed from SOC. Right- and left-hand transitions of electrons (w.r.t. magnetization) are not equal in probability, resulting in asymmetric chiral features. This asymmetry is the basis of the skew-scattering mecha-

nism, and accounting it in the Boltzmann equation shows a current perpendicular to the electric field and the magnetization. When the skew-scattering mechanism dominates, the Hall resistivity is proportional to longitudinal resistivity. The challenge for calculations of the contribution from this mechanism lies in accurate modeling of chiral asymmetry and accuracy of characterization of disorder. The Boltzmann equation and *ab initio* theory have been used for highly accurate calculations of the spin Hall angle.

In simple models, the skew-scattering contributions to the SHE or AHE are considered to arise only from the SOC in the disorder potential. This is only valid when the typical disorder broadening is larger than the splitting of the bands due to the SOC. In systems with strong SOC in the bands, such as heavy transition metals, considering the SOC only in the disorder potential would be incorrect. The reason is because, in this case, a strong contribution to the skew scattering also arises from the scattering of the spin-orbit-coupled quasiparticle from the scalar potential. In fact, the SOC of the disorder potential is typically strongly renormalized by the other nearby subbands as well, and therefore the effect of the multiband character can never be ignored in these materials.

Side jump mechanism: The side jump mechanism arises with the transverse displacement of a Gaussian wave packet after scattering off a spherical impurity with SOC. Extrinsic side jump involves the non-spin-orbit-coupled portion of the wave packet scattering off the spin-orbit-coupled disorder. Intrinsic side jump involves the spin-orbit-coupled portion of the wave packet scattering off the scalar potential. The Keldysh formula, Kubo formula and semiclassical theory have been used to evaluate the side jump contribution to SHE [40]. The reasons are the complexity of the measurements, the dephasing of spin and the lack of practical general theories that can bring one from a weak to a strong spin-orbit-coupled regime. We mentioned earlier that in certain simplified models, there exist relative cancellations, either total or partial, between the contributions that depend to zeroth order on the scattering lifetime. This is a topic that has entertained the research community of the AHE and the SHE for quite some time.

Two non-magnetic material devices have been discussed based on these effects. Firstly, the Hankiewicz H-bar device combines SHE and ISHE behavior. In the SHE portion of the device, an unpolarized electrical current generates transverse spin. This spin is then injected into the second leg, the ISHE portion of device, where electrical current is generated upon injection. The Hankiewicz device has been realized experimentally [41]. The second device was proposed by Hirsch and focuses on the oppositely polarized spin edge accumulation in NMs. Upon connection of opposite edges, the resulting spin-dependent chemical potential gradient generates a circulating spin current. This spin current is then detected by an ISHE meter. Challenges arising from the short spin lifetime compatibility with device and wire dimensions have prevented the experimental realization of the Hirsch device.

Devices combining ferromagnetic and non-magnetic materials have also been proposed. Zhang proposed the electrical detection of edge spin accumulation in NM with an FM probe. The inefficiency of spin transport across the NM/FM interface due to resistance mismatch was solved by implementing a highly resistive tunnel barrier between the FM electrode and semiconductor channel. Another hybrid device was realized [29] in which the SHE in NMs was used to inject spin currents into FMs and magnetization was electrically controlled via STT switching. It is also noted that ISHE can be used to detect spin currents in NMs and FMs. In multilayered devices, SHE induces torques in the adjacent FMs. FMR can be used to observe two types of SHE-induced torques: an antidamping torque with similar behavior to Gilbert damping and a field-like torque manifesting a shift in the FMR resonance line. These torques also induce a lateral current along the NM–FM interface.

References

1. L. Fowler. *Historical beginnings of theories of electricity and magnetism.* University of Virginia, Charlottesville, VA, 1997.

2. H. P. Vowles. *Early Evolution of Power Engineering.* University of Chicago Press, Chicago, IL, 1932.

3. S.H. Li. Origine de la boussole II. Aimant et boussole. *Isis*, 45:175–196, 1954.

4. P. Weiss. L'hypothese du champ moleculaire et de la propriete ferromagnetique. *Journal de Physics*, 6:661–690, 1907.

5. J. Stöhr. *NEXAFS Spectroscopy.* Springer Series in Surface Sciences, vol. 25. Springer, Heidelberg, 1992.

6. IBM Archives. www-03.ibm.com/ibm/history/index.html.

7. J. C. Slonczewski. Current-driven excitation of magnetic multilayers. *Journal of Magnetism and Magnetic Materials*, 159(1–2):L1 – L7, 1996.

8. L. Berger. Emission of spin waves by a magnetic multilayer traversed by a current. *Phys. Rev. B*, 54:9353–9358, 1996.

9. A. V. Kimel, A. Kirilyuk, P. A. Usachev, R. V. Pisarev, A. M. Balbashov, and Th. Rasing. Ultrafast non-thermal control of magnetization by instantaneous photomagnetic pulses. *Nature*, 435(7042):655–657, 2005.

10. W. H. Meiklejohn and C. P. Bean. New magnetic anisotropy. *Physical Review*, 102:1413, 1956.

11. H. Ohldag, P. Esquinazi, E. Arenholz, D. Spemann, M. Rothermel, A. Setzer, and T. Butz. The role of hydrogen in room temperature ferromagnetism at graphite surfaces. *New Journal of Physics*, 12:123012, 2010.

12. A. Scholl, H. Ohldag, F. Nolting, S. Anders, and J. Stöhr. Study of ferromagnet-antiferromagnet interfaces using x-ray PEEM. In H. Hopster and H. P. Oepen, editors, *Magnetic Microscopy of Nanostructures.* Springer, Berlin.

13. H. Ohldag, T.J. Regan, J. Stöhr, A. Scholl, F. Nolting, J. Lüning, C. Stamm, S. Anders, and R. L. White. Spectroscopic identification and direct imaging of interfacial magnetic spins. *Physical Review Letters*, 87:247201, 2001.

14. T. J. Regan, H. Ohldag, C. Stamm, F. Nolting, J. Lüning, and J. Stöhr. Chemical effects at metal/oxide interfaces studied by x-ray absorption spectroscopy. *Physical Review B*, 64:214422, 2001.

15. E. Arenholz, G. van der Laan, R. V. Chopdekar, and Y. Suzuki. Angle-dependent Ni2+ X-ray magnetic linear dichroism: Interfacial coupling revisited. *Physical Review Letters*, 98:197201, 2007.

16. H. Ohldag, A. Scholl, F. Nolting, S. Anders, F. U. Hillebrecht, and J. Stöhr. Spin reorientation at the antiferromagnetic NiO(001) surface in response to an adjacent ferromagnet. *Physical Review Letters*, 86:2878, 2001.

17. H. Ohldag, G. van der Laan, and E. Arenholz. Correlation of crystallographic and magnetic domains at Co/NiO(001) interfaces. *Physical Review B*, 79:052403, 2009.

18. N. Nagaosa and Y. Tokura. Topological properties and dynamics of magnetic skyrmions. *Nature Nanotechnology*, 8(12):899–911, 2013.

19. M. N. Baibich, J. M. Broto, A. Fert, F. Nguyen Van Dau, F. Petroff, P. Etienne, G. Creuzet, A. Friederich, and J. Chazelas. Giant magnetoresistance of (001) Fe/(001) Cr magnetic superlattices. *Physical Review Letters*, 61:2472–2475, 1988.

20. G. Binsach, P. Grunberg, F. Saurenbach, and W. Zinn. Enhanced magnetoresistance in layered magnetic structures with antiferromagnetic interlayer exchange. *PRB*, 39:4828, 1989.

21. Japan National Institute for Materials Science. www.nims.go.jp/apfim/GMR.html.

22. Ever Spin Technologies. www.everspin.com/.

23. N. F. Mott. The electrical conductivity of transition metals. *Proceedings of the Royal Society of London. Series A - Mathematical and Physical Sciences*, 153(880):699–717, 1936.

24. A. Fert and I. A. Campbell. Two-current conduction in nickel. *Physical Review Letters*, 21:1190–1192, 1968.

25. M. Stiles and A. Zangwill. Anatomy of spin-transfer torque. *Physical Review B*, 66(1):1–14, 2002.

26. F. J. Jedema, A. T. Filip, and B. J. van Wees. Electrical spin injection and accumulation at room temperature in an all-metal mesoscopic spin valve. *Nature*, 410(6826):345–348, 2001.

27. M. Johnson and R. H. Silsbee. Interfacial charge-spin coupling injection and detection of spin magnetization in metals. *PRL*, 55:1790–1793, 1985.

28. J. Wunderlich C. H. Back J. Sinova, S. O. Valenzuela and T. Jungwirth. Spin Hall effects. *Reviews of Modern Physics*, 87:1213, 2015.

29. D. Ralph and M. Stiles. Spin transfer torques. *Journal of Magnetism and Magnetic Materials*, 320:1190–1216, 2008.

30. M. N. Baibich, J. M. Broto, A. Fert, F. Nguyen Van Dau, F. Petro, P. Etienne, G. Creuzet, A. Friederich, and J. Chazelas. Giant magnetoresistance of (001) Fe/(001) Cr magnetic superlattices. *Physical Review Letters*, 61:2472, 1988.

31. G. Binsach, P. Grunberg, F. Saurenbach, and W. Zinn. Enhanced magnetoresistance in layered magnetic structures with antiferromagnetic interlayer exchange. *PRB*, 39, 1989.

32. M. N. Baibich, J. M. Broto, A. Fert, F. Nguyen Van Dau, F. Petro, P. Etienne, G. Creuzet, A. Friederich, and J. Chazelas. Giant magnetoresistance of (001) Fe/(001) Cr magnetic superlattices. *Physical Review Letters*, 61:2472, 1988.

33. G. Binsach, P. Grunberg, F. Saurenbach, and W. Zinn. Enhanced magnetoresistance in layered magnetic structures with antiferromagnetic interlayer exchange. *Physical Review B*, 39:4828, 1989.

34. J. Wunderlich, A. Irvine, J. Sinova, B. G. Park, L. P. Zârbo, X. L. Xu, B. Kaestner, V. Novák, and T. Jungwirth. Spin-injection Hall effect in a planar photovoltaic cell. *Nature Physics*, 5:675, 2009.

35. J. Wunderlich, B. Kaestner, J. Sinova, and T. Jungwirth. Experimental observation of the spin-Hall effect in a two-dimensional spin-orbit coupled semiconductor system. *Physical Review Letters*, 94:047204, 2005.

36. Y. K. Kato, R. C. Myers, A. C. Gossard, and D. D. Awschalom. Observation of the spin Hall effect in semiconductors. *Science*, 306:1910–1913, 2004a.

37. Y. K. Kato, R. Myers, A. Gossard, and D. D. Awschalom. Current-induced spin polarization in strained semiconductors. *Physical Review Letters*, 93:176601, 2004b.

38. H. Zhao, E. Loren, H. van Driel, and A. Smirl. Coherence control of Hall charge and spin currents. *Physical Review Letters*, 96:246601, 2006.

39. J. Železný, H. Gao, K. Výborný, J. Zemen, J. Mašek, A. Manchon, J. Wunderlich, J. Sinova, and T. Jungwirth. Relativistic Néel-order fields induced by electrical current in antiferromagnets. *Physical Review Letters*, 113:157201, 2014.

40. N. Nagaosa, J. Sinova, S. Onoda, A. H. MacDonald, and N. P. Ong. Anomalous Hall effect. *Reviews of Modern Physics*, 82:1539, 2010.

41. C. Brüne, A. Roth, E. G. Novik, M. König, H. Buhmann, E. M. Hankiewicz, W. Hanke, J. Sinova, and L. W. Molenkamp. Evidence for the ballistic intrinsic spin Hall effect in HgTe nanostructures. *Nature Physics*, 6:448, 2010.

42. J. Stöhr and H. C. Siegmann. *Magnetism, from Fundamentals to Nanoscale Dynamics*, Springer, Berlin, 2008.

18

Physics of Nanomagnets

Ralph Skomski,
Balamurugan Balasubramanian,
and D. J. Sellmyer
University of Nebraska

18.1 Introduction

Magnetic nanostructures exhibit a great diversity concerning geometry, feature size, chemistry, crystal structure, spin structure, and applications. Geometries include single-phase nanostructures, such as nanoparticles [1–3], thin films [4], nanowires [5–7], and nanotubes [8], as well as a variety of inhomogeneous structures, some of which are shown in Figure 18.1. Feature sizes of typical nanostructures are between 2 nm and a few 100 nm. Chemically, the phases are often based on iron-series ($3d$) transition-metal elements, such as Fe, Co, and Ni, but palladium-series ($4d$), platinum-series ($5d$), and rare-earth ($4f$) elements are also used. Depending on the application, the phases may be metallic or nonmetallic, elements, alloys, oxides, and other compounds being used. Concerning crystal structure, one important distinction is between cubic and noncubic structures, but other features are also important, for example, atomic packing density and the absence or presence of inversion centers.

Geometry, feature size, and chemistry depend on the application. Rare-earth transition-metal permanent magnets exploit nanoscale structural features, as exemplified by grain-boundary phases in Nd-Fe-B and cellular structures in Sm-Co [9,10]. Such high-performance permanent magnets are used in a wide range of high-tech applications,

ranging from computer hard disk drives and wind generators to magnetic resonance imaging devices, hybrid car motors, and cell phone vibrators. Traditional magnets, such as hexagonal ferrites, are structured on a somewhat bigger length scale and used in everyday applications such as loudspeakers, windscreen wipers, locks, microphones, and toy magnets [11–16]. By mass, the most widespread magnetic materials are soft magnets for low-frequency power electronics and for high-frequency applications, for example, in microwave devices. Many soft magnets are structured on a scale of several or many micrometers (microstructured), but some are nanostructured or amorphous [17].

Magnetic recording and spin electronics are another important application with a variety of nanostructures used to write, read, and store information [18–20]. The areal storage density and the corresponding effective bit size has been reduced from 250 µm in 1960 to about 25 nm in 2015. Current recording media are nanogranular, with Fe-Pt or Co-Pt as the main magnetic phase, and there is an ongoing trend towards bit patterning, where each grain forms a bit. A lower limit to the bit size, somewhat less than 5 nm, is provided by the thermal stability of the stored information (Section 18.5.4). Other applications include but are not limited to magnetic nanoparticles for clinical hyperthermia, drug administration, and catalysis, and thin-film nanostructures for sensors.

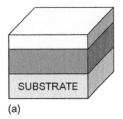
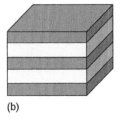
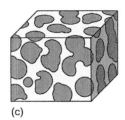

(a)   (b)   (c)

FIGURE 18.1   Some inhomogeneous nanostructures: (a) thin film on substrate, (b) multilayer, and (c) granular nanostructure. Gray and white regions denote chemically different substances.

In spite of all diversity, the physics of nanomagnets reflects a relatively small number of basics findings from quantum mechanics, relativistic physics, and statistical mechanics. It is important to distinguish between intrinsic and extrinsic magnetic properties. Intrinsic properties reflect the atomic origin of magnetism, both chemical composition and crystal structure. The following intrinsic properties are of particular interest. First, to yield a macroscopic *magnetization* $M(r)$, the atoms (subscript i) must have a magnetic moment mi. The magnetization is then defined as $M(r) = \mathrm{d}<m>/\mathrm{d}r$, where the volume element $\mathrm{d}r$ contains many atoms but is usually smaller than the nanoscale feature size. Second, the atomic magnetic moments mi must be aligned to yield a nonzero magnetization. Thermal disorder randomizes the atomic magnetic moments mi and yields $M = 0$ above the *Curie temperature* T_c. Third, it is important to control the direction of the magnetization relative to the crystal axes or to the coordinate frame of the nanostructure, a feature known as magnetic anisotropy. For example, permanent magnets must be stable with respect to mechanical torque and external magnetic fields and therefore require easy-axis anisotropy. The opposite is true for soft magnets, because magnetic anisotropy tends to suppress magnetization changes and create losses. A major source of anisotropy is the interaction of the atomic spins with the crystal lattice, known as *magnetocrystalline anisotropy*, but there are also magnetostatic contributions to the anisotropy.

Intrinsic properties are normally realized on a length scales of less than 1–2 nm, although some effects are longer range and may then interfere with nanostructuring. Extrinsic properties, such as the coercivity H_c and the remanent magnetization (remanence) M_r, are inherently nanoscale and strongly depend on the nanostructure. This is because the extrinsic behavior responsible for hysteresis (Section 18.4) involves magnetic domains and domain walls,

which contain many atomic moments. Table 18.1 lists the magnetic properties of some materials.

M_s (Section 18.2), K_1 (Section 18.3), and A (Section 18.4) are experimental room-temperature values, whereas $\mu_o H_A$, δ_B, l_{ex}, and R_{SD}, all discussed in Section 18.4, are calculated from M_s, K_1, and A [10,13,17].

The atomic moments in a solid are not necessarily parallel. Figure 18.2a shows some zero-temperature spin structures, each arrow corresponding to one atomic spin or moment. In a ferromagnet (FM), all atomic spin are parallel, which is the spin structure preferred for permanent and soft magnet, because it yields the largest magnetization. The three iron-series elements Fe, Co, and Ni are ferromagnetic, as are many alloys; and alloys, for example, PtCo, $SmCo_5$, $Nd_2Fe_{14}B$. Antiferromagnets (AFM) contain two (or more) sublattices whose magnetizations cancel. Such magnets cannot be used as permanent magnets, but there are applications where the zero net magnetization is advantageous, for example, in spin electronics. Ferrimagnets (FiM) also contain sublattices, but in contrast to AFM, the sublattices are chemically or crystallographically inequivalent. Most oxides and halides are antiferromagnetic (MnO, NiO, MnF_2) or ferrimagnetic (Fe_3O_4, $BaFe_{12}O_{19}$) [21], but CrO_2 is a ferromagnet, whereas some metals are antiferromagnets, for example, Mn.

The spins of Figure 18.2a are all collinear (parallel or antiparallel). There are also many noncollinear spin structures, one of which is shown in Figure 18.2b. Zero-temperature or 'frozen' spin structures of this type are found in spin glasses and amorphous magnets. A similar spin structure is also found above the Curie temperature, but with one important difference: above Tc, the spins are not frozen but rotate rapidly. The difference is similar between that of a glass and a liquid, where snapshots of the atomic positions look similar but rapidly change with time in the case of liquids. Figure 18.2c shows a typical micromagnetic

TABLE 18.1   Magnetic Properties of Some Ferro- and Ferrimagnetic Materials

Material	Crystal Structure	$\mu_o M_s$ T	T_c K	K_1 MJ/m^3	A pJ/m	$\mu_o H_A$ T	δ_B nm	l_o nm	R_{SD} nm
Fe	Cubic	2.15	1043	0.048	8.3	0.056	41	1.5	6.2
Co	Cubic	1.76	1388	0.53	10.3	0.76	14	2.0	34
Ni	Cubic	0.62	631	−0.005	3.4	0.014	82	3.3	15
Fe_3O_4	Cubic	0.60	858	−0.011	7.0	0.031	79	4.9	35
$BaFe_{12}O_{19}$	Hexagonal	0.48	723	0.33	6.1	1.7	14	5.8	280
$Ni_{80}Fe_{20}$	Cubic	1.0	815	−0.0005	10	0.001	440	3.5	3.2
FePt	Tetragonal	1.43	750	6.6	6.3	12	3.1	2.0	140
$SmCo_5$	Hexagonal	1.07	1003	17.0	22.0	40	3.6	4.9	760
$Nd_2Fe_{14}B$	Tetragonal	1.61	585	4.9	7.7	7.6	3.9	1.9	110

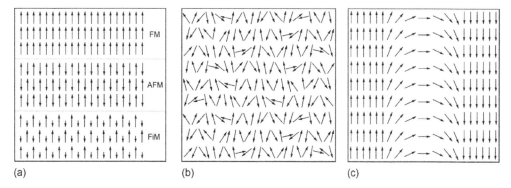

FIGURE 18.2 Some basic spin structures: (a) collinear spin structures at zero temperature, (b) noncollinear spin structure, and (c) micromagnetic spin structure (domain wall). Each arrow represents one atomic spin, and both (a) and (b) have zero net magnetization.

spin structure, namely a domain wall. The basic atomic spin structure is ferromagnetic, so that neighboring spins are nearly parallel and M(r) changes only slowly. Other noncollinear spin structures helical, characterized by a wave vector k, or canted, for example, due to Dzyaloshinskii–Moriya interactions. If the term "spin structure" is used without specification, it refers to the atomic spin structure, as contrasted to finite-temperature and micromagnetic spin structures.

18.2 Quantum-Mechanical Origin of Magnetism

Solid-state magnetism is nearly exclusively caused by electrons. The particle-wave duality of matter means that light particles, such as electrons, are predominantly wave like, described by the Schrödinger equation:

$$i\hbar \frac{\partial \psi}{\partial t} = -\frac{\hbar^2}{2m_e}\nabla^2\psi + V(r)\psi \qquad (18.1)$$

Here $V(\mathbf{r})$ the potential energy. Equation (18.1) is obtained by substituting the operators

$$\mathcal{H} = i\hbar\frac{\partial}{\partial t} \quad \text{and} \quad \mathbf{p} = -i\hbar\nabla \qquad (18.2)$$

into the classical one-particle Hamiltonian

$$\mathcal{H} = \frac{\mathbf{p}^2}{2m_e} + V(\mathbf{r}). \qquad (18.3)$$

Physical quantities have the character of quantum-mechanical averages, $<A> = \int \psi^*(\mathbf{r})\ \hat{A}\ \psi(\mathbf{r})\ d\mathbf{r}$. For example, $<V> = \int \psi^*(\mathbf{r})\ V(\mathbf{r})\ \psi(\mathbf{r})\ d\mathbf{r}$ is the average potential energy, and since $n(\mathbf{r}) = \psi^*(\mathbf{r})\ \psi(\mathbf{r})$ is the electron density, we can also write $<V> = \int n(\mathbf{r})\ V(\mathbf{r})\ d\mathbf{r}$. Wave functions also carry a phase factor, $\psi \to \exp(i\gamma)\psi$ known as the Berry phase (Section 18.6).

18.2.1 Spin, Orbital Moment, and Coulomb Interaction

The stationary solutions of Eq. (18.3) correspond to quantized eigenstates. Nuclei in atoms create an electrostatic potential $V(\mathbf{r}) \approx Ze^2/4\pi\varepsilon_o r$, which leads to eigenfunctions of the type $\psi(x,\ y,\ z) = \psi_r(r)\ \psi_\theta(\theta)\ \psi_\phi(\phi)$. This factorization is the basis for the division into atomic shells and subshells, and for the assignment of quantum numbers to these shells. The periodic table of the elements amounts to filling the orbitals, starting with $1s$, $2s$, and $2p$. Fully occupied (sub)shells have zero magnetic moment. Of particular interest in magnetism are electrons in the partially filled inner shells of transition-metal atoms, namely $3d$, $4d$, and $5d$ electrons ($L = 2$) and $4f$ and $5f$ electrons ($L = 3$). The number of orbitals per shell is $2L + 1$, meaning that there are five d orbitals and seven f orbitals, and each orbital is empty, half-filled, or filled by a ↑↓ pair of electrons.

In a crystal, $V(\mathbf{r})$ reflects the lattice periodicity, which leads to hybridizations between neighboring atomic orbitals. Applied to many electron systems, the Schrödinger equation yields intra- and interatomic exchange (Section 18.1.2), the latter being responsible for magnetic order (Section 18.1.5).

In the case of the *atomic magnetic moment*, the quantization corresponds to the *Bohr magneton* $\mu_B = e\hbar/2m_e = 9.274 \times 10^{-24}$ J/T of the electron. Neutrons and protons are much heavier than electrons and contribute very little to the magnetization, but they are important, for example, in resonance imaging. Electrons yield two contributions to the atomic moment, both quantized in terms of μ_B. First, each electron has a spin, symbolically written ↑ or ↓, whose explanation requires both quantum mechanics and relativistic physics (Section 18.3). Second, an electron's circular orbital motion corresponds to a coil or current loop with a quasiclassical magnetic moment of the order of one μ_B. In more detail, the orbital motion is quantized according to L, and $m = -L_z\ \mu_B$, where $L_z = -L, \ldots, L{-}1, L$ and L is measured in $\hbar$.

Most electrons in solids do not contribute to the net moment, because they form ↑↓ pairs with canceling spin and orbital moments. The electrons try to fill all low-lying states, but Pauli principle forbids ↑↑ (and ↓↓) occupancies of quantum states, so that one of the two parallel electrons must occupy a one-electron state of higher energy. If this one-electron energy was the only consideration, then all electrons would form ↑↓ pairs and macroscopic solids would essentially be nonmagnetic. In fact, electrostatic electron–electron interactions favor parallel spin orientations, because ↑↓ electron pairs in one orbital experience a strong Coulomb repulsion

$$\mathcal{U} = \int n_\uparrow(\mathbf{r})\, U(\mathbf{r}, \mathbf{r}')\, n_\downarrow(\mathbf{r}')\, d\mathbf{r}\, d\mathbf{r}' \qquad (18.4)$$

where

$$U(\mathbf{r}, \mathbf{r}') = \frac{e^2}{4\pi\varepsilon_0 |\mathbf{r} - \mathbf{r}'|} \qquad (18.5)$$

Parallel spin configurations are not penalized by this energy, because they are forbidden by the Pauli principle. Parallel spins are therefore favorable if the net gain in Coulomb energy, about 1 eV, outweighs the difference in one-electron energies. This indicates that the one-electron energy-level distribution is of utmost importance for the understanding of magnetism (Section 18.2.2).

Equation (18.4) is very difficult to treat quantum mechanically, because it is quadratic in $n = \psi^*\psi$ and quartic in the wave function ψ. However, it is possible to approximate the electron gas by an effective potential $V(\boldsymbol{r})$, which is known as the independent-electron or quantum-mechanical mean-field approximation and forms the basis for local spin-density (LSDA) density-functional (DFT) calculations. Simplifying somewhat, one writes

$$n_\uparrow(\mathbf{r})\, n_\downarrow(\mathbf{r}') = n_\uparrow(\mathbf{r}) <n_\downarrow> + <n_\uparrow> n_\downarrow(\mathbf{r}') + C \quad (18.6)$$

and ignores the correlation terms $C = (n_\uparrow(\mathbf{r}) - <n_\uparrow>)$ $(n_\downarrow(\mathbf{r}') - <n_\downarrow>)$. The procedure amounts to the introduction of a spin-dependent effective potential $V_{\uparrow,\downarrow} \sim \mathcal{U} <n_{\downarrow,\uparrow}>$, and this potential favors parallel spin alignment.

As a very crude rule of thumb, ferromagnets have one unpaired d or f electron per atom, which translates into a magnetization of the order of 1 tesla. In iron-series transition-metal magnets, the spin moment is usually the leading contribution. The orbital moment is largely quenched, that is, the circular orbital motion of the electrons is disrupted by the crystal field (Section 18.3.2).

18.2.2 Interatomic Hopping and Stoner Criterion

The net Coulomb interaction of about 1 eV is too weak to spin-polarized metallic free-electron gases, because this polarization would require an energy per electron of 0.6 $(2^{2/3} - 1)\, E_F$, that is, 4.1 eV for Al and 2.5 eV for Cu. In fact, the metallic magnetism of Fe, Co, and Ni involves rather narrow d bands. In this case, the one-electron energy-level distribution scales with the band width, not with the Fermi level (Figure 18.3), and ferromagnetism occurs for sufficiently narrow bands (itinerant ferromagnetism). The band width is proportional to the interatomic hopping integral $\mathcal{T}$ and therefore increases with wave function overlap. The iron-series $3d$ bands are fairly narrow, having hopping integrals of the order of 1 eV, because the d orbitals between neighboring atoms do not overlap very much. This explains the widespread occurrence of ferromagnetism in $3d$ elements

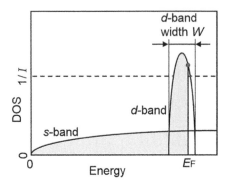

FIGURE 18.3 Stoner criterion for the onset of itinerant ferromagnetism. The solid curves show the densities of states (DOS) as a function of energy, and the dashed line shows the DOS corresponding to the Stoner criterion $D(E_F) > 1/I$.

and alloys. The 4d and 5d bands are broader, but Pd and Pt are close to the onset of ferromagnetism, and several 4d and 5d alloys are ferromagnetic.

A more detailed description involves the density of states (DOS), that is, the number of energy levels per energy interval, $D(E) = dn/dE$. Since the total number of states in a band is fixed (10 states per atom in a d band), the DOS is inversely proportional to the band width and to $\mathcal{T}$. Stability analysis [22] of the Pauli-paramagnetic or 'nonmagnetic' state yields a ferromagnetic stability when $D(E_F) > 1/I$ (Stoner criterion), where $I \sim \mathcal{U}$ is the Stoner parameter. In Figure 18.3, the Stoner criterion is satisfied, because the gray dot showing $D(E_F)$ is above the dashed line. The DOS peak positions and heights exhibit strongly depend on the crystal structure, which explains, for example, why bcc Fe and fcc Co are ferromagnetic but fcc Fe is not [16].

It is instructive to analyze metallic magnetism in terms of the Stoner factor $1/(1 - I\mathcal{D}(E_F))$. This factor describes the ability of an external magnetic field to spin-polarize the metal, quantified by the susceptibility $\chi = \chi_p/(1 - \mathcal{D}(E_F)I)$. Here χ_p is the very small paramagnetic Pauli susceptibility of noninteracting electrons ($\mathcal{T} \gg \mathcal{U}$). In simple metals, $\mathcal{D}$ is small and the Stoner factor $\chi/\chi_p \approx 1$. As the Stoner criterion is approached, the Stoner factor χ/χ_p increases, which is known as *exchange-enhanced Pauli paramagnetism*. Examples are Pd and Pt. At the Stoner transition, χ/χ_p diverges and remains quite large directly above the transition. This case is known as *very weak itinerant ferromagnetism* (VWIF) and realized in ZrZn$_2$ [23]. The ↑ and ↓ bands of VWIFs are only weakly spin-polarized, and the magnetic moment per atom is very small and exhibits a relatively strong field dependence. These materials also exhibit an intriguing Curie temperature behavior and rich physics when subjected to nanostructuring (Section 18.5). Further enhancement of $\mathcal{D}$ yields *weak ferromagnetism*, as epitomized by bcc Fe. The d bands of weak ferromagnets are not fully spin-polarized, that is, there remain a few holes in the d↑ band, but the moments are generally large and stable. Finally, magnets with fully spin-polarized bands are referred to as *strong ferromagnets*, exemplified by Co and Ni.

18.2.3 Exchange

The Stoner criterion describes the transition from metallic paramagnetism (PM) to itinerant ferromagnetism but does not address possible transitions from PM to AFM, nor does it address the competition between FM and AFM. Let us assume, for the moment, that there are stable atomic magnetic moments and compare FM order, Figure 18.1a, with AFM order, Figure 18.1b. In this case, we can use the Heisenberg model

$$\mathcal{H} = --2 \sum_{i>j} \mathcal{J}_{ij} \boldsymbol{S}_i \cdot \boldsymbol{S}_j - g\mu_o\mu_B \sum_i \boldsymbol{H}_i \cdot \boldsymbol{S}_i \qquad (18.7)$$

where the $\boldsymbol{S}_i$ are the atomic spins, $\boldsymbol{H}_i$ is the local magnetic field, and the summation is over all atoms. The g-factor is equal to 2 for spin-only magnetism but $g \neq 2$ in case of a nonzero orbital moment contribution. In the simplest case, the interatomic exchange $\mathcal{J}_{ij}$ is of the nearest-neighbor type and leads to ferromagnetism ($\mathcal{J} > 0$) or antiferromagnetism ($\mathcal{J} < 0$). The Heisenberg model, which exists in both classical and quantum-mechanical versions, describes most insulating and many metallic magnets but fails to describe very weak itinerant ferromagnets. Atomic moments in oxides and other nonmetallic compounds tend to obey Hund's rules. For example, Hund's first rule states that the total spin is maximized subject to the Pauli principle and yields 5 μ_B (↑↑↑↑↑, $S = 5/2$) for Fe^{3+} and 4 μ_B (↑↑↑↑↓, $S = 2$) for Fe^{2+}. The continuous band filling in itinerant magnets means that the moments per atom are generally noninteger, for example, 2.2 μ_B in Fe, 1.7 μ_B in Co, and 0.6 μ_B in Ni. Defining quantum-mechanical Heisenberg spins is therefore nontrivial for itinerant magnets [24].

Exchange Integral

The Heisenberg exchange is frequently equated with the exchange integral

$$\mathcal{J}_D = \int \phi^*_1(\mathbf{r})\,\phi_1(\mathbf{r}')\,U(\mathbf{r},\mathbf{r}')\,\phi^*_2(\mathbf{r}')\,\phi_2(r)\,d\mathbf{r}d\mathbf{r}' \quad (18.8)$$

where the ϕ_1 and ϕ_2 are wave functions centered around neighboring atoms. Unlike Eq. (18.4), which has a classical interpretation as a repulsion between charge clouds of the type $n(\mathbf{r}) = \phi^*(\mathbf{r})\phi(\mathbf{r})$, Eq. (8) mixes the real-space coordinates $\mathbf{r}$ and $\mathbf{r}'$, whence the name exchange. The reason for the exchange is the Pauli principle, which requires two-electron wave functions to be Slater determinants rather than simple products of the type $\phi_1(\mathbf{r})\phi_2(\mathbf{r}')$.

However, the exchange integral of Eq. (18.7) is always positive, that is, always ferromagnetic. This can be shown by introducing the fictitious particle density $n_f(\mathbf{r}) = \phi^*_1(\mathbf{r})\phi_2(\mathbf{r})$: Eq. (18.8) then describes the self-interaction of n_f, and such self-interactions are always positive. How can we explain negative Heisenberg exchange constants, which are well established through comparison of Eq. (18.7) with experiment? The answer lies in the absence or inclusion of interatomic hopping $\mathcal{T}$. The atomic wave functions used in Heisenberg's original calculation were nonorthogonal, which introduces overlap. This overlap yields hopping corrections that oppose and often overcompensate the relatively small $\mathcal{J}_D$. It is therefore common to treat the Heisenberg exchange constants $\mathcal{J}_{ij}$ as model parameters, estimated from experiments or using models. The Bethe–Slater curve plots $\mathcal{J}$ as a function of the interatomic distance and predicts antiferromagnetism for very short distances and a maximum near Fe and Co [13]. However, reality is more complicated, because the atomic neighborhood, for example, fcc-like vs. bcc-like, is at least as important as the interatomic distance.

A more elegant approach to Heisenberg exchange is to use orthogonalized (Wannier-type) wave functions. For two electrons on two atoms, the model is exactly solvable, including correlation effects [25]. For small $\mathcal{T}$, for example, in oxides, the corresponding Heisenberg exchange $\mathcal{J} = \mathcal{J}_D - 2\mathcal{T}^2/\mathcal{U}$. This equation shows that interatomic hopping tends to destroy ferromagnetism. Exchange in oxides is normally antiferromagnetic, but in some cases, for example, in CrO_2, the hopping integrals are zero by symmetry, then only $\mathcal{J}_D$ survives and the coupling is ferromagnetic (Goodenough–Kanamori–Andersen rules). The opposite limit of large interatomic hopping qualitatively reproduces the Stoner criterion, $\mathcal{J} = \mathcal{J}_D + \frac{1}{4}\mathcal{U} - \mathcal{T}$, since $\mathcal{J}_D \ll \mathcal{U}$. The model's exact solution for the onset of FM is $\mathcal{T}^2 = \mathcal{J}_D{}^2 + \frac{1}{2}\mathcal{J}_D\mathcal{U}$, which corresponds to $I \sim (\mathcal{J}_D\mathcal{U})^{1/2}$.

The neglect of the correlation term C in Eq. (18.6) is justified if the interatomic hopping energy $\mathcal{T}$ are large than the interaction energy $\mathcal{U}$. This is a reasonable assumption for itinerant electrons, for example, in iron-series elements and alloys. It is a very poor approximation in the opposite limit. Consider, for example, a hydrogen molecule having a very large interatomic distance. By definition, the neglect of correlations means that the electrons hop between the two atoms and do not care on which atom the other electron sits. In reality, the hopping energy is too small to enforce double occupancy leading to ionic states of the type H^+-H^-, because the accommodation of the second electron on the H^- ion is punished by a big Coulomb energy $\mathcal{U}$. For large interatomic distances, it is better to completely ignore ionic states and to assume full correlations, that is, each hydrogen atom has exactly one electron. This is basically the spin-1/2 Heisenberg model (Section 18.2.3).

The exchange also depends on the band filling, which is $\frac{1}{2}$ in the two-electron dimer model. Ferromagnetism is generally favored for the late transition metals, where the bands are narrow and nearly filled. Ferromagnetism in half-filled bands means that the electrons fill all available orbitals with ↑ spins. This occupancy includes both bonding and antibonding orbitals, so no net energy gain is achieved through the preferential occupancy of bonding orbitals. Mn^{2+} has a moment of 5 μ_B per atom, as compared to 2.2 μ_B in bcc iron, and the exploitation of this high magnetic moment is an ongoing challenge in magnetism research (Figure 18.4).

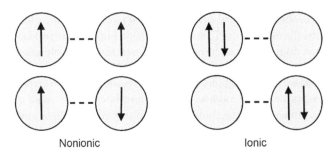

FIGURE 18.4 Correlations and ionicity. The Heisenberg limit is nonionic, that is, the strong electron correlations suppress $\uparrow\downarrow$ occupancy. The Stoner model is undercorrelated (50% ionic).

Types of Exchange

The involvement of interatomic hopping becomes even more intriguing if several types of atoms are involved, for example, in oxides. For example, in the antiferromagnetic rock-salt oxide MnO, neighboring Mn^{2+} ions interact through the O^{2+} ions (superexchange). From a fundamental viewpoint, it is necessary to distinguish intra-atomic exchange, typically of the order of 1 eV, and interatomic exchange, typically 0.1 eV or less. Intra-atomic exchange describes how the electron spins inside a given atom combine to yield an atomic moment. Since the eigenfunctions inside a given atom are (nearly) orthogonal, interatomic exchange is always positive. In itinerant magnets, the distinction between intra- and interatomic exchange blurs, especially in VWIFs (Section 18.5).

Double exchange occurs in oxides with mixed valence, for example, in Fe_3O_4, which contains both ferric (Fe^{3+}) and ferrous (Fe^{2+}) ions. In Fe_3O_4, the intra-atomic exchange is between Fe^{3+} ($\uparrow\uparrow\uparrow\uparrow$) and an extra $\downarrow$ electron: the extra electron hops between neighboring iron atoms, undergoes polarization through intra-atomic exchange, and thereby creates a ferromagnetic net exchange through "spin memory". The double exchange is of particular importance in magnetoresistive perovskites.

Ruderman–Kittel or RKKY interactions describe the exchange between localized magnetic S_i and S_j in a sea of nonmagnetic conduction electrons. A nonzero effect is obtained in second-order perturbation theory, where S_i and S_i mix conduction-electron states with different wave vector k [15]. Pictorially, S_i and S_i create concentrical waves in the electron gas, similar to those created by stones thrown into a lake. The interference of these waves translates into an exchange $\mathcal{J}_{ij}(R) = \mathcal{J}_{ij}(|R_i - R_j|)$ having the asymptotic form $\mathcal{J}_{ij}(R) \sim \cos(2k_F R)/R^3$. The Fermi vector k_F appears in this expression, because the concentrical waves must be constructed from electron states with $k \leq k_F$. The function $\cos(x)/x^3$ is oscillatory but long range and may therefore interfere with nanostructuring. For example, the fast oscillations do not average to zero but increase with the size of RKKY-interacting embedded clusters or nanoparticles. However, the increase is less pronounced than that of the magnetostatic interactions, which dominate for particles sizes larger than about 1 nm [26]. In

semiconductors and semimetals, such as Sb, the low density of carriers means that k_F is small, and the period of the oscillations is nanoscale. Note that the free-electron character of the conduction electrons is of secondary importance, and even localized electron states can produce RKKY-like interactions though second-order perturbation theory [27].

Noncollinearity

The spin structures of many materials are noncollinear (Figure 18.1). A major and arguably the most important source of atomic scale noncollineariy is competing interatomic exchange, that is, nearest-neighbor and more distant neighbors yield exchange interactions of opposite sign. Competing exchange is very common in some classes of magnetic materials, such magnetic oxides and rare-earth elements, but also occurs in itinerant magnets, especially if they contain Cr or Mn. One source of competing exchange is RKKY interactions, but there are also other mechanisms, involving for example different hopping paths.

A simple model describes the helicoidal spin structure of some rare-earth elements, $S = S_o \cos(\theta(z)) e_x + S_o \sin(\theta(z)) e_y$. It uses the classical Heisenberg expression

$$E = -\mathcal{J} \sum_n \cos(\theta_{n+1} - \theta_n) - \mathcal{J}' \sum_n \cos(\theta_{n+2} - \theta_n) \quad (18.9)$$

where θ_i is the magnetization angle of the i-th atomic layer, $\mathcal{J}$ is the exchange between neighboring layers, and $\mathcal{J}'$ is the exchange between next-nearest layers. The energy of Eq. (18.9) is minimized by $\theta_{n+1} = \theta_n + \delta$, where δ is the magnetization rotation between subsequent layers. The angle δ obeys

$$(\mathcal{J} + 4\mathcal{J}' \cos\delta) \sin\delta = 0 \quad (18.10)$$

Aside from FM ($\delta = 0$) and AFM ($\delta = \pi$) solutions, this equation has noncollinear or helimagnetic ($0 < \delta < \pi$) solutions, $\delta = \arccos(-\mathcal{J}/4\mathcal{J}')$, depending on the ratio $\mathcal{J}/\mathcal{J}'$. The noncollinearity angle d may be very large, basically covering the wholes range between 0° and 180°.

Noncollinearity may also be caused by Dzyaloshinskii–Moriya (DM) interactions, which arise in structures with broken inversion symmetry. Example are MnSi [28], α-Fe_2O_3 [21], and structurally disordered magnets [10,29]. However, DM interactions are weak relativistic corrections (Section 18.3), with nearest-neighbor noncollinearity angles of normally less than 1°. The same applies to micromagnetic noncollinearities involving magnetic anisotropy, Figure 18.1c, which are micromagnetic rather than atomic scale.

18.3 Relativistic Aspects of Magnetism

Solid-state magnetism cannot be understood without relativistic physics. For example, an external field H introduces a magnetization $M = \chi H$, where χ is the magnetic susceptibility. A naïve expectation would be that a field of about 1 tesla, created by a strong permanent magnet, can be used to align all spin in a solid, for example, in a piece of wood. This is not observed, and in Section 18.2, we have seen that magnetic materials, such as Fe, exhibit a "pre-alignment" of the spins due to interatomic exchange. Due to the exchange, Fe exhibits a spontaneous magnetization of $\mu_o M_s = 2.15$ T, which is easily realized in magnetic fields smaller than 1 T. By contrast, the very small magnetic susceptibility of Al, $\chi = 2.2 \times 10^{-5}$, means that an external magnetic field of 1 T creates a tiny magnetization of only 0.022 mT. What is the reason for inert magnetic behavior of most materials? The answer is provided by the relativistic character of magnetism. Relativistic physics is also necessary to explain many magnetotransport phenomena, such as the anomalous Hall effect and the behavior of topological insulators.

18.3.1 Pauli Expansion and Spin–Orbit Coupling

Relativistic physics follows from the four-vector (space-time) symmetry of nature, $\mathbf{X} = (ct, x, y, z)$, meaning that time coordinates obey a slightly different metrics (minus sign) but are otherwise equivalent to the spatial coordinates. For example, the propagation of light, $x^2 + y^2 + z^2 = c^2 t^2$, can be written very elegantly as $\mathbf{X}^2 = 0$. Another example of four-vector symmetry is the energy expression

$$E = \pm\sqrt{m_e^2 c^4 + p^2 c^2} \tag{18.11}$$

where $\mathbf{p} = m_e \mathbf{v}$ and the kinetic energy E combine to yield the four-vector $(E/c, \mathbf{p}) = (E/c, p_x, p_y, p_z)$. Series expansion of the root in Eq. (18.11) yields

$$E = m_e c^2 + \frac{1}{2} m_e v^2 - \frac{1}{8} \frac{v^2}{c^2} m_e v^2 \pm \ldots \tag{18.12}$$

In this expansion, $m_e c^2$ is the rest energy of the electron, $\frac{1}{2} m_e v^2$ is Newton's kinetic energy, and the last term is the lowest-order relativistic correction. In solids, v/c is of the order of $\alpha = 1/137$, where $\alpha = e^2/4\pi\varepsilon_o \hbar c$ is Sommerfeld's fine-structure constant. Four-vector symmetries also apply to quantum-mechanical expressions. For example, $-i\hbar\partial/\partial\mathbf{X} = -i\hbar\,(\bar{}\,c^{-1}\partial/\partial t, \partial/\partial x, \partial/\partial y, \partial/\partial z)$ expresses the momentum and energy (Hamilton) operators in a relativistically invariant fashion. The magnetic vector potential $\mathbf{A}$ and the electrostatic scalar potential ϕ form the four potential $(\phi/c, \mathbf{A})$, and the electromagnetic fields are four-vector derivatives of this potential:

$$E = -\nabla\phi - \partial\mathbf{A}/\partial t \quad \text{and} \quad \mathbf{B} = \nabla \times \mathbf{A} \tag{18.13}$$

The interaction of the electron with external magnetic fields is described by adding the four potential of the external fields to the four-vector energy, yielding $(E/c + e\phi/c, \mathbf{p} + e\mathbf{A})$.

Taking the square of Eq. (18.11) yields an expression quadratic in $E = i\hbar\partial/\partial t$, which needs to be factorized. This can be done by using the three-plus-one-dimensional identity

$$a^2 + b^2 = (\boldsymbol{\sigma} \cdot \mathbf{a} + ib)(\boldsymbol{\sigma} \cdot \mathbf{a} - ib) \tag{18.14}$$

where a and b form a four-vector and σ is the vector formed by the Pauli matrices

$$\boldsymbol{\sigma} = \begin{pmatrix} 0 & 1 \\ 1 & 0 \end{pmatrix} \mathbf{e}_x + \begin{pmatrix} 0 & -i \\ i & 0 \end{pmatrix} \mathbf{e}_y + \begin{pmatrix} 1 & 0 \\ 0 & -1 \end{pmatrix} \mathbf{e}_z \tag{18.15}$$

In particular, the Dirac equation, which describes relativistic quantum mechanics, can be written as

$$(\boldsymbol{\sigma} \cdot \mathbf{p} + imc)(\boldsymbol{\sigma} \cdot \mathbf{p} - imc)\psi = 0 \tag{18.16}$$

where $\pm m$ distinguishes between matter and antimatter.

Incorporating the four potential into the Dirac equation and performing perturbation theory for small $\mathbf{p}$ yields an expansion (Pauli expansion) that is a magnetic analog to Eq. (18.9). The expansion yields a small admixture of antimatter and involves the identity

$$(\mathbf{a} \cdot \boldsymbol{\sigma})(\mathbf{b} \cdot \boldsymbol{\sigma}) = (\mathbf{a} \cdot \mathbf{b})\,\mathrm{I} + i\boldsymbol{\sigma} \cdot (\mathbf{a} \times \mathbf{b}) \tag{18.17}$$

where I is the 2×2 unit matrix. This identity explains the occurrence of cross products in $3 + 1$-dimensional spaces, for example, in Eq. (18.13), a feature noticed by Hamilton as early as 1843.

Including some low-order terms of specific interest to solid-state physics and magnetism, the Pauli expansion

$$\mathcal{H} = -\left(\hbar^2/2m\right)\nabla^2 - e\phi - 2\mu_o\mu_B \mathbf{H} \cdot \mathbf{S} - \lambda\mathbf{L} \cdot \mathbf{S} \tag{18.18}$$

where $\mathbf{L} = \mathbf{r} \times \mathbf{p}$ is the orbital angular momentum and $\mathbf{S} = \frac{1}{2}\boldsymbol{\sigma}$.

The first two terms in Eq. (18.18) are non-relativistic ($\sim v^2$), whereas the Zeeman term (H) and the spin–orbit coupling (λ) are relativistic corrections ($\sim v^4$). Since $v/c \approx 1/137$, the relativistic terms are usually small, the above-mentioned Al susceptibility being a good example. The Zeeman term contains the Bohr magneton, $\mu_B = e\hbar/2m_e$, which is equal to 9.274×10^{-24} J/T or 57.88 µeV/T. This very small value explains why external magnetic fields of the order of 1 T are unable to compete against the leading energy contributions in solids, which are of the order of 1 eV (Figure 18.5).

18.3.2 Magnetic Anisotropy

The energy of a magnet depends on the magnetization angles θ and ϕ with respect to the crystal axes or to

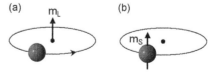

FIGURE 18.5 Atomic origin of magnetism. Both the orbital moment (a) and the spin moment (b) contribute to the magnetic moments, but in iron-series transition-metal magnets, the spin moment dominates (quenching, Section 18.3.3).

the coordinate frame of a nanostructure. This property is known as magnetic anisotropy. Lowest-order (second-order) anisotropy is described by the energy density

$$\frac{E_a(\theta, \phi)}{V} = K_1 \sin^2 \theta + K_1' \sin^2 \theta \cos(2\phi) \quad (18.19)$$

where K_1 and K_1' are first-order anisotropy constants. Minimizing this expression with respect to θ and ϕ yields one or more easy magnetization axes. For sufficiently high symmetry, especially C_3, C_4, and C_6 in the case of trigonal, tetragonal, and hexagonal crystal structures, $K_1' = 0$, and it is sufficient to consider K_1. Magnets with cubic crystal structure do not exhibit second-order anisotropy contributions, but it is possible to define K_1 for such magnets through a second-order series expansion. Typical K_1 values are listed in Table 18.1.

Depending on symmetry, there is a variety of higher-order anisotropy contributions. For example, up to fourth order, magnets with *hexagonal* symmetry are described by the uniaxial anisotropy expression

$$\frac{E_a}{V} = V(K_1 \sin^2 \theta + K_2 \sin^4 \theta) \quad (18.20)$$

whereas *tetragonal* crystals obey

$$\frac{E_a}{V} = K_1 \sin^2 \theta + K_2 \sin^4 \theta + K_2' \sin^4 \theta \cos(4\phi) \quad (18.21)$$

Hexagonal crystals exhibit a $\cos(6\phi)$ term equivalent to the $\cos(4\phi)$ in Eq. (18.20), but this term is of sixth order and therefore not included in Eq. (18.21).

Anisotropy is a higher-order relativistic effect, and typical anisotropy energies per atom are 1 μeV for very soft materials and 1 meV for very hard materials. Both the Zeeman and spin–orbit terms of Eq. (18.18) contribute to the anisotropy, as schematically shown in Figure 18.6. The Zeeman interaction is realized through *magnetostatic dipole–dipole interactions*. There are two types of dipole interactions. First, there are magnetostatic interactions between close atomic neighbors. In the example of

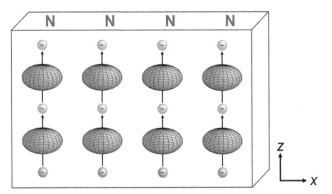

FIGURE 18.6 Magnetic anisotropy of a cuboidal nanoparticle (schematic). The leading anisotropy contribution is usually magnetocrystalline anisotropy, here corresponding between crystal-field charges (dark gray) and aspherical Nd^{3+} ions (light gray).

Figure 18.6, these interactions favor spin arrangements in form of chains in the z-direction. Their anisotropy contribution is rather small, normally much less than 0.1 MJ/m³. Second, the magnetization creates macroscopic pole distributions at the surface of the magnet, such as the north poles in Figure 18.6. This anisotropy, known as shape anisotropy, is approximately

$$K_{1,sh} = \frac{\mu_0}{4} (1 - 3D) M_s^2 \quad (18.22)$$

where D is the demagnetizing factor ($D = 0$ for long cylinders, $D = 1/3$ for spheres, and $D = 1$ for plates) [30]. In Figure 18.6, shape anisotropy favors a magnetization orientation along the x-axis, due to the aspect ratio of the cuboidal particle. Shape anisotropy is important in nanostructures made from soft magnetic materials, where it often exceeds 0.1 MJ/m³. An interesting class of nanostructured permanent-magnet materials is alnicos, which contain long nanorods ($D = 0$) of a high-magnetization material (essentially FeCo) in a nonmagnetic matrix [31,32]. Shape anisotropy becomes less important with increasing feature size, due to the formation of magnetic domains (Section 18.4).

The main contribution to the anisotropy of most materials and nanostructures, known as *magnetocrystalline anisotropy*, involves spin–orbit coupling (SOC) rather than magnetostatic interactions [33]. This anisotropy is often very high, with room-temperature values of the order of 10 MJ/m³ in some rare-earth transition-metal alloys [16]. In a nutshell, the crystal-field potential $V(\boldsymbol{r})$ affects the orbital motion of the electrons and creates, through SOC, preferential spin orientations. A conceptually very transparent version of the magnetocrystalline anisotropy mechanism is realized in rare-earth magnets. The spin–orbit coupling of the rare-earth $4f$ electrons is very strong, of the order of 500 meV, as opposed to about 50 meV for the late $3d$ elements. Combined with the location of $4f$ electrons the deep inside the rare-earth atoms, the strong SOC yields a rigid coupling between spin and $4f$ electron charge cloud. This charge cloud is given by Hund's rules, which predict that flat or 'pancake-like' oblate orbitals with large L are occupied first as the number of $4f$ electrons increases. These flat orbitals dominate in the first three elements of each half series (Ce, Pr, Nd, and Tb, Dy, Ho), whereas the tripositive ions of Pm, Sm, Eu, Er, Tm, and Yb are prolate. Gd^{3+} has a completely filled ↑ shell and therefore a spherical charge cloud. Figure 18.6 depicts the situation for the oblate $4f$ charge cloud of Nd^{3+} (light gray). The aspherical $4f$ charge clouds interacts electrostatically with the crystal charges (dark gray). The interaction is of the quadrupole type and amounts to a repulsion between the crystal-field charges and the $4f$ shell. In Figure 18.6, this mechanism yields a preferential magnetization direction along the z-axis. Turning the spins by 90° would move the ridge of the pancake close to the negative crystal-field charges and be energetically unfavorable.

A more complicated situation is encountered in $3d$ magnets, where the spin–orbit coupling is a small perturbation compared to the leading crystal-field and hopping interactions. The relatively strong crystal field is also the reason for the *quenching* of the orbital moment. For example, about 95% of the moment of bcc Fe is due to spin, with orbital contribution of only 5%. According to Eq. (18.15), the SOC favors the creation of an orbital moment L in the presence of a spin S, but the circular motion of the electrons competes against the interaction of the electrons with the crystal charges, the latter disrupting the orbital motion and suppressing L [25]. Note that the Pauli expansion responsible for the spin–orbit coupling is essentially an expansion in v/c, Eq. (18.12). Electrons inside heavy atoms, such as rare-earth $4f$ electrons, move fast due to enhanced effective nuclear charges Z and therefore exhibit a stronger spin–orbit couplings. Palladium- and platinum-series transition-metal atoms are intermediate between $3d$ and $4f$ systems regarding both SOC and crystal-field interaction.

18.3.3 Other Anisotropies Involving Spin-Orbit Coupling

It is sometimes claimed that magnetic anisotropy is caused by broken symmetry. In fact, broken symmetry is a necessary but not sufficient requirement. For example, the Heisenberg Hamiltonian of Eq. (18.5) can be used to describe arbitrary structures, including, for example, surfaces and thin films, but is describes the relative spin reorientation inside a magnet only (spin structure), not the orientation of the spin system with respect to crystalline or other symmetry axes (magnetic anisotropy). By its definition, the Heisenberg model is isotropic, and there is no point in invoking "broken exchange bonds" used to explain anisotropy. There are anisotropic extensions of the Heisenberg model, using, for example, anisotropic exchange $\mathcal{J}_{zz} = \mathcal{J}$ and $\mathcal{J}_{xx} = \mathcal{J}_{yy} = \mathcal{J}^*$, but $|\mathcal{J}^* - \mathcal{J}|$ is relativistically small and unimportant in most materials.

Magnetoelastic and surface anisotropies are predominantly magnetocrystalline as far as their physical origin is concerned. As in bulk magnets, the determination of the anisotropy amounts to the consideration of all atoms in the structure. Magnetoelastic anisotropy, which is substantial in magnets with cubic crystal structure, means that a cubic crystal is distorted to yield a tetragonal or otherwise distorted. Surfaces differ from bulk materials by a substantially modified and generally enhanced crystal field acting on the surface atoms. Surface anisotropy critically depends on the indexing of the surface in question, which determines sign, direction, and magnitude of the surface anisotropy [34]. A naïve but incorrect approach [35] is to assume a perpendicular anisotropy of the type $-K_s (M \cdot n)^2$, where n is the surface normal. In fact, cutting a nanosphere from a cubic material yields surface patches of varying indexing, and such surfaces are reminiscent of random-anisotropy magnets [10]. Surface and interface anisotropy contributions are often large in nanostructures, especially if the underlying bulk material is soft magnetic. A rule of thumb is that the

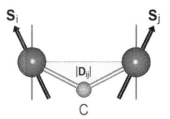

FIGURE 18.7 Atomic origin of Dzyaloshinskii–Moriya (DM) interactions.

surface-anisotropy contribution of top-layer $3d$ atoms can be as high as 5 MJ/m^3, corresponding to a surface anisotropy of about 1 mJ/m^2.

An interesting relativistic effect is Dzyaloshinskii–Moriya (DM) exchange (Section 18.2.3.3). It is described by the Hamiltonian $\mathcal{H} = -\Sigma_{ij} D_{ij} \cdot S_i \times S_j$, where the Dzyaloshinskii–Moriya vector $D_{ij} \sim (r_i - r_C) \times (r_j - r_C)$ [29]. Figure 18.7 illustrates the meaning of these expressions. Two magnetic atoms with spins S_i and S_j are connected by interatomic hopping trough a third atom (C). Atom C is not necessarily magnetic, but electrons moving from the i-th to the j-th atoms, or *vice versa*, need to change their direct of motion at C. This amounts to a temporary orbital motion around C and to some spin–orbit coupling. Ferromagnetic Heisenberg exchange favors parallel spins (dark gray vertical lines), but the DM mechanism yields a small noncollinearity. The magnitude of the effect is given by the DM vector, which corresponds to the gray triangle in Figure 18.7.

18.4 Micromagnetics

The interplay between structural and magnetic length scales is an intriguing feature of nanostructuring. Most intrinsic properties are realized on a length scale of a few a_o, where $a_o = 0.52$ Å is Bohr's hydrogen radius. For example, interatomic exchange is often dominated by nearest neighbors, which are separated by 2.5 Å = 0.25 nm in the late iron-series transition metals. Interatomic exchange is very strong and yields, ideally, long-range magnetic order. However, relativistic effects work against this order. A homogeneously magnetized macroscopic magnet of volume L^3 creates a magnetostatic field outside the magnets, but the creation of this stray field costs energy of the order of $\mu_o M_s^2 L^3$. This energy penalty is avoided by the creation of magnetic domains, which drastically reduce the net magnetization and therefore the stray field. However, domains are separated by domain walls, Figure 18.1c, which cost exchange energy due to their inhomogeneous spin distribution. Micromagnetism addresses the competition between exchange energies and other energy contributions, with the aim of explaining how extrinsic (hysteretic) properties depend on intrinsic properties and on the nanostructure of a magnet.

Before going into details, let us discuss micromagnetism from a big-picture viewpoint. Relativistic effects, such as Zeeman interaction and magnetic anisotropy, are too weak to compete against exchange on an atomic scale. However, on a macroscopic scale, they are important as one can see from the attractive force between two permanent magnets. Above what length scale are relativistic interactions no longer unimportant? Since the strength of relativistic effects is described by $\alpha = 1/137$, we expect this constant to enter the consideration. This is indeed the case, and it can be shown [36] that the corresponding length is $a_o/\alpha = 7.25$ nm. Table 18.1 shows that typical domain-wall widths are of this order.

18.4.1 Magnetic Hysteresis

Figure 18.8 shows typical *M-H* hysteresis loops (solid lines) and visualizes the meaning of coercivity (H_c) and remanence (M_r). Hysteresis strongly depends on the nanostructure or "microstructure" of the magnet and is therefore a nanoscale phenomenon. The performance of permanent magnets is described by the energy product, which is twice the usable magnetostatic energy outside the magnet divided by the magnet volume. The maximum energy product $(BH)_{max}$ of a magnet also derives from the hysteresis loop, but it is obtained from the *B-H* loop, which displays $B(H)$ rather than $M(H)$.

The field created by a magnet and therefore the energy product depends on the magnets macroscopic shape. For ellipsoidal shapes, the analysis of the problem yields

$$(BH) = \mu_o D \left(1-D\right) M^2 \qquad (18.23)$$

where D is the demagnetizing factor introduced in Section 18.3.2. Maximization of this expression yields $D = 1/2$, corresponding to magnet of compact shape, and $(BH)_{max} = \frac{1}{4}\mu_o M_2$ [16]. However, the realization of this energy product requires a sufficiently high coercivity. Low coercivity reduces the $(BH)_{max}$ area in Figure 18.8 and also means that the magnet must have a cumbersome elongated shape, for example, in form of a horseshoe. This scenario is realized in carbon–steel magnets, whose small coercivity leads to cumbersome horseshoe shapes ($D \ll 1$) and to energy

products of only about 1 kJ/m^3 [16,38]. Steel magnets have attracted renewed attention under the name of tetragonally distorted Fe-Co [39,40], but the theoretically assumed strains of $c/a = 1.23$ [39] are virtually impossible to sustain metallurgically. Modern permanent magnets have sufficiently high coercivities and compact shapes. Hexagonal ferrites (BaFe$_{12}$O$_{19}$ and SrFe$_{12}$O$_{19}$) have modest energy products (about 40 kJ/m^3) but are very cheap and dominate the permanent-magnet markets by mass. Today's high-performance permanent magnets are made from rare-earth-transition-metal intermetallics [9,16,41], some grades of neodymium–iron–boron having energy products as high as 460 kJ/m^3.

Most permanent magnet are nanostructured, because coercivity development relies on nanoscale features (Section 18.3.3). For example, sintered Sm-Co has a cellular microstructure where hexagonal 2:17 regions are surrounded by a 1:5 grain-boundary phase, whereas sintered Nd-Fe-B magnets consist of Nd$_2$Fe$_{14}$B regions embedded in an Nd-rich matrix [16].

18.4.2 Stoner–Wohlfarth Model

The simplest micromagnetic model is the *Stoner–Wohlfarth model*. The corresponding reversal mechanism is also known as coherent rotation or uniform rotation. The (free) energy per unit volume is the sum of anisotropy and Zeeman energies:

$$E/V = K_1 \sin^2\theta - \mu_o M_s H \cos\theta \qquad (18.24)$$

Here θ is the magnetization angle with respect to the field and anisotropy axis (Figure 18.9). In zero field, E has two minima, at $\theta = 0$ ($\uparrow$) and $\theta = \pi$ ($\downarrow$). In a reverse field ($H < 0$), an initial $\uparrow$ state becomes unstable at $H = -H_c$, which is referred to as nucleation or delocalized nucleation. Expanding Eq. (18.24) into powers of θ yields

$$\left(K_1(\boldsymbol{r}) + \frac{\mu_0}{2} M_s H \right) \theta = 0 \qquad (18.25)$$

In this equation, nucleation means that the term in parentheses becomes zero. This happens if the reverse field is equal to the anisotropy field $H_A = 2K_1/\mu_o M_s$. The corresponding hysteresis loop has a rectangular shape with $M_r = M_s$ and $H_c = H_A$ (Table 18.1).

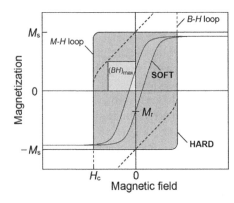

FIGURE 18.8 Typical hysteresis loop: hard or permanent magnet and soft magnet. A magnetic field normally creates clockwise hysteresis loops, but counterclockwise loops (magnetic proteresis) are observed in some nanostructures [37].

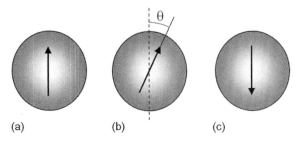

FIGURE 18.9 Stoner–Wohlfarth model. The $\uparrow$ and $\downarrow$ states are separated by an anisotropy-energy barrier, which creates magnetic hysteresis.

18.4.3 Micromagnetic Free Energy

To describe structural and magnetization inhomogeneities, it is necessary to replace Eq. (18.23) by a more general (free) energy:

$$E = \int \left\{ A \left[\nabla \left(\frac{\boldsymbol{M}}{M_s} \right) \right] - K_1 \frac{(\boldsymbol{n} \cdot \boldsymbol{M})^2}{M_s^2} - \mu_0 \boldsymbol{M} \cdot \boldsymbol{H} \right.$$
$$\left. - \frac{\mu_0}{2} \boldsymbol{M} \cdot \boldsymbol{H}_{\mathrm{d}}(\boldsymbol{M})^2 \right\} \mathrm{d}r \qquad (18.26)$$

Here $M_s(\boldsymbol{r}) = J_s(\boldsymbol{r})/\mu_0$ is the spontaneous magnetization, $K_1(\boldsymbol{r})$ denotes the first uniaxial anisotropy constant, $A(\boldsymbol{r})$ is the exchange stiffness, and $\boldsymbol{n}(\boldsymbol{r})$ is the unit vector of the local anisotropy direction. The exchange stiffness is approximately equal to J/a, and in magnets with strongly anisotropic structure, there are corrections due to $A_{xx} \neq A_{yy} \neq A_{zz}$ [42]. Interatomic exchange in antiferromagnets, which is sublattice-specific, can also be described micromagnetically [43]. The paramaters entering this equation are temperature-dependent, whence the name "free" energy. $\boldsymbol{H}$ is the external magnetic field, and $\boldsymbol{H}_{\mathrm{d}}$ is the magnetostatic self-interaction field:

$$\boldsymbol{H}_d(\boldsymbol{r}) = \frac{1}{4\pi} \int \frac{3(\boldsymbol{r} - \boldsymbol{r}')(\boldsymbol{r} - \boldsymbol{r}') \cdot \boldsymbol{M}(\boldsymbol{r}') - |\boldsymbol{r} - \boldsymbol{r}'|^2 \, \boldsymbol{M}(\boldsymbol{r}')}{|\boldsymbol{r} - \boldsymbol{r}'|^5} \mathrm{d}\boldsymbol{r}'$$
$$(18.27)$$

The exchange stiffness $A(\boldsymbol{r}) \sim \mathcal{J}/a$ describes the interatomic exchange on a continuum level and is of the order of 10 pJ/m for a broad range of ferro- and ferrimagnetic materials. Note that the ∇ operator in Eq. (18.26) penalizes rapid magnetization rotations, which are opposed by $\mathcal{J}$.

The parameters entering Eq. (18.26) are all local, that is, they depend on local chemistry, crystal structure, and crystallite orientation. In some materials, it is necessary to add additional terms, such as higher-order anisotropy constants and Dzyaloshinskii–Moriya interactions.

Hysteresis loops are obtained by tracing the local magnetization $\boldsymbol{M}(\boldsymbol{r})$ as a function of the applied (external) magnetic field H. Micromagnetic phenomena are almost exclusively realized by *rotations* of the local magnetization vector, because the intra-atomic exchange yields a robust spontaneous magnetization $M_s = |\boldsymbol{M}|$. As a consequence, only two of the three magnetization components (M_x, M_y, M_z) are independent, and one can restrict the consideration to the magnetization angles θ and ϕ, defined by $\boldsymbol{M} = M_s (\sin\theta \cos\phi \, \boldsymbol{e}_x + \sin\theta \sin\phi \, \boldsymbol{e}_y + \cos\theta \, \boldsymbol{e}_z)$.

The field may reverse the magnetization by creating domains and domain walls, such as those in Figure 18.2c, or by moving existing domain walls. The former reversal mechanism is known as domain nucleation, whereas the latter is referred to as domain-wall (de)pinning. Coercivity amounts to impeding magnetization reversal, which can be achieved by suppressing nucleation (nucleation-controlled coercivity) or creating defects that pin the domain walls (pinning-controlled coercivity).

Series expansion of Eq. (18.26) yields an approximate description of the nucleation problem in very hard and crystallographically aligned magnets ($K_{\mathrm{eff}} \approx K_1$ large and $\mathbf{n} = \mathbf{e}_z$). Treating θ as a small quantity yields

$$-A\nabla^2\theta + \left(K_{\mathrm{eff}}(\boldsymbol{r}) + \frac{\mu_0}{2} M_s H \right) \theta = 0 \qquad (18.28)$$

This equation describes localized nucleation and differs from Eq. (18.25) by the exchange term $-A\nabla^2\theta$. It is instructive to discuss the two limits of very strong and very weak exchange. In nanostructures of feature size L, for example, in nanoparticles of diameter L, magnetization inhomogeneities are penalized by an energy density of the order of A/L^2. When L is small, this penalty ensures a homogeneous magnetization $\theta(\boldsymbol{r})$ and $H_c = 2{<}K_1(\mathrm{r}){>}/\mu_0 M_s$. For large L, the exchange contribution is negligible and $H_c = 2\min(K_1(\boldsymbol{r}))/\mu_0 M_s$. In other words, nucleation starts in the softest region of structure. This is of practical importance in the processing of sintered permanent magnets, where sophisticated annealing processes are necessary to get rid of harmful soft regions [16]. In the intermediate regime, the nucleation field roughly scales as $H_A \, \delta_B^2/L^2$, with intriguing physics as the small-L limit is approached. For example, nanoscale exchange coupling of a high-magnetization soft phase to hard phase may be used to enhance the energy product of permanent magnets beyond that of the hard phase [10,38,44].

18.4.4 Domains and Domain Walls

Magnetic domains are an important aspect of magnetization reversal and hysteresis. Figure 18.10 shows a typical domain structure. Light- and dark gray regions are of opposite magnetization ($\theta = 0°$ and $\theta = 180°$), and the medium gray region is the domain wall. There are several types of domain walls such as Bloch walls and Néel walls, but the domain-wall thickness is generally many interatomic distances, as visualized in Fig. 2(c) and listed in Table 18.1.

Complicated domain structures, such as that of Figure 18.10, require numerical treatments of the micromagnetic free energy [45,46]. However, *micromagnetic scaling* provides useful information about domains and domain walls. The micromagnetic free energy contains three basic quantities, namely the exchange stiffness A, measured in J/m, the anisotropy K_1, measured in J/m³,

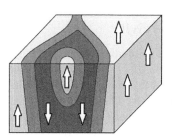

FIGURE 18.10 Magnetic domains near the surface of a permanent magnet with noncubic crystal structure.

and the magnetostatic self-energy $\mu_o M_s^2$, also measured in J/m³. Typical external fields are of the order of M_s, so that $\mu_o M_s^2$ also applies to the Zeeman energy.

The wall-width parameter $\delta_o = (A/K)^{1/2}$ determines the thickness of the domain walls that separate magnetic domains (Figure 18.2c) and the response to local imperfections. It ranges from about 1 nm for very hard magnets to about 100 nm in very soft magnets (Table 18.1). Essentially, the thickness of the walls is determined by the competition between exchange, which favors extended walls, and anisotropy, which favors narrow transition regions. A more detailed analysis shows that 180° Bloch domain walls have a width of $\delta_B = \pi \delta_o$ and a wall energy pf $\gamma = 4(AK_1)^{1/2}$. Other domain-wall types, such as Néel walls, additionally involve long-range magnetostatic interactions, which complicate the situation and mean that the wall width also depends on $(K_1/\mu_o M_s^2)^{1/2}$.

The exchange length $l_o = (A/\mu_o M_s^2)^{1/2}$ is the length below which atomic exchange interactions dominate typical magnetostatic fields. It determines, for example, the coherence radius $R_{coh} \approx 5 \, l_o$ above which the Stoner–Wohlfarth model can no longer be used to describe homogenous magnets. The exchange length l_o is of the order of 2 nm for a broad range of ferromagnets. Note that the wall-width parameter δ_o is sometimes interpreted as an exchange length. If this was a valid consideration, then ideally soft materials, where $K_1 = 0$ and $\delta_o = \infty$, would realize exchange coupling on a truly macroscopic scale. This is contradictory to experiment.

The critical single-domain radius $R_{SD} = 36 \, l_o^2/\delta_o$ determines the onset of *equilibrium* domains. It reflects the competition between magnetostatic energy (which favors domain formation) and anisotropy. R_{SD} reaches nearly 1 µm in very hard materials. It is important to emphasize that R_{SD} is an *equilibrium* property. It involves the comparison of the of single-domain and multi-domain energies and determines, for example, the initial or virgin state after thermal demagnetization. It is independent of the energy barriers that control coercivity. In particular, R_{SD} is unrelated to the applicability of the coherent-rotation theory.

Compared to domain-*wall* thickness, there are no simple estimates exist for the domain size. As analyzed by Landau nearly a century ago [47], even simple geometries yield rather complicated expression for the domain size, such as a square-root dependence on thin-film thickness. Similarly, domain tend to be smaller near surfaces that in the bulk.

18.4.5 Coercivity

Equation (18.25) predicts the coercivity to equal the anisotropy field, $H_c = H_a$, but measured coercivities are normally much smaller than H_A, which is known as Brown's paradox [48]. The paradox is solved by the involvement of inhomogeneous magnetization states $M(r) = M(\theta(r), \phi(r))$. There are several types of inhomogeneous nucleation- and pinning-controlled coercivity mechanisms, such as magnetization curling in nanoparticles [35],

thin-film patches [36], and nanowires [49], localized nucleation [44,50,51], strong pinning, and weak pinning [16]. A detailed description of these mechanisms goes beyond the scope of this chapter, but it is useful to rationalize coercivity in terms of the Kronmüller equation [52]

$$H_c = \alpha_K H_A - D_{eff} M_s \qquad (18.29)$$

where α_K is dimensionless and D_{eff} is an effective *local* demagnetizing factor. The Kronmüller factor α_K parameterizes the magnet's nanostructure and is often much smaller than one. For example, in annealed high-purity iron, $\alpha_K \approx 0.002$, because the annealing removes defects that could impede magnetization reversal through domain-wall pinning. On the permanent-magnet side, it is very difficult to achieve a values in excess of $\alpha_K = 0.3$, even in optimally processed magnets. D_{eff} is normally positive but sometimes negative, for example, in alnico magnets (Section 18.3.2), where it describes the shape anisotropy of the embedded FeCo needles.

18.4.6 Grain Boundaries and Skyrmions

Micromagnetic is not only important for the understanding of hysteresis and coercivity but also yields specific nanoscale effects of importance, for example, in spin electronics. One of these topics is grain-boundary micromagnetism. At an interface in the y-z-plane, the magnetization angle obeys the Erdmann–Weierstraß boundary condition [10,44]

$$A_{left} \frac{\partial \theta_{left}(x)}{\partial x} = A_{right} \frac{\partial \theta_{right}(x)}{\partial x} \qquad (18.30)$$

Figure 18.11a explains the meaning of this equation. If the adjacent phases have different exchange stiffnesses A, then the magnetization change is predominantly confined to the phase with the lower exchange stiffness, $A' < A$ (the light-gray regions in the figure). The total interface exchange energy $\mathcal{J}_{eff}$ is obtained by integrating the micromagnetic free energy over x. This effective interlayer exchange is different from the interatomic exchange at the interface and typically much smaller. For a thin nonmagnetic interlayer of thickness d_o and very small 'effective' exchange stiffness A', $\mathcal{J}_{ie}$ is reduced by a factor of about $2 \, \delta_B A'/\pi d_o A$ [10]. It is

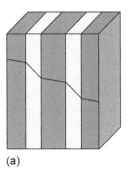 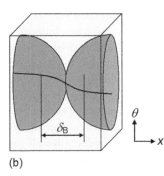

(a) (b)

FIGURE 18.11 Spin structure at interfaces: (a) multilayer and (b) granular nanocomposite.

therefore difficult to reduce intergranular exchange by using an atomically thin grain-boundary phase ($d_o \approx 0.2$ nm), because $\delta_B \gg d_o$ in this case.

When two regions have strong exchange coupling A and touch each other, Figure 18.11b, then the weakly exchange-coupled matrix (light gray) is no longer involved directly. However, since magnetization gradients are punished by an energy density proportional to $A\nabla^2$, the magnetization change is confined to the vicinity of the constriction. The corresponding length scale is approximately equal to the Bloch-wall width δ_B, irrespective of contact area and magnitude of the magnetization change.

A research area of renewed interest is magnetic *skyrmions*. By definition, skyrmions are solitonic solutions of nonlinear field equations originally used to describe nuclear matter [53]. Much of the interest in magnetic skyrmions comes from potential applications in date storage and processing, where miniaturization is a major consideration. An early example of magnetic skyrmions is magnetic bubbles, first described in 1967 [17]. Bubble skyrmions are stabilized by magnetic anisotropy and magnetostatic interactions, and DM interactions (Section 18.3.3) can be used to modify skyrmions, add functionality, and further improve stability [54]. The micromagnetic description of skyrmions continues to be a challenge, aside from simple thin-film geometries [55,56], and often the DM vector is approximated by a scalar.

One challenge is to reduce the skyrmion size below about 30 nm, to make the areal density of skyrmion systems competitive. The skyrmion diameter increases with the exchange stiffness A, because the latter favors homogeneous magnetization states and suppresses skyrmions. Materials with strongly reduced A can therefore be used to create very small skyrmions, but this strategy is limited to low temperatures, because the Curie temperature is proportional to $A \approx \mathcal{J}/a$ (Section 18.5.2). Aside from this general consideration, systems with good permanent-magnet properties are potential skyrmion materials, due to similar requirements (high magnetization, strong spin–orbit coupling favoring magnetic anisotropy and DM interactions). One example is FePt, where the theoretical limit of the bubble-skyrmion size is unusually low, below 20 nm. Co-Pt and Fe-Pt interfaces also support strong DM interactions.

An intriguing feature of skyrmions is their contribution to the topological Hall effect (Section 18.6.2). This contribution is caused by the inhomogeneous spin structure in the skyrmions and described by topological quantum number (winding number)

$$N = \frac{1}{4\pi} \int \boldsymbol{s} \cdot \left(\frac{\partial \boldsymbol{s}}{\partial x} \times \frac{\partial \boldsymbol{s}}{\partial y} \right) \mathrm{d}x\,\mathrm{d}y \qquad (18.31)$$

where $\boldsymbol{s}(\boldsymbol{r}) = \boldsymbol{M}(\boldsymbol{r})/M_s$ is the normalized magnetization vector [54,57]. Bubble skyrmions have the winding number $N = -1$ [54]. Winding numbers are topologically protected, that is, unchanged by small to moderate perturbations. Figure 18.12 illustrates topological protection in bubble domains: (a) and (b) are topologically equivalent, because the deformation does not change the integral Eq. (18.31).

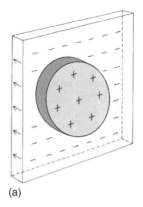 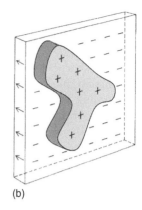

(a)   (b)

FIGURE 18.12 Magnetic bubble skyrmions. The spin structure inside the domain walls is noncollinear.

Interesting effects occur when the skyrmions are confined in space, for example, in nanoparticles [58].

18.5 Finite-Temperature and Nonequilibrium Magnetism

Up until now, we have restricted ourselves to zero-temperature equilibrium processes. Dynamic processes, occurring at zero temperature or created by thermal excitations, are important in many regards. For example, magnetic recording needs to combine long-time information storage with fast writing and reading. The read heads are typically magnetoresistive, so their understanding and improvement requires the consideration of the dynamics of conduction electrons. The distinction between intrinsic and extrinsic properties carries over to times dependent properties: intrinsic processes are very fast, extrinsic processes are often very slow, and interesting effects occur in the intermediate region. This section provides a brief outline of the physical principles involved in finite-temperature and nonequilibrium magnetism.

18.5.1 Fast Magnetization Dynamics

The time-dependent Schrödinger equation predicts $\Delta E = \hbar\omega$, that is, large energy differences correspond to fast processes. Relativistic interaction energies are smaller than the leading interactions of electrostatic origin and therefore correspond to slower processes. For example, a typical interatomic exchange of 0.1 eV is equivalent to 24 THz, whereas a soft-magnetic anisotropy energy of 1 μeV corresponds to 0.24 GHz. Expressing the Bohr magneton in terms of frequency yields almost precisely 14 GHz/T. The ENIAC computer used a clock speed of 0.1 MHz, whereas clock rates of current PC are a few GHz. Moving towards THz data processing requires new approaches, and exploiting interatomic exchange, for example, in antiferromagnets, is one strategy.

In a classical description, AFM exchange leads to ground-state sublattice magnetizations $\boldsymbol{M}_A$ and $\boldsymbol{M}_B$ that point in opposite directions ($\theta = \pi$) and cancel each other

(Figure 18.13). The high-frequency AFM mode corresponds to oscillations of the Néel vector $\boldsymbol{M}_{\mathrm{N}} = \boldsymbol{M}_{\mathrm{A}} - \boldsymbol{M}_{\mathrm{B}}$. The Néel vector is also able to rotate, but this rotation is rather slow, because it corresponds to relativistic energies, mainly in-plane magnetic anisotropy. Challenges are the reading and writing of the information, one reason being the smallness of the Néel vector. Several methods are presently being explored, such as heating above the magnetic ordering temperature (Néel temperature) and magnetoresistive reading. The quantum mechanics of antiferromagnets is actually very complicated due to quantum fluctuations, even in the ground state [15,24], which adds to the complexity of the approach.

Spin waves or magnons (Figure 18.14) are low-energy excitations in ferro- and ferrimagnets that were first considered by Bloch [59]. Ignoring damping, the dynamics (resonance) of the uniform mode ($\boldsymbol{k} = 0$) is described by Landau–Lifshitz equation $\mathrm{d}\boldsymbol{M}/\mathrm{d}t = \gamma(\boldsymbol{M} \times \mu_{\mathrm{o}}\boldsymbol{H}_{\mathrm{eff}})$, where $\gamma = e/m_{\mathrm{e}} = 2\mu_{\mathrm{B}}/\hbar$ is the electron's gyromagnetic ratio and $\mathrm{H}_{\mathrm{eff}}$ contains both external-field and magnetic anisotropy contributions [60]. This equation is based on the Heisenberg model and, as in the Pauli expansion, the cross product reflects the algebra of the Pauli matrices, Eq. (18.14). For ferromagnetic resonance in uniaxial ferromagnets, this equation yields $\omega = \mu_{\mathrm{o}}\gamma\,(H + H_{\mathrm{A}})$. The involvement of μ_{B}, H, and H_{A} shows that FM resonance is a relativistic GHz phenomenon, as opposed to THz AFM resonance.

In nanostructures, the resonance modes do not simply superpose but are coupled by interatomic exchange, similar to the nucleation mode of Eq. 18.32, which can be interpreted as the $\omega = 0$ spin-wave mode. In long and thin nanowires oriented along the z-direction [10,61]

$$-2A\frac{\mathrm{d}^2\theta}{\mathrm{d}z^2} + \left(2K_{\mathrm{eff}}(z) + \mu_0 M_s H - M_s \frac{\omega}{\gamma}\right)\theta = 0 \quad (18.32)$$

This equation describes how magnons interact with nanoscale structural features, $K_{\mathrm{eff}}(z)$ in the present example. For $\omega = 0$, this equation reduces to the static or nucleation limit of Eq. (18.24), whereas $\mathrm{d}\theta/\mathrm{d}z = 0$ reproduces $\omega = \mu_{\mathrm{o}}\gamma\,(H + H_{\mathrm{A}})$. For homogeneous systems, $K_{\mathrm{eff}} = const.$, Eq. (18.32) yields a quadratic dispersion relation, $\omega(\boldsymbol{k}) - \omega(0) \sim A\,k^2$. In this context, the exchange stiffness A is normally referred to as the *spin-wave stiffness*. While A is large, the generally very long wavelength of spin waves (Figure 18.13) corresponds to small wave vectors k and to low excitation energies. Spin waves are therefore important for the low-temperature behavior of magnets. In d dimensions, the corresponding magnetization reduction $\Delta M \sim T^{\mathrm{d}/2} \int k^{\mathrm{d}-3}\,\mathrm{d}k$. This integral diverges for $d \leq 2$, indicating the absence of ferromagnetism in low-dimensional magnets (Wagner–Mermin theorem) [15]. In three dimensions, the integral yields Bloch's law, $\Delta M \sim T^{3/2}$.

18.5.2   Equilibrium Statistical Physics

The theoretical description of magnets at nonzero temperature is conceptually very simple. It is necessary and sufficient to determine the partition function

$$\mathcal{Z} = \sum_\mu \exp\left(-E_\mu/k_{\mathrm{B}}T\right) \quad (18.33)$$

where the E_μ are the eigenvalues of the quantum-mechanical Hamiltonian and the summation includes all microstates (spin configurations). The partition function yields the free energy $\mathcal{F} = -k_{\mathrm{B}}T \ln \mathcal{Z}$, from which all thermal averaged are obtained as derivatives. The practical challenge is that the number of microstates increases exponentially with system size, so that ingenious methods are necessary to determine the partition function.

Zero-temperature ferromagnetism means that all spins are parallel, but thermal excitations tend to rotate or switch individual spins, thereby reducing the magnetization. Thermal disorder competes against interatomic exchange and causes the magnetization of ferromagnets to vanish at a well-defined sharp *Curie temperature* T_C (Figure 18.15). The magnitude of the Curie temperature is determined by interatomic exchange (Section 18.2.3). Magnetostatic dipole interactions are far too weak to create room-temperature magnetic order. That can be seen from the smallness of the Bohr magneton, $\mu_{\mathrm{B}}/k_{\mathrm{B}} = 0.672$ kelvin per tesla in temperature units. Similarly, magnetic anisotropy cannot explain long-range magnetic order [62].

The sharpness of the Curie temperature is nontrivial, because Z, Eq. (18.33), and therefore all its derivatives are smooth functions of the temperature. In fact, the Curie point implies a transition to an infinite system. This requirement interferes with nanostructuring, and low-dimensional systems, such as nanoparticles, do not have well-defined Curie temperatures.

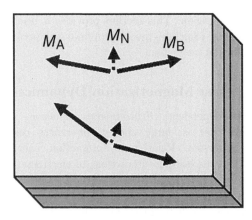

FIGURE 18.13 Antiferromagnetic resonance in thin film. Due to the dominance of the AFM interatomic exchange, the Néel vector $\boldsymbol{M}_{\mathrm{N}}$ is very small.

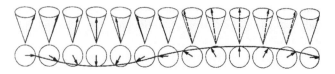

FIGURE 18.14 Magnon in a ferromagnet (schematic).

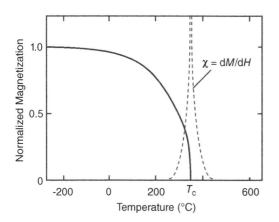

FIGURE 18.15 Temperature dependence of the magnetization (solid line) and susceptibility (dashed line, schematic) of bulk Ni in zero magnetic field.

Alternatively, Curie transitions are characterized by the divergence of the susceptibility χ, and by analyzing Z, it can be shown rigidly that this divergence is accompanied by a divergence of the correlation length ξ, defined through

$$< \mathbf{S}\,(r) \cdot \mathbf{S}\,(r+R) > - < \mathbf{S}\,(r) > \cdot < \mathbf{S}\,(r+R) >$$
$$\approx \ \exp(-R/\xi) \tag{18.34}$$

Nanostructuring on a length scale L suppresses correlations having $\xi > L$, so the c remains finite and $\chi(T)$ is a smooth peak.

The correlation length exhibits a power-law dependence on temperature [63,64], approximately

$$\xi \approx a \frac{T_c^v}{|T - T_c|^v} \tag{18.35}$$

Here $a \approx 0.2$ nm and v is a critical exponent ($v = 0.5$ in the dimensionality-independent mean-field approximation, $v \approx 0.6$ in magnets with three-dimensional crystal structure). Equation (18.35) can be used to estimate the smoothing $\Delta T = |T - T_c|$ of the Curie transition due to nanostructuring. Taking $v = 0.6$, $L = 2$, nm and $T_c = 600$ K yields $\Delta T = 13$ K, or 2.2%. For $L = 20$ nm, $\Delta T = 0.3$ K. These estimates show that effect of nanostructuring rapidly decreases with feature size. In practice, it is difficult to distinguish nanomagnetism from true ferromagnetism when the feature sizes are much larger than 1 nm: nanocomposites look like two-phase mixtures [65] and nanoparticles look like bulk magnets [66].

The analysis of the previous paragraphs shows that exchange coupling of low-Curie-temperature phase to a phase with a high temperature is no promising approach to enhance the Curie temperature of a system: nanostructuring on a length scale of 2 nm is difficulty to achieve and yields a gain of only 2.2%. Micromagnetic exchange coupling, for example, in hard–soft composites (Section 18.4.3), is also difficult to realize experimentally, but it persists on much bigger length scales. Taking $H_c = H_A\, \delta_B{}^2/L^2$ and $\delta_B = 10$ nm to estimate the length scale where H_c is reduced to 2.2%

($\alpha_K = 0.022$) yields a range of 70 nm. Such an α_K is not attractive in permanent magnetism but may be a concern in soft magnets.

18.5.3 Quantum-Phase Transitions

In the preceding subsections, we have seen how thermal fluctuations affect ($d = 3$) or destroy ($d < 2$) long-range ferromagnetic order. In the spin-wave picture, the magnetization reduction due to spin waves diverges in $d \le 2$ dimensions. Furthermore, by a more detailed analysis of the partition functions, it can be shown explicitly that critical fluctuations described by the correlation length ξ destroy ferromagnetism in zero and one dimensions ($T_c = 0$).

Quantum fluctuations yield similar effects. The corresponding transitions, known as quantum-phase transitions (QPT), are physically very different from Curie transition, but the phenomenologies of the two phenomena have much in common, so QPTs are treated in this section. For simplicity, we will restrict ourselves to the mean-field approximation. Ignoring fluctuations, mean-field theory amounts to the analysis of the Landau free-energy density

$$\mathcal{F} = \frac{1}{2}a\,\mathrm{M}^2 + \frac{1}{4}\mathrm{M}^4 - \mu_0 M\,H \tag{18.36}$$

Near the Curie point, $a = a_o\,(T - T_c)$, and for $T < T_c$, the Landau free-energy has a double-well structure with two minima corresponding to the ferromagnetic solutions $M > 0$ and $M < 0$. The spontaneous magnetization $M(T)$ and the susceptibility $\chi(T)$ (Figure 18.15) follow from the analysis of these minima.

A similar expansion can be made for ferromagnets near the Stoner transition and for other zero-temperature quantum phenomena. The main difference is that the onset of ferromagnetism is not determined by T_c but by the density of states at the Fermi level or some other quantum-mechanical parameter. The DOS can be tuned, for example, by changing the chemical composition of an alloy $A_{1-x}B_x$, with ferromagnetism occurring above or below a critical concentration x_c. The relation corresponding to Eq. (18.35) is

$$\xi \approx \frac{a}{\sqrt{|1 - \mathcal{D}(E_F)I|}} \tag{18.37}$$

In the limits of both strongly exchange-enhanced paramagnets and very weak itinerant ferromagnets (Section 18.2.2), $\xi \gg a$. To be precise, as in the Curie temperature analogy, ξ is not much bigger than about 1 nm, but this range has a big impact on the magnetization if the surface layer is ferromagnetic and the particle core paramagnetic. This is due to the large fraction of atoms near the surface. For example, a 1-nm shell in a 10-nm nanoparticle contains more than 50% of all atoms. Note that metallic surface atoms are often magnetic, because $\mathcal{D}$ is inversely proportional to the band width W and W sales as $z^{1/2}$, where z is the number of nearest neighbors [16,67,68]. Surface atoms have fewer neighbors and are therefore easier to spin-polarize. A good example of this mechanism is Co_2Si nanoparticles produced by cluster deposition [69].

18.5.4 Slow Magnetization Dynamics

Compared to intrinsic processes, which are normally very fast, extrinsic processes tend to sluggish, from milliseconds in typical superparamagnetic particles up to millions of years in magnetic rocks. This is advantageous in data storage and permanent magnets but undesired, for example, during the writing of the information. Slow relaxation involves the thermally activated overcoming of energy barriers E_a and is governed by the Boltzmann–Arrhenius law [70]

$$\tau = \tau_o \exp\left(\frac{E_a}{k_B T}\right) \qquad (18.38)$$

where τ_o is a time constant of the order of 10^{-10} s. Simplifying somewhat, τ_o is the time after which the damped precession (Section 18.s5.1) yields equilibrium on a local scale. Assuming that relaxation takes place after a few seconds and that the energy barrier is Stoner–Wohlfarth-like, $E_a = K_1 V$, yields the thermal-stability criterion $K_1 V = 25\, k_B T$.

The field dependence of the energy barrier E_a entering Eq. (18.38) is approximately

$$E_a(H) = K_1 V_o \left(1 - \frac{H}{H_o}\right)^m \qquad (18.39)$$

where V_o is a nanoscale effective activation volume. The exponent $m = 2$ for aligned Stoner–Wohlfarth particles and $m = 3/2$ for a wide range of other reversal mechanisms [10,71].

Equations (18.38 and 18.39) have two main consequences. First, the magnetization exhibits a logarithmically slow decay known as *magnetic viscosity* [70]. For example, permanent magnets and recording media loose a small fraction of their remanent magnetization per decade. In magnetic recording, it is customary to assume $K_1 V \gtrsim 60\, k_B T$, because data need to be stored for years rather than seconds and because some redundancy is necessary to minimize errors. When $K_1 V < 25\, k_B T$, relaxation is rather fast and the particles' behavior is referred to as superparamagnetism. A second consequence is that the coercivity shows a logarithmic dependence on the sweep rate dH/dt.

18.6 Spin-Dependent Transport

Electron transport is intrinsically nanoscale, due to the involvement of the mean free path l, and taking into account the electron spin yields additional rich physics. Figure 18.16

shows some basic measurement geometries. The resistivity of perfectly periodic structures, including nanostructures such as multilayers, is *exactly* zero. The reason is that resistivity requires transitions between the quantum-mechanical eigenstates labeled by wave vectors $\boldsymbol{k}$, but the corresponding matrix elements are zero in periodic solids [15]. Resistivity requires imperfections, such as atomic scale defects [72]. Surfaces are also imperfections in this sense, because the incoming electrons are no eigenfunctions of the solid. Quasi-classically,

$$\rho = \frac{2m_e v_F}{e^2 n_e l} \qquad (18.40)$$

where n_e is the electron density and v_F is the Fermi velocity. This equation is a good approximation if the mean free path exceeds a few interatomic distance. As the system approaches the "dirty-metal" limit ($l \approx a$), the description of electron states in terms of wave vectors $\boldsymbol{k}$ is no longer meaningful. The corresponding magnetoresistance is about 200 $\mu\Omega$cm [73]. Of course, Eq. (18.40) also fails to describe metal-insulator transitions (Mott transitions).

There are many magnetoresistance contributions, and a comprehensive description of these contribution goes far beyond the scope of this chapter. Ordinary magntoresistance means that a magnetic field H curves the path of conduction electrons in nonmagnetic metals and thereby creates an H^2 term in the resistivity. Anisotropic magnetoresistance (AMR) occurs in magnetic materials [74] and reflects spin-orbit coupling. Pictorially, the spin perturbs the surrounding electron cloud and makes the scattering dependent on the spin direction. For example, in Figure 18.6, the resistivity depends on whether the electrons move in the x- or z-direction. Giant magnetoresistance (GMR) and grain-boundary magnetoresistance are due to rapid magnetization variations, which act as scattering centers [75]. Tunnel magnetoresistance (TMR) means spin-dependent electron tunneling through a thin insulating layer [76]. Colossal magnetoresistance (CMR) involves double exchange (Mn^{3+}, Mn^{4+}) and means that the antiferromagnetic insulator $LaMnO_3$ ($La^{3+}Mn^{3+}O^{2-}_3$) becomes a ferromagnetic conductor in an external magnetic field H.

18.6.1 Anomalous Hall Effect and Berry Phase

The Hall effect consists in the creation of a voltage perpendicular to current and magnetic field (Figure 18.14b). The ordinary Hall effect is also found in nonmagnetic metals and

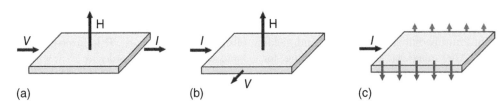

(a)   (b)   (c)

FIGURE 18.16 Basic spin-dependent transport geometries: (a) magnetoresistance, (b) Hall effect, and (c) spin Hall effect.

caused by the Lorentz force acting on the conduction electrons. The anomalous Hall effect (AHE) [78], which often involves the magnetization of a material, is closely related to the spin Hall effect (SHE), where a current leads to the separation of ↑ and ↓ electrons at different sides of the conductor, as illustrated in Figure 18.14c. Two major contributions to AHE and SHE are skew scattering and side-jump scattering by impurity atoms [77]. Both are caused by spin–orbit coupling (SOC), and the impurity is not necessarily magnetic. The difference is that skew scattering changes the *direction* of the electron velocity v, creating a velocity component Δv perpendicular to the current and a net electron displacement $\tau_R \Delta v$, where $\tau_R = l/v$ is the relaxation time. The side-jump scattering events do not create a perpendicular velocity component ($\Delta v = 0$) but a finite displacement Δr perpendicular to the current. Skew scattering in linear in the resistivity ρ and dominates in rather pure metals, whereas side-jump scattering scales as ρ^2 and dominates in moderately disordered metals. There is also an intrinsic (resistivity-independent) Hall contribution due to the Berry phase reflecting the periodic electronic structure of the material.

The *Berry phase* is a long overlooked quantum-mechanical phase factor of great importance in a variety of physical systems. It occurs in systems whose wave functions undergo parametric changes. In the example originally discussed by Berry [79], this change is realized by an external magnetic field H of constant magnitude but adiabatically rotating direction. The spin, originally pointing in the z-direction, y = (1, 0), follows the external field and changes to $\psi(t) = (\cos(\theta/2), e^{-i\phi}\sin(\theta/2))$. Since the energy $E_{spin} = -\mu_o \mu_B \sigma \cdot H$ remains unchanged during this process, E_{spin} can be chosen as the zero-point energy, so that the time-dependent Schrödinger equation becomes $i\hbar \partial \psi/\partial t = 0$. The solution of this equation is $\psi(t) = \psi(0)$, corresponding to a fixed spin and in striking contradiction to $\psi(t) = \psi(\theta, \phi)$. The reason for this unphysical result is the neglect of the *Berry phase* $e^{i\gamma}$, meaning that the correct wave function is $e^{i\gamma}\psi$ rather than ψ.

Some quantities, such as the potential energy, are not affected by the Berry phase, because $\psi^* e^{-i\gamma} e^{i\gamma}\psi = \psi^*\psi$. However, the kinetic energy involves the operator ∇^2 acting on the wave function:

$$\nabla^2(e^{i\gamma}\psi) = e^{i\gamma}(\nabla + i\nabla\gamma)^2\psi \qquad (18.41)$$

The term $e^{i\gamma}$ is now on the left, where it can be annihilated by $e^{-i\gamma}$, as in the case of the potential energy. However, this simplification is paid by a modification of $p \sim \nabla$. Comparison with the four-vector below Eq. (18.13) shows that this modification is equivalent to the introduction of a magnetic vector potential. In other words, rotating spins experience a magnetic field caused by the Berry phase. In more detail, it can be shown that this vector potential (or Berry connection) yields a nonzero "emergent magnetic field" if the spin rotation is non-coplanar [80].

18.6.2 Emergent Magnetic Field in Nanostructures

The emergent magnetic field contributes to the anomalous Hall effect. Physically, a conduction electron undergoes spin rotation due to exchange interaction with local atomic spins. When the local spins form a noncollinear structure, as the they do in the domain-wall regions of skyrmions, then the corresponding Berry phase contributes to the Hall effect. This is known as the topological Hall effect. The net contribution is determined by the winding number, so that Figure 18.11a, b yields the same Hall contribution.

Berry-phase contributions to the Hall effect are not limited to skyrmions but also occur in other nanostructures. The magnitude and direction of the emergent magnetic are given by

$$B = \frac{\hbar}{2e}\sin\theta\,(\nabla\theta \times \nabla\phi) \qquad (18.42)$$

The magnitude of the B-field scales as $\Phi_o/2\pi R^2$, where $\Phi_o = 2.067$ fTm^2 is the magnetic flux quantum and R is the distance over which the electron changes its spin directions [81]. On a macroscopic scale, the fields are weak, but they are fairly large when R is smaller than 1 μm.

18.7 Conclusions

In summary, this chapter has outlined basic physical principles governing the magnetism of nanostructures. A distinction has been made between intrinsic and extrinsic properties. Intrinsic magnetic properties are quantum-mechanical in nature, realized on an atomic scale, and very fast. Extrinsic properties are of the micromagnetic type, realized on length scales of several nanometers, and often very slow. Nanomagnetics goes beyond a superposition of atomic and macroscopic physics. One reason is the involvement of relativistically small energy contributions, specifically Zeeman interaction and spin–orbit coupling. These interactions establish a scale length of $a_o/\alpha = 7.25$ nm, which establishes nanomagnetism as a separate branch of magnetism and solid-state physics.

Acknowledgment

This chapter has benefitted from discussions with A. Kashyap, R. Pahari, R. Pathak, and G. C. Hadjipanayis. Its scientific content is partially based on original research supported by DOE BES (DE-FG02-04ER46152) and locally through NCMN and HCC.

References

1. J. L. Dormann, L. Bessais, and D. Fiorani, A dynamic study of small interacting particles: Superparamagnetic model and spin-glass laws, *J. Phys.* C **21**, 2015–2034 (1988).

2. D. J. Sellmyer, B. Balamurugan, B. Das, P. Mukherjee, R. Skomski, and G. C. Hadjipanayis, Novel structures and physics of nanomagnets (invited), *J. Appl. Phys.* **117**, 172609-1–6 (2015).

3. B. Balasubramanian, R. Skomski, X.-Zh. Li, Sh. R. Valloppilly, J. E. Shield, G. C. Hadjipanayis, and D. J. Sellmyer, Cluster synthesis and direct ordering of rare-earth transition-metal nanomagnets, *Nano Lett.* **11**, 1747–1752 (2011).

4. U. Gradmann, Magnetism in ultrathin transition metal films, In: *Handbook of Magnetic Materials*, Vol. 7, K. H. J. Buschow (ed.), Elsevier, Amsterdam (1993), pp. 1–95.

5. G. Zangari and D. N. Lambeth, Porous aluminum oxide templates for nanometer-size cobalt arrays, *IEEE Trans. Magn.* **33**, 3010–3012 (1997).

6. W. Wernsdorfer, K. Hasselbach, A. Benoit, B. Barbara, B. Doudin, J. Meier, J.-Ph. Ansermet, and D. Mailly, Measurements of magnetization switching in individual nickel nanowires, *Phys. Rev.* B **55**, 11552–11559 (1997).

7. R. Skomski, H. Zeng, M. Zheng, and D. J. Sellmyer, Magnetic localization in transition-metal nanowires, *Phys. Rev.* B **62**, 3900–3904 (2000).

8. Y. C. Sui, R. Skomski, K. D. Sorge, and D. J. Sellmyer, Nanotube magnetism, *Appl. Phys. Lett.* **84**, 1525–1527 (2004).

9. K. Kumar, $RETM_5$ and RE_2TM_{17} permanent magnets development, *J. Appl. Phys.* **63**, R13–R57 (1988).

10. R. Skomski, Nanomagnetics, *J. Phys. Condens. Matter* **15**, R841–R896 (2003).

11. R. M. Bozorth, *Ferromagnetism*, van Nostrand, Princeton (1951).

12. J. B. Goodenough, *Magnetism and the Chemical Bond*, Wiley, New York (1963).

13. S. Chikazumi, *Physics of Magnetism*, Wiley, New York (1964).

14. D. C. Mattis, *Theory of Magnetism*, Harper and Row, New York (1965).

15. N. W. Ashcroft and N. D. Mermin, *Solid State Physics*, Saunders, Philadelphia, PA (1976).

16. R. Skomski and J. M. D. Coey, *Permanent Magnetism*, Institute of Physics, Bristol (1999).

17. Ch.-W. Chen, *Magnetism and Metallurgy of Soft Magnetic Materials*, North-Holland, Amsterdam (1977).

18. D. Weller, A. Moser, L. Folks, M.E. Best, W. Lee, M.F. Toney, M. Schwickert, J.-U. Thiele, M. F. Doerner, High K_u materials approach to 100 Gbits/inz, *IEEE Trans. Magn.* **36** (l), 10–15 (2000).

19. D. Weller and T. McDaniel, Media for extremely high density recording. In: *Advanced Magnetic Nanostructures*, D. J. Sellmyer and R. Skomski (eds.), Springer, Berlin (2006), pp. 295–324.

20. R. L. Comstock, *Introduction to Magnetism and Magnetic Recording*, Wiley, New York (1999).

21. J. M. Coey, *Magnetism and Magnetic Materials*, University Press, Cambridge (2010).

22. P. Mohn, *Magnetism in the Solid State*, Springer, Berlin (2003).

23. E. P. Wohlfarth, Very weak itinerant ferromagnets; application to $ZrZn_2$, *J. Appl. Phys.* **39**, 1061–1066 (1968).

24. R. Skomski, P. Kumar, B. Balamurugan, B. Das, P. Manchanda, P. Raghani, A. Kashyap, and D. J. Sellmyer, Exchange and magnetic order in bulk and nanostructured Fe_5Si_3, *J. Magn. Magn.* Mater. **460**, 438–447 (2018).

25. R. Skomski, *Simple Models of Magnetism*, Oxford: University Press (2008).

26. R. Skomski, RKKY Interactions between nanomagnets of arbitrary shape, *Europhys. Lett.* **48**, 455–460 (1999).

27. R. Skomski, J. Zhou, J. Zhang, and D. J. Sellmyer, Indirect exchange in dilute magnetic semiconductors, *J. Appl. Phys.* **99**, 08D504-1–3 (2006).

28. P. Bak and H. H. Jensen, Theory of helical magnetic structures and phase transitions in MnSi and FeGe, *J. Phys. C* **13**, L881–L885 (1980).

29. K.-H. Fischer and A. J. Hertz, *Spin Glasses*, University Press, Cambridge 1991.

30. J. A. Osborn, Demagnetizing factors of the general ellipsoid, *Phys. Rev.* **67**, 351–357 (1945).

31. R. A. McCurrie, *Ferromagnetic Materials: Structure and Properties*, Academic Press, London (1994).

32. L.-Q. Ke, R. Skomski, T. D. Hoffmann, L. Zhou, D. D. Johnson, M. J. Kramer, I. E. Anderson, and C.-Z. Wang, Alnico coercivity, *Appl. Phys. Lett.* **111**, 022403-1–5 (2017).

33. F. Bloch and G. Gentile, Zur anisotropie der magnetisierung ferromagnetischer einkristalle, *Z. Phys.* **70**, 395–408 (1931).

34. D. Sander, R. Skomski, C. Schmidthals, A. Enders, and J. Kirschner, Film stress and domain wall pinning in sesquilayer iron films on W(110), *Phys. Rev. Lett.* **77**, 2566–2569 (1996).

35. A. Aharoni, *Introduction to the Theory of Ferromagnetism*, University Press, Oxford (1996).

36. R. Skomski, H.-P. Oepen, and J. Kirschner, Micromagnetics of ultrathin films with perpendicular magnetic anisotropy, *Phys. Rev.* B **58**, 3223–3227 (1998).

37. X.-H. Wei, R. Skomski, Z.-G. Sun, and D. J. Sellmyer, Proteresis in Co:CoO core-shell nanoclusters, *J. Appl. Phys.* **103**, 07D514-1–3 (2008).

38. R. Skomski, P. Manchanda, P. Kumar, B. Balamurugan, A. Kashyap, and D. J. Sellmyer, Predicting the future of permanent-magnet materials (invited), *IEEE Trans. Magn.* **49**, 3215–3220 (2013).

39. T. Burkert, L. Nordström, O. Eriksson, and O. Heinonen, Giant magnetic anisotropy in tetragonal FeCo alloys, *Phys. Rev. Lett.* **93**, 027203-1–4 (2004).

40. G. Andersson, T. Burkert, P. Warnicke, M. Björck, B. Sanyal, C. Chacon, C. Zlotea, L. Nordström, P. Nordblad, and O. Eriksson, Perpendicular magnetocrystalline anisotropy in tetragonally distorted Fe-Co alloys, *Phys. Rev. Lett.* **96**, 037205-1–4 (2006).

41. J. F. Herbst, $R_2Fe_{14}B$ materials: Intrinsic properties and technological aspects, *Rev. Mod. Phys.* **63**, 819–898 (1991).

42. R. Skomski, A. Kashyap, J. Zhou, and D. J. Sellmyer, Anisotropic exchange, *J. Appl. Phys.* **97** 10B302-1–3 (2005).

43. J. Richter and R. Skomski, Antiferromagnets with random anisotropy, *Phys. Stat. Sol (b)* **153**, 711–719 (1989).

44. R. Skomski and J. M. D. Coey, Giant energy product in nanostructured two-phase magnets, *Phys. Rev. B* **48**, 15812–15816 (1993).

45. T. Schrefl and J. Fidler, Finite element modeling of nanocomposite magnets, *IEEE Trans. Magn.* **35**, 3223–3228 (1999).

46. J. Fischbacher, S. Bance, M. Gusenbauer, A. Kovacs, H. Oezelt, F. Reichel, and T. Schrefl, Micromagnetics for the coercivity of nanocomposite permanent magnets, In: *Proc. REPM 2014*, Annapolis, Maryland, pp. 241–243.

47. L. Landau and E. Lifshitz, On the theory of the dispersion of magnetic permeability in ferromagnetic bodies, *Phys. Z. Sowjetunion* **8**, 153–169 (1935).

48. W. F. Brown, *Micromagnetics*, Wiley, New York (1963).

49. H. Zeng, R. Skomski, L. Menon, Y. Liu, S. Bandyopadhyay, and D. J. Sellmyer, Structure and magnetic properties of ferromagnetic nanowires in self-assembled arrays, *Phys. Rev. B* **65**, 134426-1–8 (2002).

50. A. Aharoni, Theoretical search for domain nucleation, *Rev. Mod. Phys.* **34**, 227–238 (1962).

51. B. Balamurugan, P. Mukherjee, R. Skomski, P. Manchanda, B. Das, and D. J. Sellmyer, Magnetic nanostructuring and overcoming Brown's paradox to realize extraordinary high-temperature energy products, *Sci. Rep.* **4**, 6265-1–6 (2014).

52. H. Kronmüller, Theory of nucleation fields in inhomogeneous ferromagnets, *Phys. Stat. Sol. (b)* **144**, 385–396 (1987).

53. T. H. R. Skyrme, A non-linear theory of strong interactions, *Proc. Roy. Soc.* (London) A **247**, 260–278 (1958).

54. Sh. Seki and M. Mochizuki, *Skyrmions in Magnetic Materials*, Springer, Cham (2016).

55. M. Bode, M. Heide, K. von Bergmann, P. Ferriani, S. Heinze, G. Bihlmayer, A. Kubetzka, O. Pietzsch, S. Blügel, R. Wiesendanger, Chiral magnetic order at surfaces driven by inversion asymmetry, *Nature* **447**, 190–193 (2007).

56. R. Skomski, J. Honolka, S. Bornemann, H. Ebert, and A. Enders, Dzyaloshinskii–moriya micromagnetics of magnetic surface alloys, *J. Appl. Phys.* **105**, 07D533-1–3 (2009).

57. W. Jiang, P. Upadhyaya, W. Zhang, G. Yu, M. B. Jungfleisch, F. Y. Fradin, J. E. Pearson, Y. Tserkovnyak, K. L. Wang, O. Heinonen, S. G. E. te Velthuis, A. Hoffmann, Blowing magnetic skyrmion bubbles, *Science* **349**, 283–286 (2015).

58. Bh. Das, B. Balasubramanian, R. Skomski, P. Mukherjee, Sh. R. Valloppilly, G. C. Hadjipanayis, and D. J. Sellmyer, Effect of size confinement on skyrmionic properties of MnSi nanomagnets, *Nanoscale* **10**, doi:10.1039/C7NR08864G (2018).

59. F. Bloch, Zur theorie des ferromagnetismus, *Z. Phys.* **61**, 206–219 (1930).

60. C. Kittel, *Introduction to Solid-State Physics*, Wiley, New York (1986).

61. R. Skomski, M. Chipara, and D. J. Sellmyer, Spin-wave modes in magnetic nanowires, *J. Appl. Phys.* **93**, 7604–7606 (2003).

62. J. Shen, R. Skomski, M. Klaua, H. Jenniches, S. S. Manoharan, and J. Kirschner, Magnetism in one dimension: Fe on Cu(111), *Phys. Rev. B* **56**, 2340–2343 (1997).

63. K. G. Wilson, The renormalization group and critical phenomena, *Rev. Mod. Phys.* **55**, 583–600 (1983).

64. J. M. Yeomans, *Statistical Mechanics of Phase Transitions*, University Press, Oxford 1992.

65. R. Skomski and D. J. Sellmyer, Curie temperature of multiphase nanostructures, *J. Appl. Phys.* **87**, 4756–4758 (2000).

66. R. Skomski, B. Balamurugan, P. Manchanda, M. Chipara, and D. J. Sellmyer, Size dependence of nanoparticle magnetization, *IEEE Trans. Magn.* **53** (1), 1–7 (2017).

67. M. C. Desjonquères and D. Spanjaard, *Concepts in Surface Physics*, Springer, Berlin (1993).

68. A. P. Sutton, *Electronic Structure of Materials*, Oxford University Press, (1993).

69. B. Balamurugan, P. Manchanda, R. Skomski, P. Mukherjee, Bh. Das, T. A. George, G. C. Hadjipanayis, and D. J. Sellmyer, Unusual spin correlations in a nanomagnet, *Appl. Phys. Lett.* **106**, 242401-1–5 (2015).

70. R. Becker and W. Döring, *Ferromagnetismus,* Springer, Berlin (1939).

71. E. Kneller, *Ferromagnetismus*, Springer, Berlin (1962).

72. N. F. Mott and H. Jones, *The Theory of the Properties of Metals and Alloys*, University Press, Oxford (1936).

73. J. H. Mooij, Electrical conduction in concentrated disordered transition-metal alloys, *Phys. Stat. Sol. A* **17**, 521–530 (1973).

74. R. V. Coleman and A. Isin, Magnetoresistance in iron single crystals, *J. Appl. Phys.* **37**, 1028–1029 (1966).

75. M. N. Baibich, J. M. Broto, A. Fert, F. Nguyen Van Dau, F. Petroff, P. Eitenne, G. Creuzet, A. Friederich, and J. Chazelas, Giant magnetoresistance of (001)Fe/(001)Cr magnetic superlattices, *Phys. Rev. Lett.* **61**, 2472–2475 (1988).

76. J. S. Moodera and G. Mathon, Spin polarized tunneling in ferromagnetic junctions. *J. Magn. Magn. Mater.* **200**, 248–273 (1999).

77. P. B. Allen, Electron transport. In: *Conceptual Foundations of Materials: A Standard Model for Ground- and Excited-State Properties*, S. G. Louie and M. L. Cohen (eds.), Elsevier, Amsterdam (2006), Ch. 6, pp. 165–218.

78. N. Nagaosa, J. Sinova, Sh. Onoda, A. H. MacDonald, and N. P. Ong, Anomalous hall effect, *Rev. Mod. Phys.* **82**, 1539–1592 (2010).

79. M. V. Berry, Quantal phase factors accompanying adiabatic changes, *Proc. R. Soc. Lond. A* **392**, 45–57 (1984).

80. D. Xiao, M.-Ch. Chang, and Q. Niu, Berry phase effects on electronic properties, *Rev. Mod. Phys.* **82**, 1959–2007 (2010).

81. R. Skomski and D. J. Sellmyer, Nonadiabatic Berry phase in nanocrystalline magnets, *AIP Adv.* **7**, 055802-1–4 (2017).

Magnetic Disorder at the Nanoscale

Nader Yaacoub
Le Mans University

Rodaina Sayed Hassan
Lebanese University

19.1 Introduction

Nanoscience became, in the last decades, one of the most important research areas in modern science. The general physical properties and concepts related to nano-sized objects (between 1 and 100 nm) differ greatly from the corresponding bulk materials. Among nanostructured materials, magnetic nanostructures, particularly magnetic nanoparticles, show many interesting phenomena (Baibich et al. 1998, Dormann et al. 1997a, 1997b, Morup et al. 2011, Lu et al. 2007, Nogués et al. 2005).

Magnetic nanoparticles are unique complex physical objects and their physical properties are strongly affected by the size and morphology effects. These original physical properties arise from size and confinement effects and related to the reduced number of coordination and in general to the surface effects, which became more pronounced when the size decreases. The breaking of the symmetry, the broken bonds, and the reduction of the nearest-neighbor coordination at the surface induce a big modification of the structural and physical properties. Among the relevant features of the size effect, the occurrence of non-collinear spin structures (spin canting, magnetic frustration, spin disorder) deserves a great attention. Indeed, the non-collinear spin structures could strongly modify the magnetic properties of the magnetic nanostructures (Cannas et al. 2006, Coey 1971, De Toro et al. 2017, Peddis et al. 2011). In this domain, nanoparticles with spinel structure have attracted great attention from fundamental and applications point of view because their rich crystal chemistry, and they are good model to study the correlation between cationic distribution and spin structure and consequently with magnetic behavior. We have to note that the study of the effect of spin disorder in an assembly of magnetic nanoparticles is not an easy task because of the collective behaviors (exchange interaction, dipolar interaction, etc.) (Franceschin et al. 2018, Gaudisson et al. 2016, Peddis 2014). In this context,

hollow nanoparticles appear to be very interesting from a fundamental point of view because surface effects are more pronounced and could be a good system to study the complex effect of surface anisotropy and spin disorder (Cabot et al. 2007, 2009, Khurshid et al. 2012, Sayed et al. 2016, 2018).

The aim of the present chapter is to better insight into the correlation between the magnetic structure (spin canting, spin disorder) and the magnetic properties in the rich system of spinel ferrite full nanoparticles and in the system of hollow nanoparticles (HNPs) with very high surface-to-volume ratio (R).

In the subsequent sections, a short overview on the magnetism of magnetic nanostructures followed by short description of the non-collinear spin structure and the experimental investigation of the spin structure will be reported hereafter. Then, some general features of the effect of spin disorder and the cationic distribution on the magnetic properties of spinel nano-ferrite are given. The last section is concerned by the investigation of the spin disorder in the HNPs characterized by a very big contribution of the surface, highlighting the complex relationship between the spin disorder and the macroscopic magnetic properties. All examples and results presented in this chapter have been the subject of our own research works developed in collaboration with different groups of research.

19.2 General Concepts in Magnetism of Magnetic Nanostructures

A magnetic material is characterized by the magnetic domains structure, and their magnetic properties are determined by the cooperative contribution of these magnetic domains. The main characteristic of the magnetic material is the hysteresis loop, characterized by saturation magnetization, remanence, saturation field, and the coercive field. By decreasing the size and at the nanometer scale the magnetic materials are characterized by new effects that influence its structural and magnetic properties leading to new physical properties and applications (Binns 2014, Dormann et al. 1997, Morup et al. 2011, Néel 1949, Suber and Peddis 2010). When the size reaches a critical value, the energy cost for the formation of domain walls becomes energetically unfavorable, and the formation of a single-domain configuration becomes energetically favorable. The magnetic nanoparticles become single domains and the critical diameter can be calculated by the expression:

$$D_{\mathrm{MD}} = 4\gamma/\mu_0 N_{\mathrm{MD}} M_s^2 \qquad (19.1)$$

where M_S is the magnetization at saturation, N_{MD} is the demagnetizing factor for a single domain, and γ represents the wall energy ($\gamma = 4/\sqrt{AK}$) with A and K are the exchange and magnetic effective anisotropy energies, respectively. In a single domain, all atomic moments are aligned in the same direction, they like a compressive magnetic moment of 10^3–10^5 μ_B, *superspins*, and characterized by a specific time of the magnetization reversal (Bedanta and Kleemann 2009).

The energy of a magnetic material is thus the resultant of different contributions: the exchange energy, the anisotropy energy, the dipolar energy, Zeeman energy, etc. This resulting energy depends on different parameters and varies with the temperature, the applied field, the size, and the shape.

19.2.1 Exchange Interactions

Broadly speaking, in solid state, the direct exchange arises from a direct overlap of electronic wave functions of the neighboring atoms and the Pauli Exclusion Principle. The exchange energy could be described by using the Heisenberg Hamiltonian:

$$H_{\mathrm{ex}} = -2\sum_{i>j} J_{ij}\widehat{S}_i.\widehat{S}_j \qquad (19.2)$$

where J_{ij} is the exchange integral describing the coupling between two spins of magnetic moments represented by the spin operators $\widehat{S}_i$ and $\widehat{S}_j$, respectively. We can note that, in ionic solid-like oxide, the indirect exchange interaction between non-neighboring magnetic ions, which is mediated by a non-magnetic ion, allows to superexchange and double exchange (Blundell 2001) interactions.

In the magnetic nanostructures, when the particles are in close contact, the same Heisenberg Hamiltonian for direct exchange can be used to describe the interactions between the surface atomic spins, which are responsible of superferromagnetic behavior (SFM) (Bedanta and Kleemann 2009, Peddis et al. 2014). In the case of insulating matrix included magnetic particles, an indirect exchange coupling like superexchange, mediating by atoms or ions, could be observed (Blundell 2001, Coey 2010). But, if the matrix is metallic, a RKKY (Ruderman, Kittel, Kasuya, and

Yosida) interaction between the inner de or f shells can occur through conduction electrons (Kasuya 1956, Ruderman and Kittel 1954, Yosida 1957). RKKY interaction is long range and has an oscillatory, between ferromagnetic or antiferromagnetic, dependence on the distance between the magnetic moments.

In assembly of magnetic nanoparticles, the dipolar magnetic interaction between two magnetic nanoparticles of moments μ_i and μ_j, which are separated by a distance r_{ij} is written as:

$$E_{\mathrm{dip}} = \frac{\mu_0}{4\pi r_{ij}^3}\left[\boldsymbol{\mu}_i.\boldsymbol{\mu}_j - \frac{3}{r_{ij}^2}\left(\boldsymbol{\mu}_i.\boldsymbol{r}_{ij}\right)\left(\boldsymbol{\mu}_j.\boldsymbol{r}_{ij}\right)\right] \qquad (19.3)$$

This energy is negligible in the case of atoms (bulk material), but in the case of magnetic particles (superspins having moments in the order of 10^3–$10^5 \mu_B$) becomes large enough (up to tens of K) to produce a collective behavior, leads to different magnetic behavior, like spin glass or superspin glass, etc. (De Toro et al. 2013, Morup et al. 2010) and influence the magnetic properties of the magnetic nanostructures systems.

The competition between interparticle interactions and the magnetic anisotropy, and depending on the nature and strength of the interactions, can influence deeply the magnetic properties and show different magnetic behavior of an assembly of magnetic superspins: superferromagnetism (SFM, ferromagnetic-like), superspin glass (SSG, paramagnetic-like), or superparamagnetism (SPM). The magnetism of an assembly of magnetic nanoparticles is called often supermagnetism (Bedanta and Kleemann 2009, Bedanta et al. 2014, Morup et al. 2010, Vestal et al. 2004).

19.2.2 Magnetic Anisotropy

The dependence of the internal energy of a superspin on the direction of spontaneous magnetization is known as magnetic anisotropy. Magnetic anisotropy strongly affects the shape of the hysteresis loops and controls the coercivity and remanence. This dependence creates easy and hard directions. The origin of the energetic difference between the easy and hard axes comes from different contributions and depend on the size, shape, and morphology of the magnetic particles. In the following, we will briefly discuss about these different magnetic anisotropies (Bedanta and Kleemann 2009, Blundell 2001, Coey 2010).

Spin–orbit interaction is responsible for the magnetocrystalline anisotropy. This anisotropy depends on the crystalline symmetry of the material. The simplest forms of the magnetocrystalline anisotropies are the uniaxial hexagonal or cubic anisotropies, in case of cubic crystal, for example, the energy is given by:

$$E_{\mathrm{a}}^{\mathrm{uni}} = K_1 V\sin^2\theta + K_2 V\sin^4\theta + \dots \qquad (19.4)$$

where V is the particle volume, and θ is the angle between the magnetization and the symmetry axis. By convention, $K_1 > 0$ implies an easy axis direction. It is usually noted that the anisotropy constant $K_1 >> K_2$, and thus for single-domain particles (superspin) with uniaxial anisotropy, the magnetocrystalline energy becomes,

$$E_a^{uni} = K_{eff}V\sin^2\theta \qquad (19.5)$$

with K_{eff} now the effective uniaxial anisotropy constant.

When we reduce the size of the magnetic material, surface can contribute to the magnetocrystalline anisotropy. The magnetic anisotropy, caused by the breaking of the symmetry and reduction of the nearest-neighbor coordination at the surface of the nanostructures, is called surface anisotropy and given by:

$$E_s = K_s\sin^2\theta, \qquad (19.6)$$

where θ is the angle between the magnetization direction and normal to the surface. In small nanoparticles, when the surface-to-volume ratio increases, the magnetic contribution from the surface will be more important than those from the bulk and surface anisotropy enhances the total anisotropy. For the small spherical particles the effective magnetic anisotropy is given by:

$$K_{eff} = K_v + \frac{S}{V}K_S \qquad (19.7)$$

where K_V and K_S are the volume and surface constant anisotropy, respectively.

Another source of magnetic anisotropy results from the shape of the superspin. The shape effect of the dipolar interaction can be described via an anisotropic demagnetizing field, H_d, given by $H_d = -NM$, where M is the magnetization vector and N is the shape-dependent demagnetizing tensor. In case of non-spherical sample, it will be easier to magnetize the sample along a long axis than along a short direction. As an example, the shape anisotropy energy of a uniform magnetized ellipsoid is given by:

$$E_a^{shape} = \frac{1}{2}\mu_0 V \left(N_x M_x^2 + N_y M_y^2 + N_z M_z^2\right) \quad (19.8)$$

where the N_x, N_y, and N_z represent the demagnetization factors.

A magnetostriction is a property of ferromagnetic materials that causes them to change their shape or dimensions during the process of magnetization. This kind of anisotropy is often described by a magnetostatic energy term since it is a consequence of magnetoelastic coupling.

$$E_a^{strain} = -\frac{3}{2}\lambda_s\sigma S\cos^2\theta' \qquad (19.9)$$

where λ is the saturation magnetostriction, σ the strain value by surface unit, S the particle surface, and θ' the angle between magnetization and the main principal axis.

The magnetic interactions induced at the interface between the spins of different magnetic phases, exchange bias coupling, can provide an extra source of anisotropy, leading to magnetization stability and designing new composite materials. Since the discovery of this effect by

Meiklejohn, and beam in Co/CoO core–shell nanoparticles in 1956 (Meiklejohn and Bean 1956), the exchange bias (EB) coupling has been extensively described in the different reviews published on the subject because the interest of this mechanism for spintronics (spin valve) (Berkowitz and Takano 1999, Kiwi 2001, Nogués and Schuller 1999, Nogués et al. 2005). This phenomenon occurs when the system is cooled under field at a temperature T, such as $T_N < T < T_C$, where T_N and T_C are the Néel and Curie temperatures, respectively. The net effect will be a shift in the hysteresis loop along the magnetic field axis (exchange bias field H_E) or/and a significant increase in the coercive field H_c. H_E is calculated as $H_E = -(H_{c-right} + H_{c-left})/2$. In other words, the spins in the coupled FM have only one stable configuration (i.e. unidirectional anisotropy). Note that, in systems of nanoparticles of different architectures (core/shell, hollow, etc.), different parameters are involved, such as surface and interface effects (surface anisotropy, spin disorder, etc.). Chemical inhomogeneity, dipolar interactions, and collective behavior make the study of the EB more complex in these kinds of systems.

19.2.3 Superparamagnetism

As mentioned before, the magnetic energy in the case of single-domain nanoparticles with uniaxial anisotropy is given by:

$$E_a = K_{eff}V\sin^2\theta \qquad (19.10)$$

According to Néel theory, we observe two minima energy corresponding to two antiparallel easy axis orientation. The two minima are separated by the energy barrier ($K_{eff}V$). If $K_{eff}V \geq k_BT$, the magnetic moment is stuck in an easy magnetization direction. However, in the case corresponding to $KV \leq k_BT$, the magnetic moment can freely rotate and fluctuate from easy direction to another, and Langevin theory of paramagnetism could describe the magnetization process. This thermally activated phenomenon corresponding to paramagnetic-like behavior is called the superparamagnetic relaxation. The relaxation time between magnetization reversals can be described by the Néel–Brown relation (Brown 1963, Néel 1949) as:

$$\tau = \tau_0 \exp\left(\frac{K_{eff}V}{k_BT}\right) \qquad (19.11)$$

with τ_0 is of the order $10^{-13} \sim 10^{-9}$ s (Dekkers 1997) and K_B is the Boltzmann constant. The critical size for a superparamagnetic particle in given conditions is defined by considering that the magnetization switches once a second, which corresponds to $E_a = KV \sim 25 \, k_BT$. Thus:

$$d_s \sim \left(\frac{25k_BT}{K}\right)^{\frac{1}{3}} \qquad (19.12)$$

The observation of superparamagnetic relaxation phenomenon depends on the characteristic measurement time τ_m. The particle appears static when $\tau >> \tau_m$. For a given experimental technique and at fixed measuring time, the observed behavior varies with temperature. We define the blocking temperature T_B for a particle of volume V as the temperature at which $\tau = \tau_m$. At this specific temperature the system appears as blocked. In the absence of field or presence of a weak one ($\frac{KT}{KV} \gg 1$), T_B can be defined as:

$$T_B = \frac{K_{eff} V}{k_B \ln(\tau_m/\tau)} \qquad (19.13)$$

According to Wohlfarth and Stoner model (SW) (Stoner and Wohlfarth 1948) the free energy in presence of an external magnetic field is given by.

$$E = KV (\sin\theta)^2 - M_s H V \cos\theta \qquad (19.14)$$

where θ is the angle between M_S and H. For a sufficiently low magnetic field, as $h = \frac{M_s H}{2K} < 1$, energy has two potential wells, which give rise to the superparamagnetic relaxation by crossing the energy barrier. For $h \geq 1$, there is only single well of energy in the field direction and therefore no relaxation.

19.3 Non-Collinear Magnetic Structure

The no simultaneously satisfaction of different magnetic interactions allows to a certain degree of spin disorder, in this case the magnetic material exhibits a non-collinear spin structure. This is a very large subject, and this behavior may have a different origin. Due to geometrical constraints the system can exhibit a geometrical frustration like triangular or Kagome lattices (Martinez et al. 1994, Nishimoto et al. 2016). In this system the nearest-neighbor interactions are naturally frustrated and demonstrate unusual behavior and could have degenerate ground states (Ramirez et al. 1999). The mismatch between the crystal symmetry and the desired bonding in water, for example, allows to natural frustrated system (Bramwell et al. 2009). Magnetic moments frustration could result also from the competition between exchange and dipole–dipole interactions in natural spin ice (Ramirez et al. 1999). In this general framework, and in order to better understand the nature and the ground state in disordered system, frustrated artificially structures have been studied, like artificial square (Wang et al. 2006) and artificial Kagome spin ice (Qi et al. 2008, Ladak et al. 2010, Morgan et al. 2011, Moeller and Moessner 2009, Rougemaille et al. 2011). Disorder associated with random interactions (random anisotropy, random exchange interaction, etc.) and the competition between them might lead to destruction of long-range order and important change in the magnetic properties. The speromagnetism is an example of the random anisotropy. This behavior was established in finely divided amorphous ferric gel (Coey and Readman 1973a, b, Hurd 1982) for the first time by studying their Mössbauer spectra. The spins are frozen in essentially random orientations because topological frustration, which leads to frustration of the

individual superexchange, bonds in amorphous oxide. The speromagnetic material exhibit a zero magnetization due to an isotropic moment distribution. We can cite also the asperomagnetic and sperimagnetic behavior, observed in amorphous systems. In case of A-B binary amorphous alloy, because the competition between the ferromagnetic and antiferromagnetic interactions, a sperimagnetic configuration arises (Coey and Readman 1973a, 1973b Coey 1993). When the exchange distribution is broad, an intermediate behavior is observed, the asperomagnetism, which is characterized by a non-zero magnetization with a moment distribution pointing in a preferential direction. In general, the spin system is frustrated when one cannot find a configuration of spins to fully satisfy the interactions between every pair of spins.

At nanoscale, finite size and surface effects dominate the magnetic properties of nanostructures and influence strongly the magnetic spin structure (Morup et al. 2013). The symmetry breaking at the surface induces changes in exchange integrals, related to the variation of superexchange angles and/or distances among moments, in addition to the presence of surface strain, changes in the coordination, vacancies, and lattice disorder giving rise to topological magnetic frustration. Consequently a complex non-collinear spin structure (spin canting, randomly oriented uncompensated spins, spin-like glass behavior) at the particle surface and inside the particle may occur (Kodama 1999).

19.3.1 Spin Disorder Effect in Magnetic Nanoparticles with Spinel Ferrites

General Description of Spinel Ferrites Structure

Magnetic nanostructures with spinel ferrite structure ($MeFe_2O_4$, $Me = Fe_{2+}$, Co_{2+}, Ni_{2+}, Mg_{2+}, Mn_{2+}, Zn_{2+}, etc.) have attracted a lot of attention. The variety of their chemical compositions allows to control their physical properties (magnetic, electronic, transport, etc.). This spinel structure has appeared very promising in the nanometric state for applications in biomedicine, catalysis, etc. (Kodama and Berkowitz 1999, Machala et al. 2007, Mathew and Juang 2007, Suber and Peddis 2010, Tartaj et al. 2003).

Spinel ferrite has the general chemical formula of the spinel structure $A^{2+}B_2^{3+}O_4$ where A and B are cations of a transition material. The associated space group is $Fd\bar{3}m$. Oxygen anions O^{2-}, which crystallize in an anion centered cubic stack, defining tetrahedral (A) and octahedral [B] sites into which the metal cations will be inserted. The unit cell consists of 8 formulas, we have 32 oxygen atoms generating 64 A sites usually the eighth is occupied, and 32 B sites, half of which is occupied. If the divalent cations occupy the site A and the trivalent cations occupy the site B, this leads to a direct spinel structure $(A^{2+})[B^{3+}]_2O_4$. In the case where the divalent cations occupy the site A and the trivalent cations are distributed equally between the sites A and B one speaks of an inverse spinel structure $(B^{3+})[A^{2+}B^{3+}]_2O_4$. Generally the cationic distribution in the tetrahedral and octahedral sites is defined by the degree of inversion γ, which is defined as the fraction of divalent ions in the octahedral sites.

Spin Disorder in Spinel Ferrite Structure

The spinel structure ferrites are ferrimagnetic materials, described by Néel theory, below the Curie temperature Tc. The exchange interactions between A sites (J_{AA}: exchange integral), between sites B (J_{BB}) and between sites A and B (J_{AB}) determine the magnetic order in these structures. The exchange mechanisms are mainly the superexchange and the double exchange via oxygen. The antiferromagnetic interactions between the (A) and [B] site ions dominate the J_{AA} and J_{BB} interactions and imposes an uncompensated antiferromagnetic order (ferrimagnetic). It should also be noted that the magnetic anisotropy is related to the cationic distribution. It is clear that the magnetic properties of the spinel structure can be explained by the magnetic couplings. These couplings are related to the chemical compositions and the crystallographic structure of the materials. The physical properties and the spins structure could be controlled by the choice of the chemical nature of the metal, and the control of the cationic distribution allows us to obtain different magnetic behaviors of the spinel ferrites (Liu et al. 2000a, 2000b, Sun et al. 2004). At the nanoscale, the magnetic properties of nanoparticles of spinel structure ferrites are due to a complex competition between different effects, among which the cationic distribution and the spin "canting", which play a fundamental role. In the case of non-interacting nanoparticles, the spin canting is likely attributed to the cationic inversion due to chemical disorder, which modifies superexchange interactions or to the surface effect. Both these effects give rise to magnetic topological frustration, and consequently a spin canting occurs (Kodama et al. 1997a, 1997b). The collective behavior, in assembly of interacting nanoparticles, contributes to the non-collinear magnetic structure. The spin canting may be explained, in ferrimagnetic structures, in terms of the Yafet–Kittel triangular arrangement due to the magnetic frustration resulting from the competition between *A-B* and *B-B* exchange interactions (Anhoj et al. 2003, Kodama et al. 1996, 1997, Morup 2003). In addition to these effects, the local anisotropy, surface anisotropy, and surface state can play a role in the spin disorder. All these effects should be considered to fully understand intrinsic and extrinsic physical properties of spinel ferrite magnetic nanostructures.

Experimental Investigation of Magnetic Spin Structure

Several experimental techniques were used to investigate spin disorder in magnetic nanostructures as neutron diffraction (Peddis et al. 2011, Rodriguez-Carvajal et al. 1993), polarized small-angle neutron scattering and nuclear forward scattering techniques (Herlitschke et al. 2016,

Vivas et al. 2017) and the XMCD (X-ray magnetic circular dichroism) (Brice et al. 2005, De Groot et al. 1990, Stöhr 1999). Using these techniques, we study the properties of the local magnetic moment, the spatial magnetization distribution, and the orientation of the hyperfine field experienced by the iron nuclei. The correlation between these properties and the macroscopic ones leads to a detailed description of the complex spin structure in the magnetic nanostructures. To have a quantitative description of the spin structure, Mössbauer spectrometry plays a determined role (De Toro et al. 2017, Muscas et al. 2018, Yoshida and Langouche 2012). ^{57}Fe Mössbauer spectrometry (MS) is an excellent tool to provide local information (local electron density, effective magnetic moment, etc.) of Fe-containing nanostructured materials, particularly their magnetic spin structure, through the hyperfine parameters (the isomer shift (δ), the quadrupole splitting or the quadrupole shift (2ε), and the hyperfine field (B_{hyp}). These parameters can be useful in determining ligand bonding states, electron shielding, oxidation state, electrons density, magnetic structure, magnetic moments (Muscas et al. 2018, Greenwood and Gibb 1971, Yoshida and Langouche 2012). In the case of magnetic materials, the six lines are correlated to the Fe moment configuration respect to the γ-ray direction. Indeed, the relative area ratios are given by 3:p:1:1:p:3, where $p = 4(\sin\theta)^2 / \left(2 - (\sin\theta)^2\right)$ where θ is the angle between the hyperfine field and the propagation direction of the γ-ray. Analyzing this parameter and in the case of applying an external magnetic field allows us to access to different magnetic order (paramagnetism or superparamagnetism, ferromagnetism, antiferromagnetism) and to the spin canting (non-collinear magnetic structure) or spin disorder (random orientation of spins). In addition, this methodology allows us to study the spin structure and to access to particular cases of non-collinear spin structure such as asperomagnetic (ASP), speromagnetic (SP), or sperimagnetic (SPi) structure (Coey 1971, Chappert 1974, Chappert et al. 1979, Dormann 1997, Greneche 1995). Details on the technique, hyperfine interactions, and applications are given, for example, in Muscas et al. (2018), Greenwood and Gibb (1971), Yoshida and Langouche (2012).

Influence of Spin Disorder on Magnetic Properties of Magnetic Nanoparticles with Spinel Structure: $CoFe_2O_4$ and $Co_xNi_{1-x}Fe_2O_4$

As we mentioned before, magnetic spin structure strongly affect magnetic properties of spinel ferrite nanoparticles. In this section, a short overview of some results obtained on full cobalt ferrite $CoFe_2O_4$ nanoparticles and on $Co_xNi_{1-x}Fe_2O_4$ ($0 \leq x \leq 1$) will be provided. Using these two systems, an example of the effect of magnetic structure on macroscopic magnetic properties will be given, highlighting the complex relationship between cationic distribution, spin-canting, and macroscopic magnetic properties.

$CoFe_2O_4$

The $CoFe_2O_4$ nanoparticles have been obtained by sol–gel self-combustion technique (Cannas et al. 2004, 2006b, Peddis et al. 2011). The investigation by neutron powder diffraction (NPD) at 300 K indicates only the presence of the $CoFe_2O_4$ phase crystallized in the $Fd\bar{3}m$ space group with mean particle size around 6 nm, and the cation distribution is given by $(Fe_{0.73}Co_{0.27})[Fe_{0.63}Co_{0.37}]_2O_4$ with an inversion degree of $\gamma = 0.74$. The magnetic moments obtained at the A and B sublattices are lower with respect to the values reported in literature for $CoFe_2O_4$ nanoparticles with similar particle size and cationic distribution, inducing to believe that some non-collinear spin components coexist with the ordered spins. Despite the small particle size, the value of saturation magnetization at 300 K ($M_s \cong 70$ A m^2/Kg) and at 5K ($M_s \cong 100$ A m^2/Kg) are rather close to the bulk values (at 5 K 80–93 A m^2/kg) (Cannas et al. 2006).

As shown in Figure 19.1 the spectrum obtained at 10 K under a magnetic field of 8 T applied parallel to the γ-beam is unambiguously consistent with a ferrimagnetic structure; the refinement allows to attribute clearly the two sextets to the Fe^{3+} in tetrahedral – and octahedral sites, according to the values of the isomer shift. Then, accurate value of the atomic Fe_A^{3+}/Fe_B^{3+} population ratio has been obtained (Table 19.1). The cation distribution for stoichiometric $CoFe_2O_4$ can be given as $(Fe_{0.76}Co_{0.24})[Fe_{0.62}Co_{0.38}]_2O_4$, with $\gamma_{Moss} = 0.76$, which is in good agreement with that estimated from NPD analysis ($\gamma_{NPD} = 0.74$), within the experimental error of the two techniques. The presence of second and fifth lines evidences a canted structure for Fe^{3+} magnetic moments with respect to the applied field (non-collinear magnetic structure).

TABLE 19.1   In-Field Mossbauer Fitted Parameters of the $CoFe_2O_4$: Isomer Shift (δ), Quadruple Shift (2ε), Canting Angle (θ), and the Relative Fraction of the Fe^{3+} Located in A and B Sites $\left(Fe_{A,B}^{3+}/Fe_{total}^{3+}\right)$ (Peddis et al. 2011)

	Mössbauer Hyperfine Parameters of the Co-Ferrite			
	$<\delta>$ (mm/s) ± 0.01	$<2\varepsilon>$ (mm/s) ± 0.01	$<\theta>$ (°) ± 10	$Fe_{A,B}^{3+}/Fe_{total}^{3+}$ ± 0.01
Fe_A^{3+}	0.36	−0.04	41	0.38
Fe_B^{3+}	0.47	−0.03	36	0.62

Each subspectra is characterized by broadening lines. The broadening of Mössbauer lines, due to angle and hyperfine field distributions, suggests unambiguously different atomic neighboring of Fe sites. As illustrated in Figure 19.2, the corresponding hyperfine field distribution, obtained from the magnetic effective field distributions ($P(B_{eff})$) correlated to canting angle distribution ($P(\theta)$), is ranged from 55.5 and 49.5 T, in agreement with Fe with 6 A-Fe up to 6 A-Co first neighbors (Peddis et al. 2011, Sawatzky et al. 1969). The correlation between B_{hf} and canting angle allows to relate $P(B_{hf})$ and $P(\theta)$ to the cation distribution (inversion degree, Fe^{3+} ions in B (A)-sites with nearest-neighbors Co^{2+} A (B)-site ions) and to surface effects. The mean values of hyperfine parameters are given in Table 19.1. The system is consistent with a saturated Co-rich core and a canted Fe-rich shell, suggesting non-chemical homogeneous Co-ferrite nanoparticles originating from a cationic gradient.

In many studies on ferrimagnetic nanoparticles the reduction of the saturation magnetization with the particle size is attributed to spin canting that yields magnetic disorder at the particle surface (Martínez et al. 1998). However, the freezing of surface spins (Aquino et al. 2005) allows to increasing the saturation magnetization at low temperature. Generally, the effect on magnetic disorder

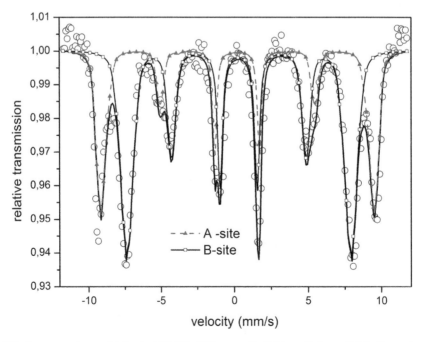

FIGURE 19.1 ^{57}Fe Mössbauer spectrum obtained at 10 K in 8 T magnetic field applied parallel to the γ- beam (Peddis et al. 2011).

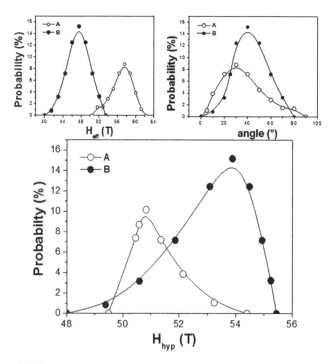

FIGURE 19.2 Left part, the distribution of the effective magnetic field $P(H_{eff})$ correlated with the distribution of canting angle $P(\theta)$; right part, obtained by fitting procedure. Below part, the distribution of the hyperfine magnetic field $P(H_{hf})$ observed for both A and B sites (Peddis et al. 2011).

on saturation magnetization, can be explained by the interplay between spin canting and inversion degree. In these magnetic nanoparticles of cobalt ferrite, the spin canting in each site A and B contributes to the decrease of the moments (measured by NPD). The high value of saturation magnetization is attributed to the decrease in inversion degree (in the bulk $\gamma_{bulck} = 0.82$). For more details, please see the following article (Peddis et al. 2011).

$Co_xNi_{1-x}Fe_2O_4$

In order to better investigate the effect of non-collinear spin structure on magnetic properties, small crystalline ferrite nanoparticles of the formula $Co_xNi_{1-x}Fe_2O_4$ ($x = 1$ sample 1; $x = 0$ sample 2) with equal size ($\cong 4.5$ nm, size obtained from XRD and TEM Table 19.2) were studied. These samples were obtained by polyol method (Muscas et al. 2015).

As shown in Table 19.2, a significant decrease in the magnetic anisotropy with the decrease in cobalt content

TABLE 19.2 Chemical Formula by ICP Analysis; Mean Diameter $\langle D_{TEM}\rangle$ Evaluated Using TEM Images, Saturation Magnetization (M_S); Coercive Field ($\mu_0 H_C$); Effective Anisotropy Constant (K_{eff}). Uncertainties on the Last Digit Are Given in Parentheses

Sample	Chemical Formula	$\langle D_{TEM}\rangle$ nm	M_s (Am^2/Kg)	$\mu_0 H_c$ (T)	K_{eff} (J/m^{-3})
Sample 1	Co$_{1.0}$Fe$_2$O$_4$	4.5(1)	130(10)	0.88(7)	$10.7(1) \times 10^{-5}$
Sample 2	Ni$_{1.0}$Fe$_2$O$_4$	4.3(1)	37(4)	0.028(1)	$0.3(1) \times 10^{-5}$

was observed at 5 K. For a cubic crystalline structure with magnetic easy axis along [100] direction, we can empirically evaluate the effective anisotropy constant according to: $K_{eff} = \frac{H_K M_S}{2}$. All the estimated values of K_{eff} are higher than those reported for bulk and nanostructured cobalt (~ 1–$4 \cdot 10^5$ J/m^3) (Laureti et al. 2010, Rondinone et al. 2000, Virden et al. 2007) and nickel ferrites ($\sim 1 \cdot 10^4$ J/m^3) (Nathani et al. 2005, Sousa et al. 2004). As listed in Table 19.2, the experimental values of saturation magnetization decrease as expected when we reduce the Co content. The value of magnetic saturation of sample 1 is relatively high compared to that of the bulk values (83–90 A m^2/kg) and of highly crystalline cobalt ferrite nanoparticles (Cruz-franco et al. 2014). But for sample 2 we obtained a relatively low value (bulk around 55 A m^2/kg). These results show that the interplay between inversion degree and magnetic spin disorder should play a key role.

The ^{57}Fe Mössbauer spectra recorded at 10 K under a magnetic field of 8 T on the sample 1 and sample 2 are shown in Figure 19.3. The two sextets correspond to the Fe^{3+}in tetrahedral and octahedral sites according to the isomer shift values. The mean values of hyperfine parameters and spin-canting angle (non-collinear magnetic structure) are presented in Table 19.3. An asymmetrical broadening lines in site B for sample 2 was observed, probably due to a wide range of possible chemical environments of Fe ions.

The inversion degree γ was estimated at 0.74(1) and 0.44(1) for sample 1 and sample 2, respectively. The first value is in agreement with values usually reported for cobalt ferrite ($\gamma \approx 0.7$–0.8) (Artus et al. 2011, Carta et al. 2009), however, the second value is significantly different from that expected for pure nickel ferrite ($\gamma \approx 0.9$–1). Using the theoretical values of magnetic moment and the canting angle, the corresponding theoretical M_S values have been calculated as 90(1) and 89(1) A m^2/kg,

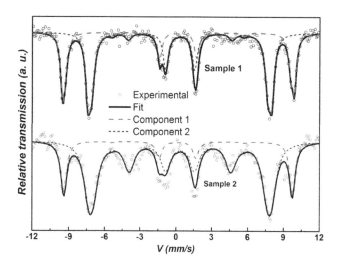

FIGURE 19.3 In-field Mössbauer spectra recorded at 10 K under 8 T, for sample 1 and sample 2. The individual components for Fe in tetrahedral and octahedral sites are reported as dotted and dashed line, respectively.

TABLE 19.3 The Isomer Shift (δ); the Hyperfine Field (B_{hyf}); the Average Canting Angle (ϑ); and the Ratio of Each Component Evaluated from In-Field Mössbauer Fitted Spectra Are Reported for Samples: Sample 1 and Sample 2. Uncertainties on the Last Digit Are Given in Parentheses

Sample	Site	δ (mm/s^{-1})	B_{hyf} (T)	θ ($°$) $\pm$ 10$°$	$Fe^{3+}_{A,B}/Fe^{3+}_{\mathrm{total}}$
Sample 1	Fe^{3+}_A	0.34(1)	51(2)	10	0.37(1)
	Fe^{3+}_B	0.48(1)	54(2)	18	0.63(1)
Sample 2	Fe^{3+}_A	0.35(1)	51(2)	12	0.22(1)
	Fe^{3+}_B	0.48(1)	53(2)	38	0.78(1)

for sample 1 and sample 2, respectively. These values are not comparable with the experimental ones, inducing to hypothesize a non-homogeneous cationic distribution for these two samples. The cationic distribution is obtained as $((Fe_{0.74}\square_{0.26})[Co_{1.00}Fe_{1.26}]O_4$ for sample 1, with vacancies ($\square$) in tetrahedral sites and overpopulation of octahedral sites. Using the effective magnetic moments and the effective inversion degree γ_{sat}, this results such elevated value of $M_S \approx 130$ A m^2/Kg^{-1} could be understanding. The result for sample 2 leads to cationic distribution $(Ni_{1.00}Fe_{0.44})[Fe_{1.56}\square_{0.44}]O_4$, explaining at the same time the low saturation and the unusual iron distribution from Mössbauer spectrometry, which leading to the observed spin canting. For more details, refer to the article Muscas et al. (2015).

The result presented above shows that the missing coordination of surface cations that induce an increase in the local anisotropy, and the cationic distribution leads to non-collinear spin structure. These effects are strongly correlated with the magnetic properties of the magnetic nanoparticles.

19.3.2 Spin Disorder Effect in Hollow Magnetic Nanoparticles

As discussed before, surface effects (spin canting, spin disorder, surface anisotropy, etc.) lead generally to a complex physical behavior. Moreover, the study of such effects remains very delicate because of the collective behavior in assembly of magnetic nanoparticles.

In this context, hollow nanoparticles (HNPs) appear to be very interesting from a fundamental point of view because surface effects are more pronounced, due to the availability of extra surface layers (inner layer) and at the interface between randomly oriented grains in the shell (Cabot et al. 2007, 2009). So, these HNPs present features of spin canting or spin disorder and appear as excellent candidates to study both the surface effects and the complex mechanism of surface anisotropy. In the case of hollow nanostructures, we defined the surface-to-volume ratio as: $R = \frac{S}{V_{\mathrm{shell}}} = \frac{S_{O-\mathrm{shell}} + S_{I-\mathrm{shell}}}{V_{\mathrm{total}} - V_{\mathrm{inner}}}$, where $S_{O-\mathrm{shell}}$ and $S_{I-\mathrm{shell}}$ correspond to the outer and inner surface, respectively. In this section, we present the investigation of the correlation between magnetic structure and magnetic properties in a system of HNPs with very high R value ≈ 1.5 and with very thin shell (~ 1.4 nm)

Correlation Between Spin Disorder and Magnetic Properties in Ultra Thin Iron Oxide Hollow Nanoparticles

The synthesis of the HNPs was carried out using the chemical protocol presented in Khurshid et al. (2011, 2012, 2013), Sayed et al. (2018) and based on the Kirkandall effect. TEM investigations clearly indicate the formation of hollow nanostructures (Figure 19.4). A detailed statistical analysis allows to measure the outer diameter (D_O, 9.4(1) nm) and the inner diameter (D_I, 6.6(2) nm), the mean shell thickness around 1.4 nm, and $R \cong 1.5$. This value is quite larger compared to previous studies (Khurshid et al. 2011) highlighting a more important role of the surface in our system.

Magnetization measurements versus temperature were carried out by means of ZFC (zero field cooled)/FC (field cooled) protocols under an applied field of 20 Oe (Figure 19.5). ZFC/FC curves give evidence for irreversibility above 45 K suggesting a superparamagnetic behavior of the magnetic entities. M_{FC} decreases with decreasing temperature, suggesting the presence of very weak interparticle interactions (Bean and Livingston 1959, Goya et al. 2003). The interesting feature in this measurement is that the ZFC curve presents two maxima: a broad one at 32 ± 2K($T_{\max\_1}$) (corresponding to blocking temperature), and a narrow one at lower temperature,

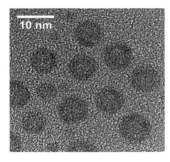

FIGURE 19.4 TEM images of hollow iron oxide nanoparticles (Sayed et al. 2018).

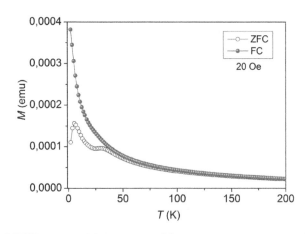

FIGURE 19.5 Magnetization M versus temperature in mode ZFC (empty symbols) and FC (full symbols) recorded at 20 Oe.

$7 \pm 0.5\text{K}$ ($T_{\text{max\_2}}$) (corresponding to freezing temperature). Thermal dependence of AC and DC magnetization reveals the "canonical" blocking of superspins below 35 K, and the freezing of a more anisotropic phase below 15 K.

ZFC/FC loops measured at 2K show an evident large horizontal shift, indicating the presence of large exchange biased coupling field (Figure 19.6). Moreover, the hysteresis loops at low temperatures exhibit also an important vertical shift, which is attributed to some spins freezing in the direction of the FC. At 20 K, there is no exchange bias coupling as ZFC and FC loops are somehow superimposed (Figure 19.6).

To better understand the macroscopic magnetic behavior and to get relevant information about the local cationic environment and magnetic structure, ^{57}Fe Mössbauer spectra were recorded at 300 K and 77 K without external field and at 11 K under intense magnetic field of 8 T.

The spectra at 300 K and 77 K result from quadrupolar doublets, not reported here, with broadened and non-Lorentzian lines of similar intensities (Sayed et al. 2018). At a first observation, this symmetrical shape of the distribution seems to indicate that our sample acquires a behavior typical of an amorphous structure. But, a more detailed analysis based on the ratio $q = <\Delta^2>/<\Delta>^2$ (1.22 at 300 K and 1.28 at 77 K, parameter quantifying the topological disorder) shows that it is somehow far from the value obtained in case of amorphous structure (Lopez-Herrera et al. 1983) and allows us to conclude that the HNPs sample can be described neither amorphous nor crystalline.

^{57}Fe Mössbauer spectrometry at 11 K and under high external magnetic field of 8 T has been carried out. As shown in Figure 19.7 the spectrum is complex. Different fitting procedures have been considered to model the in-field Mössbauer spectrum and are illustrated in Figure 19.7 (Sayed et al. 2018). Two models give rise to excellent description of the experimental spectrum and rather comparable mean hyperfine data (Table 19.4). The first method consists of two independent distributions of the hyperfine parameters, and the second method consists of two independent Gaussian distributions of Lorentzian lines sextets, the quadrupolar shift, and canting angle β values being commonly. The refined values of isomer shift are rather

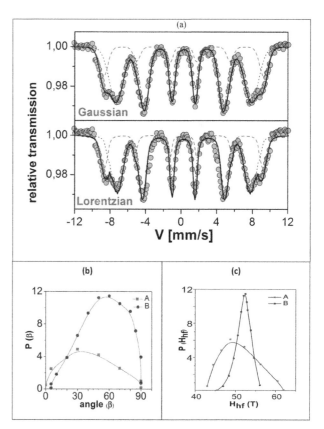

FIGURE 19.7 (a) ^{57}Fe Mössbauer spectra measured at 11 K under a magnetic field of 8 T fitted using a Gaussian distribution (upper spectrum) and fitted using a Lorentzian distribution (lower spectrum); black line stands for the total fit, dashed one for tetrahedral component, and dotted one for octahedral component. (b) and (c) Distribution of hyperfine field, $P(H_{\text{hf}})$, and angular distribution, $P(\beta)$, obtained by the fitting procedure using Lorentzian lines, respectively; A and B represent tetrahedral and octahedral sites, respectively (Sayed et al. 2018).

equal and suggest Fe^{3+} species located in tetrahedral and octahedral sites. Contrary to the collinear ferrimagnetic structures expected in spinel iron oxides, the fit methods provide evidence of a distribution of β angles, consistently with a non-collinear magnetic structure. Indeed, in this kind of hollow structure, the morphology enhances the surface

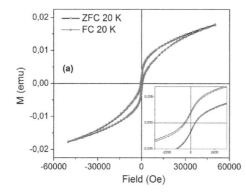

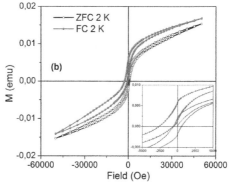

FIGURE 19.6 Hysteresis loops recorded after ZFC (empty symbols) and FC (full symbol) for cooling field of 1 T at 2 K (a) and 20 K (b).

TABLE 19.4 Summary of Obtained Values of Hyperfine Parameters (Isomer Shift, Quadrupolar Shift, Effective and Hyperfine Fields (H_{eff} and H_{hf}), Angle β, and Weight) Obtained at 11 K Under External Field of 8 T for Gaussian and Lorentzian Distributions Compared Also to Results Obtained in Sayed et al. (2018).

		$<\delta>$ mm/s ($\pm$0.01)	2ε mm/s ($\pm$0.01)	H_{eff}T ($\pm$0.5)	H_{hf}T ($\pm$0.5)	β ($\pm$10°)	% ($\pm$5%)
Gaussian 11K	First component	0.47	0.02	55.7	50.1	42	22
	Second component	0.5	−0.01	46.9	51.7	57	78
Lorentzian 11K	First component	0.45	0	56.2	50	37	24
	Second component	0.47	0	46.5	52.2	58	76
Moshex	A	0.43	−0.05	−	−48.2	−	18
	B	0.48	0.02	−	50.6	−	82

contribution and consequently the spin disorder. The last fitting procedure consists in the superimposition of two different ideal magnetic subnetworks, i.e. two speromagnetic models. In this model, the two types of Fe moments are frozen in perfectly random orientations but opposite, according to the positive and negative refined values of hyperfine fields, as reported in Table 19.4. This model allows a rather good fit to be achieved, as illustrated in Figure 19.8.

Finally, the three different and independent fitting approaches allow us to conclude that such hollow systems are characterized by high magnetic disorder, and the Fe magnetic lattice does result from two opposite non-collinear structures close to speromagnets, corresponding to octahedral and tetrahedral networks, antiferromagnetically coupled. It is worth to underline that these novel experimental features can be essentially ascribed to the high R value (i.e. very thin shell) observed in the sample under investigation with respect to what reported in literature (Belaïd et al. 2013, Cabot et al. 2001, Khurshid et al. 2015). These results allow us to understand the presence of exchange bias coupling in this system. Indeed, on decreasing the thickness, there could exist an inner layer with ferrimagnetic-like structure, which is still distinguishable from the outer layers. These results induce to believe the presence of a trilayer-like structure where some magnetic moments of the outer and inner surface layers interact with the ferrimagnetic ultrathin central layer: the local structural distortions give rise to some short superexchange paths, with antiferromagnetic interactions, originating the exchange bias coupling with the inner ferrimagnetic layer.

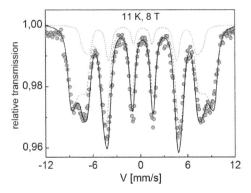

FIGURE 19.8 Mössbauer spectra obtained at 11 K under a magnetic field of 8 T fitted using "Moshex". Black line stands for the total fit, dashed line for tetrahedral component, and dotted line for octahedral component (Sayed et al. 2018).

Size and Thickness Effect on Spin Structure in Hollow Iron Nanoparticles: Monte Carlo Simulation

Monte Carlo simulations were carried out on these system of hollow structure. The aim is to understand the spin structure established from Mössbauer experiments. In this simulation, we take into account the surface magnetic anisotropy (K_S), for an individual hollow maghemite nanoparticle. In our model, as previously described in Sayed et al. (2016), the magnetic ions are represented by classical Heisenberg spins interacting according to the following Hamiltonian:

$$
H = \sum_{i=1}^{N} \left\{ -\frac{1}{2}\sum_{j\in V} J_{ij}\vec{S}_i\vec{S}_j - \mu_0\vec{\mu}_i\overrightarrow{H_{\mathrm{ext}}} \right.
$$
$$
\left. + K_s\left[1 - \left(\vec{n}_i\frac{\vec{S}_i}{S_i}\right)^2\right] + K_v\left[1 - \left(\vec{U}_i\frac{\vec{S}_i}{S_i}\right)^2\right] \right\}
$$

The first term is the nearest-neighbor (nn) exchange interaction, the second is the Zeeman energy (H_{ext} is the magnetic field and μ the atomic magnetic moment) while the third and fourth ones are surface and bulk anisotropies, respectively.

After the creation of maghemite unit cell and introducing the vacancies with the help of a certain algorithm respecting nearest-neighbor distances, we tune the inner radius considered in order to get a hollow particle with approximate external diameter of 8 nm and thickness of 2 nm respecting charge neutrality as much as possible. From literature, we took the values of the coupling constants between tetrahedral and octahedral sites as $J_{\mathrm{TO}} = -16.95$ K, between octahedral and octahedral site as $J_{\mathrm{OO}} = +3.65$ K and between tetrahedral and tetrahedral $J_{\mathrm{TT}} = -0.65$ K (Kodama and Berkowitz 1999, Restrepo et al. 2006). Starting from a random spin configuration at high temperature, the energy is minimized by simulated annealing using the Metropolis algorithm with a decreasing exponential law for temperature ($T_n = T_0 e^{\gamma n}$ with $\gamma = \log\alpha$ and $\alpha = 0.96$). The simulations were started at $T = 1,500$ K, well above T_{C}. The final temperature is normally small ~0.01K. The number of Monte Carlo Steps (MCS) per spin was 12,000 MCS. The calculation conditions have been recently discussed in Sayed et al. (2016).

The different spin structures are illustrated in Figure 19.9. Increasing gradually the value of K_{s}, the structure gets far away from collinearity until a spike structure is attained at $K_{\mathrm{s}} \sim (100$ K ~25 mJ/m^2), which is somehow a small

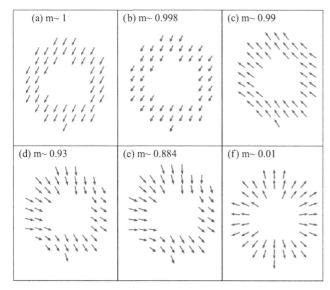

FIGURE 19.9 Different spin structures for different values of K_s: (a) $K_s = 1$, m $\sim$ 1, (b) $K_s = 10$, m $\sim$ 0.998, (c) $K_s = 25$, m $\sim$ 0.99, (d) $K_s = 75$, m $\sim$ 0.93, (e) $K_s = 87.5$, m $\sim$ 0.884, (f) $K_s = 100$, m $\sim$ 0.01; where m is the ratio of total magnetization with respective K_s in each case to the value of magnetization at $K_s = 0$. Note that only spins in the octahedral site are presented.

value compared to that needed in the case of full nanoparticles (Restrepo et al. 2006). Such a structure appears to be compatible with the results obtained from in-field Mössbauer experiments confirming that the experimental

results must be explained by the effects of surface anisotropy originating from the symmetry breaking and the local atomic disorder induced by the peculiar structure of the surface.

From Figure 19.10, which compares the angular distribution probability versus surface anisotropy, one can observe that nearly no distribution occurs for low values of K_s as the structure is almost collinear. When the value of K_s increases, some of the spins at the outer surface start to be affected and deviate from the mean direction of magnetization and this is illustrated by a distribution of 20° as a higher range. This yields a "throttled" structure in which the surface spins are oriented inward for the upper hemisphere, they reverse progressively at the equator and become oriented outward for the lower hemisphere. The peaks of the distribution refer to spin majorities and minorities corresponding to those located at octahedral and tetrahedral sites, respectively. An interesting case is represented for $K_s = 75$ K where four sharp peaks can be well discriminated in the angular distribution. These peaks are enhanced by the high symmetry and can be related to some particular zones of the HNP; these contributions appear to be localized in well-defined planes perpendicular to the mean magnetization (Nehme et al. 2015). However, the largest K_s (100) favors a rather random distribution, giving rise to a "hedgehog" or "spike" type spin structures.

This kind of numerical study was recently performed and extended to other HNPs with different sizes and thicknesses (Sayed et al. 2016). Figure 19.11 shows the values of

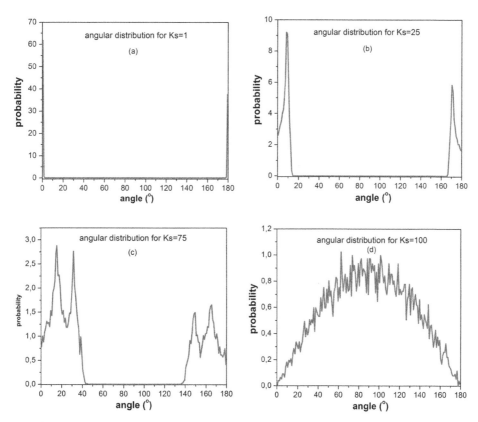

FIGURE 19.10 (a)–(d): Angular probability distribution for different cases of K_s: 1, 25, 75, and 100, respectively (Sayed et al. 2016).

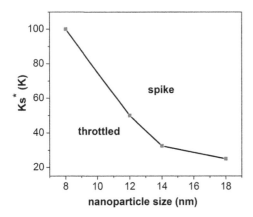

FIGURE 19.11   Variation of K_s^* as a function of the size of hollow nanoparticle with the same thickness (Sayed et al. 2016).

K_s^* function of the nanoparticle size. We observe that the spin structure depends on the size of the hollow nanoparticle assuming the same value of K_s. This dependence is given in the phase diagram that splits clearly the spike and throttled structures. The 18-nm nanoparticle attains a fully non-collinear structure for a value of $K_s = 25.0$, while the collinear spin structure is partially affected for smaller nanoparticles. As the size of the HNP increases, the misalignment of neighboring surface spins decreases, and the exchange energy is smaller.

In order to correlate the competition between the surface anisotropy and the exchange interactions through the evolution of the three different spin configurations as observed: collinear, throttled, and "hedgehog", we have thus considered the hollow nanoparticle of external diameter 12 nm and tuned the shell thickness from about 1 nm up to 3.8 nm. We showed that when the thickness of the shell increases (Figure 19.12), the value of K_s^* increases. The slope at the beginning changes slowly then increases clearly as the thickness becomes large. This large slope could be related to the fact that the number of spins at the inner part of the hollow nanoparticle becomes significantly important in comparison to those spins lying at the inner or outer surfaces. Such a

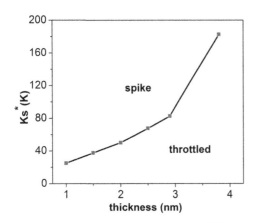

FIGURE 19.12   Variation of the critical value K_s^* as a function of the shell thickness for 12-nm-sized hollow nanoparticles (Sayed et al. 2016).

behavior could be usually attributed to a competition that arises between surface anisotropy and the exchange energy due to the decrease in the misalignment of coupled spins with increasing distance to the center. For further description and analysis of the Monte Carlo simulation results, refer to the recent article Sayed et al. (2016).

19.4   Conclusion

The investigation of the effect of spin disorder on magnetic properties in magnetic nanostructures is not an easy task. We showed through some examples that in the spinel ferrite nanostructures systems the correlation between cationic distribution, spin structure and magnetic properties should be established in order to understand and control their physical properties. As we mentioned before, the hollow architecture is a good candidate to study both the surface effects and the complex mechanism of surface anisotropy and spin disorder. We showed that the large surface area of HNPs originated from inner and outer surface contributions exhibits a large spin disorder described by two opposite speromagnetic structures. Monte Carlo simulation highlighted the role of surface anisotropy and the effect of size and thickness on the spin structure and confirm the experimental observation. The presence of large portion of disordered spins on the outer and inner surfaces of HNPs interacting with a ferrimagnetic ultrathin central layer allows the presence of strong anisotropy energy and the exchange bias effect in these kinds of nanostructures. It is important to mention that in order to explore "the no-man's land between molecular nanomagnets and magnetic nanoparticles" (Gatteschi et al. 2012) and better understand the effect of spin disorder in the physics of magnetic nanostructures, we should study the HNPs with thickness lower than ≈1 nm. But it is clear that such ultrathin objects remain *a priori* challenging to be obtained because the smaller the thickness is, the lower the stability.

References

Aquino, R., Depeyrot, J., Sousa, M. H. et al. 2005. Magnetization temperature dependence and freezing of surface spins in magnetic fluids based on ferrite nanoparticles. *Phys. Rev. B* 72: 184410–184435.

Anhoj, T. A., Bilenberg, B., Thomsen B. et al. 2003. Spin canting and magnetic relaxation phenomena in $Mn_{0.25}Zn_{0.75}Fe_2O_4$. *J. Magn. Magn. Mater.* 260: 115–130.

Artus, M., Tahar, L. B., Herbst, F. et al. 2011. Size dependent magnetic properties of $CoFe_2O_4$ nanoparticles prepared in polyol. *J. Phys. Condens. Matter* 23 (50) 506001.

Baibich, M. N., Broto, J. M., Fert, A. et al. 1988. Giant magnetoresistance of (001) Fe/(001) Cr Magnetic superlattices. *Phys. Rev. Lett.* 61: 2472–2475.

Bean, C. P., Livingston, J. D. 1959. Superparamagnetism. *J. Appl. Phys.* 30: S120–S129.

Bedanta, S., Kleemann, W. 2009. Supermagnetism, *J. Phys. D. Appl. Phys.* 42: 13001.

Bedanta, S., Petracic, O., Kleemann, W., Supermagnetism, in: K.F.H. Buschow (Ed.), *Handbook of Magnetic Materials*, Elsevier, Amsterdam, 2014, pp. 1–83.

Belaïd, S., Laurent, S., Vermeech, M. et al. 2013. New approach to follow the formation of iron oxide nanoparticles synthesized by thermal decomposition. *Nanotechnology* 24: 55705–55711.

Berkowitz, A. E., Takano, K. 1999. Exchange anisotropy: A review. *J. Magn. Magn. Mater.* 200: 552–570.

Binns, C. 2014. *Nanomagnetism: Fundamentals and Applications.* Oxford: Elsevier.

Blundell, S. 2001. *Magnetism in Condensed Matter.* New York: Oxford University Press.

Bramwell, S. T., Gibin, S. R., Aldus, R. et al. 2009. Measurement of the charge and current of magnetic monopoles in spin ice. *Nature* 461, 956–959.

Brice-Profeta, S., Arrio, M.-A., Tronc, E. et al. 2005. XMCD investigation of spin disorder in γ-Fe_2O_3 nanoparticles at the Fe $L_{2,3}$ edges. *Phys. Scr.* T115: 626–628.

Brown, W. F. J. 1963. Thermal fluctuations of single-domain particle. *Phys. Rev.* 130: 1677–1686.

Cabot, A., Puntes, V. F., Shevchenko, E. et al. 2007. Vacancy oalescence during oxidation of iron nanoparticles. *J. Am. Chem. Soc.* 129: 10358–10360.

Cabot, A., Alivisatos, A. P., Puntes, V. F. et al. 2009. Magnetic domains and surface effects in hollow maghemite nanoparticles. *Phys. Rev. B* 79: 94419.

Cannas, C., Musinu, A., Peddis, D., Piccaluga, G. 2004. New synthesis of ferrite–silica nanocomposites by a sol–gel auto-combustion. *J. Nanoparticles Res.* 6: 223–232.

Cannas, C., Falqui, A., Musinu, A. et al. 2006a. CoF_e2O_4 Nanocrystalline powders prepared by citrate-gel ethods: Synthesis, structure and magnetic properties. *J. Nanoparticle Res.* 8: 255–267.

Cannas, C., Musinu, A., Piccaluga, G. et al. 2006b. Magnetic properties of cobalt ferrite-silica nanocomposites prepared by a sol–gel autocombustion technique. *J. Chem. Phys.* 125: 164714.

Carta, D., Casula, M. F., Falqui, A. et al. 2009. A structural and magnetic investigation of the inversion degree in ferrite nanocrystals MFe_2O_4 (M = Mn, Co, Ni). *J. Phys. Chem. C* 113 (20) 8606–8615.

Chappert, J. 1974. High field mössbauer spectroscopy. *J. Phys. Colloq.* 35: C6-71–C6-88.

Chappert, J., Teillet, J.,Varret, F. 1979. Recent developments in high field Mössbauer spectroscopy. *J. Magn. Magn. Mater.* 11: 200–207.

Coey, J. M. D. 1971. Noncollinear spin arrangement in ultrafine ferrimagnetic crystallites. *Phys. Rev. Lett.* 27: 1140.

Coey, J. M. D., Readman, P. W. 1973a. Characterisation and magnetic properties of natural ferric gel. *Earth Planet. Sci. Lett.* 21: 45–51.

Coey, J. M. D., Readman, P. W. 1973b. New spin structure in an amorphous ferric gel. *Nature.* 246: 476–478.

Coey, J. M. D. 1993. Interpretation of the Mossbauer spectra of speromagnetic materials. *J. Phys. Condens. Matter.* 5: 7297–7300.

Coey, J. M. D. 2010. *Magnetism and Magnetic Materials.* New York: Cambridge University Press.

Cruz-franco, B., Gaudisson, T., Ammar, S. et al. 2014. Magnetic properties of nanostructured spinel ferrites. *IEEE Trans. Magn.* 50: 2–7.

Dekkers, M. J. 1997. Environmental magnetism: An introduction. *Geol. en Mijnb.* (*Geology Mining*) 76: 163–182.

De Groot, F. M. F., Fuggle, J. C., Thole, B. T. et al. 1990. The 2p x-ray absorption of 3d transition-metal compounds: An atomic multiplet description including the crystal field. *Phys. Rev. B: Condens. Matter* 42: 5459–68.

De Toro, J. A., Lee, S. S., Salazar, D. et al. 2013. A nanoparticle replica of the spin-glass state. *Appl. Phys. Lett.* 102: 183104.

De Toro, J. A., Vasilakaki, M., Lee, S. S. et al. 2017. Remanence plots as a probe of spin disorder in magnetic nanoparticles. *Chem. Mater.* 29: 8258–8268.

Dormann, J. L., Fiorani, D., Tronc, E. 1997a. Magnetic relaxation in fine particle systems In *Advances in Chemical Physics,* eds. I. Prigogine, S. A. Rice, Vol. 98, Hoboken, NJ: John Wiley & Sons, Inc.

Dormann, J. L., Fiorani, D., Tronc, E. 1997b. Magnetic relaxation in fine-particle systems. *Adv. Chem. Phys.* XCVIII: 283–494.

Franceschin, G., T. Gaudisson, T., N. Menguy, N. et al. 2018. Exchange-biased $Fe_{3-x}O_4$-CoO granular composites of different morphologies prepared by seed-mediated growth in polyol: From core–shell to multicore embedded structures. *Part. Part. Syst. Charact.*, 35: 1800114.

Gaudisson, T., Sayed Hassan, R., Yaacoub, N. et al. 2016. On the exact crystal structure of exchange-biased Fe_3O_4-CoO nano-aggregates produced by seed mediated growth in polyol. *CrystEngComm.* 18: 3799.

Gatteschi, D., Fittipaldi, M., Sangregorio, C. et al. 2012. Exploring the No-Man's land between molecular nanomagnets and magnetic nanoparticles. *Angew. Chemie Int. Ed.* 2012, 51: 4792–4800.

Goya, G. F., Berquo, T. S., Fonseca, F. C., Morales, M. P. 2003. Static and dynamic magnetic properties of spherical magnetite nanoparticles. *J. Appl. Phys.* 94: 3520 (1–8).

Greenwood N. N., Gibb T. C. 1971. *Mossbauer Spectroscopy.* London: Chapman and Hall.

Greneche, J. M. 1995. Noncollinear magnetic structures investigated by high-field Mossbauer spectometry. *Acta Phys. Slovaca.* 45: 45.

Herlitschke, M., Disch, S., Sergueev, I. et al. 2016. Spin disorder in maghemite nanoparticles investigated using polarized neutrons and nuclear resonant scattering. *J. Phys. Conf. Ser.* 711: 012002.

Hurd, C. M. 1982. Varieties of magnetic order in solids. *J. Contemp. Phys.* 23: 469–493.

Kasuya, T. 1956. A theory of metallic ferro- and antiferromagnetism on Zener's model. *Prog. Theor. Phys.* 16: 45–57.

Khurshid, H., Li, W., Tzitzios, V., Hadjipanayis, G. C. 2011. Chemically synthesized hollow nanostructures in iron oxides. *Nanotechnology* 22: 265605.

Khurshid, H., Li, W., Phan, M.-H. et al. 2012. Surface spin disorder and exchange-bias in hollow maghemite nanoparticles. *Appl. Phys. Lett.* 101: 22403.

Khurshid, H., Hadjipanayis, C. G., Chen, H. et al. 2013. Core/shell structured iron/iron-oxide nanoparticles as excellent MRI contrast enhancement agents. *J. Magn. Magn. Mater.* 331: 17–20.

Khurshid, H., Lampen-Kelley, P., Iglesias, Ò. et al. 2015. Spin-glass-like freezing of inner and outer surface layers in hollow γ-Fe$_2$O$_3$ nanoparticles. *Sci. Rep.* 5: 15054.

Kiwi, M. 2001. Exchange bias theory. *J. Magn. Magn. Mater.* 234: 584–595.

Kodama, R. H., Berkowitz, A. E., McNiff, E. et al. 1996. Surface spin disorder in NiFe$_2$O$_4$ nanoparticles. *Phys. Rev. Lett.* 77: 394–397.

Kodama, R. H., Berkowitz, A. E., McNiff, J. E. J. et al. 1997a. Surface spin disorder in ferrite nanoparticles (Invited). *J. Appl. Phys.* 81: 5552–5557.

Kodama, R. H., Makhlouf, S. A., Berkowitz, A. E. 1997b. Finite size effects in antiferromagnetic NiO nanoparticles. *Phys. Rev. Lett.* 79: 1393.

Kodama, R. H. 1999. Magnetic nanoparticles. *J. Magn. Magn. Mater.* 200: 359–372.

Kodama, R. H., Berkowitz, A. E. 1999. Atomic-scale magnetic modeling of oxide nanoparticles. *Phys. Rev. B* 59: 6321–6336.

Ladak, S., Read, D. E., Perkins, G. K. et al. 2010. Direct observation of magnetic monopole defects in an artificial spin-ice system. *Nat. Phys.* 6: 359–363.

Laureti, S., Varvaro, G., Testa, M. et al. 2010. Magnetic interactions in silica coated nanoporous assemblies of CoFe$_2$O$_4$ nanoparticles with cubic magnetic anisotropy. *Nanotechnology* 21: 31 315701.

Liu, C., Rondinone, A. J., Zhang, Z. J. 2000. Synthesis of magnetic spinel ferrite CoFe$_2$O$_4$ nanoparticles from ferric salt and characterization of the size-dependent superparamagnetic properties. *Pure Appl. Chem.* 72: 37–45.

Liu, C., Zou, B., Rondinone, A. J. et al. 2000. Chemical control of superparamagnetic properties of magnesium and cobalt spinel ferrite nanoparticles through atomic level nagnetic couplings. *J. Am. Chem. Soc.* 122: 6263–6267.

Lu, A.-H., Salabas, E. L., Schüth, F. 2007. Magnetic nanoparticles: Synthesis, protection, functionalization, and application. *Angew. Chem. Int. Ed. Engl. 46*: 1222–12244.

Lopez-Herrera, M. E., Greneche, J. M., Varret, F. 1983. Analysis of the mössbauer quadrupole spectra of some amorphous fluorides. *Phys. Rev. B Condens. Matter Mater. Phys.* 28: 4944–4948.

Machala, L., Zboril, R., Gedanken, A. 2007. Amorphous iron (III) oxide: A review. *J. Phys. Chem. B* 111: 4003–4018.

Martinez, B., Alberta, A., Rodriguez-Sola, R. et al. 1994. Magnetic transition in highly frustrated SrCr$_8$Ga$_4$O$_{19}$: the archetypal Kagoné system. *Phys. Rev. B* 50 (21).

Martínez, B., Obradors, X., Balcells, L. et al. 1998. Low temperature surface spin-glass transition in gamma G-Fe$_2$O$_3$ nanoparticles. *Phys. Rev. Lett.* 80: 181–184.

Mathew, D. S., Juang, R. S. 2007. An overview of structure and magnetism of spinel ferrite nanoparticles and their synthesis in microemulsions. *Chem. Eng. J.* 129: 51–65.

Meiklejohn, W. H., Bean, C. P. 1956. New magnetic anisotropy. *Phys. Rev.* 102: 1413–1414.

Moeller, G., Moessner, R. 2009. Magnetic multipole analysis of kagome and artificial spin-ice dipolar arrays. *Phys. Rev. B* 80: 14.

Morgan, J. P., Stein, A., Langridge, S. et al. 2011. Thermal ground-state ordering and elementary excitations in artificial magnetic square ice. *Nat. Phys.* 7: 75–79.

Morup, S. 2003. Spin-canting and transverse relaxation at surfaces and in the interior of ferrimagnetic particles. *J. Magn. Magn. Mater.* 266: 110–118.

Morup, S., Hansen, M. F., Frandsen, C. el al. 2010. Magnetic interactions between nanoparticles. *Nanotechnoly* 1: 182–190.

Morup, S. et al. 2011, Magnetic nanoparticle. In *Comprehensive Nanoscience and Technology*, eds. D. L. Andrews, G. D. Scholes, G. P. Wiederrecht, 437–491, New York: Elsevier.

Morup, S., Brok, E., Frandsen, C. 2013. Spin structures in magnetic nanoparticles. *J. Nanomater.* 2013: 1–8.

Muscas, G. , Yaacoub, N., Concas, G. et al. 2015. Evolution of the magnetic structure with chemical composition in spinel iron oxide nanoparticles. *Nanoscale* 7: 13576.

Muscas, G. et al. 2018. Magnetic disorder in nanostructured materials, Novel magnetic nanostructure. In *Novel Magnetic Nanostructures*. eds. N. Domracheva, M. Caporali, E. Rentshler, 127–163. New York: Elsevier.

Nathani, H., Gubbala, S., Misra, R. D. K. 2005. Magnetic behavior of nanocrystalline nickel ferrite. *Mater. Sci. Eng. B* 121: 126–136.

Nishimoto, S., Katukuri, V.M., Yushankhai, V. et al. 2016. Strongly frustrated triangular spin lattice emerging from triplet dimer formation in honeycomb Li$_2$IrO$_3$. *Nat Commun.* 18 (7) 10273.

Néel L. 1949. Théorie du trainage magnétique des ferromagnétiques en grains fins avec applications aux terres cuites. *Ann. Géophys.* 5: 99–136.

Nehme, Z., Labaye, Y., Sayed Hassan, R. et al. 2015. Modeling of hysteresis loops by monte carlo simulation. *AIP Adv.* 5:127124.

Nogués, J., Schuller, I. K. 1999. Exchange bias. *J. Magn. Magn. Mater.* 192: 203–232.

Nogués, J., Sort, J., Langlais, V. et al. 2005. Exchange bias in nanostructures. *Phys. Rep.* 422: 65–117.

Peddis, D. 2014. Magnetic properties of spinel ferrite nanoparticles: Influence of the magnetic structure. In *Magnetic Nanoparticle Assemblies*, ed. K. N. Trohidou, vol. 7, 978–981, Singapore: Pan Stanford Publishing.

Peddis, D., Yaacoub, N., Ferretti, M. et al. 2011. Cationic distribution and spin canting in $CoFe_2O_4$ nanoparticles. *J. Phys. Condens. Matter* 23: 426004 1–8.

Peddis, D., Jönsson, P., Varvaro, G. et al. 2014. Magnetic interactions: a tool to modify the magnetic properties of materials based on nanoparticles, in: C. Binns (Ed.), *Nanomagnetism Fundamentals and Applications*, Elsevier B.V, Oxford, UK, pp. 129–189.

Qi, Y., Brintlinger, T., Cumings, J. 2008. Direct observation of the ice rule in an artificial kagome spin ice. *Phys. Rev. B* 77: 9.

Ramirez, A. P., Hayashi, A., Cava, R. J. et al. 1999. Zero-point entropy in 'spin ice'. *Nature* 399: 333–335.

Restrepo, J., Labaye, Y., Greneche, J. M. 2006. Surface anisotropy in maghemite nanoparticles. *Phys. B Condens. Matter* 384: 221–223.

Rodríguez-Carvajal, J. 1993. Recent advances in magnetic structure determination by neutron powder diffraction. *Physica B: Condens. Matter* 192: 55–69.

Rondinone, A. J., Samia, A. C. S., Zhang, Z. J. 2000. Characterizing the magnetic nnisotropy constant of spinel cobalt ferrite nanoparticles. *Appl. Phys. Lett.* 76: 3624–3626.

Rougemaille, N., Montaigne, F., Canals, B. et al. 2011. Artificial kagome arrays of nanomagnets: A frozen dipolar spin ice. *Phys. Rev. Lett.* 106: 5.

Ruderman, M. A., Kittel, C. 1954. Indirect exchange coupling of nuclear magnetic moments by conduction electrons, *Phys. Rev.* 96: 99–102.

Sawatzky, G. A., Van Der Woude, F., Morrish, A. H. 1969. Mossbauer study of several ferrimagnetic spinels. *Phys. Rev.* 187: 747–757.

Sayed, F., Yaacoub, N., Labaye, Y. et al. 2018. Surface effects in ultrathin iron oxide hollow nanoparticles: Exploring magnetic disorder at the nanoscale. *J. Phys. Chem. C* 122: 7516.

Sayed, F., Labaye, Y., Sayed Hassan, R. et al. 2016. Size and thickness effect on magnetic structures of maghemite hollow magnetic nanoparticles. *J. Nanopart. Res.* 18: 279.

Sousa, E. C., Sousa, M. H., Goya, G. F. et al. 2004. Enhanced surface anisotropy evidenced by Mössbauer spectroscopy in nickel ferrite nanoparticles. *J. Magn. Magn. Mater.* 272–276: e1215–e1217.

Stöhr, J. 1999. Exploring the microscopic origin of magnetic anisotropies with X-ray magnetic circular dichroism (XMCD) spectroscopy. *J. Mag. Mag. Mat.* 200: 470–497.

Stoner, E.C., Wohlfarth, E. P. 1948. A mechanism of magnetic hysteresis in heterogeneous alloys. *Philos. Trans. R. Soc. Lond. A Math. Phys. Eng. Sci.* 240 (826) 599–642.

Suber, L., Peddis, D., 2010. Approaches to synthesis and characterization of spherical and anisometric metal oxide magnetic nanomaterials. In *Nanomaterials for Life Science*, eds. C. S. S. R. Kumar, vol. 4, 431475, Weinheim: Wiley.

Sun, S., Zeng, H., Robinson, D. B. et al. 2004. MFe_2O_4 (M=Fe, Co, Mn) nanoparticles. *J. Am. Chem. Soc.* 126: 273–279.

Tartaj, M., Del Puerto Morales, S., Veintemillas-Verdaguer, T. et al. 2003. *J. Phys. D: Appl. Phys.* 36: R182.

Vivas, L. G., Yanes, R., Michels, A. 2017. Small-angle neutron scattering modeling of spin disorder in nanoparticles. *Sci. Rep.* 7: 13060.

Vestal, C., Song, Q., Zhang, Z. 2004. Effects of inter-particle interactions upon the magnetic properties of $CoFe_2O_4$ and $MnFe_2O_4$ nanocrystals. *J. Phys. Chem.* 108: 18222–18227.

Virden, A.,Wells, S., O'Grady, K. 2007. Physical and magnetic properties of highly anisotropic cobalt ferrite particles. *J. Magn. Magn. Mater.* 316: e768–e771.

Wang, R. F., Nisoli, C., Freitas, R. S. et al. 2006. Artificial 'spin ice' in a geometrically frustrated lattice of nanoscale ferromagnetic islands. *Nature* 439, 303–306.

Yoshida Y. and Langouche G., 2012. *Mössbauer Spectroscopy*. Springer, London.

Yosida, K. 1957, Magnetic properties of Cu-Mn alloys. *Phys. Rev.* 106: 893–898.

The Study of Hexagonal Fe$_2$Si: In Terms of Its Structure and Electronic Properties

Chi Pui Tang and
Kuan Vai Tam
Macau University of Science and Technology

In this section, we use first principle calculations to show that a hexagonal structure of Fe$_2$Si is a ferromagnetic crystal. We looked at the phonon spectra which indicate that it is a stable structure. Such material exhibits a spin-polarized and half-metal-like band structure. From the calculations of generalized gradient approximation, metallic and semiconducting behaviors are observed with a direct and nearly 0 eV band gap in various spin channels. The densities of states in the vicinity of the Fermi level are mainly contributed from the d-electrons of Fe. We also calculated the reflection spectrum of Fe$_2$Si, which has minima at 275 nm and 3,300 nm with the reflectance of 0.27 and 0.49, respectively. Such results may provide a reference for the search of hexagonal Fe$_2$Si in experiments. With this band characteristic, the material may be applied in the field of novel spintronics devices.

Among the Fe-Si compounds, the β-FeSi$_2$ is a well-known compound as a Kankyo (environmentally friendly) semiconductor with a 0.8 eV indirect band gap [1].

Applying the density functional theory (DFT) [2,3], we use pseudopotential method of CASTEP (the Cambridge sequential total energy package) [4] with the ultrasoft pseudo potential method [5] and all-electron full potential electronic structure code of FHI aims (Fritz Haber Institute ab initio molecular simulations) [6] to calculate the crystal structure and the band structure. We adopt two approximations in both methods: the local density approximation (LDA) with the CA-PZ form [7] and the generalized gradient approximation (GGA) with the PBE form [8]. Because when the kinetic energy cutoff exceeds 400 eV, the change in total energy is <1 meV/atom, the cutoff energy is set to 450 eV; $12 \times 12 \times 12$ k points are taken for calculations in the Brillouin zone (BZ) and use Monkhorst–Pack method of the point distribution [9]. In all electronic calculations, the basis set and numerical grids are set to high quality as defined by the "tight" option [6]. In addition, the structural optimization is performed by using the Broyden–Fletcher–Goldfarb–Shanno (BFGS) method [10], and the convergence tolerances are set to 5.0×10^{-6} eV/atom for energy, 0.01 eV/Å for maximum force, 0.02 GPa for maximum stress, and 10^{-4} Å for maximum displacement. The phonon spectrum and the optical properties are calculated with the CASTEP code, and for phonons the finite displacement method with the cutoff radius 5.0 Å is used. The structure of hexagonal Fe$_2$Si is shown in Figure 20.1. It is a hexagonal crystal system and belongs to the P-3M1 space group. There are 4 Fe atoms and 2 Si atoms to form a primitive cell. It is composed with an uneven honeycomb sublattice of components Fe and Si, and a hexagonal sublattice of component Fe, similar to the Ni$_2$In structure. The optimized lattice parameters obtained with different methods are shown in Table 20.1. The results of experiment and calculations indicate that the symmetry constraints of the lattice parameters are $\alpha = \beta = 90°$; $\gamma = 120°$; and $a = b$. From the results, deviations of the optimized values of lattice parameters from those of experiment are <3%–6% for a and b, <5%–9% for c, and less than 3% for c/a. The results of experiment and calculations indicate that the symmetry constraints of the lattice parameters are $\alpha = \beta = 90°$; $\gamma = 120°$; and $a = b$. From the results, deviations of the optimized values of lattice parameters from those of experiment are <3%–6% for a and b, <5%–9% for c, and <3% for c/a. These deviations are probably caused by the higher temperature in experiment.

The magnetic properties of Fe$_2$Si are similar to the materials of Ni$_2$In structure. It exhibits ferromagnetism, like MnRhAs and MnCoAs, which are calculated by Korringa–Kohn–Rostoker (KKR) method and the KKR2 CPA (coherent potential approximation) with magnetic moments 3.31 and 3.25 per cell, respectively [11]. By means of Mulliken population analysis, we found that the magnetic moments of Fe atoms in the upper and lower layers of the hexagonal sublattices are 2.78 μB and 2.50 μB obtained from the GGA and LDA methods, respectively. For the Fe atoms in the middle layer of the hexagonal sublattices, the moments are 0.04 μB from GGA and 0.12 μB from LDA. For the Fe atoms in the honeycomb sublattices, the moments are 0.66 μB from GGA and 0.58 μB from LDA. The moments

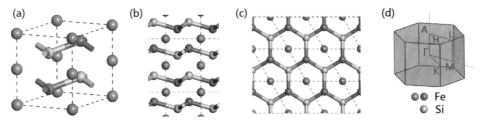

FIGURE 20.1 The crystalline structure of hexagonal Fe_2Si with its primitive cell surrounded by dashed line. (a) The hexagonal Fe_2Si belongs to P3-m1 space group. The two groups indicate a different position of Fe atoms. (b,c) Patterns viewed along the [110] and [001] directions, respectively. (d) The Brillouin zone and the lines with high symmetry.

TABLE 20.1 The Cell Parameters and Cell Moments of Primitive Cell in Hexagonal Fe_2Si Structure

	Experiment	GGA	GGA-FHI	LDA	LDA-FHI
$a = b$	4.05	3.90	3.93	3.81	3.85
C	5.09	4.80	4.82	4.69	4.72
c/a	1.26	1.23	1.23	1.23	1.23
V	72.30	63.23	64.47	58.96	60.59
Cell Moments	–	4.020	4.040	3.689	3.710

of Si atoms are 0.06 μB from GGA and 0.04 μB from LDA. This indicates that the main contribution of the magnetic moments of Fe_2Si comes from the Fe atoms in the upper and lower layers of the hexagonal sublattices. The magnetic moments of a primitive cell with the hexagonal Fe_2Si structure are shown in Table 20.1. Such a distribution of the magnetic moments is similar to that of the ferromagnetic crystal with Ni_2In structure [11]. It is also worth noting that some of Ni_2In structures exhibit magnetic-field-induced martensite transformation effects, and martensite phase transformation to orthogonal structure occurs in decreased temperature [12,13]. In order to further discuss the stability of Fe_2Si, the phonon dispersion relation is calculated and the stability of the structure is examined by the absence of imaginary-frequency phonon modes. The phonon dispersion relation for Fe_2Si is shown in Figure 20.2. It indicates that the hexagonal structure of Fe_2Si is stable.

We use the pseudopotential method together with GGA to calculate the band structure and the partial densities of states (PDOS) of hexagonal Fe_2Si. The results are displayed in Figure 20.3. One of the notable characteristics of hexagonal Fe_2Si is that the band diagram has a half-metal-like

structure. Compared with the band structures of the non-polarized β-$FeSi_2$ calculated with the augmented spherical wave method [14] and the non-polarized FeSi calculated with the linear augmented plane-wave method [15], the increase in the proportion of iron atoms in hexagonal Fe_2Si causes the spin polarization. From Figure 20.3, we can observe a zero direct band gap in spin-down band at the L point of BZ (the band gap of β-$FeSi_2$ and FeSi are 0.44 eV and 0.11 eV, respectively [14,15]). It is a half gapless semiconductor. When the band structure was calculated with LDA, we can also observe a −0.3 eV negative band gap in spin-down band at the L point. In addition, the results from the all-electronic calculation together with the method of +U (1 eV) are similar to those from the pseudopotential method. The band gaps from different methods are shown in Table 20.2.

These results indicate that this material is a semimetal with the negative gap or gapless band in spin-down subband. Because with different methods the top of the valence band (VBT) and to the bottom of the conduction band (CBB) have a very small overlap or have a little band gap near the Fermi energy, this property may need further experimental validation. On the other hand, the spin-up subband

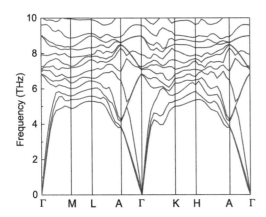

FIGURE 20.2 The phonon dispersion relation of Fe_2Si.

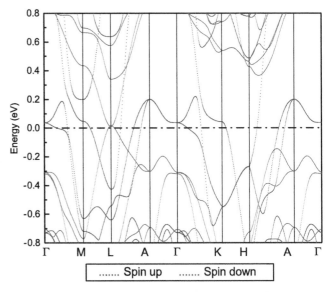

FIGURE 20.3 Electronic band structure of hexagonal Fe_2Si calculated by using GGA functional and pseudopotential method.

TABLE 20.2 The Valence Band Top (VBT), the Conduction Band Bottom (CBB) and the Band Gap of Hexagonal Fe$_2$Si Obtained from Different Calculation Methods

Methods	VBT (eV)	CBB (eV)	Band gap (eV)
GGA	0.01	0.02	−0.01
GGA-FHI	0.02	−0.06	0.08
GGA+U (1 eV)	0.38	−0.03	−0.41
LDA	−0.38	−0.10	−0.28
LDA-FHI	−0.37	−0.13	−0.24
LDA+U (1eV)	0.20	−0.08	0.28

manifests metallic behavior. As the spin-down subband shows the gapless behavior, the hexagonal Fe$_2$Si should belong to a spin gapless half-metal. Since the excitation energy in quantum dots made of a negative band gap or gapless semiconductor can be regulated with the methods of changing the size or doping [16], and some materials with a negative band gap exhibit a charge-density wave and excitonic insulator phase [17], such a spin gapless half-metal has potential application for controlling electron spins in spintronics.

The results of PDOS from different calculation methods show the similar behavior. Figure 20.4 shows the contributions to the DOS from different spins and from different atoms calculated from GGA together with the pseudopotential method. Near the Fermi energy the DOS are significantly contributed from the d-electrons of Fe atoms and the contributions from s or p states are ineffective. It also illustrates that the half-metal-like properties of the negative band gap comes from the d-electrons of Fe atoms. At the same time, the electrons from Si atoms play a main role in the covalent bonds. Corresponding to the results of the band structure near the Fermi energy that shows a small overlap of the spin-down subbands as shown in Figure 20.3, the total DOS of spin-down subbands have a nonzero minimum, which is slightly deviated from the Fermi energy. In our previous work, the PDOS of iron-containing compound – 2D pt-Fe2S also exhibit similar characteristics [18].

Similarly to the hapkeite, hexagonal Fe$_2$Si has not been found in the Earth. However, hapkeite could be synthesized under the space weathering in airless environment and on the effect of meteorite impact. Since the hapkeite was found in lunar meteorites (Dhofar 280), it also contains Fe and Si elements as FeSi and FeSi$_2$ compounds [19], the difference between hexagonal Fe$_2$Si structure and the hapkeite as well as the possibility of mutual transformation are worth further study.

Based on conventional technologies, the main methods for the study of mineral compositions on the lunar surface include reflectance spectroscopy, thermal radiation, etc. Therefore, we calculate the reflectivity of plane polarized light with polarizations in the 100, 010, and 001 directions from the surface of hexagonal Fe$_2$Si. Because of the small anisotropy of the lattice structure and the electronic bands, the results for different polarization directions are similar. The reflectance spectroscopies are shown in Figure 20.5. With the GGA calculation, the reflectivity of hexagonal Fe$_2$Si has a relative minimum at 4.51 eV (275 nm) with reflectance of 0.27 in the range of near infrared to ultraviolet (2000 to 124 nm), it also has a relative minimum at 0.37 eV ($\approx$3,300 nm) with reflectance of 0.49 and a relative maximum at 0.60 between these two minima. The results from the LDA calculation are similar. This is different from the results of β-FeSi$_2$ and FeSi, for which there are significant differences between the LDA and GGA calculations. In β-FeSi$_2$, there are a relative minimum 0.45 at 3 eV and a relative maximum 0.54 at 1 eV [20,21]. In FeSi, there are two relative minimum 0.25 and 0.12 at 4 eV and 10 eV, respectively, and a relative maximum 0.35 at 7 eV between them [22]. The differences in the reflection spectrum may be used for the search of hexagonal Fe$_2$Si on the lunar surface or other meteorites.

In conclusion, we use first principle calculation to study the structure, lattice parameter, phonon spectrum, and reflectance spectroscopy of hexagonal Fe$_2$Si. From the results of the band structure and the PDOS, it is a ferromagnetic crystal and has a spin-polarized half-metal-like band structure and a near 0 eV band gap, which is mainly contributed from the spin-down d-electrons of Fe atoms. Because of these characteristics, it may be used in spintronics. Finally, the reflection spectra of hexagonal Fe$_2$Si are calculated and compared with other Fe-Si compounds. The distinct reflection spectra provide a reference for the search of the compound in the future, because it is possible that the hexagonal Fe$_2$Si will symbiont with other Fe-Si compounds under space weathering or meteorite impacts. Whether the hexagonal Fe$_2$Si exists in the space environment or coexists with other Fe-Si compounds is worth further study.

The work is supported by the Science and Technology Development Fund of Macau (020/2014/A1,039/2013/A2), and the State Key Programs for Basic Research of China

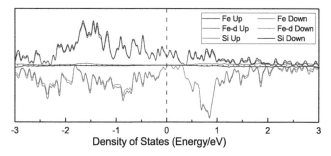

FIGURE 20.4 Partial densities of states (PDOS) of hexagonal Fe$_2$Si calculated by using GGA pseudopotential method.

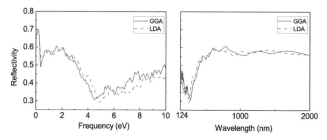

FIGURE 20.5 The reflectance spectroscopies of hexagonal Fe$_2$Si are calculated by the method of polycrystal.

(Grant No. 2011CB922102), and by National Natural Science Foundation of China of Grant Nos. 61076094 and 10874071.

References

1. NE Christensen. Electronic structure of β-FeSi$_2$. *Physical Review B*, 42(11):7148, 1990.

2. P Hohenberg and W Kohn. Inhomogeneous electron gas. *Phyical Review*, 136:B864–B871, 1964.

3. M Levy. Universal spin-orbitals and solution of the v-representability problem. *Proceedings of the National Academy of Sciences*, 76(12):6062–6065, 1979.

4. MD Segall, PJD Lindan, MJ Probert, CJ Pickard, PJ Hasnip, SJ Clark, and MC Payne. First-principles simulation: Ideas, illustrations and the CASTEP code. *Journal of Physics: Condensed Matter*, 14(11):2717, 2002.

5. D Vanderbilt. Soft self-consistent pseudopotentials in a generalized eigenvalue formalism. *Physical Review B*, 41(11):7892, 1990.

6. V Blum, R Gehrke, F Hanke, Pa Havu, V Havu, X Ren, K Reuter, and M Scheffler. Ab initio molecular simulations with numeric atom-centered orbitals. *Computer Physics Communications*, 180(11):2175–2196, 2009.

7. DM Ceperley and BJ Alder. Ground state of the electron gas by a stochastic method. *Physical Review Letters*, 45(7):566, 1980.

8. JP Perdew, KBurke, and M Ernzerhof. Generalized gradient approximation made simple. *Physical Review Letters*, 77(18):3865, 1996.

9. HJ Monkhorst and JD Pack. Special points for Brillouin-zone integrations. *Physical Review B*, 13(12):5188, 1976.

10. BG Pfrommer, M Cote, SG Louie, and ML Cohen. Relaxation of crystals with the quasi-Newton method. *Journal of Computer Physics*, 131:233–240, 1997.

11. R Zach, J Toboa, W Chajec, D Fruchart, and F Ono. Magnetic Properties of MMX (M = Mn, M = 3d or 4d Metal, X = P, As, Si, Ge) Compounds with hexagonal or orthorhombic crystal structure. In *Solid State Phenomena*, vol. 194, pp. 98–103. Trans Tech Publ, 2013.

12. V Johnson. Diffusionless orthorhombic to hexagonal transitions in ternary silicides and germanides. *Inorganic Chemistry*, 14(5):1117–1120, 1975.

13. W Jeitschko. A high-temperature X-ray study of the displacive phase transition in MnCoGe. *Acta Crystallographica Section B: Structural Crystallography and Crystal Chemistry*, 31(4):1187–1190, 1975.

14. R Eppenga. Ab initio band-structure calculation of the semiconductor β-FeSi$_2$. *Journal of Applied Physics*, 68(6):3027–3029, 1990.

15. LF Mattheiss and DR Hamann. Band structure and semiconducting properties of FeSi. *Physical Review B*, 47(20):13114, 1993.

16. N Malkova and GW Bryant. Negative-bandgap quantum dots: Gap collapse, intrinsic surface states, excitonic response, and excitonic insulator phase. *Physical Review B*, 82(15):155314, 2010.

17. H Cercellier, C Monney, F Clerc, C Battaglia, L Despont, MG Garnier, H Beck, P Aebi, L Patthey, H Berger et al. Evidence for an excitonic insulator phase in 1 T- TiSe$_2$. *Physical Review Letters*, 99(14):146403, 2007.

18. C-P Tang, S-J Xiong, W-J Shi, and J Cao. Two-dimensional pentagonal crystals and possible spin-polarized Dirac dispersion relations. *Journal of Applied Physics*, 115(11):113702, 2014.

19. M Anand, LA Taylor, MA Nazarov, J Shu, H-K Mao, and RJ Hemley. Space weathering on airless planetary bodies: Clues from the lunar mineral hapkeite. *Proceedings of the National Academy of Sciences of the United States of America*, 101(18):6847–6851, 2004.

20. VN Antonov, O Jepsen, W Henrion, M Rebien, P Stauss, and H Lange. Electronic structure and optical properties of β-FeSi$_2$. *Physical Review B*, 57(15):8934, 1998.

21. H Udono, I Kikuma, T Okuno, Y Masumoto, H Tajima, and S Komuro. Optical properties of β-FeSi$_2$ single crystals grown from solutions. *Thin Solid Films*, 461(1):182–187, 2004.

22. H Ohta, S-I Kimura, E Kulatov, SV Halilov, T Nanba, M Motokawa, M Sato, and K Nagasaka. Optical measurements and band calculations of FeSi. *Journal of the Physical Society of Japan*, 63(11):4206–4212, 1994.

21

Tunable Picosecond Magnetization Dynamics in Ferromagnetic Nanostructures

Samiran Choudhury,
Sucheta Mondal,
Anulekha De, and
Anjan Barman
S. N. Bose National Centre for Basic Sciences

21.1 Introduction

In today's world, the magnetism has traversed a long way with vast applications in a range of multidisciplinary fields in modern and future nanotechnologies with the rapid invention of several unique magnetic materials, synthesized structures, micro- and nanostructures along with the smart materials. These artificial nanostructures have received enormous attention from both the research and development communities for their promising utilization as non-volatile magnetic memory [1,2], magnetic recording heads [3], magnetic storage media [4], magnetic resonance imaging [5] as well as in the field of biology such as nano-biomedicine [6] and health science [7]. More recently, the immense progress in the nanoengineering is proposed in the fields of various spin-based logic systems [8,9], spin-torque (ST) and spin-Hall nano-oscillators (SHNO) [10,11], and magnonic crystals (MCs) [12]. But with these growing demands of inventing improved technology, new material properties are required which is not always possible to achieve from natural material. Instead, artificially structured materials may exhibit desirable material properties. A relevant example is the bit-patterned media (BPM) that utilize the ordered arrays of lithographically patterned two-dimensional (2-D) networks of magnetic bits, and the magnetic switching behaviors of such systems including the switching field distribution have been thoroughly investigated.

One of the essential criteria of BPM is to eliminate the crosstalk among the individual magnetic bits. On the other hand, similarly ordered nanomagnet arrays can be utilized to transmit collective spin waves (SWs) with long wavelength to carry information in the MCs when the nanomagnets are strongly magnetostatically coupled. For multidisciplinary technologies, exploitations of variety of new phenomena are required, which range over different temporal frames, *e.g.* from slower processes like magnetic domain wall dynamics and magnetic vortex core motion to the faster phenomena such as SW dynamics, damping to ultrafast demagnetization and relaxation. The manipulation of magnetization dynamics over this large time scale is required in ferromagnetic (FM) structures at various length scales such as magnetic nanowires and nanostripes, nanodots, antidots (ADs), bi-component or binary structures as well as the 3-D nanostructures. In the following sections, we will describe the fabrication, characterization and investigation of magnetization dynamics in such artificially patterned magnetic structures.

21.2 Fabrication Techniques

The most challenging task to investigate the physical phenomena of such patterned structures is the fabrication of high-quality magnetic materials and their accurate characterization. Current research and technology need fabrication of ultrathin films and multilayers (MLs) with very high surface and interface qualities so that in the following step, one can efficiently fabricate the nanostructures with very narrow size dispersion, arranged in an ordered array over macroscopic length scales, which are heavily required for applications in various magneto-electronics, spintronics as well as in magnonic devices. These challenges have driven the scientific community to develop a number of top-down and bottom-up approaches in the nanofabrication techniques and more recently a combination of both. The bottom-up approach heavily depends upon the solution-phase colloidal chemistry [13] and electro-chemistry using different templates, *e.g.* track-etched polymer [14], anodic alumina [15] and diblock copolymer membranes [16].

However, the top-down approach primarily deals with the physical processes, which include several kinds of lithography techniques such as photolithography [17], electron-beam lithography (EBL) [18] (see Figure 21.1), deep ultraviolet lithography (DUV) [19], ion-beam lithography (IBL) [20], shadow masking [21] and laser or ion-beam irradiation [22]. Besides these methods, X-ray lithography [23], scanning probe lithography [24], holographic or interface lithography (IL) [25], nano-imprint lithography (NIL) [26], step-growth methods [27] are also very popular methods. Recently, two-photon lithography [28] combined with the electrodeposition has shown a promising future in fabricating high-quality 3-D micro- and nanostructures.

21.3 Effect of Miniaturization

The spin structures and magnetization dynamics of the FM nanostructures differ by a great extent from their bulk counterparts. The dynamics of these systems are strongly influenced by their static magnetization profiles, which depend not only on their intrinsic material parameters like exchange stiffness constant, saturation magnetization, magnetocrystalline anisotropy, etc. but also on their physical structures, *i.e.* size and shape of the magnetic elements, lattice geometry as well as the extrinsic factors such as strength and orientation of the bias magnetic field. The magnetization reversal [29] in such nanostructures takes place coherently or incoherently through different magnetization reversal modes such as curling, buckling, fanning, corkscrew, etc. (see Figure 21.2) [30,31] depending upon the length scale of the systems.

The magnetization reversal in magnetic nanowires is dominated by domain wall movement. There are primarily

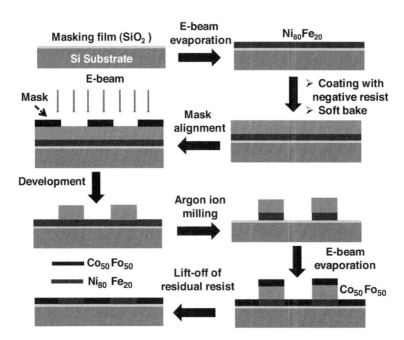

FIGURE 21.1   Schematic of the electron-beam lithography (EBL) technique used to fabricate 2-D arrays of nanostructures.

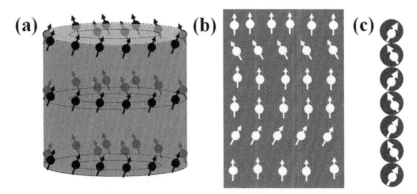

FIGURE 21.2   Schematics of (a) curling and (b) buckling modes are shown for a cylinder. (c) The fanning mode is shown for chains of spherical nanoparticles.

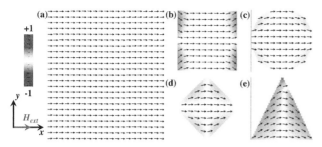

FIGURE 21.3 Simulated static magnetic configurations of 20-nm-thick (a) $10 \times 10~\mu m^2$ NiFe (permalloy) thin film. Static magnetization profiles of 20-nm-thick single (b) square-, (c) circular-, (d) diamond-, and (e) triangular-shaped nanodots each having width of 300 nm demonstrating the formation of flower-like state (square nanodot), leaf-like state (circular and diamond nanodots), and C-like state (triangular nanodot) due to the miniaturization of a continuous magnetic film into nanostructures with different shapes. The grayscale map for the static magnetization and the schematic of magnetic field direction are shown at the left side of the figure.

two different fundamental reversal modes, namely transverse wall mode and vortex wall mode [32,33] observed in magnetic nanowires depending upon their thickness. Also, 2-D nanostructures with low thicknesses regime show different quasi-single-domain states [34] such as flower, leaf, S, C, vortex and onion states (see Figure 21.3) due to the inhomogeneous internal field distributions.

21.4 Experimental Techniques

In order to investigate the magnetic properties of these systems, a range of sensitive characterization methods have been developed over the last few decades [35]. Magnetic force microscopy (MFM) [36] and Lorentz force microscopy [37] are two such examples, which have been utilized extensively for spatially mapping the gradient of the stray magnetic field and the sample magnetization, respectively. The magnetic force between the scanning magnetic tip and the gradient of the stray magnetic field coming from the magnetic surface gives rise to the magnetic contrast in MFM, while in Lorentz microscopy, this originates from the deflection of the electrons accelerated by Lorentz force after getting transmitted through the magnetic thin films. However, none of these two imaging methods provide any straightforward way to directly extract any quantitative information, whereas, electron holography [38] is another strong imaging tool based on the electron interference by which both the amplitude and phase information of the spin configurations, as well as the magnetostatic field distributions, can be easily mapped with a very high spatial resolution down to ~2 nm. On the other hand, magneto-optical Kerr effect (MOKE) microscopy [39] is yet another powerful technique that has been widely used to map the magnetization states of the magnetic microstructures with a sub-micrometer (sub-μm) spatial resolution, while the photoemission electron microscopy

(PEEM) [40] is a form of X-ray microscopy, which provides a far better spatial resolution than the visible light imaging. Spin-polarized low-energy electron microscopy (SPLEEM) [41], scanning electron microscopy with polarization analysis (SEMPA) [42], spin-polarized scanning tunneling microscopy (SP-STM) [43], and ballistic electron magnetic microscopy (BEMM) [44] are some other powerful imaging techniques, where spin-dependent transmission, scattering, or tunneling of electrons for magnetic contrast give excellent spatial freedom of 10 nm or better. However, most of the above techniques suffer from poor or moderate time resolution despite having a very high spatial resolution.

Subsequently, various experimental tools have emerged to investigate the faster magnetization dynamics [45] of magnetic micro- or nanostructures over a broad time scale. The conventional cavity-based ferromagnetic resonance (FMR) [46] and the vector network analyzer-based FMR (VNA-FMR) [47] methods (see Figure 21.4a) are very useful tools to study high-frequency magnetic response and the magnetic properties like permeability or susceptibility in MHz to tens of GHz frequency regime with a very high spectral resolution. Subsequently, the spatially resolved FMR [48] has also been developed to map the resonant SW modes profiles of the confined magnetic elements. Pulsed inductive microwave magnetometry (PIMM) [49], which is an oscilloscope-based time-domain detection tool, is later invented to probe the magnetization dynamics with a temporal resolution of tens of ps.

Another very powerful technique with the wave vector resolution is the Brillouin light scattering (BLS) technique [50] (see Figure 21.4b), which is based on the inelastic light scattering from the SWs and other quasiparticles. Here, the frequency or the energy dispersion of the SWs with wave vector can be directly measured by simply varying the angle of incidence of light with respect to the plane of the sample. Recently, both space- and time-resolved BLS methods have been developed to study and map the SWs with sub-μm spatial resolution and few ns temporal resolution [51], respectively. However, better spatio-temporal resolution can be achieved using time-resolved MOKE (TRMOKE) microscopy [52] (see Figure 21.4c) where ultrafast magnetization dynamics can be probed with tens of femtosecond (fs) temporal resolution, which is limited only by the pulse width of the laser. The magneto-resistive methods [53] and X-ray microscopy [54] have also shown great potential. Time-resolved scanning Kerr microscopy (TRSKM) [55], where the time evolution of magnetization dynamics is excited by a time-dependent magnetic field pulse, has also served well in this regard.

21.5 Time Scales of Magnetization Dynamics

Various kinds of magnetization dynamics can be observed with their characteristic time scales [56] (see Figure 21.5),

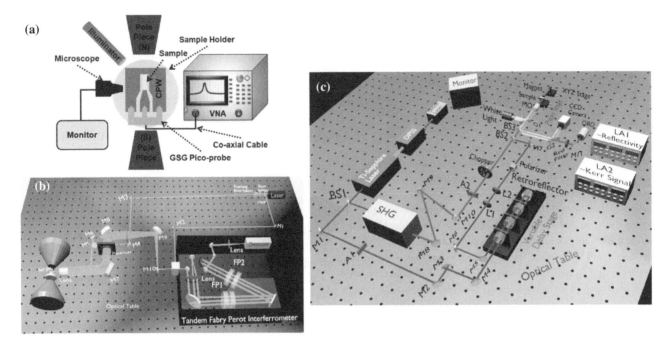

FIGURE 21.4 (a) Schematic diagram of broadband ferromagnetic resonance (FMR) spectrometer. (b) Schematic diagram of conventional Brillouin light scattering (BLS) setup. The notations for the components are given as, M, mirrors; FP, Fabry–Pérot interferometer. (c) Schematic diagram of the time-resolved magneto-optical Kerr effect (TR-MOKE) microscope. The notations for the components are given as, BS, Beam Splitter; M, Mirror; A, Attenuator; L, Lens; G, Glass Plate; MO, Microscope Objective; LA, Lock in Amplifier; OBD, Optical Bridge Detector; DPSS, Diode Pumped Solid-State laser; SHG, Second Harmonic Generator. The details can be found elsewhere [45].

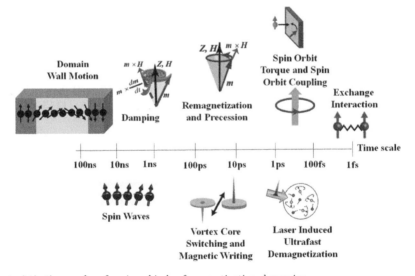

FIGURE 21.5 Characteristic time scales of various kinds of magnetization dynamics.

which are mainly governed by the interaction energies through the Heisenberg's uncertainty principle. The fastest process is the fundamental exchange interaction that takes place within ∼10 fs. This is followed by the spin–orbit (SO) coupling and related phenomena such as the spin-transfer torque (STT), which occur in the time scale between 10 fs – 1 ps. The laser-induced ultrafast demagnetization happens within few hundreds of fs. The fast remagnetization covers the time span of 1– 10 ps, which is followed by the slow remagnetization.

The magnetic writing process, which may occur *via* reversal of spins as well as the vortex core switching, has a time scale of few ps to few hundreds of ps. The precessional magnetization dynamics takes place in few ps to few ns time period, while the damping associated with the magnetization precession occurs in sub-ns to tens of ns time. The SWs in FM material generally can propagate within a time period of few hundreds of ps to few tens of ns before decaying completely. The domain wall motion is the slowest process of all and it has the time scale of few ns to μs.

21.6 Ferromagnetic Nanostructures

Artificial crystals are engineered to have properties not found commonly in nature. Such new properties are not always possible to achieve in naturally occurring materials. Instead, structuring of known materials in one-, two-, or three-dimensions (see Figure 21.6) at various length scales and exploiting dynamical magnetic properties over a broad frequency range may potentially offer the desirable material properties.

Magnonic crystals (MCs), photonic crystals, and phononic crystals are popular artificial crystals where spin waves (SWs) or magnons, electromagnetic waves (photons), and acoustic waves (phonons) are the transmission waves, respectively. However, due to the low speed of propagation compared to the electromagnetic wave, SWs with frequencies in the range of tens of GHz have wavelengths in the nanometer regime, making MCs ideal candidates for cellular nonlinear network and nanoscale on-chip data communication, including magnonic waveguides, filters, splitters, phase shifters, spin wave emitters, as well as for magnonic logic devices. As a result, a new research field named magnonics has emerged, which has the potential to build all-magnetic computation. One important problem is to tune the magnonic band structure of the MCs by varying different geometrical parameters like shape, size, lattice spacing, and lattice symmetry and also by changing the constituent ferromagnetic materials. In the following, we review existing research works in that direction.

21.6.1 One-Dimensional Ferromagnetic Nanostructures

Modulated ferromagnetic thin films with gap soliton were proposed in the early 1990s [57]. After one decade the focus shifted towards 1-D and 2-D arrays including the quasiperiodic and fractal magnetic structures [58,59]. The spectrum of magnons in closely packed 1-D arrays of magnetic nanoelements, *i.e.* nanowires and nanostripes, can also possess band structures with Brillouin zone (BZ) boundaries determined by the artificial periodicity of the arrays. The homogeneously magnetized wires can act as a spin waveguide that can channel, split, and manipulate submicrometer wide SW beams analogous to the optical fiber in photonics. In early studies [60], magnetostatic coupling in arrays of differently spaced $Ni_{80}Fe_{20}$ nanowires (NWs) was investigated. The magnetic hysteresis curves for the wires

with the smallest spacing showed a series of steps followed by plateaus corresponding to stable magnetization configurations related to the antiparallel alignment. However, the NWs with larger inter-wire spacing were non-interacting. The BLS studies showed a dispersive character of the lowest frequency mode in the magnetostatically coupled wires, instead of the dispersion-free behavior of the resonant modes in uncoupled wires. A systematic investigation [61] in arrays of $Ni_{80}Fe_{20}$ NWs with alternating width showed that the magnetization reversal process of the NWs is strongly sensitive to the wire thickness to width ratio. BLS technique was employed [62] to investigate the collective SWs in a dense array of dipolar-coupled 1-D $Ni_{80}Fe_{20}$ stripes of alternating widths. Both the saturated parallel (F) state and antiparallel (AF) state during the magnetization reversal process have been analyzed. In the F state, periodic variation in peak frequencies as a function of the wave vector (k) suggested strong dipole coupling of the stripes and formation of traveling collective excitations in the arrays. The band gaps can be substantially modified by changing the magnetic ground state from F to AF states.

FMR study [63] of a single micron-sized Co stripe under rotation of the individual stripe with respect to a static magnetic field showed magnetostatic as well as localized excitations along with quasi-uniform excitations. Recently, an all-optical TRMOKE study [64] (see Figure 21.7) in few micron long $Ni_{80}Fe_{20}$ nanostripes with varying stripe width down to 50 nm revealed a strong variation in the frequency, anisotropy, and spatial nature of SWs depending on the width of the nanostripes and the orientation of the bias field. Bias field applied along the length of the wider stripes showed uniform SW mode, a combination of backward volume (BV)- and Damon–Eshbach (DE)-like standing SW modes and pure BV-like standing SW mode, while uniform SW mode and pure BV-like standing SW mode were obtained from the 50-nm-wide stripe. For the bias field applied along the stripe width, arrow-like standing SWs for wider stripes, and only DE-like standing SWs with quantization axis deviated from the stripe axis for 50-nm-wide stripe were observed. Later, the investigation [65] of SWs in high aspect ratio single-crystal Ni nanowires revealed that the standing and uniform SW modes coalesce to form a single mode with uniform precession over the entire NW below a certain magnetic field. More recently, a combined experimental and theoretical study of the magnonic excitation was demonstrated [66] in an asymmetric saw-tooth-shaped (ASW) ferromagnetic array forming a pseudo-1-D magnonic crystal by manipulating the orientation between the wave vector (directed along ASW axes) and magnetic field. A transition in the internal magnetic field distribution, from the "S" state to the "leaf" state, was resulted by switching the magnetic field from one direction to another. A particular field orientation showed a pronounced band gap whose size was effectively modified by varying field direction. Propagation of SWs in microscopic magnetic stripes demonstrated [67] nonlinear effects resulting from the strong influence on the nonlinear wave dynamics of the

FIGURE 21.6 Schematics of (a) 1-D, (b) 2-D, and (c) 3-D magnonic crystals (MCs).

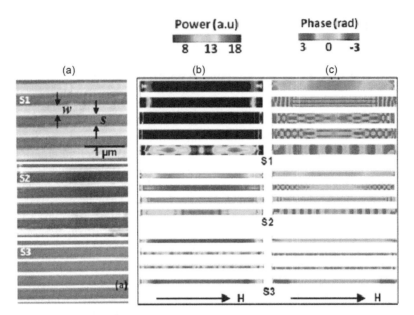

FIGURE 21.7 (a) Scanning electron micrographs of $Ni_{80}Fe_{20}$ nanostripes with different stripe widths S1 ($w = 200$ nm), S2 ($w = 100$ nm), and S3 ($w = 50$ nm). The (b) power and (c) phase profiles of the simulated SW modes for S1, S2, and S3 for $H = 1$ kOe. The direction of the external bias magnetic field (H) is shown at the bottom of the figure. The grayscale maps for power and phase distributions are shown at the top of the figure. (Adapted from reference [64].)

nonlinear magnetic damping in thin metallic ferromagnetic films. A considerable decrease in SW transmission caused by the resonant backscattering from periodical lattice was observed [68] in a micron-sized MC in the form of notched $Ni_{80}Fe_{20}$ waveguide. A plane periodical structure consisting of parallel dipolar-coupled magnetic stripes was found [69] to support propagation of Bloch waves at any angles with respect to the stripes length. Due to the one-dimensional periodicity, the gaps in the SW spectrum are partial, *i.e.* the frequency passbands for propagation along the direction of periodicity overlap with the stop bands for propagation along the stripes.

21.6.2 Two-Dimensional Ferromagnetic Nanostructures

Ferromagnetic Nanodots

The SW frequency and wavelength can both be tuned significantly by periodic modulation of exchange and dipolar field inside the nanodot arrays (see Figure 21.8). Therefore, the use of spin waves for logic devices is the natural extension of magnetic non-volatile elements for storage.

Ever since the research on confined magnetic elements started, their static and dynamic magnetic properties showed distinct differences with their bulk counterparts. Due to confinement in various dimensions, the usual magnetic domain formation gets affected and several anisotropic energies (*e.g.* shape anisotropy, configurational anisotropy) come into the picture for periodically patterned dot arrays. Initial works on magnetization dynamics in submicron-sized circular NiFe disks [70], showed multiresonance FMR spectra corresponding to nonuniform magnon modes whose positions strongly depend on the

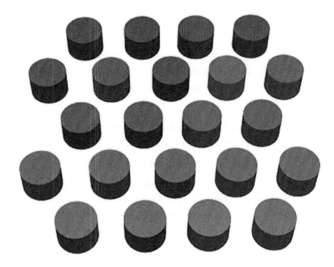

FIGURE 21.8 Schematic of 2-D nanodot array.

orientation of the external magnetic field and the inter-disk interaction. As the aspect ratio of nanodots increased, the in-plane shape anisotropy increased although the contribution from the nonuniform demagnetizing effect was reduced [71–73]. BLS investigations [72] in cylindrical $Ni_{80}Fe_{20}$ nanodots showed standing SW modes of DE origin, dipole-exchange coupled modes and reminiscent of BV mode oscillating parallel to the applied magnetic field. Additionally, a laterally confined edge mode was also observed with frequency independent of the dot radius. TRSKM study [74,75] of the precessional dynamics in square-shaped $Co_{80}Fe_{20}/Ni_{88}Fe_{12}$ bilayer nanodot arrays showed (see Figure 21.9) an increase in the precession frequency with the decrease in element size and also a crossover from one branch of modes to another below element size of 220 nm

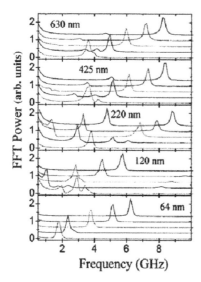

FIGURE 21.9 Fast Fourier transform (FFT) spectra of the out-of-plane component of the average dynamic magnetization for different bias magnetic field values of square arrays of square elements with lengths indicated in the figures. Curves from top to bottom correspond to bias field values of 772, 589, 405, 267, and 152 Oe, respectively. (Adapted from reference [75].)

due to the increased area of demagnetized regions. In arrays of hexagonally ordered Co elements [73], a periodic change in coercivity and remanence repetitive over every 60° mirrored the structural symmetry of the arrays.

An all-optical excitation and detection of collective magnetization dynamics in $Ni_{80}Fe_{20}$ square nanodot arrays with varying areal density showed a transition from a single uniform collective mode at a very high areal density to completely isolated dynamics of the individual nanodots at very small areal density *via* weakly collective dynamics at intermediate areal density [76]. Precessional dynamics of arrays with 50nm-sized $Ni_{80}Fe_{20}$ dots down to the single nanodot regime revealed single dominant resonant mode corresponding to the precession of the spins at the edge regions [77]. With the increase in areal density, the precession frequency increased significantly due to the enhancement in magnetostatic interactions between the nanodots. Further, the presence of anisotropy in the collective dynamics with the variation of the azimuthal angle of the bias magnetic field in similar $Ni_{80}Fe_{20}$ square nanodot arrays [78] showed a gradual transition from uniform collective regime to a non-collective regime depending upon the in-plane orientation of the bias field for closely spaced nanodots. However, for nanoelements with greater inter-element spacing, no clear trend in the transition was obtained. Another study [79] in $Ni_{80}Fe_{20}$ circular nanodots arranged in hexagonal lattice showed strongly collective dynamics for smaller inter-dot separation. At larger separation, the system entered into an isolated regime. A sharp variation in the SW spectra and mode profiles was observed as the lattice symmetry was changed from square to octagonal through rectangular, hexagonal, and honeycomb, respectively [80,81]. The single uniform collective

mode in the square lattice (edge mode of all the dots in the array) splits in two distinct modes in the rectangular lattice, which were edge mode and center mode over the entire lattice. For hexagonal and octagonal lattices, three modes were observed. For hexagonal lattice, these were uniform collective mode, a BV-like mode, and a bowtie-like mode of the entire lattice, while for the octagonal lattice the three modes corresponded to the uniform, DE-like and localized modes of the lattice. Interestingly, the honeycomb lattice showed broad and rich SW spectra, which included various localized and extended modes of the lattice. The dominant mode showed two-, six-, and eightfold anisotropy superposed with a weak fourfold anisotropy for rectangular, honeycomb, and octagonal lattices, respectively. The anisotropic dipolar interaction is more pronounced in octagonal lattice having horizontally as well as vertically coupled paired nanodots compared to honeycomb lattice with the paired nanodots placed only parallel with the field [82] (see Figure 21.10).

The SW dynamics in $Ni_{80}Fe_{20}$ nanodot lattices by varying dot shape showed [83] a single collective mode in elliptical dot lattice gets transformed into three distinct modes for the half-elliptical, rectangular, and diamond dot lattices. A drastic change, with eight modes covering a broad band for the triangular dots was observed. The dynamics under the angular variation of the applied bias field for a unique system of cross-shaped $Ni_{80}Fe_{20}$ nanodots showed [84] a significant anisotropic nature and frequencies of the SWs revealed very interesting phenomena such as mode softening, mode crossover, mode splitting, and merging of SW frequency branches (see Figure 21.11).

Further, an interesting investigation [85] revealed the direct probing of the behavior of the static dipolar field from neighboring disk affecting the dynamic behavior of a single disk by the micro-BLS technique using pairs of identical $Ni_{80}Fe_{20}$ disks with varying the inter-disk spacing. At a high magnetic field, the disks were in single-domain state, which diminishes at the low field when the disks were in a vortex state. The magnetization dynamics in 2-D diatomic MCs, where $Ni_{80}Fe_{20}$ nanodots of two different diameters, placed very close to each other to form a complex double dot unit cell [86], can be tuned by both the bias magnetic field strength and orientation. New interacting modes (get either blue or red-shifted for the two different orientations of the bias field) appeared in the diatomic unit along with a number of quantized, localized, and extended modes in the individual nanodots indicating the interaction among the larger and smaller dots within the diatomic unit.

Ferromagnetic Nanorings

One important feature of a memory device is that the magnetic hard layer should have two remnant states and the switching process must be simple and reliable. Micron- and nano-sized ferromagnetic ring structures have two kinds of magnetic states: a flux closure or "vortex" state and an "onion" state with same moment orientation in each half of a ring and they exhibit a range of different switching mechanisms including domain wall and vortex core nucleation,

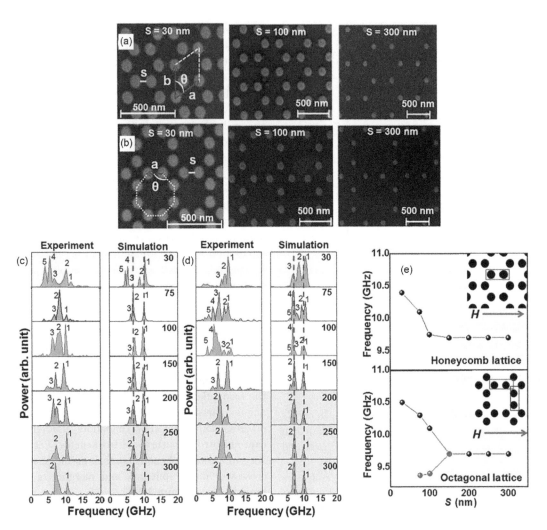

FIGURE 21.10 Scanning electron micrographs of the $Ni_{80}Fe_{20}$ nanodot array arranged in (a) honeycomb and (b) octagonal lattices with varying inter-dot separations (S), which along with length scales are mentioned in the figures. The experimental and simulated FFT power spectra of the arrays with (c) honeycomb and (d) octagonal lattice symmetries. Inter-dot separations are mentioned on the right-hand side of each panel. The dotted lines as shown in the simulated FFT spectra indicate the peak position of two SW modes of a single nanodot. Different shades of gray indicate different magnetostatic interaction regimes. (e) Variation of precessional frequencies of mode 1 obtained from simulated FFT spectra of the honeycomb lattice and mode 1, 2 for the octagonal lattice with S. Constant field (H) of 1.3 kOe is applied as shown in the insets. (Adapted from reference [82].)

annihilation, propagation, etc. Either the two orientations of the onion state or the two different chiralities of the vortex state can be deployed as two memory states in the storage layer [87,88]. Extensive research has been carried out on micro- and nanoscale square, rectangular, circular, triangular, elliptical, and off-centered asymmetrical circular magnetic rings.

Initial studies in $Ni_{80}Fe_{20}$ square nanorings [89] showed a two-step switching process as the external field is swept along both the edges and the diagonal of the ring. Magnetostatic interactions in rectangular $Ni_{80}Fe_{20}$ nanoring arrays by MFM method [90] showed that the transitions from onion-to-vortex and vortex-to-reverse onion states were strongly dependent on the edge-to-edge spacing of the rings due to dipolar interaction as well as shape anisotropy. In symmetric nanorings, magnetization reversal through the formation of a vortex state (V-process) and the rotation

of an onion state (O-process) with comparable probabilities were manifested as the competition between the exchange energy and the magnetostatic energy, and the relative probability of the two processes in symmetric nanorings cannot be altered after fabrication. Interestingly, these two processes in asymmetric nanorings fabricated with oblique angle ion-beam etching can be controlled depending on the direction of the external field [91]. 2-D $Ni_{80}Fe_{20}$ elliptical ring arrays arranged in parallel type, *i.e.* the inter-unit distance was varied along the long axis and shifted type arrays, *i.e.* where the inter-unit distance could be changed by a shifted arrangement of rows in the short axis direction confirmed [92] that more stable magnetization behavior could be achieved with the shifted-type array arrangement.

Dynamics of micron-sized circular $Ni_{80}Fe_{20}$ rings by broadband FMR technique revealed two dominant modes

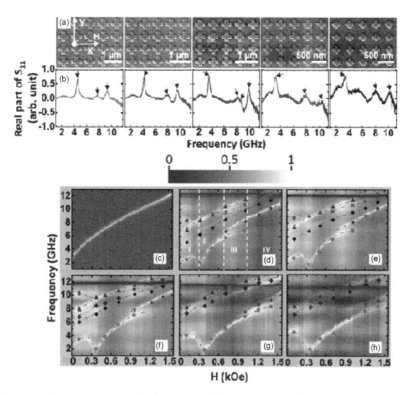

FIGURE 21.11 (a) Scanning electron micrographs for arrays of $Ni_{80}Fe_{20}$ nano-cross with the applied magnetic field direction. (b) The real part of the S_{11} parameter as a function of frequency at $H = 497$ Oe for all samples. The SW modes are marked by arrows. (c)–(h) Surface plots of bias field dependence of SW mode frequencies for (c) $Ni_{80}Fe_{20}$ thin film of 20 nm thickness and nano-cross arrays with arm lengths (d) 600 nm, (e) 500 nm, (f) 400 nm, (g) 300 nm, and (h) 200 nm. The solid line in (c) represents the Kittel fit to the experimental data. Simulated SW frequencies in (d)–(h) are shown by filled symbols. The grayscale map is shown at the top of the figure. (Adapted from reference [84].)

with uniform phase around the ring, *i.e.* a circularly symmetric mode and a rotationally antisymmetric mode [93]. BLS study [94] of the eigenmode spectrum of $Ni_{80}Fe_{20}$ circular nanorings at different values of the applied field (in both vortex and saturated states) showed localization effects for all the modes. As the applied field was switched on, each radial mode split into two and got localized at opposite arms of the ring. The influence of partial decoherence studied [95] by micro-BLS technique on $Ni_{81}Fe_{19}$ rings revealed that SW wells created due to the inhomogeneous internal field in the pole regions of the rings led to SW confinement. In the intermediate region, modes with a well-pronounced quantization in radial direction experienced a transition from partial to full coherence in the azimuthal direction as a function of decreasing ring size. An interesting investigation [96] on the dynamics of triangular-shaped $Ni_{80}Fe_{20}$ nanorings showed multiple resonance peaks with frequencies strongly dependent on the orientation of the applied magnetic field displaying sixfold anisotropy. This was in contrast to the circular rings. Recently, the internal-field-driven ultrafast magnetization dynamics of $Ni_{80}Fe_{20}$ square nanorings with varying ring width were explored [97] using TRMOKE. The SW spectrum of the widest ring showed quantized modes along the applied field direction, while as the ring width was reduced additional quantization in the azimuthal direction appeared causing mixed modes, which was followed by

SWs with quantization only along azimuthal direction for the narrowest ring (see Figure 21.12).

Ferromagnetic Antidot Array

The introduction of periodic antidots (holes) into a continuous FM thin film is a very effective way to tailor their magnetic properties. However, this novel patterning leads to less uniform and more complex magnetization dynamics. This nonuniformity can significantly contribute to the noise in the read sensor devices and may affect the ultrafast precessional switching of magnetic data storage elements induced by a pulsed magnetic field. This has a very promising future in the modern SW-based applications, *e.g.* in microwave-assisted magnetic recording (MAMR) technology. Ferromagnetic antidot (hole) lattices (ADLs), *i.e.* periodically arranged holes on a ferromagnetic thin film (see Figure 21.13) are one of the strongest candidates for reconfigurable MCs due to higher SW propagation velocity and longer propagation distance for the SWs.

They can also be considered as a potential candidate for magneto-photonic crystals due to their strong influence of the magnetic field on the light (phonon) coupling to the surface plasmons. These magnetic nanostructures have been considered as a promising candidate for ultrahigh density data storage devices. Moreover, these antidot arrays have

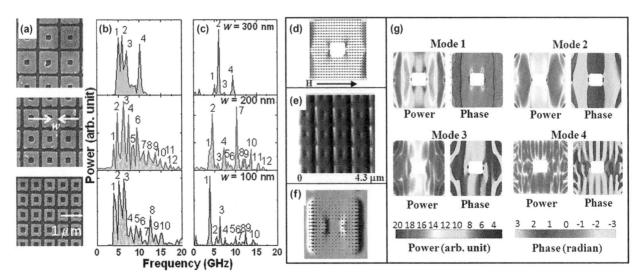

FIGURE 21.12 (a) Scanning electron micrographs of the ring arrays of different ring width. (b) Experimental and (c) simulated FFT spectra of the ring arrays with different ring widths as indicated. (d) Static magnetic configuration of a single ring with $w = 300$ nm at $H = 550$ Oe. (e) Experimental and (f) simulated MFM images of this nanoring array in presence of $H = 550$ Oe. (g) Power and phase profiles of the observed SW modes. The grayscale maps for the same are shown at the bottom of (g). (Adapted from reference [97].)

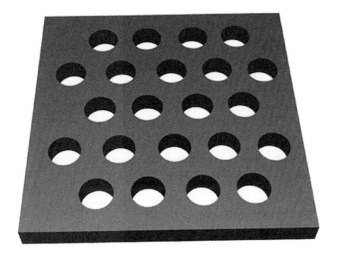

FIGURE 21.13 Schematic of 2-D antidot lattice (ADL).

some unique advantages over the array of magnetic dots at the deep nanoscale regime due to the absence of any small isolated magnetic entity. Interestingly, these antidot systems can be described as a mess of connected networks, which do not suffer from the superparamagnetic bottleneck as opposed to the magnetic dot array structures.

Consequently, extensive research has been carried out to investigate the SW dynamics in 1- or 2-D arrays of magnetic antidot structures using various methods ranging from numerical and theoretical techniques to several experimental tools. Initial studies [98,99] on the micron-sized $Ni_{80}Fe_{20}$ antidot arrays revealed pattern induced anisotropy coming from the lattice arrangement. Later, collective SW excitation in 2-D Co antidot array [100] showed a strong attenuation of the uniform precessional mode in the array unlike that in the continuous film. In another study [101],

the pattern induced splitting of surface and volume modes were modeled by the demagnetizing field confirming that edge boundary conditions of the array have negligible influence on the SW dynamics as opposed to the confined nanodots, while the magnon spectra got strongly affected by the local modulation of the demagnetizing field around the antidot array. The magnetoresistance behavior [102] of the nanoscale $Ni_{80}Fe_{20}$ antidot arrays was found to have a strong dependence on the film thickness. Again, the magneto-transport properties [103] of multilayer nanoscale antidot heterostructures with varying nonmagnetic spacer layer thickness showed a marked dependence of magnetization reversal mechanisms on the spacer layer thickness due to the interplay between different interlayer coupling phenomena. Further, a systematic study [104] in arrays of Cu/NiFe/IrMn/Cu nanoscale antidots with varying IrMn thickness [105] revealed that such magnetic system can be applied as graded-index material for which the index can be modulated by magnetic field analogous to the photonic counterpart. Later, an all-optical investigation [106] in a series of multilayer Co/Pd antidot systems with PMA fabricated using FIB demonstrated the effect of exchange coupling, tunneling as well as dipolar interactions on the observed modified magnetic properties. Several studies [107–109] on $Ni_{80}Fe_{20}$ antidot arrays with different hole sizes revealed the presence of multiple resonant SW modes. Later, the field-dependent localization of SW mode, SW confinement, and field-controlled propagation of SWs [110] showed (see Figure 21.14) that the characteristic SW eigenfrequencies can be correlated with both local inhomogeneities of the demagnetization field and specific wave vectors. This is caused by the geometry-imposed mode quantization conditions along with the presence of a preferential direction of SW propagation depending upon the direction of the applied magnetic field.

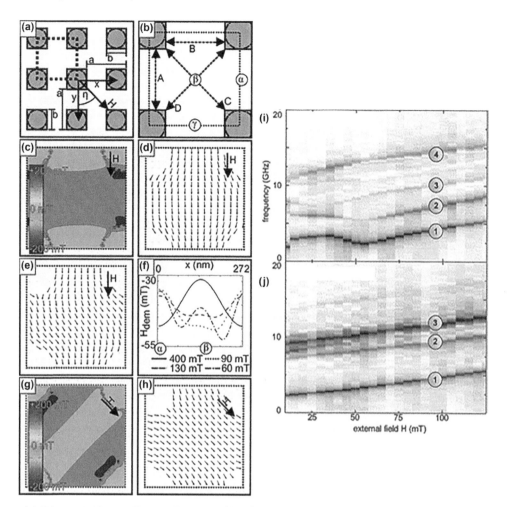

FIGURE 21.14 (a) Schematic of permalloy antidot lattice (ADL) along with the lateral parameters where dark gray regions represent antidots (holes). Here the holes of both square shape (full line) and circular shape (dashed line) have been investigated. (b) The square shown as a dotted line describes the unit cell. Relevant lattice directions A to D and specific areas α, β, and γ are indicated. (c)–(h) demonstrate the static behavior of ADL with a lateral unit cell size of 272 nm. In particular: (c) demagnetization field H_{Dem} in the butterfly state at 100 mT (grayscale code: see legend). Magnetization configuration: (d) the butterfly state at 100 mT ($\eta = 0°$) and (e) waterfall state at 30 mT ($\eta = 0°$). Small arrows indicate local orientation of spins. (f) Internal field profile through areas α and β in the x direction at $y = a/2$. For smaller fields the butterfly state becomes more pronounced. (g) Demagnetization field H_{Dem} and (h) magnetization configuration for $\eta = 45°$ and $\mu_0 H = 100$ mT. Dynamic demagnetization field changes for less than 0.1 mT). Simulated magnetic field dispersions for the same lattice at (i) $\eta = 0°$ and (j) $\eta = 45°$. (Adapted from reference [110].)

The BLS investigation [111] on 2-D sub-micrometric permalloy holes demonstrated the propagating SW modes along the edges of an effective stripe-like waveguide having a width equal to inter-hole distance. Interestingly, the anisotropic propagation and the modulation of damping characteristics of SWs in a network of interconnected nanowires were also observed [112] where the extrinsic damping was found to be dependent upon the edge roughness mediated scattering in the ADL although a considerable SW propagation velocity in such nanostructures making these interesting for SW filtering and field-controlled SW guiding. Afterwards, a number of investigations have been performed to manipulate the magnon dynamics by tuning the geometrical configurations, *i.e.* the lattice arrangement as well as the lattice periodicity of the nanoscale antidot systems. The experimental findings about the magnonic

modes in a large-area NiFe or Co antidot array have been reported [113–116], where the SWs exhibited an in-plane anisotropy induced by the lattice symmetry along with an interesting mode conversion from localized to extended mode or *vice versa* depending upon the in-plane direction of the external magnetic field for SWs with either positive or negative group velocity (see Figure 21.15).

The spin dynamics [117] of CoFeB antidot arrays with varying periodicity with one SW mode localized in the regions of strongly inhomogeneous internal field was observed, which was further confirmed by another experimental investigation [118] of magnetic normal modes in nanoscale antidot lattice. Tunable metamaterial response of $Ni_{80}Fe_{20}$ ADL has been extensively studied [119] revealing a strong dependence of the transmission coefficient on the field direction while a large reflectivity of the ADL

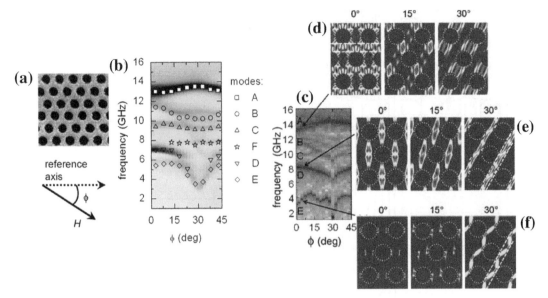

FIGURE 21.15 (a) Scanning electron micrograph of 250-nm-wide permalloy antidots arranged in rhombic geometry with periodicity 400 nm. φ is the in-plane angle defined between H and the reference axis (broken arrow). (b) Variation of spin wave (SW) dynamics with φ obtained for wave vector $q = 0$, which have been obtained by both all-electrical broadband spectroscopy (grayscale plot) and Brillouin light scattering (symbols) for $\mu_0 H = 90$ mT where dark shade encodes the excitation of eigenmodes. (c) SW excitation spectra obtained at $q = 0$ simulated as a function of φ with $\mu_0 H = 90$ mT where dark shade denotes eigenfrequencies. Local spin-precession amplitudes for SW branches (d) A, (e) D, and (f) E. The spatial spin-precession profiles have been demonstrated for $\varphi = 0°$, $15°$, and $30°$, which are labelled on the top (from left to right) where dark (bright) represents zero (maximum) precession amplitude and dashed lines highlight stripe-like precession profiles extending through the lattice unravelling the presence of mode crossover with in-plane orientation of H. (Adapted from reference [115].)

boundary was observed for a few selective frequencies at certain in-plane field direction, which can be very interesting for the development of frequency-selective mirrors in magnonics through nanopatterning. A significant artificial crystal-like behavior with magnonic minibands in antidot lattices has been described [120] where the dynamic coupling of edge modes was found to create field controllable magnonic minibands with a surprisingly large propagation velocity. This opens up interesting perspectives in the field of future nanoscale magnonic devices. The SW dynamics [121] in nanometric Co ADLs revealed significant engineering in the magnonic bands as the number of SW frequency bands were significantly modulated with periodicity providing an additional control parameter in the design of magnonic devices based upon the composite antidot systems. The realization of a repeatable evolution of domain walls (DWs) mediated by strong pinning by antidot edges indicating a frustrated flux closure configuration consisting of regions of highly ordered DWs has been found [122] in nanometer-sized antidot systems. Later, the high-symmetric magnonic modes for perpendicularly magnetized ADLs having a linear bias magnetic field dependence were also demonstrated [123] revealing a very high degree of isotropy of magnonic modes at the center of the Brillouin zone. There are few reports [124,125] on the bi-layered magnetic ADLs on Ni/Co heterostructures to investigate the magnetization reversal mechanism and the magneto-optical characteristics. The dynamic response of antidot nanostructures with alternating hole diameters, *i.e.* di-atomic

antidot array having two antidots with different sizes as the unit of the lattice revealed [126,127] stark modulation of magnonic spectra due to the existence of modulated demagnetizing field distributions. The antidot shape led to a strong variation in the internal field profile as well as the SW mode profile and its anisotropy [128]. Recently, the shape effect was utilized to manipulate the SW mode conversion controlled by bias magnetic field direction [129] in triangular-shaped $Ni_{80}Fe_{20}$ nanohole arrays. Later, alternative methods of fabricating ADLs have been successfully demonstrated [130]. FMR measurements of dynamic response in $Ni_{80}Fe_{20}$ ADLs fabricated by template-assisted method showed that the FMR field symmetry was deeply affected by the presence of defects in the ADL. The reversal mechanisms in a self-assembled iron (Fe) antidot array with high ordering fabricated by polystyrene nanosphere lithography revealed [131] a strong rotational anisotropy induced by the lattice arrangement. Similar nanosphere lithography allowing for non-closely packed voids in the in-plane magnetized Fe, Co, and NiFe, as well as out-of-plane magnetized GdFe thin films, showed [131] rich magnetic switching properties exploring the domain configurations and additional anisotropies controllable by the geometry of the ADLs. Interestingly, the antidots with certain aspect ratio were found to obey either the spin ice rules or bias field direction-dependent coercivity while the frustration effects led to significant out-of-plane magnetization contributions during the reversal process as observed in GdFe systems. Overall these effects can offer variety of possibilities to design future applications such as SW filters or artificial spin ice structures. Interestingly, crescent-shaped Co antidot systems fabricated by self-assembled anodized alumina templates unraveled [132] the presence of multi-step reversal of the hysteresis loops associated with the feature size of the array and growth morphology. The experimental realization of MC-based waveguide (MCWG) with large SW propagation velocity was reported [133] for perpendicular-to-plane magnetized 2-D CoFeB antidot lattices of varying periodicity where the frequency of MCWG relied upon the allowed bands of the MC, unlike the established photonic crystals. Later, the propagation of SWs in NiFe ADLs has been exploited [134] introducing line defects in the nanostructures by removing the holes systematically from the row or column of the arrays where the modulation of SW modes by exciting the system by a microwave current was obtained to guide the SWs at different frequencies, which find promising applications in magnonic devices as SW filters or spin-based logic devices. The universal dependence of the SW band structure on the geometrical characteristics of 2-D permalloy antidot systems was demonstrated [135] exhibiting exclusive dependence of the width and position of the frequency bandgap of the fundamental propagating SW on the aspect ratio (thickness over width) of the effective nano-channel between the adjacent rows of holes. Recently, a novel bi-structure MC in the form of embedded nanodots in a periodic NiFe antidot array namely annular antidot system has been investigated [136] showing a significant tunability in spin dynamics as compared to its antidot or dot lattice counterparts. These findings may lead to next-generation microwave devices such as magnonic filters and phase shifters.

Theoretical and numerical investigations of magnetic antidots have also led toward important findings. The effects of variation in areal density, lattice symmetry and tailored defects on magnonic spectra of Co antidot arrays revealed a strong dependence on the areal density and lattice symmetry but a weak dependence on the defect [137]. Another investigation [138] demonstrated that the magnetization pinning can facilitate the magnon band opening in nanoscale magnonic antidot waveguides (MAWs) indicating an additional functionality of such systems as SW filters with tunable forbidden and allowed bands where the pinning at the interfaces can be correlated with the surface magnetic anisotropy. This is crucial in the practical applications of 1- or 2-D MAWs. Subsequently, similar nanoscale MAWs have shown [139] the tunability of the SW band structure especially the bandgaps by varying the shape of the antidots while the inhomogeneity in the exchange fields at the antidot boundaries within the MAW was found to play a crucial role in controlling the band structure along with the observation of direct bandgap opening at the same filling fraction, *i.e.* without removing additional magnetic material during fabrication. Active control over magnonic bandgap opening by structural changes in terms of breaking the mirror symmetry of the MAWs has been demonstrated [140] by adjusting the external magnetic field.

The influence of structural changes in such MAWs was further investigated [141] where the effects of antidot size, shape, distance between the antidots, and the scale factor of MAWs on magnonic band structures and bandgaps were studied to identify the main parameters and mechanisms affecting the bandgaps in such nanoscale MAWs. Achievement of large SW velocity and filtering properties due to the existence of magnonic bandgaps along with the attainment of different mechanisms of bandgap opening resulting from Bragg scattering or anti-crossing of SW modes was shown. The observation of a complete bandgap and collective SW excitation in 2-D MCs consisting arrays of nanoscale antidots and nanodots was reported [142], where the iso-frequency contours analogous to the Fermi surfaces were obtained and also the physical origin of SW bands, partial and complete magnonic bandgaps were explained by the spatial distribution of SW energy spectral density. Further, a magnonic device consisting of the studied 2-D magnonic metamaterial exploiting the definition of small wave vector was proposed [143] for mapping the frequencies of the collective modes of the SW spectra along with a proposition for mapping the spatial profiles of the magnonic modes.

Artificial Quasiperiodic Magnonic Crystal

Recently, artificial quasicrystals have become an emerging topic of research because the most beneficial feature of quasiperiodicity is that it can combine perfectly ordered structures with purely point-diffractive spectra of arbitrarily high rotational symmetry. Recently, quasiperiodic structures have started to be explored in magnonics due to their various interesting properties like branching features in the band structure, self-similarity, and scaling properties in the transmission spectra. A theoretical analysis of the spectra of magnons in quasiperiodic MCs arranged according to the generalized Fibonacci sequences in the exchange regime revealed [144] the presence of a very rich bandpass structure with self-similar behavior as a function of the generalized Fibonacci generation number for these magnetic quasiperiodic arrays, which can be suitable as multiplex logical gates systems. Analytical calculation [145] on a 1-D Thue–Morse magnonic quasicrystals (MQCs) with arbitrary orders and layer thicknesses for arbitrary bonding materials revealed the perfect transmission resonances of the SWs, which is useful for developing ultrahigh quality, multichannel, filters, or resonators. The ability of signal storage capacity of 1-D defect-free MQC using its characteristic slow SW mode has been proposed [146] analogous to the slow light mode in optics and achievement of rich and wavelength-selective sharp resonance in such MQCs due to the unique fractal transmission spectra was shown. The plane wave method investigation [147] of SWs in 2-D planar quasicrystal having Penrose tiling structure in the form of Ni (or NiFe) disks embedded in Fe (or Co) matrix revealed the localization of the SW eigenfrequencies resulting from the quasiperiodicity of the magnetic structure. Later, the magnetic disorder on the arrays of dipolar-coupled nanostripes has been investigated [148] experimentally showing the possibility of controlled variation of the degree of the disorder by varying the applied magnetic field. The experimental and theoretical concepts of self-generation of dissipative solitons in MQCs as active ring resonator were also reported [149] where the frequency selectivity of the developed magnonic quasicrystal having quasiperiodic Fibonacci type structure leading to the self-generated dissipative soliton together with the parametric three-wave decay of the magnetostatic surface SW (MSSW) inside the MQC active ring resonator. Another investigation [150] described controlled magnetization reversal of a novel class of FM NiFe films patterned into quasiperiodic Penrose tiling-type MC with long-range order but no periodic translational symmetry that revealed a series of abrupt transitions between ordered magnetization textures culminating in a smooth evolution into a saturated state strongly influenced by the orientation of the film segments with respect to dc field. Interestingly, the FMR spectra for finite quasiperiodic segments exhibited a tenfold rotational symmetry in the nearly saturated regime, which is only expected for infinite MQCs with such Penrose tiling ordering offering a new paradigm of MQCs.

Recently, the efficient tunability magnonic spectra have been demonstrated [151] in 2-D FM antidot systems with an octagonal periodicity having a broken translational symmetry (see Figure 21.16). In addition to this, the presence of eightfold rotational symmetry was obtained in the anisotropic behavior of the SW frequency originating due to the variation of the internal field at different regions of the nanostructure.

Ferromagnetic Nanocomposites and Bicomponent Magnonic Crystal

Recently, it has been observed that the magnonic band structures of nanoscale systems can be efficiently tailored in bi-component magnonic crystals (BMCs) where one

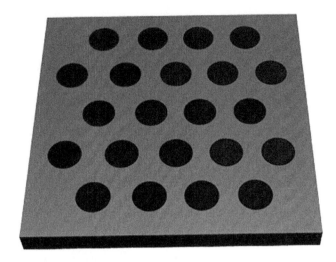

FIGURE 21.17 Schematic of 2-D bi-component magnonic crystal (BMC).

magnetic material is embedded within the continuous film of another magnetic material (nanocomposite).

In such embedded structures, where two different magnetic materials are in direct contact with each other (see Figure 21.17), the dynamic dipole coupling is maximized due to the presence of exchange coupling at their interfaces. As a result, SWs are subjected to scattering at the interface between the two materials and can easily transmit through the interface. Therefore, SWs can propagate across its entire structure with large group velocities. Initial studies on 1-D BMC in the form of laterally patterned periodic arrays of alternating Co and NiFe stripes (see Figure 21.18) with varying widths revealed [152–155] well-defined frequency bandgaps strongly dependent on their structural dimensions. Interestingly, the bandgap width and the center SW frequency were found to possess contrasting dependences on the constituent FM, *i.e.* NiFe and Co stripe widths. Such functionality can lead to the tailoring of magnonic bandgap structure and thereby

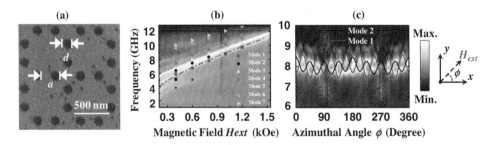

FIGURE 21.16 (a) Scanning electron micrographs of circular-shaped $Ni_{80}Fe_{20}$ (NiFe) antidots of antidot diameter $d = 140$ nm arranged in quasiperiodic octagonal lattices with variable lattice spacing $a = 300$ nm. (b) Bias field (H_{ext})-dependent spin wave (SW) absorption spectra of the lattice shown at $\phi = 0°$. Here, the dotted line represents Kittel fit to the lower frequency mode 1 of the lattice. The surface plot corresponds to the experimental result while the symbols represent the simulated data. (c) Variation of SW frequency with ϕ varying from $0°$ to $360°$ for the same system has been demonstrated at $H_{ext} = 800$ Oe. The surface plot depicts the experimental results while the solid lines describe the sinusoidal fits for the observed anisotropic SW modes. The map associated with the surface plots and the schematic of the orientation of the external applied field (H_{ext}) are given at the right side of the figure. (Adapted from reference [151].)

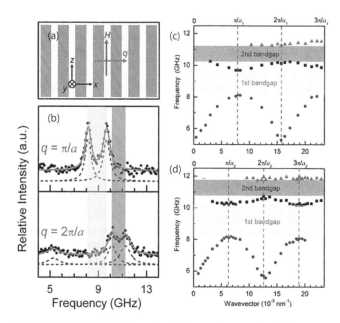

FIGURE 21.18 (a) Scanning electron micrographs of the nanostripes indicating the directions of applied magnetic field H for saturation of samples and magnon wave vector q relative to orientation of the nanostripes. (b) Brillouin light scattering spectra at $H = 0$ for 150 nm Co/250 nm $Ni_{80}Fe_{20}$ sample at various Brillouin zone boundaries. (c) Dispersion relations, showing bandgaps of spin waves in (c) 150 nm Co/250 nm $Ni_{80}Fe_{20}$ (lattice constant $a_1 = 400$), and (d) 250 nm Co/250 nm $Ni_{80}Fe_{20}$ samples ($a_2 = 500$ nm). Symbols represent the experimental data. The first and second frequency bandgaps are represented by shaded bands and the dashed lines represent the Brillouin zone boundaries. (Adapted from reference [153].)

controlling the transmission of information-carrying SWs in the nanoscale microwave devices.

Dual magnonic and phononic bandgaps in 1-D BMCs in the form of linear periodic arrays of alternating Fe (or Ni) and NiFe nanostripes has been demonstrated [156] revealing the presence of no magnon–phonon interaction for the observed modes making these periodic composites suitable for the separate simultaneous processing of information carried by hypersonic magnons and phonons along with the absence of any undesirable crosstalk between them. The nonreciprocal properties [157] of SWs in metalized 1-D BMC led to the appearance of several magnonic bandgaps located within the first BZ for propagating SWs along the metallized surface. Recently, the investigation [158] of nonlinear localized magnetic excitations in 1-D BMC under a periodic magnetic field of spatially varying strength showed the existence of excitation of magnetization in the form a soliton on the oscillatory background.

However, the exploration of magnetic aspects in 2-D BMCs has been started very recently. The magnonic spectra of 2-D periodic composite materials consisting of arrays of parallel cylinders made of an FM material embedded in another FM matrix revealed [159] the existence of magnonic bandgap, which was found to have a dependence on the physical parameters of the constituent FM materials as well

as on the lattice periodicity and filling fraction of the cylinders. Later, several experimental and theoretical investigations have been demonstrated to unravel the different properties of the SW dynamics in such nanocomposites. One such experimental study [160] on periodic arrays of lateral 1-D and 2-D bi-component exchange and magnetostatically coupled MCs consisting of alternating Co and NiFe nanowires (NWs) and nanodots lying side by side revealed two distinct switching steps corresponding to the reversal of the constituent Co and NiFe in the 1-D array in contrast to the single-step switching of individual homogeneous NWs. But for 2-D BMCs, strong dipolar fields from neighboring Co elements influencing the reversal mechanism of NiFe elements were found to result in the vortex nucleation and annihilation fields. Further, the tunability of the center frequencies as well as the width of forbidden bandgaps over tens of gigahertz (GHz) in the similar periodic composites of Fe nanodots in yttrium iron garnet (YIG) matrix was reported [161,162].

However, the spatial and field control of magnon modes in BMCs formed by periodic Co nanodisks introduced in nano-troughs etched into a thin NiFe film were investigated [163–166] (see Figure 21.19) where spatial positions of SW frequencies were dependent mainly on the Co nanodisks when the polarity of the internal field was reversed, which can allow one to optimize the transmission of signal in various spatial directions making the BMCs extremely useful in rotatable magnetic field. Recently, a very interesting investigation [167] on periodically nanostructured magnets in the form of NiFe nanodisks embedded into shallow-etched CoFeB matrix has been demonstrated as an omnidirectional grating coupler showing a giant enhancement of the amplitude of the short-wavelength SWs compared to a bare microwave antenna. Exploring the dependence of SW dynamics on the constituent FM elements, lattice periodicity, and the applied magnetic field, this magnonic grating coupler can be compared with gratings in photonics and plasmonics, which can play a key role in cellular nonlinear networks and integrated magnonics as omnidirectional emitters for shortwavelength SWs. The investigation [168] of collective SWs in the BMC consisting of a 2-D array of alternating NiFe and Co elements demonstrated that the active modes in the BZ can be characterized by the simplest mode profiles within the NiFe elements where the Co elements act as mediators or amplifiers of dipolar coupling between the NiFe elements for such excitations. Later, the mutual magnetostatic coupling between the two constituent FM elements was found [169] to modify the magnetization dynamics significantly as compared to the corresponding individual nanostructured FM elements. The difficulties regarding small propagation velocity can be potentially overcome by an increased interface exchange interaction between the two different elements of the BMC.

A recent study [170] (see Figure 21.20) on such 2-D composite nanostructures consisting arrays of NiFe elements of different shape embedded in a CoFe matrix revealed that the demagnetizing field, as well as the exchange interaction

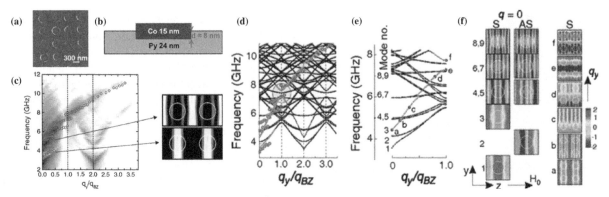

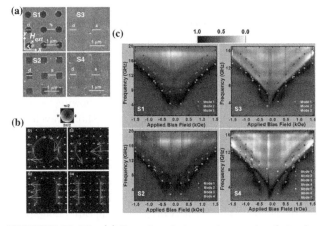

FIGURE 21.19 (a) Scanning electron micrograph of the 2-D bi-component magnonic crystal (BMC). (b) Schematic cross section of the sample showing a 15-nm-thick Co nanodisk embedded in a 24-nm-thick $Ni_{80}Fe_{20}$ film. (c) Measured (open circles) and simulated (grayscale plot) mode frequencies as a function of the SW wave vector. Right side: spatial spin-precession profiles of two modes calculated at $q_y/q_{BZ} = 0.19 \times 10^{-5}$. Bright color depicts large spin-precession amplitude. (d) Magnonic band structure of the $Co/Ni_{80}Fe_{20}$ 2-D BMC calculated by the plane wave method (PWM) (black dotted lines). The bold solid lines mark the calculated bands with the largest intensities. The BLS results from Figure 21.2 have been shown as open circles while the filled gray circles highlight the magnonic gaps opened at the 1st and 2nd BZ boundaries (indicated by vertical dashed lines). (e) The enlarged part of the calculated magnonic band structure from the 1st BZ shown in (d). The SW frequency modes have been numbered from 1 to 9 according to arrangements at the center of the BZ. (f) Left panel: x components of the magnetization vector, m_x, for modes 1–9. Right panel: evolution of mode 3 for wave vector values $q_y/q_{BZ} = 0, 0.18, 0.34, 0.8$, and 1, as indicated by filled dots and indexed with small letters from a to f in (e). In points e and f, the horizontal broken lines indicate the regions where the stationary waves have the maximum of the spin precessional motion. (Adapted from reference [166].)

FIGURE 21.20 (a) Scanning electron micrographs of circular- and square-shaped $Co_{50}Fe_{50}$ (CoFe) antidot lattices (ADLs) named by S1 and S2 and $Ni_{80}Fe_{20}$ (NiFe)-filled CoFe ADLs named by S3 and S4 arranged in square lattice symmetry are shown. The geometry of the applied magnetic field (H_{ext}) for all lattices is shown in the micrograph of S1. (b) Simulated demagnetization field distributions in S1–S4 with bias field $H_{ext} = 1$ kOe applied at $\phi = 0°$. The grayscale map is shown at the top of the figure. (c) Bias field-dependent SW absorption spectra of circular and square shaped CoFe ADLs (S1 and S2) and NiFe-filled CoFe ADLs (S3 and S4) at $\phi = 0°$. The surface plots correspond to the experimental data, while the symbols correspond to the simulated data. The grayscale for the surface plots are given at top of (c) (Adapted from reference [170].)

at the interface of the two constituent materials, can play a crucial role in manipulating the magnon dynamics, and they can also potentially overcome the shortcoming of small SW

propagation velocity in these BMCs by tailoring the filling region shape of the embedded structures due to the increase in the interfacial exchange coupling.

Vortex and Skyrmions

While dealing with the static magnetic properties and internal domain configuration of micron-sized ferromagnetic disks a highly interesting spin texture was observed, named as a magnetic vortex. A curling in-plane spin configuration is energetically favored, with an out-of-plane magnetization appearing at the core of the vortex (see Figure 21.21a). This gives rise to the clockwise (-1, CW) or counter-clockwise ($+1$, CCW) chirality (C) of the vortex. The core of this vortex, which is a few nanometres in diameter, is represented by polarity (P) $+1$ (up) and -1(down).

The first experimental observation of this core by using MFM imaging was demonstrated in 50-nm-thick $Ni_{80}Fe_{20}$ disks with the varying diameter (0.3–1.0 μm), and a possibility of observing interesting vortex dynamics was proposed [171]. By using MFM and other imaging techniques (i.e. spin-polarized scanning tunneling microscopy), researchers found the signature of the closure domain structure of vortex coexisting with antivortex pair, for particular ranges of dimensions of cylindrical and other magnetic elements [172,173]. Later, a theoretical approach followed by experimental investigation showed that the core of magnetic vortex gyrates in sub-GHz to low-GHz frequency range depending upon the aspect ratio of the element [174]. Mainly, the vortex dynamics consist of low-frequency translational mode corresponding to the movement of the core near its equilibrium position. Another high-frequency vortex

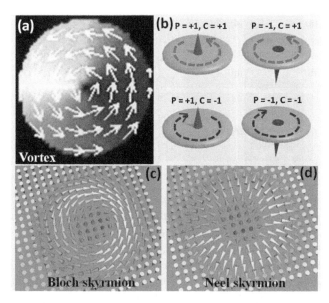

FIGURE 21.21 Schematic representation of (a) curling spin texture, (b) different combination of polarities and chiralities of the magnetic vortex. (c) skyrmion texture.

mode appears that corresponds to radially symmetric oscillations of the vortex magnetization outside the vortex core. Even vortex core switching by applying AC magnetic field or a small amount of charge current was observed in soft magnetic structures [175,176]. Direct imaging by using X-ray PEEM of the dynamics of magnetic vortices in $Ni_{80}Fe_{20}$ dot revealed that after the external perturbation the core gyrates in a self-induced magnetostatic potential [177]. Later, optical excitation and detection techniques were developed for direct observation of vortex dynamics in the time domain by using a MOKE microscopy [178]. $Ni_{80}Fe_{20}$ disks having a diameter of few micron and thickness varying from 20 to 40 nm, patterned *via* electron-beam lithography onto microwave coplanar waveguides were subjected to RF field in a VNA-FMR setup. Analogous to spin dynamics of nanodot array, the influence of the magnetostatic interaction on vortex dynamics in arrays of ferromagnetic disks was investigated [179]. By using a microwave reflection technique, the study of dynamics of magnetic soliton pairs confined in a single $Ni_{80}Fe_{20}$ ellipse created huge interest. Also, a comparison with micromagnetic simulations revealed that observed strong resonances in the sub-GHz frequency range can be originated from the translational modes of vortex pairs with parallel or antiparallel core polarization [180]. Further, it was demonstrated that gyration modes of coupled vortices can be resonantly excited by an AC current in a pair of soft magnetic disks [181]. Another work showed that persistent microwave-frequency oscillations can be induced within a vortex by a spin-polarized direct current [182]. A possibility for designing the magnonic band structure in a chain of magnetic vortex oscillators was also proposed [183]. Later, the micromagnetic simulation was exploited to develop a new way of energy transfer through a 1-D chain of vortices without having any physical

contact [184]. This unique idea was recently implemented experimentally in the construction of magnetic vortex-based logic operation [185]. By using XMCD the magnetic contrast of the vortices was studied, and the formation of XOR and OR gate was realized. Later, it was proposed that three magnetic vortices placed in a linear chain can act like a bipolar junction transistor depending upon their response to the change in relative core polarity [186]. Later, core gyration amplification with the successful fan-out operation was also demonstrated numerically [187]. Not only in soft magnetic material but also in Co-based exchange-biased nanorings, the chirality of vortex was studied in great details [188].

Recently, a different class of magnetic textures has been discovered, which also has a curling spin orientation analogous to the magnetic vortex, named as magnetic skyrmion. However, the spins inside a skyrmion rotate progressively radially inward or outward with a fixed chirality forming more like a whirl shape instead of in-plane curling like vortex [189]. Two typical types of magnetic skyrmions are observed, namely Néel and Bloch type (see Figure 21.21c). In most cases, they are induced by Dzyaloshinskii-Moriya interactions (DMI) between atomic spins in non-centro-symmetric magnetic compounds or thin films with broken inversion symmetry [190]. Recently pattering of confined nanostructures with topologically protected skyrmion was exploited at ambient condition. Theoretically, it was proposed that based upon a combination of a perpendicularly magnetized film and magnetic nanodots arranged in hexagonal lattice symmetry with closure domain structure can stabilize skyrmion in absence of DMI [191]. Following this concept, direct experimental evidence of artificial skyrmion lattices (SLs) with a stable ground state was demonstrated at room temperature [192]. The vortex-state Co nanodots of 500 nm diameter and 1-μm separation were patterned in hexagonal symmetry *via* electron-beam lithography on top of a Co/Pd thin-film multilayer with PMA and the skyrmion state was prepared by exerting a specific magnetic field sequence. To image the remanent state of the magnetic texture, magnetic imaging with MFM and SEM with polarization analysis (SEMPA) were performed after applying skyrmion lattice-setting field sequence. Magnetic transmission X-ray microscopy (MTXM) was also exploited to confirm the formation of skyrmion inside each dot through various other magnetic states. Later, the idea of exploiting only perpendicular single-domain nanomagnet with magnetic anisotropy for nucleating skyrmion at room temperature even in the absence of the Dzyaloshinskii–Moriya exchange interaction and external magnetic field was proposed [193]. It was also highlighted that the stability of skyrmion mostly depends on the nanodot size. Magnetic skyrmion can be driven by external perturbation, and they can achieve large velocity in the skyrmion track. Researchers have generated stable skyrmion lattices in transition metals and driven trains of individual skyrmions by short current pulses along a magnetic racetrack at a speed of more than 100 m/s. This aspect is highly required for applications [194]. Skyrmion crystal can achieve a size that is unreachable for conventional magnonic crystals fabricated with existing lithography techniques [195]. Making use of such functionalities will be beneficial for future spintronics devices.

Spin Ice Structure

Over long time, the possibility of the presence of magnetic monopoles has triggered the imagination of the scientific community as their existence can be reconciled with the quantum phenomena [196] where the observation of such magnetic monopoles can support the out-of-the-box models proposed in various fields of physics, *e.g.* string theory [197], grand unified theory in the particle physics. Similarly, different approaches have been initiated in the condensed matter physics to find out the presence of monopoles, which share similarities with that in quantum mechanics in terms of an artificial frustrated system [198], namely artificial spin ice (ASI) structure [199] in the form of lithographically fabricated single-domain FM islands where the dipolar interactions create a 2-D analogue to the spin ice. Here, both ice-like short-range correlations and absence of long-range correlations can be observed which are strikingly similar to the low-temperature state of spin ice with a significant fact that such artificial frustrated magnets can provide an uncharted arena where the direct visualization of the physics of frustration can be possible. However, one of the most exciting and modest approaches is to predict the monopoles occurring as fractional quasiparticles inside pyrochlore spin ice [200], which are characterized by ice rules, which in case of the tetrahedral coordination of the pyrochlore structure, two of the four Ising-type spins on the vertices point towards the tetrahedral center as opposed to the other two spins [201–204]. This resembles Pauling's ice rule [205] that determines the configuration of the proton ordering in water ice. The decay of a dipolar excitation into constituent monopoles provides the phenomenon of fractionalization [198,200,206] in 3-D, which has also been realized in 1-D as charge solitons in polyacetylene [207] or the decay of magnons into spinons in spin chains [208–210] and even in 2-D as fractionally charged excitations in the quantum Hall effect [211,212]. However, the magnetic monopoles accompanied by Dirac strings [200,213–215] in such pyrochlore systems, which are frustrated magnetic insulators, have been realized by means of neutron scattering [216,217] in the reciprocal space at sub-Kelvin (-K) temperature although their real-space observation remained inconclusive earlier. However, the direct and real-time observations of emergent monopoles associated with the Dirac strings have been successfully reported [218,219] (see Figure 21.22) at room temperature.

This is followed by the direct observation [220] of the nanoscale magnetic structure of individual monopoles in an artificially frustrated 2-D NiFe square spin ice lattice demonstrating the control of demagnetization processes [221] in such lattices. A MOKE study [222] of the collective magnetic response of NiFe square ASI revealed a nonmonotonic angular dependence of the array's coercive field indicating distinct responses by two perpendicular sublattices to the

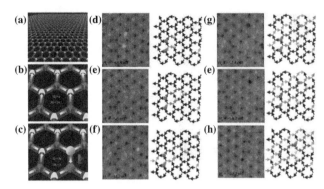

FIGURE 21.22 (a) Schematic of a magnetic charge-ordered state of alternating $Q = +q$ and $Q = -q$ vertices of a 2-D artificial spin ice (ASI) honeycomb nanoarray. (b) An in-plane field (into the page) has been applied to split the manifold and select the saturated Ising state with parallel total vertex moments $M = \Sigma m$. (c) By flipping the central moment with an applied field out of the page, two "monopole" defects are created. MFM images in zero field after the labeled conditioning fields representing (d) two 3-in monopole defects with $Q = +3q$ form (bright spots, dots), with strings of head–tail spins to opposite magnetic charges $\Delta Q = -2q$. Further bar flips are required to make the schematic (right) match the observed data; the arrows indicate a (non-unique) trial solution. (e) The $\Delta Q = +2q$ magnetic charges hop to the right, changing from $Q = +3q$ to $Q = +q$. (f–h) Another $\Delta Q = -2q$ magnetic charge appears, tracing its own light shaded string to the left, until it is blocked by the dark shaded string of the $\Delta Q = +2q$ carrier. (Adapted from reference [219].)

island edge roughness thereby influencing the magnetization reversal process. The cooperative process associated with two transverse domain walls creating the monopole defects in NiFe ASI demonstrated [223] stabilization of monopole defects whose formation within a Co honeycomb ASI has been described [224]. Later, the magnetic reversal of lithographically patterned NiFe islands in a square ASI geometry unraveled [225] the presence of a Dirac string-like flux channel between the magnetic charge monopoles.

The systematic investigations [226–231] of thermally induced properties in ASI systems (see Figure 21.23) can open a whole new direction for technological applications. The dynamic response of a NiFe square ASI revealed [232] the influence of the state change of the local orientation of nearby FM islands on the resonant mode during field hysteresis. A theoretical investigation [233] by studying the eigenmode dynamics of square NiFe ASI nanostructures with topological defects showed that these defects can play a crucial role to analyze magnetic monopoles and the associated Dirac strings. Interestingly, some experimental studies [234,235] on 2-D NiFe or Co honeycomb ASI structures demonstrated the chiral control of magnetic charge to the magnetic charge path selectivity providing an interesting tool leading to a new architecture for parallel computation.

The magnetization dynamics in ASI systems made of square or interconnected kagome NiFe nano-bars or Co cluster unraveled [236–239] the existence of a series

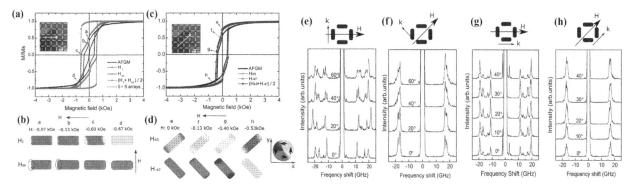

FIGURE 21.23 Hysteresis loops of the alternative gradient field magnetometry (AGFM) measurements and numerical simulation at (a) 0° and (c) 45° applied field with respect to the artificial spin ice (ASI) lattices, in which insets are SEM images showing the geometry of ASI lattices and the orientation of external field (arrows). Ground state magnetization configurations at the (b) parallel and (d) diagonal field, and gray code disk represents the direction of magnetization. Sequence of BLS spectra measured at different incidence angles θ with the external field of 3 kOe at (e) 0° and (f) 45° upon the ASI lattices. The wave vector of the incident light is perpendicular to the field orientation in Damon–Eshbach configuration. Sequence of BLS spectra measured at different incidence angles θ with the external field of 3 kOe applied at (g) 0° and (h) 45° external field at 3 kOe with respect to the ASI lattices. The wave vector of the incident light parallels to the applied field in backward configuration. (Adapted from reference [231].)

of resonances with characteristic magnetic field dependence in the spectra obtained at saturated and disordered states reflecting the spin-solid (spin ice) state and monopole-antimonopole pairs on Dirac strings. This can allow the generation of highly charged vertices in ASIs *via* microwave-assisted switching thereby opening prospective for further application of such ASIs in reprogrammable magnonics [240].

Recently, the static and dynamic properties of large spin ice and anti-spin ice NiFe structures with three different lattice configurations were investigated [241] systematically

(see Figure 21.24) exploring the intriguing static and dynamic behaviors due to the geometrical arrangement of the nanomagnets in the lattice. This may open a new possibility of application in the design of magnetically controlled tunable microwave filters. A very interesting design of NiFe ASI nanostructure producing magnetic charge ice with the tunable long-range ordering of several different configurations has been proposed [242] at room temperature. Such globally reconfigurable as well as locally writable magnetic charge ice can provide a roadmap for designing magnetic monopole defects, tailoring magnonics and also controlling

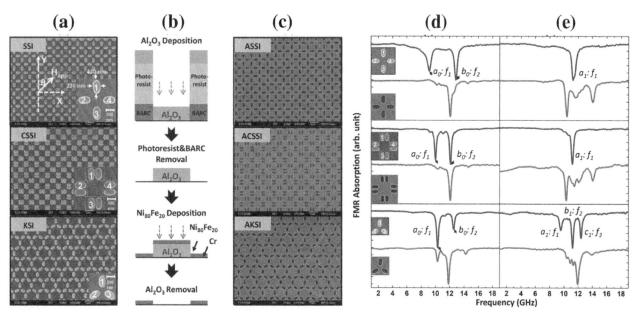

FIGURE 21.24 (a) Scanning electron images of spin ice: square spin ice (SSI), coupled square spin ice (CSSI), kagome spin ice (KSI). (b) Schematic diagrams showing detailed fabrication processes to get the reverse structures. (c) Scanning electron micrographs of anti-spin ice structures: anti-square spin ice (ASSI), anti-coupled square spin ice (ACSSI), anti-kagome spin ice (AKSI). Experimental ferromagnetic resonance (FMR) absorption curves for all the structures at saturation ($H_{\text{app}} = -1,400$ Oe) for (d) $\theta = 0°$ and (e) $\theta = 45°$. (Adapted from reference [241].)

the properties of other 2-D materials. However, few different approaches have been proposed to investigate the novel spin ice systems. A study [243] on one such unique ASI system based on a vertex-frustrated structure reminiscent of 3-D natural spin ice instead of pairwise frustrated geometry (spin ice pyrochlores) has shown that such system can exhibit a quasicritical ice phase of extensive residual entropy and more significantly algebraic correlations. Such novel realization of frustration in a vertex system can pave a new pathway to study defects in a critical manifold and to design degeneracy in such artificial magnetic nanoarrays. Later, a viable design for the realization of a 3-D ASI system demonstrated [244] that by stacking planar 2-D ASIs, one can achieve an arrangement of ice-rule-frustrated units, which is topologically equivalent to that of a pyrochlore spin ice, and it can possess a genuine ice phase in which the excitations are similar to that in natural spin ice materials, *i.e.* magnetic monopoles interacting *via* Coulomb rule. Recently, a new approach has been introduced [245], namely single interaction modifiers using mesospins in the form of disks within which the mesospin is free to rotate in the disk plane [246] enabling the exploration of spin-liquid [198] manifold.

21.6.3 Three-Dimensional Magnetic Nanostructures

An indispensable part of 21st century's nanomagnetism is 3-D magnetic structures. With the inclusion of structural freedom in the third dimension for magnetic nanostructures, more complex magnetic configurations become possible with many unprecedented properties. The most difficult part of sample fabrication and detection of high-frequency dynamics are subjected to rigorous technological advancement. Nano-patterning in the third dimension gives more control points than 2-D nanomagnets, fulfilling the demand of increasing areal density in magnetic memory devices, which is growing rapidly (schematics of 2-D and 3-D nanostructures are shown in Figure 21.25).

Not only from the application point of view but also 3-D magnetic nanostructures are important because they involve new physics and address many fundamental aspects.

FIGURE 21.25 Schematic view comparing some examples of geometries and magnetic configurations indicated by arrow for (a) 3-D and (b) 2-D nanomagnetism. (Adapted from reference [248].)

A simple example is the evolution of domain wall (DW) dynamics in nanowire (NW) in the last decade and possibly its biggest manifestation is racetrack memory [247]. Generally, a soft magnetic nanowire has shape anisotropy along the length and unsaturated magnetic charges at the edges. It can exhibit magnetization reversal through various processes (*i.e.* curling, buckling, fanning, corkscrew) within the system when it is subjected to an external magnetic field. However, in ground state, the extended domains form DWs of two types: transverse DW (TDW) (where magnetization is perpendicular to the wire surface) and vortex DW (VDW). For large lateral wire dimensions, simultaneous transverse and vortex characteristics are displayed. They may be denoted as transverse-vortex domain walls (TVDWs) [248]. This topological equivalence is valid for any NW cross-sectional shape, for instance, either square or circular (*i.e.* nanostripe with rectangular cross-section). The three-dimensional nature of the DW was revealed in micromagnetic simulation first [249]. An axial vortex, whose axis is parallel to the wire with the Bloch point lying on the axis, can be named as a Bloch point DW (BPDW). The DW dynamics associated with the nanowire or nanostripe face a bottleneck. Usually, DWs under external fields become unstable when they propagate above a critical speed resulting in Walker breakdown. Now, theoretical studies have revealed that BPDWs can move smoothly with very high speed of almost 1 km/s without having Walker breakdown [250]. Developments on building a hybrid atomistic-micromagnetic code have shown promising approach to observe dynamics of BPDWs in NWs at the atomic scale although experimental techniques still need to be improved. Recently 3-D nanomagnetic logic (NML) was achieved based on the magnetic coupling between nanowires and nanopillars in a 3-D array [251]. Vertically aligned nanopillars (NPs) actually increase the packing density. Recently, some very interesting 3-D magnetic nanostructures, *i.e.* vertical magnetic nanowires, angled magnetic nanowires, and complex 3-D tetrapod nanostructures, were fabricated using two-photon lithography and electro-deposition [28]. The domain structure within the 3-D magnetic nanostructures of complex geometry was imaged by spin-SEM and surface-sensitive magnetometry technique like MOKE. Magnetic imaging experiments provided evidence of domain wall pinning at the 3-D nanostructured junction. Unlike planar surfaces, curved surfaces do not possess space inversion symmetry. This gives rise to a new wing of nanomagnetism that is curvature-induced DMI, which consists of reducing magnetostatic effects to an effective anisotropy. An important aspect of magnetic solitons (kinks in the homogeneous magnetic plane) dynamics has now been explored in third dimension. Instead of using soliton motion in dipolar-coupled nanomagnets on the substrate plane, solitons are moved perpendicularly to the substrate plane in magnetic superlattices. 3-D magnetic nano-membrane has universal potential to be used as flexible magnetic sensors. 3-D nanomagnets could be potential candidates as sensors in the new generation of scanning probe microscopy methods.

In imaging and spectroscopy, NWs with high aspect ratio are already employed as ultra-sharp MFM tips. By the creation of complex 3-D networks of nanomagnets, a new computational paradigm can be unlocked.

Acknowledgment

We gratefully acknowledge the financial assistance from the S. N. Bose National Centre for Basic Sciences under Project Nos. SNB/AB/12-13/96 and SNB/AB/18-19/211. S. C. acknowledges the S. N. Bose National Centre for Basic Sciences for Senior Research Fellowship. S. M. and A. D. acknowledge DST, Government of India for support from the INSPIRE Fellowship.

References

1. S. Tehrani, E. Chen, M. Durlam, M. DeHerrera, J. M. Slaughter, J. Shi, and G. Kerszykowski, *J. Appl. Phys.* 85, 5822 (1999).
2. J. Åkerman, *Science* 308, 508 (2005).
3. J. R. Childress and R. E. Fontana Jr, C. R. *Physique* 6, 997 (2005).
4. T. Thomson, G. Hu, and B. D. Terris, *Phys. Rev. Lett.* 96, 257204 (2006).
5. S. H. Chung, A. Hoffmann, S. D. Bader, C. Liu, B. Kay, L. Makowski, and L. Chen, *Appl. Phys. Lett.* 85, 2971 (2004).
6. M. Arruebo, R. Fernández-Pacheco, M. R. Ibarra, and J. Santamaría, *Nano Today* 2, 22 (2007).
7. A. K. Gupta and M. Gupta, *Biomaterials* 26, 3995 (2005).
8. D. A. Allwood, G. Xiong, C. C. Faulkner, D. Atkinson, D. Petit, and R. P. Cowburn, *Science* 309, 1688 (2005).
9. A. Imre, G. Csaba, L. Ji, A. Orlov, G. H. Bernstein, and W. Porod, *Science* 311, 205 (2006).
10. S. Kaka, M. R. Pufall, W. H. Rippard, T. J. Silva, S. E. Russek, and J. A. Katine, *Nature* 437, 389 (2005).
11. V. E. Demidov, S. Urazhdin, H. Ulrichs, V. Tiberkevich, A. Slavin, D. Baither, G. Schmitz, and S. O. Demokritov, *Nat. Mater.* 11, 1028 (2012).
12. B. Lenk, H. Ulrichs, F. Garbs, and M. Münzenberg, *Phys. Rep.* 507, 107 (2011).
13. A. L. Rogach, D. V. Talapin, E. V. Shevchenko, A. Kornowski, M. Haase, and H. Weller, *Adv. Funct. Mater.* 12, 653 (2002).
14. W. D. Williams and N. Giordano, *Phys. Rev. B* 33, 8146 (1986).
15. H. Masuda, H. Yamada, M. Satoh, H. Asoh, M. Nakao, and T. Tamamura, *Appl. Phys. Lett.* 71, 2770 (1997).
16. Y. Kamata, A. Kikitsu, H. Hieda, M. Sakurai, and K. Naito, *J. Appl. Phys.* 95, 6705 (2004).
17. J. Aizenberg, J. A. Rogers, K. E. Paul, and G. M. Whitesides, *Appl. Phys. Lett.* 71, 3773 (1997).
18. P. B. Fischer and S. Y. Chou, *Appl. Phys. Lett.* 62, 2989 (1993).
19. N. Singh, S. Goolaup, and A. O. Adeyeye, *Nanotechnology* 15, 1539 (2004).
20. D. Parikh, B. Craver, H. N. Nounu, F. O. Fong, and J. C. Wolfe, *J. Microelectromech. Syst.* 17, 735 (2008).
21. F. Marty, A. Vaterlaus, V. Weich, C. Stamm, U. Maier, and D. Pescia, *J. Appl. Phys.* 85, 6166 (1999).
22. T. Devolder, C. Chappert, Y. Chen, E. Cambril, H. Bernas, J. P. Jamet, and J. Ferre, *Appl. Phys. Lett.* 74, 3383 (1999).
23. F. Rousseaux, D. Decanini, F. Carcenac, E. Cambril, M. F. Ravet, C. Chappert, N. Bardou, B. Bartenlian, and P. Veillet, *J. Vac. Sci. Technol. B* 13, 2787 (1995).
24. K. Bessho, Y. Iwasaki, and S. Hashimoto, *J. Appl. Phys.* 79, 5057 (1996).
25. E. F. Wassermann, M. Thielen, S. Kirsch, A. Pollmann, H. Weinforth, and A. Carl, *J. Appl. Phys.* 83, 1753 (1998).
26. J. Moritz et al., *IEEE Trans. Magn.* 38, 1731 (2002).
27. A. Dallmeyer, C. Carbone, W. Eberhardt, C. Pampuch, O. Rader, W. Gudat, P. Gambardella, and K. Kern, *Phys. Rev. B* 61, R5133 (2000).
28. G. Williams et al., *Nano Res.* 11, 845 (2018).
29. R. Skomski, *J. Phys. Condens. Matter* 15, R841 (2003).
30. E. H. Frei, S. Shtrikman, and D. Treves, *Phys. Rev.* 106, 446 (1957).
31. I. S. Jacobs and C. P. Bean, *Phys. Rev.* 100, 1060 (1955).
32. R. Hertel, *J. Magn. Magn. Mater.* 249, 251 (2002).
33. R. Hertel and J. Kirschner, *Phys. Condens. Matter* 343, 206 (2004).
34. M. E. Schabes and H. N. Bertram, *J. Appl. Phys.* 64, 1347 (1988).
35. M. R. Freeman and B. C. Choi, *Science* 294, 1484 (2001).
36. Y. Martin and H. K. Wickramasinghe, *Appl. Phys. Lett.* 50, 1455 (1987).
37. J. N. Chapman, *J. Phys. D Appl. Phys.* 17, 623 (1984).
38. M. S. Cohen, *J. Appl. Phys.* 38, 4966 (1967).
39. B. E. Argyle and J. G. McCord, *J. Appl. Phys.* 87, 6487 (2000).
40. J. Stohr, Y. Wu, B. D. Hermsmeier, M. G. Samant, G. R. Harp, S. Koranda, D. Dunham, and B. P. Tonner, *Science* 259, 658 (1993).
41. H. Pinkvos, H. Poppa, E. Bauer, and J. Hurst, *Ultramicroscopy* 47, 339 (1992).
42. K. Koike and K. Hayakawa, *Jpn. J. Appl. Phys.* 23, L187 (1984).
43. M. Johnson and J. Clarke, *J. Appl. Phys.* 67, 6141 (1990).
44. W. H. Rippard and R. A. Buhrman, *Appl. Phys. Lett.* 75, 1001 (1999).

45. A. Barman and J. Sinha, *Spin Dynamics and Damping in Ferromagnetic Thin Films and Nanostructures* (Springer, Berlin, 2018), 1 edn.

46. S. V. Vonsovskii, *Ferromagnetic Resonance; the Phenomenon of Resonant Absorption of a High-Frequency Magnetic Field in Ferromagnetic Substances* (Pergamon Press, Oxford; New York, 1966).

47. V. P. Denysenkov and A. M. Grishin, *Rev. Sci. Instrum.* 74, 3400 (2003).

48. S. Tamaru, J. A. Bain, R. J. M. van de Veerdonk, T. M. Crawford, M. Covington, and M. H. Kryder, *J. Appl. Phys.* 91, 8034 (2002).

49. T. J. Silva, C. S. Lee, T. M. Crawford, and C. T. Rogers, *J. Appl. Phys.* 85, 7849 (1999).

50. S. O. Demokritov, B. Hillebrands, and A. N. Slavin, *Phys. Rep.* 348, 441 (2001).

51. A. N. Slavin, S. O. Demokritov, and B. Hillebrands, in *Spin Dynamics in Confined Magnetic Structures I*, edited by B. Hillebrands, and K. Ounadjela (Springer, Berlin, 2002), p. 35.

52. A. Barman, S. Wang, J. D. Maas, A. R. Hawkins, S. Kwon, A. Liddle, J. Bokor, and H. Schmidt, *Nano Lett.* 6, 2939 (2006).

53. I. N. Krivorotov, N. C. Emley, J. C. Sankey, S. I. Kiselev, D. C. Ralph, and R. A. Buhrman, *Science* 307, 228 (2005).

54. Y. Acremann et al., *Phys. Rev. Lett.* 96, 217202 (2006).

55. W. K. Hiebert, A. Stankiewicz, and M. R. Freeman, *Phys. Rev. Lett.* 79, 1134 (1997).

56. S.-K. Kim, *J. Phys. D Appl. Phys.* 43, 264004 (2010).

57. N.-N. Chen, A. N. Slavin, and M. G. Cottam, *Phys. Rev. B* 47, 8667 (1993).

58. E. L. Albuquerque and M. G. Cottam, *Phys. Rep.* 376, 225 (2003).

59. M. Wu, B. A. Kalinikos, L. D. Carr, and C. E. Patton, *Phys. Rev. Lett.* 96, 187202 (2006).

60. G. Gubbiotti, S. Tacchi, G. Carlotti, P. Vavassori, N. Singh, S. Goolaup, A. O. Adeyeye, A. Stashkevich, and M. Kostylev, *Phys. Rev. B* 72, 224413 (2005).

61. S. Goolaup, A. O. Adeyeye, N. Singh, and G. Gubbiotti, *Phys. Rev. B* 75, 144430 (2007).

62. S. Tacchi, M. Madami, G. Gubbiotti, G. Carlotti, S. Goolaup, A. O. Adeyeye, N. Singh, and M. P. Kostylev, *Phys. Rev. B* 82, 184408 (2010).

63. C. Schoeppner, K. Wagner, S. Stienen, R. Meckenstock, M. Farle, R. Narkowicz, D. Suter, and J. Lindner, *J. Appl. Phys.* 116, 033913 (2014).

64. S. Saha, S. Barman, Y. Otani, and A. Barman, *Nanoscale* 7, 18312 (2015).

65. S. Pal, S. Saha, M. V. Kamalakar, and A. Barman, *Nano Res.* 9, 1426 (2016).

66. C. Banerjee, S. Choudhury, J. Sinha, and A. Barman, *Phys. Rev. Appl.* 8, 014036 (2017).

67. V. E. Demidov, J. Jersch, K. Rott, P. Krzysteczko, G. Reiss, and S. O. Demokritov, *Phys. Rev. Lett.* 102, 177207 (2009).

68. A. V. Chumak et al., *Appl. Phys. Lett.* 95, 262508 (2009).

69. M. Kostylev, P. Schrader, R. L. Stamps, G. Gubbiotti, G. Carlotti, A. O. Adeyeye, S. Goolaup, and N. Singh, *Appl. Phys. Lett.* 92, 132504 (2008).

70. S. Jung, B. Watkins, L. DeLong, J. B. Ketterson, and V. Chandrasekhar, *Phys. Rev. B* 66, 132401 (2002).

71. G. N. Kakazei, P. E. Wigen, K. Y. Guslienko, V. Novosad, A. N. Slavin, V. O. Golub, N. A. Lesnik, and Y. Otani, *Appl. Phys. Lett.* 85, 443 (2004).

72. G. Gubbiotti, G. Carlotti, T. Okuno, T. Shinjo, F. Nizzoli, and R. Zivieri, *Phys. Rev. B* 68, 184409 (2003).

73. S. M. Weekes, F. Y. Ogrin, and P. S. Keatley, *J. Appl. Phys.* 99, 08B102 (2006).

74. V. V. Kruglyak, A. Barman, R. J. Hicken, J. R. Childress, and J. A. Katine, *J. Appl. Phys.* 97, 10A706 (2005).

75. V. V. Kruglyak, A. Barman, R. J. Hicken, J. R. Childress, and J. A. Katine, *Phys. Rev. B* 71, 220409(R) (2005).

76. B. Rana, S. Pal, S. Barman, Y. Fukuma, Y. Otani, and A. Barman, *Appl. Phys. Express.* 4, 113003 (2011).

77. B. Rana, D. Kumar, S. Barman, S. Pal, Y. Fukuma, Y. Otani, and A. Barman, *ACS Nano* 5, 9559 (2011).

78. B. Rana, D. Kumar, S. Barman, S. Pal, R. Mandal, Y. Fukuma, Y. Otani, S. Sugimoto, and A. Barman, *J. Appl. Phys.* 111, 07D503 (2012).

79. S. Mondal, S. Choudhury, S. Barman, Y. Otani, and A. Barman, *RSC Adv.* 6, 110393 (2016).

80. S. Saha, R. Mandal, S. Barman, D. Kumar, B. Rana, Y. Fukuma, S. Sugimoto, Y. Otani, and A. Barman, *Adv. Funct. Mater.* 23, 2378 (2013).

81. S. Saha, S. Barman, S. Sugimoto, Y. Otani, and A. Barman, *RSC Adv.* 5, 34027 (2015).

82. S. Mondal, S. Barman, S. Choudhury, Y. Otani, and A. Barman, *J. Magn. Magn. Mater.* 458, 95 (2018).

83. B. K. Mahato, B. Rana, D. Kumar, S. Barman, S. Sugimoto, Y. Otani, and A. Barman, *Appl. Phys. Lett.* 105, 012406 (2014).

84. K. Adhikari, S. Choudhury, R. Mandal, S. Barman, Y. Otani, and A. Barman, *J. Appl. Phys.* 121, 043909 (2017).

85. G. Shimon and A. A. O., *Adv. Electron. Mater.* 1, 1500070 (2015).

86. A. De, S. Mondal, C. Banerjee, A. K. Chaurasiya, R. Mandal, Y. Otani, R. K. Mitra, and A. Barman, *J. Phys. D Appl. Phys.* 50, 385002 (2017).

87. V. Metlushko et al., in *Proceedings of the 2nd IEEE Conference on Nanotechnology* (2002), p. 63.

88. L. Tripp S., E. Dunin-B. R., and A. Wei, *Angew. Chem. Int. Ed.* 42, 5591 (2003).

89. P. Vavassori, M. Grimsditch, V. Novosad, V. Metlushko, and B. Ilic, *J. Appl. Phys.* 93, 7900 (2003).

90. J. Wang, A. O. Adeyeye, and N. Singh, *Appl. Phys. Lett.* 87, 262508 (2005).

91. F. Q. Zhu, G. W. Chern, O. Tchernyshyov, X. C. Zhu, J. G. Zhu, and C. L. Chien, *Phys. Rev. Lett.* 96, 027205 (2006).

92. L. J. Chang, C. Yu, T. W. Chiang, K. W. Cheng, W. T. Chiu, S. F. Lee, Y. Liou, and Y. D. Yao, *J. Appl. Phys.* 103, 07C514 (2008).

93. X. Zhu, M. Malac, Z. Liu, H. Qian, V. Metlushko, and M. R. Freeman, *Appl. Phys. Lett.* 86, 262502 (2005).

94. G. Gubbiotti, M. Madami, S. Tacchi, G. Carlotti, H. Tanigawa, T. Ono, L. Giovannini, F. Montoncello, and F. Nizzoli, *Phys. Rev. Lett.* 97, 247203 (2006).

95. H. Schultheiss, S. Schäfer, P. Candeloro, B. Leven, B. Hillebrands, and A. N. Slavin, *Phys. Rev. Lett.* 100, 047204 (2008).

96. J. Ding, M. Kostylev, and A. O. Adeyeye, *Appl. Phys. Lett.* 100, 062401 (2012).

97. C. Banerjee, S. Saha, S. Barman, O. Rousseau, Y. Otani, and A. Barman, *J. Appl. Phys.* 116, 163912 (2014).

98. C. T. Yu, H. Jiang, L. Shen, P. J. Flanders, and G. J. Mankey, *J. Appl. Phys.* 87, 6322 (2000).

99. P. Vavassori, G. Gubbiotti, G. Zangari, C. T. Yu, H. Yin, H. Jiang, and G. J. Mankey, *J. Appl. Phys.* 91, 7992 (2002).

100. O. N. Martyanov, V. F. Yudanov, R. N. Lee, S. A. Nepijko, H. J. Elmers, R. Hertel, C. M. Schneider, and G. Schönhense, *Phys. Rev. B* 75, 174429 (2007).

101. S. McPhail, C. M. Gürtler, J. M. Shilton, N. J. Curson, and J. A. C. Bland, *Phys. Rev. B* 72, 094414 (2005).

102. C. C. Wang, A. O. Adeyeye, N. Singh, Y. S. Huang, and Y. H. Wu, *Phys. Rev. B* 72, 174426 (2005).

103. C. C. Wang, A. O. Adeyeye, and N. Singh, *Appl. Phys. Lett.* 88, 222506 (2006).

104. D. Tripathy and A. O. Adeyeye, *J. Appl. Phys.* 105, 07D703 (2009).

105. C. L. Hu, R. Magaraggia, H. Y. Yuan, C. S. Chang, M. Kostylev, D. Tripathy, A. O. Adeyeye, and R. L. Stamps, *Appl. Phys. Lett.* 98, 262508 (2011).

106. S. Pal, J. W. Klos, K. Das, O. Hellwig, P. Gruszecki, M. Krawczyk, and A. Barman, *Appl. Phys. Lett.* 105, 162408 (2014).

107. M. Yu, L. Malkinski, L. Spinu, W. Zhou, and S. Whittenburg, *J. Appl. Phys.* 101, 09F501 (2007).

108. J. Ding, D. Tripathy, and A. O. Adeyeye, *J. Appl. Phys.* 109, 07D304 (2011).

109. R. Zivieri, P. Malag, L. Giovannini, S. Tacchi, G. Gubbiotti, and A. Adeyeye, *J. Phys. Condens. Matter* 25, 336002 (2013).

110. S. Neusser, B. Botters, and D. Grundler, *Phys. Rev. B* 78, 054406 (2008).

111. M. Kostylev, G. Gubbiotti, G. Carlotti, G. Socino, S. Tacchi, C. Wang, N. Singh, A. O. Adeyeye, and R. L. Stamps, *J. Appl. Phys.* 103, 07C507 (2008).

112. S. Neusser, G. Duerr, H. G. Bauer, S. Tacchi, M. Madami, G. Woltersdorf, G. Gubbiotti, C. H. Back, and D. Grundler, *Phys. Rev. Lett.* 105, 067208 (2010).

113. S. Tacchi, M. Madami, G. Gubbiotti, G. Carlotti, A. O. Adeyeye, S. Neusser, B. Botters, and D. Grundler, *IEEE Trans. Magn.* 46, 1440 (2010).

114. N. G. Deshpande, M. S. Seo, X. R. Jin, S. J. Lee, Y. P. Lee, J. Y. Rhee, and K. W. Kim, *Appl. Phys. Lett.* 96, 122503 (2010).

115. S. Tacchi et al., *Phys. Rev. B* 86, 014417 (2012).

116. R. Mandal, S. Barman, S. Saha, Y. Otani, and A. Barman, *J. Appl. Phys.* 118, 053910 (2015).

117. H. Ulrichs, B. Lenk, and M. Mnzenberg, *Appl. Phys. Lett.* 97, 092506 (2010).

118. S. Tacchi, M. Madami, G. Gubbiotti, G. Carlotti, A. O. Adeyeye, S. Neusser, B. Botters, and D. Grundler, *IEEE Trans. Magn.* 46, 172 (2010).

119. S. Neusser, H. G. Bauer, G. Duerr, R. Huber, S. Mamica, G. Woltersdorf, M. Krawczyk, C. H. Back, and D. Grundler, *Phys. Rev. B* 84, 184411 (2011).

120. S. Neusser, G. Duerr, S. Tacchi, M. Madami, M. L. Sokolovskyy, G. Gubbiotti, M. Krawczyk, and D. Grundler, *Phys. Rev. B* 84, 094454 (2011).

121. R. Mandal et al., *ACS Nano* 6, 3397 (2012).

122. V. Bhat, J. Woods, L. E. De Long, J. T. Hastings, V. V. Metlushko, K. Rivkin, O. Heinonen, J. Sklenar, and J. B. Ketterson, *Phys. C Supercond.* 479, 83 (2012).

123. R. Bali, M. Kostylev, D. Tripathy, A. O. Adeyeye, and S. Samarin, *Phys. Rev. B* 85, 104414 (2012).

124. N. G. Deshpande, M. S. Seo, S. J. Lee, L. Y. Chen, K. W. Kim, J. Y. Rhee, Y. H. Kim, and Y. P. Lee, *J. Appl. Phys.* 111, 013906 (2012).

125. N. G. Deshpande, J. S. Hwang, K. W. Kim, J. Y. Rhee, Y. H. Kim, L. Y. Chen, and Y. P. Lee, *Appl. Phys. Lett.* 100, 222403 (2012).

126. J. Ding, D. Tripathy, and A. Adeyeye, *EPL* 98, 16004 (2012).

127. M. Madami, S. Tacchi, G. Gubbiotti, G. Carlotti, J. Ding, A. O. Adeyeye, J. W. Klos, and M. Krawczyk, *IEEE Trans. Magn.* 49, 3093 (2013).

128. R. Mandal, P. Laha, K. Das, S. Saha, S. Barman, A. K. Raychaudhuri, and A. Barman, *Appl. Phys. Lett.* 103, 262410 (2013).

129. A. De, S. Mondal, S. Sahoo, S. Barman, Y. Otani, R. K. Mitra, and A. Barman, *Beilstein J. Nanotechnol.* 9, 1123 (2018).

130. F. Haering et al., *Nanotechnology* 24, 465709 (2013).

131. U. Wiedwald, J. Gräfe, K. M. Lebecki, M. Skripnik, F. Haering, G. Schütz, P. Ziemann, E. Goering, and U. Nowak, *Beilstein J. Nanotechnol.* 7, 733 (2016).

132. Y. H. Jang and J. H. Cho, *J. Appl. Phys.* 115, 063903 (2014).

133. T. Schwarze and D. Grundler, *Appl. Phys. Lett.* 102, 222412 (2013).

134. M. Madami, G. Gubbiotti, S. Tacchi, G. Carlotti, and S. Jain, *Phys. B Condens. Matter* 435, 152 (2014).

135. S. Tacchi, P. Gruszecki, M. Madami, G. Carlotti, J. W. Kłos, M. Krawczyk, A. Adeyeye, and G. Gubbiotti, *Sci. Rep.* 5, 10367 (2015).

136. N. Porwal, S. Mondal, S. Choudhury, A. De, J. Sinha, A. Barman, and P. K. Datta, *J. Phys. D Appl. Phys.* 51, 055004 (2018).

137. A. Barman, *J. Phys. D Appl. Phys.* 43, 195002 (2010).

138. J. W. Kłos, D. Kumar, J. Romero-Vivas, H. Fangohr, M. Franchin, M. Krawczyk, and A. Barman, *Phys. Rev. B* 86, 184433 (2012).

139. D. Kumar, P. Sabareesan, W. Wang, H. Fangohr, and A. Barman, *J. Appl. Phys.* 114, 023910 (2013).

140. J. W. Kłos, D. Kumar, M. Krawczyk, and A. Barman, *Sci. Rep.* 3, 2444 (2013).

141. J. W. Kłos, D. Kumar, M. Krawczyk, and A. Barman, *Phys. Rev. B* 89, 014406 (2014).

142. D. Kumar, J. W. Kłos, M. Krawczyk, and A. Barman, *J. Appl. Phys.* 115, 043917 (2014).

143. R. Zivieri and L. Giovannini, *Photonics Nanostruct.- Fundam. Appl.* 11, 191 (2013).

144. C. H. O. Costa, M. S. Vasconcelos, P. H. R. Barbosa, and F. F. Barbosa Filho, J. *Magn. Magn. Mater.* 324, 2315 (2012).

145. W. J. Hsueh, C. H. Chen, and R. Z. Qiu, *Phys. Lett. A* 377, 1378 (2013).

146. C. H. Chen, R. Z. Qiu, C. H. Chang, and W. J. Hsueh, *AIP Adv.* 4, 087102 (2014).

147. J. Rychły, S. Mieszczak, and J. W. Kłos, *J. Magn. Magn. Mater.* 450, 18 (2018).

148. J. Ding, M. Kostylev, and A. O. Adeyeye, *Phys. Rev. Lett.* 107, 047205 (2011).

149. S. V. Grishin, E. N. Beginin, M. A. Morozova, Y. P. Sharaevskii, and S. A. Nikitov, *J. Appl. Phys.* 115, 053908 (2014).

150. V. S. Bhat, J. Sklenar, B. Farmer, J. Woods, J. T. Hastings, S. J. Lee, J. B. Ketterson, and L. E. De Long, *Phys. Rev. Lett.* 111, 077201 (2013).

151. S. Choudhury, S. Barman, Y. Otani, and A. Barman, *ACS Nano* 11, 8814 (2017).

152. Z. K. Wang, V. L. Zhang, H. S. Lim, S. C. Ng, M. H. Kuok, S. Jain, and A. O. Adeyeye, *Appl. Phys. Lett.* 94, 083112 (2009).

153. Z. K. Wang, V. L. Zhang, H. S. Lim, S. C. Ng, M. H. Kuok, S. Jain, and A. O. Adeyeye, *ACS Nano* 4, 643 (2010).

154. C. S. Lin, H. S. Lim, Z. K. Wang, S. C. Ng, and M. H. Kuok, *Appl. Phys. Lett.* 98, 022504 (2011).

155. V. L. Zhang, H. S. Lim, C. S. Lin, Z. K. Wang, S. C. Ng, M. H. Kuok, S. Jain, A. O. Adeyeye, and M. G. Cottam, *Appl. Phys. Lett.* 99, 143118 (2011).

156. V. L. Zhang, F. S. Ma, H. H. Pan, C. S. Lin, H. S. Lim, S. C. Ng, M. H. Kuok, S. Jain, and A. O. Adeyeye, *Appl. Phys. Lett.* 100, 163118 (2012).

157. M. Mruczkiewicz, M. Krawczyk, G. Gubbiotti, S. Tacchi, A. F. Yu, D. V. Kalyabin, I. V. Lisenkov, and S. A. Nikitov, *New J. Phys.* 15, 113023 (2013).

158. D. Giridharan, P. Sabareesan, and M. Daniel, *Phys. Rev. E* 94, 032222 (2016).

159. J. O. Vasseur, L. Dobrzynski, B. Djafari-Rouhani, and H. Puszkarski, *Phys. Rev. B* 54, 1043 (1996).

160. A. O. Adeyeye, S. Jain, and Y. Ren, *IEEE Trans. Magn.* 47, 1639 (2011).

161. F. S. Ma, H. S. Lim, Z. K. Wang, S. N. Piramanayagam, S. C. Ng, and M. H. Kuok, *Appl. Phys. Lett.* 98, 153107 (2011).

162. F. S. Ma, H. S. Lim, Z. K. Wang, S. N. Piramanayagam, S. C. Ng, and M. H. Kuok, *IEEE Trans. Magn.* 47, 2689 (2011).

163. G. Duerr, M. Madami, S. Neusser, S. Tacchi, G. Gubbiotti, G. Carlotti, and D. Grundler, *Appl. Phys. Lett.* 99, 202502 (2011).

164. G. Duerr, S. Tacchi, G. Gubbiotti, and D. Grundler, *J. Phys. D Appl. Phys.* 47, 325001 (2014).

165. M. Krawczyk, S. Mamica, M. Mruczkiewicz, J. W. Klos, S. Tacchi, M. Madami, G. Gubbiotti, G. Duerr, and D. Grundler, *J. Phys. D Appl. Phys.* 46, 495003 (2013).

166. S. Tacchi, G. Duerr, J. W. Klos, M. Madami, S. Neusser, G. Gubbiotti, G. Carlotti, M. Krawczyk, and D. Grundler, *Phys. Rev. Lett.* 109, 137202 (2012).

167. H. Yu, G. Duerr, R. Huber, M. Bahr, T. Schwarze, F. Brandl, and D. Grundler, *Nat. Commun.* 4, 2702 (2013).

168. G. Gubbiotti, S. Tacchi, M. Madami, G. Carlotti, S. Jain, A. O. Adeyeye, and M. P. Kostylev, *Appl. Phys. Lett.* 100, 162407 (2012).

169. X. M. Liu, J. Ding, and A. O. Adeyeye, *Appl. Phys. Lett.* 100, 242411 (2012).

170. S. Choudhury, S. Saha, R. Mandal, S. Barman, Y. Otani, and A. Barman, *ACS Appl. Mater. Interfaces* 8, 18339 (2016).

171. T. Shinjo, T. Okuno, R. Hassdorf, K. Shigeto, and T. Ono, *Science* 289, 930 (2000).

172. K. Shigeto, T. Okuno, K. Mibu, T. Shinjo, and T. Ono, *Appl. Phys. Lett.* 80, 4190 (2002).

173. A. Wachowiak, J. Wiebe, M. Bode, O. Pietzsch, M. Morgenstern, and R. Wiesendanger, *Science* 298, 577 (2002).

174. K. Y. Guslienko, W. Scholz, R. W. Chantrell, and V. Novosad, *Phys. Rev. B* 71, 144407 (2005).

175. B. Van Waeyenberge et al., *Nature* 444, 461 (2006).

176. K. Yamada, S. Kasai, Y. Nakatani, K. Kobayashi, H. Kohno, A. Thiaville, and T. Ono, *Nat. Mater.* 6, 270 (2007).

177. K. Y. Guslienko, X. F. Han, D. J. Keavney, R. Divan, and S. D. Bader, *Phys. Rev. Lett.* 96, 067205 (2006).

178. J. P. Park, P. Eames, D. M. Engebretson, J. Berezovsky, and P. A. Crowell, *Phys. Rev. B* 67, 020403 (2003).

179. A. Vogel, A. Drews, T. Kamionka, M. Bolte, and G. Meier, *Phys. Rev. Lett.* 105, 037201 (2010).

180. K. S. Buchanan, P. E. Roy, M. Grimsditch, F. Y. Fradin, K. Y. Guslienko, S. D. Bader, and V. Novosad, *Nat. Phys.* 1, 172 (2005).

181. S. Sugimoto, Y. Fukuma, S. Kasai, T. Kimura, A. Barman, and Y. Otani, *Phys. Rev. Lett.* 106, 197203 (2011).

182. V. S. Pribiag, I. N. Krivorotov, G. D. Fuchs, P. M. Braganca, O. Ozatay, J. C. Sankey, D. C. Ralph, and R. A. Buhrman, *Nat. Phys.* 3, 498 (2007).

183. R. Antos, Y. Otani, and J. Shibata, *J. Phys. Soc. Jpn.* 77, 031004 (2008).

184. S. Barman, A. Barman, and Y. Otani, *IEEE Trans. Magn.* 46, 1342 (2010).

185. H. Jung, Y.-S. Choi, K.-S. Lee, D.-S. Han, Y.-S. Yu, M.-Y. Im, P. Fischer, and S.-K. Kim, *ACS Nano* 6, 3712 (2012).

186. D. Kumar, S. Barman, and A. Barman, *Sci. Rep.* 4 (2014).

187. S. Barman, S. Saha, S. Mondal, D. Kumar, and A. Barman, *Sci. Rep. 6*, 33360 (2016).

188. W. Jung, F. J. Castaño, and C. A. Ross, *Phys. Rev. Lett.* 97, 247209 (2006).

189. A. Fert, N. Reyren, and V. Cros, *Nat. Rev. Mater.* 2, 17031 (2017).

190. C. Moreau-Luchaire et al., *Nat. Nanotechnol.* 11, 444 (2016).

191. L. Sun, R. X. Cao, B. F. Miao, Z. Feng, B. You, D. Wu, W. Zhang, A. Hu, and H. F. Ding, *Phys. Rev. Lett.* 110, 167201 (2013).

192. D. A. Gilbert, B. B. Maranville, A. L. Balk, B. J. Kirby, P. Fischer, D. T. Pierce, J. Unguris, J. A. Borchers, and K. Liu, *Nat. Commun.* 6, 8462 (2015).

193. K. Y. Guslienko, *IEEE Magn. Lett.* 6, 1 (2015).

194. S. Woo et al., *Nat. Mater.* 15, 501 (2016).

195. F. S. Ma, Y. Zhou, H. B. Braun, and W. S. Lew, *Nano Lett.* 15, 4029 (2015).

196. P. A. M. Dirac, *Proc. R. Soc. London, Ser. A* 133, 60 (1931)

197. G. t. Hooft, *Nucl. Phys. B* 79, 276 (1974).

198. L. Balents, *Nature* 464, 199 (2010).

199. R. F. Wang et al., *Nature* 439, 303 (2006).

200. C. Castelnovo, R. Moessner, and S. L. Sondhi, *Nature* 451, 42 (2008).

201. M. J. Harris, S. T. Bramwell, D. F. McMorrow, T. Zeiske, and K. W. Godfrey, *Phys. Rev. Lett.* 79, 2554 (1997).

202. S. T. Bramwell and M. J. P. Gingras, *Science* 294, 1495 (2001).

203. A. P. Ramirez, A. Hayashi, R. J. Cava, R. Siddharthan, and B. S. Shastry, *Nature* 399, 333 (1999).

204. Y. Qi, T. Brintlinger, and J. Cumings, *Phys. Rev. B* 77, 094418 (2008).

205. L. Pauling, *J. Am. Chem. Soc.* 57, 2680 (1935).

206. O. Tchernyshyov, *Nature* 451, 22 (2008).

207. A. J. Heeger, S. Kivelson, J. R. Schrieffer, and W. P. Su, *Rev. Mod. Phys.* 60, 781 (1988).

208. L. D. Faddeev and L. A. Takhtajan, *Phys. Lett. A* 85, 375 (1981).

209. B. Lake, D. A. Tennant, C. D. Frost, and S. E. Nagler, *Nat. Mater.* 4, 329 (2005).

210. H.-B. Braun, J. Kulda, B. Roessli, D. Visser, K. W. Krämer, H.-U. Güdel, and P. Böni, *Nat. Phys.* 1, 159 (2005).

211. R. de-Picciotto, M. Reznikov, M. Heiblum, V. Umansky, G. Bunin, and D. Mahalu, *Phys. B Condens. Matter* 249–251, 395 (1998).

212. M. Dolev, M. Heiblum, V. Umansky, A. Stern, and D. Mahalu, *Nature* 452, 829 (2008).

213. L. D. C. Jaubert and P. C. W. Holdsworth, *Nat. Phys.* 5, 258 (2009).

214. D. J. P. Morris et al., *Science* 326, 411 (2009).

215. T. Fennell, P. P. Deen, A. R. Wildes, K. Schmalzl, D. Prabhakaran, A. T. Boothroyd, R. J. Aldus, D. F. McMorrow, and S. T. Bramwell, *Science* 326, 415 (2009).

216. A. D. Bianchi et al., *Science* 319, 177 (2008).

217. M. Laver and E. M. Forgan, *Nat. Commun.* 1, 45 (2010).

218. E. Mengotti, L. J. Heyderman, A. F. Rodríguez, F. Nolting, R. V. Hügli, and H.-B. Braun, *Nat. Phys.* 7, 68 (2010).

219. S. Ladak, D. E. Read, G. K. Perkins, L. F. Cohen, and W. R. Brandford, *Nat. Phys.* 6, 359 (2010).

220. C. Phatak, A. K. Petford-Long, O. Heinonen, M. Tanase, and M. De Graef, *Phys. Rev. B* 83, 174431 (2011).

221. J. H. Rodrigues, L. A. S. Mól, W. A. Moura-Melo, and A. R. Pereira, *Appl. Phys. Lett.* 103, 092403 (2013).

222. K. K. Kohli, A. L. Balk, J. Li, S. Zhang, I. Gilbert, P. E. Lammert, V. H. Crespi, P. Schiffer, and N. Samarth, *Phys. Rev. B* 84, 180412 (2011).

223. L. Sam, R. Dan, T. Tolek, R. B. Will, and F. C. Lesley, *New J. Phys.* 13, 023023 (2011).

224. S. Ladak, D. E. Read, W. R. Branford, and L. F. Cohen, *New J. Phys.* 13, 063032 (2011).

225. S. D. Pollard, V. Volkov, and Y. Zhu, *Phys. Rev. B* 85, 180402 (2012).

226. U. B. Arnalds et al., *Appl. Phys. Lett.* 101, 112404 (2012).

227. J. M. Porro, A. Bedoya-Pinto, A. Berger, and P. Vavassori, *New J. Phys.* 15, 055012 (2013).

228. A. Farhan et al., *Phys. Rev. Lett.* 111, 057204 (2013).

229. L. J. Heyderman, *Nat. Nanotechnol.* 8, 705 (2013).

230. Y. Li et al., *J. Appl. Phys.* 121, 103903 (2017).

231. Y. Li et al., *J. Phys. D Appl. Phys.* 50, 015003 (2017).

232. J. Sklenar, V. S. Bhat, L. E. DeLong, and J. B. Ketterson, *J. Appl. Phys.* 113, 17B530 (2013).

233. S. Gliga, A. Kákay, R. Hertel, and O. G. Heinonen, *Phys. Rev. Lett.* 110, 117205 (2013).

234. K. Zeissler, S. K. Walton, S. Ladak, D. E. Read, T. Tyliszczak, L. F. Cohen, and W. R. Branford, *Sci. Rep.* 3, 1252 (2013).

235. W. R. Branford, S. Ladak, D. E. Read, K. Zeissler, and L. F. Cohen, *Science* 335, 1597 (2012).

236. M. B. Jungfleisch et al., *Phys. Rev. B* 93, 100401 (2016).

237. V. S. Bhat, F. Heimbach, I. Stasinopoulos, and D. Grundler, *Phys. Rev. B* 93, 140401 (2016).

238. V. S. Bhat, F. Heimbach, I. Stasinopoulos, and D. Grundler, *Phys. Rev. B* 96, 014426 (2017).

239. M. Pohlit, I. Stockem, F. Porrati, M. Huth, C. Schröder, and J. Müller, *J. Appl. Phys.* 120, 142103 (2016).

240. M. Krawczyk and D. Grundler, *J. Phys. Condens. Matter* 26, 123202 (2014).

241. X. Zhou, G.-L. Chua, N. Singh, and A. O. Adeyeye, *Adv. Funct. Mater.* 26, 1437 (2016).

242. Y.-L. Wang, Z.-L. Xiao, A. Snezhko, J. Xu, L. E. Ocola, R. Divan, J. E. Pearson, G. W. Crabtree, and W.-K. Kwok, *Science* 352, 962 (2016).

243. G.-W. Chern, M. J. Morrison, and C. Nisoli, *Phys. Rev. Lett.* 111, 177201 (2013).

244. G.-W. Chern, C. Reichhardt, and C. Nisoli, *Appl. Phys. Lett.* 104, 013101 (2014).

245. E. Östman, H. Stopfel, I.-A. Chioar, U. B. Arnalds, A. Stein, V. Kapaklis, and B. Hjörvarsson, *Nat. Phys.* 14, 375 (2018).

246. U. B. Arnalds et al., *Appl. Phys. Lett.* 105, 042409 (2014).

247. S. S. P. Parkin, M. Hayashi, and L. Thomas, *Science* 320, 190 (2008).

248. A. Fernández-Pacheco, R. Streubel, O. Fruchart, R. Hertel, P. Fischer, and R. P. Cowburn, *Nat. Commun.* 8, 15756 (2017).

249. H. Forster, T. Schrefl, D. Suess, W. Scholz, V. Tsiantos, R. Dittrich, and J. Fidler, *J. Appl. Phys.* 91, 6914 (2002).

250. M. Yan, C. Andreas, A. Kákay, F. García-Sánchez, and R. Hertel, *Appl. Phys. Lett.* 99,122505 (2011).

251. M. Gavagnin, H. D. Wanzenboeck, S. Wachter, M. M. Shawrav, A. Persson, K. Gunnarsson, P. Svedlindh, M. Stöger-Pollach, and E. Bertagnolli, *ACS Appl. Mater. Interfaces* 6, 20254 (2014).

22

Nanothermodynamics: Fundamentals and Applications

Vladimir García-Morales,
Javier Cervera, and
José A. Manzanares
University of Valencia

22.1 Introduction

Atoms at the surface and, therefore, surface energies have a marked influence on the physical and chemical properties of nanoscale materials due to their high surface to volume ratio. These properties often obey relatively simple scaling equations involving a power-law dependence on the system size, which can be explained from the surface energy contributions to the free energy. An adequate description of these contributions is then essential to understand the thermodynamic behavior of nanoscale systems.

Hill's nanothermodynamics [Hill, 1994, Hill, 2001] is a framework to properly describe the equilibrium thermodynamics of small systems that was initially developed from the applications of statistical thermodynamics to polymers and biomacromolecules. Ralph V. Chamberlin realized that the small systems do not need to be separate entities and considered that a bulk material could be divided into small regions, so that the methods of Hill's theory could be applied. The consideration of inhomogeneous regions of unrestricted sizes in combination with relatively simple mean-field models or with Landau theory of phase

transitions has resulted in significant advances in the understanding of many complex systems, including non-exponential relaxation phenomena in glasses [Chamberlin, 2015].

Tsallis' thermostatistics [Tsallis, 1988, Tsallis, 2009] is a generalization of Boltzmann–Gibbs statistical mechanics to make it valid in complex nonextensive systems, including nanosized systems and systems with correlations or long-range interactions. Tsallis' entropic index q has often been considered to be intimately related to and determined by the microscopic dynamics. Many authors have contributed to the clarification of its physical foundations [Abe and Okamoto, 2001, Tsallis, 2009, Naudts, 2011], including its connection to Hill's nanothermodynamics [García-Morales et al., 2005]. In this chapter, Tsallis' theory is briefly outlined within the more general context of superstatistics [Beck, 2002, García-Morales et al., 2011].

As the system size decreases, the fluctuations in the equilibrium thermodynamic variables become more important. Temperature fluctuations, in particular, have received much attention both from theoretical and experimental points of view, especially with the recent advances in nanothermometry [Brites et al., 2012]. In this chapter, the equilibrium fluctuations are described within the context of Einstein's theory [Falcioni et al., 2011].

The attention to nonequilibrium thermodynamics and fluctuations can be traced back to Einstein, but the field underwent a revolution in the early 1990s [Evans et al.; 1993]. Since then, researchers have been proposing a growing number of fluctuation theorems (FTs) [Sevick et al., 2008, Evans et al., 2016] and fluctuation relations [Spinney and Ford, 2013, Ford, 2013]. In fact, any convex function defined over a trajectory may lead to functionals that satisfy an integral FT. The FTs seem at odds with the traditional nineteenth-century thermodynamics and have changed our understanding of equilibrium and nonequilibrium thermodynamics. The Evans–Searles FT [Evans and Searles, 1994, Evans et al., 2016] results in a generalization of the second law that applies to small systems, including those far from equilibrium. The Crooks FT [Crooks, 1999] provides a method of predicting equilibrium free energy differences from nonequilibrium paths that connect two equilibrium states. Undoubtedly, the FTs are essential for the application of statistical mechanics concepts to irreversible processes of nanoscale systems [Seifert, 2012].

The FTs can be derived within a framework of deterministic, time-reversible mechanics [Evans et al., 2016] and from stochastic dynamics (often with white noise and using the overdamped limit) [Kurchan, 1998, Lebowitz and Spohn, 1999, Spinney and Ford, 2013, Seifert, 2012]. In the deterministic framework, irreversibility finds its origins in nonlinear terms that provide a contraction of phase space. In the stochastic approach, irreversibility directly appears in the dynamical equation. In this chapter, only the stochastic approach is described, but applications to chemical and electrochemical systems are also explained in detail.

The lack of sound mechanical or quantum-mechanical foundations of thermodynamics has been seen as a major unsolved problem. Recently, there have been interesting attempts to rebuild thermodynamics from quantum mechanics, as the latter is the framework required in practical applications to nanoelectronic components and atom-sized or single-molecule machines [Brandao et al., 2015, Horodecki and Oppenheim, 2013]. They are, however, out of the scope of the present chapter, and the interested reader is referred to recent books [Mahler, 2015] and contributions presented at the quantum thermodynamics conferences [Castelvecchi, 2017].

22.2 Surface Thermodynamics of Nanomaterials

22.2.1 The Gibbs, Euler, and Gibbs–Duhem Equations in Surface Thermodynamics

Surface thermodynamics is a successful framework to describe the smooth size effects on the physicochemical properties. Consider an interfacial region, a few atomic diameters in

thickness, that separates two homogeneous phases α and β. In this region, the densities of the extensive quantities vary smoothly with position, from their values in phase α to those in β [Inzoli et al., 2010, Li and Truhlar, 2014]. This smooth variation can be replaced by an equivalent, abrupt variation so that the methods of macroscopic thermodynamics can still be used. The interfacial region, or phase σ, is then represented by an imaginary surface, the Gibbs dividing surface. In planar geometry, the extension of the interfacial region is $-x^\alpha \leq x \leq x^\beta$, where x is the distance to the interface and $-x^\alpha$ and x^β are two positions inside phases α and β close to the interface. Consider an extensive quantity Y such as energy, entropy, or the amount n_i of component i. The amount of Y in the interfacial region is $\Sigma \int_{-x^\alpha}^{x^\beta} y_V(x)dx$, where Σ is the area of the interface and $y_V(x)$ is the local density of Y. In the Gibbs description, this same amount is evaluated extrapolating the densities y_V^α and y_V^β in phases α and β. The equivalence of these descriptions requires $y_V^\alpha x^\alpha \Sigma + y_V^\beta x^\beta \Sigma + Y^\sigma = \Sigma \int_{-x^\alpha}^{x^\beta} y_V(x)dx$ where Y^σ is the surface *excess* of Y, which can be positive, negative, or zero. For instance, the surface excess entropy S^σ is the contribution of the interface to the entropy of a system formed by two homogenous phases α and β and the interface σ.

The position of the Gibbs surface is chosen so that the surface excess of the amount of component 1 (e.g., the solvent) is zero $n_1^\sigma = 0$, that is, $c_1^\alpha x^\alpha + c_1^\beta x^\beta = \Sigma \int_{-x^\alpha}^{x^\beta} c_1(x)dx$. Because different components have different tendencies to accumulate in the interfacial region, the surface excesses of the other components are usually nonzero, $n_{i\neq1}^\sigma \neq 0$ [Inzoli et al., 2010].

The area Σ describes the size of the interface, and a convenient choice of state variables is $(T, \Sigma, \mathbf{n}^\sigma)$, where $\mathbf{n}^\sigma = \{0, n_2^\sigma, \ldots, n_c^\sigma\}$. A consistent choice for phases α and β is $(T, V^\alpha, \mathbf{n}^\alpha)$ and $(T, V^\beta, \mathbf{n}^\beta)$. The Gibbs equations in the free energy representation are

$$dA^\varphi = -S^\varphi dT - p^\varphi dV^\varphi + \sum_i \mu_i dn_i^\varphi, \qquad \varphi = \alpha, \beta,$$

(22.1)

$$dA^\sigma = -S^\sigma dT + \gamma d\Sigma + \sum_{i\neq1} \mu_i dn_i^\sigma,$$ (22.2)

where $\gamma \equiv (\partial A^\sigma/\partial\Sigma)_{T,n^\sigma}$ is the interfacial free energy. Under equilibrium conditions, the temperature T and the chemical potentials μ_i do not need phase superscripts as they take the same values in phases α, β, and σ; there are, however, generalized approaches that allow for temperature differences between the phases [Schmelzer et al., 2013].

For a bulk phase $\varphi = \alpha, \beta$, the Euler equation

$$A^\varphi = -p^\varphi V^\varphi + \sum_i \mu_i n_i^\varphi$$ (22.3)

is a consequence of the system extensivity. That is, if T and the concentrations $\mathbf{c}^\varphi$ are fixed, then A^φ and $\mathbf{n}^\varphi$ scale linearly with V^φ. Similarly, A^σ and $\mathbf{n}^\sigma$ scale linearly with the surface area Σ, for fixed intensive state, and the Euler equation of A^σ is

$$A^{\sigma}(T, \Sigma, \mathbf{n}^{\sigma}) = \gamma\Sigma + \sum_{i \neq 1} \mu_i n_i^{\sigma}. \qquad (22.4)$$

Furthermore, the Euler theorem for homogeneous functions implies that the surface density of excess free energy $a^{\sigma} \equiv A^{\sigma}/\Sigma$, the chemical potentials μ_i and γ are independent of Σ. Dividing by Σ, Eq. (22.4) reduces to $a^{\sigma} = \gamma + \sum_{i \neq 1} \mu_i \Gamma_i^{\sigma}$. In one-component systems, $a^{\sigma} = \gamma$ justifies the name "interfacial free energy" for γ.

In the case of curved interfaces, the pressure can be different in phases α and β. According to the second law, a mechanical equilibration process in which the volumes of these phases vary at fixed $(T, V^{\alpha} + V^{\beta}, \mathbf{n})$, where $n_i = n_i^{\alpha} + n_i^{\sigma} + n_i^{\beta}$, ends in a state of minimum free energy $A = A^{\alpha} + A^{\sigma} + A^{\beta}$,

$$\left(\frac{\partial A}{\partial V^{\alpha}}\right)_{T,V,\mathbf{n}} = -p^{\alpha} + p^{\beta} + \gamma \left(\frac{\partial \Sigma}{\partial V^{\alpha}}\right)_{T,\mathbf{n}} = 0. \qquad (22.5)$$

When phase α is a spherical drop of radius r, $dV^{\alpha} = 4\pi r^2 dr$ and $d\Sigma = 8\pi r dr = (2/r)dV^{\sigma}$. Then, Eq. (22.5) becomes the Young–Laplace equation

$$p^{\alpha} = p^{\beta} + \frac{2\gamma}{r}. \qquad (22.6)$$

The pressure is larger in the phase α from which the interface looks concave. The instability induced by the curvature can be illustrated considering a volume transfer between the inside of two drops of different radii. Since p^{α} increases when r decreases, the larger drop would grow, and the smaller one would disappear.

22.2.2 Size Effects on the Thermodynamic Properties of Monocomponent Nanosystems

In one-component systems, the distribution equilibrium condition along the saturation curve, $d\mu^{\alpha} = d\mu^{\beta}$ or $-s^{\alpha}dT + v^{\alpha}dp^{\alpha} = -s^{\beta}dT + v^{\beta}dp^{\beta}$, leads to the generalized Clausius–Clapeyron equation

$$(s^{\beta} - s^{\alpha})dT - (v^{\beta} - v^{\alpha})dp^{\beta} + 2v^{\alpha}d(\gamma/r) = 0, \qquad (22.7)$$

where $v^{\varphi} \equiv V^{\varphi}/n^{\varphi}$ and $s^{\varphi} \equiv S^{\varphi}/n^{\varphi}$ are the molar volume and molar entropy of phase $\varphi = \alpha, \beta$. For a solid or liquid phase α in equilibrium with a vapor phase β (i.e., under saturation conditions), the chemical potential is a function of T and r, because $\mu^{\alpha} = \mu^{\beta}$ fixes the pressures $p^{\beta} = p_{\mathrm{sat}}^{\alpha\beta}(T, r)$ and $p^{\alpha} = p^{\beta} + 2\gamma/r$. Since $v^{\beta} >> v^{\alpha}$ and $d\mu = v^{\beta}dp^{\beta} \approx (v^{\beta} - v^{\alpha})dp^{\beta} = 2v^{\alpha}d(\gamma/r)$ at constant T, integration with respect to $1/r$ leads to the Gibbs–Thomson–Freundlich equation

$$\mu_{\mathrm{sat}}^{\alpha\beta}(T, r) = \mu_{\mathrm{sat}}^{\alpha\beta}(T, \infty) + \frac{2\gamma v^{\alpha}}{r}. \qquad (22.8)$$

The interfacial system $\alpha + \sigma$ satisfies $dG^{\alpha+\sigma} + S^{\alpha+\sigma}dT - V^{\alpha}dp^{\alpha} = \mu^{\alpha}dn^{\alpha} + \gamma d\Sigma = \mu^{\alpha+\sigma}dn^{\alpha}$. Contrarily to $\mu^{\alpha} = \mu_{\mathrm{sat}}^{\alpha\beta}(T, \infty)$, $\mu^{\alpha+\sigma} = \mu^{\alpha} + \gamma(\partial\Sigma/\partial n^{\alpha}) = \mu_{\mathrm{sat}}^{\alpha\beta}(T, r)$ includes

the surface free energy contribution. Note also that $G^{\alpha+\sigma} = \mu^{\alpha+\sigma}n^{\alpha} + \gamma\Sigma/3$.

The vapor pressure of metallic nanoparticles is notably higher than that of the bulk material [Nanda et al., 2003]. Indeed, if phase β is the vapor of a condensed phase α, then the integration of Eq. (22.7) at constant T leads to Kelvin's equation

$$\frac{p^{\beta}(T, r)}{p^{\beta}(T, \infty)} = \exp\left(\frac{2\gamma v^{\alpha}}{RTr}\right) \geq 1. \qquad (22.9)$$

Nanoparticles can only be in equilibrium with a supersaturated vapor because they have a greater tendency to evaporate than a flat surface of the bulk material at the same T. This effect is only noticeable if the radius is not much larger than the so-called Kelvin radius (e.g., $2\gamma v^L/RT = 1.04$ nm for water at 300 K). Thus, solid–vapor and liquid–vapor saturation curves on a p–T diagram shift to higher pressures with decreasing drop size, and in the opposite direction in the case of nanobubbles. Kelvin's equation describes an unstable equilibrium as any perturbation drives the nanoparticle away from the equilibrium state. A stable distribution equilibrium is only possible at flat interfaces or in muticomponent condensed nanophases but not in single component ones.

The melting temperature $T_m(r)$ of metallic nanoparticles is lower than that of the bulk material. William Thomson (Lord Kelvin) in 1870 and J. J. Thomson in 1888 were already interested in this phenomenon, and Pawlow established the first thermodynamic relation for $T_m(r)$ in 1909. The melting point depression is observable in nanoparticles with radius r below 100 nm and is very large for radii in the nm range. For example, a Pb nanoparticle of radius 3 nm (about 1800 atoms) has a standard melting temperature 200 K lower than bulk Pb ($T_m^b = 600$ K) [Kofman et al., 1999].

At constant p^{β}, Eq. (22.7) can be integrated to give

$$\ln\frac{T(p^{\beta}, r)}{T(p^{\beta}, \infty)} = -\frac{2\gamma v^{\alpha}}{h^{\beta} - h^{\alpha}}\frac{1}{r}, \qquad (22.10)$$

where the distribution equilibrium condition $h^{\beta} - h^{\alpha} = T(s^{\beta} - s^{\alpha})$ has been used. When applied to the liquid–vapor equilibrium, the vaporization temperature of liquid drops decreases with decreasing size, $T_{\mathrm{vap}}(p^V, r) < T_{\mathrm{vap}}(p^V, \infty)$. When applied to the (melting) equilibrium between solid nanoparticles and their liquid phase, $h^{\beta} - h^{\alpha} = \Delta H_m^b$ is the bulk molar enthalpy of melting and $T_m(r) = T_m(p^L, r) < T_m(p^L, \infty) = T_m^b$. Spherical particles then melt at lower temperatures than the corresponding bulk phase. If we use that $\ln x \approx x - 1$ when $x \approx 1$, Eq. (22.10) can be transformed to [Couchman and Jesser, 1977]

$$1 - \frac{T_m(r)}{T_m^b} = \frac{2\gamma^{SL}v^S}{\Delta H_m^b}\frac{1}{r}. \qquad (22.11)$$

Most theoretical models for the dependence of the melting point on the particle size predict similar expressions [Jiang and Wen, 2011].

The $1/r$ dependence is widely accepted for metallic nanoparticles with diameters larger than a few nanometers; group IV semiconductors show a $1/r^2$ dependence and hence lower vapor pressure increase and lower melting point depressions [Farrell and Van Siclen, 2007]. For smaller diameters, $T_m(r)$ depends nonlinearly on $1/r$ [Chushak and Bartell, 2001]. The study of the melting transition of atomic clusters is necessarily more complicated not only because of experimental difficulties, but also because the very concept of melting has no meaning for atoms and molecules. Moreover, a slush phase is observed between the freezing and melting temperatures that determine the boundaries of the solid and liquid phases [Li and Truhlar, 2014].

Although phase transitions in nanosystems are always continuous, melting of Pb nanoparticles with diameter larger than 5 nm has been considered to be first-order [Kofman et al., 1999]. The order parameter in the system (density, crystallinity, etc.) evolves continuously with radial coordinate but discontinuously with temperature. The melting process is to be understood as a nucleation and growth phenomenon of the liquid on the solid. This phenomenon is favored if the liquid wets the solid, i.e., if $\Delta\gamma \equiv \gamma^{SV} - \gamma^{SL} - \gamma^{LV} > 0$ and there is a gain in energy when a solid surface layer is replaced by a liquid one [Peters et al., 1998, Shi et al., 2004]. The surface liquid layer starts with a continuous growth and suddenly invades the cluster at $T = T_m(r)$ and $r_c = r_m$. In other words, the solid core has a radius r_c which varies from the particle radius r at low temperatures to a critical value r_m when the particle melts. At this critical value of the cluster radius, both surface melting and irreversibility in freezing disappear. For non-spherical nanoparticles, the thickness δ of the liquid surface layer depends on the local curvature so that high curvature regions enhance surface melting.

The influence of curved interfaces upon the behavior of materials is manifested primarily through the shift of phase boundaries on phase diagrams derived from the altered condition of mechanical equilibrium [Defay and Prigogine, 1966, Dehoff, 2006]. For a number of substances, the stable phase at standard conditions is not that of highest density, as the latter is usually stable at higher pressures or lower temperatures. However, a metastable high-pressure phase can be formed at ambient pressure when the material size decreases to the nanoscale. For instance, in the nucleation stage of clusters from gases during chemical vapor deposition, the phase stability is quite different from that determined at ambient pressure. The high additional internal pressure associated with the interfacial free energy through Young–Laplace equation makes it possible to observe "unusual" phases [Zhang et al., 2004, Wang et al., 2004, Wang and Yang, 2005]. Thus, for instance, diamond has been found to be more stable than graphite in small nanocrystals [Yang and Li, 2008].

22.2.3 Ostwald Ripening and Digestive Ripening in Ensembles of Nanoparticles

The chemical potential $\mu^\alpha = \mu^\beta = \mu$ increases with decreasing radius r of the condensed phase α. Since the chemical potential is a measure of the escaping tendency of the component, a larger chemical potential indicates a lower stability, e.g., with respect to phase transitions. The Ostwald ripening phenomenon is a consequence of the higher chemical potential of smaller drops. When drops of different sizes coexist, the component that forms the drops transfers from the smaller ones, where its chemical potential is larger, to the larger ones, where its chemical potential is smaller. If the drops coexist with a bulk phase of the same component, they tend to disappear due to the transfer of component from the drops to the bulk phase. Interestingly, the Ostwald ripening growth mechanism can lead to size-dependent composition in the case of nanoalloys [Alloyeau et al., 2010].

While a narrow size distribution is essential for most applications, Ostwald ripening, sintering, and coalescence induce polydispersity. These processes can be limited, e.g., in colloidal-based synthesis, but most synthetic protocols have to include a digestive ripening thermochemical step in which polydisperse nanoparticles are transformed to a narrower size distribution of ligand-stabilized metallic nanoparticles [Manzanares et al., 2017]. The size distribution after the digestive ripening process is determined by the thermodynamic stability. The free energy minimization of the colloidal solution over all possible distributions determines the relation between the Gibbs free energy of a nanoparticle made of N atoms and the chemical potential of the metal atoms. Nanoparticles of different sizes have different mole fractions in solution; that is, there is a probability distribution function for the size under equilibrium conditions. The capping ligands exert a key role, with stronger binding ligands giving rise to smaller average sizes [Manzanares et al., 2017].

22.2.4 Size Effects on the Phase Diagrams of Binary Systems

Consider the formation of a saturated liquid solution by mixing a solvent (1) and a solid solute (2), both at the same temperature T and pressure p. The molar fraction at saturation $x_2^{L,\text{sat}}(T, \infty)$ is a measure of its solubility. A supersaturated phase with $x_2^L > x_2^{L,\text{sat}}(T, \infty)$ can be formed, at the same T and p, by cooling a liquid mixture with composition x_2^L. This liquid phase can remain metastable because the nucleation of crystallites of pure solute implies the formation of a curved solid–liquid interface with radius in the nanometer scale, and the solubility limit $x_2^{L,\text{sat}}(T, \infty)$ only applies to planar interfaces.

The distribution equilibrium condition is $\mu_2^{*,S} = \mu_2^L$. Since $\mu_2^{*,S}(T, p) = \mu_2^{*,L}(T, p) - \Delta H_{m,2}(1 - T/T_{m,2})$ and $\mu_2^L(T, p, x_2^{L,\text{sat}}) = \mu_2^{*,L}(T, p) + RT \ln x_2^{L,\text{sat}}$ (if the solution is considered an ideal mixture), the equilibrium requires $T < T_{m,2}$, and the solubility

$$x_2^{L,\text{sat}}(T, \infty) = \exp\left[\frac{\Delta H_{m,2}}{R}\left(\frac{1}{T_{m,2}} - \frac{1}{T}\right)\right] \quad (22.12)$$

increases with T. Equation (22.12) is valid for planar solid–liquid interfaces. Solute nanoparticles with different radii have different solubilities because of the pressure difference induced by the interfacial free energy, $p^S = p^L + 2\gamma^{SL}/r$. The solute chemical potential in the nanoparticles increases with decreasing radius $\mu_2^{*,S}(T, p^S) = \mu_2^{*,S}(T, p^L) + (p^S - p^L)v^S = \mu_2^{*,S}(T, p^L) + 2\gamma^{SL}v^S/r$, and hence the solubility increases. The condition $\mu_2^{*,S}(T, p) = \mu_2^{*,L}(T, p, x_2^{L,\text{sat}})$ leads then to the Ostwald–Freundlich equation

$$\frac{x_2^{L,\text{sat}}(T, r)}{x_2^{L,\text{sat}}(T, \infty)} = \exp\left(\frac{2\gamma^{SL}v^S}{RTr}\right). \quad (22.13)$$

Thus, solid nanoparticles can only be in equilibrium with a supersaturated solution [Kondepudi, 2008, Defay and Prigogine, 1966, Sugimoto and Shiba, 1999].

The liquidus curve describing the distribution equilibrium of component 2 between pure solid nanoparticles and the liquid solution in a temperature–composition diagram is

$$\ln x_2^{L,\text{sat}}(T, r) = \frac{\Delta H_{m,2}}{R}\left(\frac{1}{T_{m,2}} - \frac{1}{T}\right) + \frac{2\gamma^{SL}v^S}{RTr}. \quad (22.14)$$

In general, due to the interfacial free energy contribution to the chemical potentials of the components, the equilibrium curves in the phase diagrams of binary mixtures shift significantly with the nanoscale radii of the interfaces [Park and Lee, 2008, Pohl et al., 1998, Jabbareh and Monji, 2018].

The theoretical calculation of phase diagrams of nanosystems often consider closed systems in the microcanonical ensemble [Kaszkur, 2013]. However, the different ensembles are not equivalent in the nanoscale, and the consideration of completely open nanosystems in the nanocanonical ensemble might be more realistic, as the fluctuations themselves are to govern the distribution of grain sizes [Chamberlin, 2015].

Equation (22.14) predicts that the liquidus curve shifts toward lower temperatures, and this implies that in eutectic mixtures, the eutectic temperature must decrease with decreasing nanoparticle size as observed experimentally [Chen et al., 2011]. In the case of ideal liquid mixture and solid phases which are practically pure (i.e., the case of immiscibility in solid phase), the liquidus curves are given by Eq. (22.14) and a similar one for component 1. The eutectic point belongs to both curves, and its temperature $T = T_{\text{eut}}(r)$ can be determined as a function of the radius from the condition $x_1^{L,\text{eut}} + x_2^{L,\text{eut}} = 1$. In the case of planar interfaces ($r \to \infty$), this condition would determine the bulk eutectic temperature $T_{\text{eut}}(\infty)$. A first-order series expansion of $x_1^{L,\text{eut}} + x_2^{L,\text{eut}} = 1$ in powers of $1/r$ gives

$$\frac{T_{\text{eut}}(r)}{T_{\text{eut}}(\infty)} = 1 - 2\frac{x_1^{L,\text{eut}}\gamma_1^{SL}v_1^S + x_2^{L,\text{eut}}\gamma_2^{SL}v_2^S}{x_1^{L,\text{eut}}\Delta H_{m,1} + x_2^{L,\text{eut}}\Delta H_{m,2}}\frac{1}{r}. \quad (22.15)$$

This expression is in good agreement with the measured $T_{\text{eut}}(r)$ of Ag–Pb alloy nanoparticles of different sizes (Figure 22.1). More accurate descriptions might take into account deviations from ideality and the composition of the solid phases, but the basic trend of lowering $T_{\text{eut}}(r)$ with a term proportional to $1/r$ would be reproduced as this is a surface-induced phenomenon.

22.2.5 Size Dependence of the Interfacial Free Energy

The surface free energy $\gamma\Sigma$ of gold nanoparticles with n atoms evaluated from MD simulations can be accurately fitted to $\gamma\Sigma = an^{2/3}$ with $a = (1.8765 \pm 0.0085)$ eV, and for large radii, it simplifies to $\gamma_\infty 4\pi r^2$ with $\gamma_\infty = 0.98$ J/m^2 (Figure 22.2). The interfacial free energy γ varies with the nanoparticle size, and decrease as well as increase with decreasing size have been observed [Manzanares et al., 2017]. For very small nanoparticles, the validity of Tolman's equation [García-Morales et al., 2011] has been questioned [Nanda et al., 2003, Lu and Jiang, 2014], but these apparently contradictory observations might simply reflect

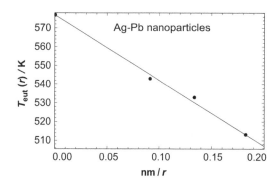

FIGURE 22.1 Measured eutectic temperatures of Ag–Pb alloy nanoparticles [Chen et al., 2011] and their comparison with the predictions by Eq. (22.15) with $\gamma_1^{SL} = \gamma_2^{SL} = 0.7$ J/m^2.

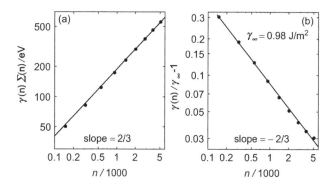

FIGURE 22.2 (a) Surface energy contribution from MD simulations [Ali et al., 2016] to the Gibbs potential of gold NPs containing n atoms. (b) Because of the increase in the fraction of edge and corner sites, $\gamma(n)$ increases with decreasing NP size [Manzanares et al., 2017]. In addition to the leading term $\propto n^{2/3}$, the surface area $\Sigma(n)$ has other size-dependent contributions, and a compensation of effects results in a reasonable accuracy of the simple expression $\gamma\Sigma = an^{2/3}$.

different conventions. When the surface atoms are compared to those in the nanoparticle core, a decrease in size reduces their energy difference. However, when the surface atoms are compared to those in bulk metal, a decrease in NP size increases their energy difference.

22.2.6 Size Effects on the Reduction Potential of Nanoparticles

Due to the interfacial free energy contribution, a metallic nanoparticle becomes less stable as its radius decreases. This also implies lower electrochemical stability. Stripping voltammetry of metallic nanoparticles on a conducting surface shows that the peak potential representing the oxidative dissolution of the metallic nanoparticle shifts negatively as r decreases. That is, the nanoparticles oxidize more easily due to the decreased stability. For the reduction of a metal cation resulting in the addition of a metal atom to the nanoparticle, the standard redox potential differs from that on a bulk metal electrode. Plieth predicted that the standard redox potential of a metal in a nanoparticle is decreased by $2\gamma v/er$, where v is the volume per atom and e is the elementary charge. Electrostatic charging effects must also be considered as they also affect the redox potential [Scanlon et al., 2015, Peljo et al., 2016, Peljo et al., 2017].

22.3 Equilibrium Nanothermodynamics

22.3.1 Hill's Subdivision Potential and the Generalized Ensemble

Compare two composite systems with the same variables (S_t, V_t, N_t) and differing in the number of subsystems (or small systems). One system has $\mathfrak{N}_1$ subsystems, each with internal energy U_1 and characterized by $S_1 = S_t/\mathfrak{N}_1$, $V_1 = V_t/\mathfrak{N}_1$ and $N_1 = N_t/\mathfrak{N}_1$. The other has $\mathfrak{N}_2$ subsystems with internal energy U_2 and variables $S_2 = S_t/\mathfrak{N}_2$, $V_2 = V_t/\mathfrak{N}_2$ and $N_2 = N_t/\mathfrak{N}_2$. In macroscopic thermodynamics, the Euler equations $U_1 = TS_1 - pV_1 + \mu N_1$ and $U_2 = TS_2 - pV_2 + \mu N_2$ imply $\mathfrak{N}_1 U_1 = \mathfrak{N}_2 U_2$, i.e., $U_{t1}(S_t, V_t, N_t, \mathfrak{N}_1) = U_{t2}(S_t, V_t, N_t, \mathfrak{N}_2)$, so that U_t is a function of (S_t, V_t, N_t) and independent of $\mathfrak{N}$. However, the Euler equation needs a correction when it is experimentally observed that $U_{t1}(S_t, V_t, N_t, \mathfrak{N}_1) \neq U_{t2}(S_t, V_t, N_t, \mathfrak{N}_2)$.

Hill's nanothermodynamics is a generalization of macroscopic thermodynamics to account for finite-size effects via the introduction of the *subdivision potential* $\mathcal{E}$. In this theory, U_t is a function of (S_t, V_t, N_t) and the number $\mathfrak{N}$ of subsystems. Therefore, the Gibbs equation of the composite system is

$$dU_t = TdS_t - pdV_t + \mu dN_t + \mathcal{E}d\mathfrak{N}, \quad (22.16)$$

where the subdivision potential is defined as

$$\mathcal{E} \equiv \left(\frac{\partial U_t}{\partial \mathfrak{N}}\right)_{S_t,V_t,N_t} \approx U_t(S_t, V_t, N_t, \mathfrak{N}+1)$$
$$- U_t(S_t, V_t, N_t, \mathfrak{N}), \quad (22.17)$$

and can be interpreted as the (positive or negative) energy required to increase in one unit the number $\mathfrak{N}$ of subdivisions of the system, while keeping constant (S_t, V_t, N_t). Thus, $(\mathcal{E}, \mathfrak{N})$ is a pair of conjugate quantities similar to (T, S_t), (p, V_t) and (μ, N_t).

Since the composite system is macroscopic and the small systems are non-interactive, U_t is a first-order homogeneous function of its extensive variables, $\mathfrak{N}$ included, and the Euler theorem implies then $U_t = TS_t - pV_t + \mu N_t + \mathcal{E}\mathfrak{N}$ and $\mathfrak{N}d\mathcal{E} = -S_t dT + V_t dp - N_t d\mu$. Division by $\mathfrak{N}$ leads to the Euler and Gibbs–Duhem equations of a small system

$$G = U - TS + pV = \mu N + \mathcal{E}, \quad (22.18)$$
$$d\mathcal{E} = -SdT + Vdp - Nd\mu. \quad (22.19)$$

Remarkably, the Gibbs equation of a small system is the same as in macroscopic thermodynamics

$$dU = TdS - pdV + \mu dN. \quad (22.20)$$

Equation (22.19) evidences that T, p, and μ can be varied independently, because the size of the small systems is an additional degree of freedom, and that the subdivision potential $\mathcal{E}$ is the thermodynamic potential whose natural variables are (T, p, μ).

The generalized, completely open, or nanocanonical ensemble [Chamberlin, 2015] considers systems with thermal, mechanical, and material interactions with its surroundings so that (T, p, μ) are environmentally-fixed variables. Equivalently, this set can be transformed to $(\beta, \beta p, \lambda)$, where $\beta \equiv 1/(k_B T)$ and $\lambda \equiv e^{\beta\mu}$. The equilibrium probability of finding the system in a microstate j with energy, volume, and number of particles (E_j, V_j, N_j) is

$$p_j = \frac{e^{-\beta(E_j + pV_j - \mu N_j)}}{\sum_j e^{-\beta(E_j + pV_j - \mu N_j)}} = \frac{1}{Y}e^{-\beta(E_j + pV_j - \mu N_j)}, \quad (22.21)$$

where $Y(\beta, \beta p, \lambda) = \sum_j e^{-\beta(E_j + pV_j - \mu N_j)}$ is the generalized partition sum. The bridge equation with thermodynamics is $\mathcal{E} = -k_B T \ln Y$. Thus, from Y and

$$d(\beta\mathcal{E}) = -d\ln Y = Ud\beta + \langle V\rangle d(\beta p) - \langle N\rangle d(\ln\lambda), \quad (22.22)$$

all equilibrium state functions can be obtained, such as

$$U = -\left(\frac{\partial \ln Y}{\partial \beta}\right)_{\beta p,\lambda}, \langle V\rangle = -\left(\frac{\partial \ln Y}{\partial(\beta p)}\right)_{\beta,\lambda},$$
$$\langle N\rangle = \lambda\left(\frac{\partial \ln Y}{\partial \lambda}\right)_{\beta,\beta p}. \quad (22.23)$$

Note that all extensive quantities, including $S/k_B = -\beta\mathcal{E} + \beta U + \beta p\langle V\rangle - \beta\mu\langle N\rangle$, $G = \mu\langle N\rangle + \mathcal{E}$, etc., are functions of the intensive variables (T, p, μ) or $(\beta, \beta p, \lambda)$ only. Intensive quantities are local fields whose gradients determine the fluxes of the extensive quantities. However, their size dependence is not the same as in classical thermodynamics. Thus, e.g., the expression $\langle N\rangle(T, p, \mu)$ can be solved for the chemical potential, which is then shown to depend on the system size, $\mu(T, p, \langle N\rangle)$.

The nanocanonical ensemble can only be used in nanosystems. Since Y does not depend on any extensive variable, the thermodynamic limit cannot be considered. This is logical, as (T, p, μ) are not independent variables and $\mathcal{E}$ is negligible in macroscopic systems.

22.3.2 Physical Interpretation of Nonlinear Corrections to the Boltzmann Factor

The Boltzmann distribution describes the occupation probability of different microstates of a system that only interacts thermally with an ideal reservoir that is able to fix the temperature by virtue of its practically infinite heat capacity and an instantaneous energy transfer between different parts of the system. However, deviations from this ideal situation may occur. For example, the "effective" number of degrees of freedom of the bath can be finite, as far as the thermal interaction with the system is concerned. Hence, corrections to the Boltzmann factor are necessary, as described in the next section.

22.3.3 Beck and Cohen's Superstatistics and Tsallis' Thermostatistics

Consider an ensemble collection of nanosystems in quasi-thermodynamic equilibrium with a thermal bath of temperature T_0. Because the temperature fluctuations of the nanosystems are important, two statistics must be superposed to describe the ensemble: the Boltzmann statistics $e^{-\beta E_j}$ and the probability distribution of $\beta = 1/(k_B T)$. The generalized probability distribution must be found by averaging over β. The effective Boltzmann factor is

$$B(E_j) = \int_0^\infty f(\beta) e^{-\beta E_j} d\beta, \qquad (22.24)$$

where $f(\beta)$ is the probability distribution function of β, and the occupation probability of microstate j is

$$p(E_j) = \frac{B(E_j)}{\int_0^\infty B(E_j) dE_j} \equiv \frac{1}{Z_B} B(E_j). \qquad (22.25)$$

This expression generalizes the Boltzmann canonical distribution, $p_j = e^{-\beta E_j}/Z$. The superstatistics approach [Beck and Cohen, 2003] is useful in many situations. For example, spatio-temporal fluctuations in temperature (or in other intensive magnitudes) may arise in driven nonequilibrium system under steady-state conditions. Spatial regions with different values of β would then play the role of different nanosystems.

The χ^2 or Γ (normalized) probability distribution

$$f(\beta) = \frac{1}{\Gamma(\gamma)} \frac{(\gamma\beta/\langle\beta\rangle)^\gamma e^{-\gamma\beta/\langle\beta\rangle}}{\beta} \propto \beta^{\gamma-1} e^{-\gamma\beta/\langle\beta\rangle}, \quad (22.26)$$

where $\Gamma(\gamma) = \int_0^\infty e^{-t} t^{\gamma-1} dt = (\gamma-1)!$ is the gamma function, is often observed in experiments. The two parameters of $f(\beta)$ are the average $\langle\beta\rangle = \int_0^\infty \beta f(\beta) d\beta$ and $\gamma \geq 0$.

The latter is the reciprocal of the squared relative fluctuation of β. Thus, fluctuations increase with increasing $1/\gamma = \langle\beta^2\rangle / \langle\beta\rangle^2 - 1$.

The bath fixes the average temperature $\langle\beta\rangle = 1/(k_B T_0)$, but temperature fluctuations can occur, and the effective Boltzmann factor of the χ^2 distribution is then

$$B(E_j) = \int_0^\infty f(\beta) e^{-\beta E_j} d\beta = e_q^{-\langle\beta\rangle E_j}, \qquad (22.27)$$

where $q \equiv 1 + 1/\gamma \geq 1$ is the entropic index and $e_q^x \equiv [1 + (1-q)x]^{1/(1-q)}$ is the Tsallis q-exponential [Tsallis, 2009, Naudts, 2011]. Although Tsallis distributions may originate from other reasons than temperature fluctuations, we have ended up with Tsallis statistics in a natural way, thus showing that the superstatistics approach contains Tsallis statistics as a particular case.

Phenomena characterized by probability distributions similar to Eq. (22.27) abound in nature. This type of statistics may arise from the convolution of the normal distribution with either a gamma or a power-law distribution, the latter being, for instance, a manifestation of the polydispersity of the system [Gheorghiu and Coppens, 2004].

The probability distribution $\{p_j\}$ can be transformed to $\{p_j^q\}$ by introducing an entropic index q as an exponent [Tsallis, 2009]. A microstate j with relatively small (large) probability p_j is a rare (frequent) event. The bias introduced by an index $q < 1$ is such that rare (frequent) events are promoted (restrained) because the probability $p_j^q/\sum_j p_j^q$ of microstate j in $\{p_j^q\}$ is larger than p_j if p_j is relatively small (smaller than p_j if p_j is relatively large). The opposite bias is introduced by an index $q > 1$. The simplest entropic form S_q that is a function of $\sum_j p_j^q$ and satisfies the conditions $\lim_{q\to1} S_q = -k_B \sum_j p_j \ln p_j = S_{Gibbs}$ and $S_q = 0$, when all microstates but one has zero probability, is Tsallis entropy [Tsallis, 1988]

$$S_q = k \frac{\sum_j p_j^q - 1}{1 - q}. \qquad (22.28)$$

The constant k is different from Boltzmann's constant but reduces to it when $q \to 1$. The entropic parameter q can be interpreted in terms of a fractal dimension for the attainable phase space [García-Morales and Pellicer, 2006]. With this interpretation, $0 \leq q \leq 1$. At the nanoscale, ions may have a reduced mobility close to highly charged interfaces (as a consequence of correlations), and a generalized Poisson–Boltzmann equation [García-Morales et al., 2004], obtained replacing the Boltzmann factor by an effective Tsallis factor with an entropic parameter $0 \leq q \leq 1$, has been shown to agree well with numerical simulations on highly charged planar interfaces.

Tsallis entropy includes the Boltzmann entropy and the Gibbs entropy equations as particular cases when $q \to 1$ as it can be transformed to

$$S_q = k \sum_j p_j \frac{(1/p_j)^{1-q} - 1}{1 - q} = k \sum_j p_j \ln_q \frac{1}{p_j}, \qquad (22.29)$$

and the q-logarithm

$$\ln_q x \equiv \frac{x^{1-q} - 1}{1 - q} \qquad (22.30)$$

becomes a natural logarithm when $q \to 1$, $\lim_{q\to 1} \ln_q x = \ln x$.

Tsallis entropy is claimed to be useful in cases where there are strong correlations between the microstates of different parts of a system [Cartwright, 2014]. Consider two systems A and B which are not independent, meaning that the probability of finding system B in microstate j_B depends on the microstate j_A in which system A is. This situation occurs, for instance, where there are interactions between the components of A and those of B. In this case, the probabilities of the microstates of $A + B$ do not factorize into those of A and B, $p_j \neq p_{j_A}^A p_{j_B}^B$, and the Gibbs entropy is not additive $S(A + B) \neq S(A) + S(B)$. According to Tsallis, the impossibility of keeping Gibbs entropy (additive and) extensive in cases like this is the crucial point of his theory [Cartwright, 2014]. In general, regardless of whether $p_j = p_{j_A}^A p_{j_B}^B$ or $p_j \neq p_{j_A}^A p_{j_B}^B$, Tsallis entropy is non-additive. However, there is one particular value q_{ent} that preserves the additivity

$$S_{q_{ent}}(A + B) = S_{q_{ent}}(A) + S_{q_{ent}}(B) \qquad (22.31)$$

in the case of non-independent systems A and B [Tsallis, 2005, Tsallis, 2009].

22.3.4 Einstein's Theory of Fluctuations

Thermal fluctuations play a dominant role in nanothermodynamics. The equilibrium fluctuations of extensive quantities can be calculated, in statistical thermodynamics, from the partition sum. The occupation probability of a microstate (X_j, Y_j) when the environment fixes the intensive variables βx and βy is $p_j = e^{\beta x X_j} e^{\beta y Y_j} / Z$, where $Z = \sum_j e^{\beta x X_j} e^{\beta y Y_j}$ is the partition sum; typical examples of pairs of conjugate quantities in entropic representation are $\beta = 1/(k_B T)$ and $-E_j$, βp and $-V_j$, and $\beta \mu$ and N_j. The variance of, e.g., X around its average value $\langle X \rangle = (\partial \ln Z / \partial(\beta x))_{\beta y}$ is $\sigma_X^2 = \langle X^2 \rangle - \langle X \rangle^2 = (\partial \langle X \rangle / \partial(\beta x))_{\beta y}$. Typical examples of these fluctuation relations are $\sigma_E^2 = k_B T^2 C_V$, $\sigma_H^2 = k_B T^2 C_p$, $\sigma_V^2 = k_B T \langle V \rangle \kappa_T$, etc. This approach cannot be used when the system is isolated, and Einstein's theory of fluctuations must then be used instead.

The second law states that, during its relaxation toward equilibrium, the entropy of an isolated system satisfies $(\partial S / \partial n)_{U,V,N} dn \geq 0$ where (U, V, N) remain fixed and the change in an internal variable n describes the advance of the process. The equilibrium condition $(\partial S / \partial n)_{U,V,N} = 0$ determines $n_{eq}(U, V, N)$ and $S_{eq}(U, V, N) = S(U, V, N, n_{eq})$. However, fluctuations in n around n_{eq} also occur at equilibrium.

The probability of observing a macrostate (U, V, N, n) is proportional to its multiplicity, $p(n) \propto W(n)$. Boltzmann's equation $S(n) = k_B \ln W$ for isolated systems then implies

$$\frac{p(n)}{p(n_{eq})} = \frac{W(n)}{W(n_{eq})} = e^{(S - S_{eq})/k_B}. \qquad (22.32)$$

The expansion of $S(n)$ around n_{eq}, truncated to second order, is

$$S = S_{eq} + \frac{1}{2}\left(\frac{\partial^2 S}{\partial n^2}\right)_{eq} (n - n_{eq})^2 \qquad (22.33)$$

because $(\partial S / \partial n)_{eq} = 0$. The symmetry of $S - S_{eq}$ around $n = n_{eq}$ implies that $\langle n \rangle = n_{eq}$. The probability distribution is Gaussian

$$p(n) = \frac{\exp\left[-(n - \langle n \rangle)^2 / (2\sigma^2)\right]}{2\int_{\langle n \rangle}^{\infty} \exp\left[-(n - \langle n \rangle)^2 / (2\sigma^2)\right] dn}, \qquad (22.34)$$

and the variance of n is $\sigma^2 \equiv -k_B/(\partial^2 S/\partial n^2)_{eq} = \langle n^2 \rangle - \langle n \rangle^2$; note that $(\partial^2 S/\partial n^2)_{eq} \leq 0$ because $S(n)$ is maximal at $n = n_{eq}$. For example, for a closed system at constant volume, and with constant isochoric heat capacity, the variance of temperature is $\sigma_T^2 = \langle T^2 \rangle - \langle T \rangle^2 = -k_B/(\partial^2 S/\partial T^2)_{V,N} = k_B T^2/C_V$. Thus, temperature measurements with nanoscale resolution [Brites et al., 2012] are limited by the need to sample a minimum isochoric heat capacity. Similarly, when C_V becomes small due to quantum effects at low T, temperature fluctuations are dominant [Mafé et al., 2000]. Einstein's theory of fluctuations can also be applied to non-isolated systems, but corrections are necessary for very small systems [Falcioni et al., 2011]. Both in classical statistical thermodynamics and in Einstein's theory, the relative fluctuations scale with the reciprocal of the square root of the system size when the interaction with the environment fixes, at least, one extensive variable. On the contrary, systems with a completely open interaction with their surroundings exhibit relative fluctuations of the order of unity [Hill and Chamberlin, 2002].

22.4 Nonequilibrium Nanothermodynamics

22.4.1 Markov Processes: The Master Equation

Chemical reactions frequently take place on nanoscale physical systems and constitute fundamental examples of microscopic stochastic events [Gillespie, 1976, Gillespie, 1977]. In macroscopic systems, the chemical kinetics is completely determined by the concentrations of chemical species and by the reaction rate constants. Starting from a known initial condition, the chemical kinetics provides deterministic evolution laws for the concentrations, yielding a system trajectory on the phase space spanned by these dynamical variables. At the nanoscale, however, fluctuations of the numbers of particles on small volumes are large, and knowledge of the concentrations alone does not suffice to describe the evolution of the system. The macroscopic, deterministic reaction kinetics breaks down in this case, and the stochastic character of chemical reactions produces significant deviations from the macroscopic average concentrations. Knowledge on the probability distribution of the number of particles and its evolution over time is, therefore, required in this case, to describe the system dynamics. The chemical master equation [Nicolis, 1972, Nicolis and Prigogine, 1977] constitutes the rigorous and general mathematical expression that allows one to address this problem. Since not all stochastic processes at the nanoscale are of chemical origin, we devote this chapter to derive and study an even more general expression, simply called the master equation from which the chemical master equation is a particular instance. Because of its major importance, the master equation is discussed in detail in now classical texts on stochastic processes [van Kampen, 2007, Gardiner, 2009], being also the subject of entire monographs [Oppenheim et al., 1977, Haag, 2017].

Let us consider a system described by a vector of stochastic variables $\mathbf{X}(t)$, and let $\mathbf{x}_0$ denote a random value of this vector measured at the 'present' time t_0. In chemical systems, these stochastic variables correspond to the numbers of particles of the chemical species, and thus they can take a discrete set of values, as we shall assume hereinafter. Let $\mathbf{x}_n$, $\mathbf{x}_{n-1}$, ..., $\mathbf{x}_1$ be random values of $\mathbf{X}(t)$ measured at future times $t_n > t_{n-1} > \ldots > t_1$ and let $\mathbf{x}_{-1}$, $\mathbf{x}_{-2}$, ... be those measured at past times $t_{-1} > t_{-2} > \ldots$. The stochastic evolution of the system is completely determined by the knowledge of the joint probability density

$$p\left(\mathbf{x}_n, t_n; \mathbf{x}_{n-1}, t_{n-1}; \ldots\right). \tag{22.35}$$

Summing over all mutually exclusive events that are possible for $\mathbf{x}_k$ at a certain time t_k in a joint probability distribution eliminates the dependence on $\mathbf{x}_k$ and t_k,

$$\sum_{\mathbf{x}_k} p\left(\ldots \mathbf{x}_{k+1}, t_{k+1}; \mathbf{x}_k, t_k; \mathbf{x}_{k-1}, t_{k-1}; \ldots\right)$$
$$= p\left(\ldots \mathbf{x}_{k+1}, t_{k+1}; \mathbf{x}_{k-1}, t_{k-1}; \ldots\right). \tag{22.36}$$

The conditional probability density

$$p\left(\mathbf{x}_n, t_n; \ldots; \mathbf{x}_1, t_1 | \mathbf{x}_0, t_0; \mathbf{x}_{-1}, t_{-1}; \ldots\right) \tag{22.37}$$

is defined as the probability of measuring values $\mathbf{x}_n$, $\mathbf{x}_{n-1}$, ..., $\mathbf{x}_1$ for the stochastic vector at the corresponding future times, provided that the values of $\mathbf{x}_0$ and $\mathbf{x}_{-1}$, $\mathbf{x}_{-2}$, ... are known. The conditional probability density is given in terms of the joint probability density by

$$p\left(\mathbf{x}_n, t_n; \ldots; \mathbf{x}_1, t_1 | \mathbf{x}_0, t_0; \mathbf{x}_{-1}, t_{-1}; \ldots\right)$$
$$= \frac{p\left(\mathbf{x}_n, t_n; \mathbf{x}_{n-1}, t_{n-1}; \ldots\right)}{p\left(\mathbf{x}_0, t_0; \mathbf{x}_{-1}, t_{-1}; \ldots\right)}. \tag{22.38}$$

These valid general expressions are of little practical value, and some assumptions are needed in order to come up with meaningful models of experimental systems. A simple assumption is *complete independence* of the stochastic

processes, in which case the probability at time t does not depend on past values (nor on future ones). In this case, we have

$$p\left(\mathbf{x}_n, t_n; \mathbf{x}_{n-1}, t_{n-1}; \ldots\right) = \prod_{k=-\infty}^{n} p\left(\mathbf{x}_k, t_k\right). \tag{22.39}$$

If the independent probabilities $p\left(\mathbf{x}_k, t_k\right)$ do not depend on t_k, one has the even more simple special case of *Bernoulli trials*. A less simple stochastic process is the *Markov process* for which the joint probability density is given by

$$p\left(\mathbf{x}_n, t_n; \mathbf{x}_{n-1}, t_{n-1}; \ldots\right) = \prod_{k=-\infty}^{n} p\left(\mathbf{x}_k, t_k | \mathbf{x}_{k-1}, t_{k-1}\right). \tag{22.40}$$

From Eq. (22.40) we have, by using Eq. (22.38),

$$p\left(\mathbf{x}_n, t_n; \mathbf{x}_{n-1}, t_{n-1}; \ldots | \mathbf{x}_0, t_0; \mathbf{x}_{-1}, t_{-1}; \ldots\right)$$
$$= p\left(\mathbf{x}_n, t_n; \mathbf{x}_{n-1}, t_{n-1}; \ldots | \mathbf{x}_0, t_0\right). \tag{22.41}$$

This is called the *Markov assumption*: knowledge of the future depends on knowledge of the present only. This is, of course, an idealization, but there may be systems whose memory time is small enough so that, if we carry out our observations on a longer time scale, it is fair to approximate them by a Markov process.

The Markov assumption allows one to reduce all considerations of the stochastic evolution to the knowledge of the initial probability density and the conditional probability density. Indeed, if one knows the initial probability distribution $p(\mathbf{x}_0, t_0)$, one has that, at a later time t_1, the following relationship holds

$$p(\mathbf{x}_1, t_1) = \sum_{\mathbf{x}_0} p(\mathbf{x}_1, t_1 | \mathbf{x}_0, t_0) p(\mathbf{x}_0, t_0), \tag{22.42}$$

i.e., the initial probability distribution at time t_0 is 'propagated' to time t_1 through the conditional probability density. Let us write $\mathbf{x} \equiv \mathbf{x}_0$ and $\mathbf{y} \equiv \mathbf{x}_1$. If we also put $t_0 \equiv t$ and $t_1 \equiv t' = t + dt$, where dt is an infinitesimally small time increment, we have

$$p(\mathbf{y}, t + dt) = \sum_{\mathbf{x}} p(\mathbf{y}, t + dt | \mathbf{x}, t) p(\mathbf{x}, t)$$
$$= \sum_{\mathbf{x}} \left(p(\mathbf{y}, t | \mathbf{x}, t) + dt \left. \frac{\partial p(\mathbf{y}, t' | \mathbf{x}, t)}{\partial t'} \right|_{t'=t} \right) p(\mathbf{x}, t)$$
$$= \sum_{\mathbf{x}} \left(\delta_{\mathbf{y}\mathbf{x}} + dt \left. \frac{\partial p(\mathbf{y}, t' | \mathbf{x}, t)}{\partial t'} \right|_{t'=t} \right) p(\mathbf{x}, t)$$
$$= p(\mathbf{y}, t) + dt \sum_{\mathbf{x}} \left. \frac{\partial p(\mathbf{y}, t' | \mathbf{x}, t)}{\partial t'} \right|_{t'=t} p(\mathbf{x}, t), \tag{22.43}$$

where we have used that

$$p(\mathbf{y}, t | \mathbf{x}, t) = \delta_{\mathbf{y}\mathbf{x}} \tag{22.44}$$

(with $\delta_{\mathbf{y}\mathbf{x}} = 1$ if $\mathbf{y} = \mathbf{x}$ and $\delta_{\mathbf{y}\mathbf{x}} = 0$ otherwise) because at time t_0, we specify that $\mathbf{X}(t_0) = \mathbf{x}$ in the conditional probability density. From Eq. (22.43), we thus obtain

$$\frac{\partial p(\mathbf{y}, t)}{\partial t} = \sum_{\mathbf{x}} \left. \frac{\partial p(\mathbf{y}, t'|\mathbf{x}, t)}{\partial t'} \right|_{t'=t} p(\mathbf{x}, t). \qquad (22.45)$$

Since we also have

$$\sum_{\mathbf{y}} p(\mathbf{y}, t + dt|\mathbf{x}, t) = \sum_{\mathbf{y}} \frac{p(\mathbf{y}, t + dt; \mathbf{x}, t)}{p(\mathbf{x}, t)} = 1 \qquad (22.46)$$

and, from Eq. (22.43),

$$\sum_{\mathbf{y}} p(\mathbf{y}, t + dt|\mathbf{x}, t) = \sum_{\mathbf{y}} \left(\delta_{\mathbf{y}\mathbf{x}} + dt \left. \frac{\partial p(\mathbf{y}, t'|\mathbf{x}, t)}{\partial t'} \right|_{t'=t} \right)$$

$$= 1 + dt \sum_{\mathbf{y}} \left. \frac{\partial p(\mathbf{y}, t'|\mathbf{x}, t)}{\partial t'} \right|_{t'=t}, \qquad (22.47)$$

necessarily

$$\sum_{\mathbf{y}} \left. \frac{\partial p(\mathbf{y}, t'|\mathbf{x}, t)}{\partial t'} \right|_{t'=t} = 0. \qquad (22.48)$$

This latter equation is equivalent to

$$\left. \frac{\partial p(\mathbf{x}, t'|\mathbf{x}, t)}{\partial t'} \right|_{t'=t} = - \sum_{\mathbf{y}(\mathbf{y}\neq\mathbf{x})} \left. \frac{\partial p(\mathbf{y}, t'|\mathbf{x}, t)}{\partial t'} \right|_{t'=t}, \qquad (22.49)$$

and, therefore, we can generally write

$$\left. \frac{\partial p(\mathbf{y}, t'|\mathbf{x}, t)}{\partial t'} \right|_{t'=t} \equiv \mathbb{W}_{\mathbf{y}\mathbf{x}}$$

$$\equiv \begin{cases} W(\mathbf{y}|\mathbf{x}, t) & \text{if } \mathbf{y} \neq \mathbf{x}, \\ W(\mathbf{y}|\mathbf{x}, t) - \sum_{\mathbf{x}'} W(\mathbf{x}'|\mathbf{x}, t) & \text{if } \mathbf{y} = \mathbf{x}, \end{cases} \qquad (22.50)$$

where we have introduced the transition (or jump) probability per unit time $W(\mathbf{y}|\mathbf{x}, t)$. We can bring the two parts of Eq. (22.50) together in a single equation as

$$\left. \frac{\partial p(\mathbf{y}, t'|\mathbf{x}, t)}{\partial t'} \right|_{t'=t} = W(\mathbf{y}|\mathbf{x}, t) - \delta_{\mathbf{y}\mathbf{x}} \sum_{\mathbf{x}'} W(\mathbf{x}'|\mathbf{x}, t). \qquad (22.51)$$

By replacing Eq. (22.51) in (22.45), we finally obtain the celebrated master equation

$$\frac{\partial p(\mathbf{y}, t)}{\partial t} = \sum_{\mathbf{x}} W(\mathbf{y}|\mathbf{x}, t) p(\mathbf{x}, t) - \sum_{\mathbf{x}'} W(\mathbf{x}'|\mathbf{y}, t) p(\mathbf{y}, t)$$

$$= \sum_{\mathbf{x}} \mathbb{W}_{\mathbf{y}\mathbf{x}} p(\mathbf{x}, t). \qquad (22.52)$$

This equation admits a straightforward interpretation. The change in time of the probability of observing a state $\mathbf{y}$ is composed of two contributions: a positive term that accounts for the incoming probability flow in which state $\mathbf{y}$ is reached from any other state $\mathbf{x}$ and a negative contribution that describes the outgoing probability flow in which state $\mathbf{y}$ is left for any other state $\mathbf{x}'$. The transition rates $W(\mathbf{y}|\mathbf{x}, t)$ contain all known information about

the stochastic system and can be often inferred from phenomenological considerations. Thanks to the master equation we can know the time evolution of the average $\langle f(\mathbf{y}) \rangle \equiv \sum_{\mathbf{y}} f(\mathbf{y}) p(\mathbf{y}, t)$ of any function f of the stochastic variable $\mathbf{y}$, since

$$\frac{d \langle f(\mathbf{y}) \rangle}{dt} = \sum_{\mathbf{y}} f(\mathbf{y}) \frac{\partial p(\mathbf{y}, t)}{\partial t} = \sum_{\mathbf{y}} \sum_{\mathbf{x}} f(\mathbf{y}) \mathbb{W}_{\mathbf{y}\mathbf{x}} p(\mathbf{x}, t). \qquad (22.53)$$

In nanoscale systems, chemical reactions make the number N_i of reacting species i fluctuate as it is produced/consumed at random, each time a specific reaction takes place. Suppose that there are s reacting species. The state of a chemical system at a time t is, thus, specified by a value of the stochastic vector $\mathbf{N} = (N_1, \ldots, N_i, \ldots, N_s)$ containing the numbers of particles present in the reaction tank at that time. Since several chemical reactions may occur, let an index ρ be used to label them. Any such reaction has the general form

$$\sum_{i=1}^{s} \nu_{<\rho}^i N_i \xrightarrow{k_\rho} \sum_{i=1}^{s} \nu_{>\rho}^i N_i. \qquad (22.54)$$

Here we have introduced the numbers for each chemical species i entering in a reaction ρ as a reactant $\nu_{<\rho}^i$ or as a product $\nu_{>\rho}^i$ of that reaction. Thus, the number of particles of species i being produced through reaction ρ is given by the stoichiometric coefficients ν_ρ^i which correspond to the difference of these numbers, i.e., $\nu_\rho^i \equiv \nu_{>\rho}^i - \nu_{<\rho}^i$. As it is the case with the number of particles, for each reaction ρ, we can put all s stoichiometric coefficients in a vector $\nu_\rho = (\nu_\rho^1, \ldots, \nu_\rho^i, \ldots, \nu_\rho^s)$. The rate at which the particles are created or annihilated is given by the reaction rate constant k_ρ. We can now ask what the probability $P(\mathbf{N}, t)$ at time t of observing a value $\mathbf{N}$ for the numbers of particles in the system is. We note that the state $\mathbf{N}$ can be reached from a state $\mathbf{N} - \nu_\rho$ and can be left to a state $\mathbf{N} + \nu_\rho$. Thus, by taking $\mathbf{y} = \mathbf{N}$ in the master equation, Eq. (22.52), and $\mathbf{x} = \mathbf{N} - \nu_\rho$, we obtain the chemical master equation

$$\frac{\partial p(\mathbf{N}, t)}{\partial t} = \sum_{\rho} [W_\rho(\mathbf{N}|\mathbf{N} - \nu_\rho) \, p(\mathbf{N} - \nu_\rho, t)$$

$$- W_\rho (\mathbf{N} + \nu_\rho|\mathbf{N}) p(\mathbf{N}, t)]. \qquad (22.55)$$

The jump probabilities $W_\rho(\mathbf{N} - \nu_\rho|\mathbf{N})$ per unit time of a chemical system are called propensities [Gillespie, 1977] and are specific for the reaction ρ. They are proportional to the reaction rate constant k_ρ and to the numbers of particles involved in the relevant reaction event. If we assume Ω to be the volume of the reaction tank, the probability of finding a molecule of species i is thus proportional to N_i/Ω. The probability of finding a different second molecule of the same species is then proportional to $(N_i - 1)/\Omega$. Since in reaction ρ there are $\nu_{<\rho}^i$ reactant molecules involved, the probability of finding all of them in a reaction event is, therefore, proportional to the product of all these probabilities, i.e., $N_i(N_i - 1) \ldots (N_i - \nu_{<\rho}^i + 1)/\Omega^{\nu_{<\rho}^i}$. Therefore, the propensities are given by the following expression

$$W_\rho(\mathbf{N} + \nu_\rho|\mathbf{N}) = \Omega k_\rho \prod_{i=1}^{s} \prod_{m=1}^{\nu_{<\rho}^i} \frac{N_i - m + 1}{\Omega}. \qquad (22.56)$$

Let us take the Brusselator as an example. The chemical reactions involving two chemical species X and Y are [Gaspard, 2002]

$$\xrightarrow{k_1} \quad X, \qquad (22.57)$$

$$X \xrightarrow{k_2} Y, \qquad (22.58)$$

$$2X + Y \xrightarrow{k_3} 3X, \qquad (22.59)$$

$$X \xrightarrow{k_4} . \qquad (22.60)$$

If we denote by X and Y the number of particles of the corresponding chemical species, the state of the system is given by the vector $\mathbf{N} = (X, Y)$, and we find that, for this system, the vectors containing the number of reactant particles $\nu_{<\rho} = (\nu_{<\rho}^1, \nu_{<\rho}^2)$ and stoichiometric coefficients $\nu_\rho = (\nu_\rho^1, \nu_\rho^2)$ are

$$\nu_{<1} = (0,0) \quad \nu_{<2} = (1,0) \quad \nu_{<3} = (2,1) \quad \nu_{<4} = (1,0), \qquad (22.61)$$

$$\nu_1 = (1,0) \quad \nu_2 = (-1,1) \quad \nu_3 = (1,-1) \quad \nu_4 = (-1,0). \qquad (22.62)$$

Therefore, for this system, the propensities are readily calculated from Eq. (22.56) as

$$W_1(\mathbf{N} + \nu_1|\mathbf{N}) = \Omega k_1 \quad W_2(\mathbf{N} + \nu_2|\mathbf{N}) = k_2 X,$$
$$W_3(\mathbf{N} + \nu_3|\mathbf{N}) = k_3 X(X-1)Y/\Omega^2 \quad W_4(\mathbf{N} + \nu_4|\mathbf{N}) = k_4 X. \qquad (22.63)$$

And, thus, by replacing Eq. (22.63) in the chemical master equation Eq. (22.55), the time evolution of the probability distribution $p(\mathbf{N}, t)$ can be numerically obtained. We shall come back to this problem in Section 22.4.4 providing explicit simulation results for the Brusselator. In the next section, we exploit the master equation to derive some further important results of interest in the statistical modeling of nanoscale systems.

22.4.2 Fluctuation Theorems and Irreversibility

The master equation has always, at least, one nontrivial stationary state for the probability distribution $p_{st}(\mathbf{y})$. This can be proved by observing that at the stationary state, one has, from Eq. (22.52),

$$0 = \sum_{\mathbf{x}} W(\mathbf{y}|\mathbf{x}, t) p_{st}(\mathbf{x}) - \sum_{\mathbf{x}'} W(\mathbf{x}'|\mathbf{y}, t) p_{st}(\mathbf{y})$$

$$= \sum_{\mathbf{x}} \mathbb{W}_{\mathbf{yx}} p_{st}(\mathbf{x}), \qquad (22.64)$$

which is a linear algebraic equation that has a nontrivial solution $p_{st}(\mathbf{x}) \neq 0$ if the determinant of the matrix $\mathbb{W}_{\mathbf{yx}}$ vanishes. But this is always the case because, from

Eq. (22.48), not all row vectors of the matrix $\mathbb{W}_{\mathbf{yx}}$ are independent.

If there is a finite number of discrete states, a probability distribution governed by the master equation will always tend to one of its stationary states in the infinite time limit [van Kampen, 2007]. Let us study in more detail some fundamental thermodynamic implications of the evolution in phase space of any system governed by a master equation. Let $P(\mathbf{y}, t)$ denote the probability that after a time interval t, such a system has undergone no transition from its original state $\mathbf{y}$. Then, after a further infinitesimal time increment τ, the probability $P(t + \tau)$ that the system still remains in the same state is

$$P(\mathbf{y}, t + \tau) = p(\mathbf{y}, t + \tau|\mathbf{y}, t) P(\mathbf{y}, t)$$

$$= P(\mathbf{y}, t) \left(p(\mathbf{y}, t|\mathbf{y}, t) + \tau \left. \frac{\partial p(\mathbf{y}, t'|\mathbf{y}, t)}{\partial t'} \right|_{t'=t} \right)$$

$$= P(\mathbf{y}, t) \left(1 - \tau \sum_{\mathbf{x}' \neq \mathbf{y}} W(\mathbf{x}'|\mathbf{y}, t) \right), \qquad (22.65)$$

where Eq. (22.49) with $\mathbf{x} = \mathbf{y}$ has been used. Note that the quantity within parentheses corresponds to the probability that no transition has occurred provided that we know with certainty that, at time t, the state is $\mathbf{y}$. Therefore, in the limit $\tau \to 0$, we obtain

$$\frac{\partial P(\mathbf{y}, t)}{\partial t} = -P(\mathbf{y}, t) \sum_{\mathbf{x}' \neq \mathbf{y}} W(\mathbf{x}'|\mathbf{y}, t), \qquad (22.66)$$

which can be integrated to give

$$P(\mathbf{y}, t) = \exp\left(-\sum_{\mathbf{x}' \neq \mathbf{y}} \int_{t_0}^{t_0+t} W(\mathbf{x}'|\mathbf{y}, t') dt' \right) \qquad (22.67)$$

because $P(\mathbf{y}, 0) = 1$.

The master equation allows us to know the probability distribution at every time, and hence, we can define the probability $p(\mathbf{y}_0 \to \mathbf{y}_n)$ that a given stochastic trajectory of duration $t = t_n - t_0$ joining the events $(\mathbf{y}_0, t_0)$, $(\mathbf{y}_1, t_1)$, ..., $(\mathbf{y}_n, t_n)$ takes place. This is given the product of the joint probability distribution that, at every discrete time t_k, an event has happened (so that the state has made a transition from $\mathbf{y}_{k-1}$ to $\mathbf{y}_k$), multiplied by all probabilities that between two consecutive discrete events no other event has taken place:

$$p(\mathbf{y}_0 \to \mathbf{y}_n) \equiv p(\mathbf{y}_n, t_n; \mathbf{y}_{n-1}, t_{n-1}; \dots; \mathbf{y}_1, t_1)$$

$$\times \prod_{k=0}^{n-1} P(\mathbf{y}_k, t_{k+1} - t_k)$$

$$= \prod_{k=1}^{n} p(\mathbf{y}_{k+1}, t_{k+1}|\mathbf{y}_k, t_k) P(\mathbf{y}_{k-1}, t_k - t_{k-1}).$$

$$(22.68)$$

Here, Eq. (22.40) has been used. Since, obviously, $\mathbf{y}_{k+1} \neq \mathbf{y}_k$, for each $k = 0, \dots, n-1$ (each single transition always proceeds from a state to a different one), we have that $p(\mathbf{y}_{k+1}, t_{k+1}|\mathbf{y}_k, t_k) = (t_{k+1}-t_k)W(\mathbf{y}_{k+1}|\mathbf{y}_k, t_k)$. Therefore, by using Eq. (22.67), we obtain

$$
p(\mathbf{y}_0 \to \mathbf{y}_n) = \prod_{k=1}^{n}(t_{k+1} - t_k)W(\mathbf{y}_{k+1}|\mathbf{y}_k, t_k)
$$
$$
\exp\left(-\sum_{\mathbf{x}'\neq\mathbf{y}_k}\int_{t_{k-1}}^{t_k} W(\mathbf{x}'|\mathbf{y}_{k-1}, t')dt'\right). \tag{22.69}
$$

The probability of observing the time-reversed trajectory, provided that we know that we start from $\mathbf{y}_n$, is given by

$$
p(\mathbf{y}_n \to \mathbf{y}_0) = \prod_{k=1}^{n}(t_{k+1} - t_k)W(\mathbf{y}_k|\mathbf{y}_{k+1}, t - t_{k+1})
$$
$$
\exp\left(-\sum_{\mathbf{x}'\neq\mathbf{y}_k}\int_{t-t_k}^{t-t_{k-1}} W(\mathbf{x}'|\mathbf{y}_k, t')dt'\right). \tag{22.70}
$$

Following [Seifert, 2005], we can now define the trajectory-dependent entropy of a path $\eta : \mathbf{y}_0 \to \mathbf{y}_1 \to \dots \to \mathbf{y}_n$ connecting any initial condition $(\mathbf{x}_0, t_0)$ to a state $(\mathbf{x}_n, t_n)$ as

$$
s(\eta) \equiv -\ln\left[p(\mathbf{y}_0, t_0)p(\mathbf{y}_0 \to \mathbf{y}_n)\right]. \tag{22.71}
$$

By noting that

$$
\sum_{\eta} p(\mathbf{y}_0, t_0)p(\mathbf{y}_0 \to \mathbf{y}_n) = 1, \tag{22.72}
$$

the Gibbs–Shannon entropy is obtained by averaging $s(\eta)$ over all possible paths

$$
\langle s(\eta)\rangle_\eta \equiv -\sum_{\eta} p(\mathbf{y}_0, t_0)p(\mathbf{y}_0 \to \mathbf{y}_n) \ln\left[p(\mathbf{y}_0, t_0)p(\mathbf{y}_0 \to \mathbf{y}_n)\right]
$$
$$
= S_{GS}. \tag{22.73}
$$

From here, we now note that the functional

$$
R(\mathbf{y}_0 \to \mathbf{y}_n) = \ln\frac{p(\mathbf{y}_0, t_0)p(\mathbf{y}_0 \to \mathbf{y}_n)}{p(\mathbf{y}_n, t_n)p(\mathbf{y}_n \to \mathbf{y}_0)} = s(\eta')-s(\eta) = \Delta s_{\text{tot}} \tag{22.74}
$$

denotes the total entropy difference between a trajectory η and its reverse counterpart η'. Therefore, by summing over all possible reverse trajectories η', we have

$$
1 = \sum_{\eta'} p(\mathbf{y}_n, t_n)p(\mathbf{y}_n \to \mathbf{y}_0)
$$
$$
= \sum_{\eta'} p(\mathbf{y}_0, t_0)p(\mathbf{y}_0 \to \mathbf{y}_n)e^{-R(\mathbf{y}_0\to\mathbf{y}_n)}
$$
$$
= \sum_{\eta} p(\mathbf{y}_0, t_0)p(\mathbf{y}_0 \to \mathbf{y}_n)e^{-R(\mathbf{y}_0\to\mathbf{y}_n)}
$$
$$
= \left\langle e^{-R(\mathbf{y}_0\to\mathbf{y}_n)}\right\rangle_\eta. \tag{22.75}
$$

By Jensen's inequality [Chandler, 1987], $\langle e^{-x}\rangle \geq e^{-\langle x\rangle}$, this latter expression necessarily implies

$$
\langle R(\mathbf{y}_0 \to \mathbf{y}_n)\rangle_\eta \geq 0, \tag{22.76}
$$

and, therefore,

$$
\langle\Delta s_{\text{tot}}\rangle_\eta \geq 0. \tag{22.77}
$$

This expression constitutes the integral FT [Seifert, 2005], and it establishes that the average over all possible nonequilibrium paths for the total entropy change is non-negative. In the thermodynamic limit, and/or trajectories spanning over a long time, this theorem implies the Second Law of Thermodynamics. Out of the thermodynamic limit, the nanoscale being a paradigmatic domain, there is a significant non-zero probability of finding paths that make a negative contribution to the total entropy, although their total average contribution will still be positive. The integral FT provides an elegant explanation of the Loschmidt paradox, showing that the Second Law is a consequence of causality [Evans and Searles, 2002] and thus elucidating how macroscopic irreversible dynamics can arise out of microscopic reversible laws.

There has arisen a plethora of FTs since their discovery [Evans et al., 1993], and there exists a vast literature on the subject. Any convex function defined over a trajectory will lead to functionals that satisfy an integral FT, analogous to the one sketched above for the trajectory entropy. This mathematical fact allows one to better understand how macroscopic thermodynamics can emerge out of statistical non-equilibrium laws and how macroscopic nonequilibrium thermodynamics can be transposed down to the nanoscale. An excellent recent review [Seifert, 2012] is strongly recommended for the interested reader.

22.4.3 Van Kampen's Volume Expansion and the Fokker–Planck Equation

The master equation can be systematically expanded in the low noise (large system size, $\Omega \to \infty$) approximation. Such expansion allows one to handle relevant, statistical corrections to macroscopic deterministic laws. In this way, the master equation provides a unified nonequilibrium picture ranging from systems consisting of a few molecules to macroscopic systems composed by an Avogadro's number of them.

Because of its great general interest for nanoscale systems, and because of the wide scope of its implications, we offer in detail the rigorous expansion of the master equation. Although we strictly adhere to van Kampen's approach [van Kampen, 2007], the full details of the expansion in [van Kampen, 2007] are there only explicitly given for an univariate chemical master equation, and only a specific example of multivariate expansion is worked out in detail. Here, we offer in full detail and generality van Kampen's expansion for the multivariate master equation. We do so in view of the interest of this approach for applications to chemical systems: the latter are usually composed of many different species and lead naturally to multivariate expressions. As magnificently elucidated by van Kampen, naive use of the popular Langevin approach or careless truncations of the Kramers–Moyal expansion, as found in an enormous number of references in the literature, leads easily to paradoxes and unphysical results, and such paths should probably better be avoided.

The relative impact of fluctuations on macroscopic equilibrium thermodynamic quantities is proportional to the reciprocal of the square root of the system size Ω. This rigorous expansion consistently yields corrections proportional to powers of $\Omega^{-1/2}$. The expansion is valid for all systems in which transition probabilities per unit time have the following canonical form:

$$W(\mathbf{y}|\mathbf{x},t) =$$
$$f(\Omega)\left[\phi_0\left(\frac{\mathbf{x}}{\Omega};\mathbf{r},t\right) + \frac{1}{\Omega}\phi_1\left(\frac{\mathbf{x}}{\Omega};\mathbf{r},t\right) + \frac{1}{\Omega^2}\phi_2 + \dots\right],$$
$$(22.78)$$

where $f(\Omega)$ is just an arbitrary function of Ω alone and the functions ϕ_j, $j = 0, 1, \dots$ depend only on the 'intensive variables' $\mathbf{x}/\Omega$ and the jumps between states $\mathbf{r} \equiv \mathbf{y} - \mathbf{x}$. Most systems of physical interest have transition probabilities per unit time with the above canonical form. This is clearly the case of chemical systems, since the propensities in Eq. (22.56) are particular instances of Eq. (22.78) where we have $\mathbf{x} = \mathbf{N}$, $\mathbf{r} = \nu_\rho$, $f(\Omega) = \Omega$,

$$\phi_0 = k_\rho \prod_{i=1}^{s} \prod_{m=1}^{\nu_\rho^i} \frac{N_i - m + 1}{\Omega},$$
$$\phi_j = 0 \qquad (j > 0). \qquad (22.79)$$

We shall assume in the following expansion that any vector quantity under discussion (let us generically denote it by $\mathbf{v}$) has a number s of components, and we shall write v_j for the jth coordinate of the vector $\mathbf{v}$, so that we have $\mathbf{v} = (v_1, \dots, v_{j-1}, v_j, v_{j+1}, \dots, v_s)$.

By replacing Eq. (22.78) in (22.52), we have

$$\frac{\partial p(\mathbf{y},t)}{\partial t} =$$
$$f(\Omega)\sum_{\mathbf{r}}\left[\phi_0\left(\frac{\mathbf{y}-\mathbf{r}}{\Omega};\mathbf{r},t\right) + \frac{1}{\Omega}\phi_1\left(\frac{\mathbf{y}-\mathbf{r}}{\Omega};\mathbf{r},t\right) + \dots\right]$$
$$\times p(\mathbf{y}-\mathbf{r},t)$$
$$- f(\Omega)\sum_{\mathbf{r}}\left[\phi_0\left(\frac{\mathbf{y}}{\Omega};-\mathbf{r},t\right) + \frac{1}{\Omega}\phi_1\left(\frac{\mathbf{y}}{\Omega};-\mathbf{r},t\right) + \dots\right]$$
$$\times p(\mathbf{y},t). \qquad (22.80)$$

The essential step in the expansion of the master equation comes from introducing an appropriate change of variables that anticipates some of the main broad features of the expected solution. This consists on splitting $\mathbf{y}$ into two contributions: one deterministic (that we shall fix through the macroscopic limit), which describes the average of the probability distribution, and the other stochastic (carrying all higher moments of the probability distribution). Therefore, the state $\mathbf{y}/\Omega = (y_1/\Omega, \dots, y_s/\Omega)$ of the system is made dependent on time through a function $\mathbf{g}(t) = (g_1(t), \dots, g_s(t))$ that is to be specified in the macroscopic

limit $\Omega \to \infty$ and which describes the motion of the peak of the probability distribution $p(\mathbf{y},t)$ in that limit. The state also depends on the stochastic variable $\mathbf{w} = (w_1, \dots, w_s)$ which evolves on the scale of the fluctuations of $\mathbf{y}/\Omega$ leading to deviations from the macroscopic average. In the thermodynamic limit, these fluctuations have a relative impact which is proportional to $\Omega^{-1/2}$:

$$\frac{\mathbf{y}}{\Omega} = \mathbf{g}(t) + \Omega^{-1/2}\mathbf{w}. \qquad (22.81)$$

In this way, the probability distribution is described in terms of these new variables. We, therefore, write

$$\mathcal{P}(\mathbf{w},t) \equiv p(\mathbf{y},t) = p(\Omega\mathbf{g}(t) + \Omega^{1/2}\mathbf{w},t). \qquad (22.82)$$

We note the following relationships involving the new and the old variables

$$\frac{\partial^k \mathcal{P}}{\partial w_{j_1} \partial w_{j_2} \dots \partial w_{j_k}} = \Omega^{k/2}\frac{\partial^k p}{\partial y_{j_1} \partial y_{j_2} \dots \partial y_{j_k}}, \qquad (22.83)$$

$$\frac{\partial \mathcal{P}}{\partial t} = \frac{\partial p}{\partial t} + \Omega\sum_{j=1}^{s}\frac{dg_j}{dt}\frac{\partial p}{\partial y_j} = \frac{\partial p}{\partial t} + \Omega^{1/2}\sum_{j=1}^{s}\frac{dg_j}{dt}\frac{\partial \mathcal{P}}{\partial w_j}. \qquad (22.84)$$

We have now all necessary ingredients to perform the expansion. By replacing Eqs. (22.81) and (22.82) in Eq. (22.52) and using Eq. (22.84), we obtain

$$\frac{\partial \mathcal{P}}{\partial t} - \Omega^{1/2}\sum_{j=1}^{s}\frac{dg_j}{dt}\frac{\partial \mathcal{P}}{\partial w_j}$$
$$= f(\Omega)\sum_{\mathbf{r}}\left[\phi_0\left(\mathbf{g}(t) + \Omega^{-1/2}(\mathbf{w} - \Omega^{-1/2}\mathbf{r});\mathbf{r},t\right)\right.$$
$$\left. + \frac{1}{\Omega}\phi_1\left(\mathbf{g}(t) + \Omega^{-1/2}(\mathbf{w} - \Omega^{-1/2}\mathbf{r});\mathbf{r},t\right) + \dots\right]$$
$$\times \mathcal{P}(\mathbf{w} - \Omega^{-1/2}\mathbf{r},t) - f(\Omega)\sum_{\mathbf{r}}\left[\phi_0\left(\mathbf{g}(t) + \Omega^{-1/2}\mathbf{w};-\mathbf{r},t\right)\right.$$
$$\left. + \frac{1}{\Omega}\phi_1\left(\mathbf{g}(t) + \Omega^{-1/2}\mathbf{w};-\mathbf{r},t\right) + \dots\right]\mathcal{P}(\mathbf{w},t). \qquad (22.85)$$

We now note that the second and third lines of Eq. (22.85) contain terms in which $\mathbf{w}$ is shifted by $-\Omega^{-1/2}\mathbf{r}$. By making a Taylor expansion of those terms around $\mathbf{w}$ in powers of $-\Omega^{-1/2}\mathbf{r}$, we find that terms of order zero are canceled by those in line 4 of Eq. (22.85). Therefore, we obtain

$$\frac{\partial \mathcal{P}}{\partial t} - \Omega^{1/2}\sum_{j=1}^{s}\frac{dg_j}{dt}\frac{\partial \mathcal{P}}{\partial w_j} = -\Omega^{-1/2}f(\Omega)$$
$$\times \sum_{\mathbf{r}}\sum_{j=1}^{s}\frac{\partial}{\partial w_j}\left[r_j\phi_0\left(\mathbf{g}(t) + \Omega^{-1/2}\mathbf{w};\mathbf{r},t\right)\mathcal{P}(\mathbf{w},t)\right]$$
$$+ \frac{1}{2}\Omega^{-1}f(\Omega)\sum_{\mathbf{r}}\sum_{j_1=1}^{s}\sum_{j_2=1}^{s}\frac{\partial^2}{\partial w_{j_1}\partial w_{j_2}}$$
$$\left[r_{j_1}r_{j_2}\phi_0\left(\mathbf{g}(t) + \Omega^{-1/2}\mathbf{w};\mathbf{r},t\right)\mathcal{P}(\mathbf{w},t)\right]$$
$$- \frac{1}{3!}\Omega^{-3/2}f(\Omega)\sum_{\mathbf{r}}\sum_{j_1=1}^{s}\sum_{j_2=1}^{s}\sum_{j_3=1}^{s}\frac{\partial^3}{\partial w_{j_1}\partial w_{j_2}\partial w_{j_3}}$$

$$\left[r_{j_1} r_{j_2} r_{j_3} \phi_0 \left(\mathbf{g}(t) + \Omega^{-1/2}\mathbf{w}; \mathbf{r}, t \right) \mathcal{P}(\mathbf{w}, t) \right]$$

$$- \Omega^{-3/2} f(\Omega) \sum_{\mathbf{r}} \sum_{j=1}^{s} \frac{\partial}{\partial w_j}$$

$$\left[r_j \phi_1 \left(\mathbf{g}(t) + \Omega^{-1/2}\mathbf{w}; \mathbf{r}, t \right) \mathcal{P}(\mathbf{w}, t) \right] + \mathcal{O}(\Omega^{-2}). \quad (22.86)$$

We define the rescaled jump moments

$$\alpha_{km}^{(j_1 \ldots j_k)}(\mathbf{x}) \equiv \sum_{\mathbf{r}} r_{j_1} r_{j_2} \ldots r_{j_k} \phi_m(\mathbf{x}; \mathbf{r}, \tau). \quad (22.87)$$

In these expressions, the indices $j_1, \ldots, j_k$ are all free. Therefore, $\alpha_{1,0}^{(j)}$ corresponds to a vector, $\alpha_{2,0}^{(j_1 j_2)}$ to a matrix, etc. In terms of a rescaled time

$$\Omega^{-1} f(\Omega)t = \tau, \quad (22.88)$$

we find

$$\frac{\partial \mathcal{P}}{\partial \tau} - \Omega^{1/2} \sum_{j=1}^{s} \frac{dg_j}{d\tau} \frac{\partial \mathcal{P}}{\partial w_j} =$$

$$- \Omega^{1/2} \sum_{j=1}^{s} \frac{\partial}{\partial w_j} \left[\alpha_{1,0}^{(j)} \left(\mathbf{g} + \Omega^{-1/2}\mathbf{w} \right) \mathcal{P} \right]$$

$$+ \frac{1}{2} \sum_{j_1=1}^{s} \sum_{j_2=1}^{s} \frac{\partial^2}{\partial w_{j_1} \partial w_{j_2}} \left[\alpha_{2,0}^{(j_1 j_2)} \left(\mathbf{g} + \Omega^{-1/2}\mathbf{w} \right) \mathcal{P} \right]$$

$$- \frac{1}{3!}\Omega^{-1/2} \sum_{j_1=1}^{s} \sum_{j_2=1}^{s} \sum_{j_3=1}^{s} \frac{\partial^3}{\partial w_{j_1} \partial w_{j_2} \partial w_{j_3}}$$

$$\left[\alpha_{3,0}^{(j_1 j_2 j_3)} \left(\mathbf{g} + \Omega^{-1/2}\mathbf{w} \right) \mathcal{P} \right]$$

$$- \Omega^{-1/2} \sum_{j=1}^{s} \frac{\partial}{\partial w_j} \left[\alpha_{1,1}^{(j)} \left(\mathbf{g} + \Omega^{-1/2}\mathbf{w} \right) \mathcal{P} \right] + \mathcal{O}(\Omega^{-1}), \quad (22.89)$$

where $\mathbf{g}$ stands for $\mathbf{g}(\tau)$. From here, expansion of the quantities $\alpha_{km}^{(j_1 \ldots j_k)}(\mathbf{x})$ in powers of $\Omega^{-1/2}\mathbf{w}$ finally gives

$$\frac{\partial \mathcal{P}}{\partial \tau} - \Omega^{1/2} \sum_{j=1}^{s} \frac{dg_j}{d\tau} \frac{\partial \mathcal{P}}{\partial w_j} = -\Omega^{1/2} \sum_{j=1}^{s} \alpha_{1,0}^{(j)}(\mathbf{g}) \frac{\partial \mathcal{P}}{\partial w_j}$$

$$- \sum_{j=1}^{s} \sum_{k=1}^{s} \frac{\partial \alpha_{1,0}^{(j)}(\mathbf{g})}{\partial g_k} \frac{\partial}{\partial w_j} (w_k \mathcal{P})$$

$$+ \frac{1}{2} \sum_{j_1=1}^{s} \sum_{j_2=1}^{s} \alpha_{2,0}^{(j_1 j_2)}(\mathbf{g}) \frac{\partial^2 \mathcal{P}}{\partial w_{j_1} \partial w_{j_2}}$$

$$+ \mathcal{O}(\Omega^{-1/2}). \quad (22.90)$$

To avoid divergent terms in the limit $\Omega \to \infty$, we take $dg_j/d\tau = \alpha_{1,0}^{(j)}(\mathbf{g})$, $j = 1, \ldots, s$. In vector form, this is the same as letting $\mathbf{g}(\tau)$ be specified by

$$\frac{d\mathbf{g}}{d\tau} = \alpha_{1,0}(\mathbf{g}), \quad (22.91)$$

with the initial condition

$$\mathbf{g}(0) = \lim_{\Omega \to \infty} \frac{\mathbf{y}(0)}{\Omega}, \quad (22.92)$$

in consistency with Eq. (22.81); i.e., the initial state is determined with certainty and evolves deterministically through Eq. (22.91). Indeed Eq. (22.91) constitutes the deterministic, macroscopic evolution law. By keeping terms of order Ω^0 in Eq. (22.90), we obtain the Fokker–Planck equation:

$$\frac{\partial \mathcal{P}}{\partial \tau} = -\sum_{j=1}^{s} \sum_{k=1}^{s} \frac{\partial \alpha_{1,0}^{(j)}(\mathbf{g})}{\partial g_k} \frac{\partial}{\partial w_j} (w_k \mathcal{P})$$

$$+ \frac{1}{2} \sum_{j_1=1}^{s} \sum_{j_2=1}^{s} \alpha_{2,0}^{(j_1 j_2)}(\mathbf{g}) \frac{\partial^2 \mathcal{P}}{\partial w_{j_1} \partial w_{j_2}}. \quad (22.93)$$

In chemical systems, we have the following identifications:

$$\tau = t, \quad (22.94)$$

$$\mathbf{g} = \lim_{\Omega \to \infty} \mathbf{N}/\Omega \equiv \mathbf{c} = (c_1, \ldots, c_s), \quad (22.95)$$

$$\mathbf{w} = \Omega^{-1/2}\mathbf{N} - \Omega^{1/2} \lim_{\Omega \to \infty} [\mathbf{N}/\Omega] = \Omega^{-1/2}\mathbf{N} - \Omega^{1/2}\mathbf{c}, \quad (22.96)$$

$$\alpha_{k0}^{(j_1 \ldots j_k)}(\mathbf{g}) = \lim_{\Omega \to \infty} \sum_{\rho} \nu_{\rho}^{j_1} \nu_{\rho}^{j_2} \ldots \nu_{\rho}^{j_k} k_{\rho} \prod_{i=1}^{s} \prod_{m=1}^{\nu_{i\rho}^i} \frac{N_i - m + 1}{\Omega}$$

$$= \sum_{\rho} k_{\rho} \nu_{\rho}^{j_1} \nu_{\rho}^{j_2} \ldots \nu_{\rho}^{j_k} \prod_{i=1}^{s} c_i^{\nu_{i\rho}^i} = \alpha_{k0}^{(j_1 \ldots j_k)}(\mathbf{c}), \quad (22.97)$$

$$\alpha_{km}^{(j_1 \ldots j_k)} = 0 \qquad (\text{if } m > 0). \quad (22.98)$$

Therefore, by making these replacements in Eq. (22.93), we obtain the multivariate Fokker–Planck equation for chemical systems with M reactions:

$$\frac{\partial \mathcal{P}(\mathbf{w}, t)}{\partial t} = \sum_{j=1}^{s} \sum_{k=1}^{s} \sum_{\rho=1}^{M} k_{\rho} \left[-\nu_{\rho}^{j} \frac{\partial \left(\prod_{i=1}^{s} c_i^{\nu_{i<\rho}^i} \right)}{\partial c_k} \frac{\partial}{\partial w_j} (w_k \mathcal{P}) \right.$$

$$\left. + \frac{1}{2} \nu_{\rho}^{j} \nu_{\rho}^{k} \left(\prod_{i=1}^{s} c_i^{\nu_{i<\rho}^i} \right) \frac{\partial^2 \mathcal{P}}{\partial w_j \partial w_k} \right], \quad (22.99)$$

where the concentrations c_i are time-dependent, their behavior being given by the macroscopic chemical kinetics Eq. (22.91) as

$$\frac{dc_j}{dt} = \sum_{\rho} k_{\rho} \nu_{\rho}^{j} \prod_{i=1}^{s} c_i^{\nu_{i<\rho}^i} \qquad (j = 1, \ldots, s). \quad (22.100)$$

For example, for the Brusselator given by Eqs. (22.57)–(22.60), the macroscopic chemical kinetics given by the time evolution of the concentrations $c_X \equiv X/\Omega$ and $c_Y \equiv Y/\Omega$ of species X and Y is

$$\frac{dc_X}{dt} = k_1 - k_2 c_X + k_3 c_X^2 c_Y - k_4 c_X, \quad (22.101)$$

$$\frac{dc_Y}{dt} = k_2 c_X - k_3 c_X^2 c_Y. \quad (22.102)$$

Therefore, in the macroscopic limit $\Omega \to \infty$, we have $\mathbf{g} = \mathbf{y}/\Omega = (c_X, c_Y)$ from Eq. (22.81) and $\alpha_{1,0}(\mathbf{g}) = (k_1 - k_2 c_X + k_3 c_X^2 c_Y - k_4 c_X, k_2 c_X - k_3 c_X^2 c_Y)$ from Eqs. (22.91), (22.101), and (22.102). By gathering all this information together with the stoichiometric numbers and numbers of reactants, the Fokker–Planck equation can then be written for the Brusselator, but we omit this here.

The general univariate Fokker–Planck equation is a particular instance of Eq. (22.93),

$$\frac{\partial \mathcal{P}(w, \tau)}{\partial \tau} = a(\tau) \frac{\partial(w\mathcal{P})}{\partial w} + b(\tau) \frac{\partial^2 \mathcal{P}}{\partial w^2} \qquad (22.103)$$

with $a(\tau) \equiv -d\alpha_{1,0}(g)/dg$ and $b(\tau) \equiv \alpha_{2,0}(g)/2$. It can be explicitly solved for a Gaussian initial condition and boundary conditions $\mathcal{P}(\pm\infty, \tau) = 0$. Indeed, since

$$\mu(\tau) = \langle w(\tau) \rangle = \int_{-\infty}^{\infty} w \mathcal{P}(w, \tau) dw, \qquad (22.104)$$

the Gaussian ansatz

$$\mathcal{P}(w, \tau) = \frac{1}{\sigma(\tau)\sqrt{2\pi}} e^{-\frac{(w - \mu(\tau))^2}{2\sigma(\tau)^2}} \qquad (22.105)$$

allows the integration by parts of the equation

$$\begin{aligned}\frac{d\mu(\tau)}{d\tau} &= \int_{-\infty}^{\infty} w \frac{\partial \mathcal{P}(w, \tau)}{\partial \tau} dw \\ &= \int_{-\infty}^{\infty} w \left[a(\tau) \frac{\partial(w\mathcal{P})}{\partial w} + b(\tau) \frac{\partial^2 \mathcal{P}}{\partial w^2} \right] dw, \quad (22.106)\end{aligned}$$

where Eq. (22.103) has been used and the boundary conditions are to be applied. Thus, we obtain

$$\frac{d\mu(\tau)}{d\tau} = -a(\tau)\mu(\tau) \qquad (22.107)$$

with solution

$$\mu(\tau) = \mu(0) e^{-\int_0^\tau a(t) dt}. \qquad (22.108)$$

Since we also have

$$\sigma^2 = \langle w^2 \rangle - \langle w \rangle^2 = \mu(\tau)^2 - \int_{-\infty}^{\infty} w^2 \mathcal{P}(w, \tau) dw, \quad (22.109)$$

by taking the time derivative, replacing Eq. (22.103), integrating by parts over w, and applying the boundary conditions, we obtain

$$\frac{d\sigma(\tau)}{d\tau} = -a(\tau)\sigma(\tau) + \frac{b(\tau)}{\sigma(\tau)}, \qquad (22.110)$$

which is readily solved as

$$\sigma(\tau) = e^{-\int_0^\tau a(t) dt} \sqrt{\sigma(0)^2 + 2 \int_0^\tau dt' b(t') e^{2\int_0^{t'} a(t) dt}} \qquad (22.111)$$

And, thus, by replacing Eqs. (22.108) and (22.111) in Eq. (22.105), one obtains the solution of the Fokker–Planck equation under the naturally imposed boundary conditions

at infinity. The conclusion of the above analysis is that a probability distribution that is initially a Gaussian function of w keeps a Gaussian function at every later time. A simplest version of this univariate Fokker–Planck equation can be used, e.g., to provide an elegant description of Brownian motion [Risken, 1989].

Newton's equation of motion for a particle of mass m and velocity v_p in a medium with damping coefficient λ is given by

$$m \frac{dv_p}{dt} = -\lambda v_p. \qquad (22.112)$$

This equation is readily integrated as $v_p(t) = v_p(0) e^{-\gamma t}$, where we have defined $\gamma \equiv \lambda/m$. If the mass of the particle is small enough, fluctuations on the velocity will be observed because of thermal agitation, and we shall have $m \langle v^2 \rangle / 2 = kT$ and, thus, a 'thermal velocity' $v_{th} \equiv \sqrt{\langle v^2 \rangle} = \sqrt{2kT/m}$. We are then interested in finding the stochastic description of the dynamics of the particle. We can view $v_p(t)$ as the average (first moment) of a probability distribution $\mathcal{P}(v, t)$, and then from Eqs. (22.91), we can take $g = v_p$, $\tau = t$, and $\alpha_{1,0}(v_p) = -\gamma v_p$. Therefore, $a(\tau) = \gamma$ in Eq. (22.103). From equilibrium statistical mechanics, we know that the stationary probability distribution is the Maxwell distribution

$$\mathcal{P}_{st}(v) = \lim_{\tau \to \infty} \mathcal{P}(v, t) = \sqrt{\frac{m}{2\pi kT}} e^{-\frac{mv^2}{2kT}}. \qquad (22.113)$$

Therefore, by replacing Eq. (22.113) in Eq. (22.103), since $\partial P_{st}/\partial t = 0$, we obtain $b(t) = \gamma kT/m$, and thus, the univariate Fokker–Planck equation that describes a Brownian particle is

$$\frac{\partial \mathcal{P}(v, t)}{\partial t} = \gamma \frac{\partial(v\mathcal{P})}{\partial v} + \frac{\gamma kT}{m} \frac{\partial^2 \mathcal{P}}{\partial v^2}. \qquad (22.114)$$

One now gets, from Eqs. (22.108), (22.111), and (22.105), the time-dependent solution of Eq. (22.114) for any Gaussian initial condition and any time. If we, e.g., take a Dirac delta peaked at $v = v_p(0)$ as initial condition (and hence $\sigma(0) = 0$), we obtain

$$\mathcal{P}(v, t) = \frac{1}{\sigma(t)\sqrt{2\pi}} e^{-\frac{(v - \mu(t))^2}{2\sigma(t)^2}}, \qquad (22.115)$$

$$\mu(t) = v_p(t) = v_p(0) e^{-\gamma t}, \qquad (22.116)$$

$$\sigma(t) = \sqrt{\frac{kT}{m}} \left(1 - e^{-2\gamma t} \right) \qquad (22.117)$$

at any other time t.

The Fokker–Planck equation has been used in the modeling of biomolecular motors [Schliwa, 2003, Bustamante et al., 2001, Wang and Oster, 1998], in which irregularity in the motion of the motors arises as a consequence of molecular noise. In general, nonequilibrium fluctuations generated externally or by chemical reactions far from equilibrium can bias the Brownian motion of a particle in an anisotropic medium without thermal gradients [Astumian, 2001]. Such fluctuation-driven transport is one mechanism by which chemical energy can directly drive the motion of particles and macromolecules.

22.4.4 Numerical Simulation of the Chemical Master Equation: Gillespie's First Reaction Method

The chemical master equation, Eq. (22.55), cannot be analytically solved in general, and it usually poses daunting mathematical problems. However, useful numerical methods exist that are able to reproduce stochastic trajectories compatible with the chemical master equation. Most celebrated algorithms are due to Gillespie and are known as the 'Direct Method' and the 'First Reaction Method' [Gillespie, 1976, Gillespie, 1977]. In this section, we discuss the latter because it easily generalizes to time-dependent propensities, as shown in [Jansen, 1995]. (We discuss this generalization in Section 22.4.6.)

Let $P(\mathbf{N}, t)$ denote the probability that, after a time t, a chemical system has been in the same state, with the number of particles $\mathbf{N}$ and no reaction taking place. This probability is provided by Eq. (22.67), with $\mathbf{y} = \mathbf{N}$ which, since the propensities do not depend on time, yields the result

$$P(\mathbf{N}, t) = e^{- \sum_\rho W_\rho(\mathbf{N}+\nu_\rho|\mathbf{N})t}. \qquad (22.118)$$

Let $P_\rho(\mathbf{N}, t)$ denote the probability that, after a time t, reaction ρ has not occurred. Clearly, if there are M reactions in the network, we have

$$P(\mathbf{N}, t) = \prod_{\rho=1}^{M} P_\rho(\mathbf{N}, t), \qquad (22.119)$$

where each $P_\rho(\mathbf{N}, t)$ is given by

$$P_\rho(\mathbf{N}, t) = e^{-W_\rho(\mathbf{N}+\nu_\rho|\mathbf{N})t} \qquad (\rho = 1, \ldots, M). \quad (22.120)$$

Thus, those reactions for which the propensities are large are more likely to happen in a time interval τ, the probability $P_\rho(\mathbf{N}, t)$ dropping to zero more strongly in those cases. By inverting Eq. (22.120), we can solve for the time τ_ρ that takes, at least, a reaction ρ to happen:

$$\tau_\rho = \frac{1}{W_\rho(\mathbf{N} + \nu_\rho|\mathbf{N})} \ln\left(\frac{1}{P_\rho}\right). \qquad (22.121)$$

Therefore, those reactions with larger propensities and/or lower probability P_ρ not to happen, need a shorter time τ_ρ to happen.

Even when the probability $P(\mathbf{N}, t)$ is different to the probability $p(\mathbf{N}, t)$ entering the chemical master equation, the above development is fully compatible with the latter, because it is consistently based on the same behavior for the conditional probability density $p(\mathbf{N}, t + \tau|\mathbf{N}, t)$ as the chemical master equation.

The idea behind Gillespie's First Reaction Method is to use M equations (Eq. (22.121)) to generate M different times τ_ρ that are all obtained by drawing a random number P_ρ from the uniform distribution in the unit interval. Then the reaction with minimal τ_ρ is selected to advance. We give, for reference, Gillespie's algorithm in detail:

- Initialize the algorithm setting $t = 0$ and $\mathbf{N} = \mathbf{N}_0$, and select a time t_{end} for the duration of the stochastic trajectory.
- Calculate all propensities $W_\rho(\mathbf{N}+\nu_\rho|\mathbf{N})$ from Eq. (22.56).
- Draw M different random numbers $P_\rho \in [0,1]$ ($\rho = 1, \ldots, M$) from the uniform distribution, and calculate M reaction times τ_ρ given by Eq. (22.121).
- Find $\mu \in [1, M]$ such that $\tau_\mu = \min \tau_\rho$, the reaction μ that has the shortest time to occur, in consistency with the random number generated. Select reaction μ to advance; i.e., put $\mathbf{N} \to \mathbf{N}+\nu_\mu$.
- If $t < t_{\text{end}}$, put $t \to t + \tau_\mu$, and go to step 1; else, finish the algorithm.

We consider in the next section an application of this algorithm.

22.4.5 Application: Chemical Clocks at the Nanoscale

Chemical oscillations have been studied at the nanoscale, theoretically [Gaspard, 2002] and experimentally [McEwen et al., 2009, Visart de Bocarmé and Kruse, 2002]. The Brusselator is a well-known abstract model of a chemical clock, since it exhibits stable periodic dynamics in its macroscopic kinetics and has been considered to reach conclusions on the minimum number of particles that is able to sustain oscillations of reasonable quality [Gaspard, 2002]. We can investigate the impact of fluctuations on the oscillatory dynamics of the Brusselator by using Gillespie's First Reaction Method. In this way, we can obtain stochastic trajectories for the system when only a small number of molecules of species X and Y are present. These trajectories, as we have seen above, are fully compatible with the chemical master equation. Simulations can be run for different values of the system size Ω.

In the limit $\Omega \to \infty$, the dynamics of the Brusselator is given by Eqs. (22.101) and (22.102). From those equations, it is clear that there is a stationary, fixed point of the dynamics at $(X_{\text{st}}, Y_{\text{st}}) = (k_1/k_4, k_2 k_4/(k_1 k_3))$. This stationary state is stable for $k_2 < k_{2,c} \equiv k_4 + k_1^2 k_3/k_4^2$. At $k = k_{2,c}$, the stationary state loses stability through a supercritical Hopf bifurcation to a stable limit cycle, which then develops in phase space for $k_2 > k_{2,c}$ [Gaspard, 2002]. We consider next this limit cycle dynamics by taking the values $k_1 = 0.5$, $k_2 = 1.5$, $k_3 = 1$, and $k_4 = 1$ for the kinetic rate constants, so that we have $k_2 > k_c = 1.25$.

In Figure 22.3, simulations of the chemical master equation obtained by means of Gillespie's algorithm are shown. The leftmost column of panels in the figure corresponds to the macroscopic limit $\Omega \to \infty$, as obtained by numerically integrating Eqs. (22.101) and (22.102). The stochastic evolution of the Brusselator is shown for finite and decreasing values of the system size Ω. Shown in each case is the limit cycle in phase space (a) and the time evolution of

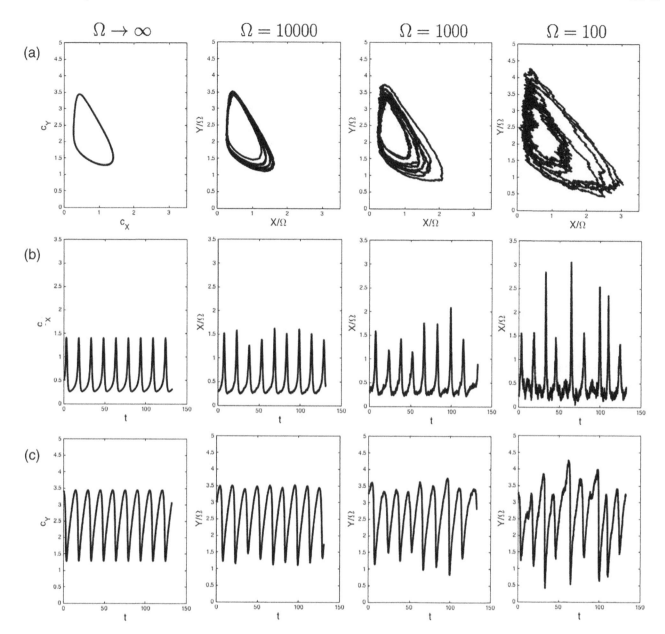

FIGURE 22.3 Phase space plots of the Brusselator dynamics obtained for the macroscopic limit $\Omega \to \infty$ (deterministic dynamics obtained integrating Eqs. (22.101) and (22.102)) and using Gillespie's algorithm with $\Omega = 10000$, $\Omega = 1000$, and $\Omega = 100$ as indicated over the panels. The limit cycle in phase space (a) and the temporal evolutions of X/Ω (b) and Y/Ω (c) are shown. In all panels, $k_1 = 0.5$, $k_2 = 1.5$, $k_3 = 1$, and $k_4 = 1$. The initial condition is $c_X(0) = X(0)/\Omega = 0.5$ and $c_Y(0) = Y(0)/\Omega = 3.41$ in all panels.

the concentrations $c_X = X/\Omega$ (b) and $c_Y = Y/\Omega$ (c). It is observed how the regular oscillations found in the macroscopic limit are affected by molecular fluctuations as the system size Ω is decreased. For $\Omega = 10000$, although apparent, the noise is weak enough so that the Fokker–Planck equation can be expected to provide a good description in this regime. However, as the system size is further decreased to $\Omega = 1000$, major deviations from regular oscillations are observed (even when their periodicity is kept), and for $\Omega = 100$, the limit cycle becomes very noisy, the oscillations being quite irregular. It can be shown that the autocorrelation function decays exponentially [Gaspard, 2002], and $\Omega = 100$ has been set as the approximate

threshold below which the quality of chemical clocks is dramatically affected by molecular noise [Gaspard, 2002]. This highlights the fact that the nanoscale dynamics of chemical systems strongly departs from the macroscopic evolution laws owing to the stochastic events taking place and their discrete nature.

22.4.6 Generalized (Electro)Chemical Master Equation and the Extension of Gillespie's Algorithm

Electrochemical systems differ from chemical systems by the fact that the rate of an electrochemical reaction depends on the electrode potential ϕ_{dl}. All information relevant to the macroscopic kinetics is provided by the Butler–Volmer electrochemical equations, which express this relationship through a preexponential factor which is analogous to the chemical rate constant and an exponential dependence on the electrode potential [Bard and Faulkner, 2004]. On a nanoelectrode, the stochastic nature of electrochemical reactions has to be considered. The problem is, however, more complex, since, in addition to the number of chemical species, there may be electron transfer to/from the electrode, and the electrode potential can become a fluctuating variable as well [García-Morales and Krischer, 2010].

In this section, we discuss the electrochemical master equation, which extends the chemical master equation to account for the effect of the stochastic variable ϕ_{dl} on the system dynamics. Therefore, we consider an extended state for the system as given now by a vector $\mathcal{N} = (\phi_{dl}, N_1, \ldots, N_s)$, where the electrode potential ϕ is incorporated to the 'chemical part' ruling the numbers of particles of the s different chemical species $\mathbf{N} = (N_1, \ldots, N_s)$. Therefore, Eq. (22.54) of a chemical system extends to

$$\sum_{i=1}^{s} \nu^i_{<\rho} N_i + n_{<\rho} e^- \xrightarrow{k_\rho(\phi_{dl})} \sum_{i=1}^{s} \nu^i_{>\rho} N_i + n_{>\rho} e^- \quad (22.122)$$

for an electrochemical system. The number of electrons transferred is a new feature compared to chemical systems and is given by $n_\rho = n_{>\rho} - n_{<\rho}$. Again, the stoichiometric numbers $\nu^i_\rho = \nu^i_{>\rho} - \nu^i_{<\rho}$ control the number of molecules of each species formed or consumed each time the reaction takes place. In the following, we shall denote by $\widetilde{\nu}_\rho = (n_\rho, \nu^1_\rho, \ldots, \nu^s_\rho)$ the vector of stoichiometric coefficients $\nu_\rho = (\nu^1_\rho, \ldots, \nu^s_\rho)$, typical of chemical systems, extended by the number of electrons transferred, n_ρ.

The macroscopic rate of electron transfer $k^{(\mathrm{mac})}_\rho$ depends on the electrode potential, ϕ_{dl}, at which the electron transfer takes place in the following form (Butler–Volmer kinetics):

$$k^{(\mathrm{mac})}_\rho(\phi_{dl}) = k^0_\rho e^{c_\rho(\phi_{dl} - \phi^0_{dl})}. \quad (22.123)$$

Here ϕ^0_{dl} is the redox potential of the reaction; the preexponential factor k^0_ρ does not depend on the electrode potential and $c_\rho = \frac{(\beta_\rho - \alpha)|n_\rho|F}{RT}$ where $\beta_\rho = 0$ for reduction reactions and 1 for oxidation reactions, α is the transfer coefficient, F the Faraday constant, R the ideal gas constant, and T the temperature [García-Morales and Krischer, 2010].

A typical electrochemical experiment is controlled by applying an external voltage U between the working and the reference electrodes [Krischer, 2003]. For nanoscale electrodes, an ohmic resistance R_e (e.g., a lipid molecule anchored to a conductive macroscopic support) is often introduced as spacer linking the nanoelectrode (e.g., a metallic nanoparticle) to the external control. Any such arrangement renders the electrode potential ϕ_{dl} a dynamic variable of the system. The evolution of ϕ_{dl} is dictated by charge conservation at the interface. The total current

flowing through the system I splits into two components, a capacitive one I_{cap} involved in the charging of the double layer and a faradaic one I_F coming from electrochemical reactions involving electron transfer to/from the electrode. Charge conservation at the interface implies $I_{\mathrm{cap}} = I - I_F$ or equivalently

$$C\frac{d\phi_{dl}}{dt} = -i_F(\phi_{dl}) + \frac{U - \phi_{dl}}{R_e A}, \quad (22.124)$$

where C is the double layer capacitance per surface area, $A = a_0\Omega$ is the area of the electrode, with a_0 denoting the density of surface sites, and $i_F = I_F/A$. Between two times, $t_0 = t$ and $t + \tau$, where no reaction event takes place at the nanoelectrode, the faradaic current is zero, the above equation can be analytically solved, yielding

$$\phi_{dl}(t + \tau) = U + (\phi_{dl}(t) - U)e^{-\frac{\tau}{R_e CA}}. \quad (22.125)$$

Let τ_ρ be the time after which a stochastic reaction event occurs, reaction ρ taking place. Then, a number n_ρ of electrons are transferred and make a contribution to the Faradaic current. Therefore, Eq. (22.125) takes the form

$$\phi_{dl}(t + \tau_\rho) = U + (\phi_{dl}(t) - U)e^{-\frac{\tau_\rho}{R_e CA}} - \frac{n_\rho e}{CA}. \quad (22.126)$$

For times $\tau_\rho \ll R_e CA$, this equation can be approximated by

$$\phi_{dl}(t + \tau_\rho) = \phi_{dl}(t) + \frac{U - \phi_{dl}(t)}{R_e CA}\tau_\rho - \frac{n_\rho e}{CA}. \quad (22.127)$$

The validity of this approximation needs to be checked for very small nanoelectrodes. The stochastic dynamics of the electrochemical system is thus given by a master equation, where the propensities are explicitly dependent on time,

$$\frac{\partial p(\mathcal{N}, t)}{\partial t} = \sum_\rho [W_\rho(\mathcal{N}|\mathcal{N} - \widetilde{\nu}_\rho, t)p(\mathcal{N} - \widetilde{\nu}_\rho, t)$$
$$- W_\rho(\mathcal{N} + \widetilde{\nu}_\rho|\mathcal{N}, t)p(\mathcal{N}, t)], \quad (22.128)$$

since they are given by

$$W_\rho(\mathcal{N} + \widetilde{\nu}_\rho|\mathcal{N}, t) = e^{c_\rho \phi_{dl}} W_\rho(\mathbf{N} + \nu_\rho|\mathbf{N})$$
$$= e^{c_\rho(\phi_{dl} - \phi^0_{dl})}\Omega k^0_\rho \prod_{i=1}^{s} \prod_{m=1}^{\nu^i_{<\rho}} \frac{N_i - m + 1}{\Omega}, \quad (22.129)$$

which explicitly depend on time through $\phi_{dl}(t)$, whose evolution we shall assume as being provided by Eq. (22.127).

Let $P(\mathcal{N}, t)$ denote the probability that, after a time t, an electrochemical system has been in the same state, with vector $\mathcal{N}$, no reaction taking place. Following the approach suggested in [Jansen, 1995, Koper et al., 1998], we can generalize Gillespie's algorithm, calculating the appropriate expressions for the waiting times of each reaction ρ to occur. In order to do this, we must take into account that the propensities now depend on time through the double layer potential. From Eq. (22.67), we now have

$$P(\mathcal{N}, \tau_\rho) = \exp\left(-\sum_\rho \int_t^{t+\tau_\rho} W_\rho(\mathcal{N} + \tilde{\mu}_\rho | \mathcal{N}, t')dt'\right),$$
(22.130)

and the probability that reaction ρ has not occurred is given by

$$P_\rho(\mathcal{N}, \tau_\rho) = \exp\left(-\int_t^{t+\tau_\rho} W_\rho(\mathcal{N} + \tilde{\mu}_\rho | \mathcal{N}, t')dt'\right)$$

$$= \exp\left(-W_\rho(\mathbf{N} + \nu_\rho | \mathbf{N}) \int_t^{t+\tau_\rho} \exp\left[c_\rho \phi_{dl}(t')\right] dt'\right)$$

$$= \exp\left(-W_\rho(\mathbf{N} + \nu_\rho | \mathbf{N}) \int_t^{t+\tau_\rho} \exp\left[c_\rho \phi_{dl}(t)\right.\right.$$
$$\left.\left.+ \frac{U - \phi_{dl}(t)}{R_e C A} c_\rho(t' - t)\right] dt'\right)$$

$$= \exp\left(-\frac{\exp\left[c_\rho \phi_{dl}(t)\right] W_\rho(\mathbf{N} + \nu_\rho | \mathbf{N}) R_e C A}{c_\rho (U - \phi_{dl}(t))}\right.$$
$$\left.\times \left[\exp\left(\frac{U - \phi_{dl}(t)}{R_e C A} c_\rho \tau_\rho\right) - 1\right]\right),$$
(22.131)

which can be inverted to give

$$\tau_\rho = \frac{R_e C A}{c_\rho (U - \phi_{dl}(t))} \ln$$

$$\times \left[1 + \frac{c_\rho (U - \phi_{dl}(t))}{\exp\left[c_\rho \phi_{dl}(t)\right] W_\rho(\mathbf{N} + \nu_\rho | \mathbf{N}) R_e C A} \ln\left(\frac{1}{P_\rho}\right)\right],$$
(22.132)

where P_ρ is a number drawn from the uniform probability distribution. When there are no electron transfer reactions involved, $c_\rho = 0$, and by applying L'Hopital's rule to Eq. (22.132), we find the waiting time of a purely chemical system, Eq. (22.121):

$$\tau_\rho = \frac{1}{W_\rho(\mathbf{N} + \nu_\rho | \mathbf{N})} \ln\left(\frac{1}{P_\rho}\right).$$
(22.133)

The extension of Gillespie's algorithm to simulate the evolution of the electrochemical master equation now proceeds as follows:

- Initialize the algorithm setting $t = 0$ and $\mathcal{N} = \mathcal{N}_0$, and select a time t_{end} for the duration of the stochastic trajectory.
- Calculate all propensities $W_\rho(\mathcal{N} + \tilde{\nu}_\rho | \mathcal{N})$ from Eq. (22.129).
- Draw M different random numbers P_ρ ($\rho = 1, \ldots, M$) from the uniform distribution, and calculate M positive reaction times τ_ρ given by Eq. (22.132)
- Find $\mu \in [1, M]$ such that $\tau_\mu = \min \tau_\rho$. The reaction μ has the shortest time to occur, in consistency with the random number generated. Select reaction μ to advance, i.e., put $\mathcal{N} \rightarrow \mathcal{N} + \tilde{\nu}_\mu$, the potential being updated by means of Eq. (22.127).
- If $t < t_{end}$, put $t \rightarrow t + \tau_\mu$ and go to step 1; else, finish the algorithm.

This algorithm has allowed to simulate electrochemical systems at the nanoscale from elementary reaction steps [García-Morales and Krischer, 2010, García-Morales and Krischer, 2011] to more complex reaction networks as the reduction of H_2O_2 on a Pt electrode [Mukouyama et al., 2001, García-Morales and Krischer, 2010], a reaction that is known to display oscillations at the macroscale [Mukouyama et al., 2001]. Because of fluctuations on the double layer potential ϕ_{dl}, electrochemical reaction steps occur faster at the nanoscale than at the macroscale [García-Morales and Krischer, 2010, García-Morales and Krischer, 2011]. Quite interestingly, the distribution of the double layer potential shows asymmetries [García-Morales and Krischer, 2011, García-Morales and Krischer, 2011] which have been related to Tsallis spectral statistics [Tsekouras and Tsallis, 2005] and superstatistics [Beck and Cohen, 2003, Beck and Cohen, 2004]. Indeed, it has been found that the stochastic kinetics of nanoscale electrochemical systems can be described by a superstatistical framework with an averaged Tsallis-like entropic parameter q_{av} that measures the impact of nanoscale correlations caused by the double layer potential and which is given by [García-Morales and Krischer, 2011]

$$q_{av} = 1 - \frac{c_\rho(U - \langle\phi_{dl}\rangle)}{R_e C A \langle W_\rho(\mathbf{N} + \nu_\rho | \mathbf{N})\rangle},$$
(22.134)

where the brackets in Eq. (22.134) denote time averages along a stochastic trajectory. We note that for a macroscopic electrode, $A \rightarrow \infty$ and $q_{av} \rightarrow 1$ so that Boltzmann–Gibbs thermostatistics is regained. However, at the nanoscale, A can be small enough so that, in general, $0 \leq q_{av} \leq 1$. This additional role of the stochastic double layer potential in electrochemical systems makes nanoscale electrochemical clocks less robust to molecular noise than purely chemical clocks [Cosi and Krischer, 2017]. Indeed, electrode sizes on the scale of 100–500 nm^2 are likely to be needed in order to observe time-correlated oscillations [Cosi and Krischer, 2017]. Note, however, that the low-noise regime of chemical and electrochemical systems display similar features: the propensities in the electrochemical master equation have the canonical form given by Eq. (22.78) in van Kampen's expansion, and a rigorous Fokker–Planck for electrochemical systems can be derived, matching the general expression Eq. (22.93). Thus, in contrast to other treatments that employ the Langevin equation in the weak noise limit [Gabrielli et al., 1993, Keizer, 1987], the approach presented in [García-Morales and Krischer, 2010] together with van Kampen's expansion of the master equation, here discussed, is able to rigorously describe the behavior of electrochemical systems from the macroscale to the nanoscale, even in regimes where noise is strong and a small number of particles are present.

22.5 Summary and Conclusions

The significant advances in the field of nanothermodynamics over the past decades have been summarized, with special attention to the foundations of nonequilibrium nanothermodynamics from stochastic processes.

The size-dependent thermodynamic properties in mono- and multicomponent systems associated with the surface energy contributions have attracted considerable attention since the late nineteenth century. Although the field is rather mature, many interesting results are still being achieved, thanks to the advances in the experimental techniques.

Hill's nanothermodynamics has proved to be successful in many fields, including the description of metastable states and complex relaxation kinetics. We have focused here on its equilibrium statistical thermodynamics basis. In the recent developments of this theory by Chamberlins group, the interactions between the nanosystems have been satisfactorily described using a mean-field approach in combination with completely open nanosystems that are allowed to adjust their size.

The range of applications of Tsallis nonextensive thermostatistics has proved to be extremely broad. This theory has been presented here as a thermodynamic formalism that can be used to study nanosystems, especially in the presence of correlations. It has been connected to the theory of superstatistics and to the fluctuations in the Boltzmann parameter.

The master equation constitutes a general and powerful tool for the stochastic modeling of nonequilibrium nanosystems beyond the linear branch of nonequilibrium thermodynamics. This equation allows the Fokker–Planck equation to be rigorously derived in the weak noise limit and, thus, it also provides an understanding on how macroscopic deterministic dynamics emerges out of an inherently probabilistic description. Although the nonlinear effects cause couplings in the collective dynamics, rigorous results of general validity can be established. These results include the integral FTs, which are of enormous interest for the understanding of both equilibrium and nonequilibrium processes at the nanoscale.

Acknowledgments

J. C. and J. A. M. thank financial support from Ministerio de Ciencia, Innovación y Universidades and the European Regional Development Funds through project PGC2018-097359-B-100.

References

Abe, S. and Okamoto, Y. 2001. *Nonextensive Statistical Mechanics and its Applications*. Berlin: Springer.

Ali, S.; Myasnichenko, V. S. and Neyts, E. C. 2016. Size-dependent strain and surface energies of gold nanoclusters. *Phys. Chem. Chem. Phys.* 18: 792–800.

Alloyeau, D.; Prevot, G.; Le Bouar, Y.; Oikawa, T.; Langlois, C.; Loiseau, A. and Ricolleau, C. 2010. Ostwald ripening in nanoalloys: When thermodynamics drives a size-dependent particle composition. *Phys. Rev. Lett.* 105: 255901.

Astumian, R. D. 2001. Thermodynamics and kinetics of a Brownian motor. *Science* 276:917–22.

Bard, A. and Faulkner, L. 2004. *Electrochemical Methods*. New York: Wiley.

Beck, C. 2002. Non-additivity of Tsallis entropies and fluctuations of temperature. *Europhys. Lett.* 57: 329–33.

Beck, C. and Cohen, E. G. D. 2003. Superstatistics. *Physica A* 322:267–75.

Beck, C. and Cohen, E. G. D. 2004. Superstatistical generalization of the work fluctuation theorem. *Physica A* 344:393–402.

Brandao, F.; Horodecki, M.; Ngc, N.; Oppenheim, J. and Wehner S. 2015. The second laws of quantum thermodynamics. *Proc. Natl. Acad. Sci.* 112: 3275–9.

Brites, C. D. S.; Lima, P. P.; Silva, N. J. O.; Mill, A.; Amaral, V. S.; Palacio, F. and Carlos, L. D. 2012. Thermometry at the nanoscale. *Nanoscale* 4:4799–829.

Bustamante, C.; Keller, D. and Oster, G. 2001. The physics of molecular motors. *Acc. Chem. Res.* 34:412–20.

Cartwright, J. 2014. Roll over, Boltzmann. Does Tsallis entropy really add up? *Phys. World* 27: 31–5.

Castelvecchi, D. 2017. Clash of the physics laws. *Nature* 543:597–8.

Chamberlin, R. V. 2015. The big world of nanothermodynamics. *Entropy* 17: 52–73.

Chandler, D. 1987. *Introduction to Modern Statistical Mechanics*. New York: Oxford University Press.

Chen, C. L.; Lee, J.-G.; Arakawa, K. and Mori, H. 2011. Quantitative analysis on size dependence of eutectic temperature of alloy nanoparticles in the Ag-Pb system. *Appl. Phys. Lett.* 98: 083108.

Chushak, Y. G. and Bartell, L. S. 2001. Melting and freezing of gold nanoclusters. *J. Phys. Chem. B* 105: 11605–14.

Cosi, F. G. and Krischer, K. 2017. Destructive impact of molecular noise on nanoscale electrochemical oscillators. *Eur. Phys. J. Spec. Top.* 226:1997–2013.

Couchman, P. R. and Jesser, W. A. 1977. Thermodynamic theory of size dependence of melting temperature in metals. *Nature* 269: 481–3.

Crooks, G. E. 1999. Entropy production fluctuation theorem and the nonequilibrium work relation for free-energy differences. *Phys. Rev. E* 60: 2721–6.

Defay, R. and Prigogine I. 1966. *Surface Tension and Adsorption*. London: Longmans.

DeHoff, R. 2006. *Thermodynamics in Materials Science*. Boca Raton, FL: Taylor & Francis.

Evans, D. J.; Cohen, E. G. D. and Morriss G. P. 1993. Probability of second law violations in shearing steady states. *Phys. Rev. Lett.* 71:2401.

Evans, D. J. and Searles, D. J. 1992. The fluctuation theorem. *Adv. Phys.* 51:1529–1585.

Evans, D. J. and Searles, D. J. 1994. Equilibrium microstates which generate second law violating steady states. *Phys. Rev. E* 50:1645–1648.

Evans, D. J.; Searles, D. J. and Williams, S. R. 2016. *Fundamentals of Classical Statistical Thermodynamics*.

Dissipation, Relaxation, and Fluctuation Theorems. Weinheim: Wiley.

Falcioni, M.; Villamaina, D.; Vulpiani, A.; Puglisi, A. and Sarracino, A. 2011. Estimate of temperature and its uncertainty in small systems. *Am. J. Phys.* 19: 777–85, 19: 980.

Farrell, H. H. and Van Siclen, C. D. 2007. Binding energy, vapor pressure, and melting point of semiconductor nanoparticles. *J. Vac. Sci. Technol. B.* 71:1441–7.

Ford, I. 2013. *Statistical Physics. An Entropic Approach.* Chichester: Wiley.

Gabrielli, C.; Huet, F.; and Keddam, M. 1993. Fluctuations in electrochemical systems. I. General theory on diffusion limited electrochemical reactions. *J. Chem. Phys.* 99:7232–9.

García-Morales, V.; Cervera, J. and Pellicer, J. 2004. Coupling theory for counterion distributions based in Tsallis statistics. *Physica A* 339:482–90.

García-Morales, V.; Cervera, J. and Pellicer, J. 2005. Correct thermodynamic forces in Tsallis thermodynamics: Connection with Hill nanothermodynamics. *Phys. Lett. A* 336:82–8.

García-Morales, V. and Pellicer, J. 2006. Microcanonical foundation of nonextensivity and generalized thermostatistics based on the fractality of the phase space. *Physica A* 361:161–72.

García-Morales, V. and Krischer, K. 2010. Fluctuation enhanced electrochemical reaction rates at the nanoscale. *Proc. Nat. Acad. Sci.* 107:4528–32.

García-Morales, V.; Cervera, J. and Manzanares, J. A. 2011. Nanothermodynamics. In K. D. Sattler (ed.) *Handbook of Nanophysics. Principles and Methods*, ch.15. Boca Raton, FL: CRC Press.

García-Morales, V. and Krischer, K. 2011. Superstatistics in nanoscale electrochemical systems. *Proc. Nat. Acad. Sci.* 108:19535–9.

García-Morales, V. and Krischer, K. 2011. Kinetic enhancement in nanoscale electrochemical systems caused by non-normal distributions of the electrode potential. *J. Chem. Phys.* 134:244512.

Gardiner, N. G. 2009. *Stochastic Methods: A Handbook for the Natural and Social Sciences.* New York: Springer.

Gaspard, P. 2002. The correlation time of mesoscopic chemical clocks. *J. Chem. Phys.* 117:8905–16.

Gheorghiu, S. and Coppens, M. O. 2004. Heterogeneity explains features of "anomalous" thermodynamics and statistics. *Proc. Natl. Acad. Sci. USA* 101: 15852–6.

Gillespie, D. T. 1976. A general method for numerically simulating the stochastic time evolution of coupled chemical reactions. *J. Comput. Phys.* 22:403–34.

Gillespie, D. T. 1977. Exact stochastic simulation of coupled chemical reactions. *J. Phys. Chem.* 81:2340–61.

Haag, G. 2017. *Modelling with the Master Equation: Solution Methods and Applications in Social and Natural Sciences.* Cham: Springer.

Hill, T.L. 1994. *Thermodynamics of Small Systems.* New York: Dover.

Hill, T.L. 2001. A different approach to nanothermodynamics. *Nano Lett.* 1:273–5.

Hill, T. L. and Chamberlin, R.V. 2002. Fluctuations in energy in completely open small systems. *Nano Lett.* 2: 609–13.

Horodecki, M. and Oppenheim, J. 2013. Fundamental limitations for quantum and nanoscale thermodynamics. *Nature Comm.* 4:2059.

Inzoli, I.; Kjelstrup, S.; Bedeaux, D. and Simon, J. M. 2010. Thermodynamic properties of a liquid-vapor interface in a two-component system. *Chem. Engn. Sci.* 65: 4105–16.

Jabbareh, M. A. and Monji F. 2018. Thermodynamic modeling of Ag-Cu nanoalloy phase diagram. *Calphad* 60: 208–13.

Jansen, A. 1995. Monte Carlo simulations of chemical reactions on a surface with time-dependent reaction rate constants. *Comput. Phys. Commun.* 86:1–12.

Jiang, Q. and Wen, Z. 2011. *Thermodynamics of Materials.* Berlin: Springer.

Kaszkur, Z. 2013. Thermodynamical properties of nanoalloys. In F. Calvo (ed.) *Nanoalloys. From Fundamentals to Emergent Applications*, chap. 5. Amsterdam: Elsevier.

Keizer, J. 1987. *Statistical Thermodynamics of Nonequilibrium Processes.* New York: Springer.

Kondepudi, D. 2008. *Introduction to Modern Thermodynamics.* New York: Wiley.

Kofman, R.; Cheyssac, P.; Lereah, Y. and Stella, A. 1999. Melting of clusters approaching 0D. *Eur. Phys. J. D* 9: 141–4.

Koper, M.; Jansen, A.; van Santen, R.; Lukkien, J.; and Hilbers, P. 1998. Monte Carlo simulations of a simple model for the electrocatalytic Co oxidation on platinum. *J. Chem. Phys.* 109:6051–62.

Krischer, K. 2003. Nonlinear dynamics in electrochemical systems. In Kolb, D. M. and Alkire, R. C. (eds.) *Advances in Electrochemical Sciences and Engineering*, Vol. 8, pp. 89–208. Weinheim: Wiley.

Kurchan, J. 1998. Fluctuation theorem for stochastic dynamics. *J. Phys. A Math. Gen.* 31:3719–29.

Lebowitz J. L. and Spohn, H. 1999. A Gallavotti-Cohen-type symmetry in the large deviation functional for stochastic dynamics. *J. Stat. Phys.* 95: 333–65.

Li, Z. H. and Truhlar, D. G. 2014. Nanothermodynamics of metal nanoparticles. *Chem. Sci.* 5: 2605–24.

Lu, H. M. and Jiang, Q. 2005. Size-dependent surface tension and Tolmans length of droplets. *Langmuir* 21:779–81.

McEwen, J. S.; Gaspard, P.; Visart de Bocarmé, T.; and Kruse, N. 2009. Nanometric chemical clocks. *Proc. Nat. Acad. Sci.* 106:3006–10.

Mafé, S.; Manzanares, J. A. and de la Rubia, J. 2000. On the use of the statistical definition of entropy to justify Planck's form of the third law of thermodynamics. *Am. J. Phys.* 68: 932–5.

Mahler, G. 2015. *Quantum Thermodynamics Processes: Energy and Information Flows at the Nanoscale.* New York: Pan Stanford.

Manzanares, J. A.; Peljo, P. and Girault, H. H. 2017. Understanding digestive ripening of ligand-stabilized, charged metal nanoparticles. *J. Phys. Chem. C* 121: 13405–11.

Mukouyama, Y.; Nakanishi, S.; Chiba, T.; Murakoshi, K. and Nakato, Y. 2001. Mechanisms of two electrochemical oscillations of different types, observed for H_2O_2 reduction on a Pt electrode in the presence of a small amount of halide ions. *J. Phys. Chem. B.* 105:7246–53.

Nanda, K. K.; Maisels, A.; Kruis, F. E.; Fissan, H. and Stappert, S. 2003. Higher surface energy of free nanoparticles. *Phys. Rev. Lett.* 91: 106102.

Naudts, J. 2011. *Generalised Thermostatistics.* Springer: London.

Nicolis, G. 1972. Fluctuations around nonequilibrium states in open nonlinear systems. *J. Stat. Phys.* 6:195–222.

Nicolis, G. and Prigogine, I. 1977. *Self-Organization in Nonequilibrium Systems.* New York: Wiley.

Oppenheim, I.; Schuler, K. E. and Weiss, G. H. 1977. *Stochastic Processes in Chemical Physics: The Master Equation.* Cambridge: The MIT Press.

Park, J. and Lee, J. 2008. Phase diagram reassessment of Ag–Au system including size effect. *Calphad* 23: 135–41.

Peljo, P.; Manzanares, J. A. and Girault, H. H. 2016. Contact potentials, Fermi level equilibration, and surface charging. *Langmuir* 32:5765–75.

Peljo, P.; Manzanares, J. A. and Girault, H. H. 2017. Variation of the Fermi level and the electrostatic force of a metallic nanoparticle upon colliding with an electrode. *Chem. Sci.* 8: 4795–803, 8: 5206.

Peters, K. F.; Cohen, J. B. and Chung, Y. W. 1998. Melting of Pb nanocrystals. *Phys. Rev. B* 57: 13430–8.

Pohl, J.; Stahl, C. and Albe, K. 2012. Size-dependent diagrams of metallic alloys: A Monte Carlo simulation study on order-disorder transitions in Pt-Rh nanoparticles. *Beilstein J. Nanotechnol.* 3:1–11.

Risken, H. 1989. *The Fokker–Planck Equation: Methods of Solution and Applications.* New York: Springer.

Scanlon, M. D.; Peljo, P.; Méndez, M. A.; Smirnov, E. and H. H. Girault 2015. Charging and discharging at the nanoscale: Fermi level equilibration of metallic nanoparticles. *Chem. Sci.* 6: 2705–20.

Schliwa, M. 2003. *Molecular Motors.* Weinheim: Wiley.

Schmelzer J. W. P.; Boltachev, G. Sh. and Abyzov, A. S. 2013. Temperature of critical clusters in nucleation theory: Generalized Gibbs' approach. *J. Chem. Phys.* 139: 034702.

Seifert, U. 2005. Entropy production along a stochastic trajectory and an integral fluctuation theorem. *Phys. Rev. Lett.* 95:040602.

Seifert, U. 2012. Stochastic thermodynamics, fluctuation theorems and molecular machines. *Rep. Prog. Phys.* 75:126001.

Sevick, E. M.; Prabhakar, R.; Williams, S. R. and Searles, D. J. 2008. Fluctuation theorems. *Annu. Rev. Phys. Chem.* 59:603–33.

Shi, Z.; Wynblatt, P. and Srinivasan, S. G. 2004. Melting behaviour of nanosized lead particles embedded in an aluminium matrix. *Acta Mater.* 52: 2305–16.

Spinney, R. and Ford, I. 2013. Fluctuation relations: A pedagogical overview. In R. Klages, W. Just and C. Jarzynski (eds.) *Nonequilibrium Statistical Physics of Small Systems. Fluctuation Relations and Beyond,* chap. 1. Weinheim: Wiley.

Sugimoto, T. and Shiba, F. 1999. A new approach to interfacial energy. 3. Formulation of the absolute value of the solid-liquid interfacial energy and experimental collation to silver halide systems. *J. Phys. Chem. B* 103: 3607–15.

Tsallis, C. 1988. Possible generalization of Boltzmann-Gibbs statistics. *J. Stat. Phys.* 52:479–87.

Tsallis, C. 2005. Is the entropy S_q extensive or nonextensive? In C. Beck, G. Benedek, A. Rapisarda, and C. Tsallis (eds.) *Complexity, Metastability and Nonextensivity. 31st Workshop of the International School of Solid State Physics,* pp. 13–32. Singapore: World Scientific.

Tsallis, C. 2009. *Introduction to Nonextensive Statistical Mechanics. Approaching a Complex World.* New York: Springer.

Tsekouras, G. A. and Tsallis, C. 2005. Generalized entropy arising from a distribution of q indices. *Phys. Rev. E* 71:046144.

van Kampen, N. G. 2007. *Stochastic Processes in Physics and Chemistry.* Amsterdam: North Holland (Elsevier).

Visart de Bocarmé, T. and Kruse, N. 2002. Kinetic instabilities during the NO(x) reduction with hydrogen on Pt crystals studied with field emission on the nanoscale. *Chaos* 12:118–30.

Wang, H. and Oster, G. 1998. Energy transduction in the F1 motor of ATP synthase. *Nature (London)* 396: 279–282.

Wang, C. X.; Yang, Y. H.; Xu, N. S. and Yang, G. W. 2004. Thermodynamics of diamond nucleation on the nanoscale. *J. Am. Chem. Soc.* 126: 11303–6.

Wang, C. X. and Yang, G. W. 2005. Thermodynamics of metastable phase nucleation at the nanoscale. *Mat. Sci. Engn. R* 49: 157–202.

Yang, C. C. and Li, S. 2008. Size-dependent temperature-pressure phase diagram of carbon. *J. Phys. Chem. C* 112: 1423–6.

Zhang, C. Y.; Wang, C. X.; Yang, Y. H. and Yang, G. W. 2004. A nanoscaled thermodynamic approach in nucleation of CVD diamond on nondiamond surfaces. *J. Phys. Chem. B* 108: 2589–93.

Characterization of Nanoscale Thermal Conductivity

Weidong Liu,
Liangchi Zhang,
Alireza Moridi,
and Mohammad Ehsan Khaled
The University of New South Wales

23.1 Introduction

Nanoscale materials and devices have shown tremendous promise as building blocks for future technologies (Balogun et al., 2016; Fu et al., 2016; Goriparti et al., 2014; Kim et al., 2017b; Lembke et al., 2015; Li and Wu, 2015; Shi et al., 2015; Thorkelsson et al., 2015; Zhao and Tan, 2014). The relentless downscaling in the size of devices and increasing in the operating speeds require new thermal theories, experimental, and numerical tools at the nanoscale (Al-Ghalith and Dumitrica, 2018; Cahill et al., 2014; Cahill et al., 2003; Chen, 2017; Cocemasov et al., 2018; Hippalgaonkar et al., 2017; Kage et al., 2016; Khan et al., 2017; Kwon et al., 2016; Mei et al., 2017; Nomura et al., 2015; Park and Prakash, 2013; Vinod and Philip, 2017; Wagner et al., 2013; Yang et al., 2012; Yeandel et al., 2018; Zhou et al., 2018). For example, transistors have followed Moore's law of continual shrinking remarkably in the last 50 years increasing the packing density (Moore, 1998). At present, we have transistors with less than 10 nm dimension and expected to reach size less than 7 nm by 2020 (VanGessel et al., 2018). However, a further shrinkage will degrade electrical characteristics as well as cause self-heating in electronic devices (Ferain et al., 2011; Majumdar, 1993; Moridi et al., 2018). In addition, phonon scattering induced by the fabrication defects, impurities, doping, and atomic mismatch at the thin film's interface can also influence the thermal conduction (Asheghi et al., 2002; Cao et al., 2017; Kovacevic and Pivac, 2014; Stevens et al., 2007).

A major challenge for the next generation semiconductor industry is finding materials with high thermal conductivity to efficiently dissipate the heat generated by the electronics (Garimella et al., 2008; Garimella and Joshi, 2002; Joshi and Garimella, 2003; Shahil and Balandin, 2012). The conventional dielectric insulation layers, such as silicon oxide (SiO_2) or alumina (Al_2O_3), have very low thermal conductivities (Cao et al., 2004; Kingon et al., 2000; Kolodzey et al., 2000; Zhang and Najafi, 2004). Hence, while functioning as good dielectric insulators, they also shield the heat induced by the self-heating of electronic devices which greatly deteriorate their performance reliability (Cao et al., 2004; Kingon et al., 2000; Kolodzey et al., 2000). To overcome this problem, bulk silicon has been utilized as the substrate material; and some dielectric materials of comparable thermal conductivity such as diamond-like carbon (DLC) (Bewilogua and Hofmann, 2014; Bullen et al., 2000) or aluminum nitride (AlN) (Alrashdan et al., 2014; Bian et al., 2015) have been deposited as the thin-film layer.

However, the applicability of Fourier's law, the foundation of thermal continuum theory at the macroscale, is questionable at the nanoscale. It is because the mean free path (MFP) of heat carriers (such as phonons, electrons) becomes comparable to the characteristic dimensions of micro/nanodevices and structures (Chen, 2005; Chen et al., 2004; Volz, 2009). When heat carriers have mean free paths (MFP) shorter than the dimension scales of the materials and devices, heat transport is diffusive; when heat carriers have MFPs comparable to, or larger than, the important dimension scales, heat transport is ballistic (Wilson and Cahill, 2014), making the continuum assumptions failure at the nanoscale.

Moreover, the complexity of nanoscale devices and structures makes the characterization of nanoscale thermal conductivity more challenging. For example, in a typical nanoscale thin-film/substrate system, there are three primary factors that could affect its thermal conductance but are very difficult to be characterized by continuum theory (Moridi et al., 2018). Firstly, the thermal resistance of the thin-film and substrate materials themselves would cause some temperature increase. Secondly, the interface thermal resistance may significantly shield the heat from the thin film into the substrate. Thirdly, any crystallographic defects such as dislocation or twinning may affect the thermal properties of the crystals and consequently could reduce the thermal conductivity. Existing theoretical models are not capable of introducing all those three effects, due to the complexity of the system. Thus, the actual temperature rise in electronic devices, such as in high-performance chips, is much more significant than theoretical predictions (Martin et al., 2008; Yang and Chen, 2004). A deep understanding of the scattering mechanisms is required to characterize the nanoscale thermal conductivity.

In this chapter, we aim to review the latest advancements in characterizing the nanoscale thermal conductivity. Section 23.2 will introduce the advances in the commonly used experimental tools, including Time-Domain Thermoreflectance, Scanning Thermal Microscopy, Coherent Optical Methods, and the 3ω Method. Special attention will be paid on a modified 3ω method recently developed for characterizing the high thermal conductivity ultra-thin-film/substrate system. Section 23.3 will summarize the progress of the key numerical methods, including First-Principles Calculation, Green's function, Boltzmann Transport Equation, and Molecular Dynamics. In addition, a recent study on the parameter selection for a reliable molecular dynamics analysis of thermal conductivity will be introduced. A brief summary and future perspective will be given in Section 23.4.

23.2 Experimental Methods

Experimentally characterizing the nanoscale thermal conductivity is challenging, due to the fact that controlling the heat penetration depth at the nanoscale is difficult (Zhao et al., 2016). Particularly, if the thermal conductivity of a nanomaterial is relatively high (such as in crystalline silicon thin films), the measurement becomes more difficult as the heat could penetrate into the substrate fast and deeply (Moridi et al., 2018). In this section, the key nanoscale thermal metrology methods, including time-domain thermoreflectance (Cahill, 2004), micro-fabricated probes (Binnig et al., 1986; Gomes et al., 2015), coherent optical methods (Mansanares et al., 1994; Merlin, 1997), and the 3ω method will be introduced (Cahill, 1990; Moridi et al., 2018). The basic concepts, the latest progress, and the advantages and drawbacks of each method will be discussed.

23.2.1 Time-Domain Thermoreflectance

Time-domain thermoreflectance (TDTR) was firstly carried out by the groups of G. Eesley and H. Maris more than 30 years ago (Cahill, 2018; Paddock and Eesley, 1986). In the last 18 years, this method was continuously developed by many researchers, particularly in D. Cahill's group (Cahill, 2018; Cahill et al., 2014; Schmidt et al., 2008; Wilson and Cahill, 2014). The main idea behind this method is that the change in surface reflectance induced by a pulse temperature increase can be used for calculating the thermal properties of the specimen including the thermal conductivity (Cahill, 2004, 2018).

A typical schematic of TDTR method implementation is demonstrated in Figure 23.1. Overall, the method consists of two parts: (i) the "pump" laser beam, which heats up the specimen surface and (ii) the "probe" beam, which is utilized to measure the reflected energy of the beam pulses for calculating the specimen temperature. A thin layer of metal with high thermoreflectance (e.g. Al) is deposited on the specimen as the transducer to absorb the heating pump beam and to convert the temperature-induced reflectance changes into the probe pulse. The thermal properties of the specimens can be determined from the temperature evolution as a function of the delay time between the pump and the probe pulses (Cahill, 2018; Kang et al., 2008). Figure 23.2 shows a typical pump pulse, the evolution of the surface temperature, the probe pulse, and the measured signal. Since the thermal penetration depth is significantly small (Capinski et al., 1999), a one-dimensional heat flow model is normally utilized to obtain the thermal properties of the specimen (Cahill et al., 2003). Recently, the surface heating problem of multi-layered structures was solved in the cylindrical coordinate and has been applied in TDTR measurements (Cahill, 2004, 2018).

The ultra-fast temperature excursion (~100 fs) and very shallow heat penetration depth (~ 100 nm) are the primary

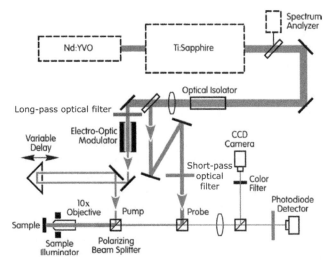

FIGURE 23.1 A typical schematic of the TDTR method implementation process (Kang et al., 2008).

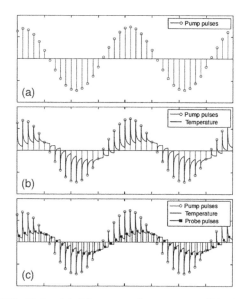

FIGURE 23.2 (a) The pump beam input, (b) the surface temperature in response to the pump input, and (c) the probe pulse (Schmidt et al., 2008).

advantages of TDTR method. Hence, the TDTR method enables the thermal conductivity characterization of high thermal conductivity and extremely thin materials (Cahill, 2018). For example, TDTR method has been successfully applied in the search of a cubic boron arsenide material with an ultra-high thermal conductivity of 1000 ± 90 W/(m·K), surpassed only by diamond and the basal-plane value of graphite (Li et al., 2018). TDTR method has also been applied to characterizing the thermal conductivity of aligned liquid crystal networks (Shin et al., 2016) and amorphous polymers (Xie et al., 2016).

It should be noted that TDTR method has specific limitations. For example, TDTR requires that the specimen surface should be smooth enough (Ra < 15 nm) to avoid any undesired modulation of the diffuse scattering produced by thermoelastic effects (Schmidt et al., 2008). The measurement results of TDTR are very sensitive to the thermal conductivity in the through-thickness direction (Schmidt et al., 2008). Thus it is still a challenging research topic to accurately characterize the thermal conductivity of anisotropic materials such as superlattices, textured polycrystalline film, and anisotropic crystals (Schmidt et al., 2008). Due to the small heat capacity of the metal transducer, the application of TDTR method at low temperature (< 30 K) is limited as well (Schmidt et al., 2008). Moreover, the initial setup cost for TDTR is expensive and the noise from the low-quality laser could affect the accuracy of the results.

23.2.2 Scanning Thermal Microscopy

The scanning thermal microscopy method (SThM) was first invented in 1986 by Williams and Wickramasinghe (Williams and Wickramasinghe, 1986). It was based on

a thermocouple sensor (see Figure 23.3), which was sensitive to local temperature and thus acts as a nanoscale thermometer. The subsequent developments moved towards cantilever-based probes (Binnig et al., 1986; Cahill et al., 2002; Gomes et al., 2015). In a SThM measurement, the localized heat transfers from the sample surface to the sensing probe and changes its temperature (see Figure 23.4). By scanning across the specimen surface, a spatial map of the temperature versus the distance between the probe and the sample could be mapped out. The thermal properties of the specimen can be obtained by fitting the temperature

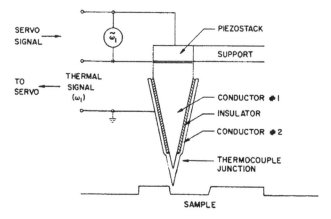

FIGURE 23.3 Schematic diagram of the thermocouple probe (Williams and Wickramasinghe, 1986).

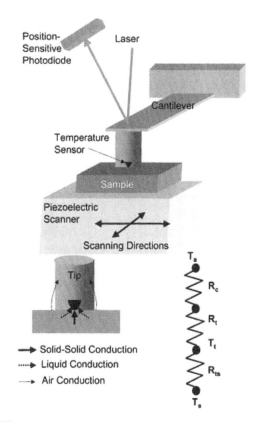

FIGURE 23.4 Schematic diagram of a scanning thermal microscope (Cahill et al., 2003).

magnitude change of the probe with the aid of a theoretical heat flux model. Utilizing this method the local temperature properties including the thermal conductivity can be calculated.

The resolution of SThM depends on the thermal probe design, the heat transfer mechanism, and the probe tip sharpness (Cahill et al., 2014; Cahill et al., 2003). Different probes have been developed and can be classified into different categories based on their underlying mechanisms including thermovoltage (Gomes et al., 2015; Kittel et al., 2005; Sadat et al., 2010), electrical resistance (Dobson et al., 2007; Hinz et al., 2008; Wielgoszewski et al., 2010; Zhang et al., 2011, 2012), fluorescence (Aigouy et al., 2005; Saidi et al., 2009), and thermal expansion (Nakabeppu et al., 1995). Thermovoltage-based methods utilize the thermoelectric voltage generated at the junction between the probe and the specimen to measure the temperature (Gomes et al., 2015). It also includes probes with a built-in thermal sensor such as a thermocouple (Kittel et al., 2005). The electrical resistance-based probe can measure the heat-induced electrical resistance changes and thus obtain the temperature. To achieve a higher resolution, smaller metallic probes (Wielgoszewski et al., 2010; Zhang et al., 2011) and doped Si resistor probes (Hinz et al., 2008) were developed. It was reported that in an ultra-high vacuum environment the SThM method with a nanofabricated probe (integrated with a nanoscale Au–Cr thermocouple) can achieve a lateral spatial resolution of 10 nm and a temperature resolution of 50 mK (Kim et al., 2012). Recently, a spatial/temperature resolution of 7 mk/sub-10 nm was achieved by using an indium arsenide nanowire (Menges et al., 2016).

It should be noted that the heat transfer between the specimen and probe is very complicated. The nanoscale tip radius comparable to phonon mean free path can induce significant error. Therefore, the major challenge of this method is to separate thermal from topographic data (Luo and Chen, 2013). Moreover, SThM is not suitable for characterizing insulators with little or no ballistic electrons, which limits its application in SOI thin-film systems (Luo and Chen, 2013).

23.2.3 Coherent Optical Methods

The optically generated coherent phonons can be utilized in thermal transport measurements, which is considered as an alternative method to determine the mean free path (Baumberg et al., 1997; Dieleman et al., 2001; Hao and Maris, 2000; Merlin, 1997). Many experimental tools have been developed based on the coherent optical beams. Scanning optical thermometry is a commonly used coherent optical method using visible or near-visible radiation with far-field optics (Anisimov et al., 2016; Goodson and Asheghi, 1997; Nitta et al., 2014). In contrast to the scanning probe methods, scanning optical thermometry directly probes the specimen without requiring the heat transfers into the sensor. Figure 23.5 shows the schematic of a typical scanning optical thermometry system (Kim et al., 2017a). The scanning optical thermometry technique is suitable for micro-machined structures thermometry due to its capability of precise focusing on a specific spot of micro-devices (Mansanares et al., 1994). However, this method has noise and calibration

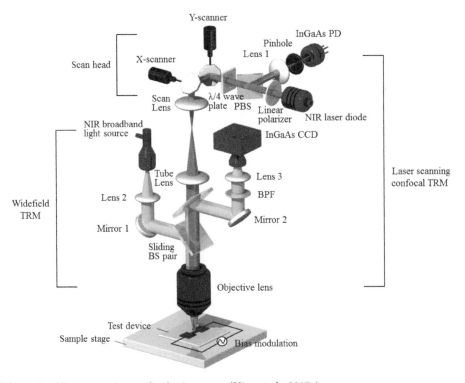

FIGURE 23.5   Schematic of laser scanning confocal microscope (Kim et al., 2017a).

complications and is not suitable for materials with relatively low transmittance.

A recently developed thermal conductivity spectroscopy technique (Minnich et al., 2011) allowed incorporating contribution of phonons with a larger mean free path in thermal conductivity. This is achieved via lowering modulation frequency and increasing penetration depth. However, this dependency limits its application in nanometer level where phonon mean free path becomes comparable to the characteristic length and propagation become ballistic rather than diffusive (Luo and Chen, 2013). Other coherent optical methods such as Raman microscopy and photoluminescence have also been developed for characterizing nanoscale thermal conductivity by measuring the shift in wavelength as a function of temperature (Balandin et al., 2008; Luo and Chen, 2013; Westover et al., 2008). However, these methods face certain limitations such as large noise and low accuracy, complicated calibration process, and inapplicability for materials with low transmittance.

23.2.4 Three-Omega (3ω) Method

The 3ω method for characterizing thermal conductivity was first intruded by Cahill et al. in 1987 (Cahill and Pohl, 1987). In the past two decades, the 3ω method gained popularity in measuring thin-film and bulk materials thermal conductivities. A great body of research is available in the literature, which has tried to implement and improve the method for various bulk and thin-film material systems (Bauer et al., 2011; Bodenschatz et al., 2013; Cahill et al., 2000; Cahill et al., 1994; Hu et al., 2006; Kaul et al., 2007; Kim et al., 1999; Lee et al., 1997; Raudzis et al., 2003; Rausch et al., 2013; Shen et al., 2013; Shenoy et al., 2010; Wang and Sen, 2009; Wang et al., 2007; Zhang and Grigoropoulos, 1995).

Figure 23.6 shows a schematic of the 3ω method measurement configuration. An AC driving current is applied through the two outer pads. A metal strip between the two pads was deposited on the surface of the specimen, acting as both a heater and a thermometer. The applied sinusoidal current, at a frequency of ω, can produce periodic Joule heating at a frequency of 2ω. This heating makes the surface temperature oscillate at the frequency of 2ω as well. On the other hand, the electrical resistance of the sensing element fluctuates proportionally with the surface temperature. Combining with the primary driving voltage,

it produces a third harmonic signal $V_{3\omega}$. The magnitude and phase of this $V_{3\omega}$ signal can be utilized to characterize the thermal properties of the underlying material including the thermal conductivity (Cahill, 2002; Cahill and Pohl, 1987).

The 3ω method has several unique advantages over the other thermometry methods described in the previous sections. One of the primary advantages of the 3ω method is its low-temperature feature, leading to three benefits. Firstly, the specimen material property does not change after applying the 3ω method. Secondly, low temperature cause lower undesirable heat loss from the sample and increase the accuracy of the measurements. In the 3ω method, the metal strip element is merely a fraction of Celsius degree hotter than the surrounding temperature. Thirdly, this technique is non-invasive, meaning that it can be utilized to carry out an in situ characterization of the specimen. In addition, the 3ω method facilities and the deposition of the metal strip on the specimen surface used for the measurement have relatively low costs.

However, the conventional 3ω method is limited in measuring only low thermal conductivity thin-film/substrate systems. It is because the low thermal conductivity of the thin-film guarantees that the heat does not pass through the film into the substrate. In that case, the effects of the substrate thermal properties in the 3ω method are minimized. In high thermal conductivity thin film, however, the amount of the heat that passed through the thin film and penetrated into the substrate may affect the accuracy of thin-film thermal conductivity measurement. This has posed a challenge for the application of the 3ω method to crystalline semiconductor on insulator systems such as AlN on Si. Recently, Moridi and Zhang et al. (Moridi et al., 2018) proposed a modified 3ω method with a nanostrip to solve this problem. The extremely narrow metal strip element provides a much better heat penetration depth control over the conventional 3ω method. The heat merely penetrates into several microns of the specimen (Moridi et al., 2018). Moreover, the narrow strip induces a much higher electrical resistance in comparison with the wide strip elements, making it easier to balance the measurement circuit. The details of this modified 3ω method and its applications will be introduced in the next section.

23.2.5 Modified 3ω Method with Nanostrip

Metal Strip Width Effect

It was found that the temperature distribution under the metal strip of the 3ω method measurement setup can be divided into three regions as shown in Figure 23.7, i.e. planar region, transition region, and linear region (Moridi et al., 2018). In the planar and transition region, the heat penetration depth is in the vicinity of the metal strip and the temperature oscillation is negligible and unstable. Thus these regions are not suitable for characterizing the thermal conductivity calculation based on the 3ω method theory. On the other hand, in the linear region the temperature

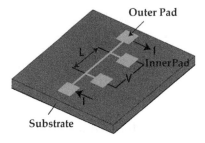

FIGURE 23.6 A schematic of the experimental configuration in the 3ω method.

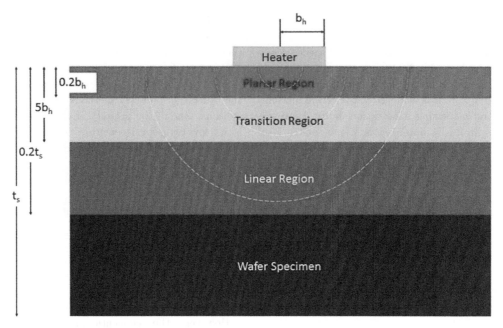

FIGURE 23.7   Different temperature region in the specimen during a 3ω method measurement.

oscillation reaches a steady-state condition required for conducting 3ω method analysis. It was reported that the inner and outer radii of the linear region are normally five times the metal strip half width and to the one-fifth of the specimen thickness, respectively, beyond which boundary effects will appear (Moridi et al., 2018). Therefore, for characterizing a high thermal conductivity thin film, one needs to have a very shallow thermal penetration depth, which can be achieved via a narrow metal strip and a high oscillation frequency.

Moridi et al. (2018) studied the effects of the metal strip half width b on the temperature oscillation ΔT in a specimen of AlN film on substrate Si over a wide range of frequencies. The results for four different metal strip elements widths are illustrated in Figure 23.8. It was found

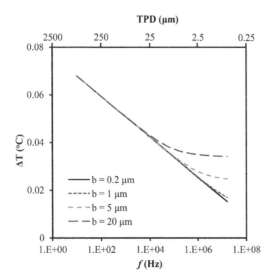

FIGURE 23.8   Effect of the metal strip width on the $\Delta T \sim f$ for a 2 µm AlN film on Si (Moridi et al., 2018).

if the measurements in small thermal penetration depth (TPD) and high frequency are of interest, the metal strip width should be accordingly narrow enough ($\sim$200 nm) (Moridi et al., 2018). Based on this idea, the authors developed a modified 3ω method with a nanostrip ($b = 200$ nm) as shown in Figure 23.9.

Algorithm Development

To apply the modified 3ω method in the high conductivity thin-film systems, Moridi et al. (2018) further developed a generalized algorithm. Considering a thin-film on substrate system with a thin film called MAT1 and the substrate MAT2 (see Figure 23.10), the total thermal penetration depth $q_{total,f}^{-1}$ under a particular oscillation frequency f can be expressed as

$$\frac{\left|q_{total,f}^{-1}\right|}{\Lambda_{total,f}} = \frac{t_{MAT1,f}}{\Lambda_{MAT1,f}} + \frac{t_{MAT2,f}}{\Lambda_{MAT2,f}} + R_{interface} \qquad (23.1)$$

where $\Lambda_{total,f}$ is the resultant thermal conductivity of all the material layers underneath the nanostrip, which can be characterized by the 3ω method measurements. $t_{MAT1,f}$ and $T_{MAT2,f}$ are the thickness of MAT1 and the heat penetration depth into the MAT2, respectively. $\Lambda_{MAT1,f}$ and $\Lambda_{MAT2,f}$ are the thermal conductivity of the thin-film and substrate materials, respectively. $R_{interface}$ is the interface thermal resistance. Based on this equation, an algorithm was developed to obtain the thermal conductivity of MAT1, as shown in Figure 23.11 (Moridi et al., 2018). Following the same principle, the algorithm for calculating the interface thermal resistance can also be developed.

Application of the Modified 3ω Method

The modified 3ω method has been applied to characterizing the high thermal conductivity AlN thin-film/Si substrate

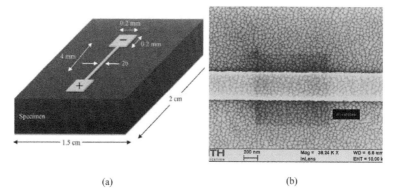

(a)   (b)

FIGURE 23.9 (a) Schematic of a modified 3ω method measurement setup with a nanostrip. (b) A typical Au nanostrip deposited on AlN (Moridi et al., 2018).

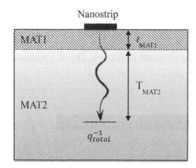

FIGURE 23.10 The schematic of a thin-film substrate system with a nanostrip deposited on the top surface (Moridi et al., 2018).

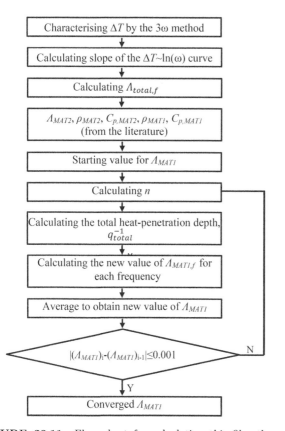

FIGURE 23.11 Flow chart for calculating thin-film thermal conductivity (Λ_{MAT1}) (Moridi et al., 2018).

system. Moridi et al. (2018) first measured the thermal conductivity of a single layer AlN thin film on a Si substrate (Figure 23.12a). Figure 23.12b shows the measurement results of the $\Delta T \sim \mathrm{Ln}(\omega)$, from which one can obtain the total heat penetration depth at that specific frequency. According to the flowchart presented in Figure 23.11, the iteration can be started with a value of 285 W/mK, which is the analytical thermal conductivity of bulk AlN (Slack et al., 1987). The AlN thin-film thermal conductivity is characterized to be 218.4 W/mK, by converging after 15 steps of iteration.

Moridi et al. (2018) further characterized interface thermal resistance between AlN and Si. To obtain a reliable measurement results, both four and ten stacks of AlN and Si thin films were used as shown in Figure 23.13. The corresponding measurement results of the $\Delta T \sim \mathrm{Ln}(\omega)$ for the two types of specimens were presented in Figure 23.14. The measured AlN-Si interface thermal resistance is 1.796×10^{-9} m²K/W from the four-stack specimen and 1.652×10^{-9} m²K/W from the ten-stack specimen, which are quite close to each other. It was found that these values here are larger compared to the calculated values (0.9×10^{-9} m²K/W) of the acoustic mismatch model, or the diffuse mismatch model 1.26×10^{-9} m²K/W (Nazari et al., 2015). However, it is significantly smaller than the experimental values (Jagannadham and Wang, 2002; Su et al., 2013). Hence, it was attributed to the good quality of specimens particular at the interface.

23.3 Numerical Methods

Alternatives to experimental methods are well-developed simulation tools for nanoscale thermal conductivity characterization, such as First-Principles Calculation (Esfarjani et al., 2011; Luo and Chen, 2013; Ward and Broido, 2010), Green's function (Mingo and Yang, 2003; Zhang et al., 2007), Boltzmann Transport Equation, and Molecular Dynamics (Khaled et al., 2018; VanGessel et al., 2018). Figure 23.15 presents the applicable range of length and timescale of different computational methods. These methods are capable of describing phonon heat transport

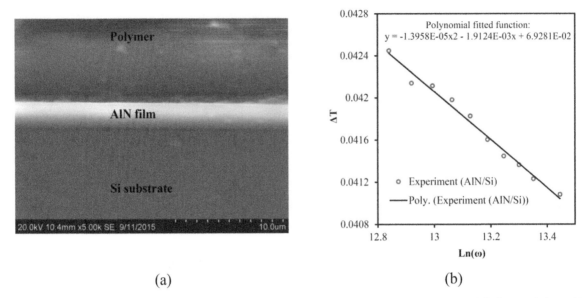

(a)   (b)

FIGURE 23.12   (a) SEM image of the cross section of the 2 μm AlN film on Si substrate specimen. (b) $\Delta T \sim Ln(\omega)$ for AlN on Si thin-film/substrate system (standard error: 0.0000575) (Moridi et al., 2018).

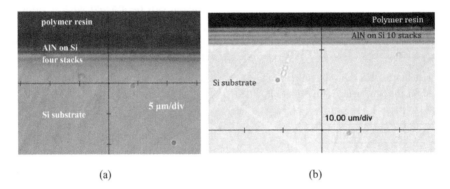

(a)   (b)

FIGURE 23.13   Optical microscope image of (a) the four-stack AlN on Si thin-film system (Moridi et al., 2018) and (b) the four-stack AlN on Si thin-film system.

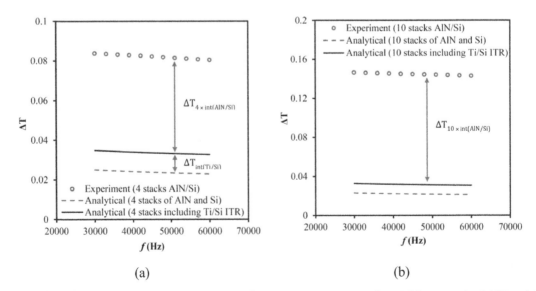

(a)   (b)

FIGURE 23.14 $\Delta T \sim f$ for (a) a four stack of AlN and Si (standard error: 0.0000668) and (b) ten stack of AlN and Si (standard error: 0.0000591) (Moridi et al., 2018).

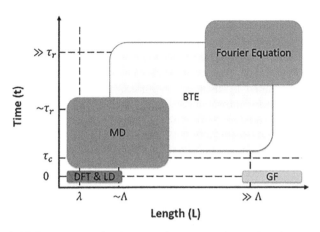

FIGURE 23.15 A schematic diagram of the applicable range of length and timescale of different computational techniques. τ_c is the phonon collision time, τ_r denotes phonon relaxation time, λ denotes phonon wavelength, and Λ is the phonon mean free path (VanGessel et al., 2018).

that cannot be characterized by traditional Fourier theory (Luo and Chen, 2013). Moreover, the advancement in the microprocessor technologies has enabled computing models and methods to solve sophisticated bulk, interface, and surface effects of thermal conduction problems with accuracy, fidelity, and reliability (VanGessel et al., 2018). This section will introduce the basic concepts, the latest progress, and the advantages and drawbacks of each numerical method.

23.3.1 First-Principles Calculation

First-principles calculation combined with lattice dynamics is a robust computational tool to predict thermal conductivity as well as phonon properties (Esfarjani et al., 2011; Luo and Chen, 2013; Ward and Broido, 2010). This method is time-independent, capable of capturing the vibrational spectrum of both unit and supercells and free of fitting to experimental data (Luo and Chen, 2013; VanGessel et al., 2018). In the first-principles calculation, a system with minimum atoms such as a primitive cell in the reciprocal space or the Brillouin zone is enough for highly symmetry points and a fewer number of energy-displacement data is required (Yin and Cohen, 1982). However, the calculation can be overwhelming when arbitrary phonon mode properties are required even for only harmonic force term calculation.

A way to avoid expensive supercell calculations to obtain harmonic force constants is to use density functional perturbation theory (DFPT), which forms a matrix of force constants in reciprocal space and requires only primitive cell calculations. Gonze (1995) used the "$2n+1$" theorem (interchange theorem) of the perturbation theory within DFPT. This method simplified the calculation by using only first-order perturbation and thus reduced the computational cost. However, such simplification does not work for complicated primitive cells such as GaN and Bi_2Te_3 (Luo and Chen, 2013). An alternative way is to fit force

constants with force-displacement data that could reduce the computational cost significantly (Esfarjani and Stokes, 2008). Esfarjani et al. (2011) used such fitting of force constants to compute thermal conductivity of silicon.

Apart from all the simplification techniques and schemes first-principles calculations are still expensive and can only be applied to systems with a small number of atoms. Moreover, first-principles calculations underestimate thermal conductivity for low-dimensional as well as three-dimensional materials, where normal phonon scattering is also significant along with Umklapp scattering of phonons (Ward and Broido, 2010). Hence, characterizing thermal conductivity of interface in thin films as well as effects of defects in thermal conductivity will be an additional challenge in complex first-principles calculations.

23.3.2 Atomistic Green's Function

Heat current is related to interatomic force constants and mostly from empirical models (Mingo and Yang, 2003; Zhang et al., 2007). There are few attempts to use DFT to obtain the interaction among atoms, but the interfacial interaction remained empirical in those cases as well (Sadasivam et al., 2017a; Sadasivam et al., 2017b). Compared to DFT, Green's function (GF) is a useful method to study thermal interface problem where heat current is expressed as a Green's function from which one can obtain phonon frequency as well as thermal conductivity.

Mingo and Yang (2004) studied the thermal conductance of a silica-coated Si nanowires by using a many-body nonequilibrium Green's function approach. New physical results were obtained including the transition from ballistic to diffusion transport, the low-temperature thermal conductance, and the influence of the wire-coated interface on thermal transport. The authors (Mingo, 2006) further included the anharmonicity in their approach and studied the thermal conductance of an anharmonic molecular junction between two solid surfaces. The Atomistic Green's Function method has also been applied to studying the interface thermal transport of graphene–metal composite (Huang et al., 2011), $Mg_2Si/Mg_2Si_{1-x}Sn_x$ (Gu et al., 2015), metal silicide–silicon (Sadasivam et al., 2017b), and Ge-Si/Ge superlattice (Latour et al., 2017; Zhang et al., 2007).

However, the Atomistic Green's Function method is highly dependent on the realistic description of phonon frequency. None of the studies with Green's function has included the imperfection at the interfaces and associated anharmonicity of phonons. Anharmonicity of phonons may affect interfacial transport and thus needed to be included in calculation (VanGessel et al., 2018). Moreover, a realistic phonon description is also required to include defect effects on thermal conductivity (VanGessel et al., 2018).

23.3.3 Boltzmann Transport Equation

Boltzmann transport equation describes the statistical behavior of a thermodynamics system (Callaway, 1959).

Thermal conductivities of material can be obtained from the solution of Boltzmann transport equation. It is regarded as the best method to study multi-length and multi-timescale thermal transport (Callaway, 1959). However, solving the Boltzmann equation can be challenging as phonon distribution in the function depends on frequency, wave vector, position, time, direction as well as polarization. Exact solution of the Boltzmann equation is thus computationally expensive and needs simplification by some approximation models such as gray, non-gray, and semi-gray models (Luo and Chen, 2013). The primary aim in these models is to approximate phonon relaxation time. With the development of molecular dynamics and first-principles calculations, accurate phonon relaxation time can be obtained and has been used as the input in solving Boltzmann transport equation (VanGessel et al., 2018).

Monte Carlo simulation is commonly used tool to solve the Boltzmann equation. However, it is computationally expensive for a whole spectrum of phonon frequency consideration (Chen et al., 2005; Jeng et al., 2008). The traditional Monte Carlo method can be sped up by using a control variant in variance reduction formulation or linearization of governing equation (Luo and Chen, 2013). Recently, McGaughey and Jain (2012) developed a method by which the thermal conductivity of a nanostructure with arbitrary geometry can be predicted through Monte Carlo sampling of the free paths. However, Peraud and Hadjiconstantinou (2012) found that such a treatment can significantly overestimate thermal conductivity in charactering a complex structure. Another challenge in the Monte Carlo simulation is to simulate phonon transmission correctly at the interface, which can be easily obtained by molecular dynamic simulation and atomic Green's function methods on the contrary (Luo and Chen, 2013).

23.3.4   Molecular Dynamics

Molecular dynamics (MD) simulations are equipped with empirical interatomic potentials as well as Newton's second law of motion to trace trajectories of all atoms in a finite simulation model. This method is capable of tracking phonon and predicting transport property, which has made it a powerful tool to study heat conduction in such materials and temperatures where phonon dominates as a heat carrier (Cahill et al., 2014; Cahill et al., 2003). Phonon relaxation time is the most critical parameter in the thermal transport study in nanostructures (Minnich et al., 2011). In MD, phonon–phonon and isotope scattering induced phonon relaxation times are successfully quantified by approximation schemes such as exponential function fit (Li et al., 1998), analytic formula (Volz and Chen, 2000), autocorrelation function decay analysis (McGaughey and Kaviany, 2005), and noise analysis (Sun and Murthy, 2006) for a wide range of materials.

Two different molecular dynamics algorithms are available for calculating the nanoscale thermal conductivity, *i.e.*, the nonequilibrium molecular dynamics (NEMD) and the

equilibrium molecular dynamics (EMD). The NEMD uses an imposed temperature gradient, $\partial T/\partial x$, that results in heat flux, q, across the simulation cell, which is analogous to the experimental situation [110]. The thermal conductivity is determined by using the Fourier's law. This method is preferable for interfacial systems but has a dependency on the type and parameters of the heat bath (Stevens et al., 2007). Moreover, the NEMD experiences nonlinear system response due to the large temperature gradient that requires large simulation time, and thus restricts its feasibility for a large system (Cahill et al., 2003; Volz and Chen, 1999).

On the contrary, the EMD with the Green–Kubo (GK) formula relies on the heat current fluctuations by applying the linear response theory in a homogeneous equilibrium system to calculate thermal conductivity (Cahill et al., 2003; Chen, 2005; Volz, 2007). Thermal conductivity κ can be obtained from the GK formula as (Zwanzig, 1964)

$$\kappa_{IJ} = \frac{V}{k_B T^2} \int_0^\infty \langle (\mathbf{J}_I(0) - \langle \mathbf{J}_I \rangle)(\mathbf{J}_J(t) - \langle \mathbf{J}_J \rangle) \rangle dt \quad (23.2)$$

where $\mathbf{J}_I$ is the heat flux in I direction, k_B is the Boltzmann constant, T is the temperature, V is volume, and $\langle \cdot \rangle$ denotes the average of the heat flux over the equilibrium ensemble and obtained by another integral of average correlation making EMD-GK a double integral system. However, the implementation of EMD-GK method is limited by three major problems. Firstly, in GK formulation the upper limits of double integration are expected to be infinite, but it is impossible in a numerical simulation. Moreover, the cut-off error of double integration is not a simple monotone function of the upper limits, specifically for heat current auto correlation function (HCACF). Secondly, EMD-GK suffers from limited size of the simulation cell and limited total simulation time due to the constraints in computational capability. Thirdly, MD assumes that all the vibrational models are equally excited and entirely a classical approach. Thus, a significant difference between experimental and MD result is observed below Debye temperature, where high-frequency phonons freeze out [85, 88-90]. MD fails to capture variation in heat flux due to the frozen phonons below Debye temperature (T_D) (Gang, 2015; Volz and Chen, 2000). Therefore, to obtain a reliable nano-conductivity prediction using EMD-GK, these issues have to be assessed thoroughly and systematically.

Great efforts have also been done to develop interatomic potentials for different materials describing the atomic interaction and lattice vibration. These interatomic potentials are empirical in nature. Consequently, large differences in the predicted thermal conductivity of crystalline materials are often found (Luo and Chen, 2013; VanGessel et al., 2018). To design reliable and transferable interatomic potential for thermal application, first-principles calculations have been used for a comparatively small system (Luo and Chen, 2013; VanGessel et al., 2018). As mentioned above, such quantum molecular systems are computationally inefficient even for simple crystalline materials let along for interfaces and defect studies (Luo and Chen, 2013). An alternative efficient way is thus needed to obtain the appropriate empirical potentials for thermal analysis using MD.

23.3.5 Critical Issues

Recently, Khaled et al. (2018) carried out a thorough investigation on the effects of the vital physical and numerical variables in the EMD-GK analysis, using the prediction of the nanoscale thermal conductivity of Si as an example. They found that a reliable prediction of EMD-GK analysis requires a crucial selection of time step, simulation time, correlation time, system size, and a proper quantum correction as well.

In brief, a suitable interatomic potential should be selected first to define the interactions among atoms in the simulation. Then time step Δt should be optimized because it directly influences the total simulation time (τ) and maximum correlation time (τ_{cmax}). After selecting Δt, the total simulation time (τ) is adjusted depending on the computational capability. The maximum correlation time is then determined by applying a cut-off where relative noise in HCACF (normalized heat current auto correlation function) is significant. It is important to determine τ_c before system size optimization as an overestimation of τ_c will include averaging error and underestimation will instigate truncation error and hence optimized system size will not be usable. Finally, simulation results below T_D are corrected by quantum correction.

With all the variables optimized, the thermal conductivities of Si at different temperature were calculated and compared with the experimental results, as shown in Figure 23.16a. Overall, the thermal conductivity of Si obtained after the parametric optimization matches well in the temperature range from 700 to 1,000 K, but deviates when the temperature is below T_D. Hence, it can be concluded that the EMD simulation is suitable for thermal conductivity analysis at a temperature above T_D. It should be noted that phonon is the predominant heat carrier for semiconductors and for Si <800 K, the contribution of heat carried by electrons is 0, 3% at 800 K, and 10% at 1,000 K, thus negligible (Glassbrenner and Slack, 1964). The quantum correction (QC) was carried out for the entire classical range of temperature as the final step to apply EMD-GK in nanoscale thermal conductivity. Figure 23.16b presents the modified κ after quantum correction (W/QC), which clearly show a better result.

23.4 Conclusions

The chapter summaries the latest advances in the characterization of nanoscale thermal conductivity, including both experimental and numerical tools. The 3ω method was recognized as the most convenient approach for experimentally characterizing the thermal conductivity of thin-film/substrate system. The recently modified 3ω method expanded its horizon to high thermal conductivity thin-film/substrate system. Time-domain thermoreflectance has been significantly improved over the last decade, but it still requires a high quality of specimen surface. Various probes have been designed and fabricated for scanning thermal microscopy, which significantly expanded the applications of this method. Many advanced methods have been developed based on the concept of coherent phonons. First-principles calculation combined with lattice dynamics is a robust numerical tool for accurately calculating the thermal conductivity of a material, but its simulation scale is limited by the computational ability. Atomistic Green's Function is a powerful method for characterizing the interface thermal resistance. The Boltzmann transport equation is a useful multi-scale tool for characterizing thermal transport problem. Different methods including Monte Carlo simulation have been developed to efficiently solve the Boltzmann transport equation. Molecular dynamics simulation method can solve various nanoscale thermal conductivity problems. However, its accuracy highly depends on the potential selection and other calculation parameters. A recent study provides a standard procedure and criteria for selecting proper simulations parameters of an equilibrium molecular dynamics.

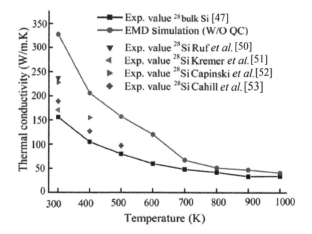
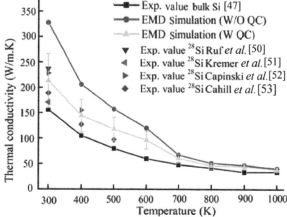

FIGURE 23.16 (Left) Simulation results without quantum correction and (Right) with quantum correction (Khaled et al., 2018).

Overall, the developments of various numerical and experimental methods enable us to explore new interesting thermal transport phenomenon at the nanoscale and reveal the underlying mechanisms. A universal, high-throughput experimental/numerical tool for measuring the nanoscale thermal conductivity and the thermal conductance of materials interfaces is yet to be developed. Experimentally, reducing the size of measuring units, such as the metal strip in the 3ω method and the probes in the scanning thermal microscopy, is a promising trend for characterizing nanoscale thermal conductivity with high accuracy. But their measurement reliability should be supported by advanced theories applicable to the nanoscale. Numerically, multiscale simulation combining the strengths of different methods is the direction, which also heavily relies on the advances in computational capacities.

References

Aigouy, L., Tessier, G., Mortier, M. et al., 2005. Scanning thermal imaging of microelectronic circuits with a fluorescent nanoprobe. *Appl Phys Lett* 87, 184105.

Al-Ghalith, J., Dumitrica, T., 2018. *Nano-Scale Heat Transfer in Nanostructures Toward Understanding and Engineering Thermal Transport Preface.* SpringerBriefs in Thermal Engineering and Applied Science. Springer, Cham, V–Vi.

Alrashdan, M.H.S., Hamzah, A.A., Majlis, B.Y. et al., 2014. Aluminum nitride thin film deposition using DC sputtering. *2014 IEEE International Conference on Semiconductor Electronics (ICSE)*, 72–75, Kuala Lumpur.

Anisimov, A.N., Simin, D., Soltamov, V.A. et al., 2016. Optical thermometry based on level anticrossing in silicon carbide. *Sci Rep-Uk* 6, 33301.

Asheghi, M., Kurabayashi, K., Kasnavi, R. et al., 2002. Thermal conduction in doped single-crystal silicon films. *J Appl Phys* 91, 5079–5088.

Balandin, A.A., Ghosh, S., Bao, W. et al., 2008. Superior thermal conductivity of single-layer graphene. *Nano Letters* 8, 902–907.

Balogun, M.S., Luo, Y., Qiu, W.T. et al., 2016. A review of carbon materials and their composites with alloy metals for sodium ion battery anodes. *Carbon* 98, 162–178.

Bauer, M.L., Bauer, C.M., Fish, M.C. et al., 2011. Thin-film aerogel thermal conductivity measurements via 3 omega. *J Non-Cryst Solids* 357, 2960–2965.

Baumberg, J.J., Williams, D.A., Kohler, K., 1997. Ultrafast acoustic phonon ballistics in semiconductor heterostructures. *Phys Rev Lett* 78, 3358–3361.

Bewilogua, K., Hofmann, D., 2014. History of diamond-like carbon films - From first experiments to worldwide applications. *Surf Coat Tech* 242, 214–225.

Bian, Y.B., Liu, M.N., Ke, G.S. et al., 2015. Aluminum nitride thin film growth and applications for heat dissipation. *Surf Coat Tech* 267, 65–69.

Binnig, G., Quate, C.F., Gerber, C., 1986. Atomic force microscope. *Phys Rev Lett* 56, 930–933.

Bodenschatz, N., Liemert, A., Schnurr, S. et al., 2013. Extending the 3 omega method: Thermal conductivity characterization of thin films. *Rev Sci Instrum* 84, 084904.

Bullen, A.J., O'Hara, K.E., Cahill, D.G. et al., 2000. Thermal conductivity of amorphous carbon thin films. *J Appl Phys* 88, 6317–6320.

Cahill, D.G., 1990. Thermal-conductivity measurement from 30-K to 750-K - the 3-omega method. *Rev Sci Instrum* 61, 802–808.

Cahill, D.G., 2002. Thermal conductivity measurement from 30 to 750 K: The 3 omega method (vol 61, pg 802, 1990). *Rev Sci Instrum* 73, 3701–3701.

Cahill, D.G., 2004. Analysis of heat flow in layered structures for time-domain thermoreflectance. *Rev Sci Instrum* 75, 5119–5122.

Cahill, D.G., 2018. Thermal-conductivity measurement by time-domain thermoreflectance. *Mrs Bull* 43, 782–789.

Cahill, D.G., Braun, P.V., Chen, G. et al., 2014. Nanoscale thermal transport. II. 2003–2012. *Appl Phys Rev* 1, 011305.

Cahill, D.G., Bullen, A., Lee, S.M., 2000. Interface thermal conductance and the thermal conductivity of multilayer thin films. *High Temp-High Press* 32, 135–142.

Cahill, D.G., Ford, W.K., Goodson, K.E. et al., 2003. Nanoscale thermal transport. *J. Appl. Phys.* 93, 793–818.

Cahill, D.G., Goodson, K.E., Majumdar, A., 2002. Thermometry and thermal transport in micro/nanoscale solid-state devices and structures. *J Heat Trans* 124, 223–241.

Cahill, D.G., Katiyar, M., Abelson, J.R., 1994. Thermal-conductivity of alpha-sih thin-films. *Phys Rev B* 50, 6077–6081.

Cahill, D.G., Pohl, R.O., 1987. Thermal-conductivity of amorphous solids above the plateau. *Phys Rev B* 35, 4067–4073.

Callaway, J., 1959. Model for lattice thermal conductivity at low temperatures. *Phys Rev* 113, 1046–1051.

Cao, S., He, H., Zhu, W.H., 2017. Defect induced phonon scattering for tuning the lattice thermal conductivity of SiO_2 thin films. *AIP Adv* 7, 015038.

Cao, X.Q., Vassen, R., Stoever, D., 2004. Ceramic materials for thermal barrier coatings. *J Eur Ceram Soc* 24, 1–10.

Capinski, W.S., Maris, H.J., Ruf, T. et al., 1999. Thermal-conductivity measurements of GaAs/AlAs superlattices using a picosecond optical pump-and-probe technique. *Phys Rev B* 59, 8105–8113.

Chen, G., 2005. *Nanoscale Energy Transport and Conversion: A Parallel Treatment of Electrons, Molecules, Phonons, and Photons.* Oxford University Press, Oxford.

Chen, G., Narayanaswamy, A., Dames, C., 2004. Engineering nanoscale phonon and photon transport for direct energy conversion. *Superlattice Microstruct* 35, 161–172.

Chen, R.K., 2017. Thermal transport in amorphous Si nanostructures. *2017 IEEE 12th International Conference on Nano/Micro Engineered and Molecular Systems (Nems)*, Los Angeles.

Chen, Y.F., Li, D.Y., Lukes, J.R. et al., 2005. Monte Carlo simulation of silicon nanowire thermal conductivity. *J Heat Trans* 127, 1129–1137.

Cocemasov, A.I., Isacova, C.I., Nika, D.L., 2018. Thermal transport in semiconductor nanostructures, graphene, and related two-dimensional materials. *Chinese Phys B* 27, 056301.

Dieleman, D.L., Koenderink, A.F., van Veghel, M.G.A. et al., 2001. Transmission of coherent phonons through a metallic multilayer. *Phys Rev B* 64, 174304.

Dobson, P.S., Weaver, J.M.R., Mills, G., 2007. New methods for calibrated Scanning Thermal Microscopy (SThM). *IEEE Sensor*, 708–711, Atlanta.

Esfarjani, K., Chen, G., Stokes, H.T., 2011. Heat transport in silicon from first-principles calculations. *Phys Rev B* 84, 085204.

Esfarjani, K., Stokes, H.T., 2008. Method to extract anharmonic force constants from first principles calculations. *Phys Rev B* 77, 144112.

Ferain, I., Colinge, C.A., Colinge, J.P., 2011. Multigate transistors as the future of classical metal-oxide-semiconductor field-effect transistors. *Nature* 479, 310–316.

Fu, K.K., Wang, Z.Y., Dai, J.Q. et al., 2016. Transient electronics: Materials and devices. *Chem Mater* 28, 3527–3539.

Gang, Z., 2015. *Nanoscale Energy Transport and Harvesting: A Computational Study*. Pan Stanford, Singapore.

Garimella, S.V., Fleischer, A.S., Murthy, J.Y. et al., 2008. Thermal challenges in next-generation electronic systems. *IEEE Trans Compon Packag Technol* 31, 801–815.

Garimella, S.V., Joshi, Y.K., 2002. Contributions from thermal challenges in next generation electronic systems (THERMES). *IEEE Trans Compon Packag Technol* 25, 567–568.

Glassbrenner, C., Slack, G.A., 1964. Thermal conductivity of silicon and germanium from 3 K to the melting point. *Phys Rev* 134, A1058.

Gomes, S., Assy, A., Chapuis, P.O., 2015. Scanning thermal microscopy: A review. *Phys Status Solidi A* 212, 477–494.

Gonze, X., 1995. Adiabatic density-functional perturbation-theory. *Phys Rev A* 52, 1096–1114.

Goodson, K.E., Asheghi, M., 1997. Near-field optical thermometry. *Microscale Therm Eng* 1, 225–235.

Goriparti, S., Miele, E., De Angelis, F. et al., 2014. Review on recent progress of nanostructured anode materials for Li-ion batteries. *J Power Sources* 257, 421–443.

Gu, X.K., Li, X.B., Yang, R.G., 2015. Phonon transmission across Mg2Si/Mg2Si1-xSnx interfaces: A first-principles-based atomistic Green's function study. *Phys Rev B* 91.

Hao, H.Y., Maris, H.J., 2000. Study of phonon dispersion in silicon and germanium at long wavelengths using picosecond ultrasonics. *Phys Rev Lett* 84, 5556–5559.

Hinz, M., Marti, O., Gotsmann, B. et al., 2008. High resolution vacuum scanning thermal microscopy of HfO(2) and SiO(2). *Appl Phys Lett* 92, 043122.

Hippalgaonkar, K., Seol, J.H., Xu, D.Y. et al., 2017. Experimental studies of thermal transport in nanostructures. *Micro Nano Technol*, 319–357. doi: 10.1016/B978-0-32-346240-2.00012-1

Hu, X.J., Padilla, A.A., Xu, J. et al., 2006. 3-omega measurements of vertically oriented carbon nanotubes on silicon. *J Heat Trans-T Asme* 128, 1109–1113.

Huang, Z., Fisher, T., Murthy, J., 2011. An atomistic study of thermal conductance across a metal-graphene nanoribbon interface. *J Appl Phys* 109, 074305.

Jagannadham, K., Wang, H., 2002. Thermal resistance of interfaces in AlN–diamond thin film composites. *J Appl Phys* 91, 1224–1235.

Jeng, M.S., Yang, R.G., Song, D. et al., 2008. Modeling the thermal conductivity and phonon transport in nanoparticle composites using Monte Carlo simulation. *J Heat Trans* 130. doi:10.1115/1.2818765

Joshi, Y.K., Garimella, S.V., 2003. Thermal challenges in next generation electronic systems. *Microelectron J* 34, 169–169.

Kage, Y., Hagino, H., Yanagisawa, R. et al., 2016. Thermal phonon transport in Si thin film with dog-leg shaped asymmetric nanostructures. *Jpn J Appl Phys* 55, 085201.

Kang, K., Koh, Y.K., Chiritescu, C. et al., 2008. Two-tint pump-probe measurements using a femtosecond laser oscillator and sharp-edged optical filters. *Rev Sci Instrum* 79, 114901.

Kaul, P.B., Day, K.A., Abramson, A.R., 2007. Application of the three omega method for the thermal conductivity measurement of polyaniline. *J Appl Phys* 101, 083507.

Khaled, M.E., Zhang, L.C., Liu, W.D., 2018. Some critical issues in the characterization of nanoscale thermal conductivity by molecular dynamics analysis. *Model Simul Mater Sc* 26, 055002.

Khan, A.I., Paul, R., Subrina, S., 2017. Thermal transport in graphene/stanene heterobilayer nanostructures with vacancies: An equilibrium molecular dynamics study. *RSC Adv* 7, 44780–44787.

Kim, D.U., Jeong, C.B., Kim, J.D. et al., 2017a. Laser Scanning confocal thermoreflectance microscope for the backside thermal imaging of microelectronic devices. *Sensors-Basel* 17, 2774.

Kim, J., Kumar, R., Bandodkar, A.J. et al., 2017b. Advanced materials for printed wearable electrochemical devices: A review. *Adv Electron Mater* 3, 1600260.

Kim, J.H., Feldman, A., Novotny, D., 1999. Application of the three omega thermal conductivity measurement method to a film on a substrate of finite thickness. *J Appl Phys* 86, 3959–3963.

Kim, K., Jeong, W.H., Lee, W.C. et al., 2012. Ultra-high vacuum scanning thermal microscopy for nanometer resolution quantitative thermometry. *ACS Nano* 6, 4248–4257.

Kingon, A.I., Maria, J.P., Streiffer, S.K., 2000. Alternative dielectrics to silicon dioxide for memory and logic devices. *Nature* 406, 1032–1038.

Kittel, A., Muller-Hirsch, W., Parisi, J. et al., 2005. Near-field heat transfer in a scanning thermal microscope. *Phys Rev Lett* 95.

Kolodzey, J., Chowdhury, E.A., Adam, T.N. et al., 2000. Electrical conduction and dielectric breakdown in aluminum oxide insulators on silicon. *IEEE Trans Electron Dev* 47, 121–128.

Kovacevic, G., Pivac, B., 2014. Structure, defects, and strain in silicon-silicon oxide interfaces. *J Appl Phys* 115, 043531.

Kwon, S., Wingert, M.C., Zheng, J.L. et al., 2016. Thermal transport in Si and Ge nanostructures in the 'confinement' regime. *Nanoscale* 8, 13155–13167.

Latour, B., Shulumba, N., Minnich, A.J., 2017. Ab initio study of mode-resolved phonon transmission at Si/Ge interfaces using atomistic Green's functions. *Phys Rev B* 96, 104310.

Lee, S.M., Cahill, D.G., Venkatasubramanian, R., 1997. Thermal conductivity of Si-Ge superlattices. *Appl Phys Lett* 70, 2957–2959.

Lembke, D., Bertolazzi, S., Kis, A., 2015. Single-Layer MoS_2 electronics. *Accounts Chem Res* 48, 100–110.

Li, J., Porter, L., Yip, S., 1998. Atomistic modeling of finite-temperature properties of crystalline beta-SiC - II. Thermal conductivity and effects of point defects. *J Nucl Mater* 255, 139–152.

Li, J.T., Wu, N.Q., 2015. Semiconductor-based photocatalysts and photoelectrochemical cells for solar fuel generation: A review. *Catal Sci Technol* 5, 1360–1384.

Li, S., Zheng, Q.Y., Lv, Y.C. et al., 2018. High thermal conductivity in cubic boron arsenide crystals. *Science* 361, 579–581.

Luo, T., Chen, G., 2013. Nanoscale heat transfer–from computation to experiment. *Phys Chem Chem Phys* 15, 3389–3412.

Majumdar, A., 1993. Microscale heat-conduction in dielectric thin-films. *J Heat Trans-T Asme* 115, 7–16.

Mansanares, A.M., Roger, J.P., Fournier, D. et al., 1994. Temperature-field determination of ingaasp/inp lasers by photothermal microscopy - evidence for weak nonradiative processes at the facets. *Appl Phys Lett* 64, 4–6.

Martin, D.M., Vallin, O., Katardjiev, I. et al., 2008. Buried aluminum nitride insulator for improving thermal conduction in SOI. *IEEE International SOI Conference*, 105–106.

McGaughey, A.J.H., Jain, A., 2012. Nanostructure thermal conductivity prediction by Monte Carlo sampling of phonon free paths. *Appl Phys Lett* 100, 061911.

McGaughey, A.J.H., Kaviany, M., 2005. Observation and description of phonon interactions in molecular dynamics simulations. *Phys Rev B* 71, 184305.

Mei, S., Foss, C.J., Maurer, L.N. et al., 2017. Boundaries, interfaces, point defects, and strain as impediments to thermal transport in nanostructures. *IEEE International Reliability Physics Symposium (IRPS)*, Monterey.

Menges, F., Mensch, P., Schmid, H. et al., 2016. Temperature mapping of operating nanoscale devices by scanning probe thermometry. *Nat Commun* 7, 10874.

Merlin, R., 1997. Generating coherent THz phonons with light pulses. *Solid State Commun* 102, 207–220.

Mingo, N., 2006. Anharmonic phonon flow through molecular-sized junctions. *Phys Rev B* 74, 125402.

Mingo, N., Yang, L., 2003. Phonon transport in nanowires coated with an amorphous material: An atomistic Green's function approach. *Phys Rev B* 68, 245406.

Mingo, N., Yang, L., 2004. Phonon transport in nanowires coated with an amorphous material: An atomistic Green's function approach (vol B 68, art no 245406, 2003). *Phys Rev B* 70, 249901.

Minnich, A.J., Johnson, J.A., Schmidt, A.J. et al., 2011. Thermal conductivity spectroscopy technique to measure phonon mean free paths. *Phys Rev Lett* 107, 095901.

Moore, G.E., 1998. Cramming more components onto integrated circuits (Reprinted from *Electronics*, pg 114–117, April 19, 1965). *Proc IEEE* 86, 82–85.

Moridi, A., Zhang, L.C., Liu, W.D. et al., 2018. Characterisation of high thermal conductivity thin-film substrate systems and their interface thermal resistance. *Surf Coat Tech* 334, 233–242.

Nakabeppu, O., Chandrachood, M., Wu, Y. et al., 1995. Scanning thermal imaging microscopy using composite cantilever probes. *Appl Phys Lett* 66, 694–696.

Nazari, M., Hancock, B.L., Piner, E.L. et al., 2015. Self-heating profile in an AlGaN/GaN heterojunction field-effect transistor studied by ultraviolet and visible micro-raman spectroscopy. *IEEE Trans Electron Devices*, 62, 1467–1472.

Nitta, J., Taguchi, Y., Saiki, T. et al., 2014. Numerical analysis of the temperature dependence of near-field polarization for nanoscale thermometry using a triple-tapered near-field optical fiber probe. *J Optics-Uk* 16, 035001.

Nomura, M., Kage, Y., Nakagawa, J. et al., 2015. Impeded thermal transport in Si multiscale hierarchical architectures with phononic crystal nanostructures. *Phys Rev B* 91, 205422.

Paddock, C.A., Eesley, G.L., 1986. Transient thermoreflectance from thin metal-films. *J Appl Phys* 60, 285–290.

Park, J., Prakash, V., 2013. Thermal transport in 3D pillared SWCNT-graphene nanostructures. *J Mater Res* 28, 940–951.

Peraud, J.P.M., Hadjiconstantinou, N.G., 2012. An alternative approach to efficient simulation of micro/nanoscale phonon transport. *Appl Phys Lett* 101, 153114.

Raudzis, C.E., Schatz, F., Wharam, D., 2003. Extending the 3 omega method for thin-film analysis to high frequencies. *J Appl Phys* 93, 6050–6055.

Rausch, S., Rauh, D., Deibel, C. et al., 2013. Thin-film thermal-conductivity measurement on semi-conducting polymer material using the 3 omega technique. *Int J Thermophys* 34, 820–830.

Sadasivam, S., Waghmare, U.V., Fisher, T.S., 2017a. Phonon-eigenspectrum-based formulation of the atomistic Green's function method. *Phys Rev B* 96, 174302.

Sadasivam, S., Ye, N., Feser, J.P. et al., 2017b. Thermal transport across metal silicide-silicon interfaces: First-principles calculations and Green's function transport simulations. *Phys Rev B* 95, 085430.

Sadat, S., Tan, A., Chua, Y.J. et al., 2010. Nanoscale thermometry using point contact thermocouples. *Nano Lett* 10, 2613–2617.

Saidi, E., Samson, B., Aigouy, L. et al., 2009. Scanning thermal imaging by near-field fluorescence spectroscopy. *Nanotechnology* 20, 115703.

Schmidt, A.J., Chen, X.Y., Chen, G., 2008. Pulse accumulation, radial heat conduction, and anisotropic thermal conductivity in pump-probe transient thermoreflectance. *Rev Sci Instrum* 79, 114902.

Shahil, K.M.F., Balandin, A.A., 2012. Graphene-multilayer graphene nanocomposites as highly efficient thermal interface materials. *Nano Lett* 12, 861–867.

Shen, B.J., Zeng, Z.G., Lin, C. et al., 2013. Thermal conductivity measurement of amorphous Si/SiGe multilayer films by 3 omega method. *Int J Therm Sci* 66, 19–23.

Shenoy, S., Barrera, E.V., Bayazitoglu, Y., 2010. Non-Fourier 3-omega method for thermal conductivity measurement. *Nanosc Microsc Therm* 14, 123–136.

Shi, Y., Peng, L.L., Ding, Y. et al., 2015. Nanostructured conductive polymers for advanced energy storage. *Chem Soc Rev* 44, 6684–6696.

Shin, J., Kang, M., Tsai, T. et al., 2016. Thermally functional liquid crystal networks by magnetic field driven molecular orientation. *ACS Macro Lett* 5, 955–960.

Slack, G.A., Tanzilli, R.A., Pohl, R.O. et al., 1987. The intrinsic thermal conductivity of AIN. *J Phys Chem Solids* 48, 641–647.

Stevens, R.J., Zhigilei, L.V., Norris, P.M., 2007. Effects of temperature and disorder on thermal boundary conductance at solid-solid interfaces: Nonequilibrium molecular dynamics simulations. *Int J Heat Mass Tran* 50, 3977–3989.

Su, Z., Freedman, J.P., Leach, J.H. et al., 2013. The impact of film thickness and substrate surface roughness on the thermal resistance of aluminum nitride nucleation layers. *J Appl Phys* 113, 213502.

Sun, L., Murthy, J.Y., 2006. Domain size effects in molecular dynamics simulation of phonon transport in silicon. *Appl Phys Lett* 89, 171919.

Thorkelsson, K., Bai, P., Xu, T., 2015. Self-assembly and applications of anisotropic nanomaterials: A review. *Nano Today* 10, 48–66.

VanGessel, F., Peng, J., Chung, P.W., 2018. A review of computational phononics: The bulk, interfaces, and surfaces. *J Mater Sci* 53, 5641–5683.

Vinod, S., Philip, J., 2017. Role of field-induced nanostructures, zippering and size polydispersity on effective thermal transport in magnetic fluids without significant viscosity enhancement. *J Magn Magn Mater* 444, 29–42.

Volz, S., 2007. *Microscale and Nanoscale Heat Transfer*. Springer, Heidelberg.

Volz, S., 2009. Thermal nanosystems and nanomaterials with 261 figures introduction. *Top Appl Phys* 118, 3–15.

Volz, S.G., Chen, G., 1999. Molecular dynamics simulation of thermal conductivity of silicon nanowires. *Appl Phys Lett* 75, 2056–2058.

Volz, S.G., Chen, G., 2000. Molecular-dynamics simulation of thermal conductivity of silicon crystals. *Phys Rev B* 61, 2651–2656.

Wagner, M.R., Chavez-Angel, E., Gomis-Bresco, J. et al., 2013. Nanoscale thermal transport and phonon dynamics in ultra-thin Si based nanostructures. *2013 19th International Workshop on Thermal Investigations of ICS and Systems (Therminic)*, 10–12, Berlin.

Wang, H., Sen, M., 2009. Analysis of the 3-omega method for thermal conductivity measurement. *Int J Heat Mass Tran* 52, 2102–2109.

Wang, Z.L., Tang, D.W., Zheng, X.H., 2007. Simultaneous determination of thermal conductivities of thin film and substrate by extending 3 omega-method to wide-frequency range. *Appl Surf Sci* 253, 9024–9029.

Ward, A., Broido, D.A., 2010. Intrinsic phonon relaxation times from first-principles studies of the thermal conductivities of Si and Ge. *Phys Rev B* 81, 085205.

Westover, T., Jones, R., Huang, J. et al., 2008. Photoluminescence, thermal transport, and breakdown in Joule-heated GaN nanowires. *Nano Letters* 9, 257–263.

Wielgoszewski, G., Sulecki, P., Gotszalk, T. et al., 2010. Microfabricated resistive high-sensitivity nanoprobe for scanning thermal microscopy. *J Vac Sci Technol B* 28, C6n7–C6n11.

Williams, C.C., Wickramasinghe, H.K., 1986. Scanning thermal profiler. *Appl Phys Lett* 49, 1587–1589.

Wilson, R.B., Cahill, D.G., 2014. Anisotropic failure of Fourier theory in time-domain thermoreflectance experiments. *Nat Commun* 5, 5075.

Xie, X., Li, D.Y., Tsai, T.H. et al., 2016. Thermal conductivity, heat capacity, and elastic constants of water soluble polymers and polymer blends. *Macromolecules* 49, 972–978.

Yang, N., Xu, X.F., Zhang, G. et al., 2012. Thermal transport in nanostructures. *AIP Adv* 2, 041410.

Yang, R.G., Chen, G., 2004. Thermal conductivity modeling of periodic two-dimensional nanocomposites. *Phys Rev B* 69, 195316.

Yeandel, S.R., Molinari, M., Parker, S.C., 2018. The impact of tilt grain boundaries on the thermal transport in perovskite SrTiO3 layered nanostructures. A computational study. *Nanoscale* 10, 15010–15022.

Yin, M.T., Cohen, M.L., 1982. Theory of static structural-properties, crystal stability, and phase-transformations - application to Si and Ge. *Phys Rev B* 26, 5668–5687.

Zhang, C.B., Najafi, K., 2004. Fabrication of thick silicon dioxide layers for thermal isolation. *J Micromech Microeng* 14, 769–774.

Zhang, W., Fisher, T.S., Mingo, N., 2007. Simulation of interfacial phonon transport in Si-Ge heterostructures using an atomistic Green's function method. *J Heat Trans* 129, 483–491.

Zhang, X., Grigoropoulos, C.P., 1995. Thermal-conductivity and diffusivity of freestanding silicon-nitride thin-films. *Rev Sci Instrum* 66, 1115–1120.

Zhang, Y., Dobson, P.S., Weaver, J.M.R., 2011. Batch fabricated dual cantilever resistive probe for scanning thermal microscopy. *Microelectron Eng* 88, 2435–2438.

Zhang, Y., Dobson, P.S., Weaver, J.M.R., 2012. High temperature imaging using a thermally compensated cantilever resistive probe for scanning thermal microscopy. *J Vac Sci Technol B* 30, 010601.

Zhao, D.L., Qian, X., Gu, X.K. et al., 2016. Measurement techniques for thermal conductivity and interfacial thermal conductance of bulk and thin film materials. *J Electron Packag* 138, 040802.

Zhao, D.L., Tan, G., 2014. A review of thermoelectric cooling: Materials, modeling and applications. *Appl Therm Eng* 66, 15–24.

Zhou, Y.G., Fan, Z.Y., Qin, G.Z. et al., 2018. Methodology perspective of computing thermal transport in low-dimensional materials and nanostructures: The old and the new. *ACS Omega* 3, 3278–3284.

Zwanzig, R., 1964. Elementary derivation of time-correlation formulas for transport coefficients. *J ChemPhys* 40, 2527–2533.

Nanothermometers: Remote Sensors for Temperature Mapping at the Nanoscale

Blanca del Rosal
Swinburne University of Technology

Dirk H. Ortgies
*Instituto Ramón y Cajal de Investigación
Sanitaria IRYCIS*
Universidad Autónoma de Madrid

24.1 Introduction—The Importance of Temperature at the Nanoscale

Temperature plays a critical role in the existence of life all the way from the cosmic to the atomic level. Due to the dependence of the aggregate state of all molecules and compounds on it, it is involved in planet formation (from gas clouds to rock planets) and is also responsible for the habitable zone around a star, where planets present temperatures that allow life to exist, mostly due to the presence of liquids and a planetary atmosphere. Going down to the macroscopic level, every living being has a sense for temperature at least in the form of a sense for hot and cold and what to do to evade the sensations of too hot or too cold.

Nevertheless, it still took until Renaissance times for the introduction of thermometers, although ancient Greeks already knew that air expanded and contracted depending on heat. Based on this knowledge, they developed thermoscopes, which could demonstrate whether it was hot or cold but lacked the necessary scale to be considered thermometers. This changed in the 17th century, when scales appeared with various forms of degrees on tubes filled with air that were open at the bottom, which was under water. These simple air thermometers could show a change of temperature through the expansion or contraction of air, but due to the fact that the system was open, it was affected by outside air pressure. The first closed thermometer (independent of air pressure) is attributed to various persons, among them the Duke of Tuscany, Ferdinando II de Medici, who used a sealed tube partially filled with a liquid. The scales at this time were all very random, and the first person attributed with

proposing a scale tied to the melting and boiling points of water was Christiaan Huygens in 1665. This paved the way to the developments of Fahrenheit and Celsius, who defined our modern understanding of temperature, as well as Kelvin, who introduced the idea of an absolute scale (McGee, 1988; Sherry, 2011).

24.1.1 Principles of Thermometry

Temperature measurements at those times were always contact measurements and therefore related to the so-called zeroth law of thermodynamics: if two bodies are in thermal equilibrium with a third body, they also must be in equilibrium with each other, meaning they must have the same temperature. Therefore, a thermometer in contact and in equilibrium with any object (air, a liquid, a body, etc.) could show its temperature. Basically, all thermometers that can directly connect the measured value to the absolute temperature via well-known simple physical equations can act as absolute thermometers and are also called primary thermometers, while empirical or secondary thermometers often offer a better resolution or more specific application but need calibration in respect to a primary thermometer (Thomsen, 1962).

Most modern thermometers work as empirical thermometers, which means that they employ a material which changes a clearly defined property with a change in temperature, and they are calibrated against well-known absolute temperature points like triple points (e.g., water: 273.16 K) or superconducting temperatures. The scales can differ, as is the case when comparing the Fahrenheit with the Celsius scale, but all require monotonic growth on the respective scale. Typical thermometric materials are liquids like alcohols and mixtures of them, employing their expansion as described above in a sealed tube, or metals and alloys, which offer a wider range of properties that scale with temperature. Although their expansion and contraction are also temperature-dependent, changes in their electric resistance or potential are commonly used for measuring temperature changes. All of these types of thermometers have in common that they need to be in contact with the subject of the measurement (Thomsen, 1962; Wang, Wolfbeis and Meier, 2013).

An absolute thermometer that also does not require contact can be obtained through radiometric thermometry, which is based on the principle of blackbody radiation and Planck's law. It connects the spectral radiance of a body to an absolute temperature value and a frequency:

$$B(v, T) = \frac{2hv^3}{c^2} \frac{1}{e^{\frac{hv}{k_b T}} - 1}. \tag{24.1}$$

Therefore, this allows the determination of the absolute temperature through the measurement of the spectral radiance (B) at a defined frequency (v) with Planck's constant (h), known speed of light (c), and the Boltzmann constant (k_b). Typically, radiometric thermometers are known as thermal cameras or infrared (IR) thermographs due to the IR wavelength of thermal radiation in relevant temperature ranges (up to 500°C). They offer a wide range of applications (fever thermometers, night vision, etc.) at good temperature resolutions of 1 K and without the necessity of contact (Corsi, 2010).

In order to evaluate the quality of any material or measuring technique as a thermometer, the most important variable is their sensitivity (S), which can be defined as the change of the measurable property (Q) with temperature (T):

$$S = \left| \frac{\delta Q}{\delta T} \right|. \tag{24.2}$$

In order to better compare sensitivities between various techniques, a normalized or relative temperature sensitivity is defined as follows:

$$S_R = \left| \frac{1}{Q} \frac{\partial Q}{\partial T} \right|. \tag{24.3}$$

In this manner, S_R can be expressed as %/K and easily indicates the response of the signal per temperature change (Jaque and Vetrone, 2012; Brites, Balabhadra and Carlos, 2018; Quintanilla and Liz-Marzán, 2018).

A second important value when talking about thermometers is its temperature resolution (ΔT_{res}). It determines the precision of the temperature measurement with the thermometer and depends therefore directly on the uncertainty and error (ΔQ) for the measurement of Q and improves with better sensitivities (S). This is expressed in the following equation:

$$\Delta T_{res} = \frac{\Delta Q}{S_R}. \tag{24.4}$$

The resulting value is usually given in °C or K and should be as small as possible for measurements in biological systems,

but larger values can often be tolerated in high-temperature applications (Brites et al., 2012; Brites, Balabhadra and Carlos, 2018; Quintanilla and Liz-Marzán, 2018).

Additionally, if the temperature is to be measured simultaneously at more than just a single point, then spatial resolution also becomes of interest. For that, a value Δx can be defined as the minimum distance at which a temperature difference higher than ΔT_{res} can be obtained. This is the spatial resolution of the measurement (Δx) (Kim et al., 2012; Brites, Balabhadra and Carlos, 2018; Quintanilla and Liz-Marzán, 2018):

$$\Delta x = \frac{\Delta T_{res}}{|\nabla T|_{max}}, \tag{24.5}$$

where ∇T_{max} is the maximum temperature gradient in the sample.

Equally, if not all the measurements occur at the same time and dynamic thermometry is to be performed, then analogously Δt can be defined as the minimum time after which the temperature difference is higher than ΔT_{res}. This is the temporal resolution (Δt):

$$\Delta t = \frac{\Delta T_{res}}{\left| \frac{dT}{dt} \right|_{max}}. \tag{24.6}$$

The challenges presented by modern microscopic-scale industrial fabrication, for example, in the electronic industry, or the advances desired for medical diagnostic techniques also require better temperature control at the micro- or nanoscale. It would be great to not just measure the body temperature or temperature of an organ but to know how various processes in cells affect the local temperature, for example, around the mitochondria. Therefore, the adaptation of macroscopic temperature-measuring techniques and the development of novel approaches are required and have led to interesting results over the last decade. In the following sections, we will discuss techniques to measure temperature at the nanoscale, beginning with methods that rely on a direct contact of the sensor or thermometer with the sample. Principally, contact-based nanothermometry techniques require very fine and small contact points and are therefore often combined with contact-based microscopy techniques, as will be explained in the next section.

24.2 Contact Nanothermometry Techniques

Most contact-based nanothermometers are based on the same techniques as their macroscopic siblings. They rely on electric variations of resistance, voltage, conductivity, or capacity influenced by the effect of temperature on the two metals in a thermocouple or thermistor. Miniaturization made it possible to add this functionality to the tip of scanning microscopes and allowed temperature measurement of 2D arrays and surfaces with the so-called scanning thermal microscopy (Kim et al., 2012). Additionally, this approach can also be integrated into devices whose temperature needs to be controlled by using micro- and nanolithographic techniques to fabricate a substrate/imprint that already includes the nanothermometer (Jaque and Vetrone, 2012; Quintanilla and Liz-Marzán, 2018).

A second microscope-based contact technique works with modified Atomic Force Microscopes (AFMs) whose cantilever has been modified to contain two metals with well-defined temperature-dependent mechanical properties. The idea to employ the deflection of a cantilever tip for thermal measurements was already demonstrated in 1995, covering a conventional AFM tip with a thin film of aluminum or gold (Nakabeppu et al., 1995). This type of probe can scan the temperature or create a temperature map when the microscope tip is passing over the sample. This way of measuring temperature has an excellent spatial resolution, basically the same as the resolution of the AFM, and thermal resolutions in the mK range can be obtained. On the other hand, the calculation of the temperature from the measured values is often noisy and complicated due to size effects on the heat flow from the sample to the tip. This problem of a heat-flux artifact, the necessity of contact, and the very complicated and expensive instruments required are the main drawbacks of these techniques. Nevertheless, some interesting applications have been demonstrated using these contact microscope approaches.

For example, Wang et al. presented in 2011 a nanothermocouple made of tungsten and platinum in the form of a 100 nm nanotip, which they applied to measure a 2 K temperature increase in a human glioblastoma cell (U251) when cell death was induced with camptothecin (Wang et al., 2011). When a different drug (doxorubicin) was employed to trigger cell death, this change in temperature did not occur, hinting at different mechanisms of action of the two different drugs. Another interesting application of a contact nanothermometer, even modeled after the classic designs of thermometry, is the study of a carbon nanotube filled with liquid gallium, whose contraction or expansion indicates the temperature (Gao and Bando, 2002). Gao and Bando demonstrated that this thermometer worked reproducibly in a range from 50 to 500°C and showed the same expansion coefficient for gallium as in the macroscopic state. Nevertheless, this type of classic nanothermometer has not yet found an application due to the fact that scanning electron microcopy is necessary to analyze the height of the liquid in the nanotube.

24.3 Remotely Sensing Temperature

Moving away from contact-based techniques, one can go back to the idea of absolute thermometers and look for a nanosystem for thermography and evaluate whether an IR thermal camera can be adjusted for operation at the nanoscale. An IR microscope with sufficient optics in the mid to far IR and a suitable detector can be employed for the rapid scanning and mapping of samples and devices.

The limitations of this technique stem, on one hand, from the optical limitations in the mid to far IR, namely the Abbe diffraction limit, which results in maximum spatial resolutions in the μm range. Therefore, this is not really a nanothermometry technique. On top of that comes the general disadvantage of IR thermography that with thermal cameras good temperature resolutions can be obtained in quick and simple measurements, but the observed temperature is only the surface temperature of the studied object. Therefore, thermographic microscopy is mainly limited to the study of electronic devices (Huth et al., 2000; Chénais et al., 2004).

Interferometry can also yield information on temperature. In an interferometer, two light beams travelling the same distance are compared after one has passed through the sample, which modifies its optical path slightly. The difference between the optical paths of both beams yields information about the refractive index of the sample, which can be correlated to temperature. For example, the refractive index of a liquid changes slightly with temperature and influences the light beam that passes through it. If the refractive index of a given liquid is known and its behavior with temperature has been characterized and/or can be modeled well enough, this type of setup can be employed to measure the temperature of and around nano-objects in liquids. Baffou et al. introduced this principle for rapid temperature measurements and heat mapping, modifying conventional optical microscopes with a wavefront analyzer and relying on a regular illumination source (Köhler illumination) (Baffou et al., 2012). They introduced the technique measuring the heating of gold nanoparticles (GNPs) (Baffou et al., 2014) and also demonstrated its applicability to the measurement of microbubble formation on a gold microwire (Bon et al., 2013).

Raman spectroscopy is a powerful and versatile tool based on the vibrations that materials, molecules, and especially their functional groups show, and it is therefore heavily employed in material characterization. But in the form of Raman scattering, it can also be employed as a remote temperature-sensing technique. Most light that reaches an object under study is scattered without any interaction, but some of it interacts with the vibrational modes (phonons) of the material and results in scattering that shows characteristic Raman spectra with shifts in the original wavelength to higher (Stokes) or lower (anti-Stokes) energy. Phonon energy is directly related to the position of the atoms in the material and therefore to temperature. Hence, the analysis of band positions, width, and intensity in the Raman scattering spectra allows the determination of temperature. Additionally, the ratio between the Stokes and anti-Stokes peak of a selected vibrational mode also allows the determination of the temperature (Jaque and Vetrone, 2012; Quintanilla and Liz-Marzán, 2018).

Based on these properties, every material can theoretically act as a Raman nanothermometer because the Raman interaction is universal, but practically it is often limited by the weak Raman signal that requires long integration times, sample transparency, and results in low thermal and spatial resolutions. Nevertheless, some interesting applications have been reported, among them indium oxide octahedrons (microparticles) which can act as a thermometer in a range from 83 to 303 K because their Raman spectrum is very well defined and the peak intensity in this range is linearly proportional to temperature (Senapati and Nanda, 2015).

24.4 Luminescent Nanothermometers as Valuable Remote Sensors of Temperature

Even though the approach presented here is only semi-contactless in that it requires the introduction of luminescent molecules or nanoparticles (NPs) as temperature sensors into the object of interest, its readout is remote. Therefore, luminescence nanothermometry is often considered a non-invasive or contactless technique, while remote nanothermometry would probably be the better and more precise term.

Luminescence is the emission of light from excited electronic states of any material, molecule, compound, ion, or atom. Although prior excitation can occur via various pathways, such as chemical reactions (chemiluminescence) or biological processes (bioluminescence), luminescence nanothermometry typically relies on optical methods (illumination with lasers or broadband illumination sources) to provide the energy required for excitation. This process is known as photoluminescence.

The population and the energy of these excited states (whether they are atomic or molecular orbitals or bands) depend on temperature and are therefore the underlying fundament for the relationship between luminescence and temperature. Changes in the local temperature around the luminescent material can then influence various aspects of the luminescence emission, and mapping and spectral analysis of the signal allows the sensing of temperature. The parameters of luminescence that can be studied to reveal temperature are illustrated in Figure 24.1 and allow an additional classification of luminescence nanothermometry (Jaque and Vetrone, 2012):

Spectral position: The spectral or peak position of a luminescence emission depends on the separation in energy between the two energy levels involved, which in turn determines the wavelength of the emission. The energetic separation is influenced by various temperature-dependent properties of the material, such as its structure and atomic distances or refractive index.

Band-shape or two intensities: The term "band-shape" is often used to describe variations in the form of a peak that is the result of emission from two energetically very close states, which are therefore thermally coupled, and changes in temperature change the population between the two close levels. Technically, these are two linked emissions or two

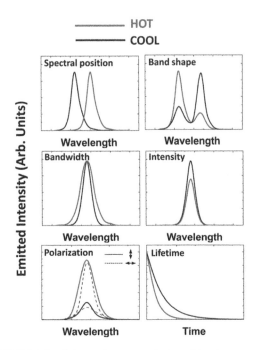

FIGURE 24.1 Schematic representation of luminescence properties that can be taken advantage of for luminescence nanothermometry. The lighter lines represent higher temperatures. (Reproduced from Jaque and Vetrone (2012) with permission from The Royal Society of Chemistry.)

intensities. Independently, this can also be the result of two separate classes of emitters inside a mixed or co-doped material.

Bandwidth: The bandwidth of a peak is directly related to properties like the order and homogeneity of the material and temperature. Increasing the temperature broadens the signal in basically any kind of spectroscopy. A linear relationship between homogenous line broadening and temperature around room temperature can be exploited for luminescence thermometry.

Intensity: A change in temperature changes the number of emitted photons per second resulting in a change of the emission intensity. Various quenching mechanisms or thermal traps, as well as an increase in non-radiative energy transfers, are activated and cause a reduction in intensity when the temperature is increased. A decrease is the more commonly studied case, but it is also possible that with an increased temperature, an emission becomes more likely and the intensity increases.

Polarization: Anisotropic luminescent materials normally emit polarized light and interact distinctively with polarized light. The ratio between the emissions at two orthogonal polarization states is defined as "polarization anisotropy" and is influenced by temperature. Therefore, it is another parameter that could be applied for thermometry.

Lifetime: The stability of an excited state and the probability of its decay are directly related to its lifetime (τ), which is defined as the necessary time for the intensity to decrease to 1/e of its initial value after excitation has been stopped. This phenomenon depends on various factors of which many are influenced by temperature, for example, the ones related to vibrations of the material (phonon-assisted energy transfer processes and multiphonon decays). This allows temperature measurements based on the changes in the lifetime of a nanomaterial.

Probably the most employed of these six principles is the intensity-based one because it is often the one where an influence of temperature can be most easily measured, even when no spectral analysis is possible. A little bit more complex are the polarization and lifetime-based measurements due to their requirement for additional instrumentation (polarization filters, pulsed excitation sources, etc.), but especially the latter is gaining more interest in general as an imaging technique (time-domain vs. frequency-domain imaging) because it is a simple way of filtering background noise and autofluorescence, which often hamper imaging in an *in vivo* biological context (Villa et al., 2015; del Rosal, Villa, et al., 2016). Nevertheless, the variety of observable phenomena that can be linked to temperature and that make thermometry possible has allowed the field of nanothermometry to increase exponentially over the last decade and led to the presentation of various luminescent nanothermometers (Brites, Balabhadra and Carlos, 2018; Quintanilla and Liz-Marzán, 2018).

24.4.1 Quality and Reproducibility of Luminescence Nanothermometry

Photoluminescence is highly dependent on various factors, both intrinsic of the luminescent material and external. In the case of the influence of temperature, this relationship is exactly the quantity to be measured as described above. But there is a whole range of other factors influencing the population of excited states, resulting in nonradiative transfers and system crossings, leading to their de-excitation or quenching. The interaction with vibrational modes of the surrounding environment, depending on their phonon energy, is especially involved in de-excitation of the luminescent sensor and reduction of the intensity of the measured signal. The excitation source or, more specifically, fluctuations in pump power and photon density also have a great influence on the final signal. On top of that comes any background noise and environmental contamination that can change the expected spectrum, and all of this turn luminescence nanothermometry at first sight into something close to roulette. Therefore, nanothermometers in general fall into the previously mentioned category of secondary thermometers that need calibration based on an established thermometer and known temperatures (Brites et al., 2012; Brites, Balabhadra and Carlos, 2018).

The calibration also needs to fulfill some requirements to turn the whole temperature measurement into not only a repeatable but also reproducible procedure. This includes measurements for the calibration as close as possible (temperature range, concentrations of sensor and/or analyte, surrounding media (fluids/solvents, tissues)) to the desired measurement or experimental

conditions. In addition, most luminescent nanothermometry methods developed employ a ratiometric approach in order to reduce the influence of external factors. At least two emissions/intensities that come from the same sensor or NP are to be measured. These emissions need to be close in energy and either thermally coupled, meaning their relative populations and as a consequence their intensities are connected, or one of them must be independent of temperature and demonstrate stable emission while the other one changes in intensity. The former case (thermally coupled emissions) is a special and highly beneficial case that will be discussed below. The latter case, on the other hand, can be compared to having an internal standard in the measurement. The temperature-independent emission, which for example can be due to a quasi-steady state in the population of a metastable excited level of the material, should be relatively close to the temperature-dependent emission, so that wavelength-dependent light–media interactions are similar (e.g., refractive index). The influence of external factors (solvent, tissue, background noise/emissions) on both emissions under observation will then, to a first approximation, be close to equal. This now allows employing the ratio between both emissions as the quantity that will be translated into temperature. Ratiometric thermometry together with a suitable calibration (as close as possible to real experimental conditions) therefore improves drastically the reproducibility of any nanothermometric approach (Brites et al., 2012; Jaque and Vetrone, 2012; Brites, Balabhadra and Carlos, 2018; Quintanilla and Liz-Marzán, 2018).

Recently, the importance of the conditions of the calibration and the experiment were highlighted by Labrador-Páez et al. (2018). They demonstrated that for the most commonly applied rare-earth-based luminescent nanothermometers (dielectric nanocrystals doped with lanthanide ions), the calibrations often do not take into account all factors that can influence the readout of the thermometric signal and might lead to artificially high temperatures. This is foremost due to self-absorption of emissions that are employed for temperature measurements in the NP. Emissions that occur via de-excitation to the ground state can easily be reabsorbed by the lanthanide ions. This is an often ignored but highly relevant property, and it becomes especially complicated when, in a ratiometric approach, one emission is self-absorbing while the other is not. This results in a drastic change in the ratio between the emissions not caused by temperature, just when the ratiometric approach is supposed to protect against this type of effect. The absorption of the emitted signal by the solvent/media or biological tissue, in which the nanothermometers are dispersed, can also alter the measurements drastically. This is especially dependent on the distance the luminescence must travel and therefore on the position (focus point) of the excitation source (laser) as well as the thickness of the surrounding media. A third often neglected problem, as discussed in Labrador-Páez et al.'s work, is the influence of the optics and experimental setup (lenses, mirrors, optical filters, gratings, detectors) itself, which often varies between calibrations and measurements, especially when moving to *in vivo* measurements. As a consequence of their findings, the authors highlight the importance of applying the same conditions during calibration and experiment. Awareness of potential sources of error like absorption or self-absorption, testing for them through additional characterization experiments (excitation vs. emission spectra), and performing calibrations under various environmental conditions (especially different media) should reveal potential artifacts.

A Primary Luminescence Nanothermometer

The availability of an absolute or primary nanothermometer (see the section "Introduction") would greatly advance the whole field of nanothermometry and its applications, knowing that even a good and thorough calibration and the ratiometric approach as described above can result in artifacts and that careful experimental design is necessary to obtain reliable and reproducible luminescence nanothermometers. Therefore, researchers have been looking into possibilities on how to link the luminescence emission of a NP directly to the absolute temperature through a simple equation where all other values are well-defined constants. This feature was achieved by Balabhadra et al. (2017) for the $SrF_2:Yb^{3+}, Er^{3+}$ NPs, which are NPs employing one of the most common lanthanide dopant pairs for nanothermometry. Yb^{3+} ions act as a sensitizer when excited with 980 nm IR laser irradiation, absorbing the excitation light and transferring their energy to the Er^{3+} ions. This triggers the

upconversion process (where short-wavelength emissions are achieved upon excitation with light of longer wavelengths) due to the long-lived excited states of the lanthanides. This means that through the sequential absorption of two photons, Er is brought to a high-energy excited state $^4F_{7/2}$, from which all following de-excitations to lower energy levels then begin, which include the highly sought after visible upconversion emissions. The luminescence thermometry in this example relies on the ratio between two well-characterized emissions of Er^{3+}, $^2H_{11/2} \rightarrow {}^4I_{15/2}$ (I_2, 510–533 nm), $^4S_{3/2} \rightarrow {}^4I_{15/2}$ (I_1, 533–570 nm) in the green (see Figure 24.2a). They are very close in energy (ΔE as difference between the barycenters of the two emissions I_2 and I_1) and hence thermally coupled. Two excited states can be considered thermally coupled if the value of ΔE is in the 200–2,000 cm^{-1} range. These empirical limits guarantee both the spectral separation of the levels and the existence of thermalization. Therefore, their coupled population is dependent on a Boltzmann distribution linking the ratio between the two corresponding emissions in the following equation:

$$R = \frac{I_2}{I_1} = Pe^{\left(\frac{-\Delta E}{k_B T}\right)}. \qquad (24.7)$$

Solving this equation for temperature results in an equation where only ΔE and P need to be determined.

ΔE can be obtained from the spectra or through calculations, and P is the extrapolated value of zero pump power or laser excitation energy in the ratio (R) vs. pump power

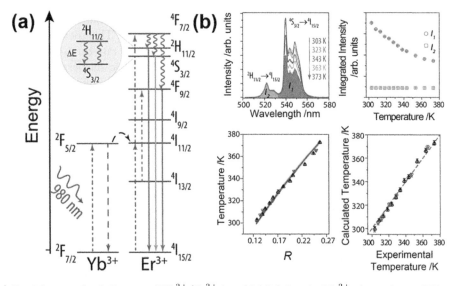

FIGURE 24.2 (a) Partial energy-level diagram of Yb^{3+}/Er^{3+} ions highlighting the Yb^{3+} absorption at 980 nm, the Yb^{3+}-to-Er^{3+} energy transfer pathway and the $^2H_{11/2} \rightarrow {}^4I_{15/2}$, $^4S_{3/2} \rightarrow {}^4I_{15/2}$, and $^4F_{9/2} \rightarrow {}^4I_{15/2}$ Er^{3+} emissions. The expansion depicts the thermally coupled $^2H_{11/2}$ and $^4S_{3/2}$ Er^{3+} levels. (b) Temperature-dependent emission spectra and thermometric performance of $SrF_2:Yb^{3+}, Er^{3+}$ NPs. Top center: emission spectra of NP powder in the 303–373 K range. Top right: Integrated emission intensities of the spectral regions depicted by I_2 (510–533 nm) and I_1 (533–570 nm). Bottom center: Temperature dependence of the experimental R values for the NPs in powder (up triangles) and water suspension (down triangles). The solid line is the theoretical prediction of temperature by solving Eq. (24.7) for T. Bottom right: Calculated temperature vs. temperature reading using a thermocouple (experimental temperature) for the NPs in powder (circles) and water suspension (squares). The dashed line is a guide for the eyes corresponding to $y = x$. (Adapted with permission from Balabhadra et al. (2017). Copyright © 2017 American Chemical Society.)

curve. This corresponds to the value of R at room temperature (T_0). In this manner, Balabhadra et al. demonstrated a very elegant way to convert luminescent nanothermometers that have just a single emission center (here Er^{3+}) into absolute thermometers, improving drastically the reliability and reproducibility for this type of nanothermometers. Figure 24.2b shows the relevant graphs going along with the determination of the temperature via this approach. For details on the whole set of equations and calculations, see the authors publications (Balabhadra et al., 2017; Brites, Balabhadra, and Carlos, 2018).

24.4.2 Luminescent Nanothermometers and Their Applications

In this section of the chapter, various types of luminescent nanothermometers will be discussed regarding their presented and potential applications as well as their contribution to the ongoing development of the field taking into consideration the aspects that were brought forward above. Of course, this can only be an incomplete overview over the variety and developments that the field has demonstrated until now. There has been an increasing interest in remotely sensing temperature, and luminescent nanothermometers have been at the forefront, as can also be seen from many reviews that have come to describe the technology in much more detail and from various angles and approaches that we cannot reproduce here in completeness. The interested reader is therefore directed to the original literature and a selection of excellent reviews in order to satiate their thirst for more knowledge (Brites et al., 2012; Jaque and Vetrone, 2012; Wang, Wolfbeis and Meier, 2013; Brites, Balabhadra and Carlos, 2018; Quintanilla and Liz-Marzán, 2018). Despite that, we will try to do the field justice and present examples and applications of luminescence nanothermometry on the following pages that demonstrate the ampleness of the field, current trends, and interesting developments.

Er,Yb — Probably the Most-Developed Rare-Earth Thermometer

As was mentioned above, the lanthanide ions Er^{3+} and Yb^{3+} deployed in various NPs are probably the most studied and developed luminescent nanothermometers. As with all lanthanide-based luminescent NPs, they present stable emissions with long lifetimes that are relatively independent of the environment or the host particle because of the partial shielding of the 4f orbitals, which are responsible for the luminescence transitions. Most often, they are employed in chemically stable fluoride hosts ($NaLnF_4$, LnF_3, etc.) that have low phonon energies, which decreases the probability of luminescence quenching via interactions with the solvent and increases the brightness of the emission. Neither their quantum yield nor their brightness can compete with that of organic dyes or quantum dots (QDs), but they present their chemical and photostability as advantage over dyes and

their often better biocompatibility, as well as the multiple emission for ratiometric thermometry as point over QDs. Furthermore, no other class of NPs is that easily modifiable via a core–shell strategy, allowing the introduction of additional functionalities and the design of multifunctional nanosensors (Jaque and Vetrone, 2012; Brites, Balabhadra and Carlos, 2018).

The temperature dependence of the green Er^{3+} emissions had been known for oxides and various glasses, but it was mentioned the first time for NPs by Alencar et al., who doped $BaTiO_3$ with Yb^{3+} and Er^{3+} and measured the fluorescence ratio for NPs of different sizes (Alencar et al., 2004). They discovered that the temperature measurements in the titanate NPs were not independent of size. Sing et al. employed the ion pair as dopants in a Gd_2O_3 phosphor for high-temperature measurements in the 300–900 K range, demonstrating the high potential for the measurements of industrially relevant temperatures with a maximum sensitivity of 0.39 %K^{-1} (Singh, Kumar and Rai, 2009).

The full potential for this pair of ions as nanothermometers became apparent when the combination was included in a fluoride host matrix. Vetrone et al. synthesized $NaYF_4{:}Er^{3+},Yb^{3+}$ NPs with an average size of 18 nm in the cubic phase (Vetrone et al., 2010). Fluorides doped with lanthanide ions show typically higher fluorescence intensities than oxides or glasses and are more easily functionalized on their surface. In this case, branched polyethyleneimine (PEI) was added during the solvothermal synthesis, resulting in water-dispersible PEI-capped nanocrystals. The surface functionalization allowed cellular uptake, and the authors were able to demonstrate the potential of this combination in biological *in vitro* applications. HeLa cells were incubated with the nanothermometers and then placed on a thermal resistor platform in a fluorescence microscope so that the temperature of the cells could be increased controlled by applying a voltage to the resistor. The NPs were excited at 920 nm to avoid the heating effect of water at 980 nm, and the luminescence was analyzed and converted to temperature based on a prior calibration. At room temperature (25°C) and at 35°C, no change in the physiology of the cell was observed, but the cell was dismantling at 45°C as a consequence of cell death.

Sedlmeier et al. investigated the effect of an inert shell on the characteristics of the Yb,Er nanothermometer by comparing hexagonal $NaYF_4{:}Er^{3+},Yb^{3+}$ NPs with hexagonal $NaYF_4{:}Er^{3+},Yb^{3+}@NaYF_4$ NPs in 2012 (Sedlmeier et al., 2012). The core–shell particles showed a better thermal resolution, on average 0.5 K, while the core-only particles showed 2.1 K. They described that this positive effect of a core–shell architecture is mainly due to a higher brightness of the core–shell NPs, which increases the intensities and therefore reduces the uncertainty in the measurement, resulting in a much improved thermal resolution. This paper demonstrated therefore the advantage that core–shell design can also have on nanothermometers.

In 2016, Brites et al. demonstrated that this type of nanothermometer shows a thermal sensitivity that, together with their sufficient spatial and temporal resolution, made the measurement of instantaneous Brownian motion possible. Specifically, they employed NaYF$_4$:Yb/Er@NaYF$_4$ core–shell NPs with a maximum thermal sensitivity of S_R of 1.15 %K^{-1} at 296 K in different concentrations (volume fractions) in an aqueous and in a chloroform dispersion. The dispersions were placed in a specially designed optical nanofluid container (see Figure 24.3), which contained a heater on one side and optics for the collection of luminescence at the opposite end, while the distance x in between could be excited by a movable laser at a 90° angle at various positions of x (x_i = 2–9 mm). Recording the time that was necessary to observe a temperature increase with the thermally coupled Er^{3+} emission in the green (onset time t_0) after turning on the heating device at the various distances and with various volume fractions, they could show a linear dependency between t_{0i} and the spatial positions x_i, resulting in a velocity (compare Figure 24.3). The authors were able to demonstrate that this measured velocity is the instantaneous Brownian velocity of a particle in a fluid. Their results were in good agreement with the literature and computational models for nanofluids taking into account the processes for heat convection applicable for a temperature gradient and the different viscosities in water and chloroform relevant in their experimental design. They corroborated this further via a range of control experiments involving different particle sizes as well as particle shapes

(nanorods instead of spherical particles). With this work, the authors set an example for the potential of nanothermometry not just for measuring temperature at a smaller scale but to apply it for the study of otherwise-difficult-to-measure physical phenomena.

The potential for multifunctionality that stems from core–shell architecture can be observed in one final representative example of an Er^{3+},Yb^{3+} thermometer that also already includes aspects that will be discussed in more detail in the following section. Wang et al. presented in 2015 a core–multishell NP with a core containing Er^{3+} ions for nanothermometry (Wang et al., 2015). In total, the NaLuF$_4$: Gd/Yb/Er@NaLuF$_4$:Yb@NaLuF$_4$:Nd/Yb@NaLuF$_4$ NPs were designed for multimodal imaging with Nd^{3+} and Yb^{3+} giving good luminescence emissions in the near-infrared (NIR), with Gd^{3+} potentially acting as contrast agent for magnetic resonance imaging (MRI) and Lu providing contrast for X-ray computer tomography (CT). Furthermore, the IR luminescence (Nd^{3+}, Yb^{3+}) as well as the visible emissions (Er^{3+}), which were of interest for nanothermometry, can be obtained with 808 nm light, which excites the Nd^{3+} ions acting as sensitizers in the penultimate shells. The energy is then transferred from Nd^{3+} to Yb^{3+} ions and passed via Yb^{3+} into the core where Er^{3+} is present (see Figure 24.4). This approach is particularly interesting for applications in aqueous media, as it avoids the potential heating of water that occurs

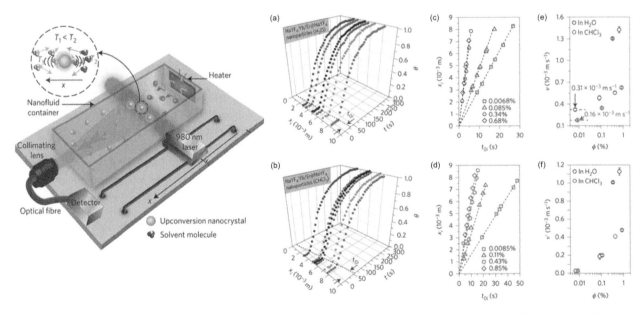

FIGURE 24.3 Left: A collimating lens collects the upconversion emissions generated at different positions by moving a 980 nm laser along the x direction, and the signals are guided to the detector by an optical fiber. The inset shows the solvent-mixing effect arising from the Brownian motion of the NP located at the interface between the cold (T_1) and hot (T_2) regions of the nanofluid. (a), (b) Reduced temperature profiles of the NPs dispersed in water (0.0068%; (a)) and chloroform (CHCl$_3$; 0.0085%; (b)), as measured by excitation from positions x_i along the x direction. The dashed line refers to the onset time t_{0i} when the change in intensity ratio is observed due to temperature variation upon turning on the heater. (c), (d) Corresponding linear correlation between x_i and t_{0i}, as measured in water (c) and chloroform (d) for NPs with different volume fractions, respectively. (e) Measured velocities of the NPs in water and chloroform as deduced from (c), (d). (f) The corresponding relative velocities in water and CHCl$_3$ obtained by subtracting the solvent effect. (Adapted with permission from Brites et al. (2016). Copyright *Springer Nature* 2016.)

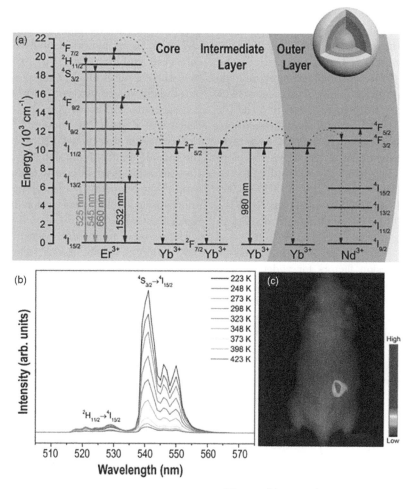

FIGURE 24.4 (a) Energy diagram showing energy transfer from Nd^{3+} to Er^{3+} via Yb^{3+} in core–multishell NPs, depicted schematically in top right corner. (b) Luminescence spectra of the NPs under 808 nm excitation in wide temperature range (223–423 K). (c) Merged optical and IR image under 808 nm excitation of a mouse after subcutaneous injection of an aqueous dispersion of multishell NPs. (Adapted from Wang et al. (2015) with permission from The Royal Society of Chemistry.)

under 980 nm excitation and adds the emission in the IR. A disadvantage is the loss of emission intensity in the thermally coupled Er^{3+} emissions and therefore a slightly lower sensitivity. Additionally, while the authors were able to demonstrate multimodal imaging *in vivo* (IR and CT) in a mouse, this was not possible for the luminescence thermometry due to the large attenuation of visible light by tissues.

Moving to the Biological Windows and In Vivo Studies

Researchers in the field of nanomedicine have probably shown the most interest in adding remote nanothermometry to their toolkit. Investigating intracellular heat phenomena not just *in vitro* (compare above) but also being able to study the temperature inside tissues and organs in living beings at a subcellular or molecular level will render new insights into metabolism, diseases, and the cellular replication processes. Furthermore, hyperthermia treatments leading to thermal ablation of malign tissues and tumors

have gained rising interest in cancer therapies and are actively investigated in the form of magnetic heating and photothermal therapy (PTT). In order to apply these therapies in a controlled manner avoiding collateral damage in healthy tissues, *in situ* thermometry is required (del Rosal et al., 2017).

Luminescence nanothermometry has set out to rise to this challenge and is presenting more and more studies with biomedical potential. The possibility of measuring temperature *in vivo* using luminescence was first demonstrated using visible-emitting nanothermometers in semi-transparent organisms. Donner et al. mapped the temperature distribution in millimeter-sized nematodes *C. elegans* genetically modified to express green fluorescence protein (GFP) (Donner et al., 2013). The polarization anisotropy of this protein changes with temperature; however, it does so in a strongly medium-dependent manner, complicating the correct interpretation of the results.

For an optical imaging/sensing technique like luminescence to achieve a broader applicability in *in vivo* studies, a shift from the visible part of the electromagnetic spectrum into the NIR needs to occur. In certain wavelength intervals in the NIR, denominated biological windows (Smith, Mancini and Nie, 2009), the absorption of biological tissues (especially skin and blood) is reduced compared to the visible (see Figure 24.5), where humans and most animals tend to be non-transparent. Additionally, scattering of light is exponentially reduced the further one moves into the IR, which enables imaging with an improved spatial resolution. The main limitation for measurement in the NIR is the presence of strong absorption bands of water, which therefore are the delimiters of the biological windows, leading to the first biological window (NIR-I) spanning from 650 to 950 nm. The second ranges from 1,000 to 1,350 nm (NIR-II), and the region from 1,500 to 1,700 nm is usually defined as NIR-IIb or NIR-III (Hong, Antaris, and Dai, 2017).

The above-discussed Er^{3+}, Yb^{3+} nanothermometers therefore present two exclusion factors for wider *in vivo* applications: their excitation via Yb^{3+} at 980 nm, the first strong absorption of water in the NIR, which leads to undesired heating *in vivo*, and the thermometry-relevant emissions in the visible, which will be absorbed or distorted

by biological tissues. Despite these limitations, Zhu et al. achieved *in vivo* temperature sensing in a mouse model of tumor using hybrid nanostructures constituted by a core–shell $NaLuF_4$: $Er^{3+}, Yb^{3+}@NaLuF_4$ NP surrounded by a carbon outer shell (Zhu et al., 2016).The role of the outer shell was to act as photothermal agent, that is, to generate sufficient heat upon light irradiation to trigger cell death and destroy the tumor. The temperature-sensitive Er^{3+} emission was used for measuring the temperature at the tumor site after 5-min-long irradiations as a function of the excitation laser intensity. However, the signal-to-noise ratio of the temperature measurements was poor due to the above-discussed low penetration of visible radiation into biological tissues.

Hence, other temperature-sensitive lanthanide ions (e.g., Nd^{3+}, Ho^{3+}), which present temperature-sensitive emission bands in the NIR, and combinations thereof are being studied. Additionally, QDs and semiconducting NPs with their emissions in the NIR-II have come into the focus as nanothermometers capable of deep-tissue temperature sensing over the last years. These IR-emitting nanothermometers can provide temperature feedback in real time during hyperthermia treatments of tumors and information about the thermal properties of biological tissues, which can be correlated to pathological processes (del Rosal et al., 2017). Continuous temperature monitoring during photothermal therapy was demonstrated by Carrasco et al. in 2015 following the experimental procedure shown in Figure 24.6a (Carrasco et al., 2015). The authors relied on the temperature-sensitive emission of intratumorally injected $LaF_3:Nd^{3+}$ to determine the laser-induced heating at the tumor site, as the ratio between the emission at 865 and 885 nm (corresponding to the $^4F_{3/2} \rightarrow {}^4I_{9/2}$ transition of Nd^{3+} ions) changed linearly with temperature. The observed difference between the *in situ* and surface (as given by IR thermography) temperatures, as can be seen in Figure 24.6b, highlights the need for temperature control in real time to optimize the outcomes of *in vivo* hyperthermia treatments. This was corroborated in a later

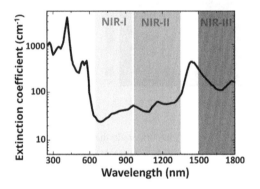

FIGURE 24.5 Extinction coefficient of a representative biological tissue. The spectral extensions of the first (NIR-I), second (NIR-II), and third (NIR-III) biological windows are indicated. (Adapted from Jaque et al. (2014) with permission of The Royal Society of Chemistry.)

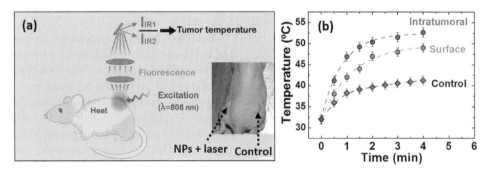

FIGURE 24.6 Real-time temperature monitoring during photothermal therapy. (a) Schematic representation of the experimental setup: upon excitation with 808 nm light, intratumorally injected $LaF_3:Nd^{3+}$ NPs trigger heat production (for therapy) and luminescence, which is spectrally analyzed to obtain the *in situ* temperature. (b) Temperature at different time points during the photothermal treatment as measured by nanothermometry (intratumoral temperature), IR thermography (surface temperature) for the treated mouse. The control case, which corresponds to the surface temperature of a mouse not inoculated with NPs, is also shown. (Reproduced with permission from Carrasco et al. (2015). Copyright 2014, Wiley-VCH.)

work where NIR-II-emitting PbS/CdS/ZnS QDs, which present a higher thermal sensitivity ($1\%\cdot K^{-1}$ as opposed to $0.25\%\cdot K^{-1}$ measured for the Nd^{3+}-doped NPs) were applied for the same purpose (del Rosal, Carrasco et al., 2016). The higher brightness of the QDs allowed temperature sensing with a higher time resolution, as shorter acquisition times were required. However, they provided an intensity-based temperature reading, which simplified the required setup as no spectral analysis of the signal was necessary, but could only yield temperature increments and not actual temperature values. This limitation can be addressed by integrating the QDs into hybrid nanostructures, where a temperature-independent emission is used as a reference to obtain a ratiometric reading, as discussed in Section 24.4.1 (Navarro Cerón et al., 2015).

Besides acting as *in situ* temperature reporters during hyperthermia cancer treatments, luminescent nanothermometers have also been used to study the thermal behavior of tissues *in vivo*, i.e., their response to external thermal stimuli. The possibility of studying the thermal response of tissues at the subcutaneous level was first demonstrated *in vivo* by Ximendes et al. using core–shell Nd^{3+}, Yb^{3+}-doped NPs (Ximendes et al., 2016). The use of Nd^{3+} ions as sensitizers allows optical excitation at 800 nm, removing the need for 980 nm excitation and minimizing any possible tissue heating. The ratio of the emissions of both rare-earth ions changed linearly with temperature, albeit with low

sensitivity ($0.41\%\cdot K^{-1}$) compared with other NIR-emitting nanothermometers. The experimental approach employed by the authors, which required the acquisition and analysis of full emission spectra, only allowed obtaining single-point temperature measurements. Two-dimensional time-resolved mapping was later achieved by changing the experimental approach (see Figure 24.7a): instead of capturing a full spectrum for each time point, a motorized filter wheel with two bandpass filters allowed capturing the intensities at two different emission wavelengths (1,000 and 1,200 nm) quasi-simultaneously (Ximendes et al., 2017). With this approach, Ximendes et al. studied the thermal dynamics of healthy tissues in mice by measuring the time evolution of the temperature after an external heating stimulus (provided by the laser labeled as "heating laser" in Figure 24.7a). The rare-earth ion content of the NPs they used as nanothermometers (core–shell $LaF_3:Er^{3+},Yb^{3+}@LaF_3:Tm^{3+},Yb^{3+}$) was optimized to maximize the temperature-dependent change in the ratio between the Yb^{3+} and Tm^{3+} emissions (centered at around 1,000 and 1,200 nm, respectively, as schematically represented in Figure 24.7b). These nanothermometers achieved the highest sensitivity ($5\%\cdot K^{-1}$at room temperature) demonstrated so far for NIR-operating nanothermometers, as shown in Figure 24.7c.

The thermal behavior of biological tissues can provide diagnostic information as it depends on several biological parameters (for instance, blood perfusion rate or metabolic

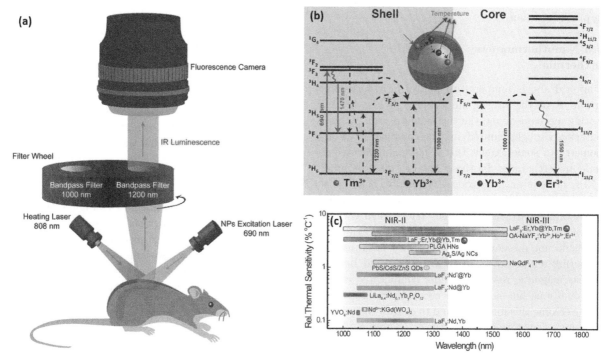

FIGURE 24.7 Real-time *in vivo* thermal dynamics recording. (a) Schematic representation of the experimental setup used by Ximendes et al. for measuring subcutaneous temperature in living mice in real time. The motorized filter wheel operated at a frequency of 0.3 Hz and allowed a quasi-simultaneous recording of the emission intensities of Tm^{3+} and Yb^{3+} ions. (b) Upon excitation with 690 nm laser light, the core–shell NPs, employed as nanothermometers, undergo different energy transfer processes, and radiative emissions occur. (c) Relative thermal sensitivity of a variety of nanothermometers operating in the NIR. (Reproduced with permission from Ximendes et al. (2017). Copyright 2017, American Chemical Society.)

heat generation) that are affected by different pathologies. Thus, an anomalous thermal behavior can indicate the presence of a disease. Ximendes et al. demonstrated that ischemic tissues returned to their baseline temperature after a heating stimulus markedly faster than their healthy counterparts using NIR-II-emitting Ag_2S NPs directly injected into the tissues as nanothermometers. On the other hand, tissues undergoing an inflammatory process showed a slower thermal relaxation than healthy and ischemic ones. The development of malignant tumors also modifies the thermal behavior of tissues due to the changes in blood perfusion that occur during tumor growth. Using Ag_2S nanothermometers in a mouse model of melanoma, Santos et al. demonstrated that tumor growth caused a significant change in the thermal dynamics of the tissue seven days before the tumor could be detected by visual inspection (Santos et al., 2018).

One of the main selling points of NIR-emitting nanoprobes as contrast agents lies in the possibility of using them for imaging hard-to-access sites, such as the brain, with high spatial and temporal resolution (Hong et al., 2014). Luminescent nanothermometers operating in the NIR allow temperature sensing through the skull, as recently demonstrated in a mouse model of coma using intracerebrally injected Ag_2S nanothermometers (del Rosal et al., 2018). The contactless nature of luminescence nanothermometry is particularly advantageous in this case, as conventional probes used for thermometry are best avoided due to the tissue damage that is likely to occur during probe insertion.

In Situ Nanothermometry in Magnetic Hyperthermia

As mentioned above, temperature control in hyperthermia treatments is of the utmost importance. While in PTT the heating mechanism and the thermometry can both rely on excitation via light, simplifying the combined PTT and thermometry to some extent, this is more complicated for magnetic hyperthermia. In these studies, superparamagnetic NPs, most often iron oxides (magnetite, maghemite) or ferrites, are exposed to alternating magnetic fields. The movement of the particles when they try to align with the rapidly changing field then produces heat. Unfortunately, these NPs do not present luminescence or a facile method for measuring temperature. In theory, this could be done via nuclear magnetic resonance under complicated experimental conditions but not at the necessary resolution. In order to overcome this limitation and to control the temperature *in situ*, a few approaches for the combination of luminescence nanothermometry with magnetic NPs have been presented.

The first iron oxide NPs that came with an attached thermometer in the form of an organic dye that was cleaved with increasing temperatures were presented by Riedinger et al. (2013). Applying an external alternating magnetic field, an azo-bond between the dye and the NPs was broken, leading to the recuperation of the luminescence of the organic molecule at 510 nm. The authors were able to correlate the increasing intensity with temperature. In 2014, Dong and Zink described the combination of Er^{3+},Yb^{3+}-doped $NaYF_4$ nanorods with superparamagnetic Fe_3O_4 nanocrystals in mesoporous silica, forming hybrid NPs containing both types of NPs (Dong and Zink, 2014). They demonstrated their heating capability in aqueous dispersions while measuring the temperature based on the typical Er^{3+} emissions. Piñol et al. presented in 2015 hybrid NPs that were based on the inclusion of maghemite and lanthanide complexes inside a block-copolymer (Piñol et al., 2015). These Fe_2O_3@P4VP-*b*-P(MPEGA-*co*-PEGA) multicore beads showed excellent heating properties and a thermal sensitivity of 5.8 %K^{-1} at 296 K. This S_R was achieved through the ratiometric measurement of the emission of a Tb^{3+}-complex emitting at 545 nm ($^5D_4 \rightarrow {}^7F_5$) and a Eu^{3+}-complex emitting at 615 nm ($^5D_0 \rightarrow {}^7F_2$), when excited with laser light at 365 nm. They were able to measure precisely the temperature change in an aqueous dispersion but also applied the particles for temperature mapping in opossum kidney (OK) cells (see Figure 24.8). Their results show a temperature distribution inside the cell, which is within the thermal resolution of 0.5 K.

Nevertheless, for *in vivo* studies, as explained in the previous section, a move to the NIR is necessary, and in 2018, Ortgies et al. presented optomagnetic hybrid nanostructures (OHMSs) that encapsulated LaF_3:Nd^{3+} NPs and Fe_3O_4 magnetic NPs inside the biocompatible polymer *poly*(lactic-co-glycolic acid) (PLGA) (Ortgies et al., 2018). The resulting OHMSs were capable of dual thermometry in the NIR-I (ratiometric, intensity-based, and simultaneously wavelength-based) and demonstrated not only magnetic heating but also photothermal heating. The dual mode thermometry was based on the ratio between the bands at 862.5 and 864 nm ($^4F_{3/2} \rightarrow {}^4I_{9/2}$), resulting in an S_R close to 0.4 %K^{-1} at 30°C. The second peak (I_{864}) interestingly also shifts to lower energies with increasing temperature by 0.002 nmK^{-1}. This is due to thermally induced strains in electron–phonon interactions of the Nd^{3+} in the LaF_3 host matrix. Therefore, two methods could be applied in the authors' *ex vivo* experiments (see Figure 24.9) to determine the heating efficacy, which were performed using an endoscopic fiber in order to measure the temperature in a magnetic setup between two coils and below a tissue (chicken breast) sample. It is easy to see that although photothermal heating is more effective, magnetic heating is not affected by the presence of biological tissues. Additionally, both thermometry techniques corroborated each other within the experimental uncertainty.

24.5 Conclusions and Outlook

In this short introduction to nanothermometry and remote temperature sensing with luminescence nanothermometry, we hope to have highlighted the potential that this

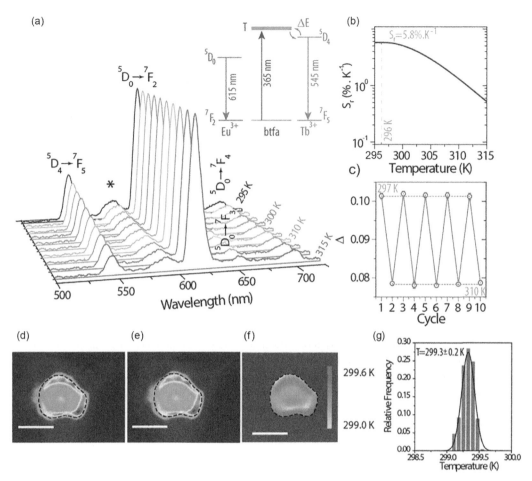

FIGURE 24.8 (a) Emission spectra of the water suspension multicore beads in the temperature range 295–315 K, excited at 365 nm. The Eu^{3+} and the Tb^{3+} transitions are identified. The inset shows a simplified energy scheme of the Eu^{3+} and Tb^{3+} ions and btfa ligand, where the most intense Eu^{3+} and Tb^{3+} transitions are presented. (b) Relative sensitivity S_R of the water suspension multicore beads in the temperature range 295–315 K. (c) Temperature cycling of the aqueous suspension between 297 and 310 K, with a repeatability better than 99.5% in the ten consecutive cycles. The solid and interrupted lines are guides for the eyes. (d, e) Temperature mapping of OK cells showing the (d) Eu^{3+} and (e) Tb^{3+} emissions. The intensity maps illustrate the co-localization of the Eu^{3+} and Tb^{3+} emissions. For better visualization, the maximum of the Tb^{3+} map was scaled by a factor of 3. The interrupted lines delimitate the nucleus of the cell, marking the region of interest where the temperature map presented in part (f) was computed. (g) The histogram of the temperature distribution near the OK nucleus follows a Gaussian distribution of mean value ($\pm$standard deviation) 299.3 $\pm$ 0.2 K ($r^2 > 0.997$), in accord with the cell culture temperature. All scale bars correspond to 10 μm. (Adapted with permission from Piñol et al. (2015). Copyright © 2015 American Chemical Society.)

technology brings to various fields from fundamental science to applications in industry and biomedicine. The demand for a technology that can ideally remotely measure temperature at a submicrometer and essentially a subcellular scale with good thermal, spatial, and temporal resolution has inspired interdisciplinary research in the nanoscience community mostly over the last decade to come forward with various types of nanothermometers in order to satiate this demand. A critical mass in proposed designs, involved investigators, and not-less-important interest in potential applications has probably been reached over the last years and has led to a partial consolidation and refocusing of ideas, for example, the focus on better reproducibility and repeatability by taking more and more external factors into consideration during the necessary calibration of most nanothermometers.

In this regard, it is a seminal advance that luminescence nanothermometers can fulfill the conditions required of an absolute thermometer when energetically close excited states form the basis of the luminescence nanothermometer. Demonstrating that thermally coupled emissions have the potential to elevate the ratiometric approach into an absolute technique under certain conditions will propel ahead the development of highly reliable nanothermometers as tools for industry and medicine. This principle will be applicable not just to the herein described Er,Yb nanothermometers but to most of the ratiometric lanthanide-based thermometers as long as the emissions stem from one type of rare earth ion (single-center emission).

Furthermore, the move toward the NIR and especially the biological windows has become more important if any

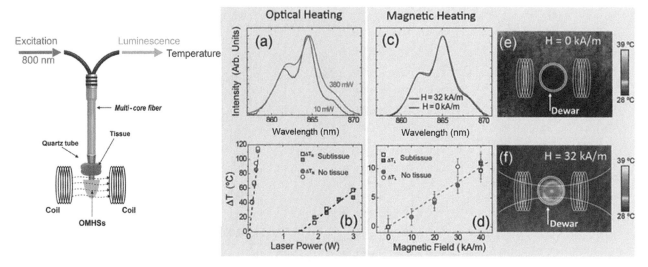

FIGURE 24.9 Left: Schematic representation of the experimental setup designed to demonstrate the ability of OMHSs to provide subtissue thermal sensing during hyperthermia induced by optical or magnetic stimuli. Right: (a) Emission spectra generated by the OMHSs as obtained for two different laser powers: 10 mW (dark line) and 380 mW (lighter line). Spectra were obtained in the absence of any tissue between the excitation/collection fiber and the solid OMHSs. (b) Temperature increment induced in the optically excited OMHSs as a function of the laser power as obtained through the ratiometric (ΔT_R) and spectral (ΔT_λ) shift calibration. Results obtained in the presence and absence of tissue are included (circles and squares, respectively). (c) Emission spectra generated by the OMHSs as obtained in the absence (dark line) and presence (lighter line) of an alternating magnetic field (H_{AC}) (100 kHz and 32 kA m^{-1}). (d) Temperature increment induced in the OMHSs as a function of the applied magnetic field in the presence and absence of tissue. Data obtained by employing the ratiometric and spectral shift calibrations are included. (e) Thermal image of the quartz tube containing the OMHSs in the absence of H_{AC}. (f) Thermal image of the tube containing the OMHSs in the presence of H_{AC} (100 kHz and 32 kA m^{-1}). (Adapted with permission from Ortgies et al. (2018). Copyright 2018, Wiley-VCH.)

future *in vivo* or medical thermometry application is desired. The field is in this regard at the point as demonstrated by the examples of proof-of-concept studies and *in vivo* nanothermometry, where real *in vivo* applications could become interesting for larger preclinical and clinical studies. This is especially the case in multifunctional nanomaterials that want to combine some form of hyperthermia treatment (magnetic or PTT) with accurate and *in situ* temperature control. Additionally, the application of luminescent nanothermometry is increasing the interest in the measurement of not just temperature but also temporal temperature gradients or transients as demonstrated in *vivo* studies. Comparing the thermal relaxation properties between healthy and potentially sick tissues has brought forward interesting applications of nanothermometers in the investigation of processes like ischemia, inflammation, and also tumorgenesis and angiogenesis.

Nevertheless, the big breakthrough in the form of an established diagnostic or treatment procedure has not yet come. But this can be said for the whole field of nanomedicine, not just nanothermometry. This is probably due to two larger problems of the design and synthesis of nanomaterials, and readers will discover this also in the other chapters of the series. So far the synthesis of most nanomaterials, whether rare-earth-doped NPs, QDs, or metallic NPs, is not well reproducible, especially not to the standards required in the pharmaceutical industry and demanded by the relevant authorities. And this is the case

for laboratory-scale synthesis of NPs; as far as we know, there is no industrial-scale synthesis established for any nanomaterial. The second problem is also the one of scaling up the available nanomaterial but this time on the user side. The amounts so far employed in typical *in vivo* studies in mice or rat in a single injection (something like 1 mg) would when translated for an 80 kg human result in a dose of 4 g. This brings a lot of additional toxicity considerations with it that will need to be addressed and still does not consider factors like a necessary switch from intravenous to oral dosing.

The good side of the above-described problem is that it affects a lot more investigations than just nanothermometry and is therefore likely going to be resolved in the next years. This means that the whole community of researchers interested in nanothermometry should build upon the achieved successes and be ready for the next big breakthrough in the synthesis/manufacturing of nanomaterials while focusing on improving reproducibility of their measurements and materials with increased thermal sensitivity, which would also mean smaller doses of the nanothermometer would be required. The recently presented Ag$_2$S nanodots seem to be promising in this regard due to their high thermal sensitivity in the NIR-II window and the low dosage that is required for their application (del Rosal et al., 2018; Santos et al., 2018). It is also very important to continue being creative and innovative and bring nanothermometry to a wider field of applications. After all, it was demonstrated nanothermometers can also be employed to measure Brownian motion. Between this type of fundamental research and the applied research of diagnostics, there is still a lot to discover and problems where a nanothermometer might deliver an unexpected solution.

References

Alencar, M. A. R. C. et al. (2004) Er^{3+}-doped $BaTiO_3$ nanocrystals for thermometry: Influence of nanoenvironment on the sensitivity of a fluorescence based temperature sensor, *Applied Physics Letters*, 84(23), pp. 4753–4755. doi:10.1063/1.1760882.

Baffou, G. et al. (2012) Thermal imaging of nanostructures by quantitative optical phase analysis, *ACS Nano*, 6(3), pp. 2452–2458. doi:10.1021/nn2047586.

Baffou, G. et al. (2014) Super-heating and micro-bubble generation around plasmonic nanoparticles under cw illumination, *Journal of Physical Chemistry C*, 118(9), pp. 4890–4898. doi:10.1021/jp411519k.

Balabhadra, S. et al. (2017) Upconverting nanoparticles working as primary thermometers in different media, *Journal of Physical Chemistry C*, 121(25), pp. 13962–13968. doi: 10.1021/acs.jpcc.7b04827.

Bon, P. et al. (2013) Three-dimensional temperature imaging around a gold microwire, *Applied Physics Letters*, 102(24), p. 244103. doi: 10.1063/1.4811557.

Brites, C. D. S. et al. (2012) Thermometry at the nanoscale, *Nanoscale*, 4(16), pp. 4799–4829. doi: 10.1039/c2nr30663h.

Brites, C. D. S. et al. (2016) Instantaneous ballistic velocity of suspended Brownian nanocrystals measured by upconversion nanothermometry, *Nature Nanotechnology*, 11(10), pp. 851–856. doi: 10.1038/nnano.2016.111.

Brites, C. D. S., Balabhadra, S. and Carlos, L. D. (2018) Lanthanide-based thermometers: At the cutting-edge of luminescence thermometry, *Advanced Optical Materials*, 2(2), p. 1801239. doi: 10.1002/adom.201801239.

Carrasco, E. et al. (2015) Intratumoral thermal reading during photo-thermal therapy by multifunctional fluorescent nanoparticles, *Advanced Functional Materials*, 25(4), pp. 615–626. doi: 10.1002/adfm.201403653.

Chénais, S. et al. (2004) Direct and absolute temperature mapping and heat transfer measurements in diode-end-pumped Yb:YAG, *Applied Physics B*, 79(2), pp. 221–224. doi: 10.1007/s00340-004-1544-0.

Corsi, C. (2010) History highlights and future trends of infrared sensors, *Journal of Modern Optics*, 57(18), pp. 1663–1686. doi: 10.1080/09500341003693011.

Dong, J. and Zink, J. I. (2014) Taking the temperature of the interiors of magnetically heated nanoparticles, *ACS Nano*, 8(5), pp. 5199–5207. doi: 10.1021/nn501250e.

Donner, J. S. et al. (2013) Imaging of plasmonic heating in a living organism, *ACS Nano*, 7(10), pp. 8666–8672. doi: 10.1021/nn403659n.

Gao, Y. and Bando, Y. (2002) Carbon nanothermometer containing gallium, *Nature*, 415(6872), pp. 599–599. doi: 10.1038/415599a.

Hong, G. et al. (2014) Through-skull fluorescence imaging of the brain in a new near-infrared window, *Nature Photonics*, 8(9), pp. 723–730. doi: 10.1038/nphoton.2014.166.

Hong, G., Antaris, A. L. and Dai, H. (2017) Near-infrared fluorophores for biomedical imaging, *Nature Biomedical Engineering*, 1(1), p. 0010. doi: 10.1038/s41551-016-0010.

Huth, S. et al. (2000) Localization of gate oxide integrity defects in silicon metal-oxide-semiconductor structures with lock-in IR thermography, *Journal of Applied Physics*, 88(7), p. 4000. doi: 10.1063/1.1310185.

Jaque, D. et al. (2014) Nanoparticles for photothermal therapies, *Nanoscale*, 6(16), pp. 9494–9530. doi: 10.1039/C4NR00708E.

Jaque, D. and Vetrone, F. (2012) Luminescence nanothermometry, *Nanoscale*, 4(15), pp. 4301–4326. doi: 10.1039/c2nr30764b.

Kim, K. et al. (2012) Ultra-high vacuum scanning thermal microscopy for nanometer resolution quantitative thermometry, *ACS Nano*, 6(5), pp. 4248–4257. doi: 10.1021/nn300774n.

Labrador-Páez, L. et al. (2018) Reliability of rare-earth-doped infrared luminescent nanothermometers, *Nanoscale*, 10(47), pp. 22319–22328. doi: 10.1039/C8NR07566B.

McGee, T. (1988) *Principles and Methods of Temperature Measurement*. 1st edn. New York: Wiley Interscience.

Nakabeppu, O. et al. (1995) Scanning thermal imaging microscopy using composite cantilever probes, *Applied Physics Letters*, 66(6), pp. 694–696. doi: 10.1063/1.114102.

Navarro Cerón, E. et al. (2015) Hybrid nanostructures for high-sensitivity luminescence nanothermometry in the second biological window, *Advanced Materials*, 37(32), pp. 4781–4787. doi: 10.1002/adma.201501014.

Ortgies, D. H. et al. (2018) Optomagnetic nanoplatforms for in situ controlled hyperthermia, *Advanced Functional Materials*, 28(11), p. 1704434. doi: 10.1002/adfm.201704434.

Piñol, R. et al. (2015) Joining time-resolved thermometry and magnetic-induced heating in a single nanoparticle unveils intriguing thermal properties, *ACS Nano*, 9(3), pp. 3134–3142. doi: 10.1021/acsnano.5b00059.

Quintanilla, M. and Liz-Marzán, L. M. (2018) Guiding rules for selecting a nanothermometer, *Nano Today*, 19, pp. 126–145. Elsevier Ltd. doi: 10.1016/j.nantod.2018.02.012.

Riedinger, A. et al. (2013) Subnanometer local temperature probing and remotely controlled drug release based on azo-functionalized iron oxide nanoparticles, *Nano Letters*, 13(6), pp. 2399–2406. doi: 10.1021/nl400188q.

del Rosal, B., Villa, I., et al. (2016) In vivo autofluorescence in the biological windows: The role of pigmentation,

Journal of Biophotonics, 9(10), pp. 1059–1067. doi: 10.1002/jbio.201500271.

del Rosal, B., Carrasco, E., et al. (2016) Infrared-emitting QDs for thermal therapy with real-time subcutaneous temperature feedback, *Advanced Functional Materials*, 26(30), pp. 6060–6068. doi: 10.1002/adfm.201601953.

del Rosal, B. et al. (2017) In vivo luminescence nanothermometry: From materials to applications, *Advanced Optical Materials*, 5(1), p. 1600508. doi: 10.1002/adom.201600508.

del Rosal, B. et al. (2018) In vivo contactless brain nanothermometry, *Advanced Functional Materials*, 28(52), p. 1806088. doi: 10.1002/adfm.201806088.

Santos, H. D. A. et al. (2018) In vivo early tumor detection and diagnosis by infrared luminescence transient nanothermometry, *Advanced Functional Materials*, 28(43), p. 1803924. doi: 10.1002/adfm.201803924.

Sedlmeier, A. et al. (2012) Photon upconverting nanoparticles for luminescent sensing of temperature, *Nanoscale*, 4(22), pp. 7090–7096. doi: 10.1039/c2nr32314a.

Senapati, S. and Nanda, K. K. (2015) Wide-range thermometry at micro/nano length scales with In_2O_3 octahedrons as optical probes, *ACS Applied Materials & Interfaces*, 7(42), pp. 23481–23488. doi: 10.1021/acsami.5b05675.

Sherry, D. (2011) Thermoscopes, thermometers, and the foundations of measurement, *Studies in History and Philosophy of Science Part A*, 42(4), pp. 509–524. Elsevier Ltd.doi: 10.1016/j.shpsa.2011.07.001.

Singh, S. K., Kumar, K. and Rai, S. B. (2009) Er^{3+}/Yb^{3+} codoped Gd_2O_3 nano-phosphor for optical thermometry, *Sensors and Actuators A: Physical*, 149(1), pp. 16–20. doi: 10.1016/j.sna.2008.09.019.

Smith, A. M., Mancini, M. C. and Nie, S. (2009) Second window for in vivo imaging, *Nature Nanotechnology*, 4(11), pp. 710–711. doi: 10.1038/nnano.2009.326.

Thomsen, J. S. (1962) A restatement of the zeroth law of thermodynamics, *American Journal of Physics*, 30(4), pp. 294–296. doi: 10.1119/1.1941991.

Vetrone, F. et al. (2010) Temperature sensing using fluorescent nanothermometers, *ACS Nano*, 4(6), pp. 3254–3258. doi: 10.1021/nn100244a.

Villa, I. et al. (2015) 1.3 µm emitting SrF_2:Nd^{3+} nanoparticles for high contrast in vivo imaging in the second biological window, *Nano Research*, 8(2), pp. 649–665. doi: 10.1007/s12274-014-0549-1.

Wang, C. et al. (2011) Determining intracellular temperature at single-cell level by a novel thermocouple method, *Cell Research*, 21(10), pp. 1517–1519. doi: 10.1038/cr.2011.117.

Wang, X., Wolfbeis, O. S. and Meier, R. J. (2013) Luminescent probes and sensors for temperature, *Chemical Society Reviews*, 42(19), pp. 7834–69. doi: 10.1039/c3cs60102a.

Wang, Z. et al. (2015) Nd^{3+}-sensitized $NaLuF_4$ luminescent nanoparticles for multimodal imaging and temperature sensing under 808 nm excitation, *Nanoscale*, 7(42), pp. 17861–17870. doi: 10.1039/C5NR04889C.

Ximendes, E. C. et al. (2016) Unveiling in vivo subcutaneous thermal dynamics by infrared luminescent nanothermometers, *Nano Letters*, 16(3), pp. 1695–1703. doi: 10.1021/acs.nanolett.5b04611.

Ximendes, E. C. et al. (2017) In vivo subcutaneous thermal video recording by supersensitive infrared nanothermometers, *Advanced Functional Materials*, 27(38), p. 1702249. doi: 10.1002/adfm.201702249.

Zhu, X. et al. (2016) Temperature-feedback upconversion nanocomposite for accurate photothermal therapy at facile temperature, *Nature Communications*. Nature Publishing Group, 7(1), p. 10437. doi: 10.1038/ncomms10437.

Luminescence Nanothermometry

Oleksandr A. Savchuk
INL - International Iberian Nanotechnology Laboratory

Joan J. Carvajal
Universitat Rovira i Virgili

25.1 Introduction

Temperature is today recognized as one of the basic variables in science. Many processes, from the chemical industry to biotechnology and medicine, or from the extraction of metals from ores to the controlled growth of microorganisms in fermentation processes, depend on temperature (Michalson et al. 2001; Leigh 1988). Thus, temperature determination is crucial in scientific research and technical development.

Temperature allows, thermodynamically speaking, determining whether two or more systems are in thermal equilibrium. It can be defined quantitatively from the thermodynamics' second law as the rate of change of entropy with energy (Childs 2001). In a qualitative way, Max Plank defined the concept of temperature as our qualitative sense of touch about the hotness or coldness of a body (Young and Freedman 2008).

To measure it, a conventional temperature measurement system comprises the following: (i) a transducer that converts a temperature-dependent phenomenon into a signal that changes with temperature; (ii) a method to transmit the signal from the transducer; (iii) some sort of signal processor to convert that signal in a parameter that can be associated with the temperature; (iv) a display to show this parameter; and (v) a method of recording the data (Childs 2016). Then, it is necessary to calibrate the system to convert the measured quantity into an actual value of temperature.

25.1.1 From "Classical" Thermometry to Nanothermometry

Since the time of the invention of the thermoscope by Galileo Galilei, consisting on a glass bulb with air and a long rod immersed in a liquid that used the expansion of air and liquids when heated as the transducing signal (McGee 1988), many other temperature measuring methods and equipment have been developed considering the field of application, the measurement accuracy, and the measurement conditions (Thyageswaran 2012). With the development of the nanotechnology, the temperature of a given system with submicrometric spatial resolution became possible to measure. This has led to the development of a new subfield of thermometry named nanothermometry, related to the temperature measurement at the nanoscale level (Lee and Kotov 2007).

The miniaturization of electronics and optoelectronics, operating at higher switching speeds that generate energy dissipation in the form of heat, increased the importance of a good knowledge of localized heat to solve manufacture and design issues and improve the reliability of such systems (Childs 2016). For example, in electronics, the reduction in size of the electrical conduction channels leads to relevant electrical resistance values, resulting in a heating of the systems caused by the Joule effect and the appearance of well-localized temperature increments, the so-called "hot spots" (Brites et al. 2012), that can affect the performance of the electronic devices and could lead to irreversible damages in them.

Biomedicine can also use the benefits offered by nanothermometry. For instance, an accurate knowledge of the temperature inside living cells has probed significant in monitoring cell health, in which metabolism and enzyme reactions are marked by temperature changes (Hildebrandt et al. 2002). An accurate and noninvasive determination of temperature is, thus, of particular importance for the investigation of cellular dynamics and heat propagation in the different intracellular compartments. Cell temperature changes with cellular activity, such as cell division or gene expression. So, pathogenesis of diseases like cancer are characterized by an increment of temperature (Weaver 2010), with cancer cells exhibiting a higher temperature than healthy cells due to the increased metabolic activity. Thus, temperature monitoring will provide not only the understanding of cellular activities but also the possibility of detection of diseases in an early stage of development. Furthermore, heat can be used as a tool to increase cells death rate if required (Pustovalov et al. 2013). Above a certain threshold (312 K), chemical reactions leading to denaturation processes are induced (Lepock et al. 1987) that damage cells and tissues and prevent them from performing their functions. Finally, at temperatures above 321 K necrosis is induced. Thus, if controlled, heat may be used to treat abnormal cells, such as cancer cells through hyperthermia treatments (Roemer 1989).

Another example of interest of temperature determination at the nanoscale includes microfluidic systems. The accurate, precise, and active temperature control inside a microfluidic system is crucial and has been demonstrated in a variety of applications, such as in the generation of localized heating, strong thermal gradients, and fast temperature cycles. In fact, the reduction in size of microfluidic devices might generate unexpected localized temperature variation when an electric field is applied across a channel containing a conducting medium because of heating caused by Joule effect (Tang et al. 2006) or the presence of photothermal effects, like laser-induced heating in optical traps (Peterman et al. 2003) or surface-enhanced Raman scattering (SERS) (Kang et al. 2010). The collapse of cavitation bubbles in microfluidic confinement is another source of important temperature variations during very short periods of time in these devices (Baffou et al. 2014). Also, when sprays are generated, the friction forces existing inside the injector might produce non-negligible changes of temperature in the droplets generated (Gibson et al. 2014). Such unwanted increase in temperature can affect the performance of the device and/or the carrying fluid, and even change some of the physical properties of the sample under study. Thus, temperature might be a perturbative parameter, whether in a positive or negative way, in the field of micro- and nanofluidics, and it should be controlled (Bergaud 2016).

Thus, nanothermometry can be understood as a multidisciplinary and challenging subject that requires the development of new approaches and new techniques, since conventional thermometry, at such scales, is no longer valid (Carlos and Palacio 2016).

Although some pioneering works about this subject were published at the very end of the past century, nanothermometry exploded during the first decade of the 21st century. However, since 2010, a clear bias towards luminescence nanothermometry and its applications, mainly in biomedicine, microelectronics, and microfluidics, has been produced.

This chapter is divided into three main parts. The first one presents the fundamentals of luminescence nanothermometry, including the different methods that can be used to determine the temperature from the different information of the emitted light that are affected by temperature changes, and also the parameters that allow us analyzing the performance of the luminescence nanothermometers. The second part presents the different families of materials that have been used up to now to develop luminescence nanothermometers, describing the main advantages and disadvantages of each kind of materials for this purpose. Finally, the last part of the chapter presents the main applications in which luminescence nanothermometry has been used up to now, distributed in two main fields: biomedical and technological applications. These last ones include the application of luminescence nanothermometry in microelectronics and microfluidics. Thus, although this chapter is not intended as an exhaustive compilation of all the work developed in this field, it will be a useful guide for the reader beginning to work in it.

25.2 Fundamentals of Luminescence Nanothermometry

25.2.1 Methods in Luminescence Nanothermometry

Luminescence is the process in which the emission of light from a given substance occurs. In general, the excitation (absorption of one or more photons in the case of photoluminescence) causes the energy of the luminescent molecule or ion to jump to higher electronic states, from where it will be emitted in the form of light or heat, returning back to the ground state or to an intermediate state (Sole et al. 2005). This process is schematically illustrated in the Jablonski energy level diagram shown in Figure 25.1.

The properties of the emitted light depend on the properties of the electronic states involved in the emission process, which in turns depend on the local temperature of the system. Thus, luminescence nanothermometry operates in base to the luminescence properties that change with temperature to achieve thermal sensing by temporal or spectral analysis of the emission. Temperature can affect the luminescence emission spectrum parameters in different ways, from which the temperature can be determined. Figure 25.2 illustrates schematically these parameters, including intensity, lifetime, bandwidth, polarization, and spectral position of the emission peak (Jaque and Vetrone 2012).

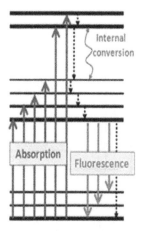

‡ - **Vibrational relaxation, quenching, or other non-radiative processes**

FIGURE 25.1 Jablonski energy level diagram showing the luminescence process, including the absorption, internal conversion, and fluorescence or luminescence phenomena.

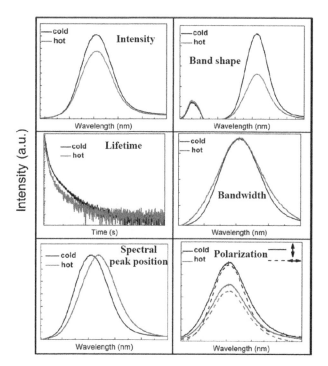

FIGURE 25.2 Schematic representation of the possible effects caused by an increase of the local temperature on the luminescence emission spectrum parameters of a given material.

Intensity-Based Luminescence Nanothermometry

Temperature-induced changes on the intensity of luminescence are caused by several factors: (i) quenching mechanisms, since temperature increments would activate mechanisms of cross-relaxation, while luminescence quenching centers would reduce the luminescence intensity; (ii) nonradiative relaxation processes, since electrons in an excited state would relax to the ground state by generating heat instead of light; (iii) population redistribution due to Boltzmann statistics, since the change of temperature would activate the population redistribution among the different

energy states; and (iv) phonon-assisted Auger conversion processes in the absence of temperature dependence (Wu et al. 2011).

Intensity-based luminescence nanothermometry has been reported in different systems, including quantum dots (QD) (Walker et al. 2003, Han et al. 2009), organic dyes (Ross et al. 2001, Löw et al. 2008), lanthanide (Ln^{3+})-doped systems (Fonger and Struck 1970, Yap et al. 2009), polymers (Uchiyama et al. 2003), and gold nanoclusters (Shang et al. 2013). Figure 25.3 illustrates the temperature dependence of the luminescence from lipoic acid-capped gold nanoclusters dispersed in phosphate-buffered saline (PBS), whose intensity decreased by 67% when the temperature increased from 283 K (10°C) to 318 K (45°C) after excitation at 580 nm.

However, the main disadvantage of this technique is that the observed emission intensity is also a function of other variables, such as the power fluctuation of the excitation source, the variation of the concentration of luminescence nanoparticles in the target in which the temperature is going to be measured, or the inhomogeneity on the distribution of the luminescent centers in the luminescent nanoparticles used as thermal probes (Jaque and Vetrone 2012).

Band-Shape Luminescence Nanothermometry

Band-shape luminescence nanothermometry is referred to as the luminescence of a system whose luminescence spectrum consists of several emission bands/lines, and the intensity at least of one of them is strongly temperature dependent (Brites et al. 2012, Jaque and Vetrone 2012). In band-shape luminescence nanothermotry, we should distinguish between two different cases: (i) when the emission bands/lines are

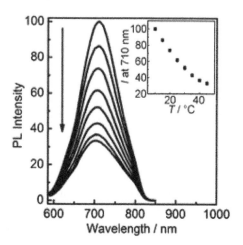

FIGURE 25.3 Temperature dependence of the fluorescence emission from lipoic acid-capped gold nanoclusters dissolved in PBS after excitation at 580 nm for various temperatures in the range from 283 K (10°C) to 318 K (45°C). Inset shows how the intensity of the emission peak centered at 710 nm decreased as the temperature increased. (Adapted with permission from Shang, L., Stockmar, F., Azadfar, N., Nienhaus, G.U. 2013. Intracellular thermometry by using fluorescent gold nanoclusters. Angew. *Chem. Int. Ed.* 52: 11154–11157. © 2013 Wiley-VCG Verlag GmbH.)

generated from a single luminescent center and (ii) when the emission bands/lines are generated by different emitting centers. In the first case, the changes in the band shape of the spectrum are generally caused by a thermally induced electronic population redistribution between the different energy levels of the emitting center. In the second case, the temperature-induced changes in the band shape of the spectrum arise from the different thermal quenching ratios of each luminescent center or from thermally induced changes in the energy transfer rates between these emitting centers.

The main advantage of this technique over the intensity-based luminescence nanothermometry is that the thermal reading is not affected by fluctuations in the concentration of the luminescence centers.

Band-shape luminescence nanothermometry is the most popular luminescence nanothermometry technique, and it has been reported in a great variety of luminescent systems, including lanthanide-doped systems (Vetrone et al. 2010, Savchuk et al. 2018b), quantum dots (Vlaskin et al. 2010, Hsia et al. 2011) and organic dyes (Peterman et al. 2003, Ebert et al. 2007).

In band-shape nanothermometry, the thermal sensing is achieved through the analysis of the changes in the emission intensity caused by temperature (Wade et al. 2003). Taking into account that the electronic population of the individual energy levels is directly proportional to the total electronic population of the system, one way to avoid the errors arising from such parameters that affect the emission intensity is to measure the emission intensity coming from two different energy levels. The particular case of the fluorescence or luminescence (depending on the source) intensity ratio (FIR or LIR) technique involves the ratio of fluorescence or luminescence from two close energy levels that can be considered to be thermally coupled, i.e., their energy difference (ΔE) is in the range of 200–2,000 cm^{-1} (Wade et al. 2003). The FIR (LIR) from two thermally coupled energy levels follows a Boltzmann-type electronic population distribution and can be written as (Wade et al. 2003)

$$FIR\ (LIR) = \frac{I_1}{I_2} = \frac{g_1 \nu_1 \sigma_1}{g_2 \nu_2 \sigma_2} exp\left(-\frac{\Delta E}{k_B T}\right)$$
$$= B exp\left(-\frac{\Delta E}{k_B T}\right) \quad (25.1)$$

where g_i, ν_I, and σ_I are the degeneracy of the electronic levels, the spontaneous emission rates, and the absorption rates, respectively, k_B is the Boltzmann constant, and T is the absolute temperature.

To illustrate the FIR (LIR) technique, Figure 25.4 shows a simplified electronic energy level diagram with the energy and transitions of interest in a general example in which the FIR (LIR) technique can be used to sense temperature. In this figure, we wanted to illustrate that the terminal level of fluorescence does not need to be the ground state, but another excited state with a lower energy than the levels from which the fluorescence (luminescence) originates.

An example of the application of the FIR (LIR) technique is shown in Figure 25.5, corresponding to the red emission

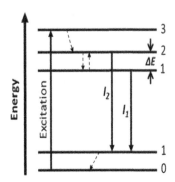

FIGURE 25.4 Simplified energy level diagram showing the energy levels and transitions of interest in a possible example in which the FIR (LIR) technique can be used to sense temperature. The dashed lines correspond to nonradiative decay processes, while solid arrows correspond to the fluorescence transitions used to calculate the fluorescence intensity ratio.

observed in Ho,Yb:KLu(WO$_4$)$_2$ nanoparticles, attributed to the $^5F_5 \rightarrow {}^5I_8$ transition of Ho^{3+} (Savchuk et al. 2015). This red band consists of two peaks assigned to the radiative transition from two different Stark sublevels of the $^5F_5$ manifold to the ground state of Ho^{3+}. As can be seen in the figure, the intensity of these two peaks evolved differently with temperature (see Figure 25.5a). Since the energy gap between these two peaks is small, of the order of 233 cm^{-1}, it can be assumed that the upper Stark sublevel has been thermally populated from the lower Stark sublevel when the temperature increased as the intensity of the peak located at 660 nm decreases faster than that of the peak located at 650 nm. Thus, one can consider that these two Stark sublevels are thermally coupled, and then, the FIR (LIR) technique can be applied. However, in this case a slight modification of Eq. 25.1 has to be introduced, since the emission peaks partially overlap, to include a term that accounts for this effect (Wade et al. 2003):

$$FIR\ (LIR) = \left(\frac{n_1}{n_2}\right) B exp\left(-\frac{\Delta E}{k_B T}\right) + \left(\frac{m_1}{m_2}\right) \quad (25.2)$$

where n_i accounts for the fraction of the total fluorescence or luminescence intensity of the transition originating from level i, considering $i = 2$ for the upper level and $i = 1$ for the lower level, and m_i corresponds to the fraction of the total intensity from level i.

Lifetime Luminescence Nanothermometry

Lifetime luminescence nanothermometry is based on the estimation of the temperature of the nanoprobe from the analysis of its fluorescence lifetime (Jaque and Vetrone 2012). Decay probabilities from electronic levels depend on a large number of parameters, and many of them are related to temperature, such as phonon-assisted energy transfer processes and multiphononic decay processes.

This technique has several advantages when compared to the intensity-based approaches. Lifetime luminescence

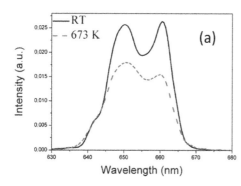
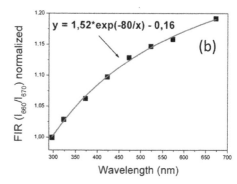

FIGURE 25.5 Application of the FIR (LIR) technique. (a) Comparison of the spectra of the emission (dashed line) observed in Ho,Yb:KLu(WO$_4$)$_2$ nanoparticles corresponding to the $^5F_5 \rightarrow {}^5I_8$ electronic transition of Ho^{3+} at two different temperature. (b) FIR (LIR) of the intensity of the 650 and 660 nm in the temperature range between 297 and 693 K, modeled according to Eq. 25.2, since there exists a clear overlapping between the two peaks observed in this emission. (Reproduced with permission from Savchuk, O.A., Carvajal, J.J., Pujol, M.C., Barrera, E.W., Massons, J., Aguilo, M., Diaz, F. 2015. Ho,Yb:KLu(WO$_4$)$_2$ nanoparticles: a versatile material for multiple thermal sensing purposes by luminescent thermometry. *J. Phys. Chem. C* 119: 18546–18558. © 2015 American Chemical Society.)

nanothermometry eliminates problems related to non-controllable spatial fluctuations of the fluorescence intensity due to the non-homogeneous distribution of nanoprobes, the uncontrolled motion of nanoprobes or biocomponents, and the light distribution and shading in the sample, among others. Since, in a first approximation, fluorescence lifetime does not depend on the local concentration of luminescent probes, all these inconveniences can be overcome. Furthermore, lifetime luminescence nanothermometry avoids the necessity of acquisition of the whole luminescence spectra, a time-consuming procedure. When this is associated with the requirement of high spatial resolution, which leads to low signal levels and long acquisition times, the short measuring times required by the lifetime technique minimize the possibility of laser-induced local heating of the system under investigation. Moreover, temperature reading is not affected by the luminescence intensity, and thus, the signal from which the temperature will be determined can be acquired at a time interval of the order of the luminescence lifetime, in the range of picoseconds to milliseconds, depending on the material in which the probe is based on, leading to a very fast temporal response of the sensor. Finally, it should also be mentioned that this technique can be used for high-temperature measurements, avoiding the undesired contributions of the blackbody radiation (Haro-Gonzalez et al. 2012).

Figure 25.6 shows one example of the application of this technique, in the comparison of the thermometric performance of Er,Yb:NaYF$_4$ and Er,Yb:NaY$_2$F$_5$O nanoparticles through the lifetime luminescence nanothermometry technique. These nanoparticles have been pumped at 980 nm with a diode laser, and the lifetime of their green emissions, generated through an upconversion mechanism (see Section 25.3.5), has been recorded as a function temperature. The results obtained clearly show that the slope of the evolution of the lifetime of the green emission arising from Er,Yb:NaY$_2$F$_5$O nanoparticles is higher than that of the Er,Yb:NaYF$_4$ nanoparticles, providing a

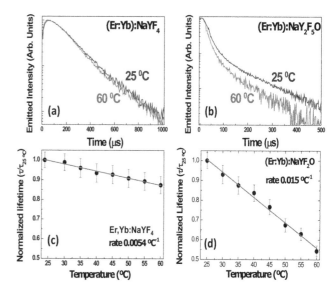

FIGURE 25.6 Luminescence decay curves of the 545 nm emission line of (a) Er,Yb:NaYF$_4$ and (b) Er,Yb:NaY$_2$F$_5$O nanoparticles at 298 K (25°C) and 333 K (60°C). Calculated lifetime values as a function of temperature for (c) Er,Yb:NaYF$_4$ and (d) Er,Yb:NaY$_2$F$_5$O nanoparticles. In all cases, dots are experimental data, and solid lines are the best linear fits.

luminescence nanothermometer that will allow to resolve smaller temperature changes (Savchuk et al. 2014).

However, this technique has a significant disadvantage, since it demands to use a pulsed excitation source, with a frequency shorter than the lifetime to be measured, that sometimes requires working in the femtosecond regime, which makes it costly and bulky.

Luminescence Nanothermometry Based on the Spectral Position of the Emission Lines

This technique, also called spectral position luminescence nanothermometry, is based on the analysis of the spectral position of the emission lines, assigned to the energy separation between the two electronic levels involved in the emission, which in turn depends on a large number of temperature-dependent parameters, such as the refractive index and the inter-atomic distances (Maestro et al. 2014).

The main advantage of this technique is that temperature reading is not affected by luminescence intensity fluctuations caused by variations in the concentration of the emitting centers, fluctuations of the power of the excitation source, and shading effects or movements of the sample.

QDs are the main class of materials that have been used for spectral position luminescence nanothermometry (Maestro et al. 2014, 2010). To illustrate this kind of luminescence thermometry technique, Figure 25.7 shows the emission spectra of CdSe QDs dispersed in phosphate-buffered saline (PBS) at two different temperatures, showing the displacement of the position of the luminescence bands and the dependence of the peak shift with the pump intensity, related to the temperature of the sample. Nevertheless, other materials, like lanthanide-doped dielectric nanoparticles, with emission bands narrower than those of QDs, have also been used for the same purpose, although the temperature-induced changes in the spectral shift were much smaller (Rocha et al. 2013).

Bandwidth-Based Luminescence Nanothermometry

For any increase in temperature above the absolute zero, higher energy phonon levels are populated. This effect leads to a broadening of the emission/absorption bands, since excited electronic sublevels are also populated and so they participate in the emission/absorption processes. The bandwidth of the emission/absorption bands ω changes with temperature according to the following expression (Henderson and Imbush 2006):

$$\omega(T) = \omega_0 \sqrt{coth\left(\frac{\hbar\Omega}{2k_BT}\right)} \qquad (25.3)$$

where ω_0 is the full width at half maximum (FWHM) of the emission/absorption band at 0 K, and $\hbar\Omega$ is the energy of the lattice vibration that interacts with the electronic transition. Thus, the higher the temperature, the wider the emission/absorption band because of the contributions of thermal vibrations of the luminescent center and its neighboring atoms/molecules.

The change in bandwidth of the emission/absorption bands is thus used in the bandwidth-based luminescence nanothermometry to get a thermal reading. However, the main disadvantage of this technique is that the magnitude of the emission band broadening caused by a temperature increment is small; thus, it can only be observed in systems with narrow emission/absorption lines, like lanthanide-doped materials (Benayas et al. 2012), although it can also be seen in QDs (Park et al. 2013). Figure 25.8 shows an example of this kind of luminescence nanothermometry technique based on the two hypersensitive luminescence lines of Nd^{3+} in Nd:YAG, located near 940 nm, and assigned to the transitions between the two Stark sublevels of the $^4F_{3/2}$ metastable state to the highest energy sublevel of the $^4I_{9/2}$ state. The bandwidths of these two transitions show a strong linear dependence with temperature (Benayas et al. 2012).

Polarization-Based Luminescence Nanothermometry

Polarization-based luminescence nanothermometry is based on the luminescence polarization anisotropy of the emitting material. In general, we can consider that the light re-emitted by a population of luminescent bodies illuminated by a linearly polarized light will be partially polarized due to the random orientation of the molecular dipoles. This polarization anisotropy of the luminescence can be expressed as (Donner et al. 2012)

$$r = \frac{I_{parallel} - I_{perpendicular}}{I_{parallel} - 2I_{perpendicular}} \qquad (25.4)$$

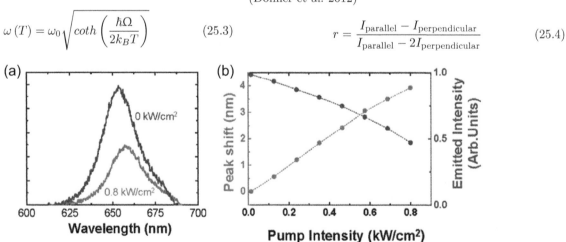

FIGURE 25.7 (a) Emission spectra of CdSe QDs dispersed in a PBS solution recorded for two different intensities of the pump beam, inducing two different temperatures in the sample, illustrating how the emission peak is shifting as the temperature changes. (b) Dependence of the peak shift of the emission of CdSe QDs with the pump intensity and thus with the temperature of the sample. The variation of the peak position is compared with the variation of the emitted intensity showing that the former is higher. (Adapted with permission from Maestro, L.M., Rodriguez, E.M., Cruz, M.C.I. et al. 2010. CdSe quantum dots for two-photon fluorescence thermal imaging. *Nano Lett.* 10: 5109–5115. © 2010 American Chemical Society.)

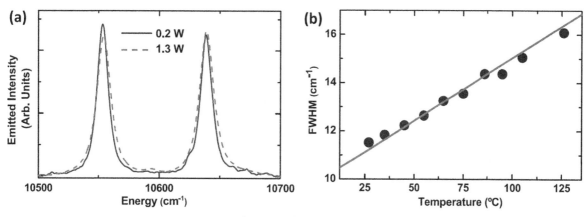

FIGURE 25.8 (a) Emission spectra from Nd:YAG $^4F_{3/2} \to {}^4I_{9/2}$ inter-Stark transitions at around 940 nm as obtained at two different absorbed pump powers of the laser with emission at 808 nm, inducing two different temperatures in the sample. The spectra reveal the different broadening of the emission lines that can be used to determine the temperature in the sample. (b) Dependence of the FWHM of the emission lines with temperature. (Adapted with permission from Benayas, A., Escuder, E., Jaque, D. 2012. High-resolution confocal fluorescence thermal imaging of tightly pumped microchip Nd:YAG laser ceramics. *Appl. Phys. B* 107: 697–701. © 2012 Springer Nature.)

where I_{parallel} and $I_{\text{perpendicular}}$ are the intensities of the luminescence polarized parallel and perpendicular to the incident polarization. This polarization anisotropy is directly related to rotational diffusion induced by molecular Brownian dynamics that is a function of temperature. In fact, the Deby–Stokes–Einstein equation correlates the rotational motion of the molecules with the temperature:

$$\tau_R = \frac{V\mu(T)}{k_B T} \quad (25.5)$$

where τ_R is the rotational lifetime, $\mu(T)$ is the viscosity, and V is the hydrodynamic volume of the luminescent molecule. From here, we can correlate the molecular rotation caused by Brownian dynamics with the polarization anisotropy (r) through the Perrin's equation (Donner et al. 2013):

$$\frac{1}{r} = \frac{1}{r_0}\left(1 + \frac{\tau_L}{\tau_R}\right) \quad (25.6)$$

where τ_L is the luminescence lifetime, and r_0 is the polarization anisotropy in the absence of any molecular motion (≤ 0.4). When the temperature increases, the Brownian rotational motion of the luminescent entities is accelerated, and the re-emitted photons will lose the memory of the incident light polarization. Thus, an increase in temperature leads to a decrease of the degree of polarization, and thus to the anisotropy, of the luminescence band.

This is probably the less used luminescence nanothermometry method, because of the complexity of the measurements, although it is insensitive to changes in intensities produced by photobleaching, variations of illumination intensity, or migration of the luminescent entities. However, for common luminescent entities in aqueous media, the rotational Brownian motion is too fast, and thus, the polarization anisotropy does not depend on temperature. To solve this problem, viscous systems, like mixtures of glycerol and water, can be used, or alternatively, the hydrodynamic

volume of the luminescent molecule can be increased, for instance using proteins like the green fluorescence protein (GFP) (Donner et al. 2012).

To illustrate how it works, Figure 25.9 shows temperature measurements performed in single transfected HeLa cells in which GFP was overexpressed as singular, nontagged proteins, so that GFP occupies the entire volume of the cells. GFP were excited with a 473 nm laser, while HeLa cells were heated through a resistive heating of the sample chamber.

25.2.2 Analysis of the Performance of the Luminescence Nanothermometers

Each of the luminescence nanothermometric techniques described in the previous section require a calibration procedure against an independent thermal probe to allow the conversion of the value recorded for the luminescence nanothermometric parameter (intensity, band shape, lifetime, etc.) into temperature.

Once this calibration is performed, we can use different parameters to analyze their performance. The main parameters used in this analysis are summarized in the following.

Absolute and Relative Thermal Sensitivities

Thermal sensitivity is one of the most important parameters to evaluate the performance of a luminescence nanothermometer. It describes the variation rate of the thermometric parameter by small changes in temperature. In luminescence nanothermometry, there are two commonly used thermal sensitivity parameters: (i) the absolute thermal sensitivity (S_{abs}) and (ii) the relative thermal sensitivity (S_{rel}).

S_{abs} can be calculated by the first derivative of the thermometric parameter (Δ) with respect to the temperature. It expresses the amount of thermally induced spectral or

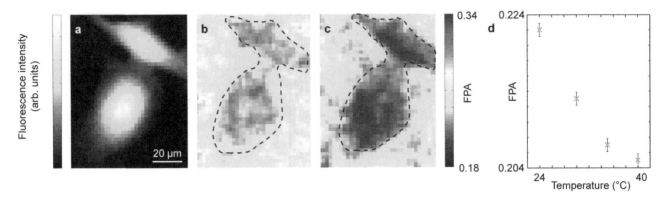

FIGURE 25.9 Luminescence polarization anisotropy measurements in HeLa cells transfected with GFP while heating the chamber temperature through resistive heating. (a) Intensity of the fluorescence of GFP after excitation at 473 nm. Polarization anisotropy (FPA) measured at (b) 296 K (23°C) and (c) 313 K (40°C). (d) Correlation between the measured polarization anisotropy of the intracellular GFP and temperature. (Reproduced from Donner, J. S., Thompson, S. A., Kreuzer, M. P., Baffou, G., Quidant, R. 2012. Mapping intracellular temperature using green fluorescent protein. *Nano Lett* 12: 2107–2111. © 2012 American Chemical Society.)

temporal changes of the luminescence nanothermometer with respect to the temperature. It can be expressed as

$$S_{\text{abs}} = \frac{\partial \Delta}{\partial T} \qquad (25.7)$$

The absolute thermal sensitivity is normally used for the comparison of luminescent nanothermometers that operates by the same mechanism, or for comparison of the material that constitutes the luminescent nanothermometer, but synthesized with a different geometry (size, shape, etc.) or with a different concentration of the luminescent center. It also depends on the experimental setup used for recording the thermometric parameter (Brites et al. 2016).

The S_{abs} in the lifetime technique has been traditionally represented by the normalized lifetime thermal coefficient (α_τ) given by the following equation (Haro-Gonzalez et al. 2012):

$$\alpha_\tau = \left| \frac{d\tau^{\text{norm}}(T)}{dT} \right| \qquad (25.8)$$

where $\tau^{\text{norm}}(T)$ is the luminescence decay time measured at a temperature T, normalized to the luminescence decay time measured at room temperature.

To facilitate the quantitative comparison among the different luminescent nanothermometers operating by different mechanisms, the use of S_{rel} has been proposed (Wade et al. 2003):

$$S_{\text{rel}} = \frac{1}{\Delta} \frac{\partial \Delta}{\partial T} \qquad (25.9)$$

The relative thermal sensitivity represents the relative change of the thermometric parameter per degree of temperature change and is usually expressed as % K^{-1}. In the last years, this parameter is becoming the most used figure of merit to compare the performance of the different luminescent nanothermometers, even if they operate by different mechanisms and are based on materials of different types. The reason for that is S_{rel} does not depend on

the nature of the nanothermometer and allows comparing the performance of the nanothermometers in a direct and quantitative way, becoming a very powerful technique when nanothermometers operating by different techniques want to be compared.

Temperature Resolution

Another important parameter that can be used to analyze the performance of the luminescence nanothermometers is the thermal resolution (δT), also called temperature uncertainty in the literature. This parameter provides information about the smallest resolvable temperature change that can be achieved by a particular luminescent nanothermometer. It depends on both the internal properties of the material that constitutes the luminescence nanothermometer, and the experimental detection setup and acquisition conditions used to record the luminescent thermometric parameter.

The thermal resolution can be determined experimentally from the evolution of the temporal fluctuation on the thermometric parameter. To do so, the temperature at each measurement point can be obtained using the previous established calibration curve. Then, the standard deviation of the resulting temperature histogram can be referred to as the thermal resolution of the luminescence nanothermometer (Savchuk et al. 2016a). Despite the high accuracy and reliability of this method, it is time-consuming.

The thermal resolution can also be calculated from S_{rel}, derived from the Taylor's series expansion of the temperature variation with the thermometric parameter and considering only the dominance of the first term (Brites et al. 2016):

$$\delta T = \frac{1}{S_{\text{rel}}} \frac{\delta \Delta}{\Delta} \qquad (25.10)$$

where $\delta \Delta$ is the standard deviation in the determination of the thermometric parameter. The $\frac{\delta \Delta}{\Delta}$ quotient depends on the experimental setup and the kind and integration time of the detector used to record the thermometric parameter. Thus, for instance, the thermal resolution increases when the signal arising from the luminescent material is low. So, it recommended the use of highly efficient luminescent materials, to work with high-performance detectors using the smallest possible integration time in order to get a faster temporal response of the thermometric sensor. For typical portable detection systems, $\frac{\delta\Delta}{\Delta}$ takes maximum values around 0.1%. This corresponds to relative thermal sensitivities of the order of 1–10% K^{-1} and temperature resolutions of the order of 0.01–0.1 K (Brites et al. 2012). This thermal resolution is similar to the one that can be achieved with thermistors (0.01 K) and better than that of noncontact infrared cameras (1.0 K), for instance. However, in general, detectors operate above that minimum value of $\frac{\delta\Delta}{\Delta}$, and the normal operation range is found in the range of 0.5–2.5%, generating worse thermal resolutions, but still in the subdegree range (Brites et al. 2016).

Spatial and Temporal Resolutions

Other relevant factors that are used to analyze the performance of a luminescence nanothermometer are the spatial (δx) and the temporal (δt) resolutions. The spatial resolution defines the minimum distance for which different temperatures can be discriminated. From its side, the temporal resolution refers to the minimum time interval for which different temperatures can be discriminated. Obviously, the values of the spatial and temporal resolutions have to be higher than those of the thermal resolution determined for that particular luminescence nanothermometer; otherwise; these temperature measurements will not be reliable.

Repeatability and Reproducibility

Repeatability is used to measure the variation among repeated measurements made under identical conditions. We can consider that a temperature measurement is repeatable when different measurements made over a certain period with the same setup or method coincide (Barlett and Frost 2008). Thus, repeatability describes the ability of the luminescent nanothermometer to generate the same results when operated under the same conditions during several heating–cooling cycles. It has to be ensured, however, that each measurement is performed when the luminescent thermometer is in thermal equilibrium with the reference temperature probe like a thermocouple, for instance.

From its side, reproducibility indicates the fluctuation of the temperature measured, determined from thermometric parameters recorded under different experimental conditions such as different equipment, measurement methods, observers, pH, or ionic strength of the surrounding environment, among others (Brites et al. 2016). Although this parameter cannot be quantified numerically, the experimenter can use statistical analysis to determine if different calibration procedures are significantly different.

25.3 Materials Used in Luminescence Nanothermometry

There exist several kinds of materials with potential applications such as luminescence nanothermometers, including QDs, organic dyes, metallic nanoparticles, luminescent polymers, and Ln^{3+}-doped dielectric nanoparticles. Organic dyes and QDs have shown the highest thermal sensitivities (Brites et al. 2012). However, most of them need to be excited using ultraviolet (UV) or visible light, thus preventing their use in biological applications due to the low penetration depth of these kinds of radiation in the biological tissues, and the background autofluorescence they can generate in this kind of samples (Sapsford et al. 2013). Ln^{3+}-doped upconversion nanoparticles (UCNPs), which emit light at higher photon energies when absorbing two or more lower energy excitation photons through sequential absorption or energy transfer processes, are a good option to overcome this situation. They can be excited in the near infrared (NIR), a radiation that causes negligible photodamage to living organisms and generates weak autofluorescence background. Also, NIR radiation exhibits deeper penetration depths in biological tissues for biomedical nanothermometry purposes when compared to visible radiation, for instance (Cheng et al. 2013).

In this section, we will briefly review the most common materials that have been used in luminescence nanothermometry.

25.3.1 Quantum Dots

A semiconductor that exhibits quantum confinement of the electrical charge carriers in the three dimensions of space due to the reduction of the particle size is called a quantum dot. This occurs when the particle size is equal to or less than the exciton size in the bulk crystal or the exciton Bohr radius (Banyai and Koch 1993). Such quantum confinement effects induce modified electronic and optical properties in the semiconductors, which are controllable to a certain degree through the flexibility in their structure design.

The fluorescence of QDs is strongly dependent on the particle size, but for a particular particle size, it is strongly dependent also on the temperature (Vlaskin et al. 2010, Li et al. 2007, McLaurin et al. 2013). This is due to several temperature-related effects, such as the thermal expansion of the crystalline lattice, temperature-induced changes in the confinement energy, and temperature-induced mechanical strength and electron–phonon coupling (Jaque and Solé 2016). With a temperature increase, the luminescence intensity decreases, a phenomenon known as luminescence quenching, which is also accompanied by a spectral shift of the position of the emission band (Dai et al. 2010), and also, in most of the cases, by a shortage of the luminescence decay time (Haro-Gonzalez et al. 2012). The luminescence quenching is attributed to thermally activated photoionization of the charge carriers (electrons and holes), so they become more delocalized and can reach a nearby trap non-fluorescent state. In QDs, the trap quencher states are normally associated with surface states, and consequently, the environment in which the QD is located influences tremendously its thermal luminescence quenching. To reduce the presence of surface trap states associated with defects and/or impurities in the surroundings of the QDs, core–shell structures can be used.

Intensity-based QDs luminescence nanothermometers often induce errors in temperature determination due to variations in the concentration of the thermal probes and their blinking, fluctuations in the excitation source, or reduced detection efficiency (McLaurin et al. 2013). These problems can be overcome by using dual emitting QDs systems and correlating the intensity of the two emission bands by a ratiometric technique (Vlaskin et al. 2010). Another important drawback of the use of QDs in luminescence thermometry is the hysteresis observed in the thermometric response observed during heating and cooling cycles if a certain temperature threshold is reached due to the creation of permanent trap states (Zhao et al. 2012). Also, due to the complexity of the thermal quenching processes in QDs, and to their dependence, not only on whether the QD is constituted by a single layer or a multilayered structure, but also on the environment hosting them, the calibration curve for thermal sensing purposes could be modified during sensing when changes in the QD environment occur (Jaque and Solé 2016). In some cases, even, especially in bare QDs, an increase in temperature might induce the spontaneous aggregation of QDs, generating a size-induced red shift of the position of their emission peak which makes that the observed spectral shift cannot be univocally related to a temperature change (Khlebtsov and Dykman 2011). Another important limitation of using QDs for luminescence thermometry is that sometimes the peak emission wavelength observed changes from dot to dot, due to the existence of a dot size distribution (Li et al. 2007). This limits the use of QDs to provide measurements of the absolute temperature, since different dot sizes could lead to different thermally induced spectral shifts, unless measurements are performed on a relatively large number of QDs (of the order of 1,200 if a temperature resolution of 1 K is regarded). This requirement also restricts the spatial resolution that can be achieved with these QDs, since it would be the area in which this number of QDs is present. These limitations might be overcome if QD suspensions with a narrower size distribution are used. Also, if the concentration of QDs changes from point to point, or with time, this could cause variations in the intensity of the emission not related to temperature. This would induce erroneous temperature measurements. To mitigate these errors, complex luminescence structures have been developed, comprising a QD and another luminescent reference, that can be another semiconductor nanostructure (Albers et al. 2012), or a metallic nanoparticle connected to the QD by a polymer acting as a molecular spring (Lee and Kotov 2007). A final point to be

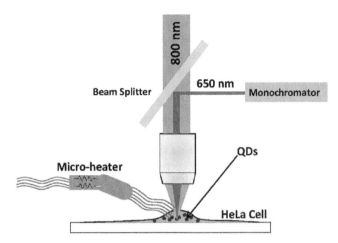

FIGURE 25.10 Experimental setup used to monitor the intracellular temperature by using the two-photon emission of internalized CdSe QDs in HeLa cells, when excited at 800 nm, emitting at 650 nm. (Adapted with permission from Maestro, L.M., Rodriguez, E.M., Cruz, M.C.I. et al. 2010. CdSe quantum dots for two-photon fluorescence thermal imaging. *Nano Lett.* 10: 5109–5115. © 2010 American Chemical Society.)

considered when QDs are used for biomedical applications is their possible toxicity, and the potential photodamage that might be generated when pumped in the UV.

QDs have been used as luminescent nanothermometers, for instance, to determine the intracellular temperature using the temperature-induced spectral shift in CdSe QDs (Maestro et al. 2010). By using a two-photon excitation scheme, a spatial resolution of 400 nm was achieved, six times larger than the one that can be obtained by one-photon excitation, due to the nonlinear nature of this procedure. CdSe QDs, internalized previously in HeLa cells, were excited with a 800 nm excitation beam that was focused by means of a 100× microscope objective. The same objective was used to collect the emission generated by the QDs, while the cell temperature was externally modified by using a micro-heater. Figure 25.10a shows the schematic diagram of the experimental setup used to monitor the intracellular heating, induced by an external micro-heater, using the spectral shift of CdSe QDs.

25.3.2 Organic Dyes

Organic dyes are aromatic organic compounds which impart color to a substrate by the selective absorption of light (Duarte 2012). These compounds exhibit a strong luminescence in the visible region when excited with UV or blue light. Their emission and absorption properties depend on their structure and chemical environment (Zollinger 2003) and, in many organic dyes, also on temperature (Sirohi 2016). These characteristics make organic dyes excellent candidates for luminescence nanothermometers.

In fact, the luminescence intensity and the emission lifetimes of organic dyes are typical parameters that are affected when the temperature changes (Ebert et al. 2007, Shah et al. 2009, Estrada-Perez et al. 2011). However, only a few organic dyes can be used in luminescence thermometry, since only those with important changes in their luminescence spectra, good photostability, and high enough luminescence quantum yields at different temperatures can be used for these purposes (Yang et al. 2016). Also, since the luminescence intensity of organic dyes is seriously affected by fluctuations in the experimental conditions due to bleaching effects, sample displacement, dye concentration, and/or light source intensity, a more common and reliable technique for temperature measurements with these materials is lifetime-based nanothermometry (Itoh et al. 2016).

Rhodamine B (RhB) is the most popular organic dye used for luminescence nanothermometry purposes, due to its good chemical stability and high luminescence efficiency. Thus, its temperature-dependent luminescence response has been extensively studied (Shah et al. 2009). Together with fluorescein, pyranine, 7-nitrobenz-2-oxa-1,3-diazol-4-yl (NBD), and 6-dodecanoyl-2-dimethylamino-naphthalene (Laurdan) have been the organic dyes most used in aqueous environments for luminescence nanothermometry, while triarylboron compounds and bispyren dyes have been used in organic solvents (Yang et al. 2016).

However, the use of free organic dyes is limited by a number of factors, like adsorption onto surfaces, molecular interactions and contamination of the surrounding material (Karstens and Kobs 1980). To solve these inconvenience, there has been significant interest in encapsulating these organic dyes. For instance, Jung et al. incorporated RhB in polydimethylsiloxane (PDMS) and SU8 polymer matrices to determine the temperature through an active layer covering a microfluidic device (Jung et al. 2011). Other examples include the use of silica matrices to encapsulate the organic dyes (Duong and Rhee 2007). Another factor that has to be taken also into account when dealing with the emission properties of organic dyes is the photobleaching effect, i.e., the photochemical alteration of the dye that reduces their luminescence quantum yield, which is a common problem with organic dyes. Also, fluctuations in the excitation source might affect the intensity of the emission generated by the dye that can be overcome if a reference dye is used, with a luminescence intensity that does not change significantly with temperature, like Rhodamine 110, for instance, or the use of a dye with a dual emission (Chandrasekharan and Kelly 2001, Schoof and Gusten 1989, Feryforgues et al. 1993).

As an example of the performance of organic dyes as luminescent thermometers, Ross et al. used RhB for mapping the temperature inside a microfluidic system (Ross et al. 2001). They used different microfluidic channel configurations, like the one with a T-shaped imprinted in acrylic that can be seen in Figure 25.11a. The pumping of the fluid containing RhB was done electrokinetically, by applying an electric field of 2,500 V along the length of the microchannel that resulted also in an increase of the temperature of the liquid by Joule heating. By changing the sign of the bias applied, the flow direction in the microchannel could be reserved. Fluorescence imaging of RhB was performed using a fluorescence microscope equipped with a Hg arc lamp for excitation at 500–550 nm and a CCD camera for detection. The measurements involved comparing the fluorescence intensity images of RhB recorded at a known and uniform temperature with those recorded at an unknown temperature. The resultant temperature distributions can be seen in Figure 25.11b and c, depending on the direction of the flow, with a temperature

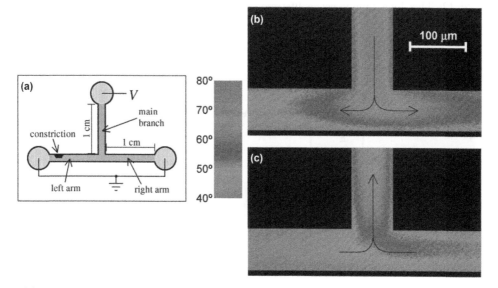

FIGURE 25.11 (a) Schematic diagram of a T-shaped microchannel used for thermal mapping with RhB using intensity-based luminescence nanothermometry. (b) and (c) Mappings of the thermal gradients observed in the fluid when pumped electrokinetically by applying a voltage of ±2,500 V, respectively, that reversed the flow direction. (Adapted with permission from Ross, D., Gaitan, M., Locascio, L.E. 2001. Temperature measurement in microfluidic systems using a temperature-dependent fluorescent dye. *Anal. Chem.* 73: 4117–4123. © 2001 American Chemical Society.)

resolution between 0.03 and 3.5 K in the range from room temperature to 363 K (90°C), a spatial resolution of 1 μm, and a temporal resolution of 33 ms.

25.3.3 Metal Nanoparticles

Metal nanoparticles, and in particular gold nanoparticles, with good biocompatibility, facile conjugation to biomolecules and fascinating luminescent properties conferred by the surface plasmon resonances they exhibit, have been used as fluorescent probes for *in vivo* and *in vitro* imaging (Lomenie et al. 2001). Moreover, due to its high photothermal conversion efficiency, they have also been investigated as photothermal agents for photothermal therapy (Goodrich et al. 2010). Also, since they present a thermosensitive luminescence response that can be evidenced by an evident luminescence quenching due to the more effective nonradiative recombination of electrons and holes as the temperature increases, they have been used as luminescence nanothermometers (Bomm et al. 2012).

In this context, they have been used to sense the intracellular temperature in HeLa cells, taking advantage of the temperature dependence of their luminescence lifetime, which changes considerably over the physiological temperature range (Shang et al. 2013). The Au nanoparticles were internalized into the HeLa cells by a simple endocytosis process, and then, their temperature was increased through a temperature-controlled stage. A thermal resolution around 0.3–0.5 K in the range of 287 K (14°C)–316 K (43°C) was achieved from the thermal response of the lifetime value of the Au nanoclusters, with little variation of the lifetime values under continuous excitations with intensities up to 2.8 kW cm^{-1} during 2 h, as can be seen in Figure 25.12. This long luminescence lifetime, of the order of 670–970 nm, could be easily separated from the autofluorescence background generated by the biological

sample under excitation at 580 nm. Furthermore, a complete reversibility, long-term stability, and minimum cell toxicity were observed for these Au nanoclusters.

However, the luminescence properties arising from Au nanoparticles are also sensitive to other local environmental parameters, including oxygen content, pH, and their concentration, which might result in inaccurate temperature measurements. To overcome these inconvenience, complex systems combining the Au nanoparticles with other luminescent materials to develop a ratiometric luminescence nanothermometer have been used, like carbon dots, encapsulated together with Au nanoparticles in TiO$_2$ microspheres (Wang et al. 2015).

25.3.4 Polymer-Based Systems

The phase transitions in polymers can be used as a response for sensing temperature, when they incorporate solvatochromic dyes that change their luminescence properties in response to changes in the local environment (Stich et al. 2010). Thus, luminescent nanothermometers based on polymers are formed by a thermoresponsive polymer that experiences a sharp phase transition when the temperature changes, and a chromophore that generates a signal that can be detected and correlated to the temperature. These thermoresponsive polymers pass from a hydrophilic molecularly dissolved state to a more hydrophobic dehydrated collapsed globule state as a response to changes in temperature (Roy et al. 2013). These changes in the structure of the polymer induce variations in the absorption and/or emission properties of the chromophore that can be detected and correlated to temperature, since solvatochromic dyes change their luminescence properties as a function of the environment polarity. There are three phenomena, occurring during the phase transition experienced by the polymers, which induce the changes in the luminescence properties of the dyes, that can be used to sense temperature (Vancoillie et al. 2016).

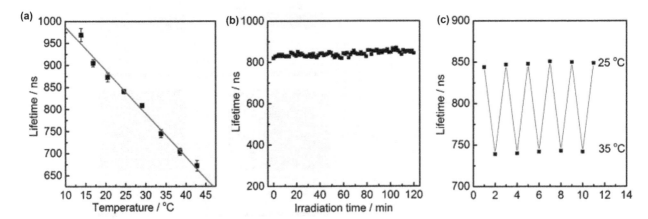

FIGURE 25.12   (a) Evolution of the fluorescence lifetime of Au nanoclusters with temperature incorporated in HeLa cells, after excitation at 580 nm, and monitoring the emission of the nanoclusters at 710 nm. Fluorescence lifetime of the Au nanoclusters (b) as a function of the irradiation time, and (c) recorded in heating and cooling cycles between 298 K (25°C) and 308 K (35°C). (Adapted with permission from Shang, L., Stockmar, F., Azadfar, N., Nienhaus, G.U. 2013. Intracellular thermometry by using fluorescent gold nanoclusters. *Angew. Chem. Int. Ed.* 52: 11154–11157. © 2013 Wiley-VCG Verlag GmbH.)

The first one is the strong dehydration of the internal part of the collapsed polymeric globules during the phase transition, inducing a decrease in the polarity of the environment. This change in polarity generates an uneven stabilization of the ground state or the excited state of the dye and, thus, a change in the energy difference between these two levels. That change can be observed in a shift in the position of the absorption or emission bands of the dye, or changes in their intensity and lifetime. An example of a luminescence nanothermometer based on this effect is that of poly(N-isopropylacrylamide) (PNIPAM) incorporating a benzoxadiazole dye (Uchiyama et al. 2012). PNIPAM is the most used thermoresponsive polymer in an aqueous environment for luminescence nanothermometry, exhibiting a phase transition at 305 K (32°C). Since this temperature is close to the human corporal temperature, and PNIPAM has an extremely low toxicity, it has been used as a luminescence nanothermometer for biomedical applications.

The second phenomenon is the decrease of the inter- and intradistances between the polymer backbone and the side-chains during the phase transition. These changes in distances can be used to play with the Förster resonance energy transfer (FRET), by incorporating two fluorophores into the same thermoresponsive polymer chain. FRET is based on a nonradiative energy transfer from an excited donor to a fluorophore acceptor that subsequently emits the transferred energy in the form of light. This nonradiative energy transfer is effective only at short distances and, thus, depends substantially on the distance between the two moieties. Thus, when the intradistances between the polymer backbone and the side-chains change during the phase transition, the FRET effect can be activated. For example, such a kind of thermometer can be made by a phenanthrene end-functionalized poly(N-decylacrylamide-b-N,N-diethylacrylamide) block copolymer with anthracene that worked by exciting phenanthrene and measuring the emission of anthracene (Prazeres et al. 2010).

Finally, the third phenomenon induced is the reduction of the mobility of the side-chains after the phase transition. This effect has been used, for instance, in PNIPAM functionalized with tetraphenylethene dye. When the polymer was precipitated, the rotation of the phenyl rings around the central double bond was restricted, which increased the intensity of the emission of the dye (Tang et al. 2009).

The formation of excimers, i.e., the formation of dimers in the excited states, can also be used for luminescence nanothermometry in polymer-based systems. Yang et al. used the temperature dependence on the intensity ratio of the fluorescence of the monomer and the excimer in the ladder-like 1,4-phenylene-bridged polyvinylsiloxane (LPPVS) for this purpose (Yang and Bai 2002). The excimer–to–monomer intensity ratio of this system exhibited a linear Arrhenius plot versus the inverse of the temperature, with a positive slope that could be used in temperature-sensing applications.

However, the main disadvantage of polymer-based luminescence nanothermometers is that they can only be used

FIGURE 25.13 Observed color shift of aqueous solutions of a copolymer formed by PEGMA and DR1 at different pH values when heated above the cloud point of the polymer (290.5 K, 17.3°C). At pH = 7, no color change was observed, although the intensity of the color increased significantly in the precipitated state. Instead, at pH = 1, a bathochromic shift was observed from 491 to 532 nm, indicating a polarity change around the dye upon copolymer precipitation. (Reproduced with permission from Pietsch, C., Hoogenboom, R., Schubert, U.S. 2009. Soluble polymeric dual sensor for temperature and pH value. *Angew. Chem. Int.* Ed. 48: 5653–5656. © 2009 Wiley-VCG Verlag GmbH.)

in a narrow temperature range of typically around 10–20 K around their phase transitions. By using polymers with a higher critical solution temperature, a broader temperature range can be explored, since they show a broader transition range. However, such polymers are quite costly (Pietsch et al. 2010a).

Despite these limitations, polymer-based luminescence nanothermometers have been used to detect the internal temperature in living cells (Gota et al. 2009), as dual sensors for temperature and pH (Pietsch et al. 2009), and as AND logic gates for developing molecular memory systems (Uchiyama et al. 2004); and for temperature and metallic ions sensing, with interest in pollution environment (Pietsch et al. 2010b). Figure 25.13 shows how a dual sensor works for temperature and pH based on polymers consisting of poly(oligoethyleneglycol methacrylate (POEGMA) to which a pH-responsive solvatochromic dye (disperse red 1, DR1) was attached. By heating aqueous solutions of this copolymer above the polymer cloud point, a change of color of the solution could be observed as a function of the pH (Pietsch et al. 2009).

25.3.5 Lanthanide-Doped Nanoparticles

Lanthanide (Ln) elements are characterized by the association of their valence electrons to the 4f orbitals, which are shielded from the environment by the outer filled 5s, and 5p orbitals, depending on the element considered, which prevent from a strong interaction of the environment. Thus, these ions can be described accurately by the free ion approximation (Henderson and Imbush 2006). The electric field created by the electrons and the effect of the spin–orbit coupling split the 4f orbital in several energy positions that the electrons will occupy. However, from Laporte's selection rule, the optical transitions between 4f states are forbidden, and they are only partially allowed when the lanthanides are incorporated in a crystalline matrix. This also makes their emissions narrow spectral lines with relatively long luminescence lifetimes (from μs to ms). The energy states on Ln ions are labeled according to the total orbital momentum (L), the total spin momentum (S), and the total angular momentum (J) of the state, respectively, according to the expression $^{2S+1}L_J$. Lanthanide ions in its most common valence state (Ln^{3+}) show emissions that lie from the ultraviolet (UV) to the near-infrared (NIR) ranges of the electromagnetic spectrum, covering also the whole visible range. The intensity of these luminescence lines depends on several parameters, among which the most critical one is the temperature.

The earliest Ln^{3+}-doped materials used for temperature-sensing applications were the thermographic phosphors (Alden et al. 2011) and the temperature-sensitive paints (Brübach et al. 2013) with applications in flame spread studies in fires, decomposition of materials, or the temperature distribution in internal combustion and aircraft engines, to name a few. In these cases, the excitation was usually performed with UV or visible light, and basically temporal and spectral methods were used to determine the temperature, allowing for discrete and two-dimensional measurements.

Another characteristic of Ln^{3+} ions is that several of them have electronic excited states that are defined by the same or a very close energy value. This match of energy allows the transfer of energy from one ion to another as long as in one of them the excited state is populated while in the other is not, and will use this energy transfer process to promote an electron from the ground state to the excited state, from which light can be emitted. These energy transfer processes can occur between Ln^{3+} of the same kind (energy migration). They can also occur between lanthanide ions of different kinds, where the ion that absorbs the energy and transfers it to the other one is called the sensitizer, and the one that gets this energy and generates the final emission is called the activator. Also, the same process can occur between populated excited states if the lifetime of the starting state is long enough to allow these processes to occur. In these cases in which the originally populated state of the activator is not the ground state, it will end up with a populated electronic state that has a higher energy than any of the starting states. Thus, a radiative relaxation from this excited state to the ground state will generate a photon of higher energy than the excitation photons, a process known as upconversion, and the mechanism that generates it is known as energy transfer upconversion (ETU) and is represented in Figure 25.14a.

Upconversion refers to nonlinear optical processes in which the sequential absorption of two or more electrons leads to the emission of light at shorter wavelengths than the excitation wavelength. Apart from the ETU process described above, this upconversion process involves other different mechanisms, mainly the ground state absorption (GSA), resulting in the promotion of electrons in an ion from its ground state (G) to an excited state (E_1); and the excited

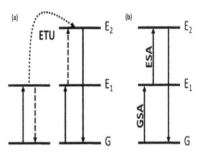

FIGURE 25.14 Upconversion mechanisms: (a) energy transfer upconversion (ETU) and (b) ground state absorption (GSA) followed by the excited state absorption (ESA). Upward arrows represent the excitation of an electron to a higher energy level, downward arrows represent radiative relaxations, dotted downward arrows represent nonradiative relaxation processes, dotted upward arrows represent the indirect excitation of an electron, and dotted curved arrows represent the energy transfer processes.

state absorption (ESA), resulting in the absorption of a photon by electrons of an ion that are already in an excited state, and their promotion to an even higher energy excited state (E_2), as is schematically represented in Figure 25.14b. However, these mechanisms are not as efficient as the ETU process, mainly because the second photon to be absorbed must be captured by an electron that populates already an intermediate energy state, where the electronic population is usually rather low. Important requirements to observe the phenomenon of upconversion are the existence of long radiative lifetimes of the excited states and a ladder-like arrangement of the electronic levels of the ion with similar energy separations, which can be found only in certain ions of the d and f elements (Bloembergen 1959). The properties of the host matrix and its interaction with the lanthanide ions have also a strong influence on the upconversion process. Thus, host matrices with low phonon energies, good chemical stability, and low lattice impurities and defects would be the perfect choice to obtain efficient upconversion processes.

Upconversion phenomena are of particular interest for luminescence nanothermometry, since through these processes, materials absorb light in the NIR region of the electromagnetic spectrum and emit light in the visible (Haase and Schäfer 2011). Thus, pumping in the NIR allows overcoming problems related to the background autofluorescence arising from biological tissues and the potential damage that UV light can generate in them. Moreover, the NIR lasers used to excite these upconversion nanoparticles are cheaper and more powerful than UV lasers. At the same time, NIR radiation preserves the operational lifetime of the phosphors used, in comparison with those illuminated with UV light.

There are plenty of examples in the literature about the use of Ln^{3+}-doped upconversion nanoparticles for luminescence nanothermometry applications. Er^{3+} is the most used Ln^{3+} ion for these purposes because of its very intense green emission consisting of two bands centered at 520 and 540 nm, arising from two thermally coupled electronic levels ($^4S_{3/2}$ and $^2H_{11/2}$, respectively), whose relative intensities are strongly temperature dependent (Fischer et al. 2011). Normally, it is sensitized by Yb^{3+}, because its absorption cross section at 980 nm is larger than that of the other Ln^{3+} ions in this region. Yb^{3+} can efficiently transfer the absorbed energy not only to Er^{3+} but also to other Ln^{3+} ions like Tm^{3+} and Ho^{3+}, with almost resonant excited energy levels. Since Yb^{3+} presents only one excited energy level in its electronic structure, the possibility of energy losses by the effect of nonradiative processes is reduced.

The use of the emissions from the two thermally coupled electronic levels $^4S_{3/2}$ and $^2H_{11/2}$ of Er^{3+} for luminescence thermometry started in 1990, when Berthout and Jörgensen compared the thermal response of different Er^{3+}-doped fluoride matrices with a fiber optics shape (Berthou and Jörgensen 1990). In a more recent example, Vetrone et al. internalized Er^{3+}, Yb^{3+}-co-doped $NaYF_4$ nanoparticles in HeLa cells and monitored their internal temperature until their death, induced by external heating (Vetrone et al. 2010). The intracellular temperature was measured by analyzing the intensity ratio between the Er^{3+} emission lines at 525 and 545 nm, arising from the $^4S_{3/2}$ and $^2H_{11/2}$ energy levels, after the excitation of Yb^{3+} at 980 nm and its efficient energy transfer upconversion to Er^{3+}. Figure 25.15 shows how this intensity ratio changed with the applied voltage that provided the heat through a metallic platform connected to a resistor and that permitted the measurement of the intracellular temperature. Figure 25.15 also shows three optical transmission images of an individual HeLa cell at three different temperatures, showing how at 318 K (45°C) the cell is destroyed.

When operating with Ln^{3+} ions, we have to take also into account if the luminescence under analysis for luminescence nanothermometry purposes is generated by a single type of lanthanide ion, or if it is generated by a combination of different lanthanide ions, as we pointed out in Section 25.2.1.2.

The examples showed in this section up to now consisted of single-center Ln^{3+}-based luminescence nanothermometry. Despite these luminescence nanothermometers based on ratiometric systems from the emission intensity arising from thermally coupled energy levels proved to be effective for temperature determination, they still suffer from a low relative thermal sensitivity. One of the ways of increasing this thermal sensitivity would be the use of lanthanide ions with thermally coupled energy levels located at larger energy differences. However, if the distance between these energy levels is too large, then thermalization is no longer observed, and even the electronic population, and hence the emission intensity of the upper level, will decrease, introducing difficulties in detecting the emission arising from it (Wade et al. 2003).

Another way to increase the thermal sensitivity of the Ln^{3+}-based luminescence nanothermometers is the use of multi-center systems. These systems are based on the incorporation of two different Ln^{3+} ions, both as emitters, in a single nanoparticle, whose emission intensities, or at least that arising from one of them, are strongly dependent on temperature (Jaque and Vetrone 2012). It has been shown that such systems are excellent candidates for noncontact luminescence nanothermometers with high thermal sensitivities. Figure 25.16 shows the S_{rel} and δT, almost 3% K^{-1} and 0.2 K, that can be achieved when optimizing the concentrations of Ho^{3+} and Tm^{3+} in a single host, $KLu(WO_4)_2$ nanoparticles. In this case, the emissions of these nanoparticles at 696 and 755 nm were used for luminescence nanothermometry, corresponding to the

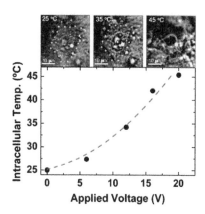

FIGURE 25.15 Optical transmission images of an individual HeLa cell at three different temperatures, and intracellular temperature as a function of the applied voltage in a metallic platform connected to a resistor in contact with the cells. The intracellular temperature was determined using the intensity ratio between the 525 and 545 nm emission lines of Er^{3+}, originated from the $^4S_{3/2} \rightarrow {}^4I_{15/2}$ and $^2H_{11/2} \rightarrow {}^4I_{15/2}$ electronic transitions of Er^{3+}, after the excitation of Yb^{3+} at 980 nm and the efficient energy transfer upconversion to Er^{3+} in Er^{3+},Yb^{3+}-co-doped $NaYF_4$ nanoparticles. (Reproduced from Vetrone, F., Naccache, R., Zamarrn, A. et al. 2010. Temperature sensing using fluorescent nanothermometers. *ACS Nano* 4: 3254–3258. © 2010 American Chemical Society.)

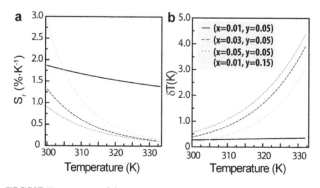

FIGURE 25.16 (a) Relative thermal sensitivity and (b) temperature uncertainty in the 300–330 K range calculated for $KLu_{1-x-y}Ho_xTm_y(WO_4)_2$ nanoparticles, by using their emissions at 696 and 755 nm generated through an upconversion process after pumping at 808 nm. (Reproduced from Savchuk, O.A., Carvajal, J.J., Brites, C.D.S., Carlos, L.D., Aguilo, M., Diaz, F. 2018. Upconversion thermometry: a new tool to measure the thermal resistance of nanoparticles. *Nanoscale* 10: 6602–6610. © 2018 The Royal Society of Chemistry.)

$^3F_{2,3} \rightarrow {}^3H_6$ and $^5S_2$, $^5F_4 \rightarrow {}^5I_7$ electronic transitions of Tm^{3+} and Ho^{3+}, respectively, generated by an upconversion process after pumping at 808 nm.

Primary Luminescence Nanothermometers

All the luminescence nanothermometers described up to now require a calibration procedure that at the same time involves the measurement of the temperature by an external and independent measurement, using, for instance, a thermocouple that allows for a correlation between the reading of the luminescence thermometer and the temperature. This is the definition of a secondary thermometer. This requires that every time that the luminescence nanothermometer is used in a different environment, the calibration has to be repeated, since parameters like ionic strength, pH, pressure, local surrounding of the luminophore, or atmosphere composition might alter the value of the thermometric parameter. Nevertheless, performing multiple calibration in different media is a time-consuming procedure, and it is not always possible to be implemented, for instance, in living cells. Thus, in general, a unique calibration is performed, and it is assumed to be valid in any media. Therefore, it is of paramount importance to develop luminescence primary thermometers, in which the thermometric parameter corresponds univocally to the absolute temperature through an equation of state, and they do not require any calibration (Bakabhadra 2017). Up to now, however, only a few luminescence primary thermometers have been reported, most of them based on Ln^{3+}-doped nanoparticles (Souza et al. 2016). There are other examples developed using QDs (Pugh-Thomas et al. 2011) and Si nanoparticles (Botas et al. 2016), by using the Varshni's law that correlates the energy bandgap of a semiconductor with temperature.

Balabhadra and co-workers developed a method to predict the temperature calibration curve of any luminescence nanothermometer based on upconversion procedures working through two thermally coupled electronic levels (Balabhadra et al. 2017). Such method is independent on the medium surround the luminescence nanothermometer, and thus, they demonstrated that these luminescence nanothermometers are intrinsically primary thermometers.

Since this method is based on the comparison of the emissions arising from two thermally coupled electronic levels, the FIR (LIR) method can be used (see Eq. 1). Analogously, the authors defined the FIR_0 (LIR_0) thermometric parameter that corresponds to the temperature T_0 corresponding to the limit of zero pump power, i.e., no laser heating:

$$FIR_0\,(LIR_0) = B exp\left(\frac{-\Delta E}{k_B T_0}\right) \qquad (25.11)$$

By determining the quotient between Eq. (25.1) and (25.11), the absolute temperature can be determined, independently of the medium (Balabhadra et al. 2017):

$$\frac{1}{T} = \frac{1}{T_0} - \frac{k_B}{\Delta E} ln\left(\frac{FIR\,(LIR)}{FIR_0\,(LIR_0)}\right) \qquad (25.12)$$

Thus, applying this equation, the authors were able to determine the absolute temperature from the experimental values of FIR (LIR), without the need to perform any previous calibration, demonstrating that these upconversion nanoparticles can work as a primary nanothermometer.

25.3.6 Transition Metal-Doped Nanoparticles

Although the luminescence arising from transition metal emitting centers is more severely quenched by an increase in temperature than that of Ln^{3+}, it can also be used for luminescence nanothermometry purposes. In fact, the use of transition metal-based materials for luminescence thermometry began earlier than that of Ln^{3+} in applications such as thermometers for cryobaric diamond anvil cells using ruby (Cr^{3+}-doped Al_2O_3 (Weinstein 1986) or thermographic phosphors using Cr^{3+}-doped $YAl_3(BO_3)_4$ (Borisov et al. 2010), despite not considering the possibility of temperature readings at the nanoscale. The interest of Cr^{3+} for these purposes is the use of its R lines, composed of two strong emission lines in the red, and due to electronic transitions between the ^2E excited and the ^1E ground states, whose intensity ratio can be used for luminescence thermometry. Similarly, Mn^{4+} in Mg_4FGeO_6, also with two emission bands in the red, can be used for the same purpose (Omrane et al. 2004). More recently, other Cr^{3+}-doped materials have been developed with the same purposes, like Cr^{3+}:$Bi_2Ga_4O_9$ mullite powders with emissions in the near infrared (Back et al. 2016).

Such ions belong to the iron group, in the fourth period of the periodic table. When embedded in solids, they form 3d transition ion centers that can generate several optical transitions originating from the multiple energy levels created by the incompletely filled d shells. 3d electrons are located outside the ion cores and thus suffer strong effects from the surrounding ions, so that their spectroscopic characteristics are a consequence of the d^n electronic configuration and the crystal field potential created by the surrounding ligands (Dramicanin 2018).

However, all these materials suffer from a low S_{rel} and also from the difficulty in determining the R lines in some host materials, particularly those exhibiting broad emission bands (Dramicanin 2018).

To solve these problems, materials with dual emission centers can be used. One of the ways of doing that is by developing transition metal-doped nanoparticles that contain also intrinsic defects that will generate a second emission, like those observed in Mn^{2+}:Zn_2SiO_4, with a green emission highly sensitive to temperature originating from Mn^{2+}, and an emission in the blue almost insensitive to temperature originating from traps in the host (Lojpur et al. 2013). Figure 25.17 shows the evolution of these two bands with temperature to illustrate the use of these transition metal-doped nanoparticles in luminescence nanothermometry.

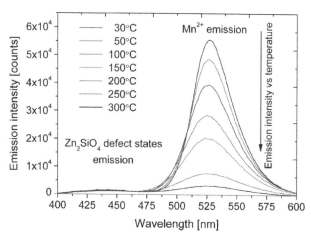

FIGURE 25.17 Emission spectra of Mn^{2+}-doped Zn_2SiO_4 powders at different temperatures recorded after excitation at 370 nm. It can be clearly seen how the Mn^{2+} emission is very sensitive to temperature changes, while the emission originating from defect states in Zn_2SiO_4 is almost insensitive to temperature. (Reproduced from Lojpur, V., Nikolic, M.G., Jovanovic, D., Medic, M., Antic, Z., Dramicanin, M.D. 2013. Luminescene thermometry with $Zn_2SiO_4:Mn^{2+}$ powder. *Appl. Phys. Lett.* 103: 141912. © 2013 AIP Publishing.)

Another option to improve the performance of transition metal-doped materials is co-doping them with Ln^{3+} ions, by combining, for instance, the highly temperature-sensitive emissions of Mn^{4+} or Cr^{3+} with those almost temperature insensitive originating from Eu^{3+}, Tb^{3+}, or Dy^{3+} (Chen et al. 2016).

25.3.7 Metal-Organic Frameworks

Metal-organic frameworks (MOFs) are solid crystalline materials formed by metal ions or clusters coordinated with organic ligands containing potential voids, and thus configuring a potential porous structure. In fact, they can be considered as a subclass of coordination polymers. Although traditionally MOFs have been of interest for the storage of gases, catalysis, and supercapacitors (Farrusseng 2011), they have also been used as luminescence nanothermometer, especially when the central metal is a lanthanide ion (Brites et al. 2016). The thermometric mechanism in this case is based on an energy transfer process between ions within the polymeric framework, through which the emission from the metal ion is sensitized by the organic ligands. They offer a clear advantage over molecular thermometers; that is, the ordered structure they present allows obtaining a highly reproducible thermal sensing response. However, their working range is determined by the triplet energy state of the organic ligands, which corresponds normally to the range from low temperatures to physiological temperatures (10–330 K), and their S_{rel} are in general smaller than those that can be obtained with Ln^{3+}-doped nanoparticles.

Cadiau et al. (2013) developed a nanometric MOF consisting of a mixture of Eu^{3+} and Tb^{3+} at a ratio of 1:100,

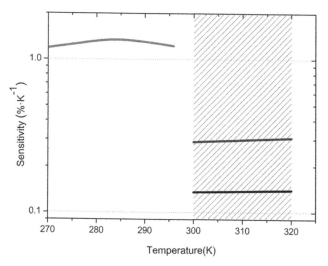

FIGURE 25.18 Relative thermal sensitivity calculated for the luminescence nanothermometer based on the comparison of the intensities of the $^5D_4 \rightarrow {}^7F_5$ emission of Tb^{3+} and the $^5D_0 \rightarrow {}^7F_2$ emission of Eu^{3+} in $Tb_{0.99}Eu_{0.01}(BDC)_{1.5}(H_2O)_2$ MOF nanorods in a solid form and in an aqueous suspension. The S_{rel} of the luminescence thermometer based on $(Eu_{0.0069}Tb_{0.9931})_2(DMBDC)_3(H_2O)_4 \cdot DMF \cdot H_2O$ and operating under the same principle is included for comparison, with a better performance when operating below 300 K. (Reproduced with permission from Cadiau, A., Brites, C.D.S., Costa, P.M.F.J., Ferreira, R.A.S., Rocha, J., Carlos, L.D. 2013. Ratiometric nanothermometer based on an emissive Ln^{3+} organic framework. *ACS Nano* 7: 7213–7218. © 2013 American Chemical Society.)

coordinated with 1,4-benzenedicarboxylate (BDC), with a shape of nanorods. This $Tb_{0.99}Eu_{0.01}(BDC)_{1.5}(H_2O)_2$ MOF was able to operate in the physiological range of temperatures as a ratiometric luminescence nanothermometer in aqueous suspension, using the $^5D_4 \rightarrow {}^7F_5$ green emission of Tb^{3+} and the $^5D_0 \rightarrow {}^7F_2$ red emission of Eu^{3+} generated by the energy transfer between these two ions after excitation at 320 nm with a Xe lamp. Importantly, its S_{rel} increased substantially when the nanorods were dispersed in water, as shown in Figure 25.18, being almost constant in the range of physiological temperatures. This increase of S_{rel} in an aqueous solution is due to a stronger quenching of the red emission of Eu^{3+} due to changes in the surrounding medium polarizability. In the figure, S_{rel} corresponding to $(Eu_{0.0069}Tb_{0.9931})_2(DMBDC)_3(H_2O)_4 \cdot DMF \cdot H_2O$, that worked through the same ratiometric parametric, is also included, since it is significantly higher when operating below 300 K.

25.3.8 Carbon-Based Materials: Nanodiamonds and Carbon Dots

Carbon-based materials, basically nanodiamonds and carbon dots, have also been used as luminescent nanothermometers with biomedical applications.

In the case of nanodiamonds, their operational principle is based on the temperature-dependent lattice strain resulting in the changes of the quantum mechanical spin properties of nitrogen color centers generated by nitrogen vacancies (Kucsko et al. 2013). This technique allowed to detect temperature variations as small as 1.8 mK. Nanodiamonds have other advantages as being chemically inert, exhibiting an excellent thermal conductivity that ensures that all defect centers within a nanocrystal are in thermal equilibrium with the local heat environment and that they can be used over a wide temperature range, extending from 200 to 600 K. However, the concept used to correlate the emitting properties of nanodiamonds with temperature was quite complex, and it required also a complex setup to determine the temperature. In fact, in its electronic ground state, each nitrogen vacancy constitutes a spin-1 system. By using microwave pulses, it is possible to coherently manipulate the spin states of these defects and detect them by laser illumination. If no external magnetic field is applied, the value of the transition frequency (Δ) between the 0 and ± 1 spin states can be determined from which the temperature sensitivity (μ) can be calculated according to the following equation (Kucsko et al. 2013):

$$\mu = \frac{1}{C\frac{d\Delta}{dT}} \frac{1}{\sqrt{T_{\mathrm{coh}}Nt}} \qquad (25.13)$$

where T_{coh} is the nitrogen vacancy coherent time, N is the number of color centers in the nanodiamond, t is the integration time, and C is a factor to account for imperfections in readout and initialization of the color centers. But to do that, it is necessary to decouple the nitrogen vacancy electronic spin from fluctuating external magnetic fields. To do that, the authors used a modified spin-echo sequence, by applying a microwave pulse at a determined frequency, to profit from the spin -1 nature of the nitrogen vacancies. After half of the total evolution time, they applied a 2π echo pulse to swap the population of the spin states, so that after another period of free evolution time, they were able to eliminate the shifts of these spin levels induced by any magnetic field, after which it was possible to determine temperature accurately. Also, nanodiamonds used had a low concentration of ^{13}C to reduce magnetic field fluctuations originating from their nuclear spin bath. Measurements were performed in a confocal microscope using high magnification ($\times 100$) oil immersion microscope objectives, and microwaves were delivered using a lithographically defined coplanar waveguide on top of the glass coverslip that contained the sample. An avalanche photodiode was used for detection to record a continuous-wave electron spin resonance (ESR) spectrum. Furthermore, the author observed a characteristic low-frequency beating of the luminescence signal that changed from defect to defect, due to local fluctuations of charge traps, although for a fixed evolution time, it did not interfere in the temperature determination, although integration times of the order of 30 s were required. However, it is complex to inject these nanodiamonds in human embryonic fibroblast SW1 cells. It was

done through silicon nanowire-mediated delivery, in which the silicon nanowires were previously treated with 3-amino-propyltrimethoxysilane so they are provided of NH_2 functionality on their surfaces to attach nanodiamonds by electrostatic binding.

A much simpler approach could be developed using carbon dots (CDs). Such dots exhibit broadband absorption, strong luminescence, good resistance to photobleaching, high chemical stability, low toxicity, and good biocompatibility. This made them a very attractive choice in fields like cellular optical and photoacoustic imaging, photothermal and photodynamic therapies, drug delivery, and biosensing (Kalytchuk et al. 2017). They have also been used as luminescence nanothermometers using intensity-based and ratiometric techniques (Chen et al. 2013, Yang et al. 2015, Wang et al. 2016). Also, nitrogen- and sulfur-co-doped CDs, with sizes $\sim$4.5 nm, have been reported to be good luminescence nanothermometers operating under the luminescence lifetime approach, with $S_{\mathrm{rel}} \approx 1.8\%$ K^{-1} and $\delta T = 0.27$ K when internalized in HeLa cells to monitor their temperature (Kalytchuk et al. 2017). These CDs exhibit excitation-independent emission; this means that their emission peak is not altered in a wide range of excitation wavelengths extending from 250 to 400 nm with maximum excitation happening at 355 nm originating from the π–π^* transitions of the aromatic sp^2 domains and the trapping of excited-state energy by surface states (Dong et al. 2013), which would suggest that their luminescence emission processes are dominated by uniform emissive states. The emission peak is relatively narrow (FWHM = 67 nm) centered at 421 nm, with an absolute luminescence quantum yield of 78%. Although the position and the intensity of this emission peak showed a weak temperature dependence, its luminescence lifetime was highly sensitive to temperature and could be used as a thermometric parameter. Furthermore, spectrally uniform single exponential decays were observed for these CDs in the range of temperatures analyzed, between 275 K (2°C) and 353 K (80°C), which would indicate that the recombination occurs through very similar and highly emissive channels across the ensemble of CDs. Figure 25.19 shows the time-resolved temperature-dependent luminescence emission of these N- and S-co-doped CDs and their luminescence lifetimes, showing that an increase in temperature leads to a decrease of the luminescence lifetime from 11.0 ns at 275 K (2°C) to 5.3 ns at 353 K (80°C), due to the activation of nonradiative relaxation channels.

25.3.9 Biomaterials

The use of bright fluorescent organic molecules as luminescent nanothermometers inside living cells has attracted the attention of researchers. In this context, green fluorescent protein (GFP) is considered as one of the most widely used biomarkers for cell imaging due to its unique optical properties. One of the first studies on the use of GFP as a luminescent thermometer involved taking advantage

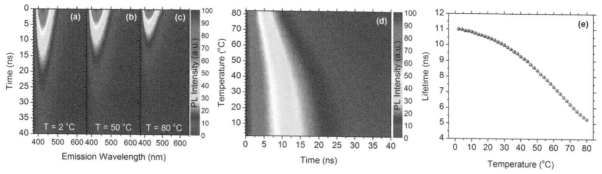

FIGURE 25.19 Time-resolved temperature-dependent luminescence emission in N,S-CDs. Normalized color plots have been used to show time-resolved luminescence emission maps at (a) 275 K (2°C), (b) 323 K (50°C), and (c) 353 K (80°C). (d) Normalized color plot of time-resolved luminescence intensity at the emission maximum of 421 nm in the range of temperatures analyzed. (e) Extracted luminescence lifetimes plotted as a function of temperature in the range of temperatures analyzed. (Reproduced from Kalytchuk, S., Polakova, K., Wang, Y., Froning, J.P., Cepe, K., Rogach, A.L., Zboril, R. 2017. Carbon dot nanothermometry: intracellular photoluminescence lifetime thermal sensing. *ACS Nano* 11: 1432–1442. © 2017 American Chemical Society.)

of the fluorescence polarization anisotropy of this protein, which decreases linearly as the temperature increases (see Figure 25.20a) (Donner et al. 2012). This could be used to obtain thermal images of HeLa cells in which local heat was delivered via a photothermal approach by dispersing gold nanorods in the extracellular medium and illuminating them with an infrared laser, as can be seen in Figures 25.20b,c. The same strategy allowed the authors to image the plasmonic heating in a leaving Caenorhabditis elegans, allowing for a fast and noninvasive method to detect subdegree temperature changes on a single neuron generated by local heating through optically excited gold nanoparticles with a Ti:sapphire laser emitting at 800 nm (Donner et al. 2013).

However, luminescence nanothermometry based on GFP relies on cellular transfection that is difficult to achieve in certain primary cellular types.

A DNA-based ratiometric luminescence nanothermometer has also been demonstrated using the fluorescence resonance energy transfer (FRET) mechanism between two organic dyes (Xie et al. 2017). The system was composed of a rigid DNA tetrahedron, where a thermally sensitive

molecular beacon was embedded in one edge of the DNA. Two organic dyes (FAM and TAMRA) were labeled at the two ends of the oligonucleotide strand, which is partially hybridized with the strand containing the molecular beacon. Temperature alters the distance between the two organic dyes, resulting in an acceptor-to-donor fluorescence intensity change. The fluorescence intensity ratio between the acceptor (TAMRA) and the donor (FAM) as a function of temperature was used as the thermometric parameter, allowing for a thermal resolution smaller than 0.5 K.

25.4 Applications of Luminescence Nanothermometry

There are many areas where temperature determination at the nanoscale is of great importance. Although this section does not want to be an exhaustive compilation of all the works in which luminescence nanothermometry has been used for practical use, it will give an overview of the most important areas in which this thermometric methodology has been applied.

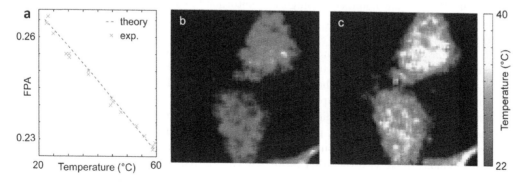

FIGURE 25.20 (a) Calibration curve that correlates the fluorescence polarization anisotropy of GFP with temperature in PBS. (b) Temperature map of HeLa cells obtained by measuring the fluorescence polarization anisotropy of GFP before heating and (c) during heating by focusing an IR laser on gold nanorods dispersed in the extracellular medium. (Adapted from Donner, J. S., Thompson, S. A., Kreuzer, M. P., Baffou, G., Quidant, R. 2012. Mapping intracellular temperature using green fluorescent protein, *Nano Lett* 12: 2107–2111. © 2012American Chemical Society.)

25.4.1   Biomedical Applications

The application of luminescence nanothermometry in biomedicine is maybe one of the most attractive ones, because of their potential use in early detection and treatment of several diseases. In this section, the use of luminescence nanothermometry in three big areas of biomedicine will be reviewed: (i) intracellular temperature determination, (ii) temperature determination inside biological tissues, and (iii) temperature monitoring during *in vivo* thermal treatments.

Intracellular Nanothermometry

Cells, as a basic unit of life, are commonly used to understand different physical and biological processes in the body. Since all processes of living organisms, including cells, are marked by temperature changes, it is of great importance to study temperature variations inside them. However, even with the crucial role of temperature in physiological phenomena, the mechanisms by which cells produce and use heat are largely unknown. Luminescent nanothermometry makes possible intracellular nanothermometry at a single-cell level, making evident spatial-temporal temperature variations and specific organelle-targeted thermogenesis (Jaque et al. 2014, Bai and Gu 2016).

To the best of our knowledge, the first intracellular temperature measurements were performed using the commercially available polarity-sensitive fluorescent 6-dodecanoy1-2-dimethylaminaphthalene (laurdan) that undergoes a gel-to-liquid transition at around 310 K (37°C), causing a red shift of the fluorescence emission, as can be seen in Figure 25.21(a) (Chapman et al. 1995). Due to its high hydrophobicity, laurdan was localized at the plasma membrane of Chinese hamster ovary cells, and intracellular temperature was determined by means of the general polarization (G.P.), given by the following equation:

$$G.P. = \frac{I_{490} - I_{440}}{I_{490} - I_{440}}$$

where I_{440} and I_{490} are the laurdan fluorescence intensities measured when the membrane is in the pure gel (440 nm) and liquid-crystalline phases (490 nm), respectively. The result of this calibration can be seen in Figure 25.21b. Through this procedure, the authors were able to demonstrate a δT in the range of 0.1–1 K that can be used to monitor the effects of pulsed lasers and optical traps on cells, following the temperature response in cells with high spatial resolution and on the time scale of normal cellular processes.

The highest thermal sensitivity (7.5%K^{-1}) in intracellular thermometry was demonstrated with the DNA ratiometric luminescence nanothermometer described in Section 3.9 (Xie et al. 2017). This luminescent nanothermometer was useful to demonstrate the photothermal effect caused by gold nanorods in HeLa cells. After 15 min of irradiation with a 980 nm laser at a power of 650 mW, the acceptor fluorescence intensity was largely reduced,

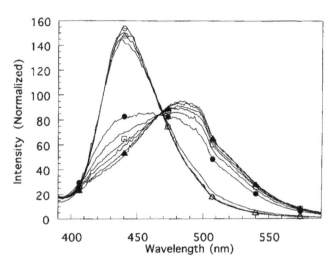

FIGURE 25.21   (a) Normalized fluorescence emission spectra of laurdan liposomes in solution at 21 (○), 25, 31 (Δ), 37, 42 (●), 48, 55 (□), 60, 66 (▲), and 72°C. (b) Generalized polarization as a function of temperature for laurdan in three Chinese hamster ovary cells. Filled symbols represent data taken in heating cycles; open symbols represent data taken in cooling cycles. (Adapted with permission from Chapman, C. F., Liu, Y., Sonek, G. J., Tromberg, B. J. 1995. The use of exogenous fluorescent probes for temperature measurements in single living cells. *Photochem. Photobiol.* 62: 416–425. © 1995 Wiley-VCG Verlag GmbH.)

suggesting that the temperature increased, while that of the control (without incubated gold nanorods) did not change (see Figure 25.22).

Nitrogen vacancy color centers in nanodiamonds have also been used for highly accurate temperature measurements inside human embryonic fibroblast WS1 cells (Kucsko et al. 2013), allowing for the detection of temperature changes as small as 44 ± 10 mK, with a response time of milliseconds. However, this system requires microinjection inside the cell, which makes its practical applications difficult.

By using genetically encoded green fluorescent proteins (GFP) acting as thermosensors, Kiyonaka and co-workers visualized the thermogenesis at a subcellular level, in the mitochondria of brown adipocytes and HeLa cells, and in the endoplasmic reticulum of myotubes (Kiyonaka et al. 2013). The luminescence nanothermometer used consisted of a thermosensitive coiled-coil protein TlpA of Salmonella that transmit conformational changes to GFP. These changes affect the GFP fluorescence, becoming a function of temperature. The thermal reading was performed by means of the temperature-dependent intensity ratio between the 400 and 480 nm bands of the GFP excitation spectrum. However, the main drawback of such luminescence nanothermometer is that it requires two excitation sources simultaneously, which complicated its operation. More recently, Nakano et al., using a temperature sensor based on two variants of GFP with different thermal sensitivities, demonstrated that carbonyl cyanide 4-(trifluoromethoxy)phenylhydrazone (FCCP) induced a temperature increase of 6–9 K in the mitochondria matrix of HeLa cells (Nakano et al. 2017).

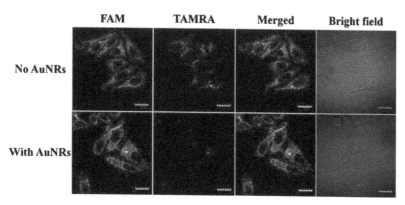

FAM TAMRA Merged Bright field

No AuNRs

With AuNRs

FIGURE 25.22 Confocal microscope images of HeLa cells containing the DNA-scallop inspired modified luminescent nanothermometers labeled with TAMRA and FAM, incubated with and without gold nanorods, after irradiation during 15 min with a 980 nm laser showing an increase in temperature generated by the photothermal effect of gold nanorods. Scale bar is 20 μm. (Adapted with permission from Xie, N., Huang, J., Yang, X. et al. 2017. Scallop-inspired DNA nanomachine: a ratiometric nanothermometer for intracellular temperature sensing. *Anal. Chem.* 89: 12115–12122. © 2017 American Chemical Society.)

Also, the authors demonstrated that it exists a difference of 2.9 ± 0.3 K between the temperature of the cytosol and that of the nucleus in the cells, as can be seen in Figure 25.23.

Despite these fascinating results, it is difficult to control the intracellular specific locations of the luminescence nanothermometers, since most of them are introduced in the cells by endocytosis or by microinjection. To avoid this limitation, Homma et al. developed a ratiometric luminescent molecular thermal probe selectively localized in the mitochondria of HeLa cells (Homma et al. 2015). The thermal probe consisted of a mixture of Rhodamine B and CS NIR dyes that possess different temperature-dependent fluorescence emission properties. The intensity ratio between these two dyes, that is temperature dependent, allowed to visualize the variation of the mitochondrial temperature in HeLa cells as a function of time induced by 10 μM of FCCP.

Another target-selective luminescence nanothermometer is developed by Arai et al., who used combinatorial synthesis to modify the side-chains of different fluorescent dye backbones that selectively targeted the endoplasmic reticulum (ER thermo yellow) (Arai et al. 2014). Using this strategy, the authors showed the temperature gradient induced by a near-infrared (NIR) laser in various cells, including HeLa cells, skeletal muscle cells, and brown adipocytes, as well as the heat produced by a Ca^{2+} gradient generated in HeLa cells. More recently, and using the same strategy, the same researchers were able to develop a targeted luminescence nanothermometer for mitochondria (Arai et al. 2015) that allowed them to demonstrate the evolution of the temperature of mitochondria during the respiration process, as can be seen in Figure 25.24. The authors also demonstrated that mitochondria were 10 K warmer than the ambient temperature of 311 K (38°C) when the respiratory chain was fully functional in human embryonic kidney 293 (HEK293) cells and primary skin fibroblasts (Chretien et al. 2018).

Internal Tissue Luminescence Nanothermometry

In vivo conditions create several obstacles for the optimum performance of luminescence nanothermometers. The main one is related to the strong extinction coefficient of the biological tissues in the visible range of the electromagnetic spectrum, restricting the use of the luminescence nanothermometers operating in this range of wavelengths.

Reduction of the absorption and scattering by biological tissues can be achieved by using specific wavelengths lying in the near-infrared or short-wavelength infrared regions: the so-called biological windows (BWs). A number of available luminescence nanothermometers operating in the BWs demonstrated the ability to sense temperature deep inside biological tissues. Mostly all of them use near-infrared light emitted by Ln^{3+} ions (Savchuk et al. 2018a). However, other types of materials have also been used for the same purpose, like quantum dots with emission in these spectral

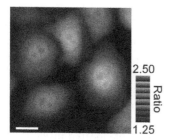

2.50

Ratio

1.25

FIGURE 25.23 Pseudo-colored ratio image of genetically encoded ratiometric fluorescent temperature indicator based on using two variants of GFP with different thermal sensitivities, ubiquitously expressed in a HeLa cells, showing an average temperature difference of 2.9 K between the cytoplasm and the nucleus regions, when the medium temperature was 310 K (37°C). Scale bar indicates 20 μm. (Reproduced from Nakano, M., Arai, Y., Kotera, I., Okabe, K., Kamei, Y., Nagai, T. 2017. Genetically encoded ratiometric fluorescent thermometer with wide range and rapid response. *PLoS One.* 12: 1–14. © 2017 PLOS.)

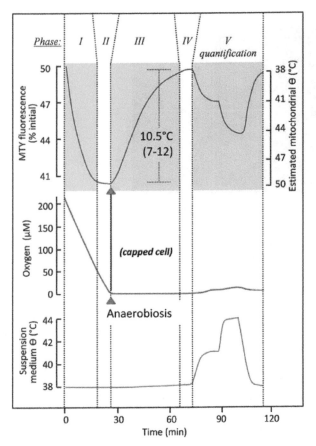

FIGURE 25.24 Evolution of the temperature of mitochondria in HEK293 cells during the different phases of the respiratory process. Phase I: cell respiration after cells are exposed to aerobic conditions in PBS. Phase II: cell respiration under aerobic conditions, maximal warming of mitochondria. Phase III: cell respiration arrested due to oxygen exhaustion. Phase IV: respiration remains stalled due to anaerobiosis. Phase V: respiration remains stalled due to anaerobiosis, while temperature of the cell suspension medium is shifted by making stepwise adjustments to water-bath temperature. (Reproduced from Chretien, D, Benit, P, Ha, H. H et al. 2018. Mitochondria are physiologically maintained at close to 50°C. *PLoS Biol.* 16:e2003992. © 2018 PLOS.)

regions, like PbS/CdS/ZnS (Del Rosal et al. 2016), and composites formed by quantum dots and metal nanoparticles, like Ag/Ag$_2$S (Ruiz et al. 2017).

Figure 25.25 shows the comparison of the performance of Nd^{3+}-based luminescence nanothermometers operating purely in the first BW, defined between 650 and 950 nm, and the second BW, defined between 1,000 and 1,350 nm (Savchuk et al. 2016b). This figure demonstrates the penetration depth that can be achieved in these BWs when testing it in chicken breast, of at least 1 cm, due to the reduced absorption and scattering of light of biological tissues at these wavelengths.

Another very interesting example of luminescence nanothermometry in *in vivo* studies with near-infrared radiation was using Nd:LaF$_3$@Yb:LaF$_3$ core–shell nanoparticles operating in the second BW (Ximendes

et al. 2016a). The spatial separation of Nd^{3+} and Yb^{3+} ions in the core and the shell of the nanoparticles allowed multiplying the thermal sensitivity by 4. This strategy permitted to unveil fundamental thermal properties of the biological tissue in living mice, like the subcutaneous thermal transients in a minimally invasive way as can be seen in Figure 25.26, from where absorption coefficient and the thermal diffusivity of living biological tissues could be determined.

In another example, by performing luminescence nanothermometry inside tissues, Ximendes et al. developed a new diagnosis tool to detect first stages of ischemia *in vivo* that could allow for the early treatment of cardiovascular diseases and accidents (Ximendes et al. 2016b). Ischemia is a disease in which a temporal or permanent restriction in blood supply to biological tissues or organs occurs that can produce a long-lasting or transient damage in the affected areas (Brown and Wilson 2004). The luminescence nanothermometer is based on PbS/CdS/ZnS/ QDs emitting light in the near infrared, at 1,200 nm after being pumped at 808 nm. By using these QDs as luminescence nanothermometers, the authors could discriminate between the ischemic and inflammatory phases that occur in live mice of the murine ischemic hindlimb model. Such discrimination is based on the faster thermal dynamics detected in ischemic tissues. This procedure also allowed monitoring the revascularization and damage recovery processes of such ischemic tissues after artery ligation, as shown in Figure 25.27.

In the last years, the possibilities of performing luminescence nanothermometry in the so-called third BW, or short-wavelength infrared region (SWIR), depending on the sources, extending from 1,350 to 2,300 nm have been analyzed (Savchuk et al. 2018a). In this region, the reduction of the light scattering compensates for the slight increase in the absorption of light with respect to the first and second BWs, so deeper penetration depths and higher resolution images can be obtained (Naczynski et al. 2013). Ln^{3+} ions emitting in this region, like Er^{3+}, Tm^{3+}, and Ho^{3+} can be used, embedded in nanoparticles, for luminescence nanothermometry, although the best performance up to date was found for combinations of Tm^{3+} and Ho^{3+} (Savchuk et al. 2018a). Figure 25.28 shows the temperatures that could be determined 2 mm below the surface of a slide of chicken breast by using the intensity ratio of the 1,480 and 1,711 nm emissions generated by Tm^{3+} Tm,Ho:KLu(WO$_4$)$_2$ nanoparticles, and their comparison with the temperatures determined by a thermocouple located inside the chicken breast, close to the luminescence nanoparticles. Heat inside the chicken breast was induced by a heating gun that was moved away from the sample, so a decrease of temperature in it was found.

Luminescence Thermometry In Vivo during Thermal Therapy

The use of heat as a tool to cure and/or improve the treatment of various diseases, including cancer, is well known in medicine. Increasing the temperature of biological tissues

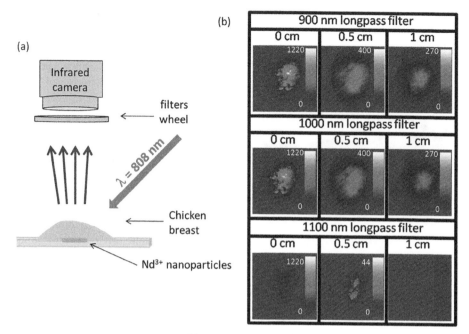

FIGURE 25.25   Near-infrared emission images of Nd^{3+}:$KGd(WO_4)_2$ nanoparticles obtained when they are covered with layers of chicken breast of different thicknesses. (a) Scheme of the setup used to determine the penetration depth of light in the first and second BWs when testing it in chicken breast. (b) Images collected from outside the biological tissue using different filters that allows passing only radiation in the first BW (900 nm longpass filter) or in the second BW (1,000 nm longpass filter). (Reproduced from Savchuk, O.A., Carvajal, J.J., De la Cruz, L.G., Haro-Gonzalez, P., Aguilo, M., Diaz, F. 2016. Luminescence thermometry and imaging in the second biological window at high penetration depth with Nd:$KGd(WO_4)_2$ nanoparticles. *J. Mater. Chem. C* 4: 7397–7405. © 2016 The Royal Society of Chemistry.)

above the limit of 312 K (39°C) induces chemical reactions leading to the denaturation of the proteins and enzymes they contain. If the temperature is increased above 321 K (48°C), then the necrosis of the biological tissue is induced. Thus, in this way malignant biological tissue can be eliminated. However, the application of heat also negatively affects the healthy tissue. With the introduction of luminescence nanothermometry, combined with the development of externally stimulated nanoheaters, it became possible to control the temperature distribution during a hyperthermia treatment at a local level, providing, therefore, an efficient and safe enough treatment of subcutaneous tumors.

Carrasco et al. (2015) performed *in vivo* efficient temperature-controlled photothermal therapy of cancer tumors induced in mice by using Nd^{3+}-doped LaF_3 nanoparticles. These nanoparticles have the ability to act at the same time as a heater, because of their high content in Nd^{3+} and the enhancement of the nonradiative transitions associated with it, and as luminescence thermal nanosensors. They sensed temperature inside the tumors from the analysis of the luminescence generated by the nanoparticles in the first BW during the hyperthermia treatment and demonstrated that it is 20% higher than the one that can be measured at the surface, as shown in Figure 25.29.

In another example, Zhu et al. used a core–shell upconversion nanocomposite, formed by Yb,Er:$NaLuF_4$@$NaLuF_4$ coated with a carbon shell, to monitor in real time the temperature variations during a photothermal therapy

process (Zhu et al. 2016). The carbon shell acted as a photothermal agent, absorbing the 730 nm laser radiation and converting it into heat, while the temperature feedback was achieved through the analysis of the green emission arising from Er^{3+} ion. Using this nanocomposite, it was possible to demonstrate efficient photothermal therapy with high spatial resolution *in vivo*, as can be seen in Figure 25.30.

Del Rosal et al. also demonstrated the use of a luminescence nanothermometer formed by PbS/CdS/ZnS quantum dots emitting in the second BW that acted simultaneously as photothermal agents, providing temperature feedback during photothermal therapy (Del Rosal et al. 2016), as can be seen in Figure 25.31.

25.4.2   Technological Applications of Luminescence Nanothermometry

Under this title, we grouped the other two important applications of luminescence nanothermometry: those related to microelectronics and microfluidics.

Luminescence Nanothermometry in Microelectronics

The continuous device miniaturization in microelectronics with the aim of increasing the functional integration density of operational devices, including data processing, energy

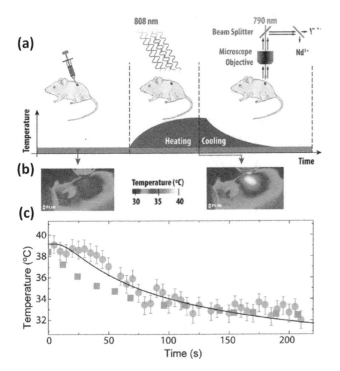

FIGURE 25.26 Determination of the subcutaneous temperature in living CD1 mice by using Nd:LaF$_3$@Yb:LaF$_3$ core–shell nanoparticles when irradiated with a 808 nm laser with a fluence of 0.35 W cm^{-1}. (a) Schematic representation of the experiments. (b) Thermal infrared images of the CD1 mouse before (left) and at the end (right) of the heating stimulus produced by the Nd^{3+} absorption and subsequent nonradiative relaxation processes that generated heat, and the residual absorption of skin at this wavelength. (c) Time evolution of the temperatures measured by the subcutaneous luminescent thermometer (circles) and the skin temperature measured with an IR thermal camera (squares). The black line corresponds to the best fitting describing the thermal relaxation of the subcutaneous tissue described by the Fourier's law. (Adapted with permission from Ximendes, E.C., Santos, W.Q., Rocha, U. et al. 2016. Unveiling in vivo subcutaneous thermal dynamics by infrared luminescent nanothermometers. *Nano Lett.* 16: 1695–1703. © 2016 American Chemical Society.)

management, sensing capability, and signal transmission and reception, has taken this technology to its limits, not only in terms of sizes but also in terms of thermal management in such devices. This increase of the functional integration density comes together with an increase in the density of heat dissipation in such devices that might generate local changes in the thermal conductivity of the materials conforming the integrated devices. Thus, thermal measurements are crucial for the design of thermal management systems in these devices so that their reliability can be ensured. Thermal monitoring allows also to detect failures in these devices, like localized hot spots, and to extract information about their electrical performance.

These thermal effects might induce the increase of electrical resistances in the different components of the devices,

as well as in the interconnects, that are detrimental for the power consumption of the chips, and might cause signal delays and line cross-talk effects when operating at high frequencies, for instance. Such thermal effects affect also the thermomechanical properties of the packaging structures and materials, inducing bond wire fatigue phenomena or cracking in the substrates. Thus, these thermally induced phenomena might affect the device performance, leading even to its complete degradation.

In optoelectronic devices, such as solid-state lasers and light-emitting diodes (LEDs), the control of temperature is even more important, especially in miniaturized devices. Changes in temperature during operation might induce shifts in their optical gain curves, broadenings or splitting in their emission lines, and jumps in their guiding modes; degrading the spectral purity of such devices; or even leading to the interruption of the laser oscillation. Also, by increasing the temperature, the threshold currents increase, affecting the power consumption of these light-emitting devices and leading, sometimes, to destructive blockages. Finally, operation above the designed temperatures might accelerate the degradation of the materials that form such structures, especially in organic-LEDs (OLEDs), in systems based on flexible and transparent substrates with poor thermal conductivities, and in high-power devices and large-area displays (Perpiñà et al. 2016).

Thus, temperature is a very important parameter to consider when designing micro- and nanoelectronics systems, and temperature measurements are crucial to monitor the average internal temperature of the device within the security thresholds established, localize defects on the devices, or detect failures in their performance, as well as to extract performance information and figures of merit. Also, by determining the temperature distribution in the devices, data are available to simulate the thermal failure phenomena that can affect their performance, providing the opportunity to analyze and solve them.

Although in microelectronics, the nanothermometers based on the electrical properties of the materials that constitute the microchips are more used (Perpiñà et al. 2016), there exist also some examples that study the thermal characteristics of such devices using luminescence nanothermometers. In this context, Brites et al. (2010) developed a luminescent molecular thermometer based on Eu^{3+} and Tb^{3+} co-doped γ-Fe$_2$O$_2$@TEOS/APTES core–shell nanoparticles, where TEOS is the acronym of tetraethyl orthosilicate and APTES stands for (3-aminopropyl)triethoxysilane, that were used for mapping the surface temperature of an integrated circuit. The strategy followed to accomplish such a task is covering the integrated circuit with a layer of the luminescent nanoparticles, and using an optical fiber to excite and collect the emission from the luminescent ions, as shown in Figure 25.32a, b. Then, the absolute temperature was calculated by using the calibration curve previously established for these luminescence nanothermometers from which the temperature profiles along the A and B lines marked in the integrated

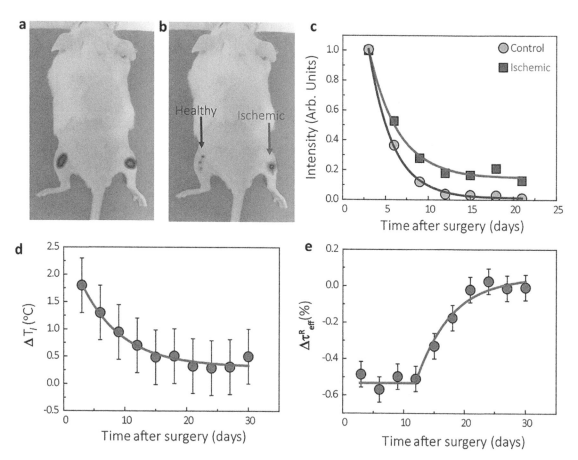

FIGURE 25.27 Near-infrared luminescence images of a representative murine ischemic hindlimb individual obtained (a) 3 and (b) 13 days after performing artery ligation to correct the ischemia problems induced in one of the legs of the mouse. (c) Time evolution of the luminescence intensity generated by the PbS/CdS/ZnS QDs injected in the ischemic and healthy limbs. (d) Temperature difference between the ischemic and healthy limbs as a function of the time passed after artery ligation, extracted from the evolution of the intensity of the PbS/CdS/ZnS QDs shown in (c). (e) Relative reduction in the characteristic thermal relaxation time of the ischemic tissue as a function of the time elapsed after artery ligation. (Reproduced with permission from Ximendes, E.C., Rocha, U., del Rosal, B. et al. 2016. In vivo ischemia detection by luminescent nanothermometers. *Adv. Healthcare Mater.* 6: 1601195. © 2016 Wiley-VCH.)

circuit, and shown in Figure 25.32b, c, respectively, could be determined and compared with the gray points corresponding to the temperature measurements performed with an infrared camera. The authors achieved a spatial resolution of 35 μm and a temperature uncertainty of 0.5 K.

In more complex example, Aigouy and collaborators (Aigouy et al. 2005) developed a scanning thermal microscope able to map the temperature of electrically excited stripes with micro/nanosizes. It was based on gluing Er^{3+}- and Yb^{3+}- co-doped fluoride nanoparticles on the scanning tip of an atomic force microscope (AFM). The particles were excited by a 975 nm laser, and the temperature-dependent fluorescence intensity ratio of the two emission bands of Er^{3+} in the green was used to determine the temperature. The scheme of the setup is illustrated in Figure 25.33.

Micro- and Nanofluidics

The main aim of measuring and controlling temperature in micro- and nanofluidic areas relies on different practical phenomena, like the capability of generating localized heat on a reduced scale. This is useful, for instance, to amplify DNA signals through polymerase chain reaction (PCR) devices, to sort and concentrate molecules, or to monitor biochemical reactions. Another interesting phenomenon is the possibility of generating strong thermal gradients in fluidic microchips and understanding how they affect the movement of fluids. This is used to separate components of a fluid whose ionic strength is a function of temperature, or by thermophoresis, for instance, or focusing the path of fluids in channels by thermally induced pH gradients in small volumes, with the advantage that it is possible to explore simultaneously a wide range of temperatures. Finally, another interesting area of research is the possibility of generating fast heating and cooling cycles in which the temperature is changing periodically but keeping an active control of these temperature changes. This can be used, for instance, to determine the protein conformation, to force the movement of droplets generated within microfluidic chips or mix solutions (Bergaud 2016).

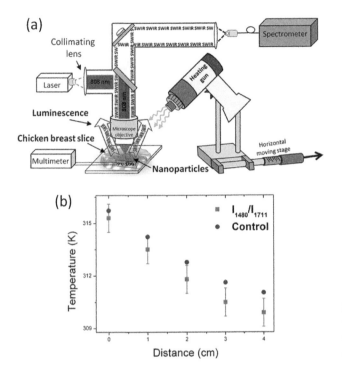

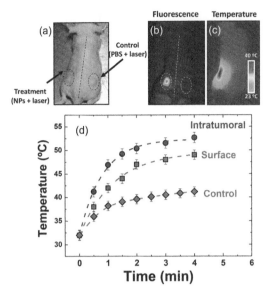

FIGURE 25.28 (a) Experimental setup used for temperature sensing inside chicken breast by using emissions in the SWIR region. Heat inside chicken breast was induced by an external heating gun that was moved away from the sample in order to decrease the temperature. (b) Temperature data determined from the intensity ratio of the 1,480 and 1,711 nm emissions generated by Tm^{3+} $Tm,Ho:KLu(WO_4)_2$ nanoparticles (squares) and a thermocouple located near the luminescence nanoparticles (control) during the experiment. (Reproduced with permission from Savchuk, O.A., Carvajal, J.J., Haro-Gonzalez, P., Aguilo, M., Diaz, F. 2018. Luminescent nanothermometry using short-wavelength infrared light. *J. Alloys Compd.* 746: 710–719. © 2018 Elsevier.)

FIGURE 25.29 (a) Optical image of a representative mouse with two induced tumors. (b) and (c) Infrared luminescence and thermal images of the same mouse irradiated with a 808 nm laser with a power density of 4 W cm^{-2}. The images show different signals depending on whether the tumor was injected with Nd:LaF$_3$ nanoparticles (only in the left-side tumor). (d) Time evolution of the temperature inside the tumor determined from the luminescence generated by the Nd:LaF$_3$ nanoparticles, and the corresponding ones determined on the surface of the tumor through and infrared camera. (Reproduced with permission from Carrasco, E., del Rosal, B., Sanz-Rodriguez, F. et al. 2015. Intratumoral thermal reading during photo-thermal therapy by multifunctional fluorescent nanoparticles. *Adv. Funct. Mater.* 25: 615–626. © 2015 Wiley-VCH.)

In fact, the reduction in size in microfluidic devices might generate unexpected localized temperature variations. For instance, when an electric field is applied across a microfluidic channel containing a conducting medium, heating caused by Joule effect can occur (Tang et al. 2006). Also, if the microfluidic channel is illuminated with a laser, the existence of optical traps in the might generate these unwanted heating spots that must alter experimental data in biophysics (Peterman et al. 2003). Other effects that might induce such localized temperature variations are photothermal conversion in surface-enhanced Raman scattering (SERS) (Kang et al. 2010), the collapse of cavitation bubbles (Baffou et al. 2014), or the friction forces generated in the injectors in sprays (Gibson et al. 2014). Such unwanted temperature changes can dramatically affect the overall performance of the microfluidic device and the surrounding fluid, which can even change some physical properties of the sample of study.

Among the different thermometers available, the need for high spatial and temporal resolution required in micro- and nanofluidics almost restricts the choice to

luminescence nanothermometry, allowing also to generate two-dimensional (2D) and three-dimensional (3D) real-time temperature maps of these devices. This, combined with the characteristics of micro- and nanofluidics devices, offers several advantages including efficiency, speed, portability, and reduced amount of reagent consumption (Shields et al. 2015).

Samy et al. reported the temperature distribution in a tapered microchannel fabricated on a poly(dimethylsiloxane) (PDMS) and heating caused by Joule effect, using a PDMS thin film with adsorbed Rhodamine B (RhB) dye (Samy et al. 2008). As can be seen in Figure 25.34a, the chip was placed in a custom-made acrylic holder that allowed holding the microfluidic chip over an inverted microscope objective. The picture obtained for the thermal map is shown in Figure 25.34b, illustrating the existence of a temperature gradient in the microchannel. In that case, spatial resolutions of the order of hundreds of micrometers were obtained, with relative temperature sensitivities between 1.3% K^{-1} and 2.3% K^{-1}.

In another example, by using two different dyes (RhB and RH110), Gibson et al. (2014) developed a ratiometric method to determine the evolution of the droplets present in an electrospray plume, by measuring the intensity ratio

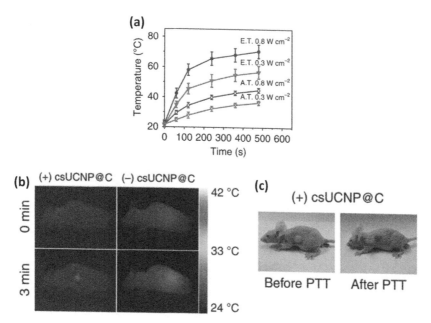

FIGURE 25.30 Photothermal treatment on mice transplanted with HeLa cells through a nanocomposite formed by Yb,Er:NaLuF$_4$@NaLuF$_4$ nanoparticles coated with a carbon shell. (a) Elevation of apparent temperature and eigen temperature of the nanocomposites in phantom tissue under irradiation with a 730 nm laser at fluences of 0.8 and 0.3 W/cm^2. (b) Thermal images of nude mice with (left panel) and without (right panel) Yb,Er:NaLuF$_4$@NaLuF$_4$ nanoparticles coated with a carbon shell. These nanoparticles were incubated in HeLa cell tumors and were irradiated at 730 nm with a fluence of 0.3 W/cm^2. The pictures show the increase in temperature observed in the tumor containing the nanoparticles of the composite after 3 min of continuous irradiation with the laser. (c) Representative pictures of the nude mice transplanted with the HeLa cells containing the luminescence thermometer/photothermal agent composite, before and after the photothermal treatment. (Adapted with permission from Zhu, X.; Feng, W.; Chang, J.; et al. 2016. Temperature-feedback upconversion nanocomposite for accurate photothermal therapy at facile temperature. *Nat. Commun.* 7: 1–10. © 2016 Nature.)

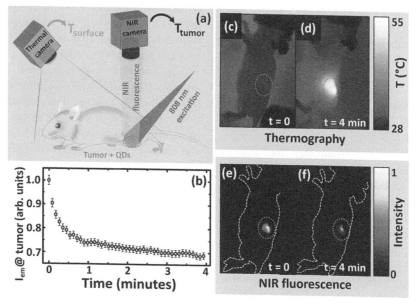

FIGURE 25.31 Real-time temperature feedback during photothermal therapy in a mice subcutaneously inoculated with A431 cells to induce tumor growth. (a) Schematic representation of the treatment procedure. (b) Time evolution of the intensity of the emission of PbS/CdS/ZnS QDs during photothermal treatment. (c) and (d) Thermographic images of the mouse before and after irradiating it during 4 min with a 808 nm laser with a power density of 1.7 W/cm^2 to treat the tumor by photothermal therapy. (e) and (f) Luminescence images of the same mouse recorded at the same moment than the thermographic images shown in (c) and (d). (Reproduced with permission from Del Rosal, B., Carrasco, E., Ren, F. et al. 2016. Infrared-emitting QDs for thermal therapy with real-time subcutaneous temperature feedback. *Adv. Funct. Mater.* 26: 6060–6068. © 2016 Wiley-VCH.)

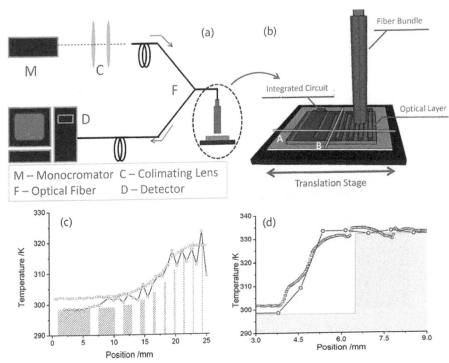

FIGURE 25.32 (a) and (b) Scheme of the setup used for the thermal mapping of an integrated circuit, whose tracks were covered with a layer of Eu^{3+} and Tb^{3+} co-doped γ-Fe_2O_3@TEOS/APTES core–shell nanoparticles. Temperature profiles recorded along the A (c) and B (d) lines shown in (b), compared with the measurements performed with an IR camera. (Reproduced from Brites, C.D.S., Lima, P.P., Silva, N.J.O. et al. 2010. A luminescent molecular thermometer for long-term absolute temperature measurements at the nanoscale. *Adv. Mater.* 22: 4499–4504. © 2010 Wiley-VCG Verlag GmbH.)

between these two dyes. Figure 25.35a shows the setup used in this experiment that allowed to translate the electrospray apparatus axially and laterally with respect to the laser beam used to excite the dyes. Figure 25.35b and c shows the axial and lateral temperature profiles, recorded for the spray at different flow rates or at different axial positions.

25.5 Conclusions

The 21st century has been testimony of the emergent interest of luminescence nanothermometry for different applications, including biological applications such as determining the intracellular and the internal tissue temperature, or monitoring it *in vivo* during thermal therapies to treat different diseases. Other applications, that we named technological, deal with temperature determination in microelectronics, microoptics, photonics, and micro- and nanofluidic devices.

Different luminescence parameters, which change with temperature, have been used for luminescence nanothermometry purposes. The self-referencing methods are the most used ones, such as the band-shape ratiometric method, followed by lifetime-based luminescence nanothermometry. However, all of them present some limitations. For instance, the lifetime-based methodology is intrinsically limited by the long acquisition times required to collect reliable data. Also, despite lifetime-based luminescence nanothermometry

allows eliminating problems related to spatial fluctuations of the luminescence intensity, for instance, the relative thermal sensitivities that can be obtained with this technique are not better than those that can be obtained using the band-shape ratiometric method that is limited by the energy difference between the emitting states used.

Also, several materials have been investigated to improve the performance of luminescence nanothermometers, including quantum dots, organic dyes, metal nanoparticles, polymers, transition metal and lanthanide-doped nanoparticles and metal-organic frameworks, nanodiamonds and carbon dots, and even biomaterials. Although almost one-third of all the publications in luminescence nanothermometry are related to lanthanide containing materials, it is not yet possible to highlight one material over the others since the most suitable one will depend on the application envisaged.

At present, the figure of merit that emerged to compare the performance of such luminescence nanothermometers seems to be the relative thermal sensitivity, with values ranging from 0.1 to 10% K^{-1}. However, up to date, only a very few number of the reported luminescence nanothermometers present values of relative thermal sensitivity that are almost constant in a wide range of temperatures, a valuable characteristic if a universal thermometer is being envisaged. Other parameters that are scarcely explored in the literature are the spatial and temporal resolutions of these thermometers, although they are critical for some of their

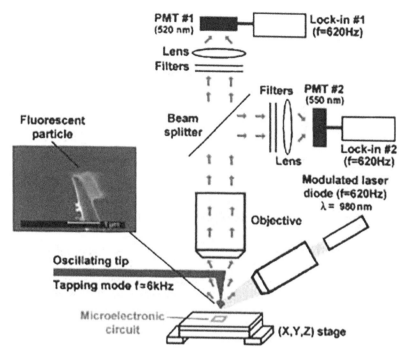

FIGURE 25.33 Scheme of the scanning thermal microscope able to map the temperature of electrically excited stripes with micro/nanosizes. A nanoparticle of Er^{3+}-and Yb^{3+}-co-doped fluoride was attached at the extremity of an atomic force microscope tip, as it is shown in the scanning electron micrograph included in the inset. The light emitted by the nanoparticle, after excitation with a 980 nm diode laser, was collected by a microscope objective and split towards two photomultiplier tubes that acted as detectors. The authors placed interferential filters, one centered at 520 nm and the other one centered at 550 nm, in front of each of the detectors, respectively, so that each of them recorded in each scan only the image corresponding to one of the emissions of Er^{3+} in the green. Finally, the thermal contrast was calculated by dividing the two images. The scanning speed was very low, only ten points per second, to allow for a good thermalization of the luminescent nanoparticle. This system allowed to measure simultaneously the topography of the sample. (Reproduced with permission from Aigouy, L., Tessier, G., Mortier, M., Charlot, B. 2005. Scanning thermal imaging of microelectronic circuits with a fluorescent nanoprobe. *Appl. Phys. Lett.* 87: 184105. © 2005 AIP Publishing.)

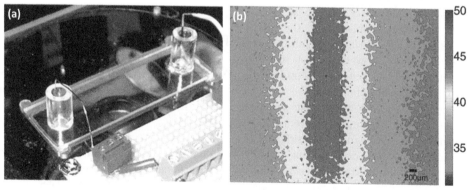

FIGURE 25.34 (a) Setup used to observe the PDMS microfluidic channel coated with a PDMS/RhB thin film on the inverted microscope for fluorescence intensity measurements. (b) Thermal map obtained in which the heating effect diffusing into the surrounding side walls can be seen. (Adapted with permission from Samy, R., Glawdel, T., Ren, C.L. 2008. Method for microfluidic whole-chip temperature measurement using thin-film poly(dimethylsiloxane)/rhodamine B. *Anal. Chem.* 80: 369–375. © 2008 American Chemical Society.)

applications, as well as their reproducibility and repeatability. Thus, in the future, novel figures of merit will need to be developed to compare their temporal response to monitor fast-developing events, the penetration depths that can be achieved in biological tissues when they want to be used for biomedical applications, or the minimum distance at which

two different temperatures can be discriminated, applicable in many different fields, just to name a few.

Other limitations are related to the distribution of the luminescence nanothermometers in the samples to be tested. Most of the reviewed applications of nanoparticles in luminescence nanothermometry rely on single-point

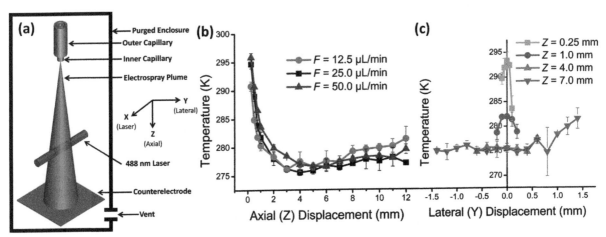

FIGURE 25.35   (a) Graphic representation of the setup used to determine the temperature distribution among the droplets of an electrospray plume by using the intensity ratio of the RhB and Rh110 dyes, after excitation at 488 nm with an Ar laser. (b) Axial temperature profiles recorded at the center of the spray at different flow rates. (c) Lateral temperature profiles recorded at different axial positions when the flow rate was kept constant at 25 μL/min. (Adapted with permission from Gibson, S.C., Feigerle, C.S., Cook, K.D. 2014. Fluorometric measurement and modeling of droplet temperature changes in an electrospray plume. *Anal. Chem.* 86: 464–472. © 2014 American Chemical Society.)

measurements. However, if they want to be used for thermal imaging purposes, for instance, a homogeneous distribution of these nanoparticles over the whole system to be imaged is required. Other possibilities for this thermal mapping can be the scan of the whole system by a single nanoparticle manipulated remotely, probably using optical tweezers or traps. This is of interest, for instance, to generate three-dimensional thermal maps of living cells, where despite the considerable amount of research developed, accurate temperature distributions within living cells have not yet been satisfactorily addressed. Also, it is clear that when these luminescence nanothermometers are envisaged for their use in *in vivo* biomedical applications, they have to be based in emissions lying in the near infrared, in the so-called biological windows. However, which of these biological windows will be the most effective is still to be determined, and probably, it will depend on the type of biological tissue being analyzed and its intrinsic composition. Furthermore, almost all the luminescence nanothermometers developed up to now can be considered as secondary thermometers. Thus, the development of self-calibrated thermometers using a well-established state equation that correlates a specific luminescence parameter with the absolute temperature is mandatory. These limitations might be overcome by combining efforts from a large number of different disciplines, including coordination and supramolecular chemistry, thermodynamics, photophysics, nanotechnology, microelectronics, micro- and nanofluidics, microscopy, materials design, assembly and integration, and biomedicine, among others.

Nevertheless, it is clear that luminescence nanothermometry is called to play a very important role in the future, being used, for instance, for disease detection and temperature control and monitoring during treatment. In this sense, multifunctional materials integrating synergistically

different functionalities in a nanometric platform (heating and thermometric elements, imaging and thermal-regulated drug delivery, or thermal and magnetic responsive materials, for instance) are envisaged to gain attention during the following years. Taking into account that temperature is a basic physical quantity that effects our daily life, and that sensors of temperature account for 80% of all sensors worldwide, we can envisage a long and healthy future for luminescence nanothermometry, whose development can be considered to be still in its infancy that would allow integrating them in commercial products.

References

Aigouy, L., Tessier, G., Mortier, M., Charlot, B. 2005. Scanning thermal imaging of microelectronic circuits with a fluorescent nanoprobe. *Appl. Phys. Lett.* 87: 184105.

Albers, A.E., Chan, E.M., McBride, P.M., Ajo-Franklin, C.M., Cohen, B.E., Helms, B.A. 2012. Dual-emitting quantum dot/quantum rod-based nanothermometers with enhanced response and sensitivity in live cells. *J. Am. Chem. Soc.* 134: 9565–9568.

Alden, M., Omrane, A., Richter, M., Särner, G. 2011. Thermographic phosphors for thermometry: A survey of combustion applications. *Prog. Energ. Combust.* 37: 422–461.

Arai, S., Lee, S.C., Zhai, D., Suzuki, M., Chang, Y.T. 2014. A molecular fluorescent probe for targeted visualization of temperature at the endoplasmic reticulum. *Sci. Rep.* 4: 6701.

Arai, S., Suzuki, M., Park, S.J. et al. 2015. Mitochondria-targeted fluorescent thermometer monitors intracellular temperature gradient. *Chem. Commun.* 51: 8044–8047.

Back, M., Trave, E., Ueda, J., Tanabe, S. 2016. Ratiometric optical thermometer based on dual near-infrared emission in Cr^{3+}-doped bismuth-based gallate host. *Chem. Mater.* 28: 8347–8356.

Baffou, G., Polleux, J., Rigneault, H., Monneret, S. 2014. Super-heating and micro-bubble generation around plasmonic nanoparticles under cw illumination. *J. Phys. Chem. C* 118: 4890–4898.

Bai, T., Gu, N. 2016. Micro/Nanoscale thermometry for cellular thermal sensing. *Small.* 34: 4590–4610.

Balabhadra, S., Debasu, M.L., Brites, C.D.S., Ferreira, R.A.S., Carlos, L.D. 2017. Upconverting nanoparticles working as primary thermometers in different media. *J. Phys. Chem. C* 121: 13962–13968.

Banyai, L., Koch, S.W. 1993. *Semiconductor Quantum Dots.* Singapore: World Scientific Publishing.

Barlett, J.W., Frost, C. 2008. Reliability, repeatability and reproducibility: Analysis of measurement errors in continuous variables. *Ultrasound Obstet. Gynecol.* 31: 466–475.

Benayas, A., Escuder, E., Jaque, D. 2012. High-resolution confocal fluorescence thermal imaging of tightly pumped microchip Nd:YAG laser ceramics. *Appl. Phys. B* 107: 697–701.

Bergaud, C. 2016. Thermometry in micro and nanofluidics. In *Thermometry at the Nanoscale: Techniques and Selected Applications*, ed. L.D Carlos, and F. Palacio, 461–492. Cambridge: The Royal Society of Chemistry.

Berthou, H. Jörgensen, C.K. 1990. Optical-fiber temperature sensor based on upconversion-excited fluorescence. *Opt. Lett.* 15: 1100–1102.

Bloembergen, M. 1959. Solid state infrared quantum counters. *Phys. Rev. Lett.* 2: 84–85

Bomm, J., Günter, C., Stumpe, J. 2012. Synthesis and optical characterization of thermosensitive, luminescent gold nanodots. *J. Phys. Chem. C.* 116: 81–85.

Borisov, S.M., Gatterer, K., Bitschnau, B., Klimant, I. 2010. Preparation and characterization of Chromium (III)-activated Yttrium Aluminum Borate: A new thermographic phosphor for optical sensing and imaging at ambient temperatures. *J. Phys. Chem. C* 114: 9118–9124.

Botas, A.M.P, Brites, C.D.S., Wu, J. et al. 2016. A new generation of primary luminescent thermometers operating in different media based on silicon nanoparticles. *Part. Part. Syst. Charact.* 33: 740–748.

Brites, C.D.S., Lima, P.P., Silva, N.J.O. et al. 2010. A luminescent molecular thermometer for long-term absolute temperature measurements at the nanoscale. *Adv. Mater.* 22: 4499–4504.

Brites, C.D.S., Lima, P.P., Silva, N.J.O. et al. 2012. Thermometry at the nanoscale. *Nanoscale* 4: 4799–4829.

Brites, C.D.S., Millán, A., Carlos, L.D. 2016. Lanthanides in luminescent thermometry. In *Handbook on the Physics and Chemistry of Rare Earths*, ed. J.C.G. Bünzli, and V.K. Pecharsky, 49: 339–427. Elsevier.

Brown, J.M., Wilson, W.R. 2004. Exploiting tumour hypoxia in cancer treatment. *Nat. Rev. Cancer.* 4: 437–447.

Brübach, J., Pflitsch, C.; Dreizler, A., Atakan, B. 2013. On surface temperature measurements with thermographic phosphors: A review. *Prog. Energ. Combust.* 39: 37–60.

Cadiau, A., Brites, C.D.S., Costa, P.M.F.J., Ferreira, R.A.S., Rocha, J., Carlos, L.D. 2013. Ratiometric nanothermometer based on an emissive Ln^{3+} organic framework. *ACS Nano* 7: 7213–7218.

Carlos, L.D., Palacio, F. 2016. Preface. In *Thermometry at the Nanoscale: Techniques and Selected Applications*, eds. L.D Carlos, and F. Palacio, vii–viii. Cambridge: The Royal Society of Chemistry.

Carrasco, E., del Rosal, B., Sanz-Rodriguez, F. et al. 2015. Intratumoral thermal reading during photo-thermal therapy by multifunctional fluorescent nanoparticles. *Adv. Funct. Mater.* 25: 615–626.

Chandrasekharan, N., Kelly, L.A. 2001. A dual fluorescence temperature based on perylene/exciplex interconversion. *J. Am. Chem. Soc.* 123: 9898–9899.

Chapman, C.F., Liu, Y., Sonek, G.J., Tromberg, B.J. 1995. The use of exogenous fluorescent probes for temperature measurements in single living cells. *Photochem. Photobiol.* 62: 416–425.

Chen, P.C., Chen, Y.N., Hsu, P.C., Shih, C.C., Chang, H.T. 2013. Photoluminescent organo-silane functionalized carbon dots as temperature probes. *Chem. Commun.* 49: 1639–1641.

Chen, D., Liu, S., Zhou, Y., Wan, Z., Huang, P., Ji, Z. 2016. Dual-activator luminescence of $RE/TM:Y_3Al_5O_{12}$ ($RE = Eu^{3+}$, Tb^{3+}, Dy^{3+}; $TM = Mn^{4+}$, Cr^{3+}) phosphors for self-referencing optical thermometry. *J. Mater. Chem. C* 4: 9044–9051.

Cheng, L., Wang, C., Liu, Z. 2013. Upconversion nanoparticles and their composite nanostructures for biomedical imaging and cancer therapy. *Nanoscale* 5: 23–37.

Childs, P. R.N. 2001. *Practical Temperature Measurement.* Boston, MA: Butterworth Heinemann.

Childs, P.R.N. 2016. Nanoscale thermometry and temperature measurement. In *Thermometry at the Nanoscale*, eds. L.D. Carlos, and F. Palacio, 3–22. Cambridge: The Royal Society of Chemistry.

Chretien, D., Benit, P., Ha, H.H. et al. 2018. Mitochondria are physiologically maintained at close to 50°C. *PLoS Biol.* 16: e2003992.

Dai, Q., Zhang, Y., Wang, Y. 2010. Size-dependent temperature effects on PbSe nanocrystals. *Langmuir* 26: 11435–11440.

Del Rosal, B., Carrasco, E., Ren, F. et al. 2016. Infrared-emitting QDs for thermal therapy with real-time subcutaneous temperature feedback. *Adv. Funct. Mater.* 26: 6060–6068.

Dong, Y., Pang, H., Yang, H.B. 2013. Carbon-based dots co-doped with Nitrogen and Sulfur for high quantum yield and excitation-independent emission. *Angew. Chem. Int. Ed.* 52: 7800–7804.

Donner, J.S., Thompson, S.A., Kreuzer, M.P., Baffou, G., Quidant, R. 2012. Mapping intracellular temperature using green fluorescent protein. *Nano Lett.* 12: 2107–2111.

Donner, J.S., Thompson, S.A., Alonso-Ortega, C. et al. 2013. Imaging of plasmonic heating in a living organism. *ACS Nano* 7: 8666–8672.

Dramicanin, M.D. 2018. *Luminescence Thermometry.* Cambridge: Woodhead Publishing.

Duarte, F.J. 2012. Tunable organic dye lasers: physics and technology of high-performance liquid and solid-state narrow-linewidth oscillators. *Prog. Quant. Electron.* 36: 29–50.

Duong, H.D., Rhee, J.I. 2007. Exploitation of thermo-effect of Rhodamine B entrapped in sol-gel matrix and silica gel for temperature detection. *Sensor Actuat. B* 124: 18–23.

Ebert, S., Travis, K., Lincoln, B., Guck, J. 2007. Fluorescence ratio thermometry in a microfluidic dual-beam laser trap. *Opt. Express* 15: 15493–15499.

Estrada-Perez, C.E., Hassan, Y.A., Tan, S. 2011. Experimental characterization of temperature sensitive dyes for lased induced fluorescence thermometry. *Rev. Sci. Instrum.* 82: 074901.

Farrusseng, D. 2011. *Metal-Organic Frameworks: Applications from Catalysis to Gas Storage.* Weinheim: Wiley-VCH.

Feryforgues, S., Fayet, J.P., Lopez, A. 1993. Drastic changes in the fluorescence properties of NBD probes with the polarity of the medium involvement of a TICT state? *J. Photochem. Photobiol.* 70: 229–243.

Fischer, L.H., Harms, G.S., Wolfbeis, O.S. 2011. Upconverting nanoparticles for nanoscale thermometry. *Angew. Chem. Int. Ed.* 50: 4546–4551.

Fonger, H. Struck, C.W. 1970. Eu^{3+} ^5D quenching to the charge-transfer states in Y_2O_2S, La_2O_2S, and LaOCl. *J. Chem. Phys.* 52:6364–6366.

Gibson, S.C., Feigerle, C.S., Cook, K.D. 2014. Fluorometric measurement and modeling of droplet temperature changes in an electrospray plume. *Anal. Chem.* 86: 464–472.

Goodrich, G.P., Bao, L., Gill-Sharp, K., Sang, K.L., Wang, J., Payne, J.D. 2010. Photothermal therapy in a murine colon cancer model using near-infrared absorbing gold nanorods. *J. Biomed. Opt.* 15: 1–8.

Gota, C., Okabe, K., Funatsu, T., Harada, Y., Uchiyama, S. 2009. Hydrophilic fluorescent nanogel thermometer for intracellular thermometry. *J. Am. Chem. Soc.* 131: 2766–2767.

Haase, M., Schäfer, H. 2011. Upconverting nanoparticles. *Angew. Chem. Int. Ed.* 50: 5808–5829.

Han, B., Hanson, W.L., Bensalah, K., Tuncel, A., Stern, J.M., Cadeddu, J.A. 2009. Development of quantum dot-mediated fluorescence thermometry for thermal therapies. *Ann. Biomed. Eng.* 37: 1230–1239.

Haro-Gonzalez, P., Martinez-Maestro, L., Martin, I.R., Garcia-Sole, J., Jaque, D. 2012. High-sensitivity fluorescence lifetime thermal sensing based on CdTe quantum dots. *Small* 8: 2652–2658.

Henderson, B., Imbush, G.F. 2006. *Optical Spectroscopy of Inorganic Solids.* London: Oxford Science Publications.

Hildebrandt, B., Wust, P., Ahlers, O. et al. 2002. The cellular and molecular basis of hyperthermia. *Critical Rev. Oncol. Hematol.* 43: 33–56.

Homma, M., Takei, Y., Murata, A., Inoue, T., Takeoka, S. 2015. A ratiometric fluorescent molecular probe for visualization of mitochondrial temperature in living cells. *Chem. Commun.* 51: 6194–6197.

Hsia, C.H., Wutting, A., Yang, H. 2011. An accessible approach to preparing water-soluble Mn^{2+}-doped (CdSSe) ZnS (core) shell nanocrystals for ratiometric temperature sensing. *ACS Nano* 5: 9511–9522.

Itoh, H., Arai, S., Sudhaharan, T. et al. 2016. Direct organelle thermometry with fluorescence lifetime imaging microscopy in single myotubes. *Chem. Commun.* 52: 4458–4461.

Jaque, D., Vetrone, F. 2012. Luminescence nanothermometry. *Nanoscale* 4: 4301–4326.

Jaque, D., del Rosal, B., Rodríguez, E.M., Maestro, L.M., Haro-González, P., Solé, J.G. 2014. Fluorescent nanothermometers for intracellular thermal sensing. *Nanomedicine* 9: 1047–1062.

Jaque, D., J.G. Solé, J.G. 2016. Quantum dot fluorescence thermometry. In *Thermometry at the Nanoscale*, eds. L.D. Carlos, and F. Palacio, 85–123. Cambridge: The Royal Society of Chemistry.

Jung, W., Kim, Y.W., Yim, D., Yoo, J.Y. 2011. Microscale surface thermometry using SU8/Rhodamine-B thin layer. *Sensor Actuat. A* 171: 228–232.

Kalytchuk, S., Polakova, K., Wang, Y., Froning, J.P., Cepe, K., Rogach, A.L., Zboril, R. 2017. Carbon dot nanothermometry: Intracellular photoluminescence lifetime thermal sensing. *ACS Nano* 11: 1432–1442.

Kang, T., Hong, S., Choi, Y., Lee, L.P. 2010. The effect of thermal gradients in SERS spectroscopy. *Small* 6: 2649–2652.

Karstens, T, Kobs, K. 1980. Rhodamine B and Rhodamine 101 as reference substances for fluorescence quantum yield measurements. *J. Phys. Chem.* 84: 1871–1872.

Khlebtsov, N., Dykman, L. 2011. Biodistribution and toxicity of engineered gold nanoparticles: A review of in vitro and in vivo studies. *Chem. Soc. Rev.* 40: 1647–1671.

Kiyonaka, S., Kajimoto, T., Sakaguchi, R. et al. 2013. Genetically encoded fluorescent thermosensors visualize subcellular thermoregulation in living cells. *Nat. Methods.* 10: 1232–1238.

Kucsko, G., Maurer, P.C., Yao, N.Y. et al. 2013. Nanometre-scale thermometry in a living cell. *Nature* 500: 54–58.

Lee, J. and Kotov, N.A. 2007. Themometer design at the nanoscale. *Nano Today* 2: 48–51.

Leigh, J.R. 1988. *Temperature Measurements and Control.* London: Peter Peregrius.

Lepock, J.R., Cheng, K.H., Al-qysi, H., Sim, I., Koch, C.J., Kruuv, J. 1987. Hyperthermia-induced inhibition of respiration and mitochondrial protein denaturation in CHL cells. *Int. J. Hyperthermia* 3: 123–132.

Li, S., Zhang, K., Yang, J.M., Lin, L., Yang, H. 2007. Single quantum dots as local temperature markers. *Nano Lett.* 7: 3102–3105.

Lojpur, V., Nikolic, M.G., Jovanovic, D., Medic, M., Antic, Z., Dramicanin, M.D. 2013. Luminescence thermometry with $Zn_2SiO_4:Mn^{2+}$ powder. *Appl. Phys. Lett.* 103: 141912.

Lomenie, N., Racoceanu, D., Gouaillard, A. 2001. *Advances in Bioimaging: From Physics to Signal Understanding Issues*, Berlin: Springer.

Löw, P., Kim, B., Takama, N., Bergaud, C. 2008. High-spatial-resolution surface-temperature mapping using fluorescent thermometry. *Small* 4: 908–914.

Maestro, L.M., Rodriguez, E.M., Cruz, M.C.I. et al. 2010. CdSe quantum dots for two-photon fluorescence thermal imaging. *Nano Lett.* 10: 5109–5115.

Maestro, L.M., Haro-Gonzalez, P., Sanchez-Iglesias, A., Liz-Marzan, L.M., Sole, J.G., Jaque, D. 2014. Quantum dot thermometry evaluation of geometry dependent heating efficiency in gold nanoparticles. *Langmuir* 30: 1650–1658.

McGee, T.D. 1988. *Principles and Methods of Temperature Measurement*. New York: John Wiley & Sons.

McLaurin, E.J., Bradshaw, L.R., Gamelin, D.R. 2013. Dual-emitting nanoscale temperature sensors. *Chem. Mater.* 25: 1283–1292.

Naczynski, D.J., Tan, M.C., Zevon, M. et al. 2013. Rare-earth-doped biological composites as in vivo shortwave infrared reporters. *Nat. Commun.* 4: 2199–2219.

Nakano, M., Arai, Y., Kotera, I., Okabe, K., Kamei, Y., Nagai, T. 2017. Genetically encoded ratiometric fluorescent thermometer with wide range and rapid response. *PLoS One.* 12: 1–14.

Omrane, A., Juhlin, G., Ossler, F., Aldén, M. 2004. Temperature measurements of single droplets by use of laser-induced phosphorescence. *Appl. Opt.* 43: 3523–3529.

Park, Y., Ko, C., Chen, H.Y., Han, A., Son, D.H. 2013. Ratiometric temperature imaging using environment-sensitive luminescence of Mn-doped core-shell nanocrystals. *Nanoscale* 5: 4944–4950.

Perpiñà, X., Vellvehi, M., Jordà, X. 2016. Thermal issues in microelectronics. In *Thermometry at the Nanoscale*, eds. L.D. Carlos, and F. Palacio, 3–22. Cambridge: The Royal Society of Chemistry.

Peterman, E.J.G., Gittes, F., Schmidt, C.F. 2003. Laser-induced heating in optical traps. *Biophys. J.* 84: 1308–1316.

Prazeres, T.J.V., Beija, M., Charreyre, M.T., Farinha, J.P.S., Martinho, J.M.G. 2010. RAFT polymerization and self-assembly of thermoresponsive poly(N-decylacrylamide-b-N,N-diethylacrylamide) block copolymers bearing a phenanthrene fluorescent α-end group. *Polymer.* 51: 355–367.

Pietsch, C., Hoogenboom, R., Schubert, U.S. 2009. Soluble polymeric dual sensor for temperature and pH value. *Angew. Chem. Int. Ed.* 48: 5653–5656.

Pietsch, C., Hoogenboom, R., Schubert, U.S. 2010a. PMMA based soluble polymeric temperature sensors based on UCST transition and solvatochromic dyes. *Polym. Chem.* 1: 1005–1008.

Pietsch, C., Vollrath, A., Hoogenboom, R., Schubert, U.S. 2010b. A fluorescent thermometer based on a pyrene-labeled thermoresponsive polymer. *Sensors.* 10: 7979–7990.

Pugh-Thomas, D., Walsh, B.M., Gupta, M.C. 2011. CdSe(ZnS) nanocomposite luminescent high temperature sensor. *Nanotechnology* 22: 185503.

Pustovalov, K., Astafyeva, L.G., Fritzche, W. 2013. Selection of thermooptical parameters of nanoparticles for achievement of their maximal thermal energy under optical irradiation. *Nano Energy* 2: 1137–1141.

Rocha, U., da Silva, C.J., Silva, W.F. et al. 2013. Subtissue thermal sensing based on neodymium-doped LaF_3 nanoparticles. *ACS Nano* 7: 1188–1199.

Roemer, R.B. 1989. Engineering aspects of hyperthermia therapy. *Annu. Rev. Biomed. Eng.* 1: 347–376.

Ross, D., Gaitan, M., Locascio, L.E. 2001. Temperature measurement in microfluidic systems using a temperature-dependent fluorescent dye. *Anal. Chem.* 73: 4117–4123.

Roy, D., Brooks, W.L.A., Sumerlin, B.S. 2013. New directions in thermoresponsive polymers. *Chem. Soc. Rev.* 42: 7214–7243.

Ruiz, D., del Rosal, B., Acebrón, M. et al. 2017. Ag/Ag_2S nanocrystals for high sensitivity near-infrared luminescence nanothermometry. *Adv. Funct. Mater.* 27: 1604629.

Samy, R., Glawdel, T., Ren, C.L. 2008. Method for microfluidic whole-chip temperature measurement using thin-film poly(dimethylsiloxane)/rhodamine B. *Anal. Chem.* 80: 369–375.

Sapsford, K.E., Algar, W.R., Berti, L. et al. 2013. Functionalizing nanoparticles with biological molecules: developing chemistries that facilitate nanotechnology. *Chem. Rev.* 113: 1904–2074.

Savchuk, O.A., Haro-González, P., Carvajal, J.J., Jaque, D., Massons, J., Aguiló, M., Díaz, F. 2014. $Er:Yb:NaY_2F_5O$ up-converting nanoparticles for subtissue fluorescence lifetime thermal sensing. *Nanoscale* 6: 9727–9773.

Savchuk, O.A., Carvajal, J.J., Pujol, M.C., Barrera, E.W., Massons, J., Aguilo, M., Diaz, F. 2015. $Ho,Yb:KLu(WO_4)_2$ nanoparticles: A versatile material for multiple thermal sensing purposes by luminescent thermometry. *J. Phys. Chem. C* 119: 18546–18558.

Savchuk, O.A., Carvajal, J.J., Cascales, C., Aguiló, M., Díaz, F. 2016a. Benefits of silica core-shell structures on the temperature sensing properties of $Er,Yb:GdVO_4$ up-conversion nanoparticles. *ACS Appl. Mater. Inlerf.* 8: 7266–7273.

Savchuk, O.A., Carvajal, J.J., De la Cruz, L.G., Haro-Gonzalez, P., Aguilo, M., Diaz, F. 2016b. Luminescence thermometry and imaging in the second biological window at high penetration depth with Nd:KGd(WO$_4$)$_2$ nanoparticles. *J. Mater. Chem. C* 4: 7397–7405.

Savchuk, O.A., Carvajal, J.J., Haro-Gonzalez, P., Aguilo, M., Diaz, F. 2018a. Luminescent nanothermometry using short-wavelength infrared light. *J. Alloys Compd.* 746: 710–719.

Savchuk, O.A., Carvajal, J.J., Brites, C.D.S., Carlos, L.D., Aguilo, M., Diaz, F. 2018b. Upconversion thermometry: A new tool to measure the thermal resistance of nanoparticles. *Nanoscale* 10: 6602–6610.

Schoof, S., Gusten, H. 1989. Radiationless deactivation of the fluorescent state of 9-methoxyanthracene and of related meso-substituted anthracenes. Ber. Bunsen. Phys. Chem. 93: 864–870.

Shah, J.J., Gaitan, M., Geist, J. 2009. Generalized temperature measurement equations for Rhodamine B dye solution and its application to microfluidics. *Anal. Chem.* 81: 8260–8263.

Shang, L., Stockmar, F., Azadfar, N., Nienhaus, G.U. 2013. Intracellular thermometry by using fluorescent gold nanoclusters. *Angew. Chem. Int. Ed.* 52: 11154–11157.

Shields IV, C.W., Reyes, C.D., Lopez, G.P. 2015. Microfluidic cell sorting: A review of the advances in the separation of cells from debulking to rare cell isolation. *Lab Chip* 15: 1230–1249.

Sirohi, R.S. 2016. *Introduction to Optical Metrology.* New York: CRC Press/Taylor & Francis Group.

Sole, J.G., Bausa, L.E., Jaque, D. 2005. *An Introduction to the Optical Spectroscopy of Inorganic Solids.* West Sussex: John Wiley & Sons.

Souza, A.S., Nunes, L.A.O., Silva, I.G.N. et al. 2016. Highly-sensitive Eu^{3+} ratiometric thermometers based on excited state absorption with predictable calibration. *Nanoscale* 8: 5327–5333.

Stich, M.I.J., Fischer, L.H., Wolfbeis, O.S. 2010. Multiple fluorescent chemical sensing and imaging. *Chem. Soc. Rev.* 39: 3102–3114.

Tang, G., Yan, D., Yang, C., Gong, H., Chai, J.C., Lam, Y.C. 2006. Assessment of Joule heating and its effects on electroosmotic flow and electrophoretic transport of solutes in microfluidic channels. *Electrophoresis* 27: 628–639.

Tang, L., Jin, J.K., Qin, A.J. et al. 2009. A fluorescent thermometer operating in aggregation-induced emission mechanism: Probing thermal transitions of PNIPAM in water. *Chem. Commun.* 2009: 4974–4976.

Thyageswaran, S. 2012. Developments in thermometry from 1984 to 2011: A review. *Recent Pat. Mech. Eng.* 5: 4–44.

Uchiyama, S., Matsumura, Y., de Silva, A.P., Iwai, K. 2003. Fluorescent molecular thermometers based on polymers showing temperature-induced phase transitions and labelled with polarity-responsive benzofurazans. *Anal. Chem.* 75: 5926–5935.

Uchiyama, S., Kawai, N., de Silva, A.P., Iwai, K. 2004. Fluorescent polymeric AND logic gate with temperature and pH as inputs. *J. Am. Chem. Soc.* 126: 3032–3033.

Uchiyama, S., Kimura, K., Gota, C. et al. 2012. Environment-sensitive fluorophores with benzothiadiazole and benzoselenadiazole structures as candidate components of a fluorescent polymeric thermometer. Chem. Eur. J. 18: 9552–9563.

Vancoillie, G., Zhang, Q. and Hoogenboom, R. 2016. Polymeric temperature sensors. In *Thermometry at the Nanoscale*, eds. L.D. Carlos, and F. Palacio, 3–22. Cambridge: The Royal Society of Chemistry.

Vetrone, F., Naccache, R., Zamarrón, A. et al. 2010. Temperature sensing using fluorescent nanothermometers. *ACS Nano* 4: 3254–3258.

Vlaskin, V.A., Janssen, N., van Rijssel, J., Beaulac, R., Gamelin, D.R. 2010. Tunable dual emission in doped semiconductor nanocrystals. *Nano Lett.* 10: 3670–3674.

Wade, S.A., Collins, S.F., Baxter, G.W. 2003. Fluorescence intensity ratio technique for optical fiber point temperature sensing. *J. Appl. Phys.* 94: 4743–4756.

Walker, G.W., Sundar, V.C., Rudzinski, C.M., Wun, A.W., Bawendi, M.G., Nocera, D.G. 2003. Quantum-dot optical temperature probes. *Appl. Phys. Lett.* 83: 3555–3557.

Wang, C., Huang, Y., Lin, H. et al. 2015. Gold nanoclusters based dual-emission hollow TiO$_2$ microspheres for ratiometric optical thermometry. *RSC Adv.* 5: 61586–61592.

Wang, C., Lin, H., Xu, Z. et al. 2016. Tunable carbon-dot-based dual-emission fluorescent nano-hybrids for ratiometric optical thermometry in living cells. *ACS Appl. Mater. Interfaces* 8: 6621–6628.

Weaver, J.B. 2010. Bioimaging: hot nanoparticles light up cancer. *Nat. Nanotechnol.* 5: 630–631.

Weinstein, B.A. 1986. Ruby thermometer for cryobaric diamond-anvil cell. *Rev. Sci. Instrum.* 57: 910.

Wu, K., Cui, J., Kong, X., Wang, Y. 2011. Temperature dependent upconversion luminescence of Yb/Er codoped NaYF$_4$ nanocrystals. *J. Appl. Phys.* 110: 053510.

Xie, N., Huang, J., Yang, X. et al. 2017. Scallop-inspired DNA nanomachine: A ratiometric nanothermometer for intracellular temperature sensing. *Anal. Chem.* 89: 12115–12122.

Ximendes, E.C., Santos, W.Q., Rocha, U. et al. 2016a. Unveiling in vivo subcutaneous thermal dynamics by infrared luminescent nanothermometers. *Nano Lett.* 16: 1695–1703.

Ximendes, E.C., Rocha, U., del Rosal, B. et al. 2016b. In vivo ischemia detection by luminescent nanothermometers. *Adv. Healthcare Mater.* 6: 1601195.

Yang, J., Bai, F. 2002. Temperature dependence of the fluorescence spectra of ladderlike polyphenylsilsesquioxane and ladderlike 1,4-phenylenen-bridged polyvinylsiloxane. *Chin. J. Polym. Sci.* 20: 15–23.

Yang, Y., Kong, W., Li, H. et al. 2015. Fluorescent N-doped carbon dots as in vitro and in vivo nanothermometer. *ACS Appl. Mater. Interfaces* 7: 27324–27330.

Yang, G., Liu, X., Feng, J., Li, S., Li, Y. 2016. Organic dye thermometry. In *Thermometry at the Nanoscale*, eds. L.D. Carlos, and F. Palacio, 3–22. Cambridge: The Royal Society of Chemistry.

Yap, S.V., Ranson, R.M., Cranton, W.M., Koutsogeorgis, D.C., Hix, G.B. 2009. Temperature dependent characteristics of La_2O_2S: Ln (Ln = Eu, Tb) with various Ln concentrations over 5–60°C. *J. Lumin.* 129: 416–422.

Young, H.D., Freedman, R.A. 2008. *Sears and Zemansky's University Physics: With Modern Physics*. 12th ed. San Francisco: Pearson Education.

Zhao, Y., Riemersma, C., Pietra, F., Koole, R., de Mello Donegá, C., Meijerink, A. 2012. High-temperature luminescence quenching of colloidal quantum dots. *ACS Nano* 6: 9058–9067.

Zollinger, H. 2003. *Color Chemistry: Syntheses, Properties and Applications of Organic Dyes and Pigments*. Weinheim: Wiley-VCH.

Zhu, X., Feng, W., Chang, J. et al. 2016. Temperature-feedback upconversion nanocomposite for accurate photothermal therapy at facile temperature *Nat. Commun.* 7: 1–10.

26

Diamond Nanothermometry

Yuen Yung Hui, Oliver Y. Chen,
and Huan-Cheng Chang
Academia Sinica

Meng-Chih Su
Sonoma State University

26.1 Introduction

Nanothermometry studies heat transfer processes in a nanoscale system using temperature probes that are of nanometric size and, typically, with a spatial resolution better than a micron. Based on the design of its temperature probe, a nanothermometer can be conveniently categorized as (i) a contacting thermal sensor where the tip of the probe directly interacts with the target as those used in scanning thermal microscopes and nanoscale thermocouples or (ii) a non-contacting thermal sensor that poses no physical interactions with the target and, hence, minimum disturbance. A luminescence nanothermometer equipped with optical thermal probes is in this non-contact category (Brites et al. 2012; Jaque and Vetrone 2012). Emerging from worldwide research activities on nanoscience and nanotechnology in the past three decades or so, the two areas that nanothermometers (contact or not) find their best use are micro/nanoelectronics and biomedicine. A result of the continuing effort in the miniaturization of electronics, from complementary metal oxide semiconductor (CMOS) to "beyond CMOS" and "more than Moore" domains (Perpina et al. 2016), is that the density of hardware construction grows at a rate exceeding the local heat dissipation rate, which has caused concerns with hot spots where highly localized temperature increments may lead to irreversible and unrecoverable defects. Monitoring temperatures through the functioning areas, as well as at hot spots, in an electronic device will provide crucial information about the thermal management necessary for the reliability and safety of the final manufactured products.

This chapter covers the research and development of luminescence nanothermometry using nanodiamonds (and their hybrids) as thermal probes. While contacting probes may have proven a sensitive tool for scanning temperatures on the surface (2D) of an electronic device, it will need the optical probe of a luminescence nanothermometer to submerge in depth (3D) of a cell or an organism when studying the energy transfer involved in biological functions at the subcellular level. A luminescence nanothermometer, whether it be dye-sensitized polymers, semiconducting quantum dots, lanthanide-doped nanoparticles, or carbon nanoparticles, has phosphors that may undergo downshifting (Stokes) emission at a longer wavelength or upconverting (anti-Stokes) emission to a shorter wavelength upon irradiation by an external light source of the absorption energy (Brites et al. 2012; Jaque and Vetrone 2012). The observed luminescence is mainly originated from the electronic states of the nanoprobe interacting with its surrounding environment, and when the property of the luminescence shows a well-defined function with temperature, the local temperature may thus be measured. These optical properties include emission intensity, lifetime, peak position and/or width, as discussed later in this chapter. Today's luminescence thermometers are small (<100 nm), sensitive (<0.1 K/Hz$^{1/2}$), and capable of providing high spatial resolution (<1 μm).

In recent years, a trend in nanothermometry research appears to shift from fabrications of more new temperature sensors to the development of technologies best utilizing the optical and magnetic properties unique with the nanothermometers, driven largely by the advancement and needs in the field of nanomedicine. In addition to the size, sensitivity, and spatial resolution requirements, a nanothermometer to be used in biomedical systems must have low cytotoxicity and high biocompatibility as well as a short response time (in milliseconds or less) to capture rapid temperature fluctuations during fast metabolic processes. Furthermore, more restrictions on the spectral response range are imposed when conducting optical imaging of the sensors where the first and second biological spectral windows, I-BW (700–950 nm) and II-BW (1,000–1,350 nm), are desired for the minimum absorption by cells and tissues (Jaque et al. 2014). As expected, after screening through the available nanothermometers with the required conditions stated above, not very many temperature sensors have actually turned out to be suited for biomedical studies. Fluorescent nanodiamonds (FNDs), containing color centers as built-in phosphors, stand out as an excellent candidate for all-in-one nanothermometry in biological systems.

In this chapter, we start with a brief overview on the magnetic and optical properties of vacancy-related color centers that pave the foundation for developing FNDs as a sensitive and versatile nanothermometer. Using the FND nanoprobe, methods of contemporary temperature sensing techniques will be presented to illustrate how diamond nanothermometry is capable of revealing heat transfer processes at the single cell level. In particular, as an example in biomedical applications of nanothermometry, we will introduce the design and synthesis of nanohybrids that are comprised of FNDs and gold nanoparticles (GNPs). The use of new GNP-FND nanohybrids as multifunctional nanothermometers for both fluorescence imaging and photothermal therapy (as hyperlocalized hyperthermia) in live human cells is an active area of research in nanomedicine with a great potential for an effective, controlled cancer treatment (Sotoma et al. 2018).

26.2 Vacancy-Related Color Centers

Diamond is a wide bandgap semiconductor with an energy gap of 5.5 eV, capable of hosting a variety of optically active crystallographic defects (Zaitsev 2001). Among over 500 types of point defects in diamond, the vacancy-related color centers deserve the closest attention. The centers were first discovered more than a century ago by the English scientist William Crookes who noticed that exposing diamond to the radiation generated by radium bromide for several months gave the crystal a bluish green color (Overton and Shigley 2008). Later studies confirmed that the color change was mainly caused by radiation damage of the crystals by radium-emitted alpha particles that knocked carbon atoms out of the diamond lattice, leaving vacancies behind. As the carbon vacancies in diamond could exist in different charged states, able to absorb light in the visible

and near-infrared wavelength regions, a blue-to-green color was resulted. Subsequent heating of the radiation-damaged diamonds above 800°C in an inert gas atmosphere could change the blue green color to yellow, brownish, or other hues due to the combination of vacancies with impurities such as nitrogen atoms and metallic ions to form vacancy-associated color centers in the crystal matrix.

Aside from alpha particles, other high-energy sources such as electrons, neutrons, protons, helium ions, or gamma rays can also change the diamond color upon exposure (Campbell et al. 2002). Particularly, vacancies near the crystal surface (within 100 nm) may be produced by ion irradiation, if the energies of the ions are properly controlled. Depending on the content of impurities in the crystal, various vacancy-related color centers form during thermal annealing. An elevation of the temperature to 500°C or higher is often required to overcome the activation energy barrier of the vacancy migration, which has been measured as 2.3 eV (Davies et al. 1992). Some of the vacancy-related centers are highly fluorescent and, therefore, useful for luminescence-based nanoscale temperature sensing. The nitrogen-vacancy (NV) center is one of such point defects, existing in two different charged states, NV^0 (neutral) and NV^- (negatively changed). Both the centers are commonly found in radiation-damaged synthetic diamond crystals (containing ~100 ppm N), given nitrogen as the most abundant impurity in diamond due to the similar size of N and C atoms.

Zooming in the nanoscale, vacancy-related color centers can also exist in diamond particles of less than 100 nm in diameter. These particles, called FNDs in short (Yu et al. 2005), are appealing nanoscale temperature sensors not only because the diamond matrix is chemically inert and biocompatible, but also because the centers are capable of emitting intense and highly stable fluorescence. Unlike other luminescence nanothermometers made of metal and semiconductor materials (Brites et al. 2012; Jaque and Vetrone 2012), the majority of the color centers in FNDs are deeply embedded inside the chemically inert matrix. Therefore, their optical properties are nearly immune to environmental changes in viscosity, pH, ion concentration, and/or other chemical factors, providing the most reliable temperature measurements at the nanoscale. The combination of these favorable properties with its high photostability and easiness of surface functionalization makes FND a superb fluorescent marker as well as an ideal temperature sensor in complex biological environments such as cytoplasm (Hsiao et al. 2016).

Table 26.1 lists seven different types of vacancy-related color centers in diamonds for luminescence-based temperature sensing reported in the literature up to date. They include NV^-, NV^0, SiV^-, GeV^-, SnV, and Ni-based S3 and NE8 centers. The detection limits of these sensors vary with their spectral features used in the measurements, including thermal shifts of the electronic transitions and spin resonance frequencies as well as thermally induced changes of the fluorescence lifetimes, fluorescence intensities, and spin coherence times. We start the discussion with NV^- centers

TABLE 26.1 Temperature Sensing with Vacancy-Related Color Centers in Diamonds

Types	ZPL (nm)[a]	Features	Sensitivity[a]	References
NV$^-$		Fluorescence lifetime	–	Plakhotnik and Gruber (2010)
		ZPL (center)	0.4 K/Hz$^{1/2}$ (nanodiamond)	Hui et al. (2019)
		ZPL (height)	0.3 K/Hz$^{1/2}$ (nanodiamond)	Plakhotnik et al. (2015)
	638	Spin resonance	0.1 K/Hz$^{1/2}$ (nanodiamond)	Kucsko et al. (2013)
			130 mK/Hz$^{1/2}$ (nanodiamond)	Neumann et al. (2013)
		Spin coherence	9 mK/Hz$^{1/2}$ (bulk diamond)	Kucsko et al. (2013)
			5 mK/Hz$^{1/2}$ (bulk diamond)	Neumann et al. (2013)
			10 mK/Hz$^{1/2}$ (bulk diamond)	Toyli et al. (2013)
			10.1 mK/Hz$^{1/2}$ (bulk diamond)	Wang et al. (2015)
NV^0	576	ZPL (center)	–	Chen et al. (2011)
SiV$^-$	738	ZPL (center)	360 mK/Hz$^{1/2}$ (bulk diamond)	Nguyen et al. (2018)
			521 mK/Hz$^{1/2}$ (nanodiamond)	
GeV$^-$	602	ZPL (center)	300 mK/Hz$^{1/2}$ (bulk diamond)	Fan et al. (2018)
SnV	619	ZPL (center)	142 mK/Hz$^{1/2}$ (bulk diamond)	Alkahtani et al. (2018)
S3	497	Fluorescence lifetime	0.1 K/Hz$^{1/2}$ (microdiamond)	Homeyer et al. (2015)
NE8	794	ZPL (center)	–	Sildos et al. (2017)

[a]Room-temperature values.

in the next section since they are the major thrusts in current development and applications of diamond nanothermometry. The other color centers are discussed separately in Section 26.4.

26.3 NV$^-$-Based Nanothermometry

26.3.1 Electronic Structure and Energy Levels

The NV center in diamond is a point defect consisting of a vacancy site adjacent to a substitutional nitrogen atom oriented along the [111] crystalline direction (Figure 26.1a). In the four "dangling" carbon bonds surrounding the vacancy site, the nitrogen atom contributes two valence electrons and each of the three adjacent carbon atoms contributes one valence electron. This five-e^- configuration is called NV^0 for its zero net charge (Figure 26.1b). If there are electron-rich sources nearby such as another substitutional N atom, the electron-deficient NV^0 can receive an extra electron to become NV$^-$, i.e., the six-e^- configuration with -1 net charge. For this particular color center, which shall be the focus of this section, an extensive review on its electronic structure and magneto-optical properties has been provided by Doherty et al. (2013) and we present only some salient features here.

When a nitrogen atom is introduced into the diamond lattice, it reduces the lattice potential symmetry from T_d to C_{3v}, which in turn alters the energy levels of the molecular orbitals. It can be shown with group theory that two of the orbitals under the C_{3v} lattice potential belong to the E irreducible representation and the other two belong to the A_1 irreducible representation. The E orbitals, being less symmetric than A_1, are at a higher energy state (Figure 26.1c). Moreover, the C_{3v} symmetry ensures that the two E orbitals are of degenerate energy while the two A_1 orbitals are not. The twofold degeneracy at the highest energy level gives the last two filled electrons a freedom to occupy different E orbitals resulting in either a triplet state with the $m_s = -1, 0, +1$ spin sublevels or a singlet state of the $m_s = 0$ spin sublevel. The triplet states with

antisymmetric orbital wave functions and symmetric spin wave functions constitute the ground state of NV$^-$. In literature, the triplet ground state is often denoted by $^3A_2$ with 3 referring to the spin state and A_2 to the orbital symmetry. A crystal field splitting (D) arises from the spin–spin interactions between the unpaired electrons, which lift the triplet degeneracy and lower the energy of $m_s = 0$ by 2.87 GHz with respect to the $m_s = \pm 1$ sublevels (Figure 26.1d). The non-axisymmetric strain in the diamond crystal can further lift the $m_s = \pm 1$ degeneracy, depending on the magnitude of the strain. External (axial or not) magnetic fields can also split the $m_s = \pm 1$ sublevels via the Zeeman effect, which is the basis behind NV magnetometry (Rondin et al. 2014).

The NV$^-$ center can be excited by promoting one of the electrons from the upper A_1 orbital to one of the E orbitals within the band gap of diamond (Figure 26.1c; Weber et al. 2010). Like the ground state, the combination of the two unpaired electron spins results in triplet states or a singlet state. Because there are two possible promotion destinations, the excited triplet states are all twofold degenerate. The six triplet excited states are often denoted by ^3E with 3 referring to the spin state and E to the orbital symmetry (Figure 26.1d). When the color center is exposed to green yellow light, the spin-conserved electronic transition from the triplet ground states to the triplet excited states is the dominant channel that produces red fluorescence from ^3E to $^3A_2$ by spontaneous emission. The zero-phonon line (ZPL) of the fluorescence emission lies in the visible spectrum at around 637 nm (or 1.945 eV) with a phonon sideband peaking around 700 nm at room temperature. Due to intricate intersystem crossing (ISC), the intensities of the spontaneous emission among the three sublevels are not equivalent. The excited $m_s = \pm 1$ sublevels have $\sim$40% chance to decay non-radiatively to the metastable states ($^1A_1$ and ^1E), thereby reducing the fluorescence intensity by $\sim$40% (Manson et al. 2006). Upon continuous excitation, the cumulative effect of the differential transition rates can polarize the spin population to be as much as 80% toward the $m_s = 0$ state (Harrison et al. 2006).

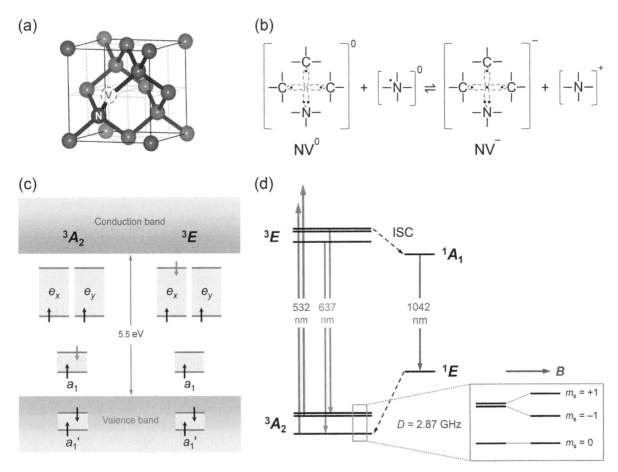

FIGURE 26.1 (a) Structure of a NV center in diamond. Carbon atoms are represented by solid spheres, while nitrogen and vacancy sites are marked in "N" and "V", respectively. (b) Lewis dot diagram of the charge state conversion reaction between NV^0 and NV^-. The dotted ovals represent the dangling bonds surrounding the vacancy site. The NV^0 and NV^- centers have five and six valence electrons, denoted by black dots, respectively. (c) Energy level diagram of NV^-, showing the relative energies (not drawn to scale) of the molecular orbitals within the band gap of diamond. At the ground state ($^3A_2$), the last two electrons are unpaired and occupying the degenerate e_x and e_y orbitals, giving rise to the triplet spin states. The excited state (^3E) corresponds to the promotion of one electron from the a_1 orbital to either the e_x or e_y orbital. (d) Electronic energy level diagram of NV^-. The dashed arrows denote the intersystem crossing (ISC) between $^3A_2$, ^3E and the metastable states, $^1A_1$ and ^1E. The crystal field splitting is $D = 2.87$ GHz and the spin sublevels $m_s = \pm 1$ split into two components in the presence of an external magnetic field (B).

The electronic energy manifold of the NV^- center is a manifestation of the physical position of its surrounding atoms. Temperature changes, for example, perturb the atomic positions in the crystal lattice, which translates to the energy level shift. In general, elongating atomic bond lengths decreases the spacing between energy levels. Such a temperature effect on NV^- can be divided into two categories: global and local. Globally, the temperature's relationship with lattice constant is well characterized by the thermal expansion coefficient of diamond, which has been experimentally measured (Sato et al. 2002). On the other hand, the temperature's local effect on atoms near the NV^- center, due to electron–phonon interactions, is less well understood. Nevertheless, it has been shown experimentally that the global effect contributes much less to the energy shift than the local one (Doherty et al. 2014). The local dominance makes the NV^- center's energy level spacing insensitive to the long-range influence such as that caused

by crystal strains. In addition, as described in the previous section, the protective diamond matrix shields the NV^- centers from exposure to the surrounding chemical changes, thereby, preventing their temperature measurements from being affected by otherwise unwanted external factors.

26.3.2 Fluorescence Lifetime

The use of FNDs in nanothermometry was first proposed in a study of the effect of temperature on NV fluorescence lifetimes, typically in the range of 10−20 ns. Plakhotnik and Gruber (2010) detected the fluorescence emission from multiple 30-nm FNDs heated in an oven and found a 2.7-fold decrease in the fluorescence lifetime from 300 K to 670 K. The observed heating–cooling cycle was reversible, suggesting that the nanomaterial is highly robust and potentially useful as a luminescence thermometer with a nanometric spatial resolution. However, compared with other approaches as discussed in the following sections (Table 26.1), the sensitivity of this method is quite limited because of the small changes of the fluorescence lifetimes with temperature. In addition, the lifetimes may vary from particle to particle depending on how the samples are prepared. For example, the fluorescence decay can be significantly accelerated if the FND surface is covered with graphitic residues, which can quench the fluorescence and alter the observed radiative rates.

26.3.3 Zero-Phonon Line

The simplest way to use the NV$^-$-based nanothermometer is to measure the shifts of the ZPL as temperature is changed. Chen et al. (2011) and Doherty et al. (2014) have studied closely the shifts of the ZPL at 638 nm from 4 K to 300 K. At room temperature, the thermal shift of the ZPL center (λ_0) is $\Delta\lambda_0/\Delta T \approx 0.015$ nm/K and the full width at half maximum (FWHM) of the band is ~ 5 nm. Such an increase in wavelength is attributed to the thermal expansion of the diamond crystal lattice together with the electron–phonon interactions. Although the thermal shift is small, it is sufficient for practical use because both the position and width of the ZPL in the fluorescence spectra are highly stable even after repeated measurements. In addition, the method is straightforward, readily implementable in any confocal fluorescence microscope system equipped with a spectrometer.

Tsai et al. (2017) realized this all-optical approach with 100-nm FNDs containing ~ 900 NV$^-$ centers per particle. A 594-nm laser was used as the light source to avoid excitation of the NV^0 centers whose fluorescence emission band (peaking at ~ 620 nm) is partially overlapped with the 638-nm band of NV$^-$, which can complicate ensuing spectral analysis. A representative result of the experiments is shown in Figure 26.2a for FNDs dispersed in water. While the overall feature of the emission band did not change much with the temperature varying over 28–75°C, the ZPL of the NV$^-$ centers was noticeably broadened and redshifted as the solution temperature was raised. To determine the sensitivity of this all-optical method, the research team spin-coated 100-nm FNDs on a glass coverslip and acquired the fluorescence spectra for the individual particles (Figure 26.2b). The values of λ_0 were then obtained by fitting the spectra over 610–660 nm to a Lorentzian function along with an exponential function to correct the baseline as

$$I(\lambda) = I_0 + B_0 \exp[b(\lambda - \lambda_0)] + \frac{A\Gamma}{\Gamma^2 + (\lambda - \lambda_0)^2}, \quad (26.1)$$

where $I(\lambda)$ is the fluorescence intensity, λ is the wavelength, Γ is the Lorentzian half-width, and I_0, B_0, b, and A are constants (Plakhotnik et al. 2015). Signal averaging over 6 s for 60 independently acquired spectra allowed the temperature measurements to achieve a sensitivity of ~ 2 K/Hz$^{1/2}$ around room temperature (inset of Figure 26.2b). Further improvement of the sensitivity to ~ 0.4 K/Hz$^{1/2}$ has recently been made by the same group for FNDs embedded densely

and uniformly in an organic polymer film (Hui et al. 2019). A limitation of this method is that its working range is confined to be $T < 120$°C due to the severe band broadening with increasing temperature; however, this temperature working range is sufficient for biosensing applications.

Exploiting the band broadening effect, Plakhotnik et al. (2015) monitored the change of the ZPL height (A) with respect to its phonon sideband background (B_0) in Eq. (26.1) as a method for the temperature measurements. They demonstrated the feasibility of this all-optical ratiometric method with 50-nm FNDs hosting more than 100 NV$^-$ centers in the crystal matrix and achieved a temperature measurement sensitivity of 0.3 K/Hz$^{1/2}$ using 590-nm laser excitation. A long-term stability of better than 0.6 K for a single FND spin-coated on a glass substrate was obtained. Although the method is intrinsically more sensitive than the detection of the ZPL shift, there is always a concern with the interference caused by background fluorescence (Hui et al. 2019). Either a lifetime gating or a magnetic modulation technique should be applied to eliminate the background signals (Hsiao et al. 2016). Taking advantage of the high sensitivity of this method, Fukami et al. (2019) have recently developed it into a nanoscale metrology platform for cryogenic temperature sensing over 100–300 K.

26.3.4 Spin Resonance

An important consequence of the optically induced spin polarization as discussed in Section 26.3.1 is that the fluorescence intensity of the NV$^-$ centers will reach its maximum after several cycles of the electronic excitation from $^3A_2$ to ^3E (Figure 26.1d). Under this condition, when a microwave radiation around 2.87 GHz is applied to the center, the electron spins will undergo a transition from $m_s = 0$ to $m_s = \pm 1$ sublevels of the ground state. Such a transition will reduce the fluorescence intensity by up to 30%, thus allowing the spin states of the individual NV$^-$ centers to be read out optically, a technique known as optically detected magnetic resonance (ODMR).

Using bulk diamonds and the ODMR technique, Acosta et al. (2010) first investigated the temperature effects on the spin resonance frequencies of the NV$^-$ centers. They measured the thermal shift of the crystal field splitting over 280–330 K for a NV$^-$ ensemble (Figure 26.3a), obtaining a value of $\Delta D/\Delta T = -74.2$ kHz/K (Figure 26.3b). Toyli et al. (2012) later studied the same effect over a wider temperature range (up to 600 K) for a single NV$^-$ center in bulk diamond and found a nonlinear frequency shift at higher temperatures. They fit the thermal shifts between 300 K and 600 K to the third-order polynomial as

$$D(T) = 2.8697 + 9.7 \times 10^{-5} T - 3.7 \times 10^{-7} T^2 + 1.7 \times 10^{-10} T^3, \quad (26.2)$$

where $D(T)$ is in unit of GHz and T in K. A similar result was also obtained with FNDs, implying that there is little variance of D as the size of diamond is reduced to the nanometer regime (Plakhotnik et al. 2014). Compared

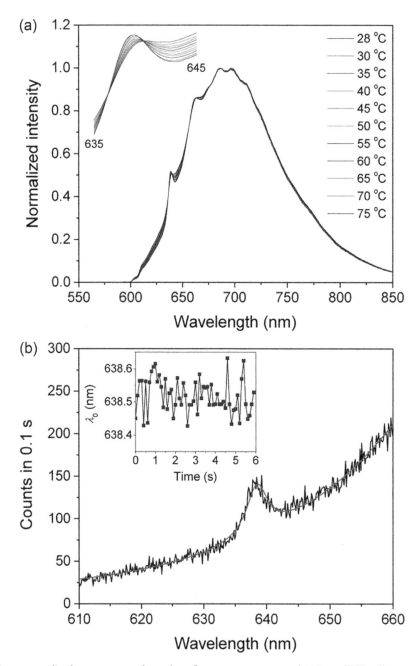

FIGURE 26.2 (a) Area-normalized temperature-dependent fluorescence spectra of 100-nm FNDs illuminated by a 594-nm laser in solution. Inset: Enlarged view of the temperature-induced shift of the ZPLs for spectra acquired at 28–75°C. (b) Typical fluorescence spectrum of a single 100-nm FND spin-coated on a glass coverslip at room temperature over 610 – 660 nm. The exposure time of the sample to the 594-nm laser with a power of 30 μW for temperature measurement is 0.1 s. Inset: Changes of λ_0 with time over 6 s, with a mean value of 638.519 ± 0.013 nm. (Reproduced with permission from Tsai et al. 2017.).

with the spin resonance width, which is typically 20 MHz (Figure 26.3a), the magnitude of the thermal shift is not substantial at room temperature. Again, thanks to the exceptional stability of the ODMR spectra, the shift can be measured with high precision. Another notable feature of the temperature-dependent spectra is that the widths of the ODMR peaks stay nearly the same over 300–600 K. This has an important implication for practical applications because the band broadening effect as found in ZPL is no longer a limiting factor here with ODMR. It allows the use of FNDs

as nanoscale thermometers over a much wider temperature range than any other nanomaterials as the sensors (Brites et al. 2012; Jaque and Vetrone 2012).

The first application of FND as a nanothermometer was reported by Kucsko et al. (2013) who used laser excitation to obtain the ODMR spectra of single 100-nm FNDs either spin-coated on a glass slide or dispersed in cells. However, instead of scanning the entire ODMR spectrum, intensities at four microwave frequencies as indicated in Figure 26.4a were measured. These frequencies (f_1, f_2, f_3,

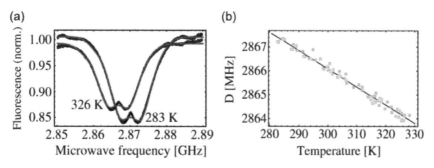

FIGURE 26.3 (a) Thermal shifts of the spin resonance of NV⁻ centers in a bulk diamond at 283 and 326 K. The splittings observed in the ODMR spectra are due to crystal strain. (b) Frequency shifts of the ODMR peak centers over 280–330 K. (Modified with permission from Acosta et al. 2010.)

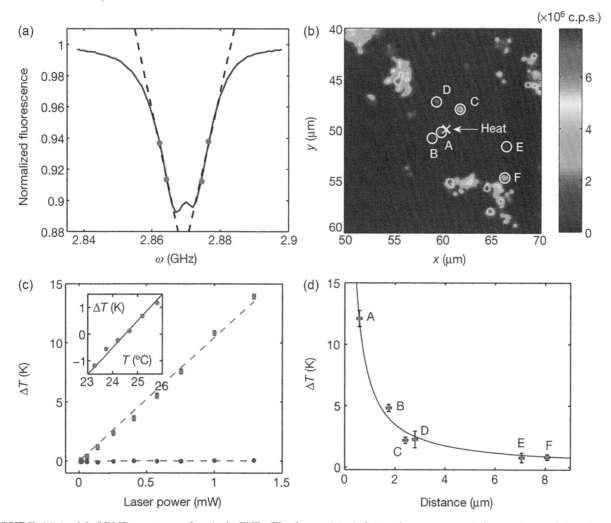

FIGURE 26.4 (a) ODMR spectrum of a single FND. The four points indicate the measurement frequencies used to extract the temperature information. (b) Two-dimensional confocal fluorescence image of FNDs (circles) and GNPs (cross) spin-coated on a glass coverslip. (c) Temperature rise of a single FND as a function of laser power at two different locations with (upper) or without (lower) laser illumination on a nearby GNP. (d) Temperature changes measured at six FND locations, indicated by white circles in (b), as a function of the distance from the illuminated GNP indicated by the white cross. (Reproduced with permission from Kucsko et al. 2013.)

and f_4) were generated by mixing radio-frequency fields (200 and 202 MHz) with two different microwave carrier frequencies around 2.8 GHz. Mathematically, it can be shown that the temperature change (ΔT) can be calculated from the fluorescence intensities measured at these four frequencies as

$$\Delta T \approx \frac{\Delta\omega}{\Delta D/\Delta T}\frac{(I_1+I_2)-(I_3+I_4)}{(I_1-I_2)-(I_3-I_4)}, \quad (26.3)$$

where $\Delta\omega = |f_1-f_2|/2 = |f_3-f_4|/2 = 1$ MHz in this specific case and I_1, I_2, I_3, and I_4 are the measured fluorescence intensities at f_1, f_2, f_3, and f_4, respectively.

The approximation holds when the temperature change is small and the fluorescence intensities at the two selected microwave frequencies around 2.8 GHz are about the same.

To induce local temperature changes, the same research team introduced GNPs of ~100 nm in diameter to the samples as the heat sources (Figure 26.4b), which were identified independently by laser excitation at 532 nm and fluorescence detection at 560 nm. With this four-point sampling method, a temperature measurement precision of better than 0.1 K was achievable within 4 s (Figure 26.4c). Also, a spatial resolution of ~100 nm was attained by measuring the separations between these two different types of nanoparticles in the same image by confocal fluorescence microscopy. Moreover, from a measurement of the temperatures of several FNDs at different locations (Figure 26.4d), the temperature rise of the laser-heated GNP was deduced based on the heat conduction equation for a sphere located in a homogeneous medium as

$$\Delta T(r) = \frac{\dot{Q}}{4\pi k r}, \qquad (26.4)$$

where $\dot{Q}$ is the heat generation rate, k is the thermal conductivity of the medium, r is the distance from the sphere center. The equation is valid at $r \geq r_0$, where r_0 is the radius of the sphere, as the thermal conductivity of gold ($k = 314$ W/m K) is much larger than that of the medium ($k = 0.6$ W/m K for water in this case) and, therefore, the temperature is uniform throughout the gold nanosphere.

The results of Kucsko et al. (2013), along with that of Neumann et al. (2013), have successfully demonstrated that it is possible to detect temperature variations in the nanometer regime with a sensitivity better than 0.1 K/Hz$^{1/2}$ if each FND contains up to 1,000 NV$^-$ centers. Further enhancement of the temperature measurement sensitivity by 1–2 orders of magnitude is achievable by manipulating the quantum coherence of single spins in ultrapure bulk diamonds (Table 26.1), a topic beyond the scope of this chapter. For readers who wish to delve in the principle and practice of these measurements, refer to the papers by Kucsko et al. (2013), Neumann et al. (2013), Toyli et al. (2013), and Wang et al. (2015).

26.4 Optical Thermometry with NV⁰, SiV⁻, GeV⁻, SnV, and Ni-Based Centers

26.4.1 NV⁰

Identical to NV$^-$ in structural configuration, the NV^0 center often appears together with NV$^-$ in synthetic type-Ib diamond, which typically contains 100 ppm atomic nitrogen (N^0) as impurities, due to charge state conversion between these two types of color centers. The process, as illustrated in Figure 26.1b, can be facilitated by either thermal annealing or laser excitation in the wavelength range of 450–610 nm, which results in a steady-state population of [NV^0] ~ 25%

out of [NV^0+NV$^-$]. When exposed to green light, the NV^0 center emits strong red fluorescence with a ZPL at 575 nm, accompanied by a broad phonon sideband peaking at ~620 nm, corresponding to the $^2A_2 \rightarrow {}^2E$ electronic transition (Manson et al. 2013). At room temperature, the thermal shift of the ZPL is $\Delta\lambda_0/\Delta T \approx 0.011$ nm/K, slightly smaller than 0.015 nm/K of NV$^-$ present in the same samples (Chen et al. 2011). However, the NV^0 center has a noticeably sharper ZPL than NV$^-$ (FWHM of ~3 nm versus ~5 nm), making it useful as a temperature sensor with the all-optical method as well.

26.4.2 SiV⁻, GeV⁻, and SnV

Carbon (C), silicon (Si), geranium (Ge), and tin (Sn) are group-IV elements (Figure 26.5a), all sharing the same crystal structure in the solid state known as the diamond cubic lattice. Because of the strong C–C bonds, diamond is a highly dense material with eight carbon atoms squeezed in a 3.57-Å lattice cube. As a result, most natural diamonds contain only N as the major impurity while the concentrations of other impurities are lower by several orders of magnitude. Silicon is a common impurity found in diamond crystals synthesized by chemical vapor deposition (CVD). A Si atom is about 1.5 times the size of a C atom and therefore can replace two neighboring carbons, situating in between two vacant sites in the diamond lattice and forming an optically active center in D_{3d} point group symmetry (inset in Figure 26.5b). The negatively charged silicon-vacancy center (SiV$^-$) is one of such defects, exhibiting a sharp ZPL at 738 nm with a weak phonon sideband in its photoluminescence spectra even at room temperature (Figure 26.5b). The radiative decay time of this center is 1.28 ns at 298 K, which increases to 1.72 ns as the temperature is lowered to 4 K (Rogers et al. 2014). Bright photoluminescence, near-infrared wavelength, and weak vibronic couplings render this center particularly appealing for quantum optics and quantum information applications (Aharonovich et al. 2011).

The ZPL at 738 nm of SiV$^-$ corresponds to the electronic transition of $^2E_g \rightarrow {}^2E_u$. A remarkable feature of this transition is that the FWHM of the ZPL at 4 K is only ~100 MHz, which is limited mostly by its spontaneous emission rate (Jahnke et al. 2015). At room temperature, the ZPL has a thermal shift of $\Delta\lambda_0/\Delta T = 0.0124$ nm/K and a FWHM of ~5 nm, both of which are similar to the characteristics of NV centers. Nguyen et al. (2018) have developed an all-optical thermometer based on an ensemble of these centers in bulk and nanoscale diamonds. They achieved a measurement precision of 70 mK at 300 K with a sensitivity of 360 mK/Hz$^{1/2}$ for bulk diamonds and 521 mK/Hz$^{1/2}$ for 200-nm diamonds. Notably, these properties varied less than 1% between samples, allowing for calibration-free temperature sensing and control in complex systems.

The GeV$^-$ centers in diamond are generated by Ge ion implantation and subsequent annealing of high-purity CVD substrates at elevated temperatures. The center has

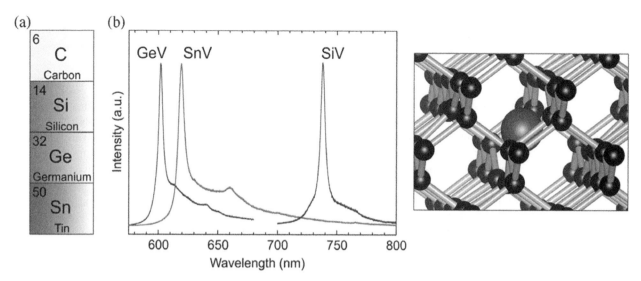

FIGURE 26.5 (a) First four elements in the group-IV periodic table. (b) Photoluminescence spectra of SiV$^-$, GeV$^-$, and SnV ensembles in bulk diamonds at room temperature. Shown to the right is an atomic structural model of these centers in diamond, where the larger gray sphere represents either the Si, Ge, or Sn atom and the smaller black spheres denote C atoms. (Modified with permission from Iwasaki et al. 2017.)

electronic energy levels similar to those of SiV$^-$, exhibiting a ZPL at 602 nm due to the $^2E_g \rightarrow {}^2E_u$ transition (Häußler et al. 2017). As temperature rises, the ZPL is red-shifted and concurrently broadened due to electron–phonon couplings as found for other vacancy-related color centers. At room temperature, the thermal shift is small, only $\Delta\lambda_0/\Delta T = 0.0083$ nm/K, with a FWHM ~ 6 nm for the ZPL. Fan et al. (2018) have developed a high-resolution, all-optical thermometer based on the GeV$^-$ ensembles and reported a sensitivity of 300 mK/Hz$^{1/2}$ for ~ 600 such color centers in bulk diamonds. Analogous to the NV$^-$ centers, the thermometer has a wide working range spanning from 4 K to 1,000 K. A more sophisticate method based on anti-Stokes excitation of the GeV$^-$ centers for nanoscale thermometry has also been developed recently by Tran et al. (2019).

In a separate study, Alkahtani et al. (2018) reported that the ZPL of SnV centers in bulk diamond is located at 619 nm with a FWHM ~ 6 nm at room temperature. The center can be effectively incorporated into diamonds via ion implantation followed by high-temperature annealing. The thermal shift of its ZPL is $\Delta\lambda_0/\Delta T = 0.0124$ nm/K, giving a sensitivity of 142 mK/Hz$^{1/2}$, well suited for practical applications. Figure 26.5b compares the ZPL positions and spectral profiles of SiV$^-$, GeV$^-$, and SnV at room temperature (Iwasaki et al. 2017), along with their performance as temperature sensors in Table 26.1.

26.4.3 Ni-Based S3 and NE8

The S3 centers are typically observed in natural diamonds or synthetic diamonds grown under high pressure and high temperature (HPHT) conditions, treated with Ni-containing catalysts, and then annealed at 2,200°C (Pereira and Santos 1990). Assigned with a structure of V-N-Ni-N-V, the center shows a ZPL at 497 nm and emits intense green fluorescence

with a lifetime of ~ 7 µs at room temperature when excited by blue light (Homeyer et al. 2015). The lifetime drastically decreases by more than two orders of magnitude upon heating the substrates from 100 K to 900 K (e.g. 277 µs at 100 K and 600 ns at 800 K), while the fluorescence intensity remains nearly constant over this temperature range. The lifetime serves well as a temperature indicator with a detection limit of 0.1 K/Hz$^{1/2}$. The sensor is unique in that it combines high sensitivity, short response time, and a wide range of use for nanoscale temperature sensing applications.

The NE8 center in diamond has a ZPL at 794 nm and is associated with a structure of a Ni atom surrounded by four N atoms (Nadolinny et al. 1999). High-temperature post-annealing (up to 1,900°C) of Ni-doped HPHT diamonds favors the formation of the NE8 centers. Sildos et al. (2017) reported a strong temperature dependence of the ZPL position above 200 K, suggesting that the NE8-containing diamonds are useful as sensitive all-optical thermometers at room temperature. Moreover, the fluorescence emission in the near-infrared range makes the centers appealing for applications in biology and medicine.

26.5 Innovative Nanothermometric Techniques

26.5.1 Time-Resolved Nanothermometry

For the luminescence nanothermometers discussed so far, the typical response time of these nanodevices is on the order of seconds if a temperature measurement precision of 0.1 K is needed (Table 26.1). Temporal resolution as such is by no means sufficient to follow the time evolution of a system under investigation at the nanoscale. Same difficulties are encountered for NV$^-$ centers because it is a time-consuming process to acquire the ZPL or ODMR spectra. Moreover, additional data processing after the experiment is required to determine the peak positions and thus the temperature changes.

A nanothermometer is far from ideal if it cannot reveal the dynamics of underlying heat transfer phenomena. To overcome this barrier, Tzeng et al. (2015) have developed a three-point sampling method to determine the temperature of a nanoscale system with a temporal resolution better than 10 μs. An important basis of this method is that the widths of the ODMR peaks of the NV$^-$ centers are nearly invariant with temperature change, as shown in Figure 26.3a. By assuming a Lorentzian signal profile of constant width (γ) and measuring the intensity changes of the fluorescence dip at three preselected frequencies (indicated in Figure 26.6a), the researchers showed mathematically that the thermal shift of the ODMR peaks could be obtained without scanning the entire spectrum according to the following equations,

$$\Delta D = \gamma \cdot \frac{1 - \sqrt{1 - 2R^2}}{R}, \tag{26.5}$$

where

$$R = -\frac{I_2 - I_1}{2I_3 - I_2 - I_1}, \tag{26.6}$$

and I_1, I_2, and I_3 are the fluorescence intensities measured at the points f_1, f_2, and f_3, respectively. The method allows real-time measurement of temperature changes at the nanoscale over ± 100 K around room temperature. More importantly, it paves the way for time-resolved temperature measurements with microwave and laser pulses.

Tzeng et al. (2015) conducted the first time-resolved temperature measurement for a solution containing gold nanorods (GNRs) using 100-nm FNDs as single-particle temperature sensors (Figure 26.6b). The GNRs employed in that work were ~10 nm in diameter and ~41 nm in length, heated by a tightly focused 808-nm laser. With the FNDs submerged in the medium and positioned only one of them near the 808-nm laser focus, the researchers

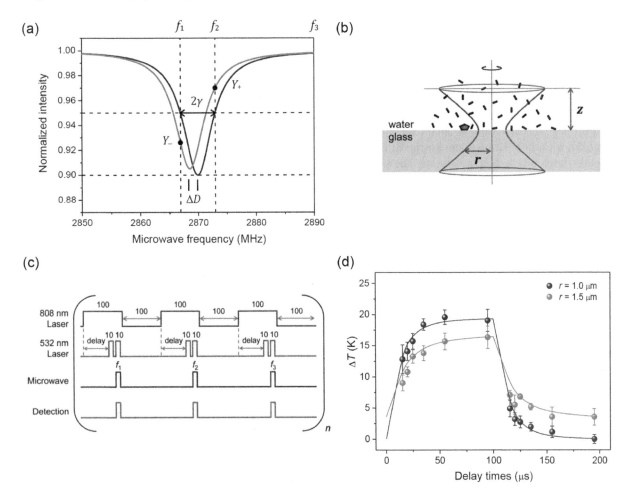

FIGURE 26.6 (a) Pictorial presentation of the three-point method based on an ODMR spectrum consisting of only one peak. The peaks before (right) and after (left) temperature change are both Lorentzian and have the same width, although their heights may vary. Three frequencies (f_1, f_2, f_3) are chosen for the intensity measurements. (b) Experimental scheme for the time-resolved temperature measurement with a 100-nm FND particle submerged in aqueous solution containing 10-nm × 41-nm GNRs (black rods) heated by an 808-nm laser (gray hyperboloid). (c) Time sequences of the laser, microwave, and detection pulses (all in μs) used in the time-resolved luminescence nanothermometry with a three-point method. (d) Time evolution of the heat dissipation in the GNR solution at the radial positions of $r = 1.0$ and 1.5 μm, as indicated in (b). (Modified with permission from Tzeng et al. 2015.)

observed superheating of the aqueous solution, achieving a measurement precision of better than ±1 K over a temperature variation range of 100 K and a spatial resolution of ∼0.5 μm limited mainly by the diffraction of light. Furthermore, by pulsing the heating laser (808 nm), the probing laser (532 nm), and the microwave radiation (Figure 26.6c), pump-and-probe type measurements based on the three-point method revealed nanoscale heat transfer dynamics with a temporal resolution of ∼10 μs in this particular experiment (Figure 26.6d).

26.5.2 Scanning Thermal Imaging

While single-particle FND nanothermometry offers an opportunity to conduct nanoscale temperature sensing, the application of the technique to map the spatial variation of temperature over an extended area proves challenging. For instance, the FND probe must be scanned through the sample in order to obtain information about the temperature distribution in 3D. Tetienne et al. (2016) tackled the challenge by grafting a single FND particle (∼100 nm in diameter containing ∼100 NV$^-$ centers) onto the tip of an atomic force microscope (AFM) for temperature sensing as well as topographic imaging. The researchers applied this integrated method, in conjunction with the ODMR technique, to map out the local temperature rise due to laser irradiation of a single 40-nm GNP particle submerged in aqueous solution with a sub-100 nm spatial resolution.

In addition to mapping local temperature changes, the FND nanothermometry can also be used for imaging the thermal conductivity (k) of a nanostructure. In a proof-of-principle experiment, Laraoui et al. (2015) attached a 50-nm FND to the apex of a silicon tip on an AFM as a local temperature sensor (Figure 26.7a). Confocal fluorescence imaging in combination with AFM confirmed the attachment of a single FND on the tip (Figure 26.7b–e). The researchers then applied an electrical current to heat up the tip and brought the FND in direct contact with a surface of varying thermal conductivities at a nanometer precision. Due to the small size and high thermal conductivity of the FND, a time response of less than 200 μs was achieved with the sensor, allowing the study of nanoscale heat transfer upon application of an electrical pulse. By measuring the temperature changes with ODMR, the research team examined in detail how the tip temperature was influenced by the contact of the tip-attached FND with the substrates having different thermal conductivities. A good agreement was obtained between the thermal conductivity image and the topographic map for a phantom microstructure made of a gold grid ($k = 314$ W/m K) on a sapphire substrate ($k = 30$ W/m K) (Figure 26.7f–h).

The two scanning thermal imaging microscopes described above promise multiple applications, from the investigation of phonon dynamics in nanostructures to the characterization of nanoscale heat transfer in solid-state systems as well as thermal imaging of material processing and microelectronic devices in operation. The methods compare well with standard scanning thermal microscopes in terms of both temperature measurement accuracy and spatial resolution, providing new perspectives in nanothermometry.

26.5.3 Magnetically Enhanced Temperature Sensing

As pointed out in the previous section, the spin resonances of NV$^-$ are relatively insensitive to the temperature change with a thermal shift of only $\Delta D/\Delta T = -74.2$ kHz/K, compared with the typical width of 20 MHz for the ODMR peaks. In an effort to enhance the sensitivity, Wang et al. (2018) took advantage of the magnetic properties of the centers and measured the magnetically induced frequency shifts due to temperature changes. They first synthesized nanohybrids comprised of FNDs and magnetic nanoparticles (MNPs) such as CuNi alloys (Figure 26.8a), which showed a critical magnetization behavior near the ferromagnetic–paramagnetic transition temperature, and then used the nanohybrids as temperature sensors. Figure 26.8b illustrates the critical magnetization behavior, where the magnetization (M) of the MNP drastically decreases with T as the temperature reaches the transition point, where the magnetization sensitivity ($\mathrm{d}M/\mathrm{d}T$) maximizes at the critical temperature (T_c). As the spin resonances of the NV$^-$ centers depend sensitively on the magnetic field generated by the MNP (Figure 26.8c), it implies that the ODMR frequencies shall vary sharply with temperature around T_c as well. For a 200-nm $Cu_{1-x}Ni_x$ MNP ($x = 0.5$–0.9) conjugated with a 100-nm FND (containing ∼500 NV$^-$ centers), the thermal shifts of the NV$^-$ spin resonances were as large as 14 MHz/K, which allowed the researchers to attain a sensitivity of 11 mK/Hz$^{1/2}$ by using a modified three-point method under ambient conditions around $T = 67°$C. The result represents a sensitivity improvement by more than two orders of magnitude. A similar high sensitivity is expected to be reached at other temperatures if the chemical compositions of MNPs in the hybrid sensors are varied (Figure 26.8d). This hybrid nanothermometer offers a novel approach to study a wide range of thermal processes at the nanoscale in high precision.

26.6 Biological Applications

26.6.1 Intracellular Nanothermometry

FNDs are carbon nanoparticles of inherent biocompatibility and non-toxicity. A natural (and logical) application of the carbon nanosensor is to use it for measuring the local temperatures of a biological system. Kucsko et al. (2013) demonstrated the feasibility by introducing FNDs and GNPs (both ∼100 nm in diameter) into human embryonic fibroblasts. They monitored the temperature changes of two FND particles within the same cell while heating a nearby GNP by a separate laser (Figure 26.9a). The particle situated closer to the heat source showed a stronger temperature dependence on the laser power than the others

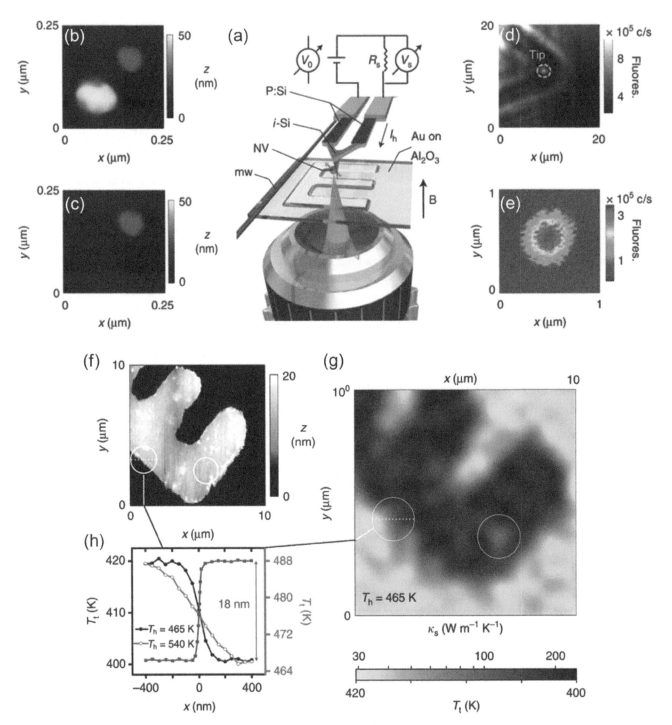

FIGURE 26.7 (a) Schematic diagram of the experimental setup used for thermal scanning microscopy. An electrical current circulates along the arms of an AFM cantilever (phosphorous-doped Si) and heats up the end section above the tip (intrinsic Si). Fluorescence from the FND attached to the AFM tip is collected by a high-numerical-aperture objective. A metal wire on the sample surface delivers microwave (mw) radiation to measure the ODMR spectra. (b) AFM image of two FNDs on a glass substrate. (c) AFM image after firmly scanning the tip on the larger FND in (b). Comparing (b) and (c) shows that the particle has been removed from the substrate and attached to the AFM tip. (d) Fluorescence image of the cantilever end and tip. Light from the tip-attached FND circled in white can be clearly separated from the background. (e) Zoom-in confocal fluorescence image of NV$^-$ centers for the FND shown in (d). (f) AFM topographic image of an 18-nm-thick gold film with an "E"-shaped pattern on a sapphire substrate. (g) FND-assisted thermal conductivity image of the same structure in (f). The heater temperature is $T_h = 465\,\mathrm{K}$. White circles in (f) and (g) indicate patches of low (high) conductivity inside (outside) the gold structure. (h) Measured tip temperatures near the edge of the gold structure at two different heater temperatures, 465 K (full black circles) and 540 K (empty gray circles). The topographic curve of the gold film (full gray squares) is also presented for reference. (Modified with permission from Laraoui et al. 2015.)

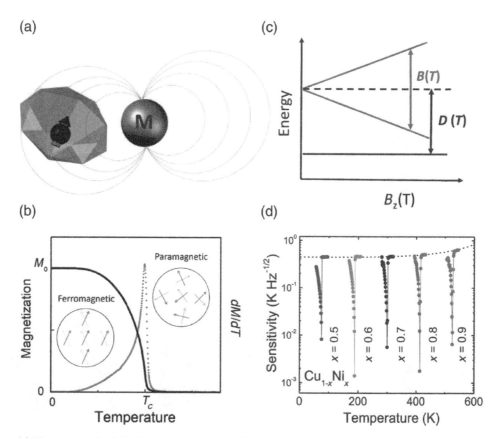

FIGURE 26.8 (a) Illustration of a hybrid nanosensor composed of a FND and a MNP. (b) Changes of magnetization (M, black solid curve) and magnetization sensitivity (dM/dT, gray dotted curve) with temperature, where T_c is the critical temperature. (c) Splitting of the spin sublevels ($m_s = \pm 1$) in the presence of an axial magnetic field, B_z. (d) Theoretical simulations for the sensitivity of the hybrid nanosensor composed of an FND and a $Cu_{1-x}Ni_x$ MNP, where x varies from 0.5 to 0.9. The simulations assume a spherical MNP with a diameter of 200 nm and a FND with 500 NV$^-$ centers, located 50 nm away from the MNP. (Reproduced with permission from Wang et al. 2018.)

(Figure 26.9b, c). At the laser power of 12 μW, a temperature change of 0.5 K was measured at the FND location, which corresponded to a change of ∼10 K at the GNP location, according to Eq. (26.4). This local temperature rise did not cause cell death. However, as the laser power increased to 120 μW, causing a temperature rise of 3.9 K for the FND (or ∼80 K at the GNP position), cell death occurred (Figure 26.9d). The cell death was confirmed by observing the fluorescence emission of ethidium homodimer-1, which is a membrane-impermeable dye whose fluorescence intensity will increase by more than tenfold once bound to nucleic acids in membrane-compromised cells. Their results demonstrate the capability of actively controlling cell viability by using FNDs and GNPs, enabling precision photothermal therapy at the single cell level, a subject to be discussed below.

26.6.2 Nanoscale Hyperthermia

Nanoscale hyperthermia, or hyperlocalized hyperthermia, is a new type of photothermal therapy. It differs from conventional local hyperthermia by creating a large temperature gradient within a region smaller than a cell with

visible light (Epperla et al. 2016). In this application, GNRs are a better heat source than gold nanospheres because its longitudinal surface plasmon resonance band can be conveniently tuned by increasing the aspect ratio of the particles from the green to the near-infrared region, where light has a better penetration depth into tissue (Huang et al. 2008). Additionally, GNRs can be readily conjugated with surface-modified FNDs by physical adsorption to form nanohybrids for various quantum sensing applications (Tsai et al. 2015).

Employing the all-optical method as presented in the previous section, Tsai et al. (2017) carried out thermometric measurements using the GNR-FND nanohybrids (containing 10-nm × 16-nm GNRs) as the dual-functional nanodevices to probe the local thermostability (or heat tolerance) of subcellular components in live cells. The research team chose to study the thermally induced rupture and retraction of membrane tunneling nanotubes (TNTs) formed between human embryonic kidney cells (HEK293T) because the events could be directly visualized by optical microscopy. To conduct the measurement, the nanohybrids were first delivered into the cells through endocytosis. Fluorescence imaging revealed that some of the endocytosed

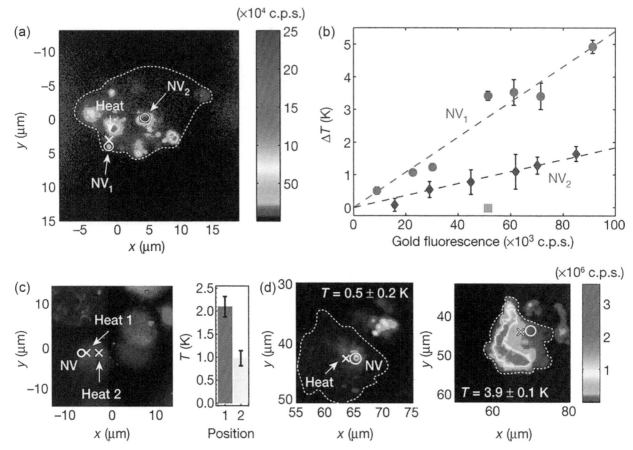

FIGURE 26.9 (a) Confocal fluorescence imaging of a single cell under laser excitation at 532 nm, with the emission collected at wavelengths greater than 638 nm. The white cross marks the position of the GNP used for heating and the locations of two FNDs are indicated by NV_1 and NV_2 for thermometric measurements. (b) Measured temperature changes at the positions of NV_1 and NV_2 relative to the incident laser power (represented by the gold fluorescence intensity) applied to the GNP. (c) Confocal fluorescence imaging of stained embryonic fibroblast cells with excitation at 494 nm or 528 nm and emission collection at 515 nm for live cells and 617 nm for dead cells. The bar graph shows the temperatures of a single FND (white circle) with the local heat applied at two different locations (white crosses). (d) Confocal fluorescence imaging of a fibroblast cell under illumination with varying laser powers. Excitation occurred at 532 nm and fluorescence was collected above 630 nm. Cell death is indicated by the penetration of ethidium homodimer-1 through the cell membrane, staining the nucleus to yield red emission (right). (Reproduced with permission from Kucsko et al. 2013.)

particles could be readily transported from one cell to the other through the TNTs, which were typically 1 μm in diameter and 30 μm in length (Figure 26.10a–c). With the use of a 594-nm laser for both heating and probing of single GNR-FNDs confined in the TNTs, the rupture temperatures of the individual nanotubes were measured (Figure 26.10d–f). For the TNT membrane, a temperature difference of up to 10°C was found between the local heating by the laser irradiation and the global heating using an on-stage incubator. The results shed new light on the nanoscale thermostability of subcellular components of a live cell, representing a significant advancement in the fields of local hyperthermia and photothermal therapy.

26.7   Conclusion and Outlook

The vacancy-related color centers were first discovered in the 1950s when scientists sought to understand the nature of

diamond color enhancement by radiation damage. However, using the color centers in the fields of nanothermometry did not start until 2013 when a benchmark experiment successfully probed the spin resonances of NV^- defects in FNDs to directly measure temperature gradients on glass slides and in single cells (Kucsko et al. 2013). After six decades of research, the NV^- centers are now finding cutting-edge applications in diverse research areas ranging from physics to biology and medicine in the 21st century (Schirhagl et al. 2014). And, yet, diamond nanothermometry is still very much in its developing stage with a lot more to offer in the foreseeable future.

In this chapter, we have first introduced some vacancy-related color centers that are good candidates of becoming sensitive temperature sensors in nanoscale systems (Table 26.1). Zooming in the NV^- centers, we have discussed their physical properties in detail, focusing on the energy transfer processes between electronic states during optical and magnetic transitions by ways of fluorescence

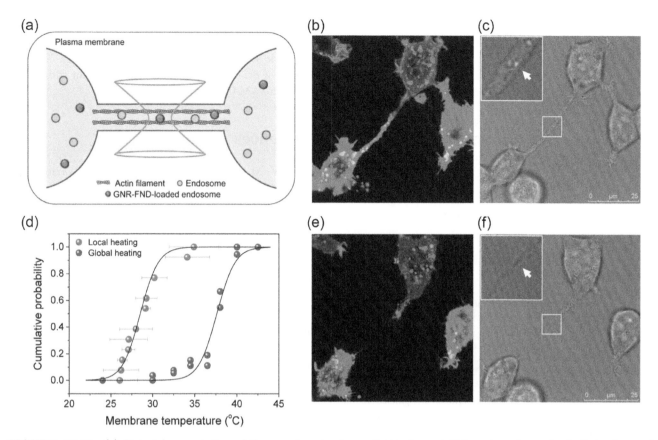

FIGURE 26.10 (a) Pictorial presentation of the experiment using a tightly focused 594-nm laser beam for both heating and temperature sensing of GNR-FNDs entrapped in the endosomes of a membrane nanotube. (b, c) Fluorescence (b) and merged bright-field/fluorescence (c) images of HEK293T cells transduced with actin-GFP fusion proteins (dark gray) and labeled with GNR-FNDs (light gray).(d) Empirical cumulative distribution plot of the membrane temperatures at which TNTs are ruptured by local heating and global heating of GFP-transduced HEK293T cells. (e, f) Fluorescence (e) and merged bright-field/fluorescence (f) images of GNR-FND-labeled, GFP-transduced HEK293T cells after exposure to the 594-nm laser with a power of 330 μW for 6 s. TNT breaking and retraction occurred upon the laser irradiation. Insets of (c) and (f): Enlarged views of the regions with the particle irradiated by the 594-nm laser in boxes. White arrows indicate the particles being irradiated. (Modified with permission from Tsai et al. 2017.)

detection. Based on the photophysical properties unique to the NV$^-$ centers, innovative nanothermometric techniques (including all-optical, magneto-optical, time-resolved, scanning probe, and magnetically enhanced methods) have been developed, and their current status has been surveyed and presented in this chapter. Finally, for practical applications in biomedical research, we have included two examples involving intracellular temperature sensing and nanoscale hyperthermia, which are arguably the two most active research areas today in photothermal therapy at the single cell level. Surface-functionalized FNDs, when properly conjugated with efficient nanoheaters like GNRs, are expected to be applicable for treating cancer cells under precisely controlled heat strokes in clinical applications (Chichel et al. 2007).

Compared with other luminescence sensors made of nanoparticles such as GNPs, semiconductor quantum dots, and rare-earth-doped oxides (Brites et al. 2012; Jaque and Vetrone 2012), FNDs turn out to be a favorable (and reasonable) choice for nanothermometric measurements in biological systems because of the following unmatchable

characteristics: (i) nontoxic and inherently biocompatible, (ii) highly fluorescent in the far-red region, (iii) excellent photostability with nearly zero photobleaching, (iv) easy functionalization with bioactive molecules through commonly used organic reactions on surface, and (v) readily conjugatable with a wide variety of nanoparticles to form multifunctional hybrids and versatile imaging modalities. For more about FNDs, refer to the recent comprehensive review by Chang et al. (2018).

As a nanoscale temperature sensor, FND is advantageous in having a broad working range and a superb single-particle detection sensitivity. Figure 26.11 compares the sensor sizes and temperature measurement accuracies between FND-based luminescence nanothermometry and other techniques reported in the literature (Kucsko et al. 2013). The comparison clearly indicates that FND outperforms other nanothermometers (such as CdSe quantum dots and green fluorescent proteins) in terms of its temperature accuracy and spatial resolution. Further improvement of the sensitivity is expected if ultrapure FNDs with long spin coherence times are available.

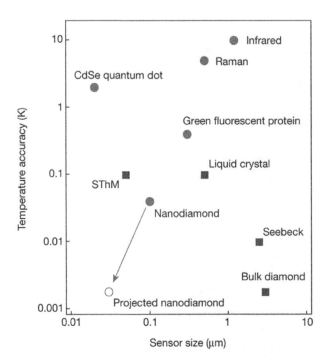

FIGURE 26.11 Comparison of sensor sizes and temperature accuracies for the NV$^-$-based quantum thermometer and other reported techniques. Biocompatible methods are labeled in full circles and the empty circle indicates the ultimate expected accuracy for FNDs. SThM: scanning thermal microscopy. (Modified with permission from Kucsko et al. 2013.)

There are challenges still ahead in the development of FNDs as a practical tool for routine nanothermal sensing operations. A major concern arises from the size of the temperature probe. At the present time, the spatial resolution of the FND-based nanothermometry is limited to the sub-micron range. In order to study, for example, the thermogenesis of intracellular organelles such as endoplasmic reticulum and mitochondria, it is crucial to use sub-20-nm FNDs capable of emitting stable fluorescence and displaying significant ZPL and ODMR signals. This requirement has posed a huge technical challenge on how such FNDs can be fabricated with high yield in today's labs. Additionally, as the particle size is reduced to 20 nm or smaller, the average number of the NV centers in each FND approaches to one and the fluorescence quantum yield drops significantly. As a result, the sensitivity in the temperature measurement with methods described in this chapter becomes so low that it would be impractical for biomedical research. We remain hopeful that with today's rapid advancement in material science and engineering, some innovative new fabrication methods may appear soon that will increase the density of color centers in FNDs as well as improve their surface properties to enhance the quantum yield. Ultimately, it is anticipated that with well-designed and synthesized FNDs, an all-in-one therapy tool will be available for tracking, heating, and thermal sensing at the nanoscale, finally becoming a general practice in nanomedicine.

Acknowledgments

This work was supported by Academia Sinica and the Ministry of Science and Technology of Taiwan with Grant No. 106-2628-M-001-004.

References

Acosta, V. M., Bauch, E., Ledbetter, M. P., Waxman, A., Bouchard, L. S., Budker, D. 2010. Temperature dependence of the nitrogen-vacancy magnetic resonance in diamond. *Phys Rev Lett* 104: 070801.

Aharonovich, I., Castelletto, S., Simpson, D. A., Su, C.-H., Greentree, A. D., Prawer, S. 2011. Diamond-based single-photon emitters. *Rep Prog Phys* 74: 076501.

Alkahtani, M., Cojocaru, I., Liu, X. et al. 2018. Tin-vacancy in diamonds for luminescent thermometry. *App Phys Lett* 112: 241902.

Brites, C. D., Lima, P. P., Silva, N. J. et al. 2012. Thermometry at the nanoscale. *Nanoscale* 4: 4799–829.

Campbell, B., Choudhury, W., Mainwood, A., Newton, M., Davies, G. 2002. Lattice damage caused by the irradiation of diamond. *Nucl Instrum Meth A* 476: 680–5.

Chang, H.-C., Hsiao, W. W.-W., Su, M.-C. 2018. *Fluorescent Nanodiamonds*. Wiley, Chichester.

Chen, X.-D., Dong, C.-H., Sun, F.-W., Zou, C.-L., Cui, J.-M., Guo, G.-C. 2011. Temperature dependent energy level shifts of nitrogen-vacancy centers in diamond. *Appl Phys Lett* 99: 161903.

Chichel, A., Skowronek, J., Kubaszewska, M., Kanikowski, M. 2007. Hyperthermia - description of a method and a review of clinical applications. *Rep Pract Oncol Radiother* 12: 267–75.

Davies, G., Lawson, S. C., Collins, A. T., Mainwood, A., Sharp, S. J. 1992. Vacancy-related centers in diamond. *Phys Rev B* 46: 13157–70.

Doherty, M. W., Acosta, V. M., Jarmola, A. et al. 2014. Temperature shifts of the resonances of the NV$^-$ center in diamond. *Phys Rev B* 90: 041201(R).

Doherty, M. W., Manson, N. B., Delaney, P., Jelezko, F., Wrachtrup, J., Hollenberg, L. C. L. 2013. The nitrogen-vacancy colour centre in diamond. *Phys Rep* 528: 1–45.

Epperla, C. P., Chen, O. Y., Chang, H.-C. 2016. Gold/diamond nanohybrids may reveal how hyperlocalized hyperthermia kills cancer cells. *Nanomedicine* 11: 443–5.

Fan, J.-W., Cojocaru, I., Becker, J. et al. 2018. Germanium-vacancy color center in diamond as a temperature sensor. *ACS Photonics* 5: 765–70.

Fukami, M., Yale, C. G., Andrich, P., Liu, X., Heremans, F. J., Nealey, P. F., Awschalom, D. D. 2019. All-optical cryogenic thermometry based on ntrogen-vacancy centers in nanodiamonds. *Phys Rev Applied* 12: 014042.

Harrison, J., Sellars, M. J., Manson, N. B. 2006. Measurement of the optically induced spin polarisation of N-V centres in diamond. *Diamond Relat Mater* 15: 586–8.

Häußler, S., Thiering, G., Dietrich, A. et al. 2017. Photoluminescence excitation spectroscopy of SiV$^-$ and GeV$^-$ color center in diamond. *New J Phys* 19: 063036.

Homeyer, E., Pailhès, S., Debord, R., Jary, V., Dujardin, C., Ledoux, G. 2015. Diamond contact-less micrometric temperature sensors. *Appl Phys Lett* 106: 243502.

Hsiao, W. W.-W., Hui, Y. Y., Tsai, P.-C., Chang, H.-C. 2016. Fluorescent nanodiamond: A versatile tool for long-term cell tracking, super-resolution imaging, and nanoscale temperature sensing. *Acc Chem Res* 49: 400–7.

Huang, X., Jain, P. K., El-Sayed, I. H., El-Sayed, M. A. 2008. Plasmonic photothermal therapy (PPTT) using gold nanoparticles. *Lasers Med Sci* 23: 217–28.

Hui, Y. Y., Chen, O. Y., Azuma, T., Chang, B.-M., Hsieh, F.-J., Chang, H.-C. 2019. All-optical thermometry with nitrogen-vacancy centers in nanodiamond-embedded polymer films. *J Phys Chem C* 123: 15366–74.

Iwasaki, T., Miyamoto, Y., Taniguchi, T. et al. 2017. Tin-vacancy quantum emitters in diamond. *Phys Rev Lett* 119: 253601.

Jahnke, K. D., Sipahigil, A., Binder, J. M. et al. 2015. Electron-phonon processes of the silicon-vacancy centre in diamond. *New J Phys* 17: 43011.

Jaque, D., Maestro, L. M., del Rosal, B. et al., 2014. Nanoparticles for photothermal therapies. *Nanoscale* 6: 9494–530.

Jaque, D., Vetrone, F. 2012. Luminescence nanothermometry. *Nanoscale* 4: 4301–26.

Kucsko, G., Maurer, P. C., Yao, N. Y. et al. 2013. Nanometer-scale thermometry in a living cell. *Nature* 500: 54–8.

Laraoui, A., Aycock-Rizzo, H., Gao, Y., Lu, X., Riedo, E., Meriles, C. A. 2015. Imaging thermal conductivity with nanoscale resolution using a scanning spin probe. *Nat Commun* 6: 8954.

Manson, N. B., Beha, K., Batalov, A. et al. 2013. Assignment of the NV^0 575-nm zero-phonon line in diamond to a ^2E-$^2A_2$ transition. *Phys Rev B* 87: 155209.

Manson, N. B., Harrison, J. P., Sellars, M. J. 2006. Nitrogen-vacancy center in diamond: Model of the electronic structure and associated dynamics. *Phys Rev B* 74: 104303.

Nadolinny, V. A., Yelisseyev, A. P., Baker, J. M. et al. 1999. A study of ^{13}C hyperfine structure in the EPR of nickel-nitrogen-containing centres in diamond and correlation with their optical properties. *J Phys Condens Matter* 11: 7357–76.

Neumann, P., Jakobi, I., Dolde, F. et al. 2013. High-precision nanoscale temperature sensing using single defects in diamond. *Nano Lett* 13: 2738–42.

Nguyen, C. T., Evans, R. E., Sipahigil, A. et al. 2018. All-optical nanoscale thermometry with silicon-vacancy centers in diamond. *Appl Phys Lett* 112: 203102.

Overton, T. W., Shigley, J. W. 2008. A history of diamond treatments. *Gems Gemology* 44: 32–55.

Pereira, E., Santos, L. 1990. Dynamics of S$_3$ luminescence in diamond. *J Luminesc* 45: 454–57.

Perpina, X., Vellvehi, M., Jorda, X. 2016. Thermal issues in microelectronics. In *Thermometry at the Nanoscale: Techniques and Selected Applications,* ed. L. D. Carlos and F. Palacio, 383–436. Royal Society of Chemistry, Cambridge.

Plakhotnik, T., Aman, H., Chang, H.-C. 2015. All-optical single-nanoparticle ratiometric thermometry with a noise floor of 0.3 K Hz$^{-1/2}$. *Nanotechnology* 26: 245501.

Plakhotnik, T., Doherty, M. W., Cole, J. H., Chapman, R., Manson, N. B. 2014. All-optical thermometry and thermal properties of the optically detected spin resonances of the NV$^-$ center in nanodiamond. *Nano Lett* 14: 4989–96.

Plakhotnik, T., Gruber, D. 2010. Luminescence of nitrogen-vacancy centers in nanodiamonds at temperatures between 300 and 700 K: Perspectives on nanothermometry. *Phys Chem Chem Phys* 12: 9751–6.

Rogers, L. J., Jahnke, K. D., Teraji, T. et al. 2014. Multiple intrinsically identical single-photon emitters in the solid state. *Nat Commun* 5: 4739.

Rondin, L., Tetienne, J.-P., Hingant, T., Roch, J.-F., Maletinsky, P., Jacques, V. 2014. Magnetometry with nitrogen-vacancy defects in diamond. *Rep Prog Phys* 77: 056503.

Sato, T., Ohashi, K., Sudoh, T., Haruna, K., Maeta, H. 2002. Thermal expansion of a high purity synthetic diamond single crystal at low temperatures. *Phys Rev B* 65: 092102.

Schirhagl, R., Chang, K., Loretz, M., Degen, C. L. 2014. Nitrogen-vacancy centers in diamond: Nanoscale sensors for physics and biology. *Annu Rev Phys Chem* 65: 83–105.

Sildos, I., Loot, A., Kiisk, V. et al. 2017. Spectroscopic study of NE8 defect in synthetic diamond for optical thermometry. *Diamond Relat Mater* 76: 27–30.

Sotoma, S., Epperla, C. P., Chang, H.-C. 2018. Diamond nanothermometry. *ChemNanoMat* 4: 15–27.

Tetienne, J. P., Lombard, A., Simpson, D. A. et al. 2016. Scanning nanospin ensemble microscope for nanoscale magnetic and thermal imaging. *Nano Lett* 16: 326–33.

Toyli, D. M., de las Casas, C. F., Christle, D. J., Dobrovitski, V. V., Awschalom, D. D. 2013. Fluorescence thermometry enhanced by the quantum coherence of single spins in diamond. *Proc Natl Acad Sci USA* 110: 8417–21.

Toyli, D. M., Christle, D. J., Alkauskas, A., Buckley, B. B., Van de Walle, C. G., Awschalom, D. D. 2012. Measurement and control of single nitrogen-vacancy center spins above 600 K. *Phys Rev X* 2: 031001.

Tran, T. T., Regan, B., Ekimov, E. A., Mu, Z, Zhou, Y., Gao, W.-b., Narang, P., Solntsev, A. S., Toth, M., Aharonovich, I., Bradac, C. 2019. Anti-Stokes excitation of solid-state quantum emitters for nanoscale thermometry. *Sci Adv* 5: eaav9180.

Tsai, P.-C., Chen, O. Y., Tzeng, Y.-K. et al. 2015. Gold/diamond nanohybrids for quantum sensing applications. *EPJ Quantum Technol* 2: 19.

Tsai, P.-C., Epperla, C. P., Huang, J.-S., Chen, O. Y., Wu, C.-C., Chang, H.-C. 2017. Measuring nanoscale thermostability of cell membranes with single gold-diamond nanohybrids. *Angew Chem Int Ed* 56: 3025-30.

Tzeng, Y.-K., Tsai, P.-C., Liu, H.-Y. et al. 2015. Time-resolved luminescence nanothermometry with nitrogen-vacancy centers in nanodiamonds. *Nano Lett* 15: 3945–52.

Wang, J., Feng, F., Zhang, J. et al. 2015. High-sensitivity temperature sensing using an implanted single nitrogen-vacancy center array in diamond. *Phys Rev B* 91: 155404.

Wang, N., Liu, G.-Q., Leong, W.-H. et al. 2018. Magnetic criticality-enhanced hybrid nanodiamond-thermometer under ambient conditions. *Phys Rev X* 8: 011042.

Weber, J. R., Koehl, W. F., Varley, J. B. et al. 2010. Quantum computing with defects. *Proc Natl Acad Sci USA* 107: 8513–8.

Yu, S.-J., Kang, M.-W., Chang, H.-C., Chen, K.-M., Yu, Y.-C. 2005. Bright fluorescent nanodiamonds: No photobleaching and low cytotoxicity. *J Am Chem Soc* 127: 17604–5.

Zaitsev, A. M. 2001. *Optical Properties of Diamond: A Data Handbook*. Springer, New York.

Index

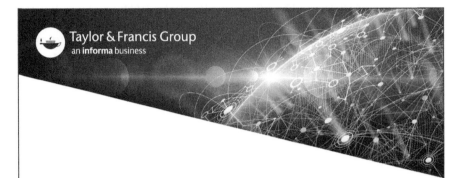

Printed and bound by CPI Group (UK) Ltd, Croydon, CR0 4YY

17/10/2024

01775672-0016